The Institute of Mathematics and its Applications Conference Series

Previous volumes in this series were published by Academic Press to whom all enquiries should be addressed. Forthcoming volumes will be published by Oxford University Press throughout the world.

NEW SERIES

1. *Supercomputers and parallel computation* Edited by D. J. Paddon
2. *The mathematical basis of finite element methods* Edited by David F. Griffiths
3. *Multigrid methods for integral and differential equations* Edited by D. J. Paddon and H. Holstein
4. *Turbulence and diffusion in stable environments* Edited by J. C. R. Hunt
5. *Wave propagation and scattering* Edited by B. J. Uscinski
6. *The mathematics of surfaces* Edited by J. A. Gregory
7. *Numerical methods for fluid dynamics II* Edited by K. W. Morton and M. J. Baines
8. *Analysing conflict and its resolution* Edited by P. G. Bennett
9. *The state of the art in numerical analysis* Edited by A. Iserles and M. J. D. Powell
10. *Algorithms for approximation* Edited by J. C. Mason and M. G. Cox
11. *The mathematics of surfaces II* Edited by R. R. Martin
12. *Mathematics in signal processing* Edited by T. S. Durrani, J. B. Abbiss, J. E. Hudson, R. N. Madan, J. G. McWhirter, and T. A. Moore
13. *Simulation and optimization of large systems* Edited by Andrzej J. Osiadacz
14. *Computers in mathematical research* Edited by N. M. Stephens and M. P. Thorne
15. *Stably stratified flow and dense gas dispersion* Edited by J. S. Puttock
16. *Mathematical modelling in non-destructive testing* Edited by Michael Blakemore and George A. Georgiou
17. *Numerical methods for fluid dynamics III* Edited by K. W. Morton and M. J. Baines
18. *Mathematics in oil production* Edited by Sir Sam Edwards and P. R. King
19. *Mathematics in major accident risk assessment* Edited by R. A. Cox
20. *Cryptography and coding* Edited by Henry J. Beker and F. C. Piper
21. *Mathematics in remote sensing* Edited by S. R. Brooks
22. *Applications of matrix theory* Edited by M. J. C. Gover and S. Barnett
23. *The mathematics of surfaces III* Edited by D. C. Handscomb
24. *The interface of mathematics and particle physics* Edited by D. G. Quillen, G. B. Segal, and Tsou S. T.

Continued overleaf

25. *Computational methods in aeronautical fluid dynamics*
Edited by P. Stow
26. *Mathematics in signal processing II*
Edited by J. G. McWhirter

Mathematics in Signal Processing II

Based on the proceedings of a conference organized by the Institute of Mathematics and its Applications on Mathematics in Signal Processing held at the University of Warwick in December 1988

Edited by

J. G. McWHIRTER
Royal Signals and Radar Establishment

CLARENDON PRESS · OXFORD · 1990

Oxford University Press, Walton Street, Oxford OX2 6DP
Oxford New York Toronto
Delhi Bombay Calcutta Madras Karachi
Petaling Jaya Singapore Hong Kong Tokyo
Nairobi Dar es Salaam Cape Town
Melbourne Auckland
and associated companies in
Berlin Ibadan

Oxford is a trade mark of Oxford University Press

Published in the United States
by Oxford University Press, New York

British Library Cataloguing in Publication Data
Mathematics in signal processing II.
1. Digital signals. Processing. Applications of mathematics
I. McWhirter, J. G. II. Series
621.38220151
ISBN 0–19–853641–0

Library of Congress Cataloging in Publication Data
(Data available)
ISBN 0–19–853641–0

Printed in Great Britain by
Bookcraft Ltd., Midsomer Norton, Avon

PREFACE

This book comprises a selection of papers presented at the second IMA Conference on Mathematics in Signal Processing which took place at the University of Warwick from the 13th to the 15th December 1988.

Signal Processing represents a major growth area for the application of mathematical concepts and techniques. For example, the modern signal processing techniques of adaptive filtering and high resolution spectral analysis encompass a broad range of mathematical concepts from linear algebra and numerical analysis to parallel processing and the solution of inverse problems. The algebra of finite fields is of fundamental importance in coding and cryptography and also pervades the literature associated with signal transformation techniques.

The aim of the conference was to bring together mathematicians and signal processing experts with a view to exploring the many areas of mutual interest, addressing the important mathematical problems which arise in signal processing and identifying fruitful avenues for further research. It was extremely successful in achieving that objective and it is hoped that the papers in this volume reflect the spirit of the meeting and the interactions which took place.

The organising committee is grateful to Dr. Keith Godfrey from the University of Warwick for helping with the organisation and to Miss C. Richards and her staff at the Institute of Mathematics and its Applications for their assistance with the organisation of the conference and these proceedings.

J.G. McWhirter (Chairman)
T.J. Shepherd
Royal Signals and Radar Establishment

A.R. Davies
University College of Wales

J.G. Hudson
P.J. Hargrave
STC Technology Ltd.

S. Hammarling
NAG, Oxford

ACKNOWLEDGEMENTS

The Institute of Mathematics and its Applications would like to thank the Office of Naval Research for their generous sponsorship of this meeting and to pay tribute to the organising committee for their dedicated service.

ACKNOWLEDGEMENTS

The Institute thanks the authors of the papers, the editor Dr. J.G. McWhirter (Royal Signal and Radar Establishment) and also Miss Pamela Irving, Miss Deborah Brown and Mrs. Anne Harding for typing the papers.

CONTENTS

VI PARALLEL ALGORITHMS

CONTRIBUTORS

G. AMMAR; *Department of Mathematical Sciences, Northern Illinois University, DeKalb, Illinois 60115, U.S.A.*

C.J. ANFINSON; *Centre for Applied Mathematics, Cornell University, Ithaca, New York 14853, U.S.A.*

A. ARTES-RODRIQUEZ; *DTC, ETSI Telecommunicacion-US, Aptdo. 62, 36280 Vigo (Pontevedra), Spain.*

M.G. BELLANGER; *T.R.T., 5 Avenue Reaumur, 92350 Le Plessis Robinson, France.*

J. BESAG; *Department of Mathematical Sciences, University of Durham, Durham, DH1 3LE.*

D. BLACKNELL; *GEC-Marconi Research Centre, Great Baddow, Chelmsford, Essex.*

D.L. BLANCHARD; *Department of Physics, University of Surrey, Guildford, Surrey, GU2 5XH.*

J.F. BOHME; *Lehrstuhl fur Signaltheorie, Ruhr Universitat, 4630 Bochum, Federal Republic of Germany.*

R.P. BRENT; *Computer Sciences Laboratory, Australian National University, Canberra, ACT 2601, Australia.*

D. CALLAERTS; *ESAT Laboratory, Department of Electrical Engineering, K.U. Leuven, Kardinaal Mercierlaan 94, B - 3030 Heverlee, Belgium.*

A. CALWAY; *Department of Computer Science, University of Warwick, Coventry, CV4 7AL.*

J.R. CASAR-CORREDERA; *DSSR, ETSI Telecommunicacion-UPM, Cuidad Universitaria, 28040, Madrid, Spain.*

J. CHEN; *ESAT Laboratory, Department of Electrical Engineering, K.U. Leuven, Kardinaal Mercierlaan 94, B - 3030 Heverlee, Belgium.*

R.J. CLARKE; *Department of Electrical and Electronic Engineering, Heriot-Watt University, 31-35 Grassmarket, Edinburgh, EH1 2HT.*

S.C. CLIPPINGDALE; *Department of Computer Science, University of Warwick, Coventry, CV4 7AL.*

C.F.N. COWAN; *Department of Electrical Engineering, University of Edinburgh, Kings Building, Mayfield Road, Edinburgh.*

D.J. CREASEY; *Department of Electronic and Electrical Engineering, University of Birmingham, P.O. Box 363, Birmingham, B15 2TT.*

M.J. CURTIS; *University of Essex, Wivenhoe Park, Colchester, Essex, CO4 3SQ.*

T.E. CURTIS; *Admiralty Research Establishment, Portland, Dorset.*

R. DE BEER; *Department of Applied Physics, Delft University of Technology, P.O. Box 5046, 2600 GA Delft, The Netherlands.*

L.M. DELVES; *Centre for Mathematical Software Research, University of Liverpool, Victoria Building, Brownlow Hill, P.O. Box 147, Liverpool, L69 3BX.*

G. DEMOMENT; *Laboratoire des Signaux et Systems, Ecole Superieure D'Electricite, Plateau du Moulon, 91190, Gif-sur-Yvette, Cedex, France.*

G. DE VILLIERS; *Royal Signals and Radar Establishment, St. Andrew's Road, Malvern, Worcs., WR14 3PS.*

D. DOCAMPO-AMOEDO; *DTC, ESTI Telecommunicacion-US, Aptdo. 62, 36280 Vigo (Pontevedra), Spain.*

P.G. EARWICKER; *Special Systems Department, Royal Aerospace Establishment, Farnborough, Hants., GU14 6TD.*

L. ELDEN; *Department of Mathematics, Linkoping University, 581 83 Linkoping, Sweden.*

D.J. EVANS; *Department of Computer Studies, University of Technology, Loughborough, Leicestershire, LE11 3TU.*

D.R. FARRIER; *Department of Electrical Engineering, The University, Southampton, SO9 5NH.*

A.R. FIGUEIRAS-VIDAL; *DSSR, ETSI Telecommunicacion-UPM, Cuidad Universitaria, 28040 Madrid, Spain.*

B. FRIEDLANDER; *Signal Processing Technology Ltd., 703 Coastland Drive, Palo Alto, CA 94303, U.S.A.*

F.M.F. GASTON; *Department of Electrical and Electronic Engineering, The Queen's University, Belfast, BT9 5AH.*

R.N. GORGUI-NAGUIB; *Department of Electrical and Electronic Engineering, University of Newcastle upon Tyne, Newcastle upon Tyne, NE1 7RU.*

Y. GOUSSARD; *Laboratoire des Signaux et Systemes, Ecole Superieure D'Electricite, Plateau du Moulon, 91190, Gif-sur-Yvette, Cedex, France.*

A. GRAHAM; *Department of Electrical Engineering, University of Southampton, Southampton, SO9 5NH.*

R.C. GREEN; *Active Memory Technology, 65 Suttons Park Avenue, Reading, RG6 1AZ.*

T. HAINSWORTH; *Department of Statistics, University of Leeds, Leeds, LS2 9JT.*

S. HAYKIN; *Communications Research Laboratory, McMaster University, 1280 Main Street West, Hamilton, Ontario, L8S 4K1, Canada.*

O.R. HINTON; *Department of Electrical and Electronic Engineering, Merz Court, University of Newcastle upon Tyne, Newcastle upon Tyne, NE1 7RU.*

A. HOUACINE; *Institut d'Electronique, Universite des Sciences et de la Technologie d'Alger, BP 32 El Alia, Bab Ezzouar, Alger, Algeria.*

J.E. HUDSON; *STC Technology Ltd., London Road, Harlow, Essex, CM17 9NA.*

G.W. IRWIN; *Department of Electrical and Electronic Engineering, The Queen's University, Belfast, BT9 5AH.*

A.J.E.M. JANSSEN; *Philips Research Laboratories, 5600 JA Eindhoven, The Netherlands.*

J.G. JONES; *Special Systems Department, Royal Aerospace Establishment, Farnborough, Hants., GU14 6TD.*

K.J. JONES; *New Systems and Techniques Laboratory, Plessey Avionics Ltd., Martin Road, West Leigh, Havant, Hants., PO9 5DH.*

M.C. JONES; *Department of Physics, University of Surrey, Guildford, Surrey, GU2 5XH.*

N.B. JONES; *Department of Engineering, University of Leicester, Leicester.*

S.C. KNOWLES; *Royal Signals and Radar Establishment, St. Andrew's Road, Malvern, Worcs., WR14 3PS.*

P.J.A. LAGO; *Grupo de Matematica Aplicada, Universidade do Porto, Porto, Portugal.*

N.K. LODGE; *Independent Broadcasting Authority, Research Laboratories, Crawley Court, Winchester, Hants., SO21 2QA.*

F.T. LUK; *School of Electrical Engineering, Cornell University, Ithaca, New York, 14853, U.S.A.*

S.P. LUTTRELL; *Royal Signals and Radar Establishment, St. Andrew's Road, Malvern, Worcs., WR14 3PS.*

R.N. MADAN; *Electronics Division, Department of the Navy, Office of the Chief of Naval Research, Arlington, Virginia 22217-5000, U.S.A.*

K.V. MARDIA; *Department of Statistics, University of Leeds, Leeds, LS2 9JT.*

S. McLAUGHLIN; *Department of Electrical Engineering, University of Edinburgh, Kings Building, Mayfield Road, Edinburgh.*

J.G. McWHIRTER; *Royal Signals and Radar Establishment, St. Andrew's Road, Malvern, Worcs., WR14 3PS.*

G.M. MEGSON; *Oriel College, Oxford.*

F. MONFRONT; *Laboratoire des Signaux et Systemes, Ecole Superieure D'Electricite, Plateau du Moulon, 91190, Gif-sur-Yvette, Cedex, France.*

B. MULGREW; *Department of Electrical Engineering, University of Edinburgh, Kings Building, Mayfield Road, Edinburgh.*

A. PARAKH; *School of Engineering and Applied Sciences, University of Sussex, Sussex House, Falmer, Brighton, BN1 9RH.*

H. PARK; *Computer Science Department, University of Minnesota, 4-192 EE/CSci Building, 200 Union Street S.E., Minneapolis, Minnesota 55455, U.S.A.*

W.W.F. PIJNAPPEL; *Department of Applied Physics, Delft University of Technology, P.O. Box 5046, 2600 GA Delft, The Netherlands.*

S.U. PILLAI; *Department of Electrical Engineering, Polytechnic University, 333 Jay Street, Brooklyn, New York, 11201, U.S.A.*

A.R. PRIOR-WANDESFORDE; *Department of Electrical Engineering, University of Southampton, Southampton, SO9 5NH.*

J.G. PROAKIS; *Department of Electrical and Computer Engineering, Northeastern University, Boston, U.S.A.*

I.K. PROUDLER; *Royal Signals and Radar Establishment, St. Andrew's Road, Malvern, Worcs., WR14 3PS.*

G.C. PRYDE; *Centre for Mathematical Software Research, University of Liverpool, Victoria Building, Brownlow Hill, P.O. Box 147, Liverpool, L69 3BX.*

B. PORAT; *Department of Electrical Engineering, Technion-Israel Institute of Technology, Haifa 32000, Israel.*

S. QUEGAN; *Department of Applied and Computational Mathematics, University of Sheffield, Sheffield, S10 2TN.*

L. REICHEL; *Bergen Scientific Centre IBM, Allegaten 36, N-5000 Bergen, Norway.*

A.P. ROCHA; *Grupo de Mathematica Aplicada, Universidade do Porto, Porto, Portugal.*

Y. ROBERT; *Laboratoire LIP-IMAG, Ecole Normale Superieure de Lyon, 69364 Lyon, Cedex, France.*

D. SALLAERTS; *Alcatel Bell, Central Hardware Division, Microelectronics Department, F. Wellesplein 1, 2018, Antwer Belgium.*

T.J. SHEPHERD; *Royal Signals and Radar Establishment, St. Andrew's Road, Malvern, Worcs., WR14 3PS.*

J. SMALLWOOD; *Department of Electrical Engineering, University of Southampton, Southampton, SO9 5NH.*

J.J. SOROGHAN; *Department of Electronic Engineering, University of Strathclyde, Glasgow.*

J.M. SPEISER; *Naval Ocean Systems Center, 271 Catalina Boulevard, San Diego, CA 92152, U.S.A.*

S.J. TABATABAIAN; *Department of Electrical and Electronic Engineering, University of Newcastle upon Tyne, Newcastle upon Tyne, NE1 7RU.*

D.M. TITTERINGTON; *Department of Statistics, University of Glasgow, Glasgow, G12 8QW.*

M. TODD; *Department of Computer Science, University of Warwick, Coventry, CV4 7AL.*

B. TOURANCHEAU; *Laboratoire LIP-IMAG, Ecole Normale Superieure de Lyon, 69364 Lyon, Cedex, France.*

S.C. TRAN; *Department of Electronic and Electrical Engineering, University of Birmingham, P.O. Box 363, Birmingham, B15 2TT.*

C.H. TRAVIS; *Department of Physics, University of Surrey, Guildford, Surrey, GU2 5XH.*

A. VAN DEN BOS; *Department of Applied Physics, Delft University of Technology, P.O. Box 5046, 2600 GA Delft, The Netherlands.*

D. VANDEPUTTE; *ESAT Laboratory, Department of Electrical Engineering, K.U. Leuven, Kardinaal Mercierlaan 94, B - 3030 Heverlee, Belgium.*

M. VANDEURZEN; *ESAT Laboratory, Department of Electrical Engineering, K.U. Leuven, Kardinaal Mercierlaan 94, B - 3030 Heverlee, Belgium.*

J. VANDEWALLE; *ESAT Laboratory, Department of Electrical Engineering, K.U. Leuven, Kardinaal Mercierlaan 94, B - 3030 Heverlee, Belgium.*

S. VAN HUFFEL; *ESAT Laboratory, Department of Electrical Engineering, K.U. Leuven, Kardinaal Mercierlaan 94, B - 3030 Heverlee, Belgium.*

D. VAN ORMONDT; *Department of Applied Physics, Delft University of Technology, P.O. Box 5046, 2600 GA Delft, The Netherlands.*

H.J. WHITEHOUSE; *Naval Ocean Systems Center, 271 Catalina Boulevard, San Diego, CA 92152, U.S.A.*

R. WILSON; *Department of Computer Science, University of Warwick, Coventry, CV4 7AL.*

B. YANG; *Lehrstuhl fur Signaltheorie, Ruhr-Universitat, 4630 Bochum, Federal Republic of Germany.*

I FINITE ALGEBRA

HIGH SPEED RECURSIVE REALISATIONS OF SMALL LENGTH DFTs

T.E. Curtis
(Admiralty Research Establishment, Portland, Dorset)

and

M.J. Curtis
(University of Essex, Colchester, Essex)

ABSTRACT

This paper develops a prime radix transform algorithm that can be used to calculate the discrete Fourier transform (DFT) of a data block of length P, a prime, using fixed coefficient second order recursive filter sections: this simple hardware structure allows compact, high performance DFT processors to be developed. Further "massaging" of the prime radix algorithm produces the zero factor transform (ZFT), implemented using first order recursive filter sections, with scaling values of -1.

The prime radix and zero factor transforms are limited to data block lengths that are prime: this constraint can be removed by re-factoring the transform equation and implementing the resultant algorithm as a two stage recursion process. This method, the composite radix Fourier transform (CRAFT), can process data blocks of any length and is particularly suitable for transforming small blocks of data of length equal to 2^n.

All of these algorithms can be implemented using first or second order recursive filter structures with fixed coefficient scaling multipliers, and high precision transforms can be produced by exploiting parallel error cancellation techniques. The simple hardware structure that results from this development has allowed high speed transform processors to be built, using both commercially available logic families and gate-arrays of moderate complexity.

1. INTRODUCTION

The discrete Fourier transform (DFT) [1] is one of the fundamental operations in digital signal processing. Many applications require compact, efficient DFT implementations and much has been published in the literature on algorithms to calculate fast Fourier transforms [2,3], often on general purpose main-frame or minicomputers. However, many real-time applications require either higher throughputs than can be obtained using "conventional" software based algorithms running on minicomputer/array processor combinations or are constrained in terms of the volume and/or power available for the complete system.

In some applications, for example "smart sensor" processing, both of these constraints appear simultaneously. Multi-channel sensor processing applications often require real-time transform systems with optimised hardware-based processing engines to cope with the processing load within the power/size budget of the system. Many of these realisations employ algorithms developed originally for software based systems and gain their speed simply by implementing the "number crunching" processors on "state-of-the-art" hardware. This approach requires that equipment manufacturers have continued access to the enabling high technology components, if they are to continue to improve system performance.

At present, a number of advanced LSI signal processing devices are available from commercial sources in Europe, the USA and Japan for implementing transform systems: however, these devices are often not ideal in specific applications, when it is often desirable or necessary to integrate the analogue sensor processing on silicon. The resulting high level of integration required for conventional (i.e. complex arithmetic FFT butterfly-based) processors, typically of the order of 80,000 gates, results in large silicon die size and high cost.

The design cycle costs of integrating the predominantly analogue sensor front end with the digital back end processing is also high, and whilst advanced technology and high performance components are being developed in various national and international programmes (e.g. VHSIC in the US, VAD in the UK), they are for predominantly military applications and are unlikely to be made widely available for commercial exploitation.

The cost of developing similar high performance components for the commercial sector is high and consequently systems companies, and other agencies that fund these developments, often prefer to market complete systems rather than individual devices. As a result, the availability of 'state-of-the-art' high performance processing elements to OEM users is limited.

The transform techniques outlined in the following sections have been developed to implement low complexity, high performance processors using low-grade, "off-the-shelf" silicon technologies, i.e. in available logic families or in gate array technologies available from commercial foundries in the UK, particularly those with a mixed analogue/digital capability, e.g. BiCMOS. The practical constraints involved in developing this type of hardware-based processor differ significantly from those in the development of software-based systems or those employing custom VLSI designs. In particular, the overall complexity of the implementation must be minimised to reduce silicon area and the number of gates used so that processors can be realised on available gate array based silicon.

The following paragraphs outline the development of two algorithms for the calculation of the DFT of a data block of length equal to a prime or the product of two or more primes, using Goods decomposition [5]. Later sections extend the method to develop a generalised algorithm for transforms with composite block length.

2. THE PRIME RADIX TRANSFORM (PRAT) [4]

The DFT of a block of complex data of length P is given by:-

$$A_r = \sum_{k=0}^{P-1} B_k \exp\{-j\Omega kr\} \tag{1}$$

where $\Omega = 2\pi/P$ and $0 \leq r \leq P-1$

The data block B_k is usually processed sequentially, with the data used in time order, but this need not be the case. The prime radix transform can be derived by expanding equation (1) and re-factoring. For example, a 7-point transform can be written:-

$$\begin{bmatrix} A_0 \\ A_1 \\ A_2 \\ A_3 \\ A_4 \\ A_5 \\ A_6 \end{bmatrix} = \begin{bmatrix} W^0 & W^0 & W^0 & W^0 & W^0 & W^0 & W^0 \\ W^0 & W^1 & W^2 & W^3 & W^4 & W^5 & W^6 \\ W^0 & W^2 & W^4 & W^6 & W^1 & W^3 & W^5 \\ W^0 & W^3 & W^6 & W^2 & W^5 & W^1 & W^4 \\ W^0 & W^4 & W^1 & W^5 & W^2 & W^6 & W^3 \\ W^0 & W^5 & W^3 & W^1 & W^6 & W^4 & W^2 \\ W^0 & W^6 & W^5 & W^4 & W^3 & W^2 & W^1 \end{bmatrix} \cdot \begin{bmatrix} B_0 \\ B_1 \\ B_2 \\ B_3 \\ B_4 \\ B_5 \\ B_6 \end{bmatrix} \tag{2}$$

It can be seen from equation (2) that all values of the coefficient vector W^0 through W^6 are used in the calculation of every frequency cell, except the first, i.e. the dc cell A_0. If we treat this as a special term, we can expand equation (2) and refactorise it so that the coefficients, rather than the data, are used sequentially.

So expanding (2) and re-factorising gives:-

$$\begin{bmatrix} A_1 \\ A_2 \\ A_3 \\ A_4 \\ A_5 \\ A_6 \end{bmatrix} = \begin{bmatrix} B_6 & B_5 & B_4 & B_3 & B_2 & B_1 & B_0 \\ B_3 & B_6 & B_2 & B_5 & B_1 & B_4 & B_0 \\ B_2 & B_4 & B_6 & B_1 & B_3 & B_5 & B_0 \\ B_5 & B_3 & B_1 & B_6 & B_4 & B_2 & B_0 \\ B_4 & B_1 & B_5 & B_2 & B_6 & B_3 & B_0 \\ B_1 & B_2 & B_3 & B_4 & B_5 & B_6 & B_0 \end{bmatrix} \cdot \begin{bmatrix} W^6 \\ W^5 \\ W^4 \\ W^3 \\ W^2 \\ W^1 \\ W^0 \end{bmatrix} \qquad (3)$$

Re-writing this in the summation form of equation (1) gives:-

$$A_0 = \sum_{k=o}^{P-1} B_k \qquad \text{for } r = 0 \qquad (4a)$$

$$A_r = \sum_{k=0}^{P-1} B_{kr*} W^k \qquad \text{for } 1 \leq r \leq p-1 \qquad (4b)$$

where r* is given by the congruence $r.r^* \equiv 1 \bmod P$.

This has not changed the transform, simply the order in which we carry out the calculation: in the general case, the coefficient vector can increment by a factor Q, with $1 \leq Q \leq P-1$, rather than as in equation (3). In this case, the summation form of the algorithm is given by:-

$$A_0 = \sum_{k=0}^{P-1} B_k \qquad \text{for } r = 0 \qquad (5a)$$

and

$$A_r = \sum_{k=0}^{P-1} B_{kr*}.W^{Qk} \qquad \text{for } 1 \leq r \leq P-1 \qquad (5b)$$

with r* given by the congruence $r.r^* \equiv Q \bmod P$

The d.c component of the transform, A_0 can be calculated by straightforward accumulation of the data samples: the rth frequency cell can be written in the general form:-

$$A_r = \sum_{k=0}^{P-1} b_k . W^{Qk} \quad \text{... where } b_k \text{ is the shuffled version of } B_{kr^*} \text{ in eq. (4)} \tag{6}$$

Equation (6) can be implemented by expanding it and factorising into partial sums:-

$$A_r = (((..(((b_{P-1}.W^Q + b_{P-2})W^Q + b_{P-3})W^Q ... \\ ... + b_2)W^Q + b_1)W^Q + b_0) \tag{7}$$

when it can be seen that equation (6) represents the first order recursive filter with complex pole W^Q, shown in Figure 1, with the z-domain transfer function:-

$$F(z^{-1}) = 1/(1 + z^{-1}.W^Q) \tag{8}$$

Equation (8) can be implemented directly, but the number of multiplications required in the recursion loop can be reduced by normalizing to generate the second order recursive structure:-

$$F(z^{-1}) = 1/(1 + z^{-1}.W^Q)$$

$$= 1/(1 + z^{-1}.[\cos\{2\pi Q/P\}-j\sin\{2\pi Q/P\}])$$

or:-

$$F(z^{-1}) = \frac{z^{-1}.[\cos\{2\pi Q/P\} + j\sin\{2\pi Q/P\}]}{1 + z^{-1}.2\cos\{2\pi Q/P\} - z^{-2}} \tag{9}$$

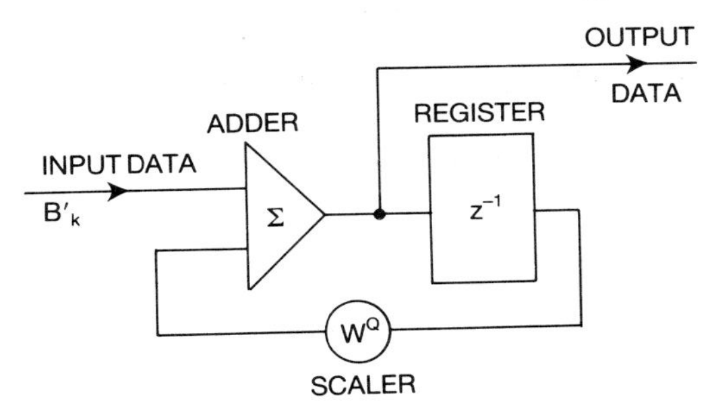

FIRST ORDER RECURSION

$$F(z^{-1}) = \frac{1}{1+z^{-1}W^Q}$$

Fig. 1 Prime radix transform

This equation can be realised using the circuit shown in figure 2: the circuit requires only one real multiplier in the recursion loop, multiplication by -1 being trivial in hardware. The recursive operations, corresponding to the demonstrator in equation (9), must be calculated as quickly as possible for high speed operation, as they limit the rate at which data can be clocked into the circuit, but the output calculations, corresponding to the numerator of equation (9) are only calculated at the transform output rate, and need to be performed much less often (in fact at 1/P times the input rate). Hence a fast combinational multiplier is required in the recursion loop, whilst much slower multipliers, for example a canonic signed serial/parallel implementation, can be used for the numerator.

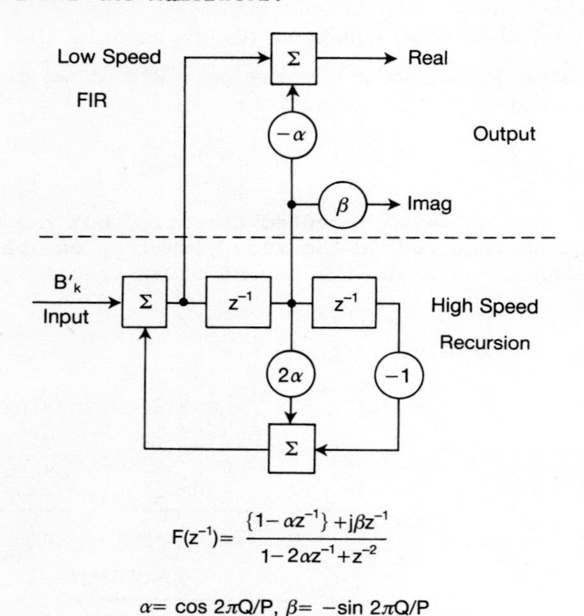

$$F(z^{-1}) = \frac{\{1-\alpha z^{-1}\} + j\beta z^{-1}}{1-2\alpha z^{-1}+z^{-2}}$$

$$\alpha = \cos 2\pi Q/P,\ \beta = -\sin 2\pi Q/P$$

Fig. 2 Second order recursion

This recursive implementation is similar to that due to Goertzel [6], but the data permutation developed in equation (5) allows us to use a scaling multiplier that is fixed in value throughout the transform calculation, rather than the multi-valued multiplier required by Goetzel. This significantly reduces the hardware complexity as we can, for example, approximate the recursive scaling, d $(= 2.\cos 2\pi Q/P)$, by some close value, D, that can be realised conveniently using a binary shift/add network, rather than a multi-valued combinational multiplier.

The error due to the use of an approximate scaling coefficient, D, rather than the exact value, d, can be corrected using a Taylor expansion:-

$$\frac{1}{1 + dz^{-1} - z^{-2}} = \frac{1}{1 + Dz^{-1} - z^{-2}} \cdot F(z)$$

Hence $$F(z) = \frac{1 + Dz^{-1} - z^{-2}}{1 + dz^{-1} - z^{-2}}$$

Writing $D = d + E$, where E is the scaling error

gives:- $$F(z) = \frac{1}{1 - Ez^{-1}/(1 + Dz^{-1} - z^{-2})}$$

and:-

$$\frac{1}{1 + dz^{-1} - z^{-2}} = \frac{1}{1 + Dz^{-1} - z^{-2}} \cdot \frac{1}{1 - Ez^{-1}/(1 + Dz^{-1} - z^{-2})} \tag{10}$$

Equation (1) can be approximated using the parallel error correction scheme in Figure 3, and high precision prime radix transforms can be produced, with low circuit complexity as in Reference 7.

The propagation delays of the scaling circuits in Figure 3 limit the rate at which the PRAT recursion can be clocked in practice, and the following section describes a method to remove the scaling multipliers from the recursion loop and hence improve transform bandwidth.

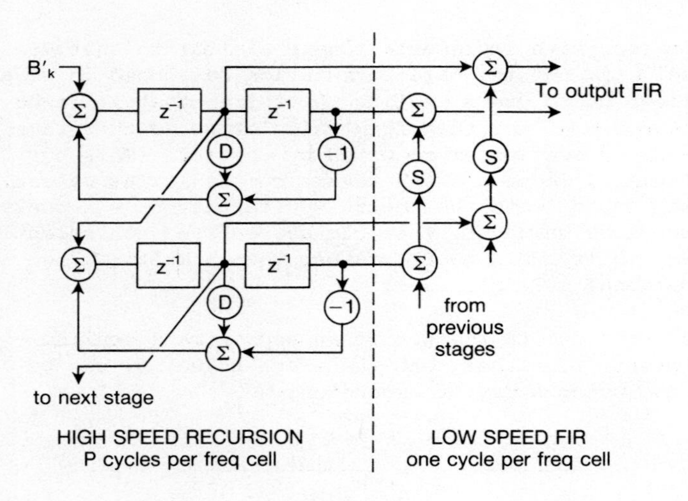

Fig. 3 Error correction

3. SIMPLIFYING THE PRIME RADIX TRANSFORM

In the previous section, scaling values were considered which could be approximated by simple binary fractions to produce minimum complexity hardware structures.

However, using the parallel error cancellation scheme shown in Figure 3, any scaling value can be approximated to arbitrary precision using the necessary number of parallel error correction stages. Consequently, for minimum complexity recursion hardware, it is convenient to consider transform lengths, P, and phase iteration factors, Q, where the scaling factor can be approximated by zero. This has the advantage that multiplication by zero can be implemented quite quickly and requires no hardware. The second order recursion then reduces to a recursion with scaling factor of -1 and a delay of z^{-2}. Filter decimation techniques can then be applied to the algorithm, allowing further simplification of the recursion structure.

3.1 The Zero Factor Transform

The value of d in the prime radix recursion is given by:-

$$d = 2.\cos\ (2\pi Q/P) \tag{11}$$

and $d \to$ zero as $2\pi Q/P \to \pi/2$.

So, to minimise d, the best values of Q for a given P are given by the integer closest to P/4, i.e.:-

$$Q = [P/4 + 0.5]$$

$$\text{or} \quad Q = P - [P/4 + 0.5]$$

where [X] denotes the integer part of X

Hence Figure 3 can be redrawn with zero multiplies as shown in FIgure 4, with the value of E in the error correction circuit given by:-

$$E = D - d$$
$$= -2.\cos(2\pi [P/4 + 0.5]/P) \qquad (12)$$

The system in Figure 4 can be implemented directly in hardware using inverting latch/adder loops for the high speed recursion and series/parallel multipliers for error correction, which can be relatively slow, since error correction is only performed at the cell output rate.

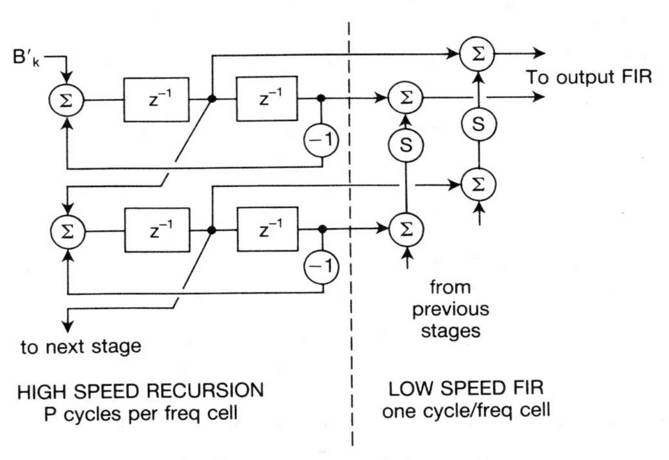

Fig. 4 "Zero Factor" transform

3.2 Input Decimation

The recursion system developed in the previous section requires P-1 recursion cycles. Since the complex plane positions of W^Q and W^{-Q} are symmetrical about the real axis, one way to increase transform throughput (and also to reduce transform recursion error) is to divide the P-1 point sequence into forward and reverse rotation recursions, each of (P-1)/2) points:-

$$A_{r+} = \sum_{k=O}^{T} b_k \, .W^{Qk}$$

$$T = (P-1)/2$$

$$A_{r-} = \sum_{k=O}^{T} b_{P-k}.W^{-Qk} \qquad (13)$$

Equation (13) can be implemented either using the same recursion hardware to generate two partial sums sequentially or using two separate but identical parallel recursions for higher throughput systems. In either case, the resulting recursion data can be used to generate a pair of frequency cells [4].

As a result of the z^{-2} delay in the PRAT recursion, only alternate input samples are combined in the process. So odd input samples form one component of the recursion output, even input samples the other. Since odd and even samples do not interact, the input sequence can be divided into odd and even parts and processed separately in parallel. This allows a number of other recursion structures to be developed.

Figure 5 shows a schematic of a system to operate on decimated input data. As can be seen from the figure, only subtracter/latch loops are required to implement the complete recursion and the increased parallelism of this architecture, together with the simpler recursion loop using only first order structures with -1 scaling, allows the transform algorithm to be realised significantly faster than when using the original PRAT system in Figure 3.

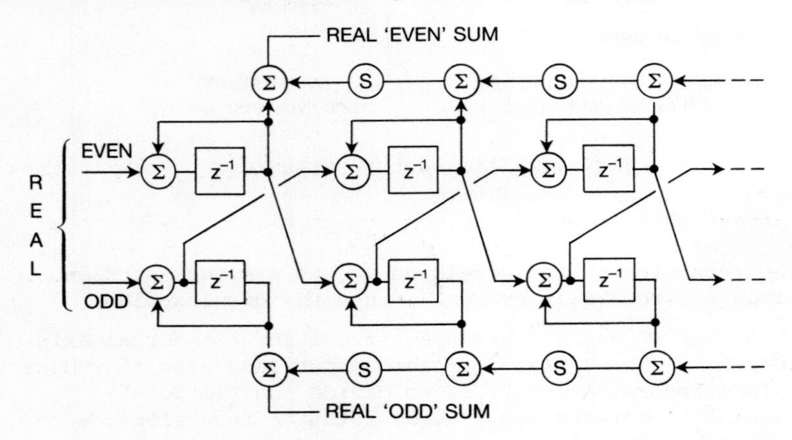

Fig. 5 Odd/even implementation

Using 74AS or 74F series logic, 29-point ZFTs can be calculated in around 2 useconds and 31-point ZFTs in around 2.5 useconds, allowing 899-point transforms to be calculated in less than 80 useconds, (using cascaded 29- and 31- point blocks).

4. HIGH SPEED ZFT PROCESSORS

The ZFT processor in Figure 6 uses only first order recursions with scaling values of -1. Since this structure is very simple, it is worth considering an FIR implementation of the basic filter structure, rather than the IIR implementation intrinsic to the PRAT algorithm. The recursion can be "unfolded" to give a systolic FIR-like system shown in Figure 6. This uses only adder/subtracters and latches and can potentially provide very high speed transforms.

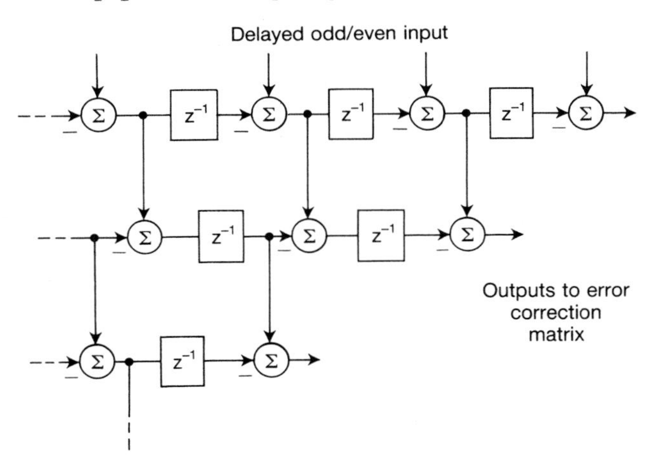

Fig. 6 Peristaltic implementation

The resultant processor, the "peristaltic ZFT" (the ultimate rubbish-in, rubbish-out machine!), differs from a systolic processor [8] in that, in the former case, the coefficient data is built into the processor structure, whilst for the systolic processor, calculation results from the interaction of separate data and coefficient streams. However both architectures have a number of common features; both have voracious appetites for data and are difficult to feed efficiently and both get rather messy when the processing pipeline gets blocked. If the peristaltic ZFT could be fed with data sufficiently quickly, then potentially it could process data at the rate of two frequency cell outputs per clock cycle.

13-point ZFTs could be calculated in 6 clock cycles, around 120 nseconds using 74AS or 74F series logic: similarly 17-point transforms requires 160 nseconds, so 221-point transforms, using cascaded 13- and 17-point ZFTs, could be calculated in just over 2 useconds, allowing transforms with bandwidths in excess of 100 MHz to be developed. Such transforms have wide application.

The regular structure of the peristaltic ZFT and the simple connectivity between the processing nodes make the system in Figure 7 attractive for VHSIC implementation.

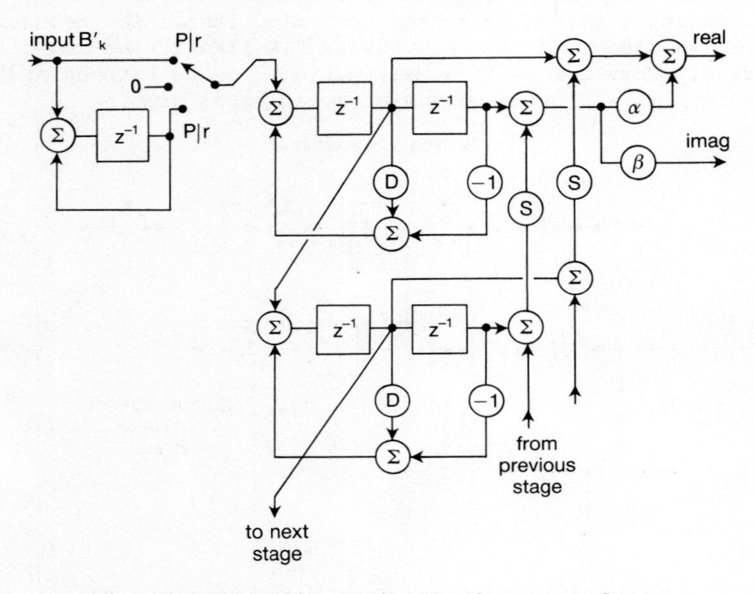

Fig. 7 Composite radix Fourier transform

5. THE COMPOSITE RADIX FOURIER TRANSFORM

The transform algorithms discussed so far have been limited to the use of block lengths equal to a prime. For new system designs, this does not create any problems, since many useful transform sizes can be realised using multiple passes through different sized small length transforms, for example, 899-point transforms using cascaded 29- and 31-point PRATs, 221-point using 13- and 17-point PRATs, etc.

However, in some applications, particularly when upgrading existing systems, this constraint to transform size equal to the product of two (or more) moderate sized primes is a problem. The following sections outline a simple modification of the prime radix algorithm to handle other data block lengths. As an example of the approach, we will consider the case when the transform length is equal to 2^n, although the method can be easily extended for any block length.

So for $N=2^n$, say $N=8$, we can write:-

$$A_r = \sum_{k=0}^{7} B_k \exp\{-j\Omega kr\} \tag{14}$$

$$\text{where } \Omega = 2\pi/8 \text{ and } 0 \leq r \leq 7$$

Proceeding as previously for the PRAT development, we can write equation (14) in matrix form as:-

$$\begin{bmatrix} A_0 \\ A_1 \\ A_2 \\ A_3 \\ A_4 \\ A_5 \\ A_6 \\ A_7 \end{bmatrix} = \begin{bmatrix} W^0 & W^0 & W^0 & W^0 & W^0 & W^0 & W^0 & W^0 \\ W^0 & W^1 & W^2 & W^3 & W^4 & W^5 & W^6 & W^7 \\ W^0 & W^2 & W^4 & W^6 & W^0 & W^2 & W^4 & W^6 \\ W^0 & W^3 & W^6 & W^1 & W^4 & W^7 & W^2 & W^5 \\ W^0 & W^4 & W^0 & W^4 & W^0 & W^4 & W^0 & W^4 \\ W^0 & W^5 & W^2 & W^7 & W^4 & W^1 & W^6 & W^3 \\ W^0 & W^6 & W^4 & W^2 & W^0 & W^6 & W^4 & W^2 \\ W^0 & W^7 & W^6 & W^5 & W^4 & W^3 & W^2 & W^1 \end{bmatrix} \cdot \begin{bmatrix} B_0 \\ B_1 \\ B_2 \\ B_3 \\ B_4 \\ B_5 \\ B_6 \\ B_7 \end{bmatrix} \tag{15}$$

It can be seen here that the coefficients W^0 through W^7 are not all used for each frequency cell calculation. In particular, the even cells use sub-sets of coefficients, the odd use them all. In the more general case with transform length N, all coefficients are used when r, the frequency index, is relatively prime to N and subsets are used when it is not.

We can expand equation (15) and again factor it to use the coefficients in any order, say sequentially:-

$$\begin{bmatrix} A_0 \\ A_1 \\ A_2 \\ A_3 \\ A_4 \\ A_5 \\ A_6 \\ A_7 \end{bmatrix} = \begin{bmatrix} 0 & 0 & 0 & 0 & 0 & 0 & 0 & (B_0+B_1+B_2+B_3+B_4+B_5+B_6+B_7) \\ B_7 & B_6 & B_5 & B_4 & B_3 & B_2 & B_1 & B_0 \\ 0 & (B_3+B_7) & 0 & (B_2+B_6) & 0 & (B_1+B_5) & 0 & (B_0+B_4) \\ B_5 & B_2 & B_7 & B_4 & B_1 & B_6 & B_3 & B_0 \\ 0 & 0 & 0 & (B_1+B_3+B_5+B_7) & 0 & 0 & 0 & (B_0+B_2+B_4+B_6) \\ B_3 & B_6 & B_1 & B_4 & B_7 & B_2 & B_5 & B_0 \\ 0 & (B_1+B_3) & 0 & (B_2+B_6) & 0 & (B_3+B_7) & 0 & (B_0+B_4) \\ B_1 & B_2 & B_3 & B_4 & B_5 & B_6 & 7 & B_0 \end{bmatrix} \cdot \begin{bmatrix} W^7 \\ W^6 \\ W^5 \\ W^4 \\ W^3 \\ W^2 \\ W^1 \\ W^0 \end{bmatrix}$$

Equation (16) can be expressed in a general form with phase rotation given by W^Q, with $(N,Q)=1$, for any transform size and can be implemented using the two stage recursion system shown in Figure 7. The main second order recursion structure uses either the PRAT or ZFT structures outlined above and a further adder/latch loop accumulates the partial sums of equation (16). Adder/latch accumulation control and data/accumulator selection are conveniently provided using extra data bits in the PROM address maps.

Expansions for a six-point and nine-point transform size are included in Appendix I: iterative implementations have been developed in this way for most small length DFTs.

6. DISCUSSION

A collection of techniques has been outlined which allow Fourier transforms to be implemented efficiently. The simple hardware structures provide transform systems with high throughput and high precision.

Low processor complexity, regular structure and simple inter-processor connectivity allow PRAT, ZFT and CRAFT systems to be developed using available LSI logic families and gate array technologies. The algorithms are sufficiently undemanding in gate count to allow them to be built in MOS, UHS bipolar or GaAs technologies as processor bandwidths and system economics dictate. Moderate throughput systems, using commercial CMOS gate arrays have been developed [6] to provide powerful, high precision DFT processors. These are designed to operate as one of the hardware macros for a high throughput, distributed architecture signal processor to implement high level signal processing graphs for sonar acoustic signal processing [9]. Similar algorithm massaging techniques and simple number theory methods are being employed to develop low complexity processors for other processing nodes in this system, e.g. for filtering, beamforming, matrix manipulation, etc.

REFERENCES

[1] Gold, B., and Radar, C.M., (1969) Digital Processing of Signals, McGraw-Hill.

[2] Cooley, J.W. and Tukey, J.W., (1965) An Algorithm for the Machine Computation of Complex Fourier Series, Math Comput., 19.

[3] Winograd, S., (1978) On Computing the Discrete Fourier Transform, Math. Comput., 32.

[4] Curtis, T.E. and Wickenden, J.T., (1983) Hardware-based Fourier Transforms: Algorithms and Architectures, IEE Proc, Pt. F, Commun., Radar and Signal Process, 130.

[5] Good, I.J., (1960) The Interaction Algorithm and Practical Fourier Series, J. Roy. Statist. Soc. Ser. B, 1958, 20, Addendum, 22.

[6] Goertzel, G., (1958) An Algorithm for the Evaluation of Finite Trigonometric Series, Am. Math. Monthly, 65.

[7] Spreadbury, D.J. and Rees-Roberts, T.M., (1984) Prime Radix Transforms- From Algorithm to Silicon Implementation, EURASIP Int. Conf. on Digital Signal Processing, Florence, Italy.

[8] Kung, H.T., (1984) Some System and Implementation Issues in Systolic Algorithm Designs, EURASIP Int. Conf. on Digital Signal Processing, Florence, Italy.

[9] Wu, Y.S. et al., (1984) Architectural Approach to Alternate Low-level Primitive Structures (ALPS) for Acoustic Signal Processing, IEE Proc, Pt. F, 131.

APPENDIX I

Examples of two stage recursion expansions for small length transforms.

1. Six-point DFT

The DFT algorithm in matrix form is:-

$$\begin{bmatrix} A_0 \\ A_1 \\ A_2 \\ A_3 \\ A_4 \\ A_5 \end{bmatrix} = \begin{bmatrix} W^0 & W^0 & W^0 & W^0 & W^0 & W^0 \\ W^0 & W^1 & W^2 & W^3 & W^4 & W^5 \\ W^0 & W^2 & W^4 & W^0 & W^2 & W^4 \\ W^0 & W^3 & W^0 & W^3 & W^0 & W^3 \\ W^0 & W^4 & W^2 & W^0 & W^4 & W^2 \\ W^0 & W^5 & W^4 & W^3 & W^2 & W^1 \end{bmatrix} \cdot \begin{bmatrix} B_0 \\ B_1 \\ B_2 \\ B_3 \\ B_4 \\ B_5 \end{bmatrix} \qquad \text{(Ia)}$$

Expanding and re-factorising:-

$$\begin{bmatrix} A_0 \\ A_1 \\ A_2 \end{bmatrix} \quad \begin{bmatrix} 0 & 0 & 0 & 0 & 0 & (B_0+B_1+B_2+B_3+B_4+B_5) \\ B_5 & B_4 & B_3 & B_2 & B_1 & B_0 \\ 0 & (B_2+B_5) & 0 & (B_1+B_4) & 0 & (B_0+B_3) \end{bmatrix} \begin{bmatrix} W^5 \\ W^4 \\ W^3 \end{bmatrix}$$

$$\begin{bmatrix} A_3 \\ A_4 \\ A_5 \end{bmatrix} = \begin{bmatrix} 0 & 0 & (B_1+B_3+B_5) & 0 & 0 & (B_0+B_2+B_4) \\ 0 & (B_1+B_4) & 0 & (B_2+B_5) & 0 & (B_0+B_4) \\ B_1 & B_2 & B_3 & B_4 & B_5 & B_0 \end{bmatrix} \begin{bmatrix} W^2 \\ W^1 \\ W^0 \end{bmatrix} \qquad \text{(Ib)}$$

2. Nine-point DFT

The DFT algorithm in matrix form is:-

$$\begin{bmatrix} A_0 \\ A_1 \\ A_2 \\ A_3 \\ A_4 \\ A_5 \\ A_6 \\ A_7 \\ A_8 \end{bmatrix} = \begin{bmatrix} W^0 & W^0 & W^0 & W^0 & W^0 & W^0 & W^0 & W^0 & W^0 \\ W^0 & W^1 & W^2 & W^3 & W^4 & W^5 & W^6 & W^7 & W^8 \\ W^0 & W^2 & W^4 & W^6 & W^8 & W^1 & W^3 & W^5 & W^7 \\ W^0 & W^3 & W^6 & W^0 & W^3 & W^6 & W^0 & W^3 & W^6 \\ W^0 & W^4 & W^8 & W^3 & W^7 & W^2 & W^6 & W^1 & W^5 \\ W^0 & W^5 & W^1 & W^6 & W^2 & W^7 & W^3 & W^8 & W^4 \\ W^0 & W^6 & W^3 & W^0 & W^6 & W^3 & W^0 & W^6 & W^3 \\ W^0 & W^7 & W^5 & W^3 & W^1 & W^8 & W^6 & W^4 & W^2 \\ W^0 & W^8 & W^7 & W^6 & W^5 & W^4 & W^3 & W^2 & W^1 \end{bmatrix} \cdot \begin{bmatrix} B_0 \\ B_1 \\ B_2 \\ B_3 \\ B_4 \\ B_5 \\ B_6 \\ B_7 \\ B_8 \end{bmatrix} \qquad \text{(Ic)}$$

Expanding and re-factorising gives:-

$$\begin{bmatrix} A_0 \\ A_1 \\ A_2 \\ A_3 \\ A_4 \\ A_5 \\ A_6 \\ A_7 \\ A_8 \end{bmatrix} = \begin{bmatrix} 0 & 0 & 0 & 0 & 0 & 0 & 0 & 0 & (B_0+B_1+B_2+B_3+B_4+B_5+B_6+B_7+B_8) \\ B_8 & B_7 & B_6 & B_5 & B_4 & B_3 & B_2 & B_1 & B_0 \\ B_4 & B_8 & B_3 & B_7 & B_2 & B_6 & B_1 & B_5 & B_0 \\ 0 & 0 & (B_2+B_5+B_8) & 0 & 0 & (B_1+B_4+B_7) & 0 & 0 & (B_0+B_3+B_6) \\ B_2 & B_4 & B_6 & B_8 & B_1 & B_3 & B_5 & B_7 & B_0 \\ B_7 & B_5 & B_3 & B_1 & B_8 & B_6 & B_4 & B_2 & B_0 \\ 0 & 0 & (B_1+B_4+B_7) & 0 & 0 & (B_2+B_5+B_8) & 0 & 0 & (B_0+B_3+B_6) \\ B_5 & B_1 & B_6 & B_2 & B_7 & B_3 & B_8 & B_4 & B_0 \\ B_1 & B_2 & B_3 & B_4 & B_5 & B_6 & B_7 & B_8 & B_0 \end{bmatrix}$$

(Id)

Other transform sizes can be expanded similarly.

GENERALISATION OF THE NUMBER THEORETIC PROPERTIES OF 120 AND 240 AND THEIR APPLICATIONS IN CRYPTOGRAPHIC SECURITY

S.J. Tabatabaian, O.R. Hinton and R.N. Gorgui-Naguib
(Department of Electrical and Electronic Engineering, University of Newcastle upon Tyne)

ABSTRACT

Recently, several attempts to break the RSA public-key cryptosystem have been studied. In this paper, a new method is introduced for determining the Euler totient function, $\varphi(n)$, where n is a product of two primes. This method has important implications to the security of the RSA system and views the factorisation problem from a different angle. Within this method it is necessary to consider the residue of the involved parameters in various fields. The different combinations of these parameters regarding their residue in various fields and their mathematical properties are analysed. The use of a modified form of the Chinese Remainder Theorem (CRT) is made. The results and consequences of using this new approach are also discussed.

1. INTRODUCTION

The most common public-key cryptosystem is the RSA in which two large primes p and q are chosen randomly to give n, where $n=p.q$. An encryption key, e is chosen at random from the interval $[2, \varphi(n)-1]$, where $\varphi(n)$ is given by $(p-1)(q-1)$ and where e and $\varphi(n)$ must be relatively prime. Both e and n are made public, but the decryption key, d, is secret and is the inverse of e mod $\varphi(n)$.

The security of the RSA system relies upon the difficulty of determining $\varphi(n)$. It is generally assumed that the best approach to break the system is to factorise n. By selecting n in the area of 200 digits, factorisation has as yet proved impractical by even the most powerful of modern computers.

Another approach to breaking the RSA system is to compute the Euler totient function $\varphi(n)$ of the composite number n where

$$\varphi(n) = \varphi(p).\varphi(q) \tag{1.1}$$

$$= (p-1).(q-1)$$

This method involves a search within the range of maximum and minimum possible values for $\varphi(n)$. It is necessary to check whether a proposed value for $\varphi(n)$ is correct, and one way of doing this is to solve the equation:

$$p,q = [n+1-\varphi(n)+\mathrm{sqrt}((n+1-\varphi(n))^2-4n)]/2 \tag{1.2}$$

Existence of an integer solution requires that the function $[(n+1-\varphi(n))^2-4n)]$ is a perfect square. Whilst fast methods for determining this may be available it is clearly very desirable to substantially reduce the search space by other means. It has been shown recently [3] that $\varphi(n)$ itself has a structure of the form

$$\varphi(n) = \frac{24.i}{n-1}, \text{ where i is an integer.} \tag{1.3}$$

A relationship such as equation (1.3) can be used as a generator of possible $\varphi(n)$. However, this reduces the number of checks by a maximum of only 24 which is not of much help since the search space is of order 10^{100} or so.

In this paper a new approach to find $\varphi(n)$ is presented which could improve on the above result. The approach is based upon fundamentals of number theory concerning properties of prime numbers, moduli, congruences, and the Chinese Remainder Theorem (CRT) [1]. Furthermore, it appears that the new approach could be combined with a fast factoring algorithm and limits determined for $\varphi(n)$ in order to further improve its efficiency.

2. OBSERVATION ON THE STRUCTURE OF $(n-1).\varphi(n)$

First, we look at some characteristics of p, q, φ and n, that provide further insight into the relationship between them. In a decimal base, since the least significant digit (LSD) of all primes greater than 5 is 1, 3, 7, or 9, therefore, there are ten possible combinations generating the LSD of n=p,q, denoted by $\langle n \rangle_{10}$, as shown in Table 1.

It can be shown [5] that the maximum factor, i, of $(n-1).\varphi(n)$, can be found simply by looking at the residue of the parameters involved in equation (1.3) mod 10.

$$(n-1)\varphi(n) = (pq-1)(p-1)(q-1) = i.k \tag{2.1}$$

A classification of the results reached with respect to the different combinations of p and q is shown in Table 1. These results imply that there are mainly two relationships which can be used in the computation of $\varphi(n)$. They are:

$$\varphi(n) = 240k/(n-1) \quad , \quad \text{when} \quad \langle n \rangle_{10}=1$$

$$\varphi(n) = 48k/(n-1) \quad , \quad \text{when} \quad \langle n \rangle_{10}=3 \vee 7 \vee 9$$

It is important to note that there will be common factors between 48, or 240, and $(n-1)$, which means that the search space is not so greatly reduced.

case	$\langle p \rangle_{10}$	$\langle q \rangle_{10}$	$\langle n \rangle_{10}$	$\mathrm{lcm}[\varphi(n),(n-1)]$	$(n-1).\varphi(n)$
1	1	1	1	$2^4.3.5^3$	6000.k
2	9	9	1	$2^4.3.5$	240.k
3	3	7	1	$2^4.3.5$	240.k
4	1	3	3	$2^4.3.5$	240.k
5	7	9	3	$2^4.3$	48.k
6	1	7	7	$2^4.3.5$	240.k
7	3	9	7	$2^4.3$	48.k
8	1	9	9	$2^4.3.5$	240.k
9	3	3	9	$2^4.3$	48.k
10	7	7	9	$2^4.3$	48.k

Table 1 Classified results for the different LSDs of n mod 10, $\langle n \rangle_{10}$

An alternative method of developing the structure for $\varphi(n)$ can be based on the fact that:

$$n^2 = 1 \pmod{120} \text{ where } n>5 \text{ and the } \langle n \rangle_{10}=1 \vee 9 \tag{2.2}$$

and

$$n^2 = 49 \pmod{120} \text{ where } n>5 \text{ and the } \langle n \rangle_{10}=3 \vee 7 \tag{2.3}$$

Also, a general equation can be used to calculate the Euler totient function of powers of n in terms of n itself [5], i.e.,

$$\varphi(n^k) = n^{k-1}\varphi(n) \quad ; \quad k > 1 \quad \text{and} \quad k \in Z \tag{2.4}$$

Therefore, there are two different expressions for $\varphi(n)$ in terms of n modulo 120, corresponding to the $\langle n \rangle_{10}$. The proof proceeds as follows:

Since $n=p.q$ and $gcd(p,q)$ is equal to 1, any power of p and q will yield coprime values. Thus, the function $\varphi(n^3)$ for the case of $<n_1>_{10}=1$ or 9,

$$\varphi(n^3) = \varphi(p^3).\varphi(q^3)$$

or $$\varphi(n^3) = p^2(p-1).q^2(q-1)$$

or $$= [1(\text{mod } 120)(p-1)].[1(\text{mod } 120)(q-1)]$$

$$= [(p-1).(q-1)(\text{mod } 120)]$$

Therefore,

$$\varphi(n^3) = \varphi(n)(\text{mod } 120) \qquad (2.5)$$

Writing congruence (2.5) in its Diophantine equation form and considering k=3 in equation (2.4), we have,

$$\varphi(n^3) = 120j + \varphi(n) \quad ; \quad j \in Z \qquad (2,6)$$

which when combined with

$$\varphi(n^3) = n^2.\varphi(n)$$

yields,

$$\varphi(n) = \frac{120.j}{n^2-1} \quad ; \quad j \in Z \qquad (2.7)$$

which is applicable when $<n>_{10}$ is 1 or 9.

If the same procedure is followed for the case where $<n>_{10}=3$ or 7, the final equation for the Euler totient function obtained is:

$$\varphi(n) = \frac{120.j'}{n^2-49} \quad ; \quad j' \in Z \qquad (2.8)$$

It can also be proved that a similar equation exists with a modulus of 240, but involving higher powers of n, i.e.,

$$\varphi(n) = \frac{240.h}{n^4-1} \quad ; \quad h \in Z \qquad (2.9)$$

However the results given in equations (2.7) and (2.8) suggest that further improvement can be achieved by considering different moduli and different residues for each modulus.

3. GENERALISATION

The arguments presented in section 2 concerning the LSD of n, p, and q and their relationship to $\varphi(n)$ suggest that there is a more general relationship between residues of n and $\varphi(n)$ in fields other than 10.

Since the product p.q is equal to n, the product of the residues of p and q in a field m_i must be equal to the residue of n in that field:

In these fields, we shall use the notation

$$\langle p\rangle_{m_i}.\langle q\rangle_{m_i} = \langle n\rangle_{m_i}$$

where "$\langle n\rangle_m$" denotes the residue of n modulo m. Hence if we determine all possible pairs of factors of $(n)_{m_i}$ in the field m_i

$$\{\langle a_{i_1}\rangle.\langle b_{i_1}\rangle \vee \langle a_{i_2}\rangle.\langle b_{i_2}\rangle \vee\ldots\vee \langle a_{i_j}\rangle.\langle b_{i_j}\rangle\}_{m_i} = \langle n\rangle_{m_i}$$

where reversed occurrences of the same pairs of $\langle a_{i_1}\rangle_{m_i}$ and $\langle b_{i_1}\rangle_{m_i}$ are neglected, then one of these pairs must be equal to $\langle p\rangle_{m_i}.\langle q\rangle_{m_i}$. We can therefore say that $\langle\varphi(n)\rangle_{m_i}$ must be given by:

$$\langle\varphi(n)\rangle_{m_i} = \{\langle a_{i_1}-1\rangle.\langle b_{i_1}-1\rangle \vee \langle a_{i_2}-1\rangle.\langle b_{i_2}-1\rangle \vee\ldots\vee \langle a_{i_j}-1\rangle.\langle b_{i_j}-1\rangle\}_{m_i}$$

$$= c_{i_1} \vee c_{i_2} \vee \ldots \vee c_{i_j}.$$

We can use this process for an arbitrary number of moduli. If these are all relatively prime, then the Chinese Remainder Theorem (CRT) can be used to combine the possible values in the different m_i so that a relationship for $\langle\varphi(n)\rangle_M$ can be derived:

$$\langle\varphi(n)\rangle_M = \{\langle d_1(n)\rangle \vee \langle d_2(n)\rangle \vee\ldots\vee \langle d_k(n)\rangle\}_M$$

where $M = \prod_{\forall i} m_i$

and $k = \prod_{\forall i} j_i$

and d1, d2 etc are all calculated according to the CRT.

To illustrate this approach, Tables 2 and 3 show the appropriate entries for the $(a_i b_i)$ for each possible value of $\langle n \rangle_{m_i}$ for just two moduli 3 and 10. These show us the possible residues of $\phi(n)$ moduli 3 and 10 for a particular value of n, and we can therefore use the CRT to construct all possible residues of $\phi(n)$ modulo 30.

case	$\langle n \rangle_3$	$\langle p,q \rangle_3$	$\langle \phi(n) \rangle_3$
1	1	(1,1) v (2,2)	(0) v (1)
2	2	(1,2)	(0)

Table 2: Possible values of n, p, q and $\phi(n)$ mod 3

case	$\langle n \rangle_{10}$	$\langle p,q \rangle_{10}$	$\langle \phi(n) \rangle_{10}$
1	1	(1,1)v(3,7)v(9,9)	(0)v(2)v(4)
2	3	(1,3)v(7;9)	(0)v(8)
3	7	(1,7)v(3,9)	(0)v(6)
4	9	(1,9)v(3,3)v(7,7)	(0)v(4)v(6)

Table 3: Possible values of n, p, q and $\phi(n)$ mod 10

For example, suppose

$$n = 113.127 = 14351$$

then $\langle n \rangle_3 = 2$ and $\langle n \rangle_{10} = 1$

hence from Tables 2 and 3:

$$\langle p,q \rangle_3 = (1,2) \quad \text{and} \quad \langle p,q \rangle_{10} = (1,1)v(3,7)v(9,9)$$

Also,

$$\langle \phi(14351) \rangle_3 = (0) \quad \text{and} \quad \langle \phi(14351) \rangle_{10} = (0)v(2)v(4)$$

Therefore,

$$\langle \phi(14351) \rangle_{30} = (0)v(12)v(24)$$

The above equation can be used to reduce the search space for $\phi(n)$.

If the field m_i is prime, then one method of determining all the pairs of factors $a_i.b_i$ is to first find the power of the primitive root of the field m_i giving $<n>_{m_i}$. Powers of the primitive root giving the factor pairs then follow readily from this. By considering this approach, it is possible to estimate the possible reduction in search space from this method by observing that for each prime m_i, the search space is reduced by approximately one half. Thus if n is around 10^{100}, and if all the m_i are relatively prime, then a $M=O(n)$ can be achieved with i approximately equal to 50. Hence the total reduction in search space will be $O(2^{50})$ or approximately $O(10^{15})$. This will be further improved upon by also using composite m_i, when a modified version of the CRT can be used.

4. CONCLUSION

In section 2 of this paper we have shown that $\varphi(n)$ has a structure related to the value of n, via equations (1.3), (2,7), (2.8) and (2.9). Unfortunately, these are of very limited use as the search space is reduced by only a small factor, and the condition to check for the correct $\varphi(n)$ requires detection of a perfect square. However, these structures have been used to point the way to a more general relationship between residues of n and $\varphi(n)$ in an arbitrary number of fields.

The major advantage of this method is that the factoring problem has effectively been decomposed into an arbitrary number of shorter finite field factorisations of the different residues of n, in much the same way as does the quadratic sieve algorithm [2]. In our case, however, it looks possible to exploit the properties of the Euler totient function in terms of the reduced number of possible combinations of residues. The computation of the above process could be readily mapped onto a parallel processing architecture. In addition, if large memory space is available, the tables could be precomputed thus saving the need to perform real-time factorisation.

REFERENCES

[1] Adams, W.W. and Goldstein, L.J., (1976) Introduction to Number Theory, Prentice-Hall, Inc.

[2] Davis, J.A. and Holdridge, D.B., (1983) Factorization using the Quadratic Sieve Algorithm, Advances in Cryptology, Proceedings of CRYPTO'83, Edited by D. Chaum, Plenum Press, pp. 103-113.

[3] Gorgui-Naguib, R.N. and Dlay, S.S., (1988) Properties of the Euler Totient Function Modulo 24 and Some of its Cryptographic Implications, presented at the Workshop on the Theory and Application of Cryptologic Technique EUROCRYPT'88, Davos, Switzerland.

[4] Rivest, R.L., Shamir, A. and Adleman, L., (1978) A Method for Obtaining Digital Signature and Public-key Cryptosystem, Communication of the ACM, Vol. 21, No. 2, pp. 120-126.

[5] Tabatabaian, S.J., (1987) Comments on the Cryptanalysis of the RSA Cryptosystem, Technical report, E.E. Eng. Dept. Un. of Newcastle upon Tyne.

THE APPLICATION OF REDUNDANT NUMBER SYSTEMS TO THE DESIGN OF VLSI RECURSIVE FILTERS

S.C. Knowles and J.G. McWhirter
(Royal Signals and Radar Establishment, Malvern, Worcs)

ABSTRACT

We show how the redundancy property of certain signed-digit number systems may be exploited to overcome a fundamental limitation in the dense pipelining of recursive digital filters. The resulting circuit can achieve a much higher throughput rate than conventional arithmetic processors for comparable hardware cost.

1. INTRODUCTION

Over the last few years there has been an enormous increase in the integration density which is economically achievable with silicon microelectronics. Consequently, the need has arisen to develop advanced architectures capable of achieving the computational performance offered by these improvements in silicon technology. This in turn has inspired a resurgence of interest in the use of novel number systems and their associated arithmetic for numerical computation. In this paper we will illustrate the use of novel arithmetic schems by considering the class of Signed-Digit Number Representations [1] and, in particular, those members of the class which exhibit the property of redundancy. These show considerable promise for a variety of applications since the redundancy enables closed arithmetic operations to be performed without the stringent need for carry propagation which is characteristic of conventional number systems.

We illustrate the potential advantage of using such a redundant number system by means of an interesting example which has arisen in the field of digital signal processing. This concerns the dense pipelining of a recursive digital filter suitable for very large scale integration (VLSI). This example is particularly illuminating because it demonstrates (i) an application where the use of a redundant number system gives a very clear advantage in overcoming what appears, at first sight, to be a fundamental problem; (ii) that the cost of implementing redundant arithmetic can be very low in terms of silicon area and speed of the associated VLSI circuits. The novel filter design described in this paper offers as much as an order of magnitude improvement in performance over comparable circuits based on conventional binary arithmetic but does not require a significant increase in hardware complexity.

We first discuss the basic problem associated with pipelining any recursive computation. Then, by analysing the micro-architecture of a bit-parallel mutiplier implemented using conventional binary arithmetic, we show how the problem of pipelining a recursive filter is dominated by the need for long range carry propagation. At this point, we digress to introduce the concept of redundancy in a number system and discuss, in particular, the Signed-Digit Number Representation. Circuits designed to implement the associated arithmetic and exploit some desirable properties such as limited carry propagation are then presented. Finally we demonstrate how the throughput rate of a recursive digital filter may be greatly increased using this type of arithmetic, and discuss the cost of implementing the solution as a VLSI silicon circuit.

2. THE PROBLEM: DENSE PIPELINING OF RECURSION

Consider the first order digital filter illustrated in figure 1. It accepts a continuously sampled input data stream x_n and produces a corresponding output stream y_n, the frequency response being determined by three programmable coefficients a_0, a_1 and b_1. The individual components perform addition (labelled Σ), multiplication (labelled X) and delay by a single sample period (labelled Δ). The filter comprises two distinct sub-sections; a non-recursive part which implements the finite impulse response (FIR) equation

$$u_n = a_0 x_n + a_1 x_{n-1}$$

and a recursive part which implements the infinite impulse response (IIR) equation

$$y_n = u_n + b_1 y_{n-1}$$

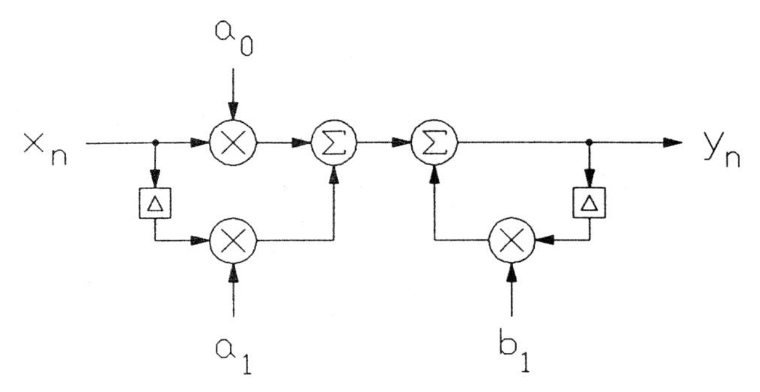

Fig. 1

Consider first the FIR- sub-section and suppose that the available circuit technology enables a multiplication such as $a_o x_n$ or $a_1 x_{n-1}$ to be performed in T seconds. Furthermore, suppose that this is much greater than the time required to perform an addition so that the multiplication units dominate the overall computation. The data throughput rate will therefore be approximately 1/T samples per second. The throughput rate may be extended beyond this "technology limit" by the architectural expedient of pipelining, which will now be described rather briefly.

A single pipeline stage may be introduced into the multiplier by splitting the computation into two sequential sub-processes separated by a storage unit which holds the intermediate results for one sample period. Two consecutive calculations can now be overlapped within the circuit. The first sub-processor generates an intermediate result from the current input data, while the second sub-processor is calculating an output value from the previous intermediate result. If the sub-processes are well balanced, approximately T/2 seconds will be required to complete each computation. Accordingly the circuit can now accept a new input sample every T/2 seconds.

The inclusion of a single pipeline stage has doubled the throughput of the multiplier at the expense of a single sample period delay, or latency, between input and output. Further levels of pipelining may be added by splitting the computation into smaller, faster sub-computations which are arranged in sequence and separated by storage registers. The pipelining analogy becomes clear. In general, n stages of pipelining will increase the throughput rate by a factor of n+1 whilst introducing a latency of n sample periods. For the majority of signal processing applications the throughput requirement is overriding whereas overall latency is of little importance.

The discrete levels of pipelining are conventionally referred to as "pipeline cuts" through the original (non-pipelined) structure. The well-known "cut theorem" [2] effectively states that the basic function of the circuit will be unaffected by pipelining (except to allow the overlapping of consecutive computations as described above) provided the following conditions are satisfied: (1) all pipeline cuts must pass through the entire circuit; and (2) all signal paths crossing each cut must do so in the same (forward) direction.

Now consider the IIR part of the circuit. Suppose that the multiplier unit is again pipelined in an attempt to increase throughput. The latency of the pipeline now becomes significant because it appears within a feedback loop. Clearly the processing loop cannot accept a new input data sample u_n until the result of the previous calculation y_{n-1} is available at the output of the multiplier block. Hence, the input delay is determined by the multipler latency. Furthermore, any cut through the overall circuit will be intersected by the feedback signal, so called because it passes in the opposite direction to all other signal paths. The cut theorem states that the basic operation of the circuit will be affected by such a cut and so it is not possible through pipelining, to overlap consecutive steps in the recursion and thereby increase the throughput rate. The fundamental problem is that pipelining introduces latency, and latency within a feedback loop decreases the throughput rate. By examining the micro-architecture of a bit-parallel multiplier, we will now show how the problem of pipelining an IIR filter based on conventional binary arithmetic is dominated by the need to propagate carries from the least significant bit (lsb) to the most significant bit (msb) position.

3. THE MICROSTRUCTURE OF MULTIPLICATION

Figure 2a indicates the structure of an electronic circuit for multiplying two 4-bit binary numbers. It constitutes a direct mapping of the conventional pencil and paper multiplication technique. An input word ($B = b_3, b_2, b_1, b_0$) is multiplied by each individual bit b_i of a second input word ($A = a_3, a_2, a_1, a_0$) and the resultant partial products are then summed. A single bit shift between consecutive partial products allows for the relative significance of consecutive digits of the second input word. Thus each cell of the array forms the product of two bits a_i and b_i, and adds the result to an accumulating sum bit which passes vertically downwards. Carries propagate from right to left along each row of the array. At the left hand end, the emerging carry bit is simply passed down to a spare input of correct significance on the row below.

The critical path, which determines the throughput rate of the multiplier, has two components. Firstly, carries may have to propagate along the entire length of any row before the outputs which it passes on to the row beneath become stable. Secondly, the accumulating sum of partial products has to pass down through four rows before a stable result is formed at the bottom of the array. As a result, the time required to form each 4-bit product could be as much as ten cell delays and accordingly, the maximum throughput of this "ripple-through" array is one input sample every ten cell delays.

The array may be densely pipelined in two simple stages as illustrated by figures 2b and 2c. In figure 2b rows of synchronously clocked latches (indicated by heavy black dots) are placed along horizontal "cuts" between rows of cells, enabling four consecutive multiplications to be overlapped within the array. The maximum throughput is now determined by the settling time within a single row rather than that of the complete array. Figure 2c illustrates a fully pipelined (so called "bit-level systolic") array obtained by adding further "cuts" parallel to the vertical axis. As a result, signals only need to propagate between adjacent cells during each clock cycle and the throughput of the entire array is determined by a single cell delay. Note that the pipelining imposes an lsb-first skew on the parallel input and output data words. By bit-level pipelining the parallel ripple-through multiplier circuit we have reduced the critical path from ten cells to a single cell, thus increasing the possible throughput rate by a factor of ten. As a penalty, we have introduced a four cycle latency and paid in hardware terms by the provision of 48 latches.

a)

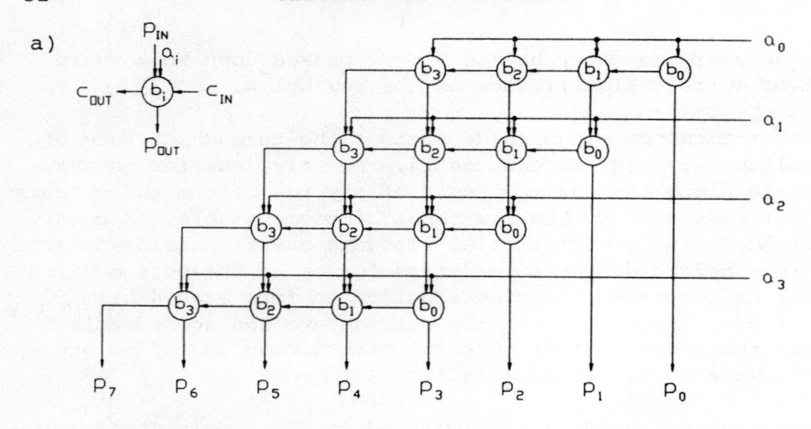

P_{IN}
a_i
C_{OUT}
b_i
C_{IN}
P_{OUT}
a_0
a_1
a_2
a_3
b_3
b_2
b_1
b_0
p_7
p_6
p_5
p_4
p_3
p_2
p_1
p_0

b)

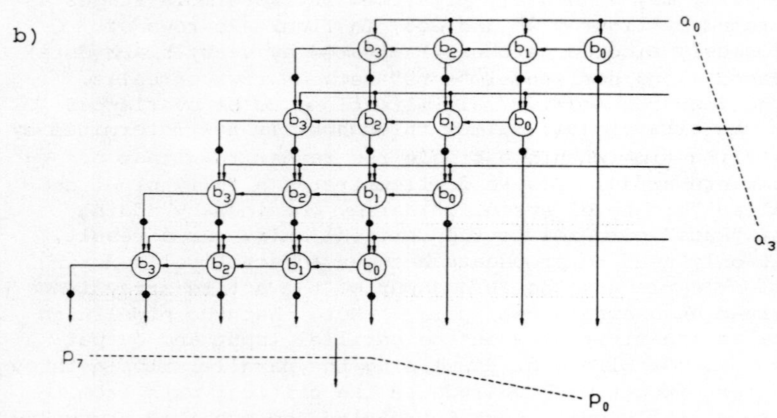

a_0
a_3
b_3
b_2
b_1
b_0
p_7
p_0

c)

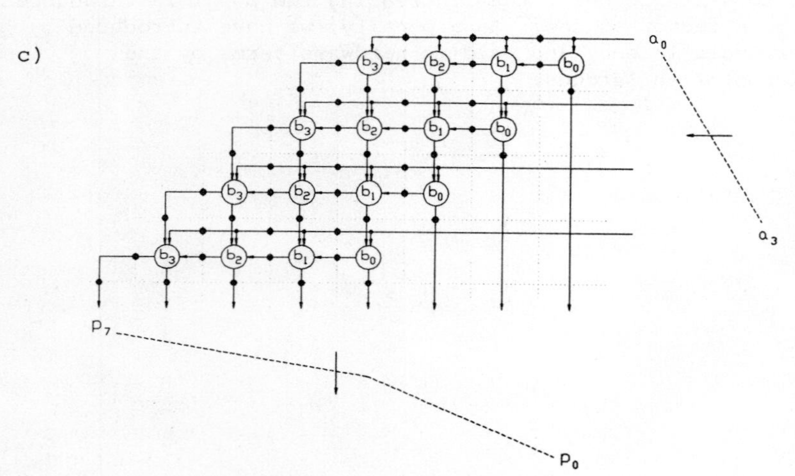

a_0
a_3
b_3
b_2
b_1
b_0
p_7
p_0

Fig. 2 a,b,c

If we are to solve the recursive filter problem we must devise a way of pipelining the multiplication process (to increase the throughput rate) without introducing significant latency. Referring back to figure 2b we note that the lsbs of the result are produced with a lower latency than the msbs. Unfortunately, this feature cannot be exploited because the recursion requires that the result of the multiplication be rounded or truncated to prevent wordgrowth and hence the lsbs are discarded. Only the msbs are fed back into the computation. However, the observation suggests that if we could reverse the order of partial product accumulation so that the msbs of the result are produced first, then they could be fed back with reduced latency. Unfortunately such a scheme cannot be implemented using conventional arithmetic because of the requirement for carries to propagate from the lsb to the msb position. Hence, conventional arithmetic is implicitly constrained to operate in an lsb-first manner. Certain alternative number systems allow arithmetic to be devised which is free from this constraint and therefore permit msb-first computation. In the next section, we introduce one such class of number systems and describe some of the corresponding arithmetic circuits.

4. SIGNED-DIGIT ARITHMETIC AND THE REDUNDANCY CONCEPT

Recall that conventional "fixed radix positional" numbers have a representation

$$\text{rep}\{X\} = x_m\ x_{m-1}\ x_{m-2}\ \ldots.\ x_1\ x_0\ .\ x_{-1}\ x_{-2}\ \ldots.\ x_{-(k-1)}\ x_{-k}$$

which corresponds to the algebraic value

$$\text{val}\{X\} = \sum_{i=-k}^{m} x_i r^i$$

where r is the radix and $x_i \in D_r$, the digit set. The digit set $D_r = \{0,1\ldots r-1\}$ is normally used, its cardinality (the number of elements) being equal to the radix r. For example in the decimal system, the digits of a number are usually chosen from the set $\{0,1\ldots9\}$.

These conventional number systems can be regarded as one subset of a broader class known as Signed-Digit Number Representations (SDNRs) in which the individual digits may be positive or negative; ie. $D_r = \{\bar{\alpha}\ldots\beta\}$ where $\bar{\alpha} = -\alpha$ denotes a negative digit. Generally, α and β are chosen to be less than r so that there is a unique representation of zero. However, the set D_r must contain at least r (consecutive integer) elements to ensure that every algebraic value may be represented.

SDNRs may have cardinality greater than the radix, in which case they are termed "redundant" because there may be more than one valid representation for a given algebraic value. It is this property which leads to the useful arithmetic behaviour described below. Quantitatively, the redundancy of a digit set is defined by

$$R = \frac{\text{Cardinality of } D_r}{r}$$

Consider, for example, the following decimal digit sets and their corresponding representations of the value 469:

Conventional non-redundant (R=1)
$D = \{0...9\}$: 469

SDNR non-redundant (R=1)
$D = \{\bar{2}...7\}$: $47\bar{1}$

SDNR symmetric minimally-redundant (R=1.1)
$D = \{\bar{5}...5\}$: 531, $1\overline{531}$

SDNR symmetric maximally-redundant (R=1.9)
$D = \{\bar{9}...9\}$: 469, $47\bar{1}$, $5\overline{31}$, $1\overline{531}$, $1\bar{6}69$, $1\bar{6}$

It should be clear from these examples that the greater the the redundancy of the chosen number system, the more freedom there is in the choice of representation for any given value. Now consider the addition of two numbers. The choice of representation means that it is not necessary to know every digit of the two operands in order to deduce a valid digit of the result. In fact a knowledge of the operand digits in one or two adjacent significance positions is adequate to make a valid selection for any given result digit and so the computation of all result digits can be performed in parallel.

We now show how this implicitly parallel, and hence inherently fast, addition may be achieved in practice. The resulting circuits may, of course, be applied to a wide range of computational tasks since all higher arithmetic functions are ultimately defined in terms of additions and subtractions and the latter may be treated as addition with a negative operand.

Figure 3a illustrates one section of a generalised adder taking two multi-digit numbers A and B as operands. The circuit consists of a chain of identical digit adder cells, each provided with a pair of input digits, a_i and b_i, and producing a corresponding sum digit s_i. Cells communicate with their neighbours by means of the signals t_{in} and t_{out}. For a conventional adder these signals correspond to carries and because, in this case, t_{out} always depends on t_{in}, the operation is implicitly sequential.

An implicitly parallel addition requires that t_{out} be independent of t_{in}. This independence is achieved by partitioning the addition process into two stages as indicated in figure 3b. We refer to the stages as "transfer generation" and "transfer recombination". The term "transfer" rather than carry is used in signed-digit arithmetic because in general these signals may be positive or negative; thus the transfer digit may act as a carry or a borrow digit. The function of each sub-cell is still the addition of its inputs and the generation of appropriately weighted outputs corresponding to the result of that addition. The important point to note is that the transfer generation stage produces a transfer digit which is dependent only on the pair of incoming operand digits. This means that transfer digits cannot propagate along the entire row of cells as is possible in a conventional adder. The sum output from each cell is generated by combining the incoming transfer digit and an intermediate value produced by the transfer generation stage.

In order the avoid long range transfer propagation, care must be taken in selecting the range of values which each digit can represent; figure 3c indicates a suitable choice for radix-4 addition. The key feature in the operation of this cell is that the tranfer generation stage limits the output tranfer digit to the range $[\bar{1}...1]$, and its output to the recombination stage to the range $[\bar{2}...2]$. This allows the incoming transfer digit to be absorbed into the recombination stage without exceeding the permitted output number range of $[\bar{3}...3]$. This "scope" for recombination is provided by the redundancy which exists within the number representation. Figure 3d provides a specific example concerning the radix-4 addition of $2\bar{3}\bar{1}2$ ($=78_{10}$) and $2\bar{2}33$ ($=111_{10}$) to give $1\bar{1}\bar{1}31$ ($=189_{10}$).

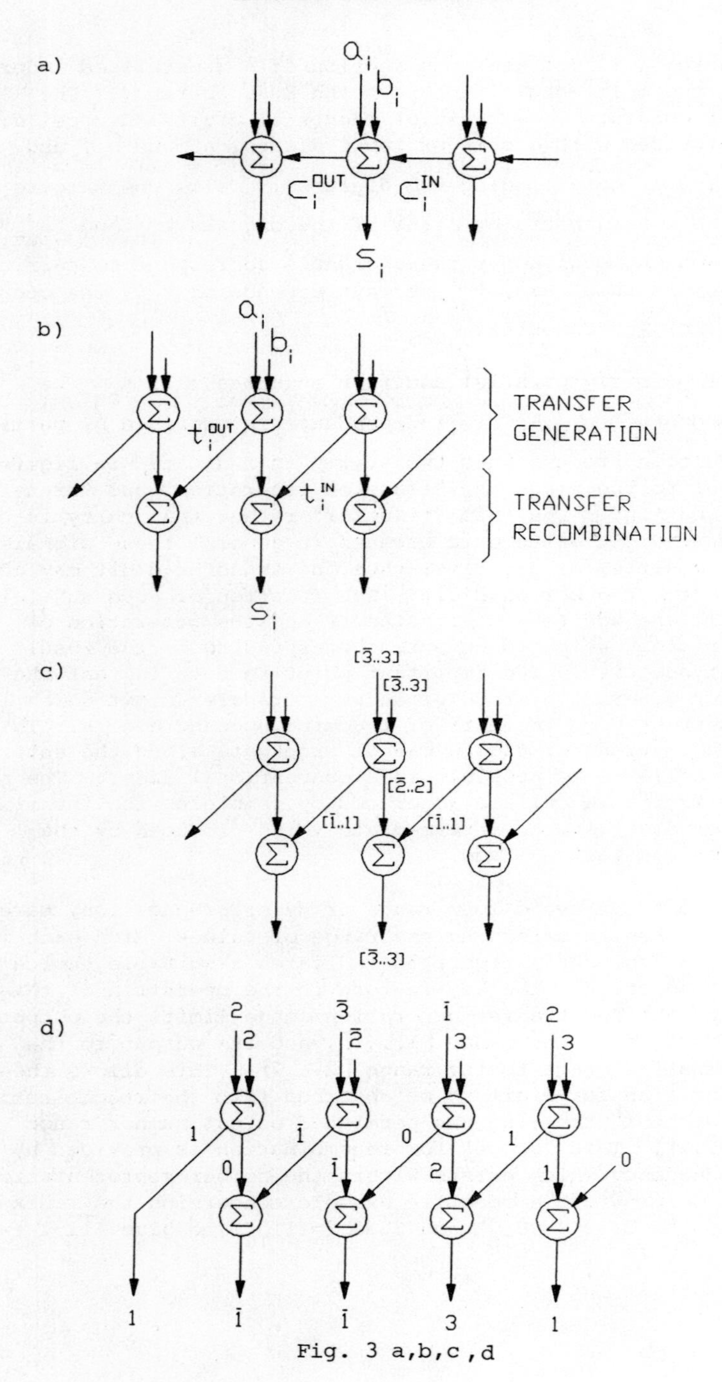
a)
a_i
b_i
c_i^{OUT}
c_i^{IN}
s_i
b)
a_i
b_i
t_i^{OUT}
t_i^{IN}
TRANSFER GENERATION
TRANSFER RECOMBINATION
s_i
c)
[3̄..3]
[3̄..3]
[2̄..2]
[1̄..1]
[1̄..1]
[3̄..3]
d)
2
2
3̄
2̄
1̄
3
2
3
1
1̄
0
1
0
0
1̄
2
1
1
1̄
1̄
3
1

Fig. 3 a,b,c ,d

We have chosen to illustrate the case of radix-4 addition in order to demonstrate the generality of these ideas and because the radix-2 case is somewhat exceptional. In radix-2 there is only one possible redundant digit set, $\{\bar{1}, 0, 1\}$, corresponding to the signed-binary number representation (SBNR). In this case, the low radix value does not provide enough scope to perform the transfer recombination in a single stage. A similar situation occurs for higher radices if digit sets with limited redundancy are used. A modified adder structure is required which has three levels and allows transfers to propagate across two digit positions rather than one. One section of a limited-carry-propagation radix-2 adder is shown in figure 4, the transfers in this case being single binary digits. Despite this additional complication, the SBNR is of great practical interest because, as with conventional binary arithmetic, it leads to very compact hardware realizations.

Finally, it is worth pointing out that the conversion from, a conventional non-redundant number system to a signed-digit representation can be performed in a manner which avoids long range transfers, but that conversion in the opposite direction does require full transfer propagation. Thus it is not worthwhile converting into a redundant number system and back again to perform a single arithmetic operation. However the recursive filtering application requires many consecutive, identical calculations to be performed on the data which is recirculated. The reconversion (with full carry propagation) need only be performed outside the recursive loop, where the latency which results from pipelining this operation incurs no penalty in terms of throughput rate.

5. AN ARCHITECTURE FOR FAST RECURSIVE FILTERING

An alternative multiplier which uses the SBNR and is suitable for application to the IIR filter problem is illustrated schematically in Fig. 5. for ease of comparison, it is represented in a similar form to figure 2. We have assumed that all signal values are fractional, denoting the most significant digit (msd) by superscript $\bar{1}$, the next most significant digit by superscript $\bar{2}$ and so on. Note that the order of summation of the partial products has been reversed so that the calculation effectively proceeds msd first. Thus each row is now right-shifted by one cell with respect to the row above. The reason for re-arranging the order of summation is to allow the msds of the result to emerge from the array with the minimum possible latency. To this end, the most significant partial products from each row are tapped off the main array and summed in the row of diamond-shaped adder cells at the left hand edge. Note that if such a circuit were constructed using conventional binary arithmetic it would be necessary for carries to propagate from the least significant to the most significant stages of the adder chain at the left hand edge of the array. Any propagation of carries in this direction would however, violate the cut theorem and render invalid the pipelining which is shown in figure 5. The circuit has not been pipelined in the horizontal direction since this would introduce additional latency into the circuit and, given that the range of carry propagation within each row has been restricted to two cells, the extra pipelining would achieve very little in terms of throughput rate.

The architecture is shown in more detail in figure 6 which includes an additive input at the top and depicts the feedback connections required for a first order IIR filter. The sub-cells labelled A, B and C are identical to those required for the radix-2 redundant adder shown in figure 4. The input q and output y both have an msd-first, skew-parallel format with a single clock cycle delay between consecutive digits of the same word. Note that the latency (only two clock cycles in this case) does not depend on the data or coefficient wordlength. Accordingly, the msds of the output y_{n-1} may be fed back to the input q_n to yield a first order recursive filter with a throughput of one sample every two clock cycles, whatever the wordlength.

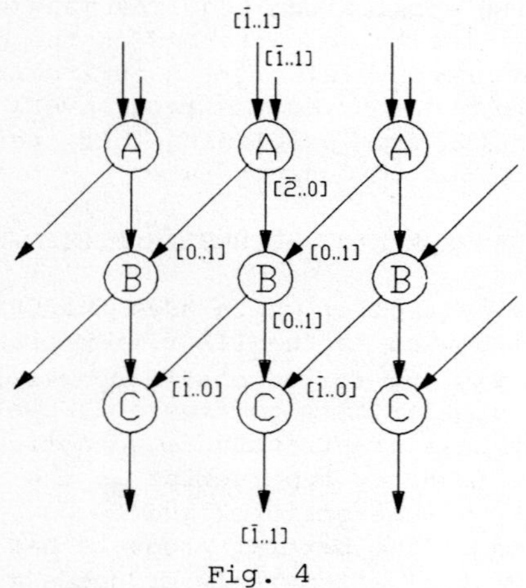

Fig. 4

The circuit architecture in figure 6 is clearly based on the 3-level SBNR adder of figure 4. The only additional sub-cell required is a signed-binary digit multiplier, which has similar complexity to the adder subcells. Note that in comparison to figure 5, some cells at the bottom right corner of the circuit in figure 6 have been omitted. This is a consequence of the msd-first operation of the circuit in that these cells would only contribute to the least significant output bits which are subsequently truncated to prevent wordgrowth in the recursion.

6. DISCUSSION

The circuit presented in this paper is described in more detail in reference [3] where the application to higher order filters is also discussed. References [4] and [5] describe significantly improved architectures which have been designed using the same basic philosophy but can achieve reduced hardware complexity and increased throughput. Comparing the latched SBNR multiply and add cell used here with an equivalent latched, binary, gated-full-adder indicates that the redundant cell is ~35% slower and carries a silicon area overhead of ~65%. However, due to the truncation of some lsd circuitry, the circuit in figures 5 and 6 requires fewer cells than an equivalent binary array and so the overall increase in silicon area is typically ~30%. References [4] and [5] show how the speed and area penalties may be completely eliminated, resulting in a very simple structure with identical harware complexity to that of a conventional binary multiplier. Thus the improvement in throughput offered by a redundant number system may be achieved at no cost.

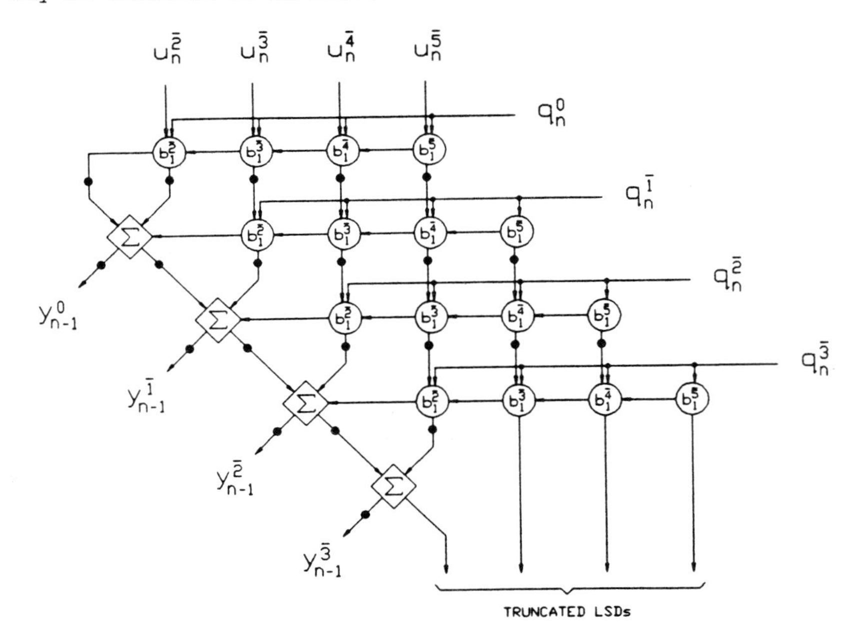

Fig. 5

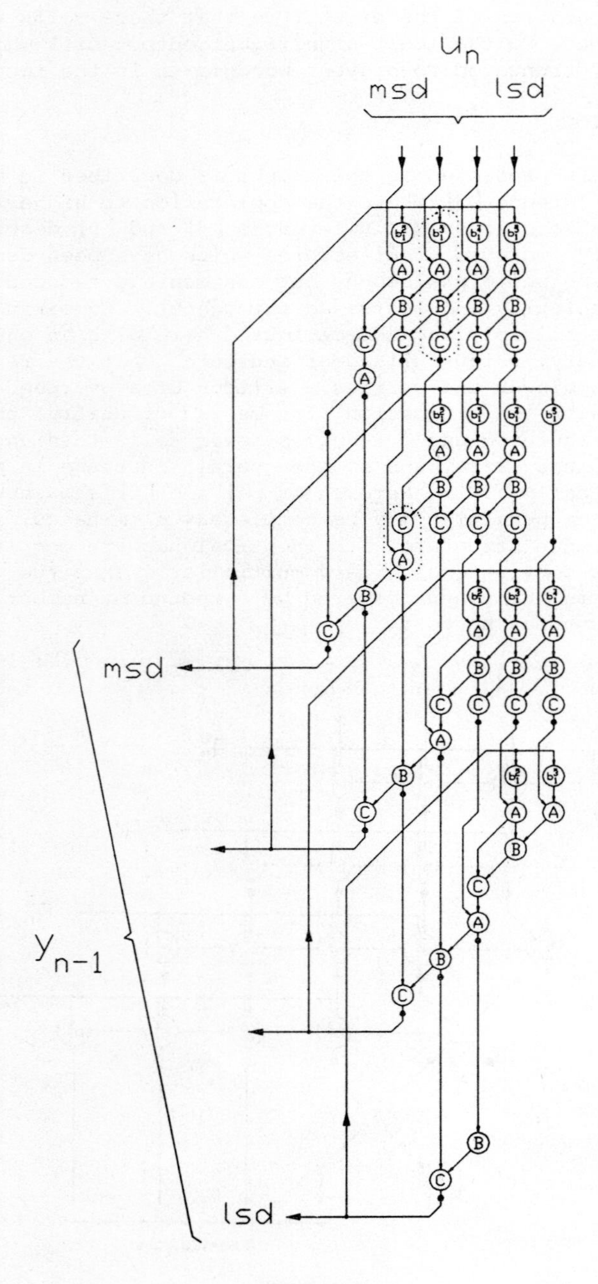

Fig. 6

The circuit shown in figure 6 is only 50% efficient in the sense that it produces one output every two clock cycles. 100% efficiency can be achieved by "iterating" the recursion once so that y_n depends on y_{n-2} rather than y_{n-1}. Thus

$$y_n = b_1^2 y_{n-2} + b_1 y_{n-1} + u_n$$

The array is now capable of producing a result on every clock cycle, but an additional multiply and add circuit is required to compute the non-recursive term $b_1 u_{n-1}$.

In section 4, we referred briefly to the reconversion of signed-binary numbers to conventional binary form. Several simple techniques are known (see for example [1]) most of which required full carry propagation. However, as stated previously, this procedure should only be applied outside the feedback loop where it can be pipelined in order to maximise the throughput rate of the filter without violating the cut theorem. Reference [5] describes an alternative technique based on conditional selection rather than carry propagation. This allows the format conversion to be achieved simultaneously with de-skewing of the output words.

A considerable body of work in the recent past has been devoted to the development of "on-line" computation, which exploits redundant arithmetic to reduce the latency of digit-serial computations (see for example [6]). The problem of latency in recursive systems such as IIR filters is much less severe for digit-serial systems because a pipeline delay of many clock cycles may only correspond to one or two word delays. In this and related papers [3, 4, 5] we have proposed an msd first skew-parallel data format which enables principles similar to those of on-line computation to be applied efficiently to digit-parallel circuits, with consequent advantages in terms of throughput rate.

In the interest of clarity, the filter circuit described in this paper employs simple truncation of the lsds as required to prevent word growth at each stage of the recursion. This has subtly different implications for signed-digit numbers than for conventional binary numbers because the part which is discarded may have different sign from the part which is retained. If this is the case then truncation will increase the magnitude of the number. In a recursive application, rounding towards zero would generally be preferable from a stability viewpoint. Modification to force such behaviour in SBNR circuits with a skew-parallel output format, involves examining each output digit as it is produced, until the first non-zero digit is detected. This digit carries the overall sign of the output word. Subsequently, when the final significant output digit is produced this sign is used in combination with the overall sign of the part to be discarded in order to determine the polarity of the rounding correction. If the sign of the part to be discarded is not clear from its first digit (because that digit is zero) then the safely pessimistic assumption that it has opposite sign to the retained part avoids having to look at lower significance digits, at the expense of a slightly increased rounding error. Such a scheme requires very little hardware and does not significantly affect the performance of the circuit.

7. REFERENCES

[1] Avizienis, A., (1961) Signed-Digit Number Representations for Fast Parallel Arithmetic, IRE Trans. on Electronic Computers, Vol. EC-10, pp. 389-400.

[2] Kung, H.T. and Lam, M.S., (1983) Fault-Tolerant VLSI Systolic Arrays and Two-Level Pipelining, Proc. SPIE, Vol. 431 (Real Time Signal Processing VI), pp. 143-158.

[3] Knowles, S.C., Woods, R.F., McWhirter, J.G. and McCanny, J.V., (1988) Bit-Level Systolic Arrays for IIR Filtering, Proc. IEEE Int. Conf. on Systolic Arrays, San Diego, pp. 653-663.

[4] Knowles, S.C., McWhirter, J.G., Woods, R.F. and McCanny, J.V., A Bit-Level Systolic Architecture for Very High Performance IIR Filters, to be published in Proc. IEEE Int. Conf. on Acoustics, Speech and Signal Processing, Glasgow.

[5] Knowles, S.C. and McWhirter, J.G., (1989) An Improved Bit-Level Systolic Architecture for IIR Filtering, to be published in Proc. IEEE Int. Conf. on Systolic Arrays, Killarney.

[6] Ercegovac, M.D. and Lang, T., (1988) On-Line Arithmetic: A Design Methodology and Applications in Digital Signal Processing, VLSI Signal Processing III, IEEE Press.

P-ADIC TRANSFORMS IN DIGITAL SIGNAL PROCESSING

R.N. Gorgui-Naguib
(Department of Electrical and Electronic Engineering, University of Newcastle upon Tyne)

ABSTRACT

This paper describes a new set of orthogonal transforms which possess the convolution property and which are suited for error-free signal processing computations. This involves the study of a relatively new type of coding called Hensel codes. The presentation of a variable-length p-adic number representation based on properties of periodicity is given. This is followed by the detailed presentation of p-adic transforms in their general form and also with their Mersenne and Fermat variations. Finally, complex p-adic transforms are briefly described.

1. INTRODUCTION

The operation performed by a linear filter on an input signal, x, is to convolve it with a quantity, h, known as the impulse response of the filter. This holds for a continuous-time signal where the output is:

$$y(t) = \int_{-\infty}^{\infty} x(t-u)\, h(u)\, du$$

or for a discrete-time signal,

$$y(nT) = \sum_{k=0}^{N-1} x[(n-k)T]\, h(kT)$$

where T is the interval between sampling instants.

The software implementation of such cyclic convolutions can be time consuming. To solve this problem, transforms having special properties have been used; the most common being the

Fourier transform. In general, such transforms have the following properties [9]:

a) they are orthogonal, i.e., the multiplication of the transform by its inverse is, formally, the identity, and,

b) they possess the convolution property:

$$T[x*y] = T[x] \ . \ T[y]$$

where * denotes the convolution operation.

These transforms, however, suffer from many drawbacks when used with discrete signals. For instance, implementing a DFT on quantised data introduces some rounding errors due to finite wordlength. The net effect is a decrease in the signal-to-noise ratio. New transforms have been proposed which solve this problem, the most widely used being the Number Theoretic Transform (NTT). The NTT involves some modular arithmetic modulo a given prime number (usually a Fermat number) thus permitting the execution of an exact integer transform. It is also possible to extend the NTT to rationals by scaling them to integers (through multiplying them by their least common multiplier). However this operation is long and tedious and may involve the usage of high magnitude numbers occupying a large memory space. As a direct consequence, it is possible indeed to lose the ability of implementing an otherwise fast and simple transformation on microprocessors.

In this paper we shall explain how to derive a new exact transform on certain rational numbers. For that, we shall study a relatively new type of coding called Hensel codes. An explanation of the methods of performing basic arithmetic operations on Hensel codes will be presented, together with some of the possible problems affecting the closure of such operations.

The various operations in the transform domain will rely on p-adic number theoretic notions. These will be investigated and the presentation of a variable-length p-adic number representation based on properties of periodicity will be given. This is an extension of the conventional Hensel coding scheme, with the result that this new representation will allow the performance of closed arithmetic operations in the p-adic number domain.

This will be followed by the detailed presentation of p-adic transforms on 1-dimension sequences, which could be systematically extended to include multidimensional applications. The analysis given here consists of presenting the Mersenne and Fermat p-adic transforms along with their complex variations.

2. THE p-ADIC FIELD $\underset{\sim}{Q}_p$ AND THE SEGMENTED p-ADIC FIELD $\hat{\underset{\sim}{Q}}_p$

The concept of p-adic field was originated by Hensel [7,8]. Hensel states that every p-adic number can be uniquely represented by an infinite series of the form:

$$\alpha = \sum_{n=-m}^{\infty} a_n p^n$$

where $a_n \in I_p$, p being a prime, and $m \in Z$, the set of integers. This infinite series converges to α with respect to the p-adic norm [2].

However, in order to perform arithmetic using p-adic numbers, a finite segment p-adic number has to be defined. A finite segment p-adic field $\hat{\underset{\sim}{Q}}_p$ is thus introduced [5, 10] by truncating the infinite p-adic expansion series of a p-adic number to a fixed number of digits, r. This representation of a finite segment p-adic expansion of a p-adic number α is also known as a Hensel code and is denoted by $H(p,r,\alpha)$. The conditions for the construction of such codes are [3]:

1. For the rational number $\alpha = a/b$ to be represented, the numerator and denominator have the prescribed bounds:

$$-[(p^r-1)/2]^{1/2} \leq a,b \leq [(p^r-1)/2]^{1/2}.$$

2. The p-adic expansions are terminated at the right such that r is even. Thus, any rational number α within the prescribed range, given above, can be represented by a Hensel code such that:

$$H(p,r,\alpha) = \sum_{i=-m}^{r} a_i p^i, \text{ where } a_i \in I_p.$$

The four basic arithmetic operations are valid in $\hat{\underset{\sim}{Q}}_p$ provided that no overflow occurs, even within the computation. Consider two rational numbers α and β. Then

$$H(p\ r,\alpha \odot \beta) = H(p,r,\alpha) \odot H(p,r,\beta) \tag{2.1}$$

where $\odot$ represents an arithmetic operation.

Expression (2.1) is valid provided that the absolute values of the numerator and denominator of α, β and $\alpha \odot \beta$ do not exceed $[(p^r-1)/2]^{1/2}$. The arithmetic operations in $\hat{\underset{\sim}{Q}}_p$ are almost identical to the p-ary arithmetic, since it is essentially modulo p^r arithmetic realised as a simple recursion of modulo p operations. Also, p-adic arithmetic operations are error-free. For a detailed presentation of p-adic arithmetic operations and algorithms, the reader is referred to [3].

From our discussion of Hensel codes it becomes apparent that the use of finite p-adic number systems confined to a fixed length r involves a constant overflow trace and a constant analysis of the structure of the operands prior to, and within, any finite p-adic operations. For this reason, the use of variable-length p-adic number systems for digital signal processing implementation can be considered.

In this context, it has been shown [4] that the canonical p-adic expansion of a rational number is periodic (or recurring). An algorithmic approach to computing the period, λ, has been given [3] together with an algorithm for the determination of the aperiodic elements and the elements comprised in one recurring cycle. It is sufficient to say here that, if $\alpha = a/b$ where $(a,b) = 1$, then α can be written in the form:

$$\alpha = a \,/\, (d.p^k)$$

where $p \nmid d$ and $k \in Z$. Therefore, the period λ must satisfy the following relation:

$$p^\lambda = 1 \pmod d \tag{2.2}$$

It is then clear that it is indeed possible to obtain a deterministic variable-length representation of an otherwise infinite p-adic expansion. This representation may suffer from the need of larger storage capability than it is required for a Hensel code. However, it provides the ability to perform error-free arithmetic in the p-adic domain.

3. THE GENERAL p-ADIC TRANSFORM

In this section a number-theoretic-like transform is defined in the segmented p-adic field $\hat{Q}_p$. The p-adic transform (PAT) of an N-point sequence x(n), represented by the corresponding Hensel codes H(p,r,x(n)) for a given prime p and even length r, is defined by [9, 12]:

$$H(p,r,X(k)) = \sum_{n=0}^{N-1} H(p,r,x(n)) \, \{H(p,r,\gamma)\}^{nk} \tag{3.1}$$

for $0 \leq k \leq N-1$

where $X(k)$ is the transformed sequence and $H(p,r,\gamma)$ is the Hensel equivalent of the Nth root of unity in $\hat{Q}_p$. The inverse transform is given by:

$$H(p,r,x(j)) = \sum_{k=0}^{N-1} H(p,r,X(k)) \{H(p,r,\gamma)\}^{-jk} \qquad (3.2)$$

for $0 \leq j \leq N-1$

To derive the conditions of orthogonality (or existence of the inverse transform), we substitute equation (3.1) into (3.2). Hence,

$$H(p,r,x(j)) = H(p,r,1/N) \sum_{k=0}^{N-1} \sum_{n=0}^{N-1} H(p,r,x(n)) \{H(p,r,\gamma)\}^{k(n-j)} \qquad (3.3)$$

and then, reordering, we obtain

$$H(p,r,x(j)) = H(p,r,1/N) \sum_{n=0}^{N-1} H(p,r,x(n)) \sum_{k=0}^{N-1} \{H(p,r,\gamma)\}^{k(n-j)} \qquad (3.4)$$

Let

$$H(p,r,s) = H(p,r,1/N) \sum_{k=0}^{N-1} \{H(p,r,\gamma)\}^{k(n-j)} \qquad (3.5)$$

We thus get

$$H(p,r,s) = H(p,r,1) \quad \text{if } (n-j) \equiv 0 \pmod N \qquad (3.6)$$

Similarly,

$$\{H(p,r,\gamma)\}^{N(n-j)} = H(p,r,1) \quad \text{in } \hat{Q}_p \qquad (3.7)$$

Alternatively, if $(n-j) \not\equiv 0 \pmod N$, then

$$\{H(p,r,\gamma)\}^{(n-j)} \neq 1$$

thus

$$\{H(p,r,\gamma)\}^{(n-j)} - 1 \neq 0 \qquad (3.8)$$

Multiplying $H(p,r,s)$ by expression (3.8), we get

$$H(p,r,s)\ [\{H(p,r,\gamma)\}^{(n-j)}-1] = H(p,r,1/N)\ [(H(p,r,\gamma)\}^{(n-j)}-1]\ \sum_{k=0}^{N-1}\{H(p,r,\gamma\}^{k(n-j)}$$

$$= H(p,r,1/N)\ [\{H(p,r,\gamma\}^{N(n-j)}-1]$$

$$= 0 \tag{3.9}$$

and since $\{H(p,r,\gamma)\}^{(n-j)} - 1 \neq 0$, then equation (3.9) implies that

$$H(p\ r,s,) = 0 \tag{3.10}$$

In conclusion

$$H(p,r,s) = \begin{cases} H(p,r,1) & \text{if } (n-j) \equiv 0 \pmod N \\ 0 & \text{if } (n-j) \not\equiv 0 \pmod N \end{cases} \tag{3.11}$$

Thus the necessary and sufficient conditions for the PAT to have a DFT structure, or the properties of cyclic convolution are:

1. $H(p,r,\gamma)$ is a root of unity of order N in the field $\hat{Q}_p$:

$$\{H(p,r,\gamma)\}^N = H(p,r,1) \tag{3.12}$$

2. $H(p,r,1/N)$ should exist, or be representable, in $\hat{Q}_p$

3 $\{H(p,r,\gamma)\}^{-1}$ should be representable

4. THE EXISTENCE AND DERIVATION OF $H(p,r,\gamma)$ IN $\hat{Q}_p$

We have shown in the previous section that one of the conditions for the existence of the PAT is that a root of unity, $H(p,r,\gamma)$, of order N exists in $\hat{Q}_p$. Bachman [2] showed that the equation $x^{p-1} = 0$ has exactly p-1 distinct roots in $\hat{Q}_p$. Thus, the maximum transform length which can be used is

$$N_{max} = p - 1 \tag{4.1}$$

which is determined by the chosen prime p [6,13]. Moreover, the root of unity of order N_{max} is called the primitive p-adic root $H(p,r,\gamma_\phi)$; the powers of this primitive root will generate all roots of unity in the field $\hat{Q}_p$. Table 1 gives the roots of unity corresponding to different primes p, with r = 4 [9, 12].

p	r	H(p,r,γ)
5	4	(3,3,2,3)
7	4	(3,4,6,3)
17	4	(3,13,2,3)
257	4	(3,127,178,130)

Table 1 Roots of unity $H(p,r,\gamma)$ in $\hat{Q}_p$ for given p and r

The number-theoretic properties are satisifed, i.e. the number of primitive roots is given by $\phi(\phi(p))$, the number of roots of order N is given by $\phi(N)$ and the Nth root γ_N can be obtained from the primitive root γ_ϕ as given by

$$\gamma_N = \gamma_\phi^{(p-1)/N} \tag{4.2}$$

In order to find $H(p,r,\gamma_\phi)$ we have to find the solution for

$$\{H(p,r,x)\}^{p-1} = H(p,r,1) \quad \text{in } \hat{Q}_p \tag{4.3}$$

or, in general,

$$\{H(p,r,x)\}^{N} = H(p,r,1) \quad \text{in } \hat{Q}_p$$

5. MERSENNE AND FERMAT p-ADIC TRANSFORMS

The basic arithmetic operations in a segmented p-adic field $\hat{Q}_p$ are performed modulo p where p is a prime number. However, operations mod p are very costly; thus Mersenne and Fermat primes are usually chosen in order to reduce the complexity of the processor's structure [11].

5.1 Mersenne p-adic Transform

Consider the segmented p-adic field $\hat{Q}_p$ with p a Mersenne prime, $p = 2^q-1$, where q is a prime number. Thus, the PAT in this field has the maximum length

$$N_{max} = p - 1 = 2^q - 2 = 2(2^{q-1} - 1) \tag{5.1}$$

which is not highly composite. Thus a simple radix 2 FFT algorithm cannot be used. However, more complex and efficient techniques, such as prime factorisation or Winograd algorithms, can be used. In these algorithms, the transform length is put into prime factorisation form. Then, by using the Chinese Remainder Theorem, this 1-D PAT is mapped into a multidimensional transform such that it is cyclic, with prime transform length in every dimension [15]. This multidimensional PAT is then implemented by the Rader algorithm [14], which is a conversion of prime length p_i transforms into circular convolutions of p_i-1 points which, in turn, are implemented by FFT algorithms, short convolutions, or polynomial product algorithms [1].

5.2 Fermat p-adic Transform

Unlike the Mersenne p-adic transform, the Fermat p-adic transform has a transform length which is highly composite; thus a radix-2 FFT algorithm can be employed. This is due to the fact that, given a Fermat prime $F_t = 2^{2^t} + 1$ for t = 1,2,3,4, the maximum transform length Nmax is given by

$$N_{max} = F_t - 1 = 2^{2^t} \tag{5.2}$$

and consequently the Fermat p-adic transform can be directly implemented by FFT algorithms.

6. COMPLEX p-ADIC TRANSFORM

Consider a complex p-adic sequence $H(p,r,u_n)$ to be filtered by a complex sequence having N terms $H(p,r,b_n)$, in which $H(p,r,y_m)$ is the output p-adic sequence given by

$$H(p,r,y_m) = \sum_{n=0}^{N-1} H(p,r,b_n)\, H(p,r,u_{m-n}) \tag{6.1}$$

$$\text{for } 0 \le n,m \le N-1$$

where

$$H(p,r,u_n) = H(p,r,x_n) + j\,H(p,r,\hat{x}_n)$$
$$H(p,r,b_n) = H(p,r,h_n) + j\,H(p,r,\hat{h}_n) \qquad (6.2)$$
$$H(p,r,y_m) = H(p,r,Z_m) + j\,H(p,r,\hat{Z}_m)$$

and $j = \sqrt{-1}$.

The complex convolution (6.1) can be decomposed into four real convolutions as follows:

$$H(p,r,y_m) = \sum_{n=0}^{N-1} [H(p,r,h_n)\,H(p,r,x_{m-n}) - H(p,r,\hat{h}_n)\,H(p,r,\hat{x}_{m-n})]$$
$$+ j \sum_{n=0}^{N-1} [H(p,r,\hat{h}_n)\,H(p,r,x_{m-n}) + H(p,r,h_n)\,H(p,r,\hat{x}_{m-n})] \qquad (6.3)$$

Thus

$$H(p,r,Z_m) = \sum_{n=0}^{N-1} [H(p,r,h_n)\,H(p,r,x_{m-n}) - H(p,r,\hat{h}_n)\,H(p,r,\hat{x}_{m-n})] \qquad (6.4)$$

and

$$H(p\ r\ \hat{Z}_m) = \sum_{n=0}^{N-1} [H(p,r,\hat{h}_n)\,H(p,r,x_{m-n}) + H(p,r,h_n)\,H(p,r,\hat{x}_{m-n})] \qquad (6.5)$$

It is seen that expression (6.3) consists of four real convolutions which can be implemented separately by a fast Mersenne or Fermat p-adic transform.

CONCLUSION

From the outline given of p-adic number transforms it is seen that, for digitial signal processing purposes, they have very attractive features due to the possibility of undertaking completely error-free computations over a p-adic field. Furthermore, they provide a larger dynamic range and a longer transform length, which makes their use more advantageous than the NTT. Their most attractive feature, however, lies in the ability to map, exactly, the field of real numbers onto the integer-like p-adic field and carry out different operations within this domain prior to the final conversion to the real number equivalent.

p-adic number algorithms are currently being practically investigated as viable tools for implementing circular convolution and as a means of eliminating noise sources in digital filters.

REFERENCES

[1] Agarwal, R.C. and Cooley, J.W., (1977) New Algorithms for Digital Convolution, *IEE Trans. Acoust., Speech, Signal Process.*, ASSP-25, 392-410.

[2] Bachman, G. (1964) Introduction to p-adic Numbers and Valuation Theory, Academic Press.

[3] Gorgui-Naguib, R.N., (1986) p-adic Number Theory and its Applications in a Cryptographic System, PhD. Thesis, Imperial College of Science and Technology, University of London.

[4] Gorgui-Naguib, R.N., (1988) p-adic Number Theory, IEE Proc., Vol. 135, Pt. G, No. 3, 107-115.

[5] Gorgui-Naguib, R.N. and King, R.A., (1986) Comments on Matrix Processors Using p-adic Arithmetic for Exact Linear Computations, *IEEE Trans. Comput.* C-35, 928-930.

[6] Gorgui-Naguib, R.N. and Leboyer, A., (1985) Comment on Determination of p-adic Transform Bases and Lengths, Electron. Lett., Vol. 21, No. 20, 905-906.

[7] Hensel, K., (1908) Theorie der Algebraischen Zahlen, Teubner, Leipzig.

[8] Hensel, K., (1913) Zahlentheorie, Goschen, Berlin and Leipzig.

[9] King, R.A., Ahmadi, M., Gorgui-Naguib R.N., Kwabwe A.S. and Azimi-Sadjadi, M., (1988) Digital Filtering in One and Two Dimensions - Design and Applications, Plenum Press

[10] Krishnamurthy, E.V., (1977) Matrix Processors Using p-adic Arithmetic for Exact Linear Computations, IEEE Trans. Comput. C-26, 633-639.

[11] McClellan, J.H. and Rader, C.M., (1979) Number Theory in Digital Signal Processing, Prentice-Hall Signal Processing Series.

[12] Nasrabadi, N.M., (1984) Orthogonal Transforms and their Applications to Image Coding, PhD. Thesis, Imperial College of Science and Technology, University of London.

[13] Pei, S.-C. and Wu., J.-L., (1985) Determination of p-adic Transform Bases and Lengths, Electron. Lett., Vol. 21, No. 10, 431-432.

[14] Rader, C.M., (1968) Discrete Fourier Transforms when the Number of Data Samples is Prime, Proc. IEEE 56, 1107-1108.

[15] Reed I.S. and Truong, T.K., (1978) Fast Mersenne-Prime Transforms for Digital Filtering, Proc. IEEE 125, 433-440.

ON NON-SEPARATE ERROR-CORRECTING ARITHMETIC CODES

I. K. Proudler
(Royal Signals and Radar Establishment, Worcestershire)

ABSTRACT

It is shown that a non-separate arithmetic error-correcting code that commutes with both addition and multiplication must be an AN code where the generator A is an idempotent element of the ring being used. Given this type of code, its ability to detect errors in arithmetic expressions is explored and shown to be poor, due to error masking in multipliers.

The constraints placed on a non-separate multiplication-preserving arithmetic code that avoids such problems are discussed. The simplest code satisfying these conditions turns out to be an AN+B code where both A and B are idempotent elements. Conditions for the existence of this type of code are given.

1. INTRODUCTION

There has been a lot of interest, in recent years, in the design of fault-tolerant arithmetic circuits. This is primarily due to various side-effects of the increasing component densities to be found in modern integrated circuits (VLSI, WSI etc.) Fault-tolerant arithmetic [6] is one of the many different approaches to the topic of fault-tolerance.

First Diamond [2] and Brown [1] and then others (see [8] p 86) investigated the use of error-correcting AN-codes to detect and correct errors. The majority of this work has been directed towards the protection of the addition operation and the theory is extensive [8]. Multiplication can, of course, be thought of as a series of additions and thus, in theory, be protected by the same techniques. This however means that only one of the operands can be in encoded form, for the other operand is used to control the sequence of additions. Hence errors can only be detected in one of the two operands using this method. More significantly, this approach limits the application of AN codes to individual adder or multiplier circuits. The fact that the encoding and decoding circuitry for arithmetic codes may itself be prone to errors means, however, that such an approach is unlikely to work. Most of the work done to date has tended to ignore this rather fundamental problem.

One solution is to build fault-tolerant encoders and decoders using the techniques of self-checking circuits [11]. Another option would be to use a replicating technique like TMR [5]. A third scheme is to ensure that the encoder and decoder circuitry is relatively small compared to the circuit being protected. In which case, although it is still possible for the encoder and/or decoder to become faulty, in all probability it will be the arithmetic circuits that fail first.

In this paper we consider the latter approach and address the problem of classification of error-detecting codes that can protect large arithmetic expressions involving only addition and multiplication. Such a scheme would be eminently suitable for Digital Signal Processing applications where most of the computation reduces to series of multiplications and additions. In section 2 we derive the form of a non-separate multiplication/addition preserving arithmetic code and show that the error protection ability of such a code is poor due to error masking in multipliers. In section 3 we consider the requirements placed on a non-separate arithmetic error detecting code by the need to avoid error masking in the multiplication process. The types of error to be found in a multiplier are discussed and a set of constraints thus imposed on the code is found. The theoretical properties of the simplest code to satisfy these constraints are studied in Section 4. The results of some computer simulations of the fault tolerance provided by this type of code are discussed briefly in Section 5.

2. MULTIPLICATION/ADDITION PRESERVING ARITHMETIC CODES

A multiplication/addition preserving arithmetic code is a redundant encoding of a finite ring [9] of integers (Z_n say) into some integer ring (Z_m, $m > n$, say) such that the arithmetic structure is preserved. In the appendix (subsection 8.1) we show that this leads to the result that the encoding function, for an integer x, has to be of the form Ax, where A is such that $A^2 = A \text{ MOD } m$ (i.e. an idempotent: see subsection 8.2).

Following standard coding theory [8], we define an "error" to be a non-zero difference between a variable's actual value and the correct one. The term "fault" is used to mean a physical failure which then results in an error, in one or more variables.

The ability of AN codes to detect, and correct, errors in addition is well documented [8], so here we only consider the process of multiplication. If each input to the multiplier consists of a valid codeword and an error (e.g. Ax, e_x respectively) then the output is

$$(Ax + e_x)\ (Ay + e_y) + e_{xy} = A^2xy + A(xe_y + ye_x) + e_xe_y + e_{xy}$$
$$= A(xy + xe_y + ye_x) + e_xe_y + e_{xy} \quad (2.1)$$

where the term e_{xy} represents any error in the multiplication process. The basic principle of error detection using AN codes is that valid codewords are multiples of the generator (A). An error can be detected if it changes a codeword into a value that is not a multiple of A (not a codeword). Consideration of equation (2.1) thus reveals the following facts:

a) faults in the multiplier (e_{xy}) are detectable (provided the associated errors are not multiples of A).

b) if only one input is in error (i.e., one of e_x, e_y is zero) then the error in the other input is masked since it appears as a multiple of A.

c) if both inputs are in error this fact can be detected through the term e_xe_y. Error correction will not be possible because the individual errors (e_x, e_y) cannot be recovered from their product.

It is easy to see that the above coding scheme will mask (the effects of) a single fault at the input to a multiplier. Adders do not mask faults so if this multiplier is part of a network consisting of just additions and multiplications then any single fault occurring before the last multiplier will be masked. Only single faults occurring in this multiplier or in any of the subsequent adders will be detectable. It is usual to assume that a single fault is more probable than many faults. Thus the coding scheme described here is somewhat poor, in general, as it masks nearly all of the most probable faults (single faults).

3. ERROR DETECTING CODES FOR MULTIPLICATION

In section 2 we have showed that the only non-separable code that commutes with both addition and multiplication is somewhat poor because of error masking in the multipliers. Thus it is not possible to have a single code that works well for both adders and multipliers. The AN-codes work well for addition, so it is interesting to ask what form a non-separable multiplication-preserving code must take if it is not to mask errors in this way. In this way it may be possible to either protect a multiplier, which is a more complex circuit than an adder, against errors or construct some sort of dual code scheme for arithmetic circuits.

Suppose the coded form of x is c(x) and consider two values x, y. The product of the two codewords c(x), c(y) has to be the encoded form of the product of the two initial values:

$$c(x)c(y) = c(xy) \ \forall \ x,y. \tag{3.1}$$

If one of the inputs was in error then ideally this ought to be detectable at the output of the multiplier i.e., the output ought not to be a valid codeword. Thus another condition on the code is

$$c(x)v \notin \mathbf{C} \text{ if } v \notin C, \ \ \forall \ x. \tag{3.2}$$

where **C** is the set of codewords. The necessity of this condition may, at first, appear to be founded solely on the requirement that the code be able to check for input errors as well as for errors generated by the multiplier itself. Condition (3.2) can, however, be viewed as representing the ability to detect systematic errors within the multiplier. If a fault occurs in the control circuitry of a 'shift and add' type multiplier then the effect of this fault will be the same as if the relevant input (usually designated the multiplier or coefficient) was in error. Hence if the code does not satisfy condition (3.2), this type of error will be undetectable and the code's error-detecting ability correspondingly weak.

It is relatively easy to see that the AN code does not satisfy condition (3.2). The next, least-complex form of encoding function to try is, perhaps, $c(x) = Ax + B$ (i.e., an AN+B code [8]).

4. AN+B CODES

We use the following notation:

if a divides b we write $a|b$,

the greatest common divisor of two integers a, b is denoted (a,b),

the least positive residue of a modulo b is $((a))_b$.

Consider the usual setting for an arithmetic code where:

the data is taken from the ring Z_n,

and the codewords belong to the ring Z_m, where $n|m$.

Let the encoding function be $c(x) = Ax + B$,

so that $$c(x)c(y) = A^2xy + AB(x+y) + B^2 \qquad (4.1)$$

whereas $$c(xy) = Axy + B \qquad (4.2)$$

In order to satisfy the condition (3.1) we require that A and B satisfy the following equations:

$$A^2 = A, \qquad B^2 = B, \qquad AB = 0 \qquad (4.3)$$

Clearly $A = 1$ is a possibility, but this then implies that $B = 0$ which does not lead to a useful code. This is not the only solution, however, since in Z_m there exists an idempotent element A (i.e., $A^2 = A$) whenever m is factorizable as $m = ab$, where $(a,b) = 1$ (see appendix). In this case it turns out that not only is

$$A = ((a^{-1}))_b a \qquad (4.4)$$

an idempotent but also that

$$B = ((b^{-1}))_a b \qquad (4.5)$$

is one as well. Clearly

$$AB = ((a^{-1}))_b ((b^{-1}))_a ab \equiv 0 \text{ MOD } m, \qquad (4.6)$$

since $ab = m$. So that provided m is factorizable in this fashion we can construct a code that satisfies condition (3.1) of section 3.

The ability of an AN+B code to detect errors, in a particular variable, is exactly the same as that of the AN code ([8] section 4.1). Its ability not to mask errors at the input to a multiplier (condition (3.2)) can be seen as follows. Let e be an error:

$$(Ax+B + e)(Ay+B) = A(x+e)y + B + Be \qquad (4.7)$$

where we have used the facts that A and B are idempotents and that $AB = 0$. Clearly the right hand side of the above equation will not be a codeword provided that

$$Be \not\equiv 0 \text{ MOD } m \qquad (4.8)$$

or, since $m = ab$ and $B \equiv 1$ MOD a,

$$e \not\equiv 0 \text{ MOD } a. \tag{4.9}$$

Hence all detectable input errors (i.e., $e \not\equiv 0 \bmod a$) are still detectable at the multiplier output. See [7] for a list of idempotent AN+B codes that can be derived from the first few, known, useful AN codes.

5. FAULT TOLERANT MULTIPLIER

A fault tolerant multiplier was simulated by computer program, however, due to lack of space it is not possible to fully discuss the results here (see [7] for further details). The simulations showed that this type of code does surprisingly well in detecting errors. Indeed they can detect errors well past their minimum distances with a high degree of success. As a result the fault tolerant multiplier based on this method could detect upwards of 95% of all active faults even if a significant number of components have failed. In comparison to an established technique like DMR, however, the use of an error-detecting code to achieve fault-tolerance was shown not to be cost-effective mainly due to the square-law increase in circuit area.

It is interesting to note that the simulation showed that the detection rate rapidly approaches 100% as the minimum distance increases. Indeed a minimum distance of 3 would appear to be quite adequate.

6. CONCLUSION

We have characterised non-separate codes that preserve the operations of both multiplication and addition. The result is found to be consistent with the previous result about addition preserving codes. A study of the error protection ability of this type of code shows that they are in fact rather bad at detecting errors in arithmetic expressions involving only additions and multiplications.

We have also derived two conditions which an arithmetic code must satisfy in order to be used to protect a multiplier from errors. The simplest code that satisfies these two conditions is shown to be an AN+B code where both A and B are idempotents. The necessary conditions for the existence of these idempotents was given. Such a code was shown to be able to detect all errors that are not a multiple of the idempotent A.

7. REFERENCES

[1] Brown, D.T., (1960) Erro Detecting and Error Correcting Binary Codes for Arithmetic Operations, IRE Trans. Electron. Comput., EC-9, 333-337.

[2] Diamond, J.L., (1955) Checking Codes for Digital Computer, Proc. IRE, 43, 487-488.

[3] Hartley, B. and Hawkes, T.O., (1980) Rings Modules and Linear Algebra, Chapman and Hall, London.

[4] MacWilliams, F.J. and Sloane, N.J.A., (1977) The Theory of Error Correcting Codes, North-Holland.

[5] von Neumann, J., (1956) Probabilistic Logics and the Synthesis of Reliable Organisms form Unreliable Components, Automata Studies, 34, 43-99, Princeton University Press.

[6] Pradhan, D.K., (Ed.), (1986) Fault Tolerant Computing: Theory & Techniques Vols. I & II, Prentice Hall.

[7] Proudler, I.K., (1988) Non-separate Arithmetic Codes, Internal Memo 4215, Royal Signals and Radar Establishment.

[8] Rao, T.R.N., (1974) Error Coding for Arithmetic Processors, Academic Press, New York.

[9] Stone, H.S., (1975) Discrete Mathematical Structures and Their Applications, Science Research Associates, Chicago.

[10] Szabo, N.S. and Tanaka, R.I., (1967) Residue Arithmetic and its Applications to Computer Technology, McGraw-Hill, New York.

[11] Wakerley, J., (1978) Error Detecting Codes Self Checking Circuits and Applications, Elsevier North Holland.

8. APPENDIX

8.1 Multiplication/Addition Preserving Arithmetic Codes

A multiplication/addition preserving arithmetic code is a redundant encoding of a finite set of integers (z_n say) such that their arithmetic structure is preserved. It is usual [8] to consider the arithmetic to be modulo the integer n ($n=2^b$ for two's complement, $n=2^b-1$ for one's complement) in which case the set Z_n can be considered to be a ring [9]. As the encoded integers are to be manipulated by a computer, or similar dedicated hardware, they too will be elements of some integer ring (Z_m, $m>n$, say). The encoding function of the code will thus be a one-to-one and into ring homomorphism (a ring monomorphism).

To prove our main result (theorem 8.1.3) we require two well known results:

Lemma 8.1.1. (see Hartley & Hawkes [3] p 20).

If ϕ: $R \to S$ is a ring homomorphism then

a) $\phi(R)=\{\phi(r)\,|\,r\varepsilon R\}$ is a subring of S.

b) let $\text{Ker}\ \phi = \{r\varepsilon R\,|\,\phi(r)=0_S\}$.

ϕ is a monomorphism iff $\text{Ker}\ \phi = \{0_R\}$. □

Theorem 8.1.2

If R is a subring of Z_n then R is a principal ideal.

Proof:

As R+ is an additive subgroup, for $q,g\varepsilon Z_n$ we have

$$((qg))_n = ((\sum_{i=1}^{q} g))_n \ \varepsilon\ R \qquad (8.1.1)$$

and hence R is an ideal of Z_n.

Let g be the smallest non-zero member of R. For $x\varepsilon R$ let

$$x = qg + r \qquad 0\leq r<g. \qquad (8.1.2)$$

Then $$((r))_n = ((x - qg))_n\ \varepsilon\ R$$

$$\text{since } x,\ ((qg))_n\ \varepsilon\ R. \qquad (8.1.3)$$

Because g is, by definition, the smallest non-zero element of R we must have that r = 0 and hence R is generated by g. □

With the aid of these results we may now prove

Theorem 8.1.3.

If ϕ: $Z_n \to Z_m$, where $m>n$, is a ring monomorphism then $\phi(x) = Ax$ and A is an idempotent in Z_m.

Proof:

As ϕ is a ring homomorphism, by virtue of lemma 8.1.1(a) and theorem 8.1.2, $\phi(Z_n)$ is a subring of Z_m and hence a principal ideal.

Let $a \varepsilon Z_m$ be a generator of the ideal:

i.e. $$\phi(Z_n) = \{ax \mid x \varepsilon Z_m\} \qquad (8.1.4)$$

Now ϕ is actually a monomorphism hence $\phi(0) = 0$ (lemma 8.1.1(b)), thus $\phi(1) = ay$ for some non-zero $y \varepsilon Z_m$. Then as ϕ is addition preserving we have

$$\phi(x) = \phi\left[\sum_{i=1}^{x}\right] = \sum_{i=1}^{x} \phi(1) = x\phi(1) = xay \qquad (8.1.5)$$

i.e. $\phi(x) = Ax$

where $A = ay = \phi(1)$. (8.1.6)

But ϕ is also multiplication preserving

i.e. $\phi(x)\phi(z) = \phi(xz) \qquad \forall\ x, z \varepsilon Z_n,$

or $A^2xz = Axz \qquad \forall\ x, z \varepsilon Z_n.$ (8.1.7)

In particular let $x=z=1$, then

$$A^2 = A.$$

Hence A is an idempotent.

□

By construction, the encoding function for an addition/multiplication preserving arithmetic codes satisfies the conditions of theorem 8.1.3. Thus we have shown that such codes have to be idempotent AN codes. This is not a totally unexpected result as it has long been known ([10] section 14.3) that a non-separate addition preserving code must be an AN code.

8.2 Idempotents

Definition:

Let **R** be a ring, then $e \in \mathbf{R}$ is an idempotent if

$$e^2 = e.$$

Notation:

The (principal) ideal generated by an element j of a ring **R** is $\langle j \rangle$, i.e., $\langle j \rangle = \{rj \mid r \varepsilon \mathbf{R}\}$.

In a ring **R**, the residue class congruent to $r \varepsilon \mathbf{R}$, modulo an ideal **J**, is denoted by $[r]$, i.e., $[r] = \{s | s-r \varepsilon \mathbf{J}\}$.

The ring of residue classes (or quotient ring) of a ring **R** modulo an ideal **J** is denoted by **R/J**.

Theorem 8.2.1

Let **K** be a Euclidean domain. If $j,g \varepsilon \mathbf{K}$ are such that $g|j$ and $(g,j/g) = 1$ then, in the quotient ring $\mathbf{R} = \mathbf{K}/<j>$, the ideal generated by the residue class $[g]$ can be generated by a unique idempotent.

Proof: (cf. [4] ch.8)

Define $h = j/g$. If $(h,g) = 1$ then, by the Euclidean division algorithm, there exists $p, q \varepsilon \mathbf{K}$ such that

$$pg + qh = 1 \qquad (8.2.1)$$

and hence, in **R**,

$$[pg] + [qh] = [1] . \qquad (8.2.2)$$

Consider $[\gamma] \doteq [pg] \varepsilon \mathbf{R}$,

$$[\gamma]([\gamma] + [qh]) = [\gamma]$$

i.e.

$$[\gamma]^2 + [pg][qh] = [\gamma] \qquad (8.2.3)$$

Now

$$[pg][qh] = [pgqh] = [pqgh] = [pqj] \qquad (8.2.4)$$

since multiplication is commutative in a Euclidean domain.

But by definition of an ideal, $pqj \varepsilon <j>$

hence

$$[pqj] = [0]. \qquad (8.2.5)$$

so that

$$[\gamma]^2 = [\gamma] \qquad (8.2.6)$$

and $[\gamma]$ is an idempotent.

Now as

$$[\gamma] = [p][g] \qquad (8.2.7)$$

then

$$<[\gamma]> \subseteq <[g]>, \qquad (8.2.8)$$

but

$$[g][\gamma] = [g]([1] - [qh])$$

$$\doteq [g] - [qj] = [g] \qquad (8.2.9)$$

i.e. $$<[g]> \subseteq <[\gamma]>$$

Thus $$<[g[> = <[\gamma]>, \qquad (8.2.10)$$

and $[\gamma]$ generates the ideal.

The idempotent $[\gamma]$ is unique, for if $[f] \varepsilon \mathbf{R}$ is another idempotent that generates $<[g]>$ then $[\gamma] \varepsilon <[f]>$

i.e. $$[\gamma] = [a][f] \quad \text{for some } [a] \varepsilon \mathbf{R} \qquad (8.2.11)$$

Thus $$[\gamma][f] = [a][f]^2 = [a][f] = [\gamma]. \qquad (8.2.12)$$

Similarly $$[f] \varepsilon <[\gamma]>$$

i.e. $$[f] = [b][\gamma] \quad \text{for some } [b] \varepsilon \mathbf{R} \qquad (8.2.13)$$

thus $$[\gamma][f] = [b][\gamma]^2 = [b][\gamma] = [f] \qquad (8.2.14)$$

Hence $$[\gamma] = [f]$$ □

As an example let $\mathbf{K}$ be the ring of integers, if $j = 65$ the ideal $<j>$ is the set of integer multiples of 65 and the quotient ring $\mathbf{K}/<j>$ is the ring of integers modulo 65 (i.e., Z_{65}). Now consider the ideal of Z_{65} generated by $g = 13$ (i.e., all integers that are multiples of 13 modulo 65):

$$<g> = \{13,26,39,52,65 \equiv 0\}.$$

We have $65/13 = 5$, and as 13 and 5 are relatively prime we can find integers p, q such that $13p + 5q = 1$.

In fact $p = 2$, $q = -5$ and hence $\gamma = 2*13 = 26$, and we have

$$<26> = \{26,52,78 \equiv 13,104 \equiv 39,130 \equiv 0\} = <g>,$$

and $$(26)^2 = 676 \equiv 26 \text{ MOD } 65$$

8.3 Idempotent Generators for AN Codes

An AN code is an ideal in $\mathbf{Z}_m$, generated by A', a divisor of m. Thus provided that $(A',m/A') = 1$, the code can be generated by the idempotent

$$A = ((pA'))_m$$

where $p = ((A'^{-1}))_{m/A'}$.

The condition that A' and m/A' be relatively prime does not hold for all useful AN codes. It is, however, true for a high percentage of published values (see [7]).

GENERALISED NUMBER THEORETIC TRANSFORMS IN DIGITAL SIGNAL PROCESSING

O.R. Hinton
(Department of Electrical and Electronic Engineering, University of Newcastle upon Tyne)

ABSTRACT

This paper describes one method of overcoming the practical shortcomings of the Fermat Number Transform: namely, very restricted transform lengths, and slow arithmetic using diminished-1 representation. The method described uses a Generalised Number Theoretic Transform (GNTT) in which a (non-Fermat) field is chosen together with a root of unity of order 2^n. It is shown that for each power-of-2 transform length up to 4096 it is possible to find sufficient relatively prime fields $M \leq 2^{16}$ to provide a combined dynamic range of $>10^8$ by making use of the Chinese Remainder Theorem. The lookup table based hardware implementation methods described in the paper show that it is feasible to implement a complete NTT butterfly on a single ASIC using current CMOS technology. The estimated computation times are fast: for example, less than 30 μs for a 256 length transform with a dynamic range of 10^8 and using only two butterfly blocks for each of the three moduli required.

1. INTRODUCTION

The Number Theoretic Transform (NTT) has been used to implement a variety of signal processing operations, such as cyclic convolution, error correction, and the computation of other tranforms [4]. The advantages over the Discrete Fourier Transform (DFT) are error-free calculations, and the possibility of reducing computational complexity. This reduction is due partly to the avoidance of complex arithmetic, but more significantly to the choice of parameters used in the transform. The NTT is defined as:

$$X(k) = \sum_{n=o}^{N-1} x(n)\, \alpha^{nk} \bmod M \tag{1.1}$$

where x(n) are the data samples being transformed, and α is the generator of order N, where N is the length of the transform and is equal to the length of the cyclic sequence generated by α in the field M.

A well known choice for these parameters results in the Fermat Number Transform (FNT) in which α is 2, and M is a Fermat number. This reduces all "twiddle factor" multiplications in the transform to simple shift operations, and ensures that N is also a power of 2 thus enabling the use of fast transform methods based on decimation in time and frequency. However, in practice it is found that the FNT has serious drawbacks. Firstly, the generator α=2 is always of low order which implies the fundamental constraint of long wordlengths (large M) and short transform lengths (small N). For example, with $M=2^{32}$ and α=2, we have N=64. Secondly, efficient implementations of arithmetic modulo a large Fermat number invariably use diminished-1 representation with end-around-carry [5], which is intrinsically slower than conventional arithmetic. These disadvantages have generally prevented the widespread use of the FNT in practice.

This paper reports on an investigation into one approach to overcome these disadvantages. This is based upon the use of Generalised NTTs (GNTTs) in which the generator and field are given no constraints other than that they support power of 2 transform lengths and that they be sufficiently small for the implementation of modulo functions using look-up tables. This enables the choice of higher order field generators and allows faster implementation methods to be used. In order to achieve sufficient dynamic range with these smaller fields, it is necessary to use an n-tuple of relatively prime fields, but each supporting the same power of 2 transform length, as shown in figure 1. The final output sample can be constructed from the n-residues by making use of the Chinese Remainder Theorem.

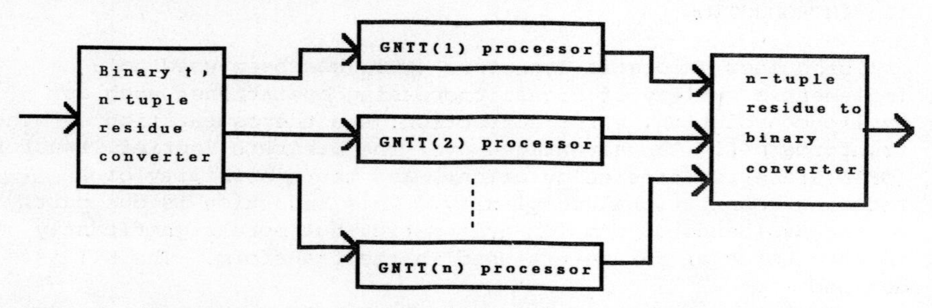

Fig. 1 An n-tuple of Number Theoretic Transform processors.

The feasibility of such an approach depends firstly upon the existence of NTTs with these characteristics, and secondly upon finding efficient look-up table based implementations for arithmetic modulo a variety of unusual numbers. The remainder of the paper reports on results which suggest that these are both possible in present day technology and lead to very high speed throughput systems.

2. DETERMINING SUITABLE NTT PARAMETERS

The NTT is defined in equation (1.1). We are interested in identifying which fields M support power of two transform lengths. This can be determined by reference to the theorem by Agarwal [3] which states that "a NTT of length N, defined modulo M, exists if and only if N divides O(M), where M is composite and given by:

$$M = p_1^{m1} . p_2^{m2} .... p_i^{mi} \tag{2.1}$$

and O(M) denotes the greatest common divisor (gcd) of the (p_{i-1})". For M prime, which in our application turns out to be the case for nearly all values of interest, this reduces to the statement: "a NTT of length N, defined modulo a prime M, exists if and only if N divides (M-1)". The fundamental theorem upon which this is based is that of Euler which states that for every α relatively prime to M:

$$\alpha^{\phi(M)} = 1 \bmod M \tag{2.2}$$

where $\phi(M)$ is the Euler totient function and is equal to (M-1) for M prime.

These theorems provide a means of determining fields that support composite length transforms. In order to implement these transforms, it is also necessary to determine the generator α. The method used requires first a search for the primitive root, which is the value of α in equation (1.1) which results in a generated sequence of length $\phi(M)$ (the maximum possible). For large values of M, checking if a particular α is primitive can become computationally intensive. To avoid this we have made use of the theorem by Pollard [6]: "the element α in $F=GF(M^n)$ generates F*, the multiplicative group of F, if and only if for each prime factor p_i of (M^n-1), with $p_i < (M^n-1)$, we have:

$$\alpha^{((M-1)/p_i)} \neq 1 \tag{2.3}$$

Hence a particular value for α will be a primitive root (α_p)

provided that equation (2.3) is satisfied for all divisors $d_i = (M-1)/p_i$.

Having found a primitive root, it is straightforward to find a generator for the length N required by using:

$$\alpha_N = \alpha_p^{\phi(M)/N} \tag{2.4}$$

The simplest (smallest) generator for N can then be determined from this.

Extensive tables have been generated using these methods, and some of the more useful values are shown in Table 1. For example, from this table it can be seen that it is possible to achieve a transform length of 128 using fields (M) of 257, 641, and 769 with generators (α) of 9, 243, and 554 respectively. The combined dynamic range of this 3-tuple is (257 x 641 x 769), or approximately 1.4 x 10^8. The two rightmost columns in the table give estimation of hardware complexity and are discussed in more detail in the following section.

It can be concluded that power of 2 transform lengths up to 4096 are indeed supported by sufficient relatively prime fields of reasonably small wordlength (≤ 16) to provide a dynamic range of 10^8 or greater.

Modulus M	Primitive root	Root of unity of order							Word length (bits)	Total LUT size (bits x 10^3)
		64	128	256	512	1024	2048	4096		
193	5	11	-	-	-	-	-	-	8	11
257	3	11	9	3	-	-	-	-	9	16
449	3	24	-	-	-	-	-	-	9	28
641	3	2	243	-	-	-	-	-	10	45
769	11	12	554	562	-	-	-	-	10	54
1153	5	43	1096	-	-	-	-	-	11	89
3329	3	56	1915	3061	-	-	-	-	12	280
7681	17	330	3449	2028	7146	-	-	-	13	700
7937	3	151	2458	2805	-	-	-	-	13	720
10753	11	136	5606	4305	4894	-	-	-	14	1054
11777	3	503	8539	4482	7795	-	-	-	14	1154
12289	11	563	12149	8340	3400	10302	1945	1331	14	1204
13313	3	1152	6105	9968	838	10076	-	-	14	1305
18433	5	83	11421	5329	18360	7673	17660	-	15	1940
19457	3	413	15718	4541	19170	17029	-	-	15	2043
40961	3	14529	19734	36043	8603	8091	20237	18088	16	4590
61441	17	22450	12270	18868	54359	421	16290	39003	16	6881

Table 1 Moduli and roots which support power-of-two length NTTs, including lookup table (LUT) sizes for implementation of a butterfly unit in the specified modulus.

3. ARCHITECTURE FOR IMPLEMENTING THE GNTT

Decimation in time and frequency techniques used in formulating the FFT may also be applied to the GNTT. Hence a radix 2 implementation based on the well known pipelined structure takes the form shown in figure 2 in which $\log_2 N$ Butterfly units (BU) are required. This number of BUs is excessive for large N, and so in [2] and [7] it is shown how the pipeline approach may be modified to form a recirculating structure. The number of BUs may then be equal to any value which is a factor of $\log_2 N$ (including 1), with a consequent reduction in the data throughput rate.

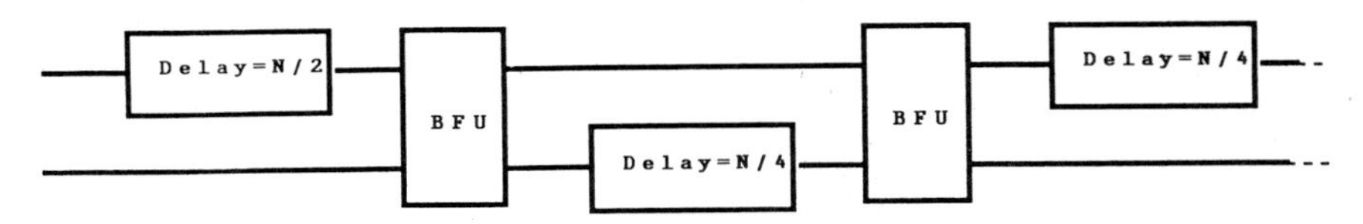

Fig. 2 Pipeline architecture for the implementation of radix-2 transforms.

Figure 3 shows the recirculating structure in which only one BU is used. Consideration of the control for this structure shows that the number of clock cycles required to compute a complete GNTT is given by [7]:

$$C = N/2 \log_2 N + 5 \log_2 N + (N/2 - 1) \tag{3.1}$$

If n butterfly units are used then the number of clock cycles required becomes approximately C/n.

For the architecture shown in figure 3, the main hardware blocks required are: (i) binary to residue, and residue to binary converters; (ii) BU with a programmable sequence of twiddle factors; (iii) programmable commutator for data crossover; (iv) fixed delay line for the recirculating buffer; (v) programmable delay line; and (vi) a control unit. Circuits for (iii), (iv), (v) and (vi) can be implemented using well established designs which may be realised either with discrete MSI components or as ASIC devices. The binary/residue conversion circuits in (i) have been shown to be feasible using techniques such as those described in [1]. It is the BU implementation in (ii) which requires innovation to achieve efficiency and speed.

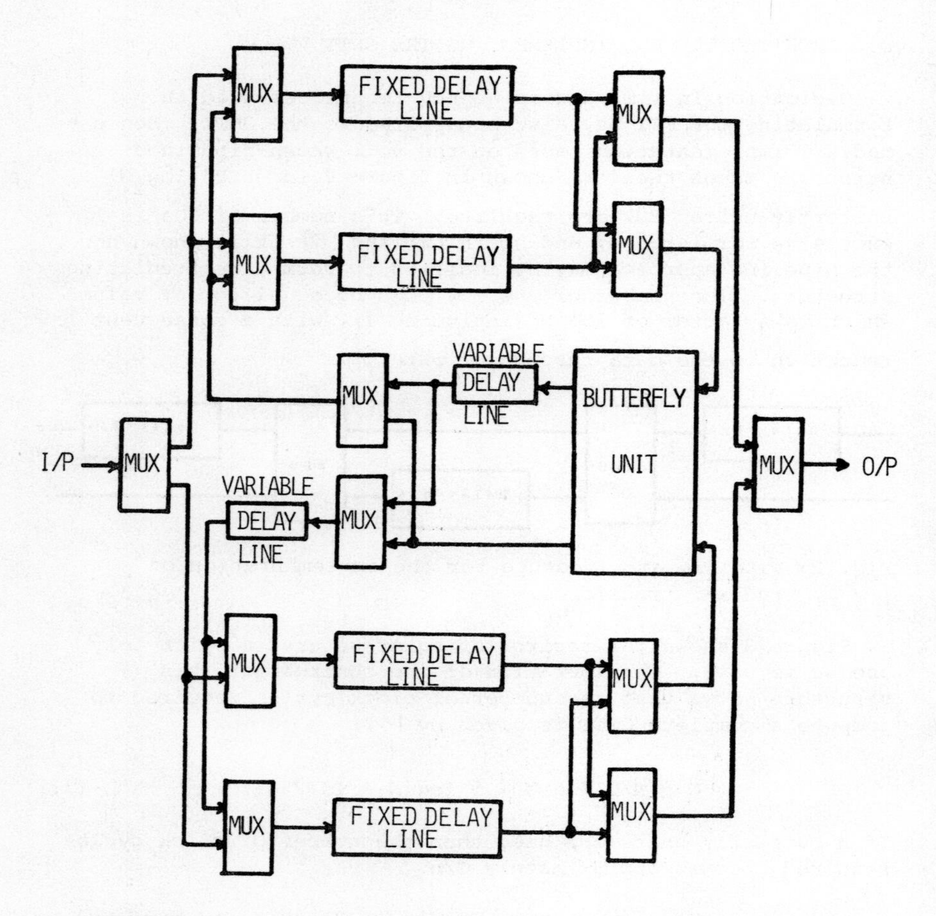

Fig. 3 A pipelined architecture for the implementation of a NTT using only one butterfly unit.

4. HARDWARE IMPLEMENTATION OF BU USING LOOKUP TABLES

The complete BU has as inputs two operands A and B, and for the Decimation in Frequency approach is required to compute two outputs: (A+B) mod M, and $(A-B)\alpha^{c}$ mod M.

4.1 *Modulo Addition*

Modulo addition requires two steps: addition, followed by modulo reduction. The addition step can use a conventional carry lookahead adder, whilst the modulo reduction step can most easily be achieved with a lookup table (LUT). If the modulo used is M, then the required LUT size is $2M \times |1+\log_2 M|$ bits.

For example, with M=769, a LUT of 1538 x 10 is required. The LUT size clearly increases exponentially with wordlength, and an alternative to lookup based modulo reduction for large M is to follow the addition step by a conventional subtraction of M, the result to be selected depending on the sign of the result.

To maximise clocking rates, either of the methods adopted can be divided into two stages of a pipeline separated by latches.

4.2 Modulo Multiplication

The method adopted here uses index mapping to convert a multiplication modulo M to an addition of indices modulo (M-1) [8]. The mapping of the value y to its index x must be of the form:

$$y = g^x \rightarrow x \qquad (4.2.1)$$

where g is the primitive root of the field M. Addition of indices can use a conventional adder as above. A neat solution for modulo reduction and inverse index mapping can be achieved by combining these steps together in a single lookup table. For a modulo M, the LUTs required are:

$M \times |1+\log_2 M|$ bits for index mapping (two required)
$2M \times |1+\log_2 M|$ bits for inverse index mapping

For example, with M=769 two LUTs of 769 x 10 bits, and one of 1538 x 10 bits are required.

4.3 Butterfly Unit (BU)

The complete BU can be realised by combining the addition and multiplication techniques described above. The block diagram of the implementation is shown in figure 4. It should be noticed that the right hand datapath contains 4 latches which form a pipeline of 4 stages; this enables high clocking speeds and hence high data throughput. The approach taken has made it possible to achieve great efficiency in the use of the LUT used. Specifically; LUT(2) carries out both modulo M reduction for the subtraction (A-B), and index mapping in preparation for the subsequent multiplication by α^c; LUT(4) stores the values of α^c as indices, so no further mapping is necessary; and LUT(3) carries out modulo (N-1) reduction and inverse index mapping. The total LUT requirements for a modulo M BU are therefore:

LUT(1)	$2M \times \lvert 1+\log_2 M \rvert$ bits
LUT(2)	$2M \times \lvert 1+\log_2 M \rvert$ bits
LUT(3)	$2M \times \lvert 1+\log_2 M \rvert$ bits
LUT(4)	$M \times \lvert 1+\log_2 M \rvert$ bits
Total	$7M \times \lvert 1+\log_2 M \rvert$ bits

The rightmost column in Table 1 shows the total LUT size in bits required for a BU for each modulus, calculated according to the above "total" formula.

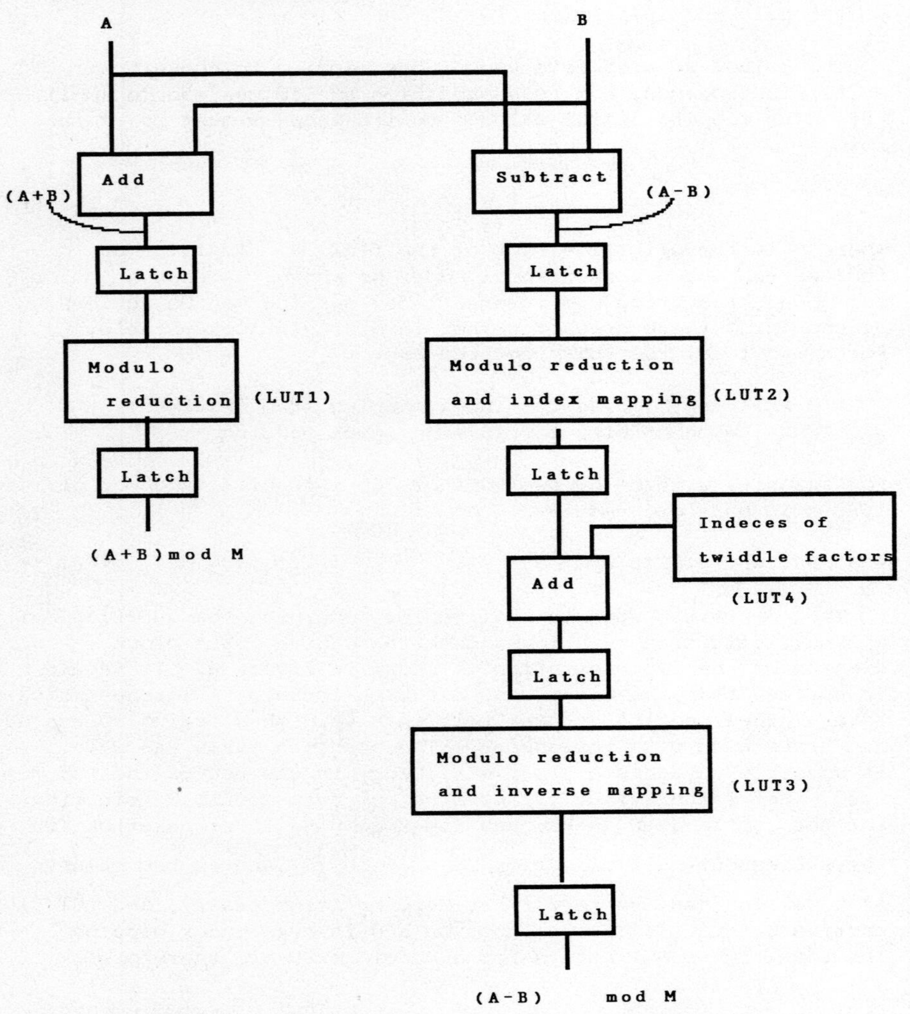

Fig. 4 Block diagram of the implementation of a NTT butterfly unit using lookup tables.

5. HARDWARE REALISATION AND PERFORMANCE

The above techniques have been used as the basis for hardware realisation using both discrete MSI components and VLSI ASIC designs.

Using MSI components a complete BU has been constructed on a single 6in x 9in board that can handle a modulus up to 257. Since all available RAM chips are of size $2^n \times m$, all the required LUT sizes need to be rounded up to the nearest power of 2. This prototype system uses slow memories with 100ns access time and can be clocked at only 5 Mhz so that a 64 point transform takes about 50 μs. Nevertheless, it has served to prove functional correctness.

It is in the realisation using ASICs that the non power-of-two size of the LUTs can be exploited. A number of CAD tools have been developed that calculate the required data values for the LUTs and pass these to the symbolic VLSI editor, VIVID (from the Microelectronics Center of North Carolina) for the automatic generation of masks for a CMOS process. The LUTs are implemented as ROMs using one active device per bit, and precharging of the output bit lines to reduce delays. The CAD tools have been integrated into an entire suite that takes as input the required modulus and transform length, and generates the masks for a complete BU. The floor plan for the BU is fixed in the sense of relative positioning of blocks, although the block sizes, and hence the chip sizes vary according to the modulus in use. The floor plan of a 5-bit pipelined BU is shown in figure 5. Switch level simulation has verified the functionality of these designs, and the feasibility of realising a GNTT butterfly unit as a single ASIC is clearly demonstrated.

Since the whole GNTT architecture is heavily pipelined, the maximum clocking rate is determined only by the access time of the largest lookup table and by the adder delays. Assuming a conservative clock rate of 20 MHz, the time taken to compute various transform lengths using a varying number of BUs is shown in Table 2.

In real systems using NTTs, it will be necessary to also compute the inverse NTT. It can readily be shown that this can be achieved with identical circuits to those for the forward NTT, but using different values in the LUTs.

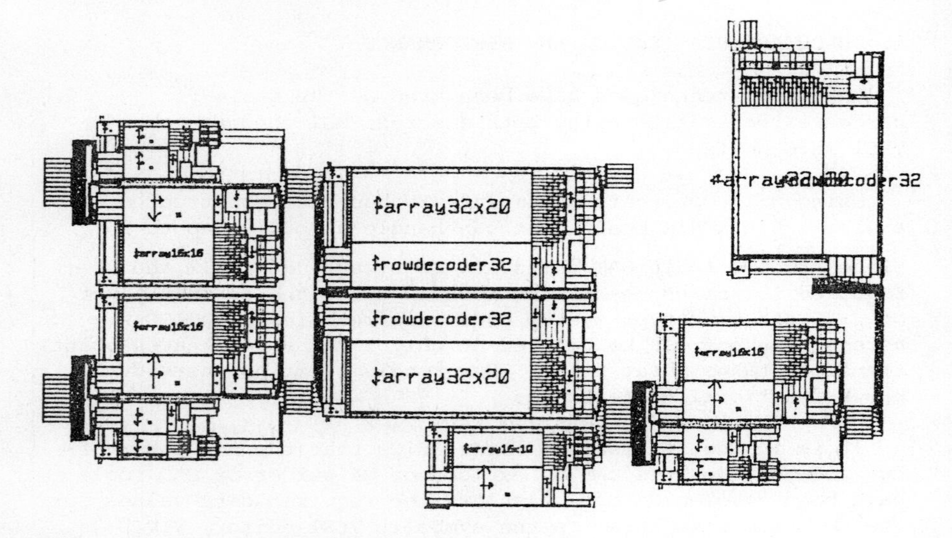

Fig. 5 Floor plan of a 5-bit pipelined NTT butterfly unit using lookup tables.

Transform length N	Number of clock cycles with only one BU	Transform computation time for a clock rate of 20 MHz and number of BUs equal to:				
		1	2	3	4	5
64	253	12 μs	6 μs	4 μs	-	-
128	546	27 μs	-	-	-	-
256	1191	60 μs	30 μs	-	15 μs	-
512	2604	130 μs	-	43 μs	-	-
1024	5681	284 μs	142 μs	-	-	57 μs
2048	12342	617 μs	-	-	-	-
4096	26683	1334 μs	667 μs	445 μs	334 μs	-

Table 2 Estimated evaluation times for a complete NTT using the architecture and ASICs described.

6. CONCLUSIONS

The methods described in this paper provide a means of implementing a generalised number theoretic transform that circumvents the length and computation time problems of the Fermat number transform. These implementations have the following characteristics: power-of-two transform lengths up to 4096; dynamic range greater than 10^8 using a 2 or 3-tuple residue system; high-speed computation with the use of lookup tables; and acceptable ASIC complexity in current VLSI technology. As an example, a 256 point transform could be implemented with 3 parallel NTTs based on moduli of 257, 769 and 3061 providing a dynamic range of 6.10^8; the butterfly units for these three moduli have lookup tables of sizes 16.10^3, 54.10^3 and 280.10^3 respectively; the estimated computation time using only one butterfly unit per modulus is approximately 60 μs assuming a conservative clock rate of 20 MHz; this time could be reduced to 30 μs or 15 μs if the number of butterfly units per modulus were increased to 2 or 4 respectively.

Because of the overhead of the residue/binary converters required, it seems that these methods are most likely to find application in problems involving extensive computation between input and output. An example of such is the evaluation of very long DFTs, or multidimensional DFTs, where the NTT has been shown [4] to be valuable in reducing computational complexity.

REFERENCES

[1] Alia, G. and Martinelli, E., (1984) A VLSI Algorithm for Direct and Reverse Conversion from Weighted Binary Number System to Residue Number System", IEEE Trans on Circuits and Systems, CAS-31, 12, 1033-1039.

[2] Amir-Alikhani, H., Hinton, O.R. and Saleh, R.A., (1986) A VLSI Pipeline Number Theoretic Transform Processor, IEEE Custom Integrated Circuits Conference, 74-77.

[3] Agarwal, R.C. and Burrus, C.S., (1975) Number Theoretic Transforms to Implement Fast Digital Convolution, Proc. IEEE, 63, 550-560.

[4] Hinton, O.R. and Saleh, R.A., (1984) Two-Dimensional Discrete Fourier Transform with Small Multiplicative Complexity Using Number Theoretic Transforms, IEE Proc. Pt. G, 131, 6, 234-236.

[5] Leibowitz, L.M., (1976) A Simplified Binary Arithmetic for Fermat Number Transforms, IEEE Trans. on ASSP, ASSP-24, 5, 356-359.

[6] Pollard, J.M., (1971) The Fast Fourier Transform in a Finite Field, *Math. Comput.* **25**, 114, 356-374.

[7] Saleh, R.A., (1985) Algorithms and Architectures Using the Number Theoretic Transform for Digital Signal Processing, PhD Thesis, University of Kent.

[8] Szabo, N.S. and Tanaka, R.I., (1967) Residue Arithmetic and its Application to Computer Technology, McGraw-Hill, New York.

II INVERSE PROBLEMS AND IMAGE PROCESSING

TOWARDS BAYESIAN IMAGE ANALYSIS

J. Besag
*(Department of Mathematical Sciences, University of Durham)**

ABSTRACT

Many of the tasks encountered in image processing can be considered as problems in statistical inference. In particular, they fit naturally into a subjectivist Bayesian framework. In this paper, we describe the Bayesian approach to image analysis. Numerical examples are not included but can be found among the references and elsewhere. It is argued that the Bayesian approach, still in its infancy, has considerable potential for future development.

1. THE BAYESIAN PARADIGM

Bayesian Image Analysis makes use of explicit probability models to incorporate general and scene-specific prior knowledge into the processing of images and aims to provide a novel unified framework within which many different tasks can be considered. Its beginnings can be found in a short note of Besag (1983) and, much more significantly, in Grenander (1983) and Geman and Geman (1984). In attempting to give an overview of the subject, this paper leans heavily on research at Brown University and at the University of Massachusetts by these last three authors and their co-workers.

We begin with a general introduction to Bayesian Image Analysis. We suppose throughout that a set of records $y = \{y_k : k \varepsilon T\}$ is generated by (stochastic) degradation D of the true pixel image $x^P = \{x_i^P : i \varepsilon S\}$; thus, $D: x^P \rightarrow y$. The finite sets S and T may be identical or only loosely related, as in tomography. T is generally determined by the sensing device but S is usually under the control of the user, though often there is a natural choice based on T. The most obvious task is

* present address: Department of Statistics GN-22, University of Washington, Seattle, WA 98195, U.S.A.

to produce an accurate restoration of x^P from y but commonly other concerns are more important. Thus, we may wish to consider conceptual image attributes that do not contribute directly to the records (Grenander, 1983; Geman and Geman, 1984). For example, satellite images are used to produce crop inventories: here the y_ks are multivariate, with observations in several spectral bands, and the goal is to provide a classification $x^L = \{x_i^L : i \in S\}$ in which x_i^L labels the land usage in pixel i, chosen from a known finite set. Another important class of problems involves segmentation of images into unlabelled homogeneous regions: here we might associate an "edge variable" $x_{ij}^E = 0$ or 1 with each edge site (i,j) between directly adjacent pixels i and j. If $x_{ij}^E = 1$, an edge is deemed present, otherwise absent: thus, the end product is an edge map $x^E = \{x_{ij}^E : i \wedge j \in S\}$, where $i \wedge j$ indicates adjacent pixels. In summary it is often useful to define an underlying image x, consisting of various constituent sub-images; thus, we shall write $x = (x^P, \ldots)$.

Note that, in considering any image attributes, scale can be an important factor. Thus, although it is notationally convenient above to attach a variable to each pixel edge, there may be advantages in restricting edge sites to a coarser array (Geman, Geman, Graffigne and Dong, 1989). Similarly, labels might be attached to small blocks, rather than to individual pixels. Indeed, even when the aim is simple restoration, with the natural choice S = T, it may be preferable to use a coarser or finer array for S, or even vary the scale during processing, as in Gidas (1988). Note also that, although it is conventional to choose a rectangular array for S, this is often no more than a computational convenience, in which case there are grounds for preferring a hexagonal array. In fact, we include the possibility that T and/or S are irregular arrays: for example, in processing geographical information, T may refer to administrative districts.

With the above ground rules, the Bayesian paradigm for image processing can be described as four successive stages.

(a) Construct a prior probability distribution $\{p(x)\}$ for the underlying image x. This distribution should capture general and scene-specific knowledge about x, though in practice we cannot expect to model the global features of the true image, and typical realizations of $\{p(x)\}$ need not resemble reality. We concentrate instead on modelling the local characteristics of x. For example, nearby pixel values tend to be similar; adjacent labels are usually the same; and boundaries around objects are generally continuous. We describe the choice of prior distribution in some detail in Section 2, though there remain many open problems. Note that there will usually be one or two unknown (hyper-) parameters in the prior: we briefly discuss their estimation at the end of Section 4.

(b) Form the joint probability density $\ell(y|x^P)$ for the observed records y given any particular pixel image x ; we use the symbol ℓ because this is the likelihood function of x^P for fixed data y. Whereas the specification of $\{p(x)\}$ is something of an art, that of ℓ is governed by more standard physical considerations concerned with the sensing device etc. Thus, in nuclear medicine, the records are generated by radio-active decay and therefore have independent Poisson distributions with means that can be closely approximated for any given x^P. Note that complications such as blur must be taken into account at this stage but not any inverse transformations. The likelihood function may also depend on unknown parameter values that need to be estimated from the data. Finally, in some applications, such as the interpolation of a partially observed surface, the degradation is deterministic.

(c) Combine the prior density and the likelihood by Bayes theorem to form the posterior density $P(x|y)$ of any image x given y; that is,

$$P(x|y) \propto \ell(y|x^P)\,p(x), \tag{1.1}$$

where the constant of proportionality is a function of y and is not generally required. Mathematically, (1.1) is no more than a use of the simple conditional probability statement,

$$P(A|B) = P(B|A)P(A)/P(B),$$

for events A and B, but philosophically it requires us to combine belief probabilities in the prior distribution with frequency probabilities in the likelihood, using the usual probability axioms: this lies at the heart of all Bayesian methods (for a brief review, see Smith, 1984).

(d) Base any inferences about x on the posterior distribution $\{P(x|y)\}$ of x given y. The exact manner in which this should be done is context dependent and will be discussed at length in Section 4. If we ignore the existence of unknown parameter values, then the most obvious method of point estimation is first to find the maximum a posteriori (MAP) estimate $\hat{x}$ of x and then to extract from $\hat{x}$ the features of prime interest. Of course, such a procedure maximizes

$$\ln \ell(y|x^P)+\ln p(x)$$

and hence resembles more traditional forms of regularization: faith in the data is balanced against the regularities supported by the prior. In fact, we shall later argue against the general use of MAP-based estimates.

A major strength of the Bayesian approach is that it is not concerned merely with point estimates. For example, in restoring a grey-level image x^P, we can attach an interval estimate (Bayesian confidence interval) to each pixel, reflecting the precision of the restoration; in detecting tumours or in prospecting for a mineral deposit, we can ascribe a posterior probability to each element of area or volume associated with x, rather than mere presence or absence; in classifying land usage, we can similarly assign probability vectors, rather than unique labels; and, in estimating image functionals $t = g(x)$, we can attach standard errors to the point estimates. Of course, confidence intervals etc. must not be interpreted too rigorously because of known defects in the prior, but nevertheless they can provide broad guidance.

2. PRIOR DISTRIBUTIONS FOR IMAGES

We describe some current views regarding choices of prior distributions for images but this is an area that requires much additional research.

2.1 Pixel Priors

We assume initally that our sole concern is with a pixel image: thus, $x = x^P$ and we omit superscripts to ease the notation. We also assume for the moment that each x is a continuous variable taking values on the whole real line $\mathbb{R}$, perhaps after suitable (e.g. logarithmic) transformation. Then a useful class of prior distributions, involving only pairwise differences, is defined by

$$p(x) \propto \exp\{-\sum_{i\sim j} \phi(x_i - x_j)\}, x \in \mathbb{R}^n, \tag{2.1}$$

where ϕ is a specified function satisfying

$$\phi(z) = \phi(-z), \phi(z) \text{ increasing with } |z|,$$

$i \sim j$ means i and j are "neighbours", as defined below, and n is the number of pixels. Note that there can be computational advantages if ϕ is convex and differentiable; see Section 4.

It follows from (2.1) that the conditional density of x_i occurring at pixel i, given all other pixel values $x_{S\setminus i}$, is

$$p_i(x_i \mid x_{S\setminus i}) \propto \exp\{-\sum_{j\in\partial i} \phi(x_i - x_j)\}, \quad x_i \in \mathbb{R}, \qquad (2.2)$$

where ∂i denotes the neighbours of pixel i; thus, we can replace $x_{S\setminus i}$ by $x_{\partial i}$ on the left-hand side of (2.2). We now have a means of defining the neighbours of each pixel i, by including pixel j if and only if we believe that its value x_j contributes (substantially) to the conditional distribution of x_i. For a rectangular array, the most common choice of ∂i is the set of pixels that are directly and diagonally adjacent to i and it is then preferable to modify (2.1) and (2.2) so that diagonal interactions are appropriately downweighted: from this viewpoint, hexagonal pixels, with six equally weighted neighbours, have a clear advantage. For irregular arrays of "pixels", it is again natural to choose neighbours according to adjacency, perhaps with different weights according to size, length of common boundary, etc. It is evident from (2.1) that such neighbourhood criteria produce priors that promote locally smooth images. Finally, note that i and j not being neighbours does not imply that x_i and x_j are stochastically independent, only that x_i and x_j are conditionally independent given the neighbouring values of one or both.

Before considering some simple examples, we must face the slight embarrassment that the right-hand side of (2.1) is non-integrable and therefore cannot be normalized: that is, $\{p(x)\}$ is an improper prior. Fortunately, with any plausible choice of ϕ, conditional distributions such as (2.2) are perfectly proper, as is the posterior distribution $\{p(x\mid y)\}$ on which we base our inferences. The impropriety arises because (2.1) only addresses differences and not the overall level of the process, which is arbitrary: it could be easily removed by restricting one or more x_is to any finite interval, or otherwise, but this is hardly worthwhile.

We shall merely note here that it is sometimes preferable to work with discrete versions of continuous pixel priors, in accord with usual grey-level representations: this is easily done. For some self-contained examples of discrete priors and a speculative discussion, see Besag (1986, Section 4.1.3).

2.1.1 Gaussian pixel prior

The simplest choice of Φ is a quadratic, so that

$$p(x) \propto \exp\{-\frac{1}{2\kappa} \sum_{i \sim j} (x_i - x_j)^2\}, x \in \mathbb{R}^n, \tag{2.3}$$

where κ is a positive parameter. This is an intrinsic Gaussian autoregression (see Künsch, 1987) and is "just" improper in that the exponent is negative semi-definite rather than negative definite. It follows that the conditional distribution of x_i, given all other values, is Gaussian with mean $\bar{x}_i$ and variance κ/ν_i, where $\bar{x}_i$ denotes the mean value of the neighbouring x_js and ν_i is the number of neighbours of i: note that, for regular arrays, the conditional variance is increased at the boundary because less neighbouring information is available. Various embellishments of (2.3) can be made but are not considered here. Also, we do not consider the more usual forms of conditional Gaussian autoregression (Besag, 1974), as we prefer to moderate the influence of neighbours, when necessary, in a different way; see Section 2.1.3.

Gaussian priors have appeal if x is known to be a smooth surface, masked by noise, but are unsatisfactory in the presence of real discontinuities, which they will smear out.

2.1.2 Median pixel prior

An alternative to (2.3) is

$$p(x) \propto \exp\{-\frac{1}{\kappa} \sum_{i \sim j} |x_i - x_j|\}, x \in \mathbb{R}^n, \tag{2.4}$$

where κ is a positive scale parameter. Then,

$$p_i(x_i | x_{\partial i}) \propto \exp\{-\frac{1}{\kappa} \sum_{j \in \partial i} |x_i - x_j|\}, x_i \in \mathbb{R}, \tag{2.5}$$

where the constant of proportionality can be evaluated explicitly but is uninstructive. Note that the density (2.5) has its mode at the median of the neighbouring x_js and this makes (2.4) an attractive alternative to (2.3) for surfaces with discontinuities. It can be interpreted as a stochastic version of the median filter that is commonly applied to satellite and other images.

Green (1989) introduces an attractive family of "log-cosh" prior distributions whose range includes (2.3) and (2.4) at its opposite extremes.

2.1.3 Convolution priors

In some applications, a convolution of two or more prior distributions may be appropriate. Consider, for example, the mapping of log relative risks x_i for a rare disease of unknown aetiology, such as childhood leukaemia, over administrative regions i, such as wards. A map based only on the raw incidence rate in each region is uninterpretable because of chance fluctuations in the low counts. Two possible strategies for smoothing the data are: (a) if there is generally close agreement in the rates for adjacent regions, use a prior of the form (2.3) or (2.4); (b) if not, shrink the individual incidence rates towards the overall mean rate. Usually it will be unclear, from the raw data, whether (a), (b), or something intermediate is most appropriate. A family of priors that covers this range is obtained by writing $x = u+v$, where u has a density of the form (2.3) or (2.4), and v is Gaussian white noise with variance λ, say. The values of κ and λ are estimated as part of the analysis; $\lambda/\kappa\downarrow 0$ yields (a), and $\lambda/\kappa\uparrow\infty$ gives (b). This type of convolution has been used successfully in the analysis of agricultural field trials (Besag and Kempton, 1986, Section 3), as well as in epidemiology (Besag and Mollié, 1989).

2.2 Priors for labels

Label variables x^L take on a finite set of states and sometimes generate the records y directly, with $x_i^P = \mu(x_i^L)$, where μ is a deterministic function. For example, a common simulation model (e.g. Geman and Geman, 1984; Besag, 1986) is

$$y = A\mu(x^L)\oplus z, \tag{2.6}$$

where A is the identity or a blurring matrix and $\oplus z$ represents additive or multiplicative noise. The simplest useful prior for c unordered labels (or "colours"), conveniently identified with the integers $\mathbb{C} = \{1,2,\ldots,c\}$, is

$$p(x^L)\propto\exp\{-\beta \sum_{1\le k<\ell\le c} n_{k\ell}\}, x^L\in\mathbb{C}^n, \tag{2.7}$$

where $n_{k\ell}$ denotes the number of distinct neighbour pairs coloured (k,ℓ) and β is a positive interaction parameter. It follows that the conditional probability of colour x_i^L occurring at i, given the colouring $x_{S\backslash i}^L$ elsewhere, is

$$p_i(x_i^L | x_{\partial i}^L) \propto \exp\{\beta m_i(x_i^L)\}, x_i^L \in C, \tag{2.8}$$

where $m_i(x_i^L)$ denotes the number of neighbours of pixel i having colour x_i^L. The two colour version of (2.7) on the infinite rectangular lattice, with direct adjacency defining neighbourhood, is the celebrated Ising model of statistical physics. This warns us that typical realizations of (2.7), over a finite array, will consist of a single colour, apart from a few isolated pixels, for quite moderate values of β. Thus, we should ensure that subsequent processing emphasizes the relatively attractive local charateristics of the prior distribution, exemplified by (2.8), and not the large scale properties. Note here that the records y act as an external magnetic field: long range order in $\{P(x|y)\}$ on the infinite lattice is thereby prevented but, on a finite array, problems may still persist. This fact often militates strongly against the MAP estimate of x because of its concentration on global characteristics; see Marroquin, Mitter and Poggio (1987) for a two-colour example.

In real applications, it seems more usual for x^P to be a degraded version of x^L. For example, in compiling a crop inventory from remotely sensed data, "wheat" is a likely label, but wheat in different fields, and indeed within fields, will vary in its maturity, creating a range of pixel values. These values are then subject to further degradation, caused by blur etc., in forming the records y. Thus, we might replace (2.6) by

$$x^P = \mu(x^L) \oplus z, \; y = Ax^P \oplus z', \tag{2.9}$$

where z has spatial structure and z' is additional noise. Note that, even when z' is negligible, (2.9) is generally distinct from (2.6), except when A is the identity, a special case discussed with examples by Geman et al. (1989) in the context of texture discrimination but with labels at a coarser scale than pixels. Note that we need to slightly modify the above formulae for y to accommodate other degradations of x^P, such as Poisson records in nuclear medicine. The important general point is that we are working with an hierarchical framework, in which we require (a) a discrete prior distribution for x^L, (b) a secondary prior distribution for x^P given x^L, and (c) a likelihood function for the degradation $D:x^P \to y$. Incidentally, the distribution (2.7) can be easily modified to allow for differing levels of inhibition between different pairs of colours and for neighbour interactions that depend on distance and orientation; see Besag (1986).

Other situations arise for ordered categories. For example, in studying a surface, we might define $x_i^L = 0$ or 1 according to whether $x_i^P \leq h$ or $x_i^P > h$, where h is specified: the extension to contour plots is obvious. Since x^P now determines x^L, a prior for x^L is not required. Otherwise, versions of (2.8) for ordered categories can be easily constructed; see Besag (1986).

2.3 Inclusion of edge variables

We now incorporate edge sites, so that the image becomes $x = (x^P, x^E)$. The only novelty in the description below is that we adopt a hexagonal array of pixels, rather than the conventional rectangular array in Geman and Geman (1984, 1986). The advantages are (a) edges have three rather than two available orientations and (b) the generally pathological case of four (or more) edges meeting at a point is precluded.

We choose the neighbourhood structure of Figure 1, motivated by a simple conditional probability argument. First consider any interior pixel i: if j is an adjacent pixel, we expect x_i^P to be similar to x_j^P unless an edge separates the pixels (i.e. $x_{ij}^E = 1$). Thus, pixels are assigned six pixel and six edge-site neighbours. Next consider the edge variable x_{ij}^E between adjacent pixels i and j: if x_i^P and x_j^P are highly dissimilar, we favour $x_{ij}^E = 1$, otherwise $x_{ij}^E = 0$. However, our views are also coloured by the states of adjacent edge sites: if no edges are present, we strongly favour $x_{ij}^E = 0$; if just one at each end is present, we expect $x_{ij}^E = 1$; and other configurations suggest an intermediate stance. Thus, edge sites are allotted two pixel and four edge-site neighbours.

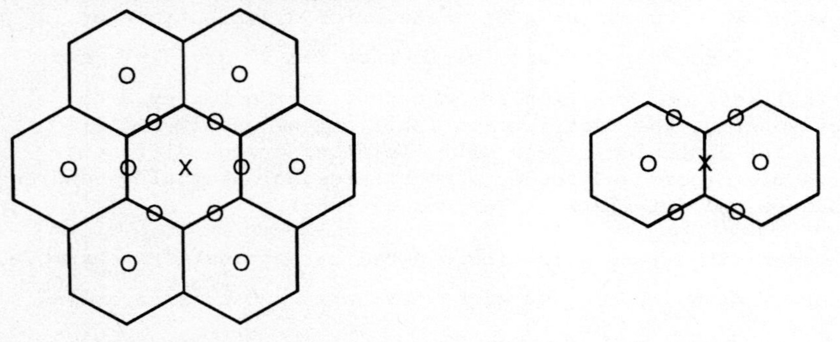

Fig. 1 Neighbours o of a pixel x and of an edge site x.

A corresponding prior distribution for x is defined by

$$p(x) \propto \exp\{-\sum_{i\sim j}(1 - x_{ij}^{E})[\varphi(x_i^P - x_j^P) - \gamma] - \delta\Sigma V(x_{ijk})\}, \tag{2.10}$$

where the final summation is over all distinct triples (i,j,k) indexing mutual pixel neighbours, and

$$x_{ijk} = x_{ij}^{E} + x_{ik}^{E} + x_{jk}^{E} = 0,\ 1,\ 2 \text{ or } 3.$$

The function φ corresponds to that in Section 2.1, while the value of $\gamma > 0$ is chosen so that

$\varphi(x_i^P - x_j^P) < \gamma$ indicates a preference for $x_{ij}^E = 0$, and vice versa.

Finally, we choose

$$0 = V(0) < V(2) < V(3) < V(1) = 1 \text{ and } \delta > 0.$$

In assigning values to the constants, it can be helpful to examine the explicit conditional distributions for pixel and edge-site variables: these are easily written down.

Geman et al (1989) define edge sites and neighbourhoods to exist at a coarser scale than that of the (rectangular) pixels. This device helps in preventing the profusion of "micro-edges" that may be suggested by apparently fuzzy boundaries. These authors ban certain configurations but bring in the taboos gradually, using simulated annealing (Section 4.1).

2.4 Other considerations

There is wide scope for further image attributes to be introduced in a similar fashion; for example, line segments to identify fractures in medical scans or roads in SAR images. Related but distinct methodologies are also being developed, including those in metric pattern theory (Grenander, 1983; Chow, Grenander and Keenan, 1988) and in stochastic geometry (Baddeley and Møller, 1989).

3. THE GIBBS SAMPLER

In Section 4, we discuss various estimators of image attributes, based on the posterior distribution $\{P(x|y)\}$ for fixed records y. As it is rarely practicable to obtain the estimates directly, we usually resort to Monte Carlo implementations and, in anticipation of this, we here describe how sample images can be generated from a given posterior distribution. Not even this can be generally achieved directly and, instead, we adopt an algorithm very close to Metropolis' method; note here that any $\{P(x|y)\}$ is a Gibbs distribution, a fact that motivates use of the term "Gibbs sampler", introduced by Geman and Geman (1984).

The procedure is to construct a discrete-time Markov chain, with state space the space of all valid images x and limit distribution $\{P(x|y)\}$. A simulation of this Markov chain then produces, after an initial evolutionary phase, a sequence of (stochastically dependent) images sampled from $\{P(x|y)\}$. The actual construction is very simple: it merely requires that each site is visited in turn and that the current value there is replaced by one sampled randomly from the associated conditional distribution, given the current states of all other image attributes. If this distribution is discrete, it is usually most convenient to generate the new value by the inverse distribution function method; if continuous, by rejection sampling, unless the distribution has a convenient form, such as gamma or Gaussian.

We consider the algorithm in more detail for a pure pixel image $x = x^P$; this simplifies the notation. Thus, when at pixel i, we generate a new x_i from the univariate conditional distribution $\{P(x_i|x_{S\backslash i}, y)\}$. Viewed at the end of each cycle we now have a time-homogeneous Markov chain whose limit distribution must be consistent with the individual conditional distributions and hence with $\{P(x|y)\}$ that they determine: this last fact follows, under very mild regularity conditions, from the Brook expansion (Brook, 1964; Besag, 1974). Note that

$$P(x_i|x_{S\setminus i},y) \propto \ell(y|x)\ p_i(x_i|x_{\partial i}) \tag{3.1}$$

and that only the dependence on x_i in $\ell(y|x)$ is relevant: at one extreme, a large number of records may depend on x_i, as happens in tomography; at the other, only y_i, so that $\ell(y|x)$ in (3.1) can be replaced by the conditional density $f_i(y_i|x_i)$ of y_i.

Although the programming itself is quite straightforward, it is clear that the CPU time involved in creating a useful sample of images can easily become excessive, particularly if approximately independent realizations are required, so that images from well-separated cycles must be used (note however that the presence of the records can greatly reduce the dependence). We therefore describe the enormous savings that can be made with the availability of parallel processing. The idea here is to group sites into "coding sets" (Besag, 1974), such that no two site variables within the same set can influence one another during a single cycle of the Gibbs sampler. It follows that sites within the same coding set can be simultaneously updated without disturbing the convergence properties of the algorithm. As one admittedly extreme example, consider a hexagonal pixel array with $x = (x^P, x^E)$, where x^E is at the same resolution as x^P and the neighbourhood structure conforms with Figure 1. Then if $S = T$ and there is no blur, we may have

$$\ell(y|x^P) = f_1(y_1|x_1^P)\ f_2(y_2|x_2^P)\ \ldots\ f_n(y_n|x_n^P) \tag{3.2}$$

in which case only three coding sets are required, as in Figure 2. With the addition of local blurring of the records, (3.2) fails and more coding sets are required but the corresponding parallelism would still produce a dramatic reduction in CPU time. Incidentally, note that simultaneous updating of all sites is never valid because it produces the wrong limit distribution.

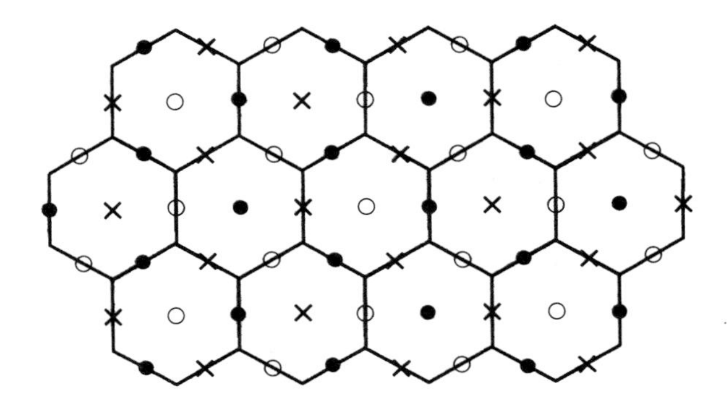

Fig. 2 The three coding sets for parallel processing of pixels and edge sites.

Modifications of the standard Gibbs sampler are sometimes advantageous. For example, suppose $x = u + v$, where u and v are stochastically independent processes with known densities, as in Section 2.1.3. Then, instead of working directly with the x_is, it is generally easier to simulate from the conditional posterior distributions of the u_is and v_is. However, note that the marginal independence of u and v does not extend to conditional independence given y.

4. BAYESIAN IMAGE ESTIMATES

We first describe some plausible point estimates of image attributes and then briefly consider the estimation of posterior probabilities. In this context, use of the Gibbs sampler, or versions thereof, originates in Grenander (1983) and Geman and Geman (1984).

4.1 Maximum a posteriori (MAP) estimate

The MAP estimate of x, defined by

$$\hat{x} = \arg\max P(x|y), \tag{4.1}$$

has been mentioned already in Sections 1 and 2, where we noted its links with more traditional forms of regularization and the possible exaggeration of undesirable global features of the prior distribution for x. Often, the behaviour of $P(x|y)$ is impossible to visualize, because of the very high dimensionality of x and the existence of multiple local maxima. In such circumstances, it may be unwise to use an estimate determined only by the position of the global maximum, assuming that this can be found! Furthermore, there is an inconsistency in that the MAP estimate of particular components of the image does not generally coincide with the components of the MAP estimate of x itself. Lastly, note that for discrete images, the MAP estimate is optimal, in a decision theoretic sense, under a 0-1 loss function; that is, if zero loss is ascribed to the true image and unit loss to any incorrect image, whether it fails in one or every component!

However, there is at least one situation for which the MAP estimate is both plausible and relatively easy to find. This concerns restoration of a pixel image $x = x^P$, where $\ln P(x|y)$ is a strictly concave differentiable function of x and therefore has a single maximum that can be located by simple hill-climbing methods, such as iterated conditional modes (ICM), described below. This condition is quite common in practice, since $\ln \ell(y|x)$ and $\ln p(x)$ are often concave differentiable functions of x and one or other is strictly so. Note that $\ln p(x)$ in (2.1) is concave when $\phi(z)$ is a convex function of z.

ICM (Besag, 1983) is a simple deterministic algorithm closely related to the Gibbs sampler. The difference is that, instead of sampling randomly at each stage, ICM selects the mode of the relevant conditional distribution. It is easily shown that this procedure cannot decrease $P(x|y)$ at any stage; hence the claim in the previous paragraph. However, in a more general setting, the algorithm will be trapped on reaching any local maximum of $P(x|y)$, though ICM may still be useful as an easy means of providing local regularity to an otherwise satisfactory initial estimate. Thus, in restoring a pixel image, the maximum likelihood estimate will often be an adequate starting point; for some (artificial) examples, see Besag (1986). For images that contain other attributes; such as edge sites, maximum likelihood is not applicable, as it does not produce a complete image; alternative methods of initialization must then be sought as described elsewhere in this Section. Finally, note that ICM usually converges rapidly, often in ten or less cycles.

In principle, the exact MAP estimate can always be obtained by simulated annealing (SA), as proposed by Geman and Geman (1984). SA is a stochastic algorithm, using the Gibbs sampler, and is therefore able to escape from local maxima. Instead of simulating from $\{P(x|y)\}$, successive samples are drawn from a density

$$q_T(x|y) \propto [\ell(y|x)p(x)]^{1/T}, \qquad (4.2)$$

where $T > 0$ is a control parameter ("absolute temperature"). In the limit $T \downarrow 0$, $\{q_T(x|y)\}$ assigns unit probability to the MAP image, so it is plausible that this will be the end product if T is decreased sufficiently slowly towards zero during the simulation; Geman and Geman (1984) provide a formal proof of convergence. In practice, it is not possible to adhere to the required temperature schedule and SA will result in an approximation to the MAP estimate; Greig, Porteous and Seheult (1989) provide an exact MAP algorithm for binary images and show some interesting comparisons with SA estimates. Three further points: first, the availability of parallel processing would make SA a more effective tool; second, ICM corresponds to immediate zero-temperature SA; third, Geman et al (1989) obtain impressive results for boundary detection and texture discrimination, using an initial phase of SA, followed by ICM.

4.2 Marginal posterior modes (MPM) estimate

Suppose for convenience that $x = (x^P, x^L, x^E)$, with labels and edges at the pixel scale. Then the MPM estimate $\hat{x}$ of x is defined to have components,

$$\hat{x}_i^P = \arg\max P(x_i^P|y),\ \hat{x}_i^L = \arg\max P(x_i^L|y),\ \hat{x}_{ij}^E = \arg\max P(x_{ij}^E|y). \tag{4.3}$$

MPM is particularly attractive for x^L, since it minimizes the expected number of mislabelled pixels under the posterior distribution; to prove this, attach an indicator variable to each pixel, with value 0 if the label is correct, and 1 otherwise. Note also that, by design, MPM passes the consistency test of Section 4.1. However, MPM is unsatisfactory for x^E, because the marginal maximizations do not seek to enforce continuity of boundaries; see instead the final paragraph of Section 4.1.

Although MPM only requires maximization of univariate functions, it is very rare for the marginal conditional distributions in (4.3) to be calculable in closed form; for partial exceptions, see Haslett (1985) and Devijver and Dekesel (1987). For a general solution, we again resort to a single run of the Gibbs sampler and assume that a large number m of (dependent) realizations from the posterior distribution $\{P(x|y)\}$ are available. Then, for example, $\hat{x}_i^L$ is closely approximated by the most frequently occurring label at pixel i, the approximation improving as m increases. For continuous variables appropriate binning must be used (or see Section 4.3). Note that simulation of $\{P(x|y)\}$ is useless for MAP, since it is necessary to estimate the most frequently occurring complete image in an infinite sequence of realizations.

4.3 Posterior mean and median

In estimating continuous or near-continuous functionals $t = g(x)$ of x, the posterior mean $\hat{t} = E(t|y)$ of t is an appealing choice and is optimal under quadratic loss. It can be estimated by the corresponding sample mean, again obtained from a single run of the Gibbs sampler. An alternative to the posterior mean is the posterior median, estimated in similar fashion.

4.4 Probability estimates

An important benefit of using the Gibbs sampler to produce images from $\{P(x|y)\}$ is that approximate posterior probability statements can be made. Some examples were quoted in the final paragraph of Section 1. In each case, the interval or posterior probability can be calculated, either on the fly or from a sample of images collected at regular intervals after an initial running-in period.

4.5 Parameter estimation

Parameter estimation has been the subject of much recent research in image analysis. Many different proposals have been made, based on methods such as maximum likelihood, moments, pseudo-likelihood, coding, empirical Bayes, full Bayes, etc. Results are generally encouraging, though no clear conclusions have yet emerged and the engineering literature has sometimes produced more heat than light. For example, there has been unfortunate confusion between the definitions of coding estimators (Besag, 1974) and the simpler and more efficient pseudo-likelihood versions (Besag, 1975, 1986; Geman and Graffigne, 1986). Classical statistical analysis generally favours maximum likelihood estimation but, in the present context, this usually requires an enormous amount of computation, even for a single parameter, though a once-and-for-all calibration curve can sometimes be constructed off line (Geman and McClure, 1987). Here we shall focus on a fully Bayesian approach but this should not be taken as a suggestion of superiority; rather it fits most easily into the context of the present paper.

For definiteness, we suppose that it is the prior density of x that contains unknown (hyper-) parameters β. Thus we re-write the prior density as $p(x|\beta)$ to signify the conditional dependence on β: for example, in the convolution prior of Section 2.1.3, $\beta = (\kappa, \lambda)$. A fully Bayesian approach requires that a prior distribution $\{q(\beta)\}$ be specified for β and that inferences about x and β be based on the corresponding posterior density

$$P(x,\beta|y) \propto \ell(y|x^P)p(x|\beta)q(\beta), \qquad (4.4)$$

assuming the usual situation in which the likelihood is independent of β. Of course, the choice of $\{q(\beta)\}$ places an additional responsibility on the user. Unless there is substantial prior knowledge about β, a rather diffuse choice is appropriate but some care must be exercised in adopting the usual "ignorance" priors, since, when improper, these can lead to bizarre singularities in the posterior (4.4).

A much more severe problem can occur, particularly when x is discrete. For example, consider the label prior (2.7): when β is known, the constant of proportionality is irrelevant but for (4.4) its dependence on β is required. However, this cannot be written down in any closed form, except in the simplest case c = 2, for which Ising model calculations provide an approximation. Thus, (4.4) is often unavailable. Fortunately, methods of estimation exist that do not require (4.4), particularly those based on pseudo-likelihood (see, for example, Besag, 1986; Chalmond, 1988), though these do not concern us here. Note that the problem does not arise in (2.3) and (2.4), where the constants of proportionality produce factors $\kappa^{-\frac{1}{2}n}$ and κ^{-n}, respectively; nor does it occur for the convolution priors suggested in Section 2.1.3.

When (4.4) is available in closed form, the simultaneous estimation of x and β can be carried out neatly using the Gibbs sampler, with each component of β providing an additional "site" to be visited on each cycle. When visiting such a site, a new value of the relevant component is generated from the corresponding conditional distribution in the usual way: note that this conditional distribution does not depend on the records y. For convolution priors (Section 2.1.3), the discussion at the end of Section 3 extends easily to include unknown parameters.

ACKNOWLEDGEMENT

The author gratefully acknowledges support from the Complex Stochastic Systems Initiative of the UK Science and Engineering Research Council.

REFERENCES

Baddeley, A. and Moller, J. (1989) Nearest-neighbour Markov point processes and random sets. Int. Statist. Rev. 57, 89-121.

Besag, J. E. (1974) Spatial interation and the statistical analysis of lattice systems (with Discussion) J. R. Statis. Soc. B, 36, 192-236.

Besag, J.E. (1975) Statistical analysis of non-lattice data. The Statistician, 24, 179-195.

Besag, J.E. (1983) Discussion of paper by P. Switzer. Bull. Int. Statist. Inst., 50, (Bk.3), 422-425.

Besag, J.E. (1986) On the statistical analysis of dirty pictures (with Discussion). J.R. Statist. Soc. B, 48, 259-302.

Besag, J. E. and Kempton, R. A. (1986) Statistical analysis of field experiments using neighbouring plots. Biometrics, 42, 231-251.

Besag, J. E. and Mollié, A. (1989) Spatial estimation of incidence rates in epidemiology. Bull. Int. Statist. Inst. (to appear).

Brook, D. (1964) On the distinction between the conditional probability and the joint probability approaches in the specification of nearest-neighbour systems. Biometrika, 51, 481-483.

Chalmond, B. (1988) An iterative Gibbsian technique for simultaneuous structure estimation and reconstruction of M-ary images. Université de Paris-Sud, Mathématiques, # 88-28.

Chow, Y., Grenander, U. and Keenan, D. M. (1988) Hands: a pattern theoretic study of biological shapes. Brown University, Division of Applied Mathematics.

Devijver, P. A. and Dekesel, M. M. (1987) Learning the parameters of a hidden Markov random field image model: a simple example. In NATO ASI Series, vol. F30, Pattern Recognition Theory and Applications (P. A. Devijver and J. Kittler, eds) Berlin: Springer-Verlag.

Geman, D. and Geman, S. (1986) Bayesian image analysis. In NATO ASI Series, vol. F20, Disordered Systems and Biological Organization (E. Bienenstock et al, eds.) Berlin: Springer-Verlag.

Geman, D., Geman, S. Graffigne, C. and Ping Dong (1989) Boundary detection by constrained optimization. I.E.E.E. Trans. Pattern Anal. Machine Intell. (to appear).

Geman, S. and Geman, D. (1984) Stochastic relaxation, Gibbs distributions and the Bayesian restoration of images. I.E.E.E. Trans. Pattern Anal. Machine Intell., 6, 721-741.

Geman, S. and Graffigne, C. (1987) Markov random field image models and their applications to computer vision. Proceedings of the International Congress of Mathematicians, 1986, Berkeley, California, USA (A. M. Gleason, ed.), 1496-1517.

Geman, S. and McClure, D. (1987) Statistical methods for tomographic image reconstruction. Bull. Int. Statist. Inst., 52, Bk. 4, 5-21.

Gidas, B. (1988) A renormalization group approach to image processing problems. I.E.E.E. Trans. Pattern Anal. Machine Intell.,

Green, P. J. (1989) Bayesian reconstruction from emission tomography data using a modified EM algorithm IEEE Trans. Medical Imaging (to appear).

Greig, D. M., Porteous, B. T. and Seheult, A. H. (1989) Exact m.a.p. estimation for binary images. J. R. Statist. Soc. B, 51, 271-279.

Grenander, U. (1983) Tutorial in Pattern Theory., Brown University, Division of Applied Mathematics.

Haslett, J. (1985) Maximum likelihood discriminant analysis on the plane using a Markovian model of spatial context. Pattern Recognition, 18, 287-296.

Künsch, H. R. (1987) Intrinsic autoregressions and related models on the two-dimensional lattice. Biometrika, 74, 517-524.

Marroquin, J., Mitter, S. and Poggio, T. (1987) Probabilistic solution of ill-posed problems in computational vision. J. Amer. Statist. Assoc., 82, 76-89.

Smith, A. F. M. (1984) Bayesian statistics. In Handbook of Applicable Mathematics, Vol. VI, Pt. B, Ch. 15 (E. Lloyd, ed.) Wiley.

MODELLING AND RESTORATION OF IMAGE SEQUENCES

D.M. Titterington
(Department of Statistics, University of Glasgow)

ABSTRACT

Statistical modelling techniques, recently developed for single frames, are extended for application to sequences of frames. With the help of plausible assumptions, restoration algorithms result, based on the Kalman filter or discrete analogues thereof. Parameter estimation is also discussed.

1. INTRODUCTION

There has recently been an arousal of interest, among statisticians, in problems of image analysis and restoration. The distinctive feature of the statisticians' approach is the rôle of modelling, as expressed in the seminal papers of Geman and Geman (1984) and Besag (1986). The main objective of this paper is to describe extensions of this approach for the treatment of sequences of images, each of which is represented by a frame of pixels. It has to be emphasised that, although the paper offers a framework, many details of practical implementation on real sequences still remain to be resolved. The main problems in this context are computational, however, and it is envisaged that their resolution will become feasible thanks to the availability of parallel and other novel computer-architectures.

There is no shortage of applications. Three fertile areas are remote sensing (where sequences of overlapping frames are often available), biophysical imaging (where, for instance, pulsations of organisms are examined by ultrasonic imaging), and more general processing of movie-film or videotape.

The plan of the paper is as follows. Section 2 describes general aspects of statistical image modelling and introduces the idea of recursive processing, which reflects the desirability of real-time analysis. Section 3 establishes a general probabilistic structure for the problem and particular cases of this structure are described in Sections 4 and 5 for "discrete" and "continuous" images, respectively. These Sections will be comparatively free of computational details, which are available already in Green and Titterington (1988). Section 6 provides a miscellaneous collection of further ideas and remarks.

2. GENERAL ASPECTS AND THE NOTION OF RECURSIVE METHODS

Our attack on the problem concentrates on three general activities:

(i) Modelling;

(ii) Restoration;

(iii) Estimation (or Identification).

To clarify these aspects, we introduce some notation. We assume a sequence of T frames, each of which is made up of N pixels, normally arranged in a rectangular array of m rows and n columns (so that $N = m \times n$). The T <u>true scenes</u> are denoted by

$$\{x_t : t = 1, \ldots, T\},$$

where, for each t, x_t is the N-vector of true pixel labels, grey levels or intensities on frame t. The $\{x_t\}$ are not available to us. Instead we have data in the form

$$\{y_t : t = 1, \ldots, T\},$$

where y_t is the observed record corresponding to x_t. Usually, y_t is also an N-vector and is a blurred, noisy version of x_t. We shall concentrate on this case but it is also possible to deal easily, in principle, with the case where the vector y_t is much longer than x_t, incorporating information such as multi-channel data.

The subscript t indicates the correct sequential order of the frames. Often this will correspond to time, and we shall refer to it thus, but sometimes it indicates a third spatial dimension, in examples where the T frames correspond to a set of three-dimensional sections of a solid body.

We shall also employ the following useful notation.

$$x_{\leqslant t} = \{x_s : s = 1, \ldots, t\}$$

$$x_{<t} = \{x_s : s = 1, \ldots, t-1\},$$

with corresponding versions based on y_t.

We now outline what we mean by the three general aspects named at the beginning of this section.

(i) Modelling corresponds to the proposal of some model, M, relating $x_{\leqslant T}$ and $y_{\leqslant T}$. This generally takes the form

$$M = M(x_{\leqslant T}, y_{\leqslant T}, \theta),$$

where M on the right-hand side means some (probabilistic) structural form and θ denotes a set of parameters whose values pin down the specific model uniquely.

(ii) Restoration means the drawing of inferences about $x_{\leqslant T}$, given $y_{\leqslant T}$ and assuming that the model M is correct. This normally requires that θ be known, or be estimated by some reliable procedure. Usually, the "inferences" take the form of "point" estimates, $\hat{x}_{\leqslant T}$, for $x_{\leqslant T}$, but the statistician would also like to create more general inferences that include, at least, some idea of the precision to be associated with the "point" restorations.

(iii) Estimation means the drawing of inferences about the parameters, θ, given $y_{\leqslant T}$ and, again, assuming that M is correct.

Steps (ii) and (iii) identify our version of the image processing problem as a statistical "missing data" problem, in which the observed data are $y_{\leqslant T}$ and the missing data are $x_{\leqslant T}$. In other manifestations of such problems, such as the problem of non-response in sample surveys, what we have called point restoration is termed imputation of the missing values. Furthermore, it is typical that the estimation of parameters would be very much easier if $x_{\leqslant T}$ were also available than it is if only $y_{\leqslant T}$ is given: see Little and Rubin (1987) for a recent, authoritative account. Our missing-data problem is particularly complex as a result of the need to acknowledge, within the model, spatial associations among the pixels in a given frame, and "temporal" associations among frames that are close together in "time". However, it will undoubtedly be revealing to see how far we can progress by recognising the parallel between our problem and this large class of missing-data problems.

The final objective of this section is to introduce the notion of <u>recursive</u> methods, the development of which would permit real-time processing of the stream of frames. The general recursive step for simultaneous restoration and estimation would have the following form: at stage (frame) t and given $(\hat{x}_{<t}, y_{\leq t}, \hat{\theta}_{t-1})$, where $\hat{x}_{<t}$ are the restorations already provided for $x_{<t}$ and $\hat{\theta}_{t-1}$ is the current estimate for θ (assumed time-invariant), to obtain $\hat{x}_t$ and to update $\hat{\theta}_{t-1}$ to $\hat{\theta}_t$.

3. PROBABILISTIC MODELS, GENERAL AND MARKOVIAN

In the sequel the letter "p" will be used generically to denote probability function, either the probability mass function if the underlying random variables are discrete-valued or the probability density function in the continuous case. Thus, a stochastic model, M, for the sequences of true scenes and recorded images is represented by the probability function

$$p_\theta(x_{\leq T}, y_{\leq T}), \tag{1}$$

where the subscript reminds us of the dependence on parameters. When considering only the first t frames we must, correspondingly, deal with

$$p_\theta(x_{\leq t}, y_{\leq t}). \tag{2}$$

In the simplest case of t = 1 we have

$$p_\theta(x_1, y_1) = p_\theta(y_1 | x_1)\, p_\theta(x_1),$$

where $p_\theta(\cdot|\cdot)$ denotes a conditional probability function, and the factorisation on the right-hand side is the foundation stone of the single-frame models of Geman and Geman (1984) and Besag (1986).

In the non-recursive approach, (1) provides the basis for procedures of restoration and estimation. If θ is known, then it is appropriate to base restoration on

$$p_\theta(x_{\leq T} | y_{\leq T}) = p_\theta(x_{\leq T}, y_{\leq T}) / p_\theta(y_{\leq T}), \tag{3}$$

where

$$p_\theta(y_{\leq T}) = \int p_\theta(x_{\leq T}, y_{\leq T})\, dx_{\leq T}, \tag{4}$$

in which the integral is replaced by a summation if $x_{\leq T}$ is discrete-valued. So far as estimation of θ is concerned, the standard statistical procedure known as maximum likelihood estimation would seek the maximum of the likelihood function corresponding to the observed data. In other words $p_\theta(y_{\leq T})$, as defined by (4), should be maximised with respect to θ. Unfortunately, (4) is typically a very complicated object, and this has the following consequences.

(i) Maximum likelihood estimation from incomplete data almost always requires iterative algorithms. One procedure that is particularly relevant to missing data problems is the EM algorithm of Dempster, Laird and Rubin (1977), whose appearance in image restoration problems has been noted in various contexts; see Vardi, Shepp and Kaufman (1985) and Titterington and Rossi (1985).

(ii) Largely because of (i), various slight and not-so-slight variations on maximum likelihood estimation have been developed for image restoration problems; see Besag (1986, Section 5.1), Possolo (1986), Younes (1988a, 1988b) and Chalmond (1988).

(iii) The appearance of the complicated function $p_\theta(y_{\leq t})$ on the right-hand side of (3) led to the decision to deal with modes, rather than means, as the source of restorations in Geman and Geman (1984) and Besag (1986).

So far, model (1) is totally general and we now impose simplifying assumptions, as follows.

(a) $p_\theta(y_t | x_{\leq T}, y_{<t}) = p_\theta(y_t | x_t)$

(b) $p_\theta(x_t | x_{<t}) = p_\theta(x_t | x_{t-1})$.

Under (a), y_t depends only on the corresponding x_t, whereas (b) is an assumption of first-order Markovian structure. As a consequence of (a) and (b),

$$p_\theta(x_{\leq t}, y_{\leq t}) = \prod_{s=1}^{t} \{p(y_s | x_s) p(x_s | x_{s-1})\} \, p(x_o).$$

For clarity, the θ subscript has been dropped; also, $p(x_o)$ represents some initialising stage. It follows, after a little calculation, that

$$p(x_t | y_{\leq t}) \propto p(y_t | x_t) \int p(x_t | x_{t-1}) p(x_{t-1} | y_{<t}) \, dx_{t-1}. \tag{5}$$

In principle, (5) shows how to generate recursively the $\{p(x_t|y_{\leq t})\}$, which are the key quantities so far as restoration is concerned. Practical difficulties arise, however, in that the right-side is usually unmanageable. (Section 5 describes an example in which these difficulties evaporate.) To deal with these computational problems, it is necessary to approximate to $p(x_{t-1}|y_{<t})$ as we go along, the simplest approach being to replace it by a degenerate distribution, concentrated on some $\hat{x}_{t-1}$. In that case, (5) becomes

$$p(x_t|y_{\leq t}) \propto p(y_t|x_t)p(x_t|\hat{x}_{t-1}). \tag{6}$$

This approach amounts, therefore, to imputing $\hat{x}_{t-1}$ and, at the next stage, arguing as if $\hat{x}_{t-1}$ were the correct, true scene on frame t-1. One would then create a restoration $\hat{x}_t$ on the basis of $p(x_t|y_{\leq t})$, and so on. There are two ways of obtaining $\hat{x}_t$, either as a summary parameter from $p(x_t|y_{\leq t})$ (such as the mean or the mode) or as a simulated value from $p(x_t|y_{\leq t})$. These approaches correspond, respectively, to the <u>decision-directed</u> (DD) and <u>probabilistic-teacher</u> (PT) techniques well-known in the literature on unsupervised-learning problems; see for instance Titterington, Smith and Makov (1985, Chapter 6). The relative merits of DD and PT methods are discussed in Section 6.4.

Although mention of unknown parameters has been dropped from (5) and (6), the need to deal with them must not be forgotten. Titterington (1984a) discusses recursive versions of the EM algorithm, mentioned earlier, to cope with a sequence of incomplete observations, and Green and Titterington (1988) extend them to cover the present context. It has to be said that the convergence properties of the resulting, complex recursions still require considerable investigation.

In the following two sections we look at two particular, Markovian models.

4. GIBBS DISTRIBUTIONS

Suppose S denotes the set of all pixels in the frame and that $x_{t,c}$ denotes the values of x_t corresponding to a subset c of pixels within S. In particular, $x_{t,i}$ denotes the value of x on pixel i in frame t. Our assumption in this section is that

$$p(x_t \mid x_{<t}) \propto \exp\{-\sum_c V_c(x_{t,c};\ x_{<t})\}, \tag{7}$$

where $\{V_c\}$ is a set of potential functions indexed by cliques $\{c\}$. (A clique is a set of pixels such that every pair of distinct vertices in the set are neighbours according to a prescribed neighbourhood system.) Relationship (7) defines a Gibbs distribution, a type that has arisen in the modelling of single true scenes by Markov random fields (Geman and Geman, 1984; Besag, 1986). As a consequence of (7).

$$p(x_{t,i} \mid x_{<t},\ x_{t,S\setminus i}) = p(x_{t,i} \mid x_{t-1,\pi i}, x_{t,\partial i}), \tag{8}$$

where πi and ∂i are, typically, small neighbourhoods of pixel i. Equation (8) reflects, therefore, both the Markovian assumption about the sequence of true scenes and the assumption of local, spatial dependence among x-values on mutually neighbouring pixels.

Figure 1 indicates just one possible choice for πi and ∂i. For such a choice and for the case of binary (black/white) scenes, a simple version of (8) is given by

$$p(x_{t,i} \mid x_{<t}, x_{t,S\setminus i}) \propto \exp\{\beta_1 \sum_{j\in\partial i} \nu(x_{t,i}, x_{t,j}) + \beta_2 \sum_{j\in\pi i} \nu(x_{t,i}, x_{t-1,j})\} \tag{9}$$

where

$$\beta_1,\ \beta_2 > 0 \text{ and } \nu(z_1, z_2) = 1 \quad \text{if } z_1 = z_2$$
$$= 0 \quad \text{otherwise.}$$

Clearly, (9) is very special, and more flexible models involving more than two parameters could be proposed.

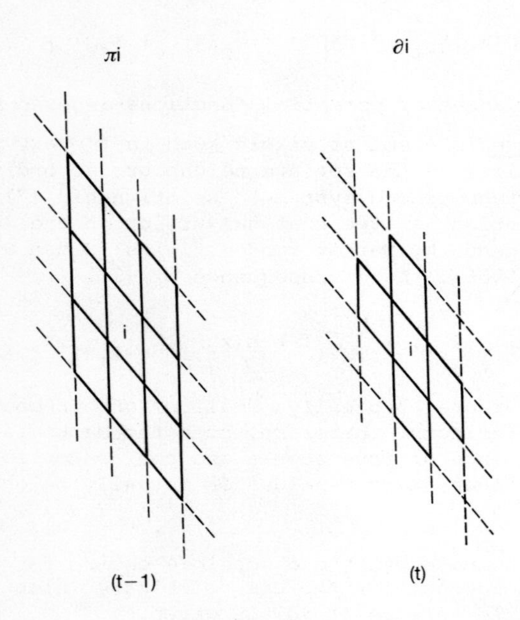

Fig. 1

5. THE LINEAR GAUSSIAN MODEL

5.1 The model and the Kalman Filter

This model, applicable when all variables in x and y are continuous-valued, provides one situation in which the analysis of Section 3 can be carried out exactly, in that the recursion implied by (5) is feasible. We assume a time-invariant, linear, additive formulation in which

$$x_t = Gx_{t-1} + \eta_t$$

$$y_t = Hx_t + \varepsilon_t,$$

where $\eta_t \sim N(\mu_\eta, V_\eta)$, $\varepsilon_t \sim N(\mu_\varepsilon, V_\varepsilon)$, the $\{\eta_t\}$ and $\{\varepsilon_t\}$ are all mutually independent and, to initialise the system, $x_o \sim N(m_o, V_o)$. Thus, the parameters are

$$\theta = \{G, H, \mu_n, V_n, \mu_\varepsilon, V_\varepsilon, m_o, V_o\}.$$

Some of these may be known: for instance, usually $\mu_\varepsilon = \mu_\eta = 0$. The consequence of this structure is that, for all $t > 0$,

$$x_t | y_{\leq t} \sim N(m_t, V_t), \tag{10}$$

where the $\{m_t\}$ and $\{V_t\}$ can be computed recursively according to the Kalman filter equations, as follows.

$$V_t^{-1} = (GV_{t-1}G^T + V_\eta)^{-1} + H^T V_\varepsilon^{-1} H,$$

and

$$m_t = Gm_{t-1} + \mu_\eta + V_t H^T V_\varepsilon^{-1}\{y_t - \mu_\varepsilon - H(Gm_{t-1} + \mu_\eta)\}.$$

The natural Decision-Directed recursive restoration for x_t is therefore $\hat{x}_t = m_t$. Alternatively, a PT-restoration could be obtained by simulating from (10).

There are, of course, problems in implementing the procedure, even in the case where θ is known exactly. The sizes of the matrices involved lead to potentially heavy computations, although assumptions of only local associations leave us with very sparse matrices. In addition, sometimes assumptions can be made of special symmetries and structure that lead to particularly simple eigenstructure.

One such special case, that of toroidal stationarity, is described in detail in Green and Titterington (1988). If in this situation the pixels are numbered sequentially by a suitable lexicographic ordering, and M is a correspondingly ordered mnxmn matrix, with elements denoted in a natural way by $M_{ijk\ell}$, $i,k=1, \ldots, m$, $j, \ell = 1, \ldots, n$, then it turns out that $M_{ijk\ell}$ depends only on $(k-i) \bmod m$ and $(\ell-j) \bmod n$. From a practical point of view, the facts that the matrices involved are block circulant and are therefore diagonalised by the Discrete Fourier Transform, imply that the Kalman Filter can be constructed in the Fourier domain using only scalar recursions. In other cases, the matrices involved are naturally block Toeplitz, but the above comments may be relevant because of the existence of circulant approximations to Toeplitz matrices. Hunt (1973) exploits these comments in the context of the analysis of a single frame and the relationships between circulant and Toeplitz matrices are discussed in Sherman (1985) and references therein.

5.2 Parameter estimation within the Kalman Filter

As remarked earlier, the fact that we are dealing with a missing-data problem means that estimation of any unknown components of θ requires iterative procedures. For instance, Shumway and Stoffer (1982) derive a version of the EM algorithm of Dempster et al (1977) that is appropriate for the Kalman Filter.

So far as recursive estimation is concerned, Green and Titterington (1988) derive the following procedure for estimating G, H, V_η and V_ε. It is assumed that m_o and V_o are specified and that $\mu_\eta = \mu_\varepsilon = 0$. Suppose the estimates at stage t are denoted by $\hat{G}_t$, $\hat{H}_t$, $\hat{V}_{\varepsilon,t}$ and $\hat{V}_{\eta,t}$. Then

$$\left.\begin{aligned} \hat{G}_t &= E_t F_t^{-1}, \\ \hat{H}_t &= B_t C_t^{-1}, \\ \hat{V}_{\varepsilon,t} &= t^{-1}\{A_t - B_t C_t^{-1} B_t^T\} \\ \text{and} \qquad \hat{V}_{\eta,t} &= t^{-1}\{C_t - E_t F_t^{-1} E_t^T\}, \end{aligned}\right\} \qquad (11)$$

where

$$A_t = \sum_{s=1}^{t} y_s y_s^T, \quad B_t = \sum_{s=1}^{t} y_s m_s^T, \quad C_t = \sum_{s=1}^{t} (m_s m_s^T + V_s)$$

$$E_t = \sum_{s=1}^{t} (m_s \hat{m}_{s-1}^T + V_{s,s-1})$$

and

$$F_t = \sum_{s=1}^{t} (\hat{m}_{s-1} \hat{m}_{s-1}^T + \hat{V}_{s-1}).$$

Within these expressions,

$$\begin{aligned} m_s &= \mathbf{E}(x_s | y_{\leqslant s}; \hat{\theta}_{s-1}) \\ V_s &= \operatorname{cov}(x_s | y_{\leqslant s}; \hat{\theta}_{s-1}), \\ \hat{m}_{s-1} &= \mathbf{E}(x_{s-1} | y_{\leqslant s}; \hat{\theta}_{s-1}), \\ \hat{V}_{s-1} &= \operatorname{cov}(x_{s-1} | y_{\leqslant s}; \hat{\theta}_{s-1}) \end{aligned}$$

and

$$V_{s,s-1} = \mathrm{cov}(x_s, x_{s-1} | y_{\leqslant s}; \hat{\theta}_{s-1});$$

all of these quantities are computable.

As explained in Green and Titterington (1988) the above algorithm is a special case of a recursive analogue of the EM algorithm proposed by Titterington (1984a). As in the case of implementation of the Filter itself, the practicalities of the above algorithm are daunting except in special situations. In the case of toroidal stationarity, a considerably simplified version of (11) exists, in the Fourier domain, where each matrix is now interpreted as the diagonal part of its Fourier transform. A specific illustration of the resulting simplicity, taken from Section 4.5 of Green and Titterington (1988), is as follows.

Suppose V is the unitary matrix $U_{ijk\ell} = (mn)^{-\frac{1}{2}} \nu^{-ik} \omega^{-j\ell}$.

where ν and ω are, respectively, the mth and n^{th} complex roots of unity. Suppose also that v_i, a_i, b_i, c_i and h_i are, respectively, the ith diagonal elements of UMU*, for $M = V_\varepsilon$, A_t, B_t, C_t and H, respectively, where "*" denotes complex conjugate. Then, estimates of $\{h_i\}$ are given by the minimisers of

$$\sum_i v_i^{-1}(a_i - c_i^{-1}|b_i|)^2 + \Sigma v_i^{-1} c_i |h_i - c_i^{-1} b_i|^2.$$

If, for instance, H is assumed to be symmetric, then h_i is estimated by $(2c_i)^{-1}(b_i + b_i^*)$.

In spite of the above remarks it is still the case that, in more general examples involving more than a few parameters, parameter estimation remains as a severe, computational problem.

6. FURTHER REMARKS

6.1 The use of the Kalman Filter in image restoration

A considerable literature already exists on the use of the Kalman Filter in image analysis, but related to the restoration of a single frame. The analogue of our sequence of frames is to regard the single frame as a sequence of individual pixels or a sequence of complete rows and columns. Of particular relevance to the present paper is the work of Biemond, Rieske and Gerbrands (1983), although there are many other related papers, from Nahi (1972) to Wu (1985).

6.2 Motion analysis

Other methodologies have also been developed for the analysis of the motion of an object over a sequence of frames; see, for instance, Sethi and Jain (1987) and Bresler and Merhav (1987) for some recent material. The work of these authors and others concentrates on the motion of specific features of the scene and, if motion is known to be present, it is often best to model it directly.

In current collaborative work with J.W. Kay and A.J. Gray, the author is developing statistical procedures for estimating, in terms of frame-to-frame shift, the steady motion of an object across a background.

6.3 More on the approximation of $p(x_t|y_{\leq t})$

It was noted in Section 3 that the updating of $p(x_t|y_{\leq t})$ as t increases, is a crucial feature of our general approach. For the Linear Gaussian model of Section 5, exact calculations are feasible. Otherwise, this normative procedure is not available, and some approximation is necessary in which the data $\{y_{\leq t}\}$ are not used efficiently.

One approach, inspired by current trends in Statistics, would be to store empirical estimates of $p(x_t|y_{\leq t})$, $t=1, \ldots$ in the form of a set of random samples from the distribution. If, for instance, a set of R such sample scenes are available for frame $t-1$, they yield an empirical estimate of $p(x_{t-1}|y_{<t})$ which, from (5), provides a probability function from which simulations of frame t can be generated as an empirical estimate of $p(x_t|y_{\leq t})$ and so on. As yet, this notion has not been properly investigated.

A somewhat related approximation is to combine the updating procedure with restoration by, say, the Probabilistic-Teacher (PT) approach. Suppose $\{\hat{x}_t\}$ is a sequence of images that are simulated in this way, and that we envisage a comparison among the following three methods.

(a) (Ideal): Simulate $\hat{x}_t$ from $p(x_t|y_{\leq t})$.

(b) (Frame-by-frame): Simulate $\hat{x}_t$ from $p(x_t|y_t)$.

(c) (Intermediate): Simulate $\hat{x}_t$ from

$$p(x_t | y_t, \hat{x}_{t-1}) \propto p(y_t | x_t) p(x_t | \hat{x}_{t-1}).$$

It is intuitively plausible that (c), which, at stage t, uses the past to at least some degree, should be somewhat superior to (b), which does not. Although (c) will not be as efficient as (a), it may well be workable when (a) is prohibitive.

To find out about the relative merits of (a), (b) and (c), a trivial case was examined, based on the Linear Gaussian model. (This was chosen in order that (a) could be assessed, although in practice one would not normally need to use method (c) if that model is correct.) The triviality of our example refers to the fact that we considered, for clarity, a scalar version of the problem with one pixel per frame.

The measure of performance on which (a), (b) and (c) were compared was a quadratic risk criterion

$$\lim_{t\to\infty} \mathbf{E}(\hat{x}_t - x_t)^2,$$

where the expectation operator averages out all random variability, namely, that caused by the noise sequences $\{\eta_t\}$ and $\{\varepsilon_t\}$ and by the inherent random variability in x_t. The limit operator ensures that our results apply to the equilibrium version of the Kalman Filter, in which the posterior variance parameter has settled down to $V_\infty = V$, where V satisfies

$$V^{-1} = (GVG^T + V_\eta)^{-1} + H^T V_\varepsilon^{-1} H.$$

For methods (a), (b) and (c), the values of the asymptotic quadratic risk turn out to be given essentially by the following quantities, whose evaluation is sketched in the Appendix.

(a) $\rho_1 = \{(WA-1)^2 + W^2A\}/\{1-G^2(WA-1)^2\} + W$

(b) $\rho_2 = 2/\{A + (1-G^2)\}$

(c) $\rho_3 = 2(A+1)^{-1}\{1-G^2/(A+1)^2\}^{-1}$,

in which $A = H^2\sigma_\eta^2/\sigma_\xi^2$ and $W = V/\sigma_\eta^2$.

Table 1 provides a comparison of the three methods for $G^2 = 0.1$ (0.2) 0.9 and $A = 2^r$, for $r = -4$ (2) 4. The quantities quoted are the inefficiency ratios ρ_2/ρ_1 and ρ_3/ρ_1, along with percentage reduction in inefficiency obtained by using the intermediate method instead of the frame-by-frame, non-recursive procedure. This last measure is π, where

$$\pi = 100\ (\rho_2 - \rho_3)/(\rho_2 - \rho_1).$$

The Table shows that method (c) can achieve substantial improvement in performance over that of method (b) and it suggests that it would be worthwhile to explore methods similar to (c) in cases where there is available no "ideal" method such as (a).

Table 1

Efficiency comparisons for methods of Section 6.3

G^2	r $(A=2^r)$	ρ_2/ρ_1	$\rho 3/\rho 1$	$\pi=100(\rho_2-\rho_3)/(\rho_2-\rho_1)$
	-4	1.04	1.04	14.5
	-2	1.15	1.13	13.2
0.1	0	1.39	1.36	9.1
	2	1.70	1.67	3.9
	4	1.90	1.89	1.2
	-4	1.09	1.07	27.6
	-2	1.26	1.19	28.5
0.3	0	1.54	1.41	23.2
	2	1.78	1.69	11.1
	4	1.92	1.89	3.5
	-4	1.19	1.13	30.8
	-2	1.47	1.29	37.1
0.5	0	1.72	1.48	34.1
	2	1.86	1.71	17.6
	4	1.95	1.90	5.7
	-4	1.47	1.32	31.7
	-2	1.85	1.47	44.2
0.7	0	1.97	1.55	48.1
	2	1.95	1.73	23.6
	4	1.97	1.90	7.9
	-4	2.66	2.00	39.4
	-2	2.73	1.80	53.6
0.9	0	2.31	1.64	51.2
	2	2.05	1.75	29.1
	4	2.00	1.90	10.0

6.4 *Decision-Directed vs Probabilistic-Teacher*

We have remarked that a sequence of restorations might be generated by DD use of, say, posterior modal values or by PT simulations from the posterior distribution. It seems likely that DD will produce more helpful restorations, particularly if the scenes consist of patches of fairly constant intensity of colour. If, however, restoration is to be combined with parameter estimation by undertaking the latter as if the restored images were the truth, then the DD restorations are much more likely to lead to biased estimates. For an intercourse about these points in the context of image segmentation, see Titterington (1984b) and Sclove (1984).

6.5 *The use of three-dimensional methods*

Instead of regarding the problem as involving a sequence of T, two-dimensional frames, one could treat the structure as a block of mxnxT pixels and restore the scenes as a three-dimensional image. It is straightforward to extend the Markov Random Field models, with associated Gibbs distributions and neighbourhood systems, to the three-dimensional case, and to use corresponding versions of, say, the Iterated Conditional Modes (ICM) algorithm of Besag (1986) to create restorations. Alternatively, one can generalise, to three dimensions, the causal Markov Mesh models developed in the two-dimensional case (Abend, Harley and Kanal, 1965; Kanal, 1980; Lacroix, 1987). This development is described in Qian and Titterington (1989), which also includes numerical comparison with three-dimensional ICM.

6.6 *The potential for parallel computing*

It is inevitable that parallel computing will have an increasing influence on image restoration and it is to be hoped that algorithms that appear to be computationally prohibitive at present will become practicable in the very near future. In our context, such developments should permit the treatment of ever larger applications, and should also enable us to implement variations of our methods, particularly of recursive approaches. In our earlier analysis, we envisaged incorporating one observed image at a time and then simply moving on to the next. In the case of models where exact calculations are not possible (that is, in cases other than the Linear Gaussian models!), the data could be used more efficiently if, during the stage of a recursive procedure at which x_t is being considered, a "window" of several observed frames, on either side of y_t, could be considered in the procedure. Such a proposal clearly escalates the computing aspects but, one hopes, not fatally in the long term.

ACKNOWLEDGEMENT

It is a pleasure to acknowledge the fact that most of this work is derivative of collaborative research with Dr. P.J. Green of the University of Durham.

REFERENCES

Abend, K., Harley, T.J. and Kanal, L.N., (1965) Classification of binary random patterns. *IEEE Trans. Inform. Th., IT-11,* 538-544.

Besag, J., (1986) On the statistical analysis of dirty pictures (with discussion). *J.R. Statist. Soc. B,* **48**, 259-302.

Biemond, J., Rieske, J. and Gerbrands, J.J., (1983) A fast Kalman filter for images degraded by both blur and noise. *IEEE Trans. Acoust. Speech Signal Proc., ASSP-31,* 1248-1256.

Bresler, Y., and Merhav, S.J., (1987) Recursive image registration with application to motion estimation. *IEEE Trans. Acoust. Speech Signal Proc., ASSP-35,* 70-85.

Chalmond, B., (1988) An iterative Gibbsian technique for simultaneous structure estimation and reconstruction of M-ary images. Preprint 88-28, Univ. Paris-Sud, Mathématiques.

Dempster, A.P., Laird, N.M. and Rubin, D.B., (1977) Maximum likelihood from incomplete data via the EM algorithm (with discussion). *J.R. Statist. Soc. B,* **39**, 1-38.

Geman, S. and Geman, D., (1984) Stochastic relaxation, Gibbs distributions and the Bayesian restoration of images, *IEEE Trans. Pattern Anal. Machine Intell., PAMI-6, 721-741.*

Green, P.J. and Titterington, D.M., (1988) Recursive methods in image processing. *Bull. Int. Statist. Inst.,* **52-4**, 51-67.

Hunt, B.R., (1973) The application of constrained least squares estimation to image restoration by digital computer. *IEEE Trans. Computers, C-22,* 805-812.

Kanal, L.N., (1980) Markov mesh models. In Image Modelling. New York: Academic Press.

Little, R.J.A. and Rubin, D.B., (1987) Statistical Analysis with Missing Data. Wiley.

Nahi, N.E., (1972) Role of recursive estimation in statistical image enhancement. *Proc. IEEE,* 60, 872-877.

Possolo, A., (1986) Estimation of binary random Markov fields. Tech. Rep. 77, Univ. Washington, Dept. Statist.

Qian, W. amd Titterington, D.M., (1989) Image restoration for three-dimensional scenes based on Markov Mesh models. Submitted for publication.

Sclove, S.C., (1984) Reply to Titterington (1984b) *IEEE Trans. Pattern Anal. Machine Intell., PAMI-6,* 657-658.

Sethi, I.K. and Jain, R., (1987) Finding trajectories of feature points in a monocular image sequence. *IEEE Trans. Pattern Anal. Machine Intell., PAMI-9,* 56-73.

Sherman, P.J., (1985) Circulant approximations to Toeplitz matrices and related quantities with application to stationary random processes. *IEEE Trans. Acoust. Speech Signal Proc., ASSP-33,* 1630-1632.

Shumway, R.H. and Stoffer, D.S., (1982) An approach to time series smoothing and forecasting using the EM algorithm. *J. Time Series Anal.,* **3**, 253-264.

Titterington, D.M., (1984a) Recursive parameter estimation using incomplete data. *J.R. Statist. Soc. B,* **46**, 257-267.

Titterington, D.M., (1984b) Comment on a paper by S.C. Sclove. *IEEE Trans. Pattern Anal., Machine Intell., PAMI-6,* 656-657.

Titterington, D.M. and Rossi, C., (1985) Another look at a Bayesian direct deconvolution method. *Signal Processing,* **9**, 101-106.

Titterington, D.M., Smith, A.F.M. and Makov, U.E., (1985) Statistical Analysis of Finite Mixture Distributions. Wiley.

Vardi, Y., Shepp, L.A. and Kaufman, L., (1985) A statistical model for positron emission tomography (with discussion). *J. Amer. Statist. Assoc.,* **80**, 8-37.

Wu, Z., (1985) Multidimensional state-space model Kalman filtering with applications to image restoration. *IEEE Trans. Acoust. Speech Signal Proc., ASSP-33,* 1576-1592.

Younes, L., (1988a) Estimation and annealing for Gibbsian fields. *Ann. Inst. H. Poincaré,* **24**, to appear.

Younes, L., (1988b) Parametric inference for imperfectly observed Gibbsian fields. Preprint 88-17, Univ. Paris-Sud, Mathématiques.

APPENDIX

In this Appendix we compute the measures of risk discussed in Section 6.3. It is assumed that x_t and η_t are scalar, that the linear Gaussian model is valid and that $|G| < 1$.

Thus

$$x_t \sim N(A_t, B_t)$$

where $A_t = G^t m_o$ and

$$B_t = G^{2t} V_o + \{(1-G^{2t})/(1-G^2)\}\sigma_\eta^2 .$$

(a) Ideal case (Equilibrium version)

Here, $\hat{x}_t \sim N(m_t, V)$, where

$$V^{-1} = (G^2 V + \sigma_\eta^2)^{-1} + H^2/\sigma_\xi^2$$

$$m_t = Gm_{t-1} + VH\sigma_\xi^{-2}(y_t - HGm_{t-1}).$$

Thus

$$\mathbf{E}\{(\hat{x}_t - x_t)^2 | x_t\} = (m_t - x_t)^2 + V,$$

from which, if E_t denotes $\mathbf{E}(m_t - x_t)^2$, it follows that

$$E_t = G^2 R^2 E_{t-1} + R^2\sigma_\eta^2 + V^2 H^2/\sigma_\xi^2$$

$$\rightarrow (R^2\sigma_\eta^2 + V^2H^2/\sigma_\xi^2)/(1-G^2R^2), \text{ as } t \rightarrow \infty,$$

where

$$R^2 = V^2/(G^2 V + \sigma_\eta^2)^2 .$$

Thus

$$\lim_{t\rightarrow\infty} \mathbf{E}(\hat{x}_t - x_t)^2 = (R^2\sigma_\eta^2 + V^2H^2/\sigma_\xi^2)/(1-G^2R^2) + V = R_1,$$

say.

(b) Frame-by-frame

With the help of Bayes' Theorem, we obtain

$$E\{(\hat{x}_t - x_t)^2 | x_t\} = (H^2/\sigma_\xi^2 + B_t^{-1})^{-1}[\{(A_t - x_t)^2/B_t^2 + (H^2/\sigma_\xi^2)\}(H^2/\sigma_\xi^2 + B_t^{-1})^{-1} + 1]$$

Averaging out x_t and letting $t \to \infty$, we obtain

$$\lim_{t\to\infty} \mathbf{E}(\hat{x}_t - x_t)^2 = 2/\{H^2/\sigma_\xi^2 + (1-G^2)/\sigma_\eta^2\} = R_2, \text{ say},$$

(c) Intermediate

In this case, $\hat{x}_t$ is generated from

$$\hat{x}_t \sim N(\{(Hy_t/\sigma_\xi^2) + (G\hat{x}_{t-1}/\sigma_\eta^2)\}[(H^2/\sigma_\xi^2) + \sigma_\eta^{-2}]^{-1}, [(H^2/\sigma_\xi^2) + \sigma_\eta^{-2}]^{-1}).$$

From this we obtain

$$\mathbf{E}(\hat{x}_t - x_t)^2 = 2(H^2/\sigma_\xi^2 + \sigma_\eta^{-2})^{-1} + (G^2/\sigma_\eta^4)\mathbf{E}(x_{t-1} - \hat{x}_{t-1})^2/(H^2/\sigma_\xi^2 + \sigma_\eta^{-2})^2,$$

which leads to

$$\lim_{t\to\infty} \mathbf{E}(\hat{x}_t - x_t)^2 = 2(H^2/\sigma_\xi^2 + \sigma_\eta^{-2})^{-1}[1 - (G^2/\sigma_\eta^4)(H^2/\sigma_\xi^2 + \sigma_\eta^{-2})^{-1}]^{-1} = R_3$$

say.

If we define $A = H^2\sigma_\eta^2/\sigma_\xi^2$ and $W = V/\sigma_\eta^2$, then

$$\sigma_\eta^{-2}R_1 = \{(WA-1)^2 + W^2A\}/\{1-G^2(WA-1)^2\} + W = \rho_1, \text{ say.}$$

$$\sigma_\eta^{-2}R_2 = 2/\{A + (1-G^2)\} = \rho_2, \text{ say}$$

and

$$\sigma_\eta^{-2}R_3 = 2(A+1)^{-1}\{1-G^2/(A+1)^2\}^{-1} = \rho_3, \text{ say.}$$

MAXIMUM A POSTERIORI DETECTION-ESTIMATION OF BERNOULLI-GAUSSIAN PROCESSES

Y. Goussard, G. Demoment and F. Monfront
*(Laboratoire des Signaux et Systèmes,
Ecole Supérieure d'Electricité, France)*

ABSTRACT

This paper deals with the restoration of Bernoulli-Gaussian (B-G) processes observed through linear systems. This operation generally involves estimation of the input process as well as identification of *hyperparameters* that characterize the system, and is referred to as myopic deconvolution. Maximum *a posteriori* (MAP) techniques have been used for myopic deconvolution with apparent success. However, a few points still require clarification. Two of them are addressed here.

On a methodological standpoint, consistency of the MAP approach for hyperparameter estimation has yet to be proven. Results presented here indicate that the MAP estimator does not always converge toward any meaningful value, which makes its use questionable. Alternative approaches should be sought.

On a practical standpoint, there is a need for efficient methods able to cope well with tight operational constraints. In these conditions, particular attention is paid to the most time-consuming task of the procedure, i.e.; simple deconvolution of the B-G process when the hyperparameters are given a set value. The emphasis is then placed on on-line processing abilities, computational load and ease of implementation of the method.

We propose two methods that provide satisfactory answers in the situation described above. One of them has a fully recursive structure suited to on-line data processing, while the other operates iteratively. Simulations are presented in order to illustrate the respective advantages and disadvantages of the two methods, and the results are briefly compared with those provided by other approaches.

1. INTRODUCTION

The subject of this communication is the restoration of sparse spike trains distorted by linear systems and corrupted by noise. This problem arises in a number of fields such as seismic exploration, medical ultrasonic imaging, nondestructive evaluation or more generally when the structure of a layered propagation medium has to be identified from measurements performed at its boundaries. Under some assumptions, the observed signals can be considered as the convolution product of a wavelet (incident waveshape) and of the reflectivity of the medium (first derivative of its acoustic impedance). Assuming the homogeneity of the layers, the reflectivity vanishes everywhere except at the boundaries of the layers and appears as a sparse spike train. A schematic representation of the phenomena is given in figure 1. The problem consists of estimating the reflectivity from the observations and from the knowledge available on the wavelet and on a set of *hyperparameters* (typically parameters of the probability density functions of the observation noise and of the reflectivity sequence). Generally the wavelet and the hyperparameters are not known precisely, and restoration of the reflectivity is a *myopic deconvolution* problem in which reflectivity, wavelet and hyperparameters have to be estimated. Restoration of the reflectivity when all other quantities are known will be referred to as *simple deconvolution*.

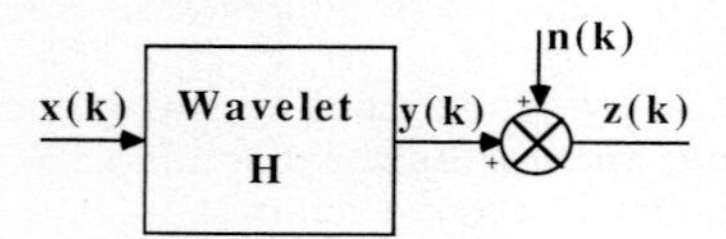

Fig. 1 Representation of the phenomena under investigation. The reflectivity x(k) characterizes the propagation medium, and is modelled as a Bernoulli-Gaussian process. It is observed through a linear system (wavelet) that accounts for the dynamic distortions introduced by the investigation device. n(k) represents the measurement noise and other unmodelled phenomena.

Simple deconvolution is an inverse *ill-posed* problem [14] which, in general, has to be solved with limited computing power an sometimes on-line. A few methods based upon the use of L_p norms [16], minimum entropy criteria [15] or multi-pulse coding techniques [5] have been proposed. None of those explicitly accounts for both the ill-posed nature of deconvolution and for the particular nature of the input. Mendel *et al* [8,11,4] have proposed an elegant and rather complete solution to myopic deconvolution. It is based on a Maximum *a posteriori* (MAP) approach for regularization of the simple deconvolution problem, on an explicit description of the reflectivity as a Bernoulli-Gaussian (B-G) random process and on the use of a unique generalized likelihood criterion for the estimation of all unknown quantities. Later, refinements have been introduced by others [10,7]. However, a few points have yet to be addressed. In particular, identifiability of hyperparameters using the MAP approach has not been fully established. On a practical standpoint, these MAP methods are often not easy to implement and computationally expensive. Therefore, the goal of the paper is twofold: On a methodological level, results on the identifiability of hyperparameters using the MAP approach are presented. On a practical level, two simple deconvolution algorithms, suited to different operational constraints, are proposed and compared on synthetic examples.

2. PROBLEM FORMULATION

With our assumptions, the input-output relationship of the system can be written as the following convolution equation

$$z(k) = \sum_{i=0}^{n} h(i)x(k-i)+n(k) \qquad 1 \leq k \leq P \tag{2.1}$$

where z,x and n denote the observation, the unknown reflectivity and the observation noise respectively. In order to simplify subsequent derivations, wavelet h is assumed to be time-invariant with support of length n+1. Note that most results still hold with a time-varying wavelet.

Concatenating all observed samples z(k), $1 \leq k \leq P$, all noise samples n(k), $1 \leq k \leq P$ and all unknown reflectivity samples x(k), $1 \leq k \leq N$ respectively into vectors z, n and x, equation (2.1) can be rewritten in a compact matrix form

$$z = H\,x + n \tag{2.2}$$

where H contains the shifted samples of wavelet h. The MAP approach used here requires a statistical description of the unknown quantities. Observation noise n is assumed to be white, Gaussian, with zero mean and variance r^n. Reflectivity x is modelled as a B-G process

$$x(k): \text{Gaussian random variable with zero mean and covariance } r^x t(k)$$

$$t(k): \text{Bernoulli random variable} \begin{cases} p\{t(k) = 1\} = \lambda \\ p\{t(k) = 0\} = 1-\lambda \end{cases} \tag{2.3}$$

Equations (2.3) indicate that $x(k)$ takes a zero value with probability $1-\lambda$ and follows a Gaussian distribution with zero mean and covariance r^x with probability λ. It is readily seen that for small values of λ model (2.3) defines a white process consisting of sparse impulses. Note also that x is a nonstationary gaussian process conditionally to the knowledge of Bernoulli sequence t.

The problem of estimating the wavelet along with all other quantities is not addressed here. h is assumed to be known *a priori*, or estimated by other means. Therefore, the signal restoration problem consists of estimating t, x and the set of hyperparameters $\theta = (r^n, r^x, \lambda)$ form observation vector z. Since some of these quantities are random while others are deterministic, we need to define a generalized likelihood function. Following [8], we use the criterion

$$J \stackrel{\Delta}{=} p\{x,t,z|\theta\} = p\{z|x,t,\theta\}\ p\ \{x|t,\theta\}p\{t|\theta\} \qquad (2.4)$$

Other functionals can be chosen (e.g. $p\{x,t|z,\theta\}$) but their evaluation generally involves computation of a normalizing factor $p\{z\}$ that is a function of θ. This makes the estimation of θ a difficult task. On the other hand, J as defined in (2.4) has been used for myopic deconvolution of seismic signals with apparent success [11] and is straightforward to compute. When θ and t are known, x is nonstationary Gaussian and maximisation of J yields an explicit solution which is linear with respect to the observations, and which can be computed using standard techniques such as Kalman filtering. For the simple deconvolution problem, θ is known and restoration of x is a detection-estimation problem that cannot be solved optimally in real size problems. Suboptimal schemes have to be designed, and the level of suboptimality has a strong influence on the final quality of the results. Two approaches to this problem will be presented in section 4. For myopic deconvolution, i.e. estimation of $\{t,x,\theta\}$ through maximization of J, interesting results have been obtained on synthetic and real world examples. Of course, no convergence result can be established for t or x since the dimension of these vectors and the number of observations grow accordingly. However, even though no such restriction applies to θ, the properties of the estimator have not been established yet. This is due to the complexity of the mathematical derivations involved. The aim of section 3 is to begin to clarify this problem by using a simplified model of the phenomena.

3. HYPERPARAMETER ESTIMATION USING THE MAP APPROACH

First, it should be noted that regularization (or ridge regression) parameters which play a crucial role in the solution of inverse problems can be considered as hyperparameters. Hence their estimation is of great practical importance, and no fully satisfactory solution is available yet, with the possible exception of the linear and Gaussian case. Questions remain regarding the choice of a proper estimation criterion and the related optimization technique. The point addressed here is whether the MAP approach introduced in section 2 can yield satisfactory solutions to our problem. In order to avoid intractable mathematical derivations a simplified model, in which H is the identity matrix, is used. The rationale is that a study conducted in this oversimplified framework will help gain insight on the nature of the complete problem and hopefully open the way to new results. Lack of space prevents us from reporting all derivations, and most results will be stated without proof. For more details, see [13].

Let J_N denote the MAP criterion when the number of observations is equal to N. With H = Id, J_N takes the simplified logarithmic form

$$J_N = -\operatorname{Ln}(\sqrt{2\pi r^n}) - \frac{1}{2r^n}\sum_{k=1}^{N}\frac{(z(k)-t(k)x(k))^2}{N} - \sum_{k=1}^{N}\frac{t(k)}{N}\operatorname{Ln}(\sqrt{2\pi r^x})$$

$$-\frac{1}{2r^x}\sum_{k=1}^{N}\frac{x(k)^2}{N} + \sum_{k=1}^{N}\frac{t(k)}{N}\operatorname{Ln}(\lambda) + \left(1-\sum_{k=1}^{N}\frac{t(k)}{N}\right)\operatorname{Ln}(1-\lambda) \tag{3.1}$$

Let $\hat{t}_N$, $\hat{x}_N$ and $\hat{\theta}_N$ respectively denote the values of t, x and θ that maximize J_N. Questions are: i) Do such estimates exist and are their values meaningful? ii) In case of positive answer to i), what is the behaviour of $\hat{\theta}_N$ when N goes to infinity? It is easily seen on (3.1) that for arbitrary values of the variables (e.g. $\hat{x}_N = 0$, $\hat{t}_N=1$, $\hat{r}^n_N=1$, $\hat{\lambda}_N=0.5$ and $\hat{r}^x_N \to 0$) J_N goes to infinity. Therefore, in order to obtain meaningful estimates, the range of variation of θ should be restricted to a set Θ such as

$$\Theta = \{(r^n,r^x,\lambda)\,|\,r^n \in [\varepsilon_n, A_n],\ r^x \in [\varepsilon_x, A_x],\ \lambda \in [\varepsilon_\lambda, 1-\varepsilon'_\lambda]\} \tag{3.2}$$

where ε takes a small positive value and where A is arbitrary. Since estimates of θ located on the boundary of Θ are meaningless, the question is that of the existence of local maxima of J_N and of their asymptotic behaviour. These two points can be studied simultaneously by virtue of the following theorem [13]:

Let J_∞ and $\hat{\theta}_\infty$ denote the asymptotic MAP criterion: $J_\infty = \lim_{N\to\infty} J_N$ and a global maximizer of J_∞ on Θ respectively.

If $\hat{\theta}_\infty$ exists, then for any positive integer N large enough, $\hat{\theta}_N$ exists; any accumulation point of $\hat{\theta}_N$ is a global maximizer of J_∞ on Θ. If $\hat{\theta}_\infty$ is unique, then $\lim_{N\to\infty} \hat{\theta}_N = \hat{\theta}_\infty$ almost surely.

In addition, if $\hat{\theta}_\infty$ does not lie on the boundary of Θ, then $\sqrt{N}(\hat{\theta}_N - \hat{\theta}_\infty)$ is asymptotically normal.

The consequence of the theorem is that a necessary condition for the MAP estimator to have acceptable properties is that J_∞ exhibits local maxima. We now examine this question.

We first rewrite the expression of J_N. It can be shown that for any value of θ the optimal detector of the Bernoulli sequence is a threshold detector on z, and that the estimate of the imput takes the form

$$\hat{x}_N(k) = \frac{r^x}{r^x + r^n}\,\hat{t}_N(k)\; z(k) \tag{3.3}$$

The expression of J_N can be rewritten as function of the threshold value T

$$J_N = -\mathrm{Ln}(\sqrt{2\pi r^n}) - \frac{1}{2r^n}\sum_{k=1}^{N}\frac{z(k)^2\, 1_{z^2(k)<T}}{N} - \sum_{k=1}^{N}\frac{1_{z^2(k)\geq T}}{N}\mathrm{Ln}(\sqrt{2\pi r^x})$$

$$- \frac{1}{2(r^x+r^n)}\sum_{k=1}^{N}\frac{z(k)^2\, 1_{z^2(k)\geq T}}{N} + \sum_{k=1}^{N}\frac{1_{z^2(k)\geq T}}{N}\mathrm{Ln}(\lambda) + \sum_{k=1}^{N}\frac{1_{z^2(k)<T}}{N}\mathrm{Ln}(1-\lambda) \tag{3.4}$$

The threshold value $\hat{T}_N$ which correspond to a local maximum of J_N depends on $\hat{\theta}_N$ through the following equation

$$T_N = \frac{\hat{r}^n_N\left(\hat{r}^n_N + \hat{r}^x_N\right)}{\hat{r}^x_N} \mathrm{Ln}\left(\sqrt{2\pi\hat{r}^x_N}\,\frac{1-\hat{\lambda}_N}{\hat{\lambda}_N}\right) \tag{3.5}$$

Since all signals are ergodic, making N tend toward infinity yields the expression of the asymptotic criterion J_∞.

$$J_\infty = -\mathrm{Ln}(\sqrt{2\pi r^n}) - \frac{E[z^2 1_{z^2<T}]}{2r^n} - p\{z^2 \geq T\}\,\mathrm{Ln}(\sqrt{2\pi r^x})$$
$$- \frac{E[z^2\, 1_{z^2\geq T}]}{2(r^x+r^n)} + p\{z^2 \geq T\}\,\mathrm{LN}(\lambda) + (1 - p\,\{z^2 \geq T\})\mathrm{Ln}(1-\lambda) \tag{3.6}$$

It should be noted that quantities $E[z^2\, 1_{z^2<T}]$, $E[z^2\, 1_{z^2\geq T}]$ and $p\{z^2 \geq T\}$ are fully determined by the actual values r^n_T, r^x_T and λ_T of the hyperparameters and by the value of threshold T.

Necessary conditions for J_∞ to have local maxima are obtained by taking the gradient of (3.6) with respect to θ and setting it to zero, and by the additional constraint on T derived from (3.5). We obtain the following set of nonlinear equations

$$\mathrm{grad}_\theta\, J_\infty = 0 \tag{3.7a}$$

$$T = \frac{r^n(r^n + r^x)}{r^x}\,\mathrm{Ln}\left(\sqrt{2\pi r^x}\,\frac{1-\lambda}{\lambda}\right) \tag{3.7b}$$

We have not been able to solve (3.7) analytically. However an analytic solution to (3.7a) for a given value of T and θ_T can be computed rather easily. Substituting this solution into (3.7b) defines an implicit equation of the form $T = \Phi(T)$ for which a solution can be sought numerically. Thus, a numerical study of the asymptotic criterion has been carried out in the following conditions: A range of variation of the true hyperparameters $\theta_T = \left\{r^n_T, r^x_T, \lambda_T\right\}$ has been chosen according to usual values of these quantities, and has been sampled on a rectangular grid. For each sample of θ_T the numerical solution of (3.7) has been sought and when a solution was found, a numerical check was performed in its neighbourhood to make sure it corresponded to a local maximum. An example of the results is given in figure 2. λ_T is set to 0.1 while $\sqrt{r^n_T}$ and $\sqrt{r^x_T}$ vary linearly on [0.01, 0.865] and [0.05, 1.95] respectively. Black areas correspond to samples of θ_T for which (3.7) does not have a solution, grey areas to samples for which a solution of (3.7) exists but is not a local maximum and white areas to samples for which a solution to (3.7) was found and corresponds to a local maximum of J_∞. As in figure 2, it was observed that no local maximum exists for small values of r_T^n and for small signal-to-noise ratios. In addition the extent of the areas with no solution tends to increase with λ_T. Of course these results only have the limited value of any numerical study. However, they indicate that the MAP estimator of the hyperparameter may not even converge toward any meaningful value, and there is no way of assessing its behaviour ahead of time. In addition, it can be shown that maximization of J_N or J_∞, whenever possible, always yields biased estimates. Therefore this is strong evidence of the risky character of the MAP approach to hyperparameter estimation. Note that our results are in agreement with those reported in [6] in a different framework, Maximization of generalized likelihood functionals depending on several quantities of different natures is also encountered in other areas such as the restoration Markov random fields. According to Besag [1], the use of marginal probabilities instead of a global MAP criterion yields better results. In our situation, this amounts to maximising the usual likelihood of θ $p\{z|\theta\}$. This approach is currently under investigation and preliminary results are encouraging.

4. INPUT RESTORATION USING THE MAP APPROACH: TWO ALGORITHMS

We go back to the original problem described in section 2. θ is assumed to be known and we propose two simple deconvolution algorithms. Both of them present a low arithmetic count in order to meet our operational constraints, and one of them has a fully recursive structure suited to on-line data processing. As noted earlier, restoration of a B-G sequence is a detection-estimation problem, and the detection step must be carried out in some suboptimal manner. The recursive approach that we now present is a way of performing such a suboptimal detection.

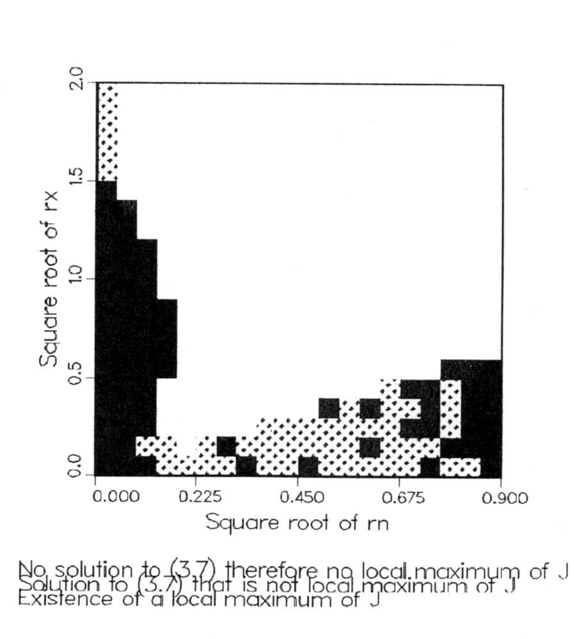

Fig. 2 Behaviour of the asymptotic criterion J_∞ for different values of the true hyperparameters. λ_T is set to 0.1 while $\sqrt{r_T^n}$ and $\sqrt{r_T^x}$ vary linearly on [0.01, 0.865] and [0.05, 1.95] respectively. Black areas correspond to samples of θ_T for which (3.7) does not have a solution, grey areas to samples for which a solution of (3.7) exists but is not a local maximum and white areas to samples for which a solution to (3.7) was found and corresponds to a local maximum of J_∞. As illustrated here, it was generally observed that no local maximum exists for small values of r_T^n and for small signal-to-noise ratios.

4.1 Recursive algorithm

The recursive algorithm is based upon a recursive approximation to MAP criterion J defined by (2.4): At time sample k, detection of t(k) and estimation of vector x are performed through maximization of J_k defined by

$$J_k(t(k),x) \triangleq p\{z^{k+\ell} | \hat{x}_{k-1}, \hat{t}^{k-1}, t(k)\} p\{\hat{x}_{k-1} | \hat{t}^{k-1}, t(k)\} p\{t(k)\} \tag{4.1}$$

where for any vector v, v^m denotes the subvector of v whose coordinate indices range between 1 and m. $\hat{x}_{k-1}$ is the estimate of x obtained at time-sample k-1, and $\hat{t}^{k-1}$ denotes the estimated Bernoulli sequence up to sample k-1. Both quantities are available at time k. ℓ is a time-shift whose function is to make the detection procedure more robust. Equation (4.1) is obtained by replacing z,x and t in the expression of J by the corresponding observations or estimates available at time k. Detection is achieved by maximizing J_k over the two possible values of t(k). Then x is estimated using its Gaussian nature conditionally to the knowledge of t.

The mathematical derivations that yield the exact expression of the algorithm are quite involved and, due to lack of space, they will not be reported here. The interested reader is referred to [3] where they appear in their entirety. On a practical standpoint, the functioning of the method is the following. Detection is achieved by maximisation of J_k over the two possible values of t(k), while all previous decisions are frozen. Therefore the decision on t(k) is made using all knowledge available at time sample k, but it will never be reversed afterwards. As noted earlier, when t is known, x can be estimated using Kalman filtering techniques. Since x is a white process, it is easily shown that computation of $\hat{x}_k$ requires the knowledge of the Bernoulli sequence up to time-sample k only. Since past detection results are never reversed, the detection step can be inserted in the estimation loop by Kalman filtering, which gives a globally recursive structure to the algorithm. An interesting point is that J_k can be evaluated from by-products of the Kalman filter used for the estimation, and the recursive detector can be interpreted as an adaptive test on a pseudo-innovation process. This contributes to achieving a low arithmetic count for the algorithm, and further computational savings are obtained by taking advantage of the sparse nature of the normal matrix of the problem

4.2 Iterative algorithm

In this approach, criterion J as defined by (2.4) is not approximated. The suboptimality comes from the maximization technique which is essentially similar to the SMLR detector proposed by Kormylo and Mendel [8]. Starting from an initial configuration (t_0, x_0, J_0) we introduce Bernoulli sequences t_k that are identical to t_0 except for the k-th coordinate t(k) whose value is changed. For each t_k, $1 \leq k \leq N$, the corresponding values of x_k and J_k are computed. Then the sequence that maximizes J_k is selected as a new initial configuration. The procedure is repeated until convergence is reached. Note that such a method guarantees an increase of J at each iteration, but can get trapped in a local maximum.

The algorithm we propose brings the following improvements over Kormylo and Mendel's SMLR detector: i) In our method detection and estimation are carried out simultaneously. ii) The detection-estimation algorithm has a much simpler structure. iii) The computational burden is low, However, the memory requirements can be quite high. These three points are a consequence of the MA representation that is used here, which allows us to express x_k as a function of x_o through RLS-type equations. The technique for obtaining such equations is outlined below.

First, we detail the expression of J. Using equation (2.4), we obtain

$$J \propto -\frac{(z - Hx)^T (z - Hx)}{2r^n} - \frac{x^T T x}{2r^x} - \frac{N_e}{2} \text{Ln}(2\pi r^x) + N_e \text{Ln}(\lambda) + (N-N_e)\text{Ln}(1-\lambda) \tag{4.2}$$

where T is a diagonal matrix defined by $T = \text{Diag}\{t(k)\}$, $1 \leq k \leq N$, and where N_e is the number of events of the Bernoulli sequence. Superscript ^{T} denotes the transpose operator. For a given Bernoulli sequence t, N_e and T are known and maximisation of J yields the following estimate of x

$$x = P\, H^T (r^n)^{-1} z \tag{4.3a}$$

$$P = \Pi - \Pi\, A\, \Pi \tag{4.3b}$$

$$A = H^T (H\, \Pi\, H^T + r^n\, \text{Id})^{-1} H \tag{4.3c}$$

$$\Pi = r^x T \tag{4.3d}$$

Note that Π and P represent the *a priori* and *a posteriori* estimation covariance matrix of x respectively. To obtain the relationship betweem x_k and x_o, the influence of a change in one diagonal element of Π is evaluated. Derivations make extensive use of the matrix inversion lemma, which cannot be applied to P since this matrix is singular. It is applied to matrix A instead, resulting in the following set of equations [2]

$$x_k = x_o - (a_o - v_k)\rho^{-1} v_k^T H^T (r^n)^{-1} (z - Hx_o) \quad (4.4a)$$

$$\rho = \pm (r^x)^{-1} + v_k^T A_o v_k \quad (4.4b)$$

$$a_o = \Pi_o A_o v_k \quad (4.4c)$$

$$A_k = A_o - A_o v_k \rho^{-1} v_k^T A_o \quad (4.4d)$$

In the above equations, all quantities related to x_o (resp. x_k) are denoted by subscript $_o$ (resp. $_k$). v_k is a N-dimensional vector whose all coordinates $v_k(i)$ are equal to zero except $v_k(k)$ which is equal to 1. Initial conditions of (4.4) are given by (4.3) for an arbitrary Bernoulli sequence t. The expensive part of the algorithm is equation (4.4d) which updates a NXN matrix. However, (4.4d) needs to be computed only once per iteration, when a new sequence is selected as initial conditions. Finally, a complete iteration of the algorithm (i.e. test of N different Bernoulli sequences t_k), can be carried out with $O(N^2)$ floating point multiplications or divisions. For usual values of N, this numerical complexity is compatible with the use of small (workstation-type) computer systems.

4.3 Simulation results

In order to evaluate the efficiency of both algorithms, tests were carried out on synthetic data. In all simulations, the signal-to-noise ratio, defined as the ratio between the mean power of the noiseless observations and the variance of the observation noise r^n, was set to 10 dB. For the sake of easy comparison with results previously published, we first used the reflectivity sequence explicitly defined by Mendel and Kormylo in [12] and used later on by Kollias and Halkias [7], along with a fourth-order Kramer wavelet [9]. Source wavelet, observed signal and deconvolution results are shown in figure 3. Both metnods perform well, with an advantage to the iterative approach. These results compare favourably with those reported in the literature.

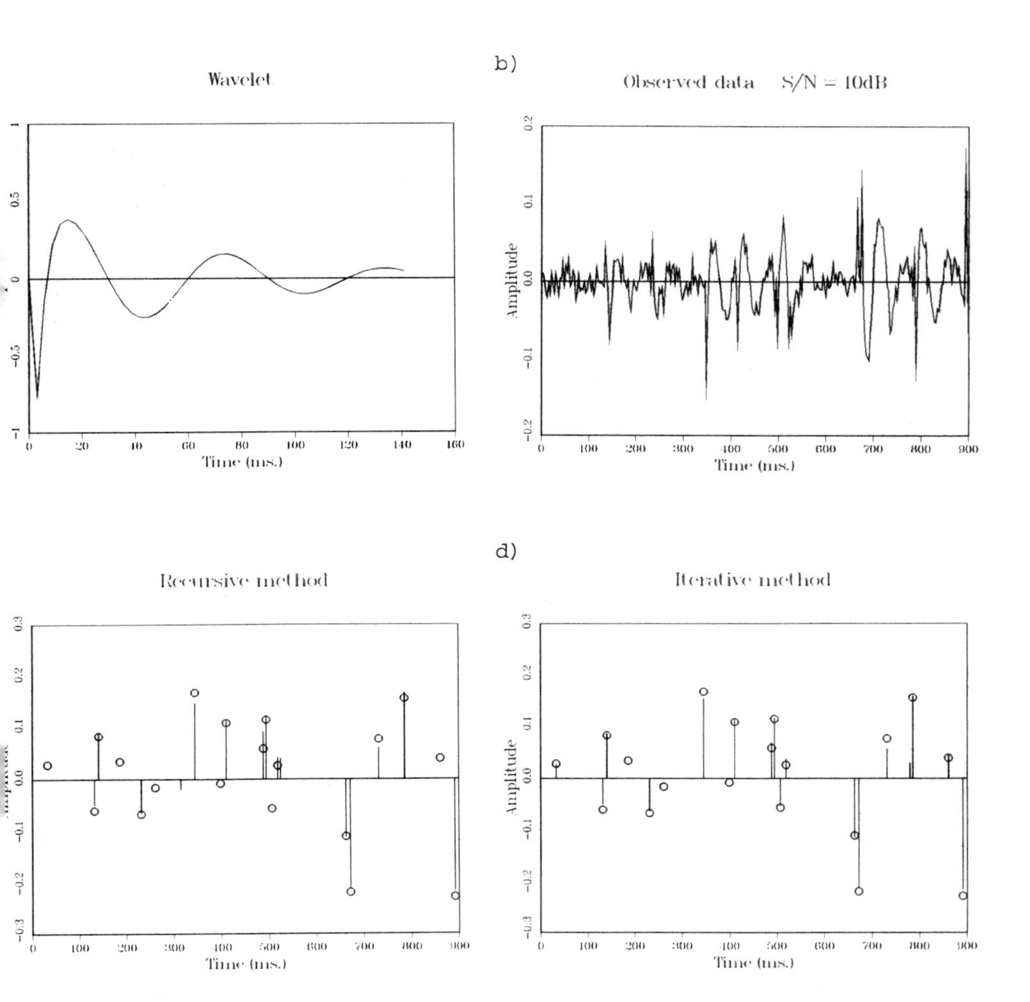

Fig. 3 a) Fourth order source wavelet (from [9]). b) Observed signal (S/N) = 10 bB). c) Deconvolution results obtained with the recursive method. d) Deconvolution results obtained with the iterative method. The circles depict the actual reflectivity, and the estimated signal is shown in solid lines. Both methods perform well, with an advantage to the iterative approach. These results compare favourably with those reported in the literature.

The recursive detector was found to be sensitive to the shape of the wavelet. This effect is illustrated in figure 4. An actual seismic wavelet was used with a simple reflectivity sequence. While the iterative method still performs well, the recursive algorithm exhibits instabilities that are typical of *decision-directed*-type of approaches in difficult situations. This is certainly the major drawback of the recursive approach, to which more robust methods should be preferred when conditions allow it.

Figure 4 was extracted from the set of simulated seismic traces shown in figure 5. Here too, the difference between the two algorithms appears clearly. The iterative method produces better results and seems to be more robust. However, a two-dimensional model of the propagation medium should bring further improvements. This illustrates the limits of 1-D processing of 2-D signals.

5. CONCLUSION

In this paper results regarding myopic deconvolution on B-G signals using a MAP approach have been presented. Estimation of hyperparameters was investigated first. It was shown that the MAP approach to this problem is very questionable since the corresponding estimator may not even converge toward any meaningful value, and since there is no way of assessing its properties beforehand. Even though the approach has already been used with apparent success, other and more reliable methods should be developed. Then, simple deconvolution was addressed and two algorithms designed to fulfil various operational constraints were presented. A recursive method suited to on-line data processing was found to provide interesting results, but exhibited a lack of robustness. An iterative method produced better results at the expense of a slight increase of computations and of a high memory requirement. Both methods could be improved. The recursive algorithm could be expanded into a *stack algorithm* in order to reverse previous decisions and improve the robustness of the method. Equations (4.4) could be used to perform an optimal maximisation of criterion J using stochastic methods. Development of 2-D models of layered propagation media should bring substantial improvements, as long as the data can be processed in a reasonable time.

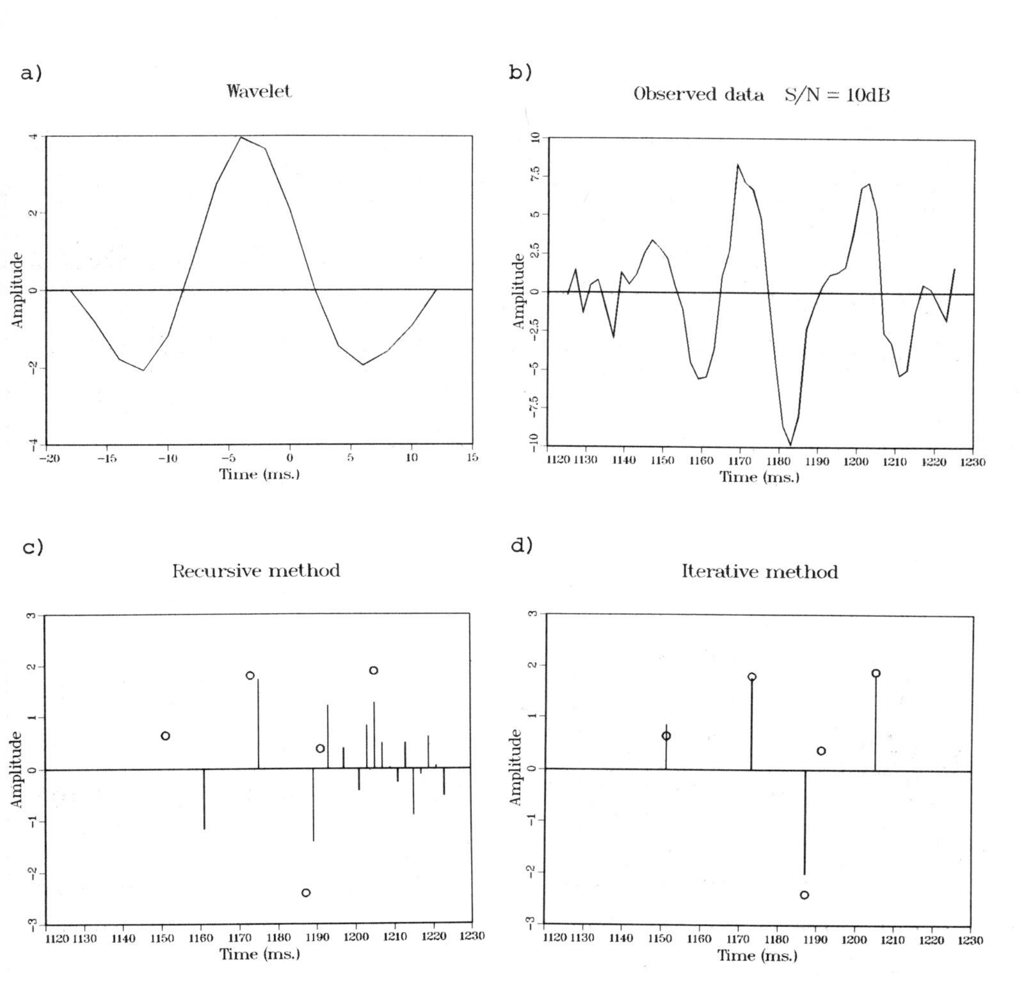

Fig. 4 a) Actual seismic wavelet. b) Observed signal (S/N=10 dB) c) Deconvolution results obtained with the recursive method. d) Deconvolution results obtained with the iterative method. The circles depict the actual reflectivity, and the estimated signal is shown in solid lines. The iterative method still performs well, while the recursive algorithm exhibits instabilities that are typical of *decision-directed*-type of approaches in difficult situations. This illustrates the sensitivity of the recursive detector to the shape of the wavelet.

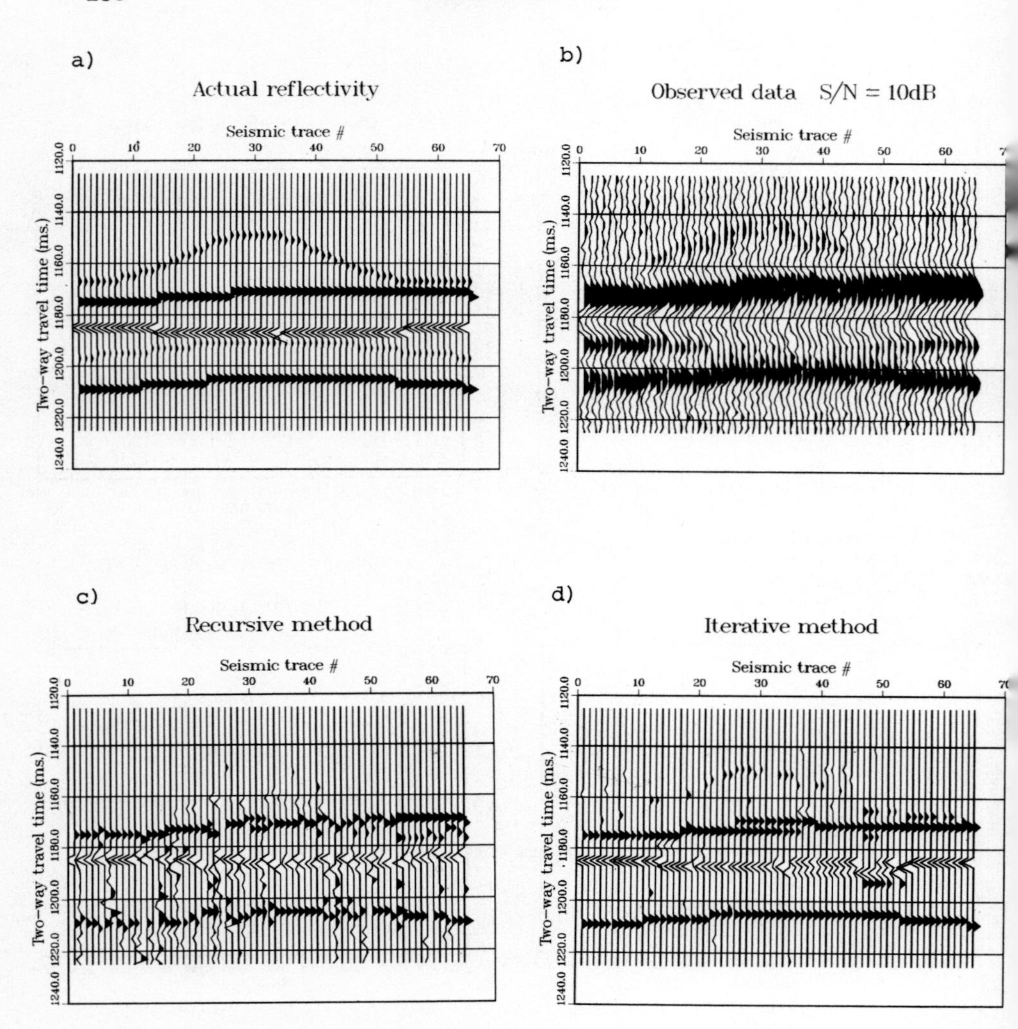

Fig. 5 Deconvolution of a set of simulated seismic traces. a) Actual reflectivity. b) Observed data (S/N = 10 dB). c) Deconvolution results obtained with the recursive method. d) Deconvolution results obtained with the iterative method. The iterative method produces better results and seems to be more robust. However, a two-dimensional model of the propagation medium should bring further improvements.

6. REFERENCES

[1] Besag, J., (1986) On the Statistical Analysis of Dirty Pictures, *J.R. Statist. Soc. B*, **48**, 259-302.

[2] Goussard, Y., (1989) Déconvolution de processus Bernoulli-gaussiens. Un algorithme itératif., *Rapport interne LSS GPI-89/02.*

[3] Goussard, Y. and Demoment, G., (1989) Recursive deconvolution of Bernoulli-Gaussian processes using a MA representation, To appear in IEEE Trans. Geoscience and Remote Sensing, GE-27.

[4] Goutsias, J. and Mendel, J.M., (1986) Maximum-likelihood deconvolution: An optimization theory perspective, *Geophysics*, **51**, 1206-1220.

[5] Grenier, Y., Fabre, P. and Omnes, M.C., (1985) Utilisation des techniques de modélisation en profil sismique vertical, Proc. ATP CNRS Géophysique appliquée à la prospection, 30-31.

[6] Harvey, A.C. and Peters, S., (1985) A note on the estimation of variance in state-space models using the maximum a posteriori procedure, *IEEE Trans. Automatic Control*, **AC-30**, 1048-1050.

[7] Kollias, S.D. and Halkias, C.C., (1985) An instrumental variable approach to minimum-variance seismic deconvolution, *IEEE Trans. Geoscience and Remote Sensing*, **GE-23**, 778-788.

[8] Kormylo, J. and Mendel, J.M., (1982) Maximum-likelihood detection and estimation of Bernoulli-Gaussian processes, *IEEE Trans. Information Theory*, **IT-28**, 482-488.

[9] Kramer, F.J., Peterson, R.W. and Walter, W.C., (Editors) (1968) Seismic energy sources 1968 handbook, 38th Ann. Meeting of the Soc. Of Exploration Geophysicists, Denver, Colorado.

[10] Mahalanabis, A.K., Prasad, S. and Mohandas, K.P., (1982) Recursive decision-directed estimation of reflection coefficients for seismic data deconvolution, *Automatica*, **18**, 721-726.

[11] Mendel, J.M., (1983) Optimal seismic deconvolution, Academic Press, New York.

[12] Mendel, J.M. and Kormylo, J., (1978) Single-channel white-noise estimators for deconvolution, *Geophysics*, **43**, 102-124.

[13] Monfront, F., (1988) Déconvolution de signaux Bernoulli-gaussiens par maximum de vraisemblance, Rapport de stage de DEA, Université de Paris-Sud, Orsay.

[14] Nashed, M.Z., (1981) Operator-Theoretic and Computational Approaches to Ill-Posed Problems with Applications to Antenna Theory, *IEEE Trans. Antennas and Propagation,* **AP-29**, 220-231.

[15] Wiggins, R.A., (1977) Minimum Entropy Deconvolution, Proc. IEEE Intern. Symp. Computer-Aided Seismic Analysis and Discrimination, Falmouth, MA, 7-14.

[16] Yarlagadda, R., Bednar, J.B. and Watt, T.L., (1985) Fast Algorithms for Lp Deconvolution, *IEEE Trans. Acoustics, Speech and Signal Processing,* **ASSP-33**, 174-182.

CORRELATION AND SMOOTHNESS CONSTRAINTS IN THE LINEAR INVERSE PROBLEM

D.L. Blanchard, C.H. Travis and M.C. Jones
(Department of Physics, University of Surrey)

ABSTRACT

This paper describes methods of applying smoothness constraints to the linear inverse problem, and shows how the smoothness constraint matrix may be partitioned to produce a more accurate estimate of a signal from some noisy measurements. We also give examples of introducing discontinuities in the approximation of the mean of a noisy signal to improve the estimate of the original signal.

The application of smoothness constraints to image processing problems has been extensively researched by Blake & Zisserman (1986a, 1986b, 1987), Grimson (1982), Terzopoulos (1983, 1984, 1986, 1988) and others. Here we show a related but conceptually and implementationally distinct method of applying smoothness/correlation constraints to the linear inverse problem.

For this paper, we restrict ourselves to the one dimensional case, for which the property of "smoothness" can be expressed in terms of the first derivative: a "smooth" function f(x) is one for which the functional $\int \left(\frac{d}{dx} f(x) \right)^2 dx$ is small.

In the linear inverse problem we are attempting to find the signal f from the measurements g where

$$g = Lf + n \qquad (1)$$

n is (zero mean) noise and L is a linear operator representing the measurement system.

The link between correlation and smoothness constraints arises when we are dealing with a class of signals f which we know to have some finite range correlation - "smooth" signals.

This is formalized in the linear minimum variance estimate for f (Sage and Melsa 1971), given by

$$\hat{f} = \mu + PL^{\dagger}(LPL^{\dagger} + R)^{-1}(g - L\mu) \tag{2}$$

which may alternatively be written (Jones and Travis 1988)

$$\hat{f} = \mu + \left(L^{\dagger}BL + \frac{\sigma_n^2}{\sigma_s^2}Q\right)^{-1} L^{\dagger}B(g - L\mu) \tag{3}$$

where μ is the mean and P the covariance of the signal (with itself), R is the covariance of the noise (with itself), σ_s^2 and σ_n^2 are the signal and noise variances respectively, with $Q = \sigma_s^2 P^{-1}$ and $B = \sigma_n^2 R^{-1}$.

For P and R diagonal (2) and (3) reduce to the Tikhonov regularized solution (Tikhonov and Arsenin 1977).

A standard model for the signal covariance is

$$P(x,x') = \sigma_s^2 \frac{\alpha}{2} e^{-\alpha|x-x'|} \tag{4}$$

from which we can obtain

$$\left(-\frac{d^2}{dx^2} + \alpha^2\right) P(x,x') = \alpha^2 \delta(x-x')\sigma_s^2 \tag{5}$$

and hence $P^{-1}(x,x') = \dfrac{1}{\sigma_s^2}\left(-\dfrac{1}{\alpha^2}\dfrac{d^2}{dx^2} + 1\right)\delta(x-x')$ (6)

Thus $\langle f,P^{-1}f\rangle = \dfrac{1}{\sigma_s^2}\left(\dfrac{1}{\alpha^2}\langle f',f'\rangle + \langle f,f\rangle\right)$ (7)

where $\langle .,. \rangle$ is the inner product.

The size of α determines the range of correlation. In the discretized case, if $x_{i+1} - x_i = \Delta$, and $\rho = e^{-\alpha\Delta}$, P is

$$(1-\rho^{1/2})\Delta^{-1}\sigma_s^2 \begin{bmatrix} 1 & \rho & \rho^2 \dots & & \\ \rho & 1 & \rho & \rho^2 & \dots \\ \rho^2 & \rho & 1 & \rho & \rho^2 \dots \\ \dots & & & & \end{bmatrix} \tag{8}$$

with inverse

$$\begin{bmatrix} 1 & -\rho & 0 & \dots & & & \\ -\rho & 1+\rho^2 & -\rho & \dots & & & \\ 0 & -\rho & 1+\rho^2 & -\rho & \dots & & \\ & & & & & & -\rho \\ & & & & & -\rho & 1 \end{bmatrix} \frac{\Delta}{\sigma_s^2(1-\rho^2)(1-\rho)} \tag{9}$$

Another way of expressing the application of a smoothness constraint to the solution of (1) can be derived from the (constrained) least squares solution:

$$\underset{\hat{f}}{\text{Minimize}}\ \beta(g - L\hat{f})^\dagger(g - L\hat{f}) + \hat{f}^\dagger A\hat{f} \tag{10}$$

The extra term $f^\dagger Af$ is the smoothness constraint. The factor β depends on the signal to noise ratio and the "strength" of the smoothness constraint. For a first order smoothness constraint the $f^\dagger Af$ is equivalent to $\langle f', f'\rangle$ (c.f.7) and a matrix representation of A may be written as

$$A = \begin{bmatrix} 1 & -1 & & & & & \\ -1 & 2 & -1 & & & & \\ & -1 & 2 & -1 & & & \\ & & \cdot & \cdot & \cdot & & \\ & & & \cdot & \cdot & \cdot & \\ & & & & \cdot & \cdot & \cdot \\ & & & & -1 & 2 & -1 \\ & & & & & -1 & 2 & -1 \\ & & & & & & -1 & 1 \end{bmatrix}$$

which is the limiting form of (9), with $\alpha \to 0$, $\beta = \dfrac{\sigma_s^2}{\sigma_n^2}\Delta\alpha^2$ (11)

Differentiating (10) above with respect to f, setting the result to zero and rearranging, we arrive at

$$\hat{f} = (\beta L^{\dagger} L + A)^{-1} L^{\dagger} g \tag{12}$$

which is of the same form as (3), with $\mu = 0$.

So far we have been applying the smoothness/correlation constraint across the entire range of x. However in many cases of interest we may wish to "relax" the constraint at an "edge" between two regions in the reconstructed $\hat{f}$ in order to reproduce a discontinuity in f of which we have some a priori knowledge. This relaxation of the constraint may be simply achieved by a suitable partitioning of the A or P^{-1} matrices, in which the off-diagonal blocks are set to zero. An alternative method of introducing edges, some results of which we shall show later in this paper, is to introduce a non-uniform μ.

For a particular partitioning of A, the smoothness constraint matrix, the matrix $(\beta L^{\dagger} L + A)$ can be inverted using standard techniques, then a value of $\hat{f}$ calculated.

How do we decide where to partition the matrix? The problem is that although a particular partition of A may lead to a minimum in expression (10) it may not be the <u>global</u> minimum. A brute force solution may be applied if number of edges is known, or if it is known that the edges are no closer together than a certain minimum distance.

The brute force approach is to solve (12) for all possible positions of a single edge, evaluate (10) for each $\hat{f}$ then look for minima in the plot of the value of (10) against edge position to decide where the edges "really" are. This process need not be as computationally costly as it first appears.

Having inverted $(\beta L^{\dagger} L + A)$ first with an unpartitioned A, the stucture of A allows the inverse of $(\beta L^{\dagger} L + A')$, where A' is a partitioned version of A, to be calculated with little extra computational effort. This because the structure of A allows A' to be written thus

$$A' = A - D \tag{13}$$

where, for example for a first derivative smoothness constraint the matrix representation of A is

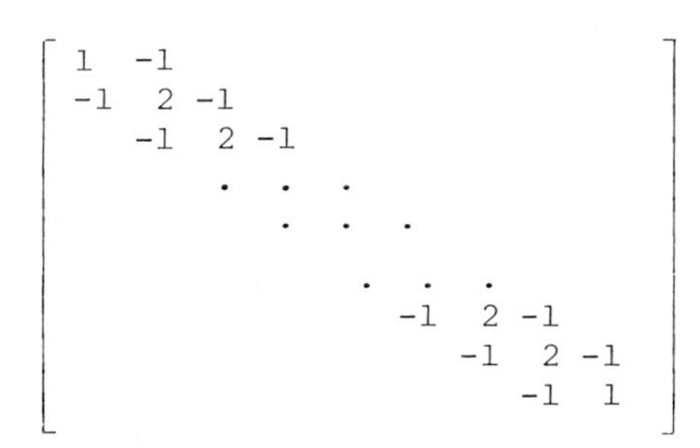

and D is simply

$$\begin{bmatrix} & & & & \\ & & 1 & -1 & \\ & & -1 & 1 & \\ & & & & \end{bmatrix}$$

Since $(\beta L^{\dagger}L + A')^{-1} = (\beta L^{\dagger}L + A - D)^{-1}$ can be rearranged as $(\beta L^{\dagger}L + A)^{-1}(I - D(\beta L^{\dagger}L + A)^{-1})^{-1}$ and D is so sparse, once the initial computational effort required to find $(\beta L^{\dagger}L + A)^{-1}$ has been made it requires very little extra computation to find $\hat{f}$ if A is changed to A'. In the case of the linear minimum variance estimate, not only may p^{-1} be partitioned as above, but the mean μ of the signal may be discontinuous. Obviously, however, considerably more a priori knowledge of the signal f is required to choose a "good" μ than is required when simply applying smoothness constraints.

Figs. 1 to 26 show the results of this process for a variety of L, f and signal to noise ratios.

Fig. 1 shows the original one-dimensional f distribution, and fig. 2 shows the (noise-free) measurement g created by using a Gaussian-weighted local averaging sampling operator for L. This local averaging operator is moved in increments of twice the discretization used in fig. 1, and so the L operator alone is singular. Thus one <u>must</u> apply some extra constraints in order to find a unique value of $\hat{f}$ from g.

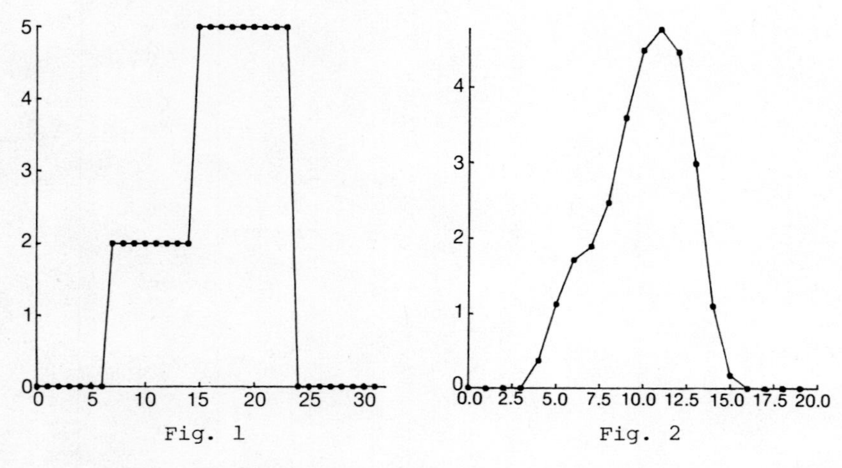

Fig. 1 Fig. 2

Fig. 3 shows $\hat{f}$ calculated from the "edge-free" inversion of $(\beta L^{\dagger} L + A)$ where A corresponds to a first derivative or "flatness" constraint. The value of β used here was 1000, corresponding to small σ_n^2 in (11). Fig. 4 shows the value of the functional (10) plotted against the position of a single edge.

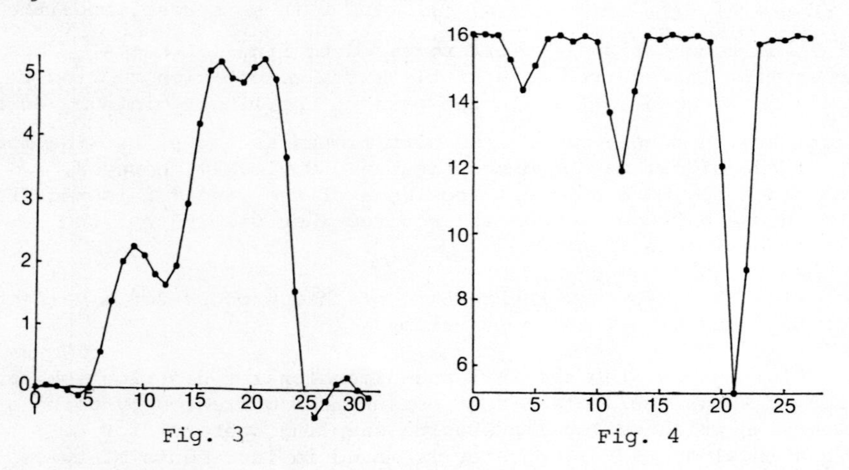

Fig. 3 Fig. 4

Figs. 5 and 6 show the values of $\hat{f}$ obtained when one (fig. 5) or two (fig. 6) edges are "inserted" into the constraint matrix at the positions indicated by the minima in fig. 4. If all three edges indicated by the minima of fig. 4 are inserted, the resultant $\hat{f}$ is indistinguishable from fig. 1, shown at this scale.

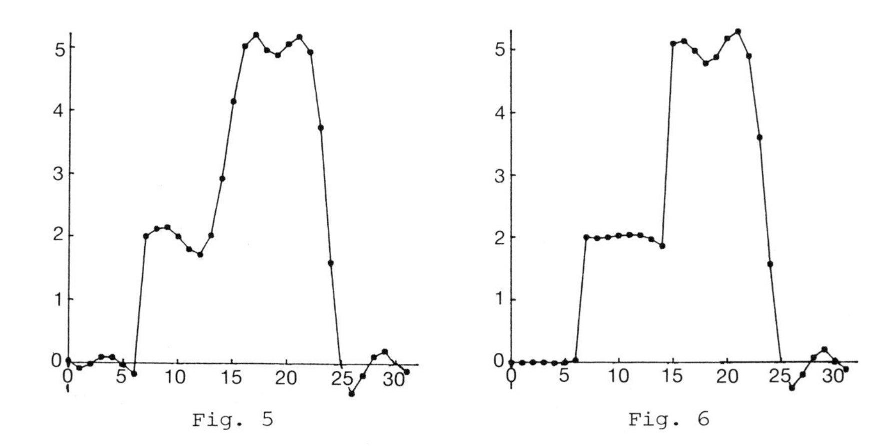

Fig. 5 Fig. 6

Fig. 7 is the g obtained from sampled the data of fig. 2 with a different operator: The Gaussian local averaging operator in this case is moved in increments of five times the discretization interval used in fig. 1. β is given a value of 1000 as in the previous example. Fig. 8 shows the value of the functional (10) plotted against the position of a single edge as before.

Fig. 9 shows the value of $\hat{f}$ from fig. 7 with no edges, and figure 10 shows $\hat{f}$ from fig. 7 with edges in the positions indicated by the minima of fig. 8.

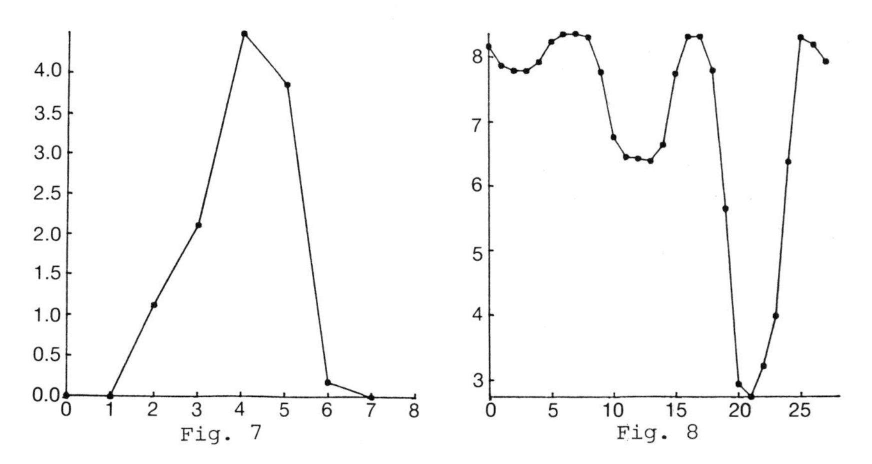

Fig. 7 Fig. 8

It is worth noting that although the edges found by the minima of fig. 8 are not exactly in the right places, they are only 1 discretization interval out, which is better than one would have expected from an initial sampling of five times that size.

Figs. 11 - 21 show the results of attempting to reconstruct a simple step edge from an extremely noisy sample g: the signal to noise ratio used (σ_s^2/σ_n^2) here is 2. The sampling operator L was simply the identity operator. In order to demonstrate the robustness of this technique, the position of the "best" step edge found was recorded for 100 different noisy g samples (all with the same signal to noise ratio), using different values for β corresponding to "erroneous" a priori knowledge of the signal to noise ratio and range of correlation. Fig. 11 shows the original f distribution prior to the addition of noise which we will attempt to reconstruct. Fig. 12 shows a "g" sample of f with the added noise. Fig. 13 is the $\hat{f}$ reconstruction from the g of fig. 12 with no edges inserted using a value of β derived from the "correct" values of the signal to noise ratio and correlation distance. Fig. 14 shows the value of the functional (10) plotted against all edge positions for the g of fig. 12. Fig. 15 shows the $\hat{f}$ reconstructic with the "best" single edge as indicated by fig. 14. Figs. 16, 17 and 18 show histograms of the number of times a particular edge position was chosen as the "best" for a variety of β values. Fig. 16 corresponds to β chosen to be the "correct" value, fig. 17 has β at twice the "correct" value, and fig. 18 has β at four times the "correct" value. Figs. 19, 20 and 21 are sample $\hat{f}$ reconstructions with the β values corresponding to figs. 16, 17 and 18 respectively, each reconstructed from the same g with the edge inserted in the same place.

Figs 22-26 show the results of using the linear minimum variance estimate to invert a noisy Laplace transform. In these examples, the covariance matrices of the signal and noise were taken to be diagonal (i.e. uncorrelated signal and uncorrelated noise), and various values of μ were used.

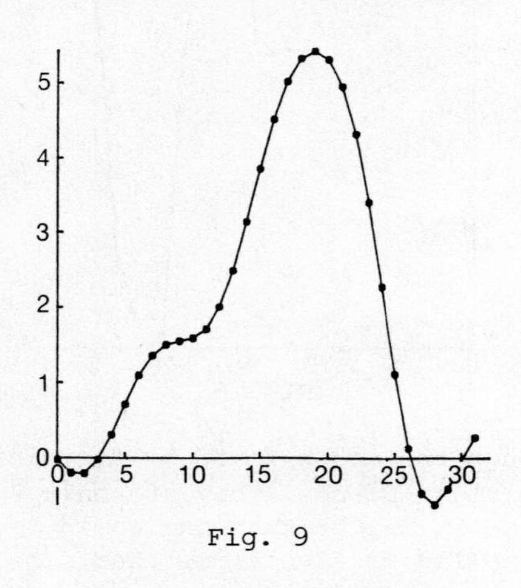

Fig. 9

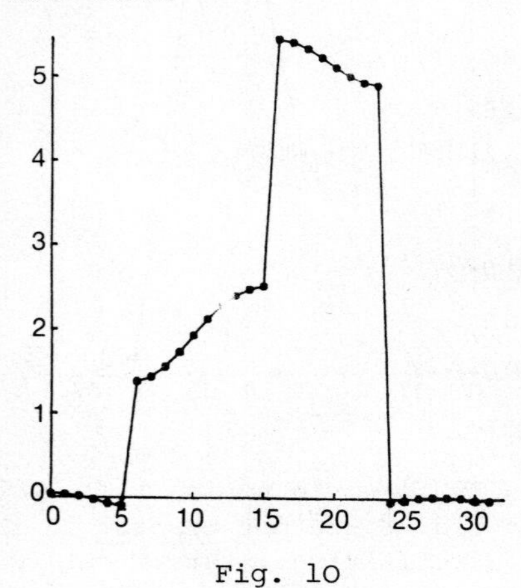

Fig. 10

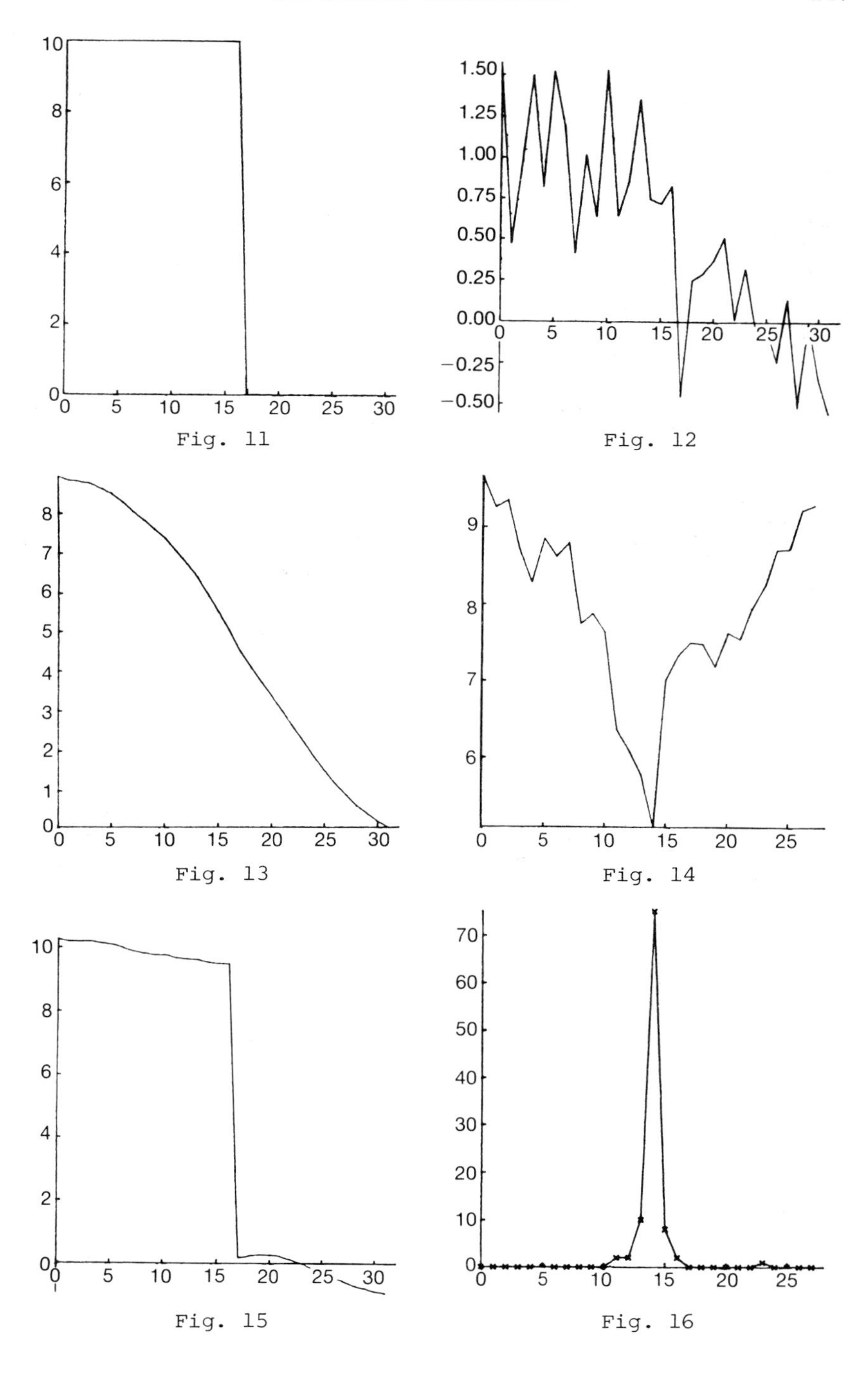

Fig. 11

Fig. 12

Fig. 13

Fig. 14

Fig. 15

Fig. 16

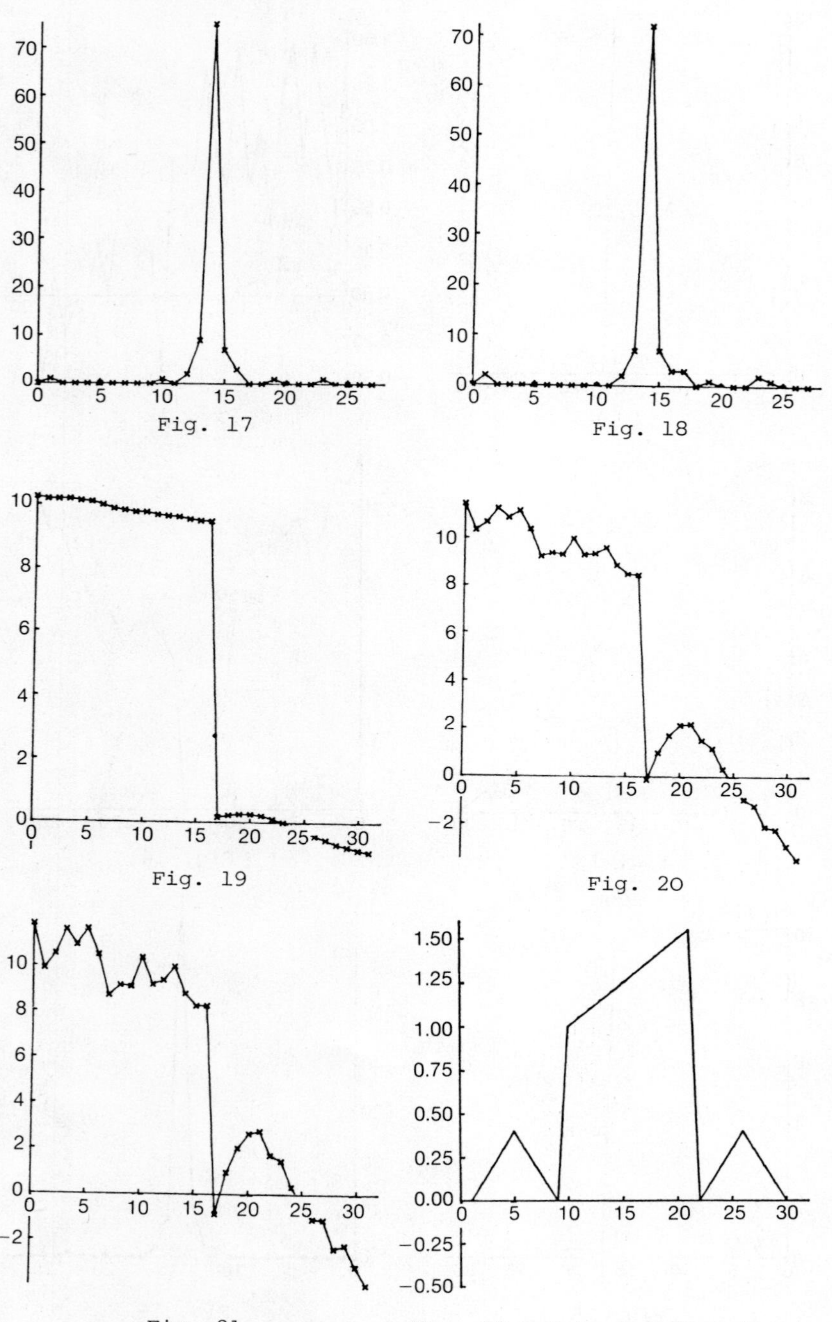

Fig. 17

Fig. 18

Fig. 19

Fig. 20

Fig. 21

Fig. 22 Original f distribution

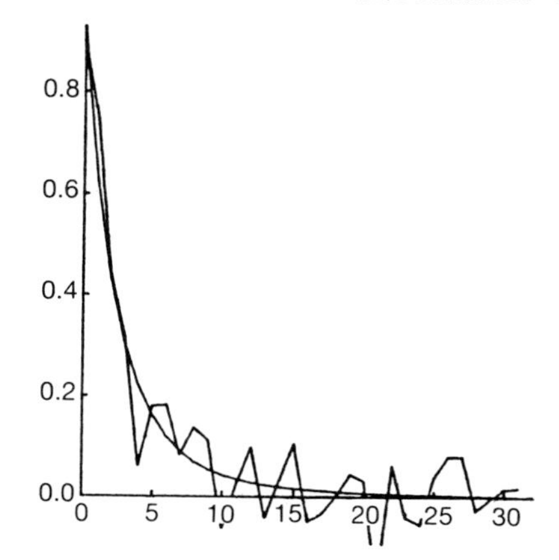

Fig. 23 g - real Laplace transform of fig. 22, with and without added noise

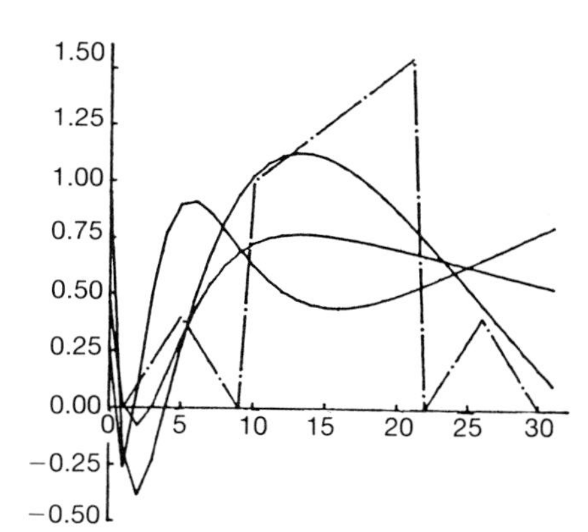

Fig. 24 Solid lines - linear minimum variance estimate of f of fig. 22 from Laplace transforms with various noise samples, all with same signal/noise ratio. $\mu = 0$ was used for all of the above. Chained line - original f.

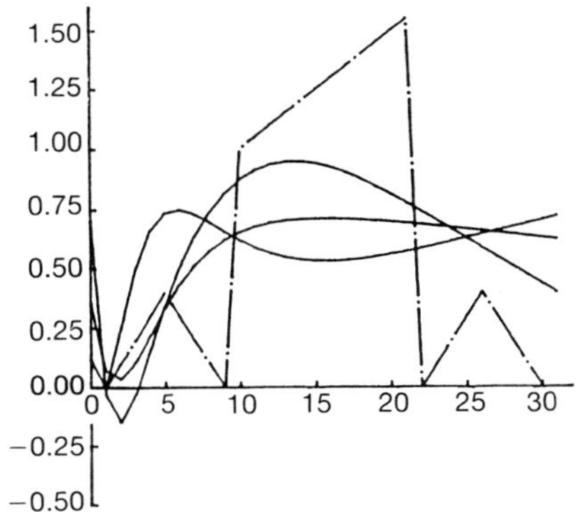

Fig. 25 As fig. 24, but with μ constant, equal to mean value of fig. 22

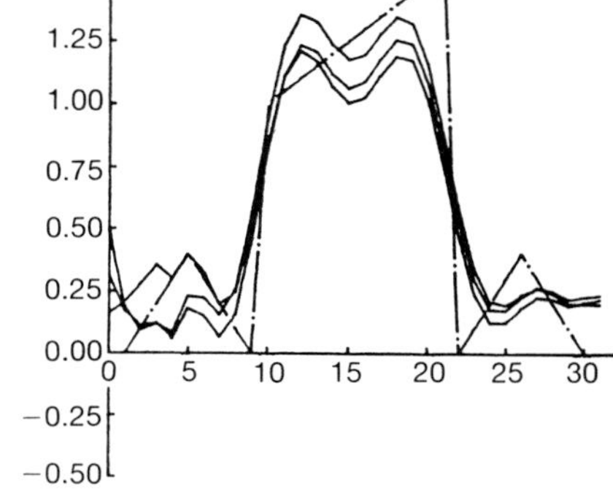

Fig. 26 As fig. 24 but with $\mu = 0$ for $x < 9$, $\mu = 1.25$ for $9 < x < 22$ and $\mu = 0$ for $x > 22$.

REFERENCES

Blake, A. and Zisserman, A., (1986a) Invariant surface reconstruction using weak continuity constraints. *Proc. CVPR-86 Miami*, pp. 62-67.

Blake, A. and Zisserman, A., (1986b) Some properties of weak continuity constraints and the GNC algorithm. *Proc. CVPR-86, Miami*, 656-661.

Blake, A. and Zisserman, A., (1987) Visual Reconstruction, MIT press, Cambridge, Massachusetts, London.

Grimson, W.E.L., (1982) A computational theory of visual surface interpolation. *Phil. Trans. Roy. Soc. Lond. B298*, 395-427.

Jones, M.C. and Travis, C.H., (1988) Bidiagonalization for inverse problems. *J. Opt. Soc. Am. A*, Vol. 5, 660-665.

Sage, A.P. and Melsa, (1971) Estimation Theory, McGraw-Hill.

Terzoploulos, D., (1983) Multilevel computational processes for visual surface reconstruction. *Computer Vision, Graphics and Image Processing*, Vol. 24, 52-96.

Terzopoulos, D., (1984) Multilevel reconstruction of visible surfaces. Variational principles and finite-element representations. From: Multiresolution Image processing and analysis. Ed. A. Rosenfeld, Springer-Verlag.

Terzopoulos, D., (1986) Regularization of inverse visual problems involving discontinuities. *IEEE Trans. PAMI-8, No. 4.*

Terzopoulos, D., (1988) The computation of visible surface representations. *IEEE Trans. PAMI-10, No. 4*, 417-38.

Tikhonov, A.N., and Arsenin, V.Y., (1977) Solution of Ill-posed Problems. Winston, Washington D.C.

APPLYING CATASTROPHE THEORY TO NONLINEAR MODEL FITTING

A. van den Bos
*(Department of Applied Physics,
University of Technology, Delft, The Netherlands)*

ABSTRACT

Elementary catastrophe theory is used to detect structural change of model fitting criteria under the influence of errors in the observations. Detection is desirable since structural change may influence the model fitting solution in an adverse and unexpected way. In particular, structural change may limit the parametric resolution of the model fitting method involved. The usefulness of these considerations is demonstrated in practical examples.

1. INTRODUCTION

The purpose of this paper is to show the usefulness of elementary catastrophe theory for the analysis and solution of a broad class of nonlinear model fitting problems. Catastrophe theory describes structural changes of a parametric function under the influence of changes of the parameters and makes it possible to compute the parameter values for which the structural change occurs [3],[4]. Here, the concept structure is to be understood as the number and nature of the stationary points of the function. In model fitting the structure of the criterion of goodness of fit for error corrupted observations may differ from that for hypothetical exact observations. Then, if the absolute minimum of the latter criterion is involved in the change, the solution with error corrupted observations may become questionable. It is, therefore, important to detect structural changes of the criterion used. For that purpose catastrophe theory provides the required tools.

The paper is organized as follows. Section 2 describes how to reduce the expression for a model fitting criterion to a <u>catastrophe.</u> This is a relatively simple polynomial whose structural changes exactly describe the structural changes of the criterion concerned. In Section 3 these ideas are applied to model fitting criteria for the class of weighted sum models. These are weighted sums of functions of the same nonlinearly parametric family. To this class belong compartmental models, diffraction patterns, multiple sonar echoes, etc. The usefulness of the results of Section 3 is demonstrated in a number of practical examples in Section 4. Conclusions are presented in Section 5.

2. CRITERIA AS CATASTROPHES

Catastrophe theory deals with structural change of a function of one or more independent variables under the influence of its parameters. For a particular value of the parameters the Hessian matrix in a stationary point of a parametric function may become degenerate. Then one or more of the eigenvalues of the Hessian matrix are equal to zero. For the purposes of this paper it is sufficient to suppose that only one of the eigenvalues is equal to zero. The set of all parameter values leading to a degenerate Hessian matrix is called the bifurcation set. If, in parameter space, the bifurcation set is crossed, a one-saddle point may, for example, become a minimum. Therefore, for parameters in the bifurcation set the function is called structurally unstable.

Catastrophe theory shows that in a parameter neighbourhood of a degenerate stationary point, the function is structurally equivalent to a Taylor polynomial described by

$$\lambda_1 x_1^2 + \ldots + \lambda_{K-1} x_{K-1}^2 + \text{Taylor polynomial in } x_K \qquad (2.1)$$

where the λ_k are scalars proportional to the eigenvalues that can not vanish under the influence of the parameters, and the x_k are curvilinear transformations of the K original independent variables of the function. The Taylor polynomial in x_K is described by

$$\mu_2 x_K^2 + \mu_3 x_K^3 + \ldots + \mu_L x_K^L. \qquad (2.2)$$

In this expression the coefficient μ_L is the first coefficient that has been found nowhere to vanish in the parameter domain concerned.

From (2.1) and (2.2) it follows that the terms in the variables $x_1, \ldots, x_{K-1}$ can, in an investigation of structural change, be left out of consideration. Their coefficients are and remain strictly positive or negative everywhere in the parameter domain. These variables are, therefore, called inessential. So, the structural changes are a consequence of changes in $\mu_2, \ldots, \mu_L$ as a result of changes in the parameters. Therefore, all that needs to be studied is the polynomial (2.2) in the essential variable x_K.

The above results may be used for the analysis of model fitting problems by replacing the function, its independent variables and its parameters by the criterion of goodness of fit, the model parameters and the errors in the observations, respectively. The advantages of this approach are by now clear: the analysis of the behaviour of the stationary points of the, often complicated, criterion as a function of the K model parameters is replaced by a similar analysis of a Taylor polynomial in the essential variable. Moreover, in practical problems, the degree L of the Taylor polynomial is often as low as four.

The problem left is the search for degenerate stationary points or, more precisely, bifurcation sets in error space. If the smallest distance from the bifurcation set to the origin is small compared with the Euclidean norm of the observations there is a reason for caution: this means that relatively small errors may give rise to structural change of the criterion.

In the next section, the usefulness of these ideas will be demonstrated for a particular class of nonlinearly parametric models. They are, however, not necessarily limited to this class.

3. APPLICATION TO WEIGHTED SUM MODELS

In many fields of applied science the process response has the form

$$f(x;\underset{\sim}{\gamma}) = \alpha_1\, g(x;\underset{\sim}{\beta}_1) + \ldots + \alpha_K\, g(x;\underset{\sim}{\beta}_K) \qquad (3.1)$$

In this expression $f(x;\underset{\sim}{\gamma})$ is the process response, x is the independent variable, $\underset{\sim}{\gamma} = (\underset{\sim}{\alpha}^T \mid \underset{\sim}{\beta}^T)^T$ is the vector of unknown parameters, $\underset{\sim}{\alpha} = (\alpha_1 \ldots \alpha_K)^T$ are the scalar linear parameters, $\underset{\sim}{\beta} = (\underset{\sim}{\beta}_1^T \ldots \underset{\sim}{\beta}_K^T)^T$ are the nonlinear parameters with $\underset{\sim}{\beta}_k \neq \underset{\sim}{\beta}_\ell$ for $k \neq \ell$ while K and g(x;.) are the order and the generic function respectively. The model (3.1) is called a weighted sum model of order K. The independent variable may be vector-valued. Throughout it will be assumed that N observations are available described by

$$w_n = f_n(\underset{\sim}{\gamma}) + v_n \tag{3.2}$$

where $f_n(\underset{\sim}{\gamma}) = f(x_n;\underset{\sim}{\gamma})$ and the values x_n are supposed known, while the v_n are errors.

Examples of generic functions are $\exp(-\sigma x)$ with $\sigma > 0$ found in compartmental models and in radioactive decay, $\{\sin(\omega x)/\omega x\}^2$ occurring in optical diffraction, $1/[1 + \{(x-\mu)/\sigma\}^2]$ called Lorentz lines in molecular spectroscopy where $\underset{\sim}{\beta} = (\mu\ \sigma)^T$, and Gabor sonar pulses used by aquatic mammals as dolphins and described by $\exp[-\{(x-\mu)/\sigma\}^2/2]\cos\omega(x-\mu)$ where $\underset{\sim}{\beta} = \mu$ or $\underset{\sim}{\beta} = (\mu\ \sigma)^T$. For simplicity in the discussion to follow $\underset{\sim}{\beta}$ will be supposed to be scalar. For vector-valued $\underset{\sim}{\beta}$, see [10].

Intuitively, difficulties in fitting (3.1) to observations described by (3.2) are not to be expected if the β_k are not too closely located and the norm of the errors v_n is small compared to any of the α_k. In the above examples this corresponds with clearly distinct decays or non-overlapping lines or pulses. Difficulties, however, arise if two or more β_k are closely located. To simplify the analysis, it will be assumed that only two of the β_k are close [5]. Clustering of three or more β_k is discussed in [9]. Suppose that β_{K-1} and β_K are the β_k that are close. Then they could, in principle, always be resolved from hypothetical exact observations using any fitting procedure and ideal computational facilities. At the solution the criterion would be equal to zero. A model of order K-1 would also nicely fit. Then the solution for the first K-2 linear and nonlinear parameters would approximately be equal to $\alpha_1, \ldots, \alpha_{K-2}$ and $\beta_1, \ldots, \beta_{K-2}$ and that for the last ones $\alpha_{K-1} + \alpha_K$ and a value somewhere in between β_{K-1} and β_K, respectively. The criterion value would, of course, no longer be equal to zero. Now suppose that errors are present whose sum of squares exceeds that criterion value. Then the signs and magnitudes of the errors may be such that observations may occur that are more or less similar to hypothetical, error corrupted observations generated by a (K-1)th order model. The surprising fact is that under these conditions a (K-1)th order solution not only fits well, it also usually fits best! Thus a (K-1)th order solution results from the Kth order observations.

The remainder of this section is devoted to an explanation of this phenomenon in terms of structural change of the criterion of goodness of fit employed. This will be confined to the least squares criterion. For other criteria including not everywhere differentiable ones, see [6] and [8].

Suppose that the Kth order weighted sum model

$$f(x;\underset{\sim}{c}) = a_1\, g(x;b_1) + \ldots + a_{K-2}\, g(x;b_{K-2}) + \eta\, a\, g(x;b_{K-1}) + (1-\eta)\, a\, g(x;b_K) \tag{3.3}$$

is fitted to the available observations with respect to $(a_1 \ldots a_{K-2}\ a\ b_1 \ldots b_K)^T$ and that this is done for a number of selected values of the scalar η. Notice that this implies fitting under the restriction that the ratio of the (K-1)th to the Kth linear parameter is $\eta/(1-\eta)$. Furthermore, let α_{K-1} and α_K have the same sign, as is the case in most applications. Next assume that a (K-1)th order model is fitted to the same observations and that the resulting (K-1)th order solution is $\hat{\underset{\sim}{c}} = (\hat{a}_1 \ldots \hat{a}_{K-1}\ \hat{b}_1 \ldots \hat{b}_{K-1})^T$. Then substitution shows that

$$\bar{\underset{\sim}{c}} = (\hat{a}_1 \ldots \hat{a}_{K-2}\ \hat{a}\ \hat{b}_1 \ldots \hat{b}_{K-2}\ \hat{b}_{K-1}\ \hat{b}_{K-1})^T \tag{3.4}$$

satisfies the normal equations for fitting the Kth order model (3.3). Notice that the stationary point $\bar{\underset{\sim}{c}}$ is fully defined by the (K-1)th order minimum $\hat{\underset{\sim}{c}}$.

Next the point $\underset{\sim}{c}$ is chosen as origin of a Taylor expansion of the Kth order criterion. The resulting series is subsequently transformed into a Taylor polynomial as described by (2.1) and (2.2). The result is

$$\lambda_1 e_1^2 + \ldots + \lambda_{2K-2}\, e_{2K-2}^2 + \mu_2 b^2 + \mu_3 b^3 + \mu_4 b^4 \tag{3.5}$$

where $b = b_K - b_{K-1}$, the λ_k may be shown to be strictly positive, and the coefficients μ_k are functions of the errors v_n, the solution $\hat{\underset{\sim}{c}}$, and the constant η. Furthermore, the coefficients μ_2 and μ_3 are linear combinations of the deviations of the (K-1)th order model from the observations at $\hat{\underset{\sim}{c}}$. They are, therefore, small compared with μ_4 which has an essentially different structure and is relatively large and positive as long as the selected ratio of a_{K-1} to a_K is not excessively large or small.

Now an elementary, first order analysis shows that (3.5) has at the origin either a single minimum if $\mu_2 \geq 0$ or a one-saddle point and two further minima on either side with $b \neq 0$. In the former case, $\bar{c}$, defined by (3.4), is the least squares solution. Then the solutions for β_K and β_{K-1} coincide and, consequently, this solution is (K-1)th order. The solutions for β_K and β_{K-1} can not be resolved from the available observations. In the latter case, one of both minima is the least squares solution and this minimum is located in the region where $b_{K-1} \neq b_K$. Hence, the solutions for β_K and β_{K-1} are distinct.

The conclusion is that the equation $\mu_2 = 0$ along with the 2K-2 normal equations defining $\hat{c}$ constitute 2k-1 equations in the N errors and the 2K-2 elements of $\hat{c}$. Eliminating these elements yields one equation in the N errors. In error space this equation defines a hypersurface, the <u>bifurcation set</u>, separating the errors or, equivalently, the observations from which both last parameters can be resolved from those from which they can not.

4. EXAMPLES

4.1 Resolution in model fitting

Using the considerations of Section 3 a simple and objective definition of the concept (parametric) resolution can be given: two terms of equal amplitude of a weighted sum model are defined as resolvable from the available error corrupted observations if the latter are situated on the same side of the bifurcation set as exact observations.

This definition removes two difficulties encountered in resolution criteria as the classical Rayleigh criterion [1]. First, this criterion is based on presumed limits to the ability of the human visual system to perceive spatial intensity variations. The criterion is, therefore, subjective. Second, in the Rayleigh theory the observations are exact observations made on a second-order weighted sum model. Errors of any kind are absent. Hence, using modern photometry and computing facilities, limits to resolution would be virtually non-existent. In practice, errors of all kinds constitute the limit. The theory of Section 3 shows how. For a detailed description of applying these considerations to optical resolution, see [7].

4.2 Maximum allowable noise-to-signal ratio

In the processing of overlapping reflected sonar pulses, one may wish to know what, for various degrees of overlap, the maximum allowable noise energy over the width of the pulses is if resolution is to be guaranteed. The question arose in experiments intended to assess the limits to the otherwise remarkable abilities of dolphins to resolve overlapping reflections to their Gabor-like sonar pulses [2]

The procedure followed to answer this question was as follows. The minimum of the sum of the squares of the errors (noise samples), that is, their energy is computed under the constraint that the errors are in the bifurcation set separating distinct solutions for the locations of the pulses from coinciding ones. Thus, the minimum energy and the corresponding errors are found required to make resolution impossible. The ratio of the error energy thus found to the energy of the transmitted pulse is the maximum noise-to-signal ratio guaranteeing resolution.

4.3 Hierarchical model fitting

The specification of initial parameter values for the numerical fitting of weighted sum models is often difficult. In particular this is so if two or more nonlinear parameters are close and/or the order is unknown. Moreover, even if the order is known, the order of the solution may be lower than that of the model underlying the observations. This has been discussed in the previous section.

A possible solution would be the following. First fit a lower-order model. Next compute for each of the solutions for the nonlinear parameters thus obtained the quantity μ_2 discussed in the previous section. Then, if this quantity is negative, try to replace the term concerned by two new terms having different nonlinear parameters using a further model fitting procedure. The solutions for the parameters can be tightly controlled if the procedure is carried out for a number of credible values of η using the model (3.3). This prevents the procedure from producing spurious solutions, for example, a very small linear parameter combined with a highly improbable solution for the corresponding nonlinear one. The procedure could then be repeated until no further splitting of any terms appears reasonable. This idea is currently under investigation.

5. CONCLUSION AND DISCUSSIONS

The analysis presented concerns structural change of model fitting criteria and the accompanying consequences for the model fitting solution under the influence of the errors in the observations. No particular assumptions have been made on the errors. These may be stochastic, systematic or both. Also, no assumptions have been made with respect to the number of observations. In both respects, this analysis differs from the model fitting literature in which errors are usually assumed to be independent and identically distributed stochastic variables while, in addition, asymptoticity is assumed. Although the results of such an asymptotic statistical analysis are not contradictory to the results presented, they do, by their very nature, not reveal them.

Central in the results concerning weighted sum models is that from lower-order model fitting solutions conclusions may be drawn with respect to higher order ones without the need to compute the latter. Since lower order solutions are usually better-conditioned and, therefore, easier to compute, this considerably contributes to practical feasibility of this theory.

REFERENCES

[1] Born, M. and Wolf, E., (1980) Principles of Optics, Pergamon, London.

[2] Kamminga, C., (1982) Temporal difference perception by tursiops truncatus, Aquatic Mammals, 9, 41-47.

[3] Poston, T. and Stewart, I.N., (1978) Catastrophe Theory and its Applications, Pitman, London.

[4] Saunders, P.T., (1980) An introduction to catastrophe theory, Cambridge University Press, Cambridge.

[5] Van den Bos, A., (1981) Degeneracy in nonlinear least squares, IEE Proceedings 128, Part D, 109-116.

[6] Van den Bos, A., (1984) Resolution of model fitting methods, International Journal of Systems Science, 15, 825-835.

[7] Van den Bos, A., (1987) Optical resolution: an analysis based on catastrophe theory, Journal of the Optical Society of America A 4, 1402-1406.

[8] Van den Bos, A., (1988) Least-absolute-values and minimax model fitting, Automatica 24, 803-808.

[9] Van den Bos, A., and Van der Werff, T.T., (1989) Degeneracy in nonlinear least squares II - Coincidence of more than two parameters. Submitted paper.

[10] Van der Werff, T.T., (1987) The nonlinear least squares criterion for weighted sum models, M.Sc. Thesis, Systems and Signals Laboratory, Department of Applied Physics, Delft University of Technology (in Dutch).

THE APPLICATION OF SIGNAL PROCESSING TECHNIQUES TO THE DIGITAL CODING OF STUDIO QUALITY TELEVISION SIGNALS

N. K. Lodge
(Independent Broadcasting Authority, Research Laboratories, Crawley Court, Winchester, Hants)

and

R. J. Clarke
(Department of Electrical and Electronic Engineering, Heriot-Watt University, Grassmarket, Edinburgh)

ABSTRACT

The dramatic impact of digital signal processing techniques upon telecommunications is nowhere better exemplified than in the manipulation of television signals in digital format, and there is significant interest at the present time in reducing the standard coding rate of 216 Mb/s to 34 Mb/s for picture exchange within Europe. The present paper describes a technique which has the advantages of rapid recovery from errors, high subjective quality and simple decoding. It employs a combination of interpolative and predictive coding, with an explicitly conveyed control signal, and uses a model of the visual response of the viewer to allow compensation for the effects of any significant coding errors.

1. INTRODUCTION

Digital signal processing has had a major effect upon all fields of telecommunication operations, and one of the most challenging of these is that of television, where signals, originally manipulated in analogue (composite) form, are now processed by high speed digital algorithms applied to the separate luminance and chrominance components at the internationally agreed total rate of 216 Mb/s.

For picture exchange between countries the signal will not be required to withstand multiple processing operations (as would original material produced by the studio) but can be conveyed directly to the transmitter to be broadcast to the viewer. It is possible, therefore, to reduce the above rate significantly and use a lower level of the ISDN hierarchy. Since the quality requirements for broadcast video information are extremely stringent, the use of a very low rate is not possible (at least at present), but the rate of 34 Mb/s, whilst still presenting a challenge to system designers, is capable of satisfactorily supporting the digital transmission of broadcast quality television image sequences.

The algorithm described here combines predictive and interpolative coding, and employs an explicitly conveyed control signal to obviate difficulties which would occur as a result of channel errors in systems where predictor choice is made independently by the receiver on the basis of previously decoded samples. A further refinement is the use of a visual model which tests the locally decoded image (at the transmitter), identifying significant errors in reconstruction and allowing the communication of a correction signal to the receiver. An additional advantage of the scheme described here is that it concentrates system complexity at the coder, and as such it is well suited to 'single coder/many decoder' situations.

2. THE CODING OPERATION

The basic sampling pattern [1] is shown in Figure 1, where the sample distribution over four consecutive fields is indicated. Samples occurring in any given field (field 1, say) will form the set to be predicted, and those midway horizontally and vertically between the predicted set will be interpolated. Thus in the "same" field of the next frame (field 3), the locations of predicted and interpolated samples are reversed. A similar operation takes place in fields 2 and 4, etc. For each interpolated sample a choice of horizontal, vertical or temporal interpolation is available (see below).

A three-state signal interpolator selection signal is present at each location and is transmitted to the receiver to enable the decoder to perform the correct interpolation and also to aid in the description of the neighbourhood of a sample to be predicted. Predictor choice at the decoder is thus explicitly defined instead of relying upon previously decoded data, ensuring rapid recovery from channel errors. A further stage in the coding operation is to locally decode (i.e. at the transmitter) the signal to be transmitted and to use a visual model to detect poorly interpolated samples which would otherwise result in visible errors in the reconstructed picture. These errors can then be compensated for by the transmission of a correction signal.

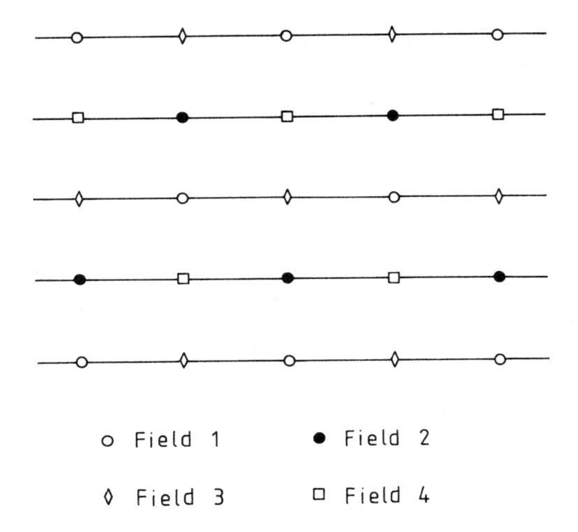

Fig. 1 Sampling Structure

3. THE INTERPOLATION PROCESS

For convenience, we use indices i, j and k to denote successive horizontal, vertical and temporal sample locations in the three-dimensional expansion of Figure 1 along the temporal axis. In static areas of the picture, temporal interpolation will reproduce full source resolution i.e.

$$I^{T}_{i,j,k} = \frac{1}{2}\left(S_{i,j,k-2} + S_{i,j,k+2}\right) \qquad 3.1$$

This is satisfactory, provided that the segmentation of the image into moving/static areas is reliable. Since the frame difference signal is susceptible to noise in areas of nearly constant luminance, and this noise will cause uncertainty in the decision if simple thresholding of single frame-difference signals is used as a moving/static indicator, an averaging operation is carried out on frame difference signals over a small area surrounding the sample to be predicted using available samples in the previous, present and following frames. This will significantly improve the reliability of the segmentation operation.

When motion is present interframe interpolation is not satisfactory and the operation must be carried out on an intra-field basis. The sampling structure of Figure 1 produces two-dimensional spectral repeats about the locations shown in Figure 2, and the prefiltering operation necessary to avoid aliasing will result in the ability to retain bandwidth only over a region such as the one shown shaded in the Figure if fixed inter-polation is used. By choosing between horizontal and vertical interpolation at each location the bandwidth which can be retained overall increases to that shown in Figure 3. Note that, at any given location, only the horizontal or vertical bandwidth extension is available, and the spectral response of Figure 3 does not tessellate in the manner of that of Figure 2. Naturally, a prefilter with the corresponding two-dimensional response must be employed prior to the sampling operation.

Horizontal and vertical intrafield interpolators are

$$I^H_{i,j,k} = \frac{1}{2}\,(S_{i-1,j,k} + S_{i+1,jk}) \qquad 3.2$$

$$I^V_{i,j,k} = \frac{1}{2}\,(S_{i,j-1,k} + S_{i,j+1,k}) \qquad 3.3$$

respectively. Each is tested against $S_{i,j,k}$ to determine which will produce the lower interpolation error. At the receiver reconstructions of the surrounding samples (which formed the interpolations at the coder) are present and so $I^H_{i,j,k}$ and and $I^V_{i,j,k}$ are likewise available. By indicating to the receiver whether the better interpolation is above or below the average, $(I^H_{i,j,k} + I^V_{i,j,k})/2$, i.e. by indicating the sign of

$$S_{i-1,j,k} + S_{i+1,j,k} + S_{i,j-1,k} + S_{i,j+1,k} - 4S_{i,j,k} \qquad 3.4$$

the appropriate interpolator may be selected and, in addition, information about the location of picture edges may be extracted from the zero crossings of Equation 3.4, the discrete Laplacian operator. In many low detail image regions no single inter-polator will have a decided advantage and so selection may be made on the basis of continuing to employ a previous choice. This improves the efficiency of the run-length coding operation performed on the interpolator selection signal for its trans-mission to the receiver.

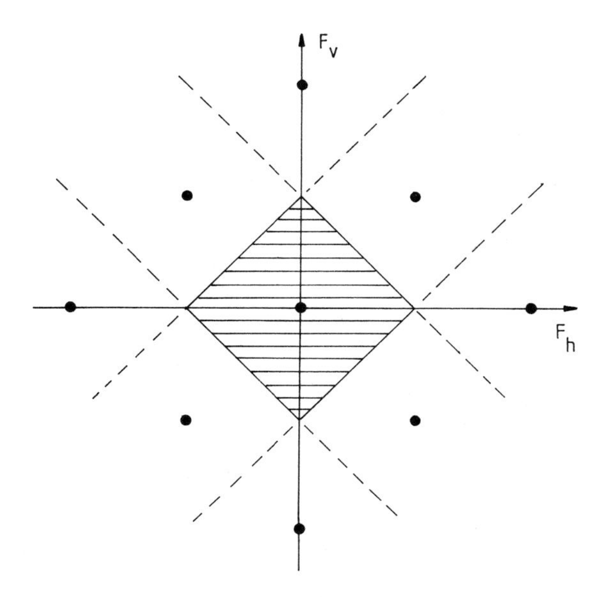

Fig. 2 Locations of spectral repeats and baseband spectrum

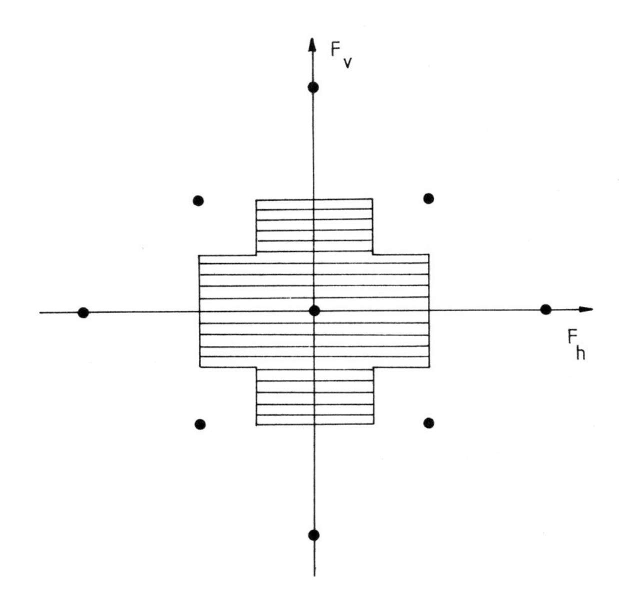

Fig. 3 Bandwidth retained using switched interpolation

4. ADAPTIVE PREDICTION

The interpolator selection signal sent to the receiver will eventually determine the nature of the interpolation to be carried out. It also provides a description of the region surrounding a sample to be predicted and will allow the selection of an appropriate prediction function. Since this selection is dependent upon an explicitly transmitted control signal and not recursively upon previously decoded samples there is a high degree of protection from the occurrence of errors in transmitted data.

5. VISUAL CONSIDERATIONS

The third stage in the coding operation is to test the quality of reconstruction of the image from decoded data by the use of a receiver model at the transmitter [2]. This model detects the more serious impairments due to defective interpolation and/or prediction, and allows the generation of a compensation term for transmission to the receiver. Such a model should incorporate the influence of the spatio-temporal frequency response of the human eye, luminance threshold effects and the visual masking of errors by rapid (spatial and temporal) luminance changes. In this context it is important to note that, subjectively, error visibility is lower if the frequency range of the error signal and the background against which it is viewed are similar [3], and this fact justifies an approach which adapts the visual filter response according to the masking function [4].

6. CONCLUSIONS

The paper has described an algorithm which will allow the distribution of broadcast quality digital television signals at 34 Mb/s. It employs a hybrid combination of predictive and interpolative coding in which a three-state descriptor signal is transmitted instead of alternate source samples, and this signal determines the selection of temporal interfield, or horizontal or vertical intrafield interpolations, and also controls the adaptive prediction operation. In its latter role, it ensures rapid recovery from the effects of channel errors and also allows simple implementation of the decoding operation. The coding scheme also includes a model of the human visual response to impairments in sample reconstruction, and thus allows the transmission of a correcting signal to the decoder in the case of significant errors. Satisfactory decoded picture quality at the receiver is thus assured.

7. REFERENCES

[1] Dubois, E., 1985, 'The Sampling and Reconstruction of Time-Varying Imagery with Applications in Video Systems', (Figure 20), Proc IEEE, 73, 502-522.

[2] Limb, J., 1973, 'Picture Coding: The Use of a Viewer Model in Source Encoding', BSTJ, 52, 8, 1271-1302.

[3] Stromeyer, C. and Julesz, B., 1972, 'Spatial-Frequency Masking in Vision: Critical Bands and Spread of Masking', JOSA, 62, 10, 1221-1232.

[4] Netravali, A., 1977, 'Interpolative Picture Coding Using a Subjective Criterion', IEEE Transactions on Communications, 25, 5, 503-508.

STATISTICAL ASPECTS OF MOMENT INVARIANTS IN IMAGE ANALYSIS

K.V. Mardia and T.J. Hainsworth
(Department of Statistics, University of Leeds)

ABSTRACT

In image analysis, the shape of an object in a 2-dimensional image is usually summarized by some suitable feature descriptors. One method is to use the image moments for which we introduce area corrections since images are only observed on a discrete lattice. In particular, moment invariants ([5], [1]) are constructed which are invariant under translation, rotation and dilation of spatial variables and rescaling of response variables. We critically examine the low order moments. In particular, it is known that any function of the invariants could be used and we propose invariants which stabilize the variance. Variance stabilizing transformations are derived and their behaviour is examined for pure and segmented objects. It is shown that unbiased estimates of the population size are obtained under naive thresholding when the object and background are of equal size. Our results indicate how an improved feature vector can be obtained by utilizing a suitable combination of variance stabilized moment invariants and the moment invariants themselves. The synthetic examples given here are motivated by real infra-red images containing objects of interest and background. We also examine other moments and approaches including the use of the shape space itself.

1. INTRODUCTION

Let $\{Z(\mathbf{x})\}$ be a non-stationary random field with $E\{Z(\mathbf{x})\} = g(\mathbf{X})$ where $g(.)$ is the response variable and $\mathbf{x}^T = (x,y) \in R^2$. In images, $Z(\mathbf{x})$ is the grey level at pixel $\mathbf{x}$. Let $\mathbf{x}_i$, $i = 1,..,n$ be the positions of n pixels on a regular lattice at which $\{Z(\mathbf{x})\}$ is observed. We define the $(r,s)^{th}$ population moment of the spatial variables (x,y) as

$$\mu'_{rs} = \int_{R^2} x^r y^s g(x,y)\,dxdy \tag{1.1}$$

with the corresponding 'process-moment'

$$m'_{rs} = \int_{R^2} x^r y^s Z(x,y)\,dxdy. \tag{1.2}$$

Note that

$$E(m'_{rs}) = \mu'_{rs}.$$

The most important case in practice is where

$$g(x,y) = \mu_1, (x,y) \in \bar{D}; \; = \mu_2, (x,y) \in D, \tag{1.3}$$

where $D+\bar{D}$ is a finite subset of R^2. If D is a silhouette image of an object, then $\mu_1 = 0$ and $\mu_2 = 1$ say. The boundary of D is the outline of a shape. Invariance under scale changes in intensity, i.e. in g(.) or Z(.), may be achieved by replacing μ'_{rs} by standardized moments,

$$\mu'_{rs}/\mu'_{00}. \tag{1.4}$$

If g(.) in (1.1) can be viewed as the bivariate probability density function of (X,Y), then the μ'_{rs} represent the usual bivariate moments. In particular, we know that under certain conditions g(.) is uniquely specified by $\{\mu'_{rs}\}$. Higher moments, such as

$$\int_{R^2} x^r y^s [Z(x,y)]^t dxdy,$$

are not invoked.

Initially, following the usual convention in Image Analysis, we will not use (1.4).

In addition to the scaling of g(.), the shape is comprised of those aspects of the image which are invariant under

(i) translation, (ii) rotation and (iii) dilation of $\mathbf{x}$.

Therefore one requires summary statistics which satisfy these invariances.

Section 2 gives area corrections required for real images and these have been used in subsequent studies in this paper. Section 3 examines moment invariants critically and Section 4 provides variance stabilizing transformations for the lower order moment invariants. Section 5 examines the moment invariants of objects segmented by naive thresholding and Section 6 utilizes a feature vector based on moment invariants for classifying pentomino shapes. Finally, in Section 7 we examine some other approaches, including the use of the shape space itself.

2. AREA CORRECTIONS

In practice, continuous images are pixelated so that the moments m'_{rs} of the spatial variables $\mathbf{x}$ are calculated over points $\mathbf{x}$ with values $Z(\mathbf{x})$, but we require the moments over the values for the squares $(x\pm\frac{1}{2}, y\pm\frac{1}{2})$ where $\mathbf{x} = (i,j)$ is a point on a regular lattice. Averaging $Z(i,j)$ uniformly over $(i\pm\frac{1}{2}, j\pm\frac{1}{2})$, we get the corrected value $\tilde{m}'_{rs}$ of m'_{rs} as

$$\tilde{m}'_{rs} = [\,(r+1)(s+1)^{-1} \sum_{i=0}^{\left[\frac{r}{2}\right]} \sum_{j=0}^{\left[\frac{s}{2}\right]} \binom{r+1}{2i+1}\binom{s+1}{2j+1}\left(\frac{1}{4}\right)^{i+j} m'_{r-2j,s-2j}. \qquad (2.1)$$

(In (2.1) it is assumed understood that the pixels are of unit area.) In particular, with $m'_{00} = 1$, we have

$$\tilde{m}'_{10} = m'_{10}, \quad \tilde{m}_{20} = \tilde{m}'_{20} + \frac{1}{12}, \quad \tilde{m}'_{11} = m'_{11},$$

$$\tilde{m}'_{30} = m'_{30} + \frac{1}{4}m'_{10}, \quad \tilde{m}'_{21} = m'_{21} + \frac{1}{12}m'_{01}$$

and so on. Note that these are the "inverse" of the Sheppard corrections, since here we are extending the value at a point to each square pixel. These corrections were applied in our simulation studies of Section 4, 5 and 6.

There are other possible sources of error; for example, see [10] for a general study of quantization and undersampling errors. However, the area-corrections (2.1) are new, and quite important in practice.

3. MOMENT INVARIANTS

We can construct a set of invariants from $\{\mu'_{rs}\}$. Obviously, for <u>translation invariance</u>, we should work on the central moments, $\{\mu_{rs}\}$. For <u>rotational invariance</u>, we require invariance under $(x,y)^T \to C(x,y)^T$ where

$$C = \begin{pmatrix} \cos\theta, -\sin\theta \\ \sin\theta, \cos\theta \end{pmatrix} \tag{3.1}$$

is an orthogonal matrix. Hu [5] obtained such invariant moments, but these are analogous to the statistical moments given in Barton and David [1]. A list of the "first seven" rotational invariants $\eta_1,\ldots,\eta_7$, is given in [5]. The first rotational invariant is

$$\eta_1 = \mu_{20}+\mu_{02} = \int_{R^2}(x^2+y^2)g(x,y)\,dxdy, \tag{3.2}$$

assuming that $\mu'_{10} = \mu'_{01} = 0$. It can easily be verified that

$$\eta_1 = \lambda_1+\lambda_2, \quad \eta_2 = (\lambda_1-\lambda_2)^2,$$

where λ_1 and λ_2 are the eigenvalues of the matrix

$$\begin{bmatrix} \mu_{20} & \mu_{11} \\ \mu_{11} & \mu_{02} \end{bmatrix}.$$

Note that η_7 is not algebraically independent of $\eta_1,\ldots,\eta_6$ as we have the relationship (see for example [7])

$$\eta_5^2+\eta_7^2 = \eta_3\eta_4^3. \tag{3.3}$$

Further, the first six rotational invariants $\eta_1,\ldots,\eta_6$ are also invariant under reflection, i.e. under $g(x,y)\to g(-x,-y)$, whereas η_7 is not. Note that sometimes, invariance under full rotations may not be needed, e.g. in discriminating between the digits 6 and 9. Note that $\{\eta_i\}$ forms a larger set than the affine invariant moments (see [9]). In particular, Mardia's skewness coefficient β_{12} (under unit covariance matrix) is

$$4\beta_{12} = \eta_3+3\eta_4.$$

Under <u>dilation</u>, $g^{(1)}(\lambda x,\lambda y) = g^{(2)}(x,y)$, giving $\mu_{rs}^{(1)} = \lambda^{r+s+2}\mu_{rs}^{(2)}$ so that we could define dilation invariants by

$$\mu_{rs}/\mu_{00}^{(r+s+2)/2}. \tag{3.4}$$

We can write down a set of <u>shape invariants</u>, $\{\eta_i^*\}$ which are invariant under translation, rotation and dilation, by using (3.4) in place of μ_{rs} in the expressions for the η_i. For example, from (3.2) and (3.4),

$$\eta_1^* = (\mu_{20}+\mu_{02})/\mu_{00}^2. \tag{3.5}$$

We are indirectly working on the shape space in forming these invariants (see also Section 7, Remark 1), but various other normalizations are possible.

The dilation invariants (3.4) can be adjusted so as to also give <u>scale invariance</u> in the response variable, e.g. we can replace μ_{rs} by the quantity

$$\mu_{rs}(\mu_{00})^{\frac{1}{2}(r+s-2)}/(\mu_{20}+\mu_{02})^{\frac{1}{2}(r+s)}. \tag{3.6}$$

Thus, we may define a set of <u>total invariants</u> $\{\eta_i^{**}\}$, having shape invariance and scale invariance, by using (3.6) in place of μ_{rs} in the expressions for the η_i. For example,

$$\eta_2^{**} = \left[(\mu_{20}-\mu_{02})^2+4\mu_{11}^2\right]/(\mu_{20}+\mu_{02})^2, \tag{3.7}$$

has total invariance. Note that η_1^* is absorbed in producing total invariance. Scale invariance is irrelevant for binary (0-1) images; we may replace (3.6) by the dilation invariants (3.4) or $\mu_{rs}/(\mu_{20}+\mu_{02})^{(r+s+2)/4}$.

Sample counterparts of these moment invariants can similarly be written down. In particular, let M_i^* denote the sample counterpart of η_i^*, $i=1,\ldots,7$.

Note that any function of the moment invariants is also invariant and so the question arises as to which functions of the invariants should be used; we give priority to those functions which stabilise the variances. This point is crucial when the object consists of a moderately small number of pixels.

4. VARIANCE STABILIZING TRANSFORMATIONS

Let Z_i, $i=1,\ldots,n$ be the observed values at n pixels, x_i, $i=1,\ldots,n$. Initially, we assume that these are a realisation of white noise with mean μ and variance σ^2. The sample counterpart of the first shape invariant, n^*_1 given at (3.5), is

$$M_1^* = \left\{\sum_{i=1}^{n} (x_i^2+y_i^2) Z_i\right\} \Big/ \left\{\Sigma Z_i\right\}^2 = \frac{1}{n} R, \text{ say} \tag{4.1}$$

where

$$R = \bar{Z}_1/\bar{Z}^2, \quad \bar{Z}_1 = \frac{1}{n}\Sigma (x_i^2+y_i^2) Z_i, \quad \bar{Z} = \frac{1}{n}\Sigma Z_i. \tag{4.2}$$

We can evaluate the mean and variance of M_1^* using the Taylor expansion assuming that σ is small and μ is large. That is, if m_1 and m_2 are the two sample moments (ie. $\bar{Z}_1$ and $\bar{Z}$), then approximately

$$E[\psi(m_1,m_2)] \simeq [\psi(m_1,m_2)]_p \tag{4.4}$$

$$\operatorname{var}[\psi(m_1,m_2)] \simeq \Big[[\psi_1(m_1,m_2)]_p^2 \operatorname{var}(m_1) + [\psi_2(m_1,m_2)]_p^2 \operatorname{var}(m_2) + 2[\psi_1(m_1,m_2)]_p [\psi_2(m_1,m_2)]_p \operatorname{cov}(m_1,m_2)\Big] \tag{4.5}$$

where $\psi_1(.)$ and $\psi_2(.)$ are the first partial derivatives of $\psi(.)$ with respect to m_1 and m_2 respectively, and the subscript p indicates substitution for m_1 and m_2 by their expectations. Using

$E(\bar{Z}) = \mu$, $E(\bar{Z}_1) = D_1\mu$, $D_1 = \frac{1}{n}\Sigma(x_i^2+y_i^2)$, $\text{var}(\bar{Z}) = \frac{\sigma^2}{n}$,

$\text{var}(\bar{Z}_1) = \frac{\sigma^2}{n}D_2$, $D_2 = \frac{1}{n}\Sigma(x_i^2+y_i^2)$, $\text{cov}(\bar{Z},\bar{Z}_1) = \frac{\sigma^2}{n}D_1$,

we find from (4.4) and (4.5) that

$$E(R) \simeq D_1/\mu = \theta \text{ say}, \ \text{var}(R) \simeq [\sigma^2 D_2/(n\mu^4)] = (\sigma^2 D_2/nD_1^4)\theta^4.$$

Hence, the variance stabilizing transformation for M_1^* is defined by $c\int\theta^{-2}d\theta = -c/\Theta$, so that the transformation for M_1^* is

$$1/M_1^*. \tag{4.5}$$

Note that this transformation also applies to the continuous version for a generalized Gaussian process $\{Z(x)\}$ with

$$\frac{1}{n} \to \iint dxdy, \quad D_1 \to \iint(x^2+y^2)\,dxdy, \quad D_2 \to \iint(x^2+y^2)^2 dxdy.$$

Similarly, we can show that

$$E(M_2^*) \simeq k_1\mu^{-2}, \quad \text{var}(M_2^*) \simeq k_2\mu^{-8}.$$

Thus, M_2^{*-1} is the required transformation. Other transformations obtained by this method are $M_3^{*-2/3}$, $M_4^{*-2/3}$, $M_5^{*-2/3}$, and $M_6^{*-3/4}$. However, except for M_1^{*-1} we have found no real gain in practice using these transformations, partly because the other moment invariants do not suffer from large skewness and the errors are small in the images we have encountered (see below).

We now indicate that the same result applies for a stationary process $\{Z(X)\}$. Define the (r,s)th complex moment by

$$C_{rs} = \int_{R^2} (x+iy)^r(x-iy)^s Z(x,y)\,dxdy.$$

For σ small and μ large, it can be shown that

$$E(C_{rs}) \propto \mu, \ var(C_{rs}) = O(\sigma^2), \ cov(C_{rs}, C_{tu}) = O(\sigma^2),$$

$$var(C_{rs}C_{tu}) = O(\mu^2).$$

Now, using the relations between C_{rs} and m'_{rs}, we can write M_i, $i=1,\ldots,6$ in terms of the C_{rs}; hence the respective approximations to the means and variances of M_i^*, $i=1,\ldots,6$ can be written down as before. Hence, the above results are valid even for stationary processes provided that μ is large and σ^2 is small. We now give two simulation studies for small images (i) to assess the transformations and (ii) to investigate the behaviour when the boundary of the object is perturbed. These examples are constructed after experience with real infra-red images.

Simulation study 1.

Suppose that the grey-levels Z_i on a 4x4 square object are distributed as $N(0.4,(0.05)^2)$ white noise. Note that $\mu=0$ is a special value, representing the 'background' mean. Table 1 gives summary statistics for M_1^* and M_1^{*-1} for 100 runs of 100 simulations The coefficients of skewness (b_1) and kurtosis (b_2) are small for both M_1^* and M_1^{*-1}, but the variability is much less for M_1^{*-1}. Further, the skewness is relatively small for M_1^{*-1}. Table 1 also gives simulation results for a Normal mixture with equal weights of $N(0.4,(0.05)^2)$ and $N(0.45,(0.05)^2)$; again M_1^{*-1} is "more Normal" than M_1^*.

Simulation study 2.

We consider an image of 'pentomino' W (step shape) constructed of five squares of 4x4 pixels (see Figure 1). We will consider the other pentominoes in Figure 1 later. To see the effect of boundary perturbation of M_1^*, M_1^{*-1} and M_2^*, we randomly add/remove a pixel at points on the object boundary. Table 2 gives these moment invariants for the whole silhouette.

		Mean	S.D.	b_1	b_2
	M_1^*	0.416	0.0147	0.0925	2.936
		(0.0013)	(0.0012)	(0.150)	(0.520)
(i)	$1/M_1^*$	2.403	0.0847	0.0556	2.864
		(0.0075)	(0.0064)	(0.0775)	(0.373)
	M_2^*	0.0002	0.0002	5.405	9.437
		(0.00002)	(0.00004)	(3.938)	(5.723)
	M_1^*	0.381	0.0135	0.149	3.031
		(0.0014)	(0.0011)	(0.224)	(0.712)
(ii)	$1/M_1^*$	2.629	0.0925	0.0723	2.884
		(0.0095)	(0.0070)	(0.106)	(0.472)
	M_2^*	0.00016	0.00018	5.114	9.173
		(0.00002)	(0.00003)	(3.365)	(4.923)

Table 1. Summary statistics for 100 runs of 100 simulations of a 4x4 square with grey-levels from white noise of (i) $N(0.4, (0.05)^2)$ and (ii) an equal mixture of $N(0.4, (0.05)^2)$ and $N(0.45, (0.05)^2)$; S.E. is given in brackets.

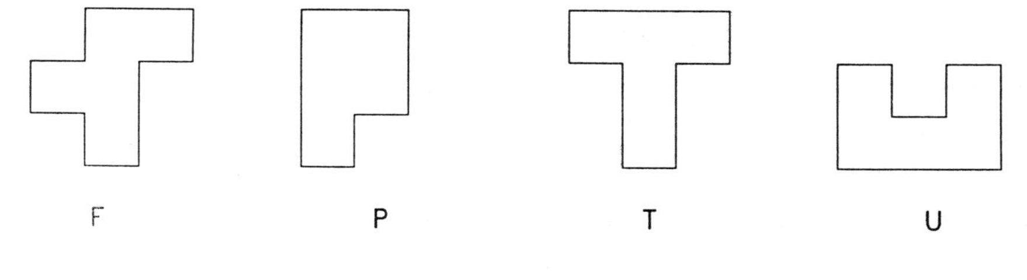

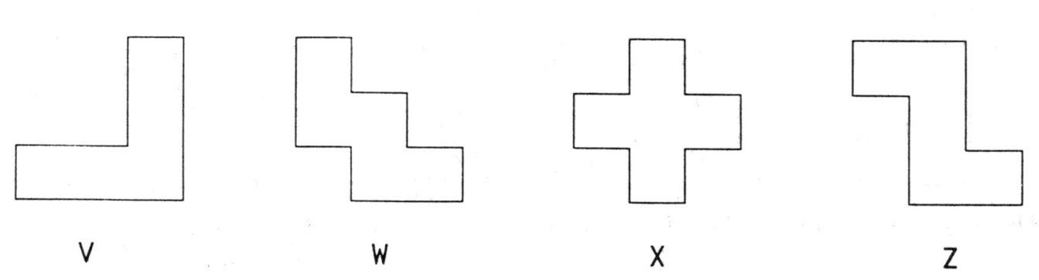

Fig. 1. Eight pentominoes consisting of 5 squares of 7x7 pixels

	Number of pixels perturbed (out of 41)	Mean	S.D.	Confidence Interval	b_1	b_2
	0	0.257				
	10	0.263	0.011	(0.241,0.282)	0.010	2.374
M_1^*	20	0.270	0.015	(0.240,0.300)	0.014	3.837
	30	0.277	0.020	(0.237,0.317)	0.222	2.433
	40	0.285	0.023	(0.239,0.331)	0.099	2.368
	0	3.89				
	10	3.81	0.157	(3.496,4.124)	0.004	2.295
$1/M_1^*$	20	3.71	0.209	(3.292,4.128)	0.102	3.533
	30	3.62	0.253	(3.114,4.126)	0.046	2.266
	40	3.54	0.284	(2.972,4.108)	0	2.281
	0	0.031				
	10	0.032	0.004	(0.023,0.040)	0.051	2.59
M_2^*	20	0.032	0.006	(0.020,0.041)	0.321	3.82
	30	0.034	0.009	(0.016,0.051)	0.908	3.18
	40	0.035	0.001	(0.016,0.054)	0.289	3.31

Table 2 Summary statistics for M_1^*, ${M_1^*}^{-1}$ and M_2^* under random boundary perturbations of ± 1 pixel for pentomino shape W (step shape).

Note that the true values of M_1^*, M_1^{*-1} and M_2^* (i.e. without perturbation) lie inside the confidence intervals for the perturbed case. These simulations show that the moment invariants M_1^*, M_1^{*-1} and M_2^* are not sensitive to fuzzy boundaries, which is desirable since the silhouette image is produced by some (imperfect) segmentation method. Fourier descriptors should be better for objects with sharper boundaries with inherent "jaggedness". Also note from the values of b_1 and b_2 that M_1^* and M_1^{*-1} are approximately Normal even under perturbation. Again, M_1^{*-1} is usually less skew than M_1^*.

5. MOMENT INVARIANTS OF SEGMENTED OBJECTS

We now consider the case of a grey-level image containing an object and background. In practice, by using some segmentation method, we can work on either

(i) a binary image, or (ii) a grey-level segmented object.

Further, we have the choice to work on either

(i) the boundary, or (ii) the silhouette.

We consider the behaviour of moments when the object is formed by the naive thresholding method, which we now describe. Let us assume that

$$\Pi_1 \text{ (background)}: Z(\mathbf{x}) \sim N(\mu_1, \sigma^2), \quad \Pi_2 \text{ (object)}: Z(\mathbf{x}) \sim N(\mu_2, \sigma^2),$$

where $\{Z(\mathbf{x})\}$ is a white noise process. The binary image $\{B(\mathbf{x})\}$ is then formed as follows:

$$\text{if } Z(\mathbf{x}) > t \text{ then } B(\mathbf{x}) = 0, \quad \text{otherwise } B(\mathbf{x}) = 1,$$

where $t = (\mu_1+\mu_2)/2$, with μ_1 and μ_2 known. Let $\{\hat{m}'_{rs}\}$ denote the corresponding values of $\{m'_{rs}\}$ for $\{B(\mathbf{x})\}$. Note that under Π_1 and Π_2, $B(\mathbf{x})$ forms independent Bernoulli trials with probabilities of success ε and $(1-\varepsilon)$ respectively, where

$$\varepsilon = \Phi[(\mu_1-\mu_2)/2\sigma]$$

and $\Phi(.)$ is the N(0,1) distribution function. Let n, n_1 and n_2 denote the number of pixels in the image, background and object respectively. We can now write down expectations of $\{\hat{m}'_{rs}\}$. For example,

$$E[\hat{m}'_{00}] = n_1\varepsilon + n_2(1-\varepsilon). \tag{5.1}$$

Thus, the size of the object is estimated without bias under naive thresholding if $n_1 = n_2$: this result is not surprising.

For the higher moments, the true shape of the object becomes relevant. Suppose we have a 2kax2ka ($k>1$, $a>0$) image containing a centralized 2ax2a square. Then we have

$$E[\hat{m}'_{20}+\hat{m}'_{02}] = \{n^2\varepsilon+(1-2\varepsilon)n_2^2\}/6$$

and so on. From (4.1), we have for the first (binary) shape invariant

$$\begin{aligned} E[\hat{M}_1^*] &\simeq \frac{1}{6}[\, n^2\varepsilon+(1-2\varepsilon)n_2^2] /[\,\varepsilon(1-\varepsilon)+\{n\varepsilon+(1-2\varepsilon)n_2\}^2] \\ &\simeq 1/6(1-\varepsilon) \quad \text{as} \quad n_2 \to n. \end{aligned} \tag{5.2}$$

Similarly, we can obtain higher moments. We now give two simulation studies: to see (i) how good is approximation (5.2) and (ii) how the shape invariants behave under segmentation.

We will call Gaussian white noise with background mean $\mu_1 = 100$, object mean $\mu_2 = 120$ and common variance $\sigma^2 = 25$ our "basic noise model". This gives grey-levels in the range 0-255, reflecting the convention in image processing.

Simulation study 1

Consider a 32x32 pixel image containing an object (centralized square) of n_2 = mxm pixels, subject to our basic noise model. Here we have $\varepsilon = 0.023$. Table 3 summarizes results for "silhouette" moments based on only a limited number of simulations (20) of segmented objects under naive thresholding. It can be seen that the approximation (5.2) to $E(M_1^*)$ depends on the size of the object within the image, as expected: the larger the object, the better the approximation.

m	$E(\hat{M}_1^*)$	Sample statistics based on 20 simulations	
		Mean	S.D.
1	6.509	6.99	1.61
2	5.249	5.48	1.21
3	3.837	3.86	0.681
4	2.660	2.52	0.380
5	1.814	1.72	0.256
10	0.394	0.371	0.049
15	0.212	0.209	0.012
20	0.179	0.179	0.003
21	0.177	0.177	0.003
22	0.175	0.175	0.002
23	0.174	0.174	0.002
24	0.173	0.173	0.002
25	0.172	0.172	0.001

Table 3 Approximation to the mean of $\hat{M}_1^*$ and simulated values of mean and variance for a 32x32 binary image obtained by thresholding an image containing an mxm object under the basic noise model.

Simulation study 2

We now consider a 32x32 square image containing a centralized pentomino Z (5 squares of 7x7 pixels) of Figure 1. The moment invariants are again calculated for the silhouette image obtained by naive thresholding under our basic noise model. Figures 2 to 4 give histograms of M_1^*, M_1^{*-1} and M_2^* based on 1000 simulations. Note from Figures 2 and 3 that

M_1^{*-1} is less skew than M_1^*, but from Figure 4 M_2^* itself is quite Normal. Note that the object size is about one third the size of the background.

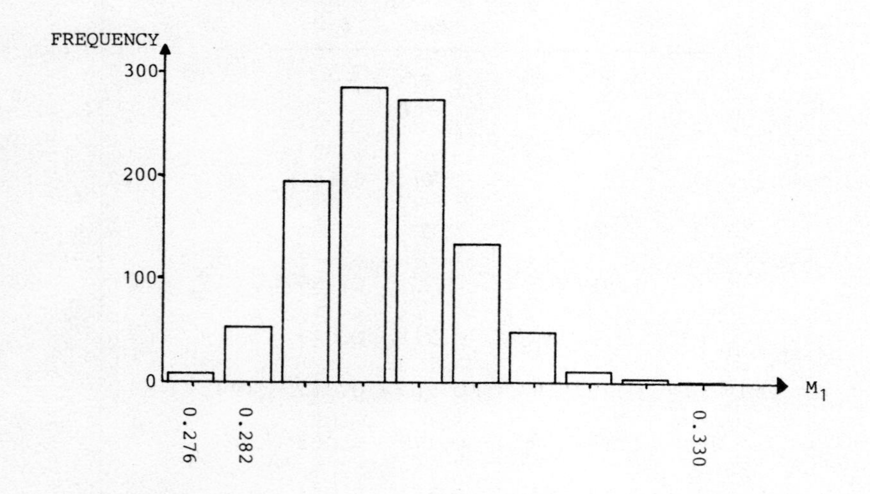

Fig. 2 Histogram of M_1^* for 1000 simulations of pentomino Z in a square background after naive thresholding

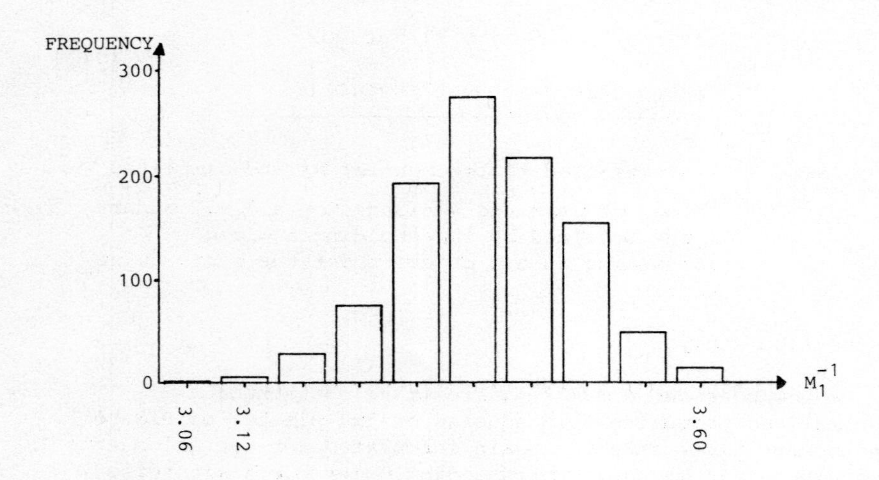

Fig. 3 Histogram of M_1^{*-1} for 1000 simulations of pentomino Z in a square background after naive thresholding.

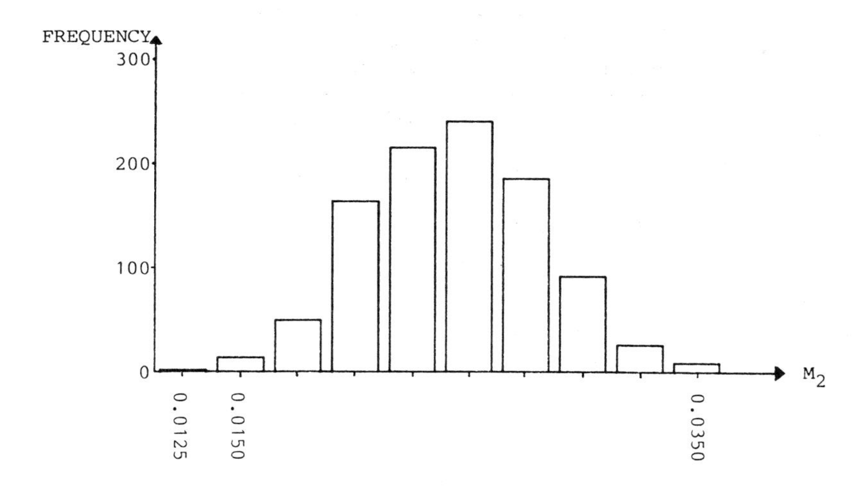

Fig. 4 Histogram of M_2^* for 1000 simulations of pentomino Z in a square background after naive thresholding

6. CLASSIFICATION

Once the image has been thresholded to give an object, the next stage is to assign it to one shape from a library of shapes. Consider images of the eight pentomino shapes, F, P, T, U, V, W, X, Z given in Figure 1, each composed of five squares of 7x7 pixels in a square of 32x32 pixels.

We can calculate the true values $\{\mu_{rs}\}$, but it is more useful to obtain $\{m'_{rs}\}$ under a large number of simulations, with modelling as in the previous sections. We calculated ν^*_{ij} and σ^{*2}_{ij}, i=1,...,6 as the mean and variance of M^*_i, i = 1,...,6

Feature Vector	δ	Angle of rotation	Error Rate
	0	0°	4.29%
	0.01	0°	2.57%
$\underset{\sim}{M}^*$	0.01	40°	4.9%
	0.01	random	4.25%
	0.05	random	77.5%
	0	0°	4.41%
	0.01	0°	2.59%
$\underset{\sim}{M}^*_{-1}$	0.01	40°	4.0%
	0.01	random	4.3%
	0.05	random	75.6%

Table 4 Average classification error rate based on the Karl Pearson distance for the feature vectors $\underset{\sim}{M}^*$ and $\underset{\sim}{M}^*_{-1}$ for eight pentominoes (5 squares of 7x7 pixels) of Figure 1, within a 32x32 square under our basic noise model contaminated by Cauchy noise.

for 1000 simulations of each of the eight objects (pentominoes) $j = 1,\dots,8$. Now we can classify an unknown object with shape invariants $M_1^*,\dots,M_6^*$ as being shape j giving a minimum to the Karl Pearson distance

$$K_j = \sum_{i=1}^{6} (M_i^* - \nu_{ij}^*)^2 / \sigma_{ij}^{*2}. \tag{6.1}$$

Let K_j^* denote the corresponding value of (6.1) on replacing M_1^* by M_1^{*-1}. Results of our study, based on an average of 4000 simulations of 32x32 images with Π_1 (background) $\sim N(100,25)$ and Π_2 (object) $\sim N(120,25)$, are given in Table 4.

The expected error rate is 2.3%. We also used (i) contamination by a fixed proportion δ of Cauchy white noise, i.e.

$$Z(x) \rightarrow (1-\delta) Z(x) + \delta Z_1(x).$$

where $\{Z_1(x)\}$ is Cauchy white noise with median 0 and semi-interquartile range 1, and (ii) rotation of the object within the 32x32 square. Let

$$\underset{\sim}{M}^* = (M_1^*, M_2^*, \dots, M_6^*) \quad \text{and} \quad \underset{\sim}{M}^*_{-1} = (M_1^{*-1}, M_2^*, \dots, M_6^*).$$

Note that the error rates are almost identical except when the contamination is large when $\underset{\sim}{M}^*_{-1}$ is slightly better.

We found that using the Mahalanobis distance does not improve the results. Further, the covariance matrices of $\underset{\sim}{M}^*_{-1}$ were slightly nearer to independence than those of $\underset{\sim}{M}^*$ for the pentomino shapes. Recent studies show that one requires a large number of moments rather than only 6 moment invariants to recreate a shape (see [11]). However, the higher moments are not reliable if there are outliers, etc.

7. OTHER MOMENTS AND APPROACHES

1. Shape space

We have concentrated on moment invariants, but there are other routes. The simplest is to work on $\{m'_{rs}\}$ themselves when the effects of translation, rotation and dilation have been removed. For example we can take

$$(x_i', y_i') = cC^T(x_i - \bar{x}^*, y_i - \bar{y}^*),$$

where $\bar{x}^* = m_{10}'$, $\bar{y}^* = m_{01}'$, $c = (m_{20}+m_{02})^{-1/2}$ and, in the orthogonal matrix C, Θ is given by

$$\tan 2\Theta = 2m_{11}/(m_{20}-m_{02}),$$

where we select the rotation Θ so that $m_{20}>m_{02}$. The latter choice fixes the first principal component with the largest eigenvalue to make angle Θ with the x-axis. Also, for invariance under intensity rescaling, we take $Z_i' = Z_i/m_{00}$. Hence, we may define a new set of totally invariant moments $\{\tilde{m}'_{rs}\}$. However, there is some degeneracy, namely,

$$\tilde{m}'_{00} = 1, \quad \tilde{m}'_{10} = \tilde{m}'_{01} = \tilde{m}'_{11} = 0 \quad \text{and} \quad \tilde{m}_0' + \tilde{m}_{20}' = 1.$$

We can now work with any subset of $\{\tilde{m}_{rs}\}$ as we like and have avoided constructing invariant functions. This method implies that we are working directly on the shape space, but the normalization is identical to that used to derive (3.7). See [9] for a partial normalization study.

Note that there has been considerable advancement in studies of shape space when a shape is characterized by a finite number of points (landmarks) on its boundary (see [6] and [2]). Another approach is through Procrustes rotation (see [4]). We could use moment invariants to summarize the shape if the number of landmarks is large, but this approach has not been used.

2. Polar moments

Another method is to use polar moments on a unit disc

$$D_{n\ell} = \int_0^1 \int_0^{2\pi} r^n e^{-il\Theta} Z(r\cos\Theta, r\sin\Theta)\, r\,dr\,d\Theta;$$

see [3], which gives an algorithm for constructing total invariants $\{\Psi_i\}$ from $D_{n\ell}$ up to any given order, n. For example, 2 complex and 3 real invariants (cf. the η_i are all real) can be formed from polar moments up to order $n = 3$:

$$\Psi_1 = D_{22}D_{2,-2}/D_{20}^2, \quad \Psi_2 = D_{31}D_{3,-1}/D_{20}^3, \quad \Psi_3 = D_{33}D_{3,-3}D_{00}/D_{20}^3,$$

$$\Psi_4 = D_{31}^2 D_{2,-2}D_{00}/D_{20}^4, \quad \Psi_5 = D_{33}D_{3,-1}D_{2,-2}D_{00}/D_{20}^4.$$

The $\{D_{n\ell}\}$ can be written in terms of the $\{\mu_{rs}\}$; in fact it can be shown that $\Psi_1,\ldots,\Psi_5$ are related to the shape invariants $\eta_1^*,\ldots,\eta_7^*$ as follows:

$$\Psi_1 = \eta_2^*/\eta_1^{*2}, \quad \Psi_2 = \eta_4^*/\eta_1^{*3}, \quad \Psi_3 = \eta_3^*/\eta_1^{*3},$$

$$\Psi_4 = [\eta_6^*+i\sqrt{(\eta_6^{*2}-\eta_2^*\eta_4^{*2})}]/\eta_1^{*4},$$

$$\Psi_5 = (\eta_5^*-i\eta_7^*)[\eta_6^*+i\sqrt{(\eta_6^{*2}-\eta_2^*\eta_4^{*2})}]/(\eta_1^{*4}\eta_4^{*2}),$$

so that the $\{\Psi_i\}$ do not contain any further independent invariant quantities to the $\{\eta_i\}$. However, the higher order Ψ_i exhibit greater stability than the corresponding η_i against their information content being swamped by noise. Perhaps a set of stabilized invariants could take the form

$$\log\Psi_i, \quad \Psi_i \quad \text{real}; \quad \text{Im}(\log\Psi_i), \quad \Psi_i \quad \text{imaginary}.$$

Note that $\text{Im}(\log\Psi_i)$ is a directional variable on $(0,2\pi)$.

We could also use a set of orthogonal polynomials in (x,y) or polar (r,Θ) as used in regression analysis, rather than the m'_{rs} (see [11]).

3. *Aspect Invariance*

We have not considered "aspect invariance", but in such cases the imposition of affine invariance rather than only dilational, translational and rotational invariance could be more fruitful (cf. [9]). Thus, moment invariants on the lines of [8] could be useful.

8. ACKNOWLEDGEMENT

We are grateful to David Lloyd, John Kent and John Haddon for their helpful comments. This work was supported by the Procurement Executive, Ministry of Defence.

9. REFERENCES

[1] Barton, D.E. and David, F.N., (1962), "Randomisation bases for multivariate tests: I. The bivariate case, randomness of N points in a plane", *Bull. Int. Statist. Inst.* **37**, (1) 158-9, (2) 455-67.

[2] Bookstein, F.L., (1986), "Size and shape spaces for landmark data in two dimensions", *Statist. Science* **1**, 181-242.

[3] Boyce, J.F. and Hossack, W.J., (1983), "Moment invariants for pattern recognition", *Pattern Recognition Letters* **1**, 451-6.

[4] Goodall, C.R. and Green, P.J., (1986), "Quantitative analysis of surface growth", *Botanical Gazette* **147**, (1) 1-15.

[5] Hu, M-K., (1962), "Visual pattern recognition by moment invariants", *IRE Trans. Inform. Theory IT*-**8**, 179-87.

[6] Kendall, D.G., (1984), "Shape manifolds, procrustean metrics and complex projective spaces", *Bull. London Math. Soc.* **16**, 81-121.

[7] Lloyd, D.E., (1985), "Automatic target classification using moment invariants of image shapes", Farnborough, UK, Rep. RAE IDN AW126.

[8] Mardia, K.V., (1970), "Measures of multivariate skewness and kurtosis with applications", *Biometrika* **57**, 519-530.

[9] Reeves, A.P., Prokop, R.J., Andrews, S.E. and Kuhl, F.P., (1988), "Three dimensional shape analysis using moments and Fourier descriptors", *IEEE Trans. Patt. Anal. Mach. Intell. PAMI*-**10**, 937-43.

[10] Teh, C.-H. and Chin, R.T., (1986), "On digital approximation of moment invariants", *Comput. Vision Graphics, Image Processing* **33**, 318-26.

[11] Teh, C.-H. and Chin, R.T., (1988), "On image analysis by the methods of moments", *IEEE Trans. Patt. Anal. Mach. Intell.* **10**, 496-513.

A CLASS OF NONSTATIONARY IMAGE MODELS AND THEIR APPLICATIONS

R. Wilson and S.C. Clippingdale
(University of Warwick, Coventry)

ABSTRACT

A new class of linear image models, based on a discrete lattice structure called a scale-space, is introduced. It is shown that such models, based on local interactions within the scale space, are capable of representing image structure across a full range of scales in the image. In other words, they capture an important property of natural images - such images typically contain structure at all scales, from global to local. Moreover, they are easily and naturally extended to cope with inhomogeneity or nonstationary behaviour at a given scale, without sacrificing the basic locality of the model. Such modifications can be used to represent boundaries, for example. Finally and most importantly, such models lead to recursive forms of estimators and predictors, which are again based on local processing within the scale space and are therefore highly efficient computationally. After a brief discussion of some of these ideas, the paper is concluded with examples of the application of these models to image enhancement and data compression. Some extensions and generalisations are also considered.

1. INTRODUCTION

For many years, it has been recognised that natural images contain structure which is not well modelled by linear stationary processes. At its most naive level, this is because images are composed of a number of more or less homogeneous regions separated by 'sharp' boundaries. Consequently, there have been numerous attempts to capture such 'non-stationarities' within image models, leading to a variety of largely heuristic modifications of the stationary linear models to incorporate adaptivity to localised image features such as region edges (e.g. [1]-[3]). Typically such models have been 'two-level' models, with an upper level process selecting the parameters defining the (linear) lower level model. While such models have met with some success, they may be criticised for failing to tackle the question of image structure 'head-on', but rather resorting to ad-hoc modifications of an inadequate model. The computational burden associated with such models is a further source of discomfort.

Over recent years, however, there has been a considerable amount of interest in so-called quadtree or pyramidal representations of image (e.g. [4]-[7]). Such representations consist of a number of levels, each of which is produced by smoothing and subsampling of the level below. On the face of it, such a representation seems merely to be an overcomplete and therefore highly redundant version of the original data. Despite their undoubted success in applications, their use has been motivated largely on heuristic grounds: there have been few attempts to place them on a more rigorous footing.

It is the main purpose of this paper to attempt to overcome this deficiency in the theoretical basis for 'pyramidal' or 'multiresolution' techniques. It will be shown that a class of linear multiresolution models is the *natural* extension to the non-causal 2-d world of images of the recursive causal models which have proved so successful in 1-d signal processing. To this end, the paper may be summarised as follows. A lattice structure called a discrete scale space is introduced for one and two dimensional data. Upon this structure are constructed the linear image models which form the basis of the investigations. Both homogeneous and inhomogeneous (non-stationary) models are presented. Efficient forms of estimator and predictor, which are *local* and *recursive* are defined and their optimality discussed. Examples are then given of the application of such estimators to the problems of image enhancement and data compression. The paper is concluded with a discussion of some of the issues raised by the new approach.

2. DISCRETE SCALE SPACES AND LINEAR SIGNAL MODELS

To begin with, it is simplest to illustrate the ideas with a 1-d scale-space, which may then be extended to 2-d. A 1-d scale-space $\Sigma\rho$ is a 2-d lattice, Σ_ρ. More precisely, to each point $(i,j) \in \Sigma_\rho, i,j \in Z$, there is associated a point $(x_i, y_j) \in R^2$ with

$$x = i\rho^j, y = j \qquad i,j = \ldots,-1,0,1,2,\ldots \tag{2.1}$$

where ρ is the *scale constant* of the lattice. In typical applications, $\rho = \frac{1}{2}$ and the resulting structure is as in figure 1, with the subsets $\{\bigcup_i (x_i, y_j)\} \subseteq R$ becoming denser in R as j increases.

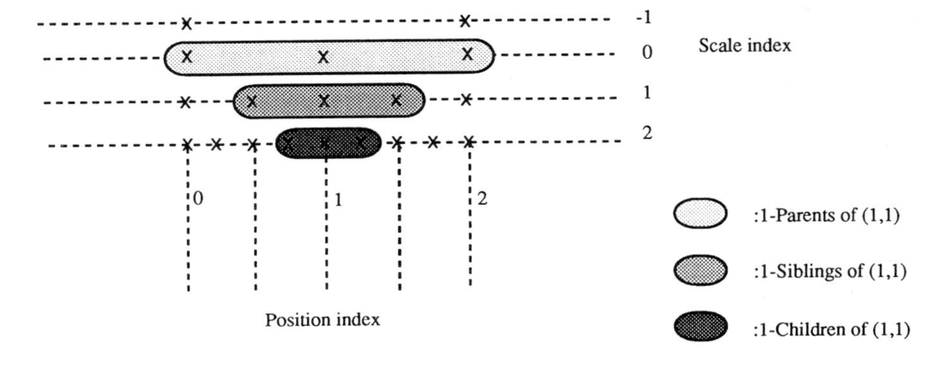

Fig. 1 One dimensional Scale-space lattice structure

Now a key property of the new models is that, in the context of the lattice Σ_ρ, they are *local*. In order to make this locality explicit, it is helpful to define the following subsets of Σ_ρ. A set of r-siblings of the point $(i,j) \in \Sigma\rho$ is a set of points $\sigma_{ij}(r), r>0$,

$$\sigma_{ij}(r) = \bigcup_{|k-i| \leq r} (k,j) \tag{2.2}$$

i.e. those points within a radius r of the point (i,j). A set of r-children of the point (i,j) is a set of point $\gamma_{ij}(r), r>0$.

$$\gamma_{ij}(r) = \bigcup_{|k-i\rho^{-1}| \leq r} (k,j+1) \tag{2.3}$$

Correspondingly, a set of r-parents of a point (i,j) is a set $\pi_{ij}(r), r>0$

$$\pi_{ij}(r) = \bigcup_{(i,j) \in \gamma_{k(j-1)}(r)} (k,j-1) \tag{2.4}$$

The r-neighbourhood of the point (i,j) is just the set of points

$$\nu_{ij}(r) = \sigma_{ij}(r) \cup \gamma_{ij}(r) \cup \pi_{ij}(r) \qquad (2.5)$$

and the class of processes with which we are concerned is those for which the interactions are confined to the r-neighbourhood of each point $(i,j) \in \Sigma_\rho$. These sets are illustrated in figure 1.

Extension of these ideas to 2-d is readily accomplished. The resulting 3-d lattice $\Sigma_\rho \in R^3$ is the set of points $\{(i,j,k)\}$ where to each (i,j,k) corresponds a point (x_i, y_j, z_k) with

$$x_i = ip^k \qquad y_j = jp^k \qquad z_k = k \qquad (2.6)$$

the r-siblings being defined by

$$\sigma_{ijk}(r) = \bigcup_{(i-l)^2 + (j-m)^2 \leq r^2} (l,m,k) \qquad (2.7)$$

and the children, parents and neighbourhood in a similar fashion.

A recursive linear signal model defined on the 2-d scale-space Σ_ρ is then expressible in the form

$$s_{ij}(k) = \sum_{(\ell,m,k-1) \in \pi_{ijk}(r)} A_{ijlm}(k) s_{\ell m}(k-1) + \sum_{(l,m,k) \in \sigma_{ijk}(r)} B_{ijlm}(k) w_{\ell m}(k) \qquad (2.8)$$

for some $p>0$ and radius $r>0$, or in linear o erator form

$$\mathbf{s}(k) = \mathbf{A}(k)\, \mathbf{s}(k-1) + \mathbf{B}(k)\, \mathbf{w}(k) \qquad (2.9)$$

The 'innovations' $w_{ij}(k)$ defining the signal model will be assumed to be samples from a stationary white normal process

$$E w_{ij}(k) w_{\ell m}(n) = \delta_{i\ell} \delta_{jm} \delta_{kn} \qquad (2.10)$$

The similarity of equation (2.9) to the conventional state-space signal models (e.g. [8]) should be readily apparent. Indeed, the well-known causal linear models can also be expressed in the form of equation (2.9). In that case, however, the index k represents not a *scale* index, but a *scanning* index, so that the equivalent lattice to the 1-d scale space has the appearance of figure 2 rather than figure 1. In other words, if $\rho = \frac{1}{2}$, then to generate an image of dimension $2^N x 2^N$ pixels, it requires only (N+1) steps of the scale space model, but 2^{2N} steps of a conventional state-space model. Nonetheless, the analogy between the scale space and state-space models is a useful one, for it underlies the recursive structure of the computationally efficient estimators for such models.

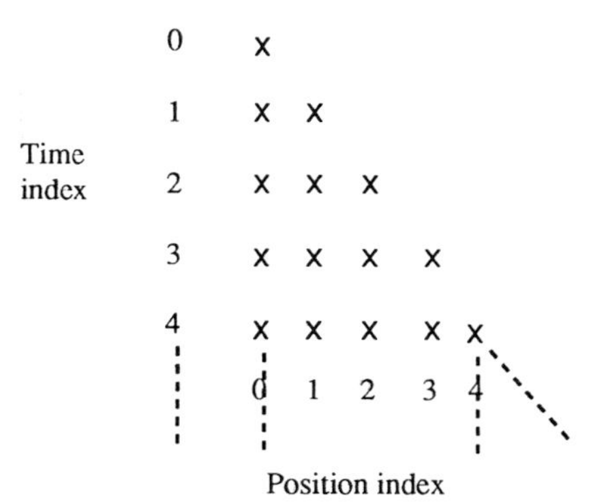

Fig. 2 Causal Recursive Lattice

A model of the form of equations (2.8)-(2.10) will be called *homogeneous* if for each k

$$A_{ij\ell m}(k) = A_{(i+r\rho^{-1})(j+s\rho^{-1})(\ell+r)(s+m)}(k) \qquad \forall i,j,\ell,m \in Z$$

$$B_{ij\ell m}(k) = B_{(i+r)(j+s)(\ell+r)(m+s)}(k) \qquad (2.11)$$

where only the integer part of each index is taken.

In other words, a homogeneous model is one for which the operators A(k), B(k) have a shift-invariant structure: if S_x, S_y are the spatial shift operators

$$(S_x s(k))_{ij} = s_{(i+1)j}(k) \qquad (S_y s(k))_{ij} = s_{i(j+1)}(k)$$

$$(2.12)$$

then from equation (2.11)

$$S_x^{i}\rho^{-1} S_y^{j}\rho^{-1} A(k) S_y^{-j}\rho^{-1} S_x^{-i}\rho^{-1} = A(k)$$
$$S_x^i S_y^j B(k) S_y^{-j} S_x^{-i} = B(k) \tag{2.13}$$

Thus homogeneous scale space models have a 'quasi-stationary' property, although the process governing s(k) is not in general stationary for any k.

Perhaps the simplest such model is the quadtree model, defined for $\rho = \frac{1}{2}$ by

$$\begin{aligned} A_{ij\ell m}(k) &= 1 \qquad i = 2\ell+p, j = 2m+q \quad 0 \leq p,q \leq 1 \\ &= 0 \qquad \text{else} \\ B_{ij\ell m}(k) &= b(k)\,\delta_{i\ell}\,\delta_{jm} \end{aligned} \tag{2.14}$$

The properties of this class of model are explored in [10].

A second example of a homogeneous scale-space process is the binary interpolative model, again defined for $\rho = \frac{1}{2}$, by

$$\begin{aligned} A_{ij\ell m}(k) &= 1 \qquad i = 2\ell,\ j = 2m \\ &= \tfrac{1}{2} \qquad i = 2\ell\pm1\ j = 2m,\ i = 2\ell\ j = 2m\pm1 \\ &= \tfrac{1}{4} \qquad i = 2\ell\pm1\ j = 2m\pm1 \\ &= 0 \qquad \text{else} \\ B_{ij\ell m}(k) &= b(k)\,\delta_{i\ell}\,\delta_{jm} \qquad i = 2\ell\pm1\ j = 2m\pm1 \\ &= 0 \qquad \text{else} \end{aligned} \tag{2.15}$$

Note that, unlike the quadtree model, the binary interpolative model has the property that the signal at all points in the lattice at level k is propagated unaltered to the next level - only the 'new' lattice points are updated by interpolation and addition of an *innovation*. The reader who is familiar with image data compression may appreciate the utility of such a model in the context of image data compression. This is indeed the application where such models are of greatest utility ([11], [12]).

There are several modifications of the above model which are useful in applications. The first of these concerns the lattice structure Σ_p. Since any digitised image is an array of, say, NxN points, in practice only a finite subset of the lattice, covering scale indices 0 to $\log_\ell N$, need be considered. Since such a subset is bounded, the homogeneity of the model cannot be preserved - 'boundary conditions' generally need to be taken into account. This is not necessary in all cases, however. The two above models, for example, avoid such problems because of their inherent block structure. Other structures than that of equation (2.6) are possible within the same general framework of an increasing sampling density with increasing scale index. In such cases, it may be necessary to alter the definitions of neighbourhood accordingly (e.g. [12].

The most significant modification, however, is to introduce an *adaptive* or 'nonstationary' element into the model, by making the coefficients $A_{ij\ell m}(k), B_{ij\ell m}(k)$ depend explicitly on some parameter $\theta_{ij}(k)$ defined on the lattice, i.e.

$$s_{ij}(k) = \sum_{(\ell,m,k-1)\in\pi_{ijk}(r)} A_{ij\ell m}(\theta_{ij}(k))\,s_{\ell m}(k-1) + \sum_{(\ell,m,k)\in\sigma_{ijk}(r)} B_{ij\ell}(\theta_{ij}(k))\,w_{\ell m}(k) \qquad (2.16)$$

The scalar (or vector) parameter field $\theta(k)$ represents a simple method of introducing *local* structure at a given scale into the model. In applications, one particularly useful such parameterisation is in terms of a local orientation representation, such as that described in [3]. This allows the model to represent the local anisotropy observed in many natural images, in which edge features show a consistency of orientation over some range of scales. It does this in a way which avoids the artefacts which tend to accompany similar causal models (e.g. [2]).

3. SCALE-SPACE PREDICTION AND ESTIMATION

The recursive structure of the above models, equation (2.9), suggests an obvious form of predictor for level k given level (k-1), namely

$$\hat{s}(k|k-1) = A(k)s(k-1) \qquad (3.1)$$

and this is clearly optimal in a m.s.e. sense (e.g. [8]). At this point, however, it is necessary to consider the major obstacle to the application of scale-space models: in general the signal s(k) is not directly accessible from the image, which corresponds to some signal level s(n), say, of the model. Thus, unlike conventional causal prediction, scale-space prediction requires in addition a retrodiction - an estimation of s(k) from s(k+1), for k<n. Denote this estimate by $\hat{s}^u(k)$. Then

$$\hat{s}^u(k) = C(k)\hat{s}^u(k+1) \qquad n_0 \le k < n \tag{3.2}$$

and the retrodiction-prediction is a *two-pass* process. On the first pass, the upward estimates of s(k) are formed and on the second, the predictions are based on the upward estimates, viz

$$\hat{s}(k|k-1) = A(k)\hat{s}^u(k-1) \tag{3.3}$$

Now the retrodiction operator C(k) will generally be restricted to be a neighbourhood operator, so that

$$\hat{s}_{ij}{}^u(k) = \sum_{(\ell,m,k+1)\in\gamma_{ijk}(r)} C_{ij\ell m}(k)\hat{s}_{\ell m}{}^u(k+1) \tag{3.4}$$

This choice is made for reasons of computational efficiency, but within this constraint, the coefficients $C_{ij\ell m}(k)$ can be chosen to minimise the m.s.e. $\in(k)$ on level k

$$\in(k) = \sum_{i,j} E(s_{ij}(k)-\hat{s}_{ij}{}^u(k))^2 \tag{3.5}$$

In any event, it remains true that $\hat{s}^u(n) = s(n)$ and so in the most obvious application of prediction, data compression [4], [12], little is lost in terms of efficiency. Moreover, for one important class of models, the retrodiction is *exact*

$$\hat{s}^u(k) = s(k) \qquad n_0 \le k \le n \tag{3.6}$$

This is the class of interpolative models, such as that of equation (2.15) and its generalisations [12]. (Note incidentally that the lower scale limit, n_0, is typically chosen so that only a 'few' (from 1 to 16x16) points cover an image of 512x512 pixels).

As in causal predictive coders, coders based on the prediction of equation (3.3) function by quantising the prediction error e(k)

$$e(k) = \hat{s}^u(k) - \hat{s}(k|k-1) \tag{3.7}$$

Compression ratios of up to 100 have been achieved using such methods [12].

The estimation problem is in many respects similar - it consists of both a retrodiction and a recursive smoothing step. In this case, the data are assumed to be a noisy version of the signal at level n

$$x(n) = s(n)+v \tag{3.8}$$

A localised upward estimate of the form of equation (3.2) is again used, i.e.

$$\hat{s}^u(k) = C(k)\hat{s}^u(k+1) \qquad n_0 \leq k \leq n$$
$$\hat{s}^u(n) = x(n) \tag{3.9}$$

where the coefficients $C_{ij\ell m}(k)$ are again chosen to minimise the m.s.e. subject to a locality constraint. From the retrodicted estimates, a final estimate is formed by a downward recursion

$$\hat{s}(k) = D(k)\hat{s}(k-1)+F(k)\hat{s}^u(k) \qquad n \geq k > n_0 \tag{3.10}$$

in which the coefficients $D_{ij\ell m}(k)$ are non-zero only within the parent set $\pi_{ijk}(r)$ and those $F_{ij\ell m}(k)$ only in the sibling set. Within this constraint they can be selected for minimum m.s.e. Although in general the resulting estimate of the image, $\hat{s}(n)$, is not the linear m.m.s.e. estimte, in one important special case - the quadtree model - it can be shown to be the m.m.s.e. estimate [10]. Even in the general case, the loss of optimality associated with the restrictions on the coefficients is no more significant than is the case for the so-called 'reduced update' Kalman filter [2].

The inhomogeneous forms of the estimator and predictor follow exactly the same pattern, except for the necessary complication of estimating the local parameter values $\theta_{ij}(k)$. These are also computed efficiently, if sub-optimally, within the same general scale-space framework, leading to a unified and rather elegant computational framework. Such matters are dealt with quite comprehensively in [10], which also describes modifications of the simple quadtree process designed to overcome blocking artefacts at minimal computational cost. On the subject of computational cost, it is worth emphasising that despite (or rather because of) the apparent redundancy of the scale-space models, the cost of achieving 'very large' smoothing operations is ridiculously cheap: the homogeneous quadtree estimator costs on the order of 5/3 multiplications per pixel in the image. Inhomogeneous estimators are more costly, but not prohibitvely so.

Examples of quadtree model based data compression and enhancement are shown in figures 3-5. Figure 3 shows the result of using a predictive coder on a familiar test image 512x512 8-bit pixels [4]. The resulting picture costs only 0.25 bpp, a compression ratio of ~30. More impressive results are described in [12]. Figures 4 and 5 show the input and output images from the estimation process, based on an input m.s. SNR of 0dB. The result compares favourably with other methods reported in the literature, both numerically (the output SNR is 13.6dB) and subjectively.

Fig. 3 Result of coding 'girl' image using scale-space predictor

Fig. 4 'Girl' image with additive white Gaussian noise (SNR = 0dB)

Fig. 5 Result of restoration of figure 4 using Scale-space recursive estimator

4. CONCLUSIONS

The purpose of this paper was to attempt to define a class of signal models which retain the obvious computational advantages of linear recursive models, but are freed from the artificial constraints imposed by image-plane causality. The practical results achieved with such methods are sufficiently encouraging to indicate their utility. Furthermore, one or two interesting theoretical results have been noted, although space permits only a reference to their proofs.

Nonetheless, much work remains to be done if they are to take their place in the theory of image processing. More results are needed on the optimality of various forms of the estimator, both in general and in particular cases. One interesting speculation is that a multipass estimator (rather than a two-pass one) may be optimal for a wider class of signal models. Furthermore, other generalisations of the lattice structure are no doubt possible and may be useful. This paper will have achieved its goal if it stimulates others to take up the challenge of this new area.

ACKNOWLEDGEMENT

This work was supported in part by UK SERC.

REFERENCES

[1] Abramatic, J.F. and Silverman, L.M., (1982), "Nonlinear Restoration of Noisy Images", *IEEE Trans. PAMI,* **4**, pp. 141-149.

[2] Woods, J.W., Dravida, S. and Mediavilla, R., (1987), "Image Estimation Using Doubly Stochastic Gaussian Random Field Models", *IEEE Trans. PAMI,* **9**, pp. 245-253.

[3] Knutsson, H., Wilson, R. and Granlund, G.H., (1983), "Anisotropic Nonstationary Image Estimation and its Applications", pts. I, II, *IEEE Trans., COM.,* **31**, pp. 388-406.

[4] Wilson, R., (1984), "Quadtree Predictive Coding", Proc. ICASSP-84, San Diego.

[5] Burt, P.J. and Adelson, E.H., (1983), "The Laplacian Pyramid as a Compact Image Code", *IEEE Trans. COM,* **31**, pp. 532-540.

[6] Burt, P.J., Hong, T.H. and Rosenfeld, A., (1981), "Segmentation and Estimation of Image Region Properties through Cooperative Hierarchical Computation", *IEEE Trans. SMC*, **11**, pp. 802-809.

[7] Wilson, R. and Spann, M., (1987), "Image Segmentation and Uncertainty", Letchworth, Research Studies Pr.

[8] Candy, J.V., (1987), "Signal Processing - the Model Based Approach", New York, McGraw-Hill.

[9] Clippingdale, S.C. and Wilson, R., (1987), "Quadtree Image Estimation: a New Image Model and its Application to MMSE Image Restoration", Proc. 5th Scand. Conf. on Image Anal., pp. 699-706, Stockholm.

[10] Clippingdale, S.C., (1988), "Multiresolution Image Modelling and Estimation", University of Warwick, Ph.D. Thesis.

[11] Beaumont, J.M., (1988), "The RIBENA Algorithm - Recursive Still Picture Coding", BT Memo. no. RT4343/88/14, March.

[12] Todd, M. and Wilson, R., (1988), "Image Data Compression Combining Multiresolution, Feature Orientation and Arithmetic Entropy Coding", Proc. IMA Conf., Warwick, December.

QUANTITATIVE ANALYSIS OF 2-D MAGNETIC RESONANCE TIME DOMAIN SIGNALS

R. de Beer, D. van Ormondt and W.W.F. Pijnappel
(Delft University of Technology, The Netherlands)

ABSTRACT

2-D Magnetic Resonance is a powerful spectroscopic tool. The signals are usually measured in the time domain. For analytical purposes, it is often required to quantify a signal in terms of physically relevant model parameters. As a rule this is done in the frequency domain, after correcting for artefacts associated with transformation from the time domain to the frequency domain. In this work quantification is brought about by fitting a 2-D time domain model function directly to the data. Two approaches are used, one based on singular value decomposition, the other on linear prediction and the variable projection method.

1. INTRODUCTION

The method of choice for studying molecular structure in solution is 2-D high resolution Nuclear Magnetic Resonance (NMR). This technique enables one to infer, among other things, the structure of proteins and nucleic acids. See e.g. A. Bax and L. Lerner [1].

The NMR phenomenon is induced by applying a sequence of pulsed radio frequency waves to the molecules under study, and is subsequently observed and recorded in the time domain. A typical 2-D NMR signal comprises 512 rows of 1024 complex-valued data points. Frequently, the data acquisition time is of the order of ten hours.

Traditionally, 2-D NMR signals are transformed to the frequency domain using FFT in both dimensions. Subsequently, the spectra thus obtained are studied and quantified. As many as several thousand spectral features (peaks) may be distinguishable. In most 2-D experiments one obtains a spectrum which is symmetric with respect to the bisector of the two frequency axes. However, 2-D experiments yielding asymmetric spectra exist also. See [1] for details.

The aforementioned determination of molecular structure hinges on the ability to detect and parametrize weak peaks, which lie off the above mentioned bisector and 'connect' strong peaks on the bisector in the sense that the respective projections on the frequency axes coincide (see Fig. 1). These connecting peaks are called cross-peaks in the 2-D NMR jargon. Unfortunately, various experimental conditions pose practical limits to the measuring time. In practice, one may therefore be forced to reduce the number of rows, which entails distortion of the spectrum as obtained by FFT. This in turn hampers the search for, and analysis of, weak peaks in a contour plot of a 2-D spectrum.

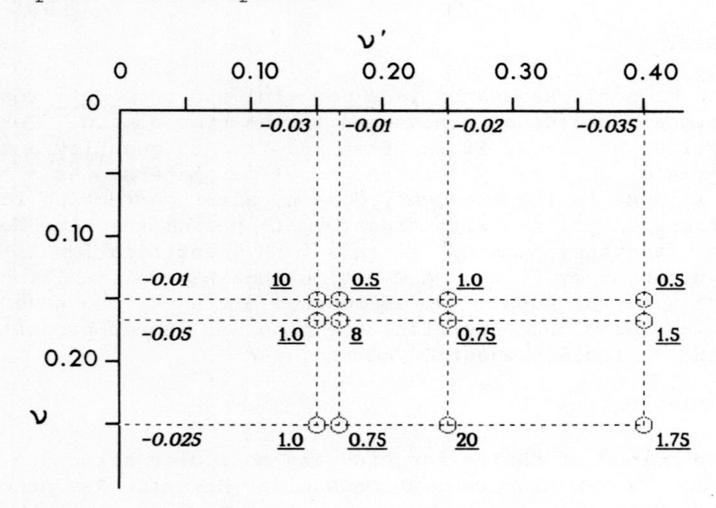

Fig. 1 Model parameters of the simulated signal of Eq.1, with K=3, K'=4, N=N'=50, and Δt=1. The 2-D sinusoids are indicated by small circles. The frequencies can be read from the projections on the axes, in units of the Nyquist frequency; the damping factors are given in italic script, and the absolute values of the amplitudes defined via $c_{kk'}=|c_{kk'}|\exp(i\phi_{kk'})$, are underlined. All phases $\phi_{kk'}$ were chosen zero.

An alternative to using FFT and processing in the frequency domain is to introduce a model function for the 2-D NMR time domain signal, and fit this function directly to the data. In the present work several such methods are proposed and tested on simulated signals. Our methods are based on previous work by Kung et al. [10], Kumaresan et al. [7,8], Bresler and Macovski [4], and Park and Cordaro [12].

The remainder of this paper is organized as follows. In Sec. 2 we introduce the model function that reasonably describes 2-D NMR signals, and then we treat two fundamentally different approaches to retrieve the wanted model parameters from the data. In Sec. 3 results of quantification of noise corrupted simulated data are given and discussed.

2. THEORY

2.1 The model function

In this subsection we introduce a model function that best approximates actual 2-D NMR time domain signals obtained from liquid or liquidlike samples. Experimentally, it is observed that such signals consist of a linear combination of damped 2-D sinusoidal 'waves'. In order to model this signal we make the following assumptions. 1) Each 2-D sinusoid can be written as a product of two 1-D sinusoids. 2) The damping of the sinusoids is *exponential*. Thus, a model function that reasonably approximates a typical 2-D NMR signal can be written as

$$\hat{x}_{nn'} = \sum_{k=1}^{K} \sum_{k'=1}^{K'} c_{kk'} \, z_k^{n-1} \, z'^{\,n'-1}_{k'} , \tag{1}$$

in which the symbols have the following meaning:

$n = 1,2, \ldots ,N$, N being the number of samples per column,

$n' = 1,2, \ldots ,N'$, N' being the number of samples per row,

$c_{kk'}$, $k=1, \ldots ,K$, and $k'=1, \ldots ,K'$, are the, generally complex-valued, amplitudes of the sinusoids, i.e. $c_{kk'}=|c_{kk'}|\exp(i\phi_{kk'})$, $\phi_{kk'}$ being the phases,

$z_k= \exp[(\alpha_k+i\omega_k)\Delta t]$, $k=1, \ldots ,K$, are the signal poles in the unprimed (column) space, α_k being damping factors, ω_k angular frequencies ($\omega_k=2\pi\nu_k$), and Δt the sampling interval,

$z'_{k'}=\exp[(\alpha'_{k'}+ i\omega'_{k'})\Delta t']$, $k'=1, \ldots , K'$, are the signal poles in the primed (row) space, the primed symbols having the same meaning as their unprimed counterparts. K *need not be equal* to K'.

A possible inadequacy of the above model function is that the actual damping of the signal can be more complicated than $\exp(\alpha t)$. The latter situation may arise, for instance, when the frequencies within a group of sinusoids happen to be so near to each other that this group is to be modelled as a single sinusoid. In such a case the overall damping of the group may be $\exp(\beta t^2)$ rather than $\exp(\alpha t)$. Another possible source of deviations is that the magnetic field used in the experiments is not sufficiently homogeneous over the physical dimensions of the sample.

2.2 Decomposition of the data matrix

One way to fit Eq. (1) to the data, $x_{nn'}$, n=1, ... , N, n'=1, ... ,N', is to seek a decomposition of the NxN' data matrix X, formed from the 2-D data, into a product of three matrices, namely an NxK Vandermonde matrix $\zeta_{N,K}$, a KxK' amplitude matrix C, and the transpose of another Vandermonde matrix $\zeta'_{N',K'}$, according to

$$X = \begin{bmatrix} 1 & \cdot & \cdot & \cdot & 1 \\ z_1^1 & \cdot & \cdot & \cdot & z_K^1 \\ \cdot & & \cdot & & \cdot \\ \cdot & & \cdot & & \cdot \\ \cdot & & \cdot & & \cdot \\ z_1^{N-1} & \cdot & \cdot & \cdot & z_K^{N-1} \end{bmatrix} \begin{bmatrix} c_{11} & \cdot & \cdot & \cdot & c_{1K'} \\ \cdot & & \cdot & & \cdot \\ \cdot & & \cdot & & \cdot \\ c_{K1} & \cdot & \cdot & \cdot & c_{KK'} \end{bmatrix} \begin{bmatrix} 1 & z'^1_1 & \cdot & \cdot & \cdot & z'^{N'-1}_1 \\ \cdot & \cdot & & & \cdot & \cdot \\ \cdot & \cdot & & & \cdot & \cdot \\ \cdot & \cdot & & & \cdot & \cdot \\ 1 & z'^1_{K'} & \cdot & \cdot & \cdot & z'^{N'-1}_{K'} \end{bmatrix} \tag{2}$$

$$= \zeta_{N,K} \qquad C \qquad \tilde{\zeta}'_{N',K'}, \tag{2'}$$

where $\tilde{}$ indicates transposition.

In absence of noise, and provided Eq. (1) adequately describes the signal, Eq. (2) is exact. In the presence of noise, Eq. (2) can be satisfied in the least squares sense. Once the decomposition has been accomplished, the model parameters are immediately available. In the present contribution we distinguish two approaches, namely non-iterative ones based on SVD [10], and iterative ones based on Variable Projection plus Linear Prediction [7,8,4,12]. A common advantage of all methods mentioned is that the model function of Eq.(1) is fitted to the data without the need to provide starting values. A common disadvantage is that spectroscopic prior knowledge other than the number of sinusoids cannot be imposed.

2.3 Decomposition by SVD (Method 1)

We start by considering the rank of the data matrix, R(X), and this in absence of noise so that the decomposition of Eq.(2) is exact. Perusing Eq.(2), one can see that R(X)=min[K,K'], provided the rank of the amplitude matrix, R(C), is full. Kung et al. have shown [10] that the wanted decomposition can be brought about by SVD if K=K' and C is diagonal. Very briefly sketched, their method is as follows. SVD of the data matrix is executed and then the rows and columns that do not contribute, are truncated,

$$X = U\Lambda V^{\dagger} = U_K \Lambda_K V_K^{\dagger} \quad , \tag{3}$$

where the subscript K indicates truncation and the symbol † Hermitian conjugation. Under the conditions quoted, it is now possible [10] to find nonsingular KxK' transformation matrices $\underset{\sim}{Q}$ and $\underset{\sim}{Q}'$, such that

$$U_K \underset{\sim}{Q} = \zeta_{N,K} \, , \quad \underset{\sim}{Q}' V_K^{\dagger} = \tilde{\zeta}'_{N',K'} \, , \text{ and } \quad \underset{\sim}{Q}^{-1} \Lambda_K \underset{\sim}{Q}'^{-1} = C, \tag{4}$$

whereby all wanted model parameters have been retrieved.

We have found that restriction of the amplitude matrix in Eq.(2) to diagonal form is not necessary. In fact, various simulated 2-D signals with full, nonsingular amplitude matrices were successfully processed using the above method. The latter result is not sufficient for the purpose of NMR, however. Firstly, signals with K≠K' exist, and secondly, even if K=K' and R(C)=K, Monte Carlo experiments show that the variance of the resulting model parameters are unnecessarily far above the Cramér-Rao lower bounds. In the following we remedy these two points.

We propose to rearrange the data in the following manner. First, we form Hankel matrices H_n (n=1, ... ,N) from the N' entries of each of the N rows of X, i.e.

$$H_n = \begin{bmatrix} x_{n1} & x_{n2} & \cdots & x_{nM} \\ x_{n2} & x_{n3} & \cdots & x_{nM+1} \\ \cdot & \cdot & & \cdot \\ \cdot & \cdot & & \cdot \\ \cdot & \cdot & & \cdot \\ x_{nL} & x_{nL+1} & \cdots & x_{nN'} \end{bmatrix}, \quad n=1, \ldots, N, \tag{5}$$

where $L+M=N'+1$. It is important to note that each H_n, when processed according to Eqs.(3) and (4), would yield the same left and right Vandermonde matrices $\zeta'_{L,K'}$ and $\zeta'_{M,K'}$. Subsequently, we form an $L\times(MN)$ compound matrix

$$H \overset{\text{def}}{=} (H_1|H_2| \ldots |H_N), \tag{6}$$

and subject this to SVD. Since the H_n are arranged in row form and each comprises the same Vandermonde matrices, the left Vandermonde matrix associated with H is identical to that of the H_n. It follows that one obtains the wanted signal poles $z'_{k'}$, $k'=1, \ldots, K'$, irrespective of the values of K and R(C) and using all data simultaneously. The SVD can be efficiently executed using the normal equations approach which amounts to diagonalizing the Hermitian matrix $HH^\dagger$, whose size is only LxL [2]. The unprimed signal poles can be found by transposing the data and repeating the procedure.

Until now, the effect of noise on the data was ignored. In the presence of noise, the method remains essentially the same, but the transformation matrix Q is to be determined in the least squares sense. As for the precision of the ensuing results, it was found by Kot et al. [6] for the case of a single, undamped 1-D sinusoid, that the variance of the frequency attains a minimum for $L \approx N/3$ (resp. $N'/3$), while rising steeply for $L \to N$ (resp. N'). We were able to derive a similar dependence for the 2-D case, subject to the same conditions as in [6].

As a last step, the complex-valued amplitudes are obtained by fitting Eq.(1), with signal poles fixed, to the data with a linear least squares procedure.

An advantage of SVD is that good estimates of K and K' can be inferred from perusing the singular values as a function of their index. In addition, the approach is noniterative. The computational load of the entire calculation is dominated by that of the SVD and is therefore proportional to resp. N^3 and N'^3, which entails high costs if the data set is large. The methods treated in the next subsection are iterative, but the computational load per cycle may be much smaller.

Finally, it should be mentioned that an alternative method to deal with the aspect of rank was very recently proposed in Ref. [11].

2.4 Decomposition by VARPRO (Methods IIa and IIb)

The starting point of VARPRO [5] is to write down the sum of squared deviations between the model function and the data, i.e.

$$E_2 = \sum_{n=1}^{N} \sum_{n'=1}^{N'} |\hat{x}_{nn'} - x_{nn'}|^2 . \tag{7}$$

E_2 is to be minimized as a function of the complex-valued amplitudes, frequencies, and damping factors, using an iterative nonlinear least squares procedure. The special feature of VARPRO is that the amplitudes entering in $\hat{x}_{nn'}$ are eliminated from Eq.(7) so that, at least initially, minimization is to be achieved only with respect to the frequencies and damping factors. Recently, it has been pointed out [7,8,4,12] that further simplification of Eq.(7) is possible by invoking Linear Prediction (LP). E_2 can then be minimized with *respect to the LP coefficients* rather than the frequencies and damping factors. A very important advantage of this is that in spite of the iterative character of the minimization procedure, starting values of the LP coefficients need not be supplied. Subsequent calculation of the wanted signal poles from the LP coefficients is a routine task.

Refs. [7,8,4,12] are primarily aimed at processing 1-D signals with the LP/VARPRO method. In this work, an extension to 2-D is made. Our strategy is described in Ref. [3], and is only very briefly indicated here for reasons of space.

First, X is decomposed, in the least squares sense, as (see e.g. [7])

$$X = \zeta_{N,K} A , \tag{8}$$

where $\zeta_{N,K}$ is defined in Eq.(2) and A is a KxN' matrix that carries the information about the signal poles in the primed dimension and the amplitudes. Subsequently, A is decomposed, using the same decomposition method once more which leads to

$$A = C\,\tilde{\zeta}'_{N',K'}\ , \tag{9}$$

the right-hand side of which is defined in Eq.(2). At this point all model parameters in Eq.(1) have been quantified. Note that the number of 1-D components in each dimension had to be known beforehand. Estimates of the latter can be obtained either from previous studies or from perusing the spectrum obtained by FFT. Since an NMR signal is usually contaminated by noise, choosing K and K' somewhat too large merely results in a limited number of insignificant amplitudes which can be discarded.

An important aspect of the LP/VARPRO method, not yet mentioned, is the potential to reduce the computational load relative to SVD-based methods, for *large* data sets. Our 2-D implementation of the LP-approach of Refs. [7,8,4] requires $N'N^2$ complex-valued multiplications in the initial stage and subsequently in each iteration cycle approximately $K(K+4)(N-K)^2$ complex-valued multiplications for executing Eq. (8), and $K(N'^2+3N'K')$ for Eq. (9) (Method IIa). Note that for large data sets, these expressions depend on the squares of the number of rows and columns of X. The appealing feature of the approach of Ref. [12] is that for a 1-D signal the computational load is proportional to the *first* power of N. Our 2-D implementation requires approximately 3KNN' complex-valued multiplications for Eq.(8), 3KK'N' for Eq.(9), and NN' for the initial stage (Method IIb).

Finally, we point out that Kumaresan and Shaw have devised [8,13] two alternative 2-D implementations of the LP/VARPRO method. One aspect of their method is that the signal poles of both dimensions are retrieved *simultaneously,* either in consecutive iteration cycles or in the same one. Conceptually, this approach is more appealing than ours, but in NMR practice the matrices involved may be rather large. In addition, these authors have so far restricted their method to the case of a diagonal amplitude matrix.

3. RESULTS AND DISCUSSION

In this section we present results of the SVD- and LP/VARPRO-based methods treated in Secs. 2.3 and 2.4. Our aims were: 1) to ascertain that retrieval of all model parameters entering Eq.(1) is feasible, 2) to investigate to what extent the standard deviations of the quantified parameters approach the theoretical (Cramér-Rao) lower bounds, 3) to investigate whether or not there is evidence for a significant bias in the values of the quantified parameters. To this end we have analyzed 50 noise-contaminated versions of the same simulated signal comprising 12 2-D exponentially damped sinusoids. The phases of all sinusoids were set to zero; the values of the frequencies, damping factors, and amplitudes are given in Fig. 1. From these model parameters, 50x50 2-D data points were generated, using Eq.(1). The real and imaginary parts of each data point were contaminated with normally distributed white noise of standard deviation 1. Each of the 50 noise-contaminated signals was analyzed using the methods described in Sec.2. The wanted standard deviations followed from the spread in the resulting 50 values of each parameter, and the biases from the average values. Note that two sinusoids had amplitudes smaller than the noise.

The space to present all results for each of the 38 model parameters involved in this paper is lacking. However, the relevant numbers for method IIa have already been published in [3]. At least for the signal-to-noise ratio (SNR) chosen here, the results can be adequately characterized qualitatively by the following points:

1) All three methods were indeed able to retrieve the 38 model parameters of the signal without posing problems. The SVD-based method, I, easily found the number of sinusoids involved in each dimension. In methods IIa and IIb these numbers were assumed to be known *a priori*.

2) Methods IIa and IIb approached the Cramér-Rao bounds to within several tens of percents. This result is to be expected, since LP/VARPRO is a maximum likelihood method. Method I (SVD) also performed well, but *occasionally* the standard deviation of some parameter (mostly a phase or an amplitude) was of the order of twice the standard deviation. Since SVD-based methods are not maximum likelihood, this result is satisfying.

3) There is no evidence for significant bias in the quantified parameters, at least at the given SNR.

4) Quantification of one noise-contaminated version of the signal presently investigated required about 4.5, 6.7, and 4.8 s. for method I, IIa, and IIb, respectively, using an IBM 3083-JX1 Mainframe computer .

From the above results we conclude that all three methods are candidates for quantifying 2-D signals. However, further tests are required. The following aspects should be investigated. 1) The robustness under possible inadequacy of the model function of Eq.(1). 2) The resolving power. 3) The feasibility of analyzing large real-world data sets comprising many sinusoids. 4) The SNR at which each of the methods ceases to function properly.

ACKNOWLEDGEMENT

This work was carried out in the program of the Foundation for Fundamental Research on Matter (FOM), and was supported (in part) by the Netherlands Technology Foundation (STW). The authors thank Th.J.L. Bosman, C.W. Hilbers, and F. van de Ven for useful discussions.

REFERENCES

[1] Bax, A. and Lerner, L., (1986) Two-Dimensional Nuclear Magnetic Resonance Spectroscopy, *Science*, **232**, 960-967.

[2] Barkhuysen, H., de Beer, R. and van Ormondt, D., (1985) Aspects of Computational Efficiency of LPSVD, *J. Magn. Reson.*, **64**, 343-346.

[3] de Beer, R., van Ormondt, D., Pijnappel, W.W.F., Bosman, Th. J.L. and Hilbers, C.W., (1988) Processing of Two-Dimensional Nuclear Magnetic Resonance Time Domain Signals, *Signal Processing*, **15**, 293-302.

[4] Bresler, Y. and Macovski, A., (1986) Exact Maximum Likelihood Parameter Estimation of Superimposed Exponentials in Noise, *IEEE Trans. Acoust., Speech, Signal Processing*, **ASSP-34**, 1081-1089.

[5] Golub, G.H. and Pereyra, V., (1973) The Differentiation of Pseudo-Inverses and Non-Linear Least Squares Problems whose Variables Separate, *SIAM J. Numer. Anal.*, **10**, 413-432.

[6] Kot, A.C., Parthasarathy, S., Tufts, D.W. and Vaccaro, R.J., (1987) The Statistical Performance of State-Variable Balancing and Prony's Method in Parameter Estimation, *Proc. ICASSP*, 1549-1552.

[7] Kumaresan, R. and Shaw, A.K., (1985) High Resolution Bearing Estimation without Eigen Decomposition, *Proc. ICASSP*, 576-579.

[8] Kumaresan, R., Scharf, L.L. and Shaw, A.K., (1986) An Algorithm for Pole-Zero Modelling and Spectral Analysis, *IEEE Trans. Acoust., Speech, Signal Processing*, **ASSP-34**, 637-640.

[9] Kumaresan, R. and Shaw, A.K., (1986) An Exact Least Squares Fitting Technique for Two-Dimensional Frequency Wavenumber Estimation, *Proc. IEEE*, **74**, 606-607.

[10] Kung, S.Y., Arun, K.S. and Bhaskar Rao, D.V., (1983) State Space and Singular Value Decomposition-Based Approximation Methods for the Harmonic Retrieval Problem, *J. Opt. Soc. Am.*, **73**, 1799-1811.

[11] Liu, Q.-G. and Zou, L.-H., (1988) A New Separable Eigenstructure Algorithm for Parameter Estimation of 2-D Sinusoids in White Noise, in "Signal Processing IV: Theory and Applications", J.L. Lacoume et al., Eds., Elsevier, pp. 443-446.

[12] Park, S.-W. and Cordaro, J.T., (1987) Maximum Likelihood Estimation of Poles from Impulse Response Data in Noise, *Proc. ICASSP,* 1501-1504.

[13] Shaw, A.K. and Kumaresan, R., (1988) Some Structured Matrix Approximation Problems, *Proc. ICASSP,* 2324-2327.

CONTRAST AS A MEASURE OF FOCUS IN SYNTHETIC APERTURE RADAR IMAGES

D. Blacknell*
(GEC-Marconi Research Centre, Chelmsford)

and

S. Quegan
(The University of Sheffield)

ABSTRACT

Synthetic aperture radar (SAR) is an all-weather, high resolution imaging system which can be carried on an aircraft or a spacecraft to provide a radar map of the Earth's surface. In order to produce high quality imagery, the motion of the SAR platform must be known accurately. For airborne SAR systems this is often not the case allowing image degradations, in particular defocussing of the image, to occur. By modifying the processing of the signal received by the SAR, a focussed image can be produced but to be able to do this automatically a measure of focus based on image statistics is needed. A possible candidate is the contrast of the image and in this paper the correspondence between contrast and focus is discussed for various images. The paper raises a number of questions about the meaning of focus.

1. INTRODUCTION

Synthetic aperture radar (SAR) is an all-weather, high resolution imaging system which can be carried on an aircraft or a spacecraft to provide a map of the varying radar cross-section of a portion of the Earth's surface. The high resolution is achieved in range by compression of a chirp pulse and in azimuth (cross - range) by coherent addition of the radar returns from a particular ground position as it is illuminated by successive parts of the radar beam. A collection of papers discussing SAR and its applications can be found in Kovaly [1].

Nominal SAR processing in the azimuth direction assumes that the SAR platform moves in a straight line and so will not necessarily produce high quality imagery when significant across-track motions occur. Since such motions are inevitable in

*Present Address: Department of Applied and Computational Mathematics, University of Sheffield, Sheffield, S10 2TN.

airborne SAR systems, methods have been developed to remove their effects. Autofocus is an automatic method of removing one of the effects of uncorrected across-track motions of the SAR platform, namely defocussing of the image.

A number of autofocus algorithm exist but in this paper only the contrast optimisation autofocus algorithm as implemented by Finley and Wood [2] will be considered. This algorithm uses the assumption that the image is in focus when the contrast of those features in the image which naturally produce large contrast values is maximised. In this paper, the limitations of this assumption will be discussed. It will be shown that certain high contrast features exist which have maximum contrast when the image is significantly out of focus. This would seem to indicate that, under certain circumstances, the contrast optimisation algorithm could fail dramatically.

In Section 2 the equations describing SAR processing and autofocus will be derived. In Section 3, target arrangements which cause the contrast to peak when the image is not in focus will be produced. In Section 4 the consequences of these observations for the successful operation of the contrast algorithm will be discussed.

2. SAR PROCESSING

The SAR imaging geometry in the slant range plane for a SAR platform trajectory, $r(x)$, is shown in Figure 1. The intended flight - path is a straight line which has been taken as the x-axis. A point target has been shown which lies at a range R_o and at an azimuth position $x = x_o$. The range from the SAR platform to the point target is $R(x)$ so that, assuming uniform illumination over the aperture, the signal received from the point target can be written as

$$s(x) = \exp[2ikR(x)], \tag{2.1}$$

where $k = 2\pi/\lambda$, and λ = wavelength.

By using the geometry, and making suitable approximations, Equation (2.1) becomes

$$s(x) = \exp i\frac{k}{R_o}[(x - x_o)^2 - 2R_o r(x)]. \tag{2.2}$$

An image of the point target, i.e. the point spread function (PSF), is formed by correlating the received signal with a 'reference function'

$$u(x) = \exp\ (i\phi(x)), \tag{2.3}$$

where $\phi(x)$ is some suitably chosen phase variation. If $\phi(x)$ is chosen to match the phase variation of the received signal then the PSF will have the narrowest possible main lobe. An image of a general scene, which can be represented as the convolution of the scene with the PSF, will then be defined as being in focus.

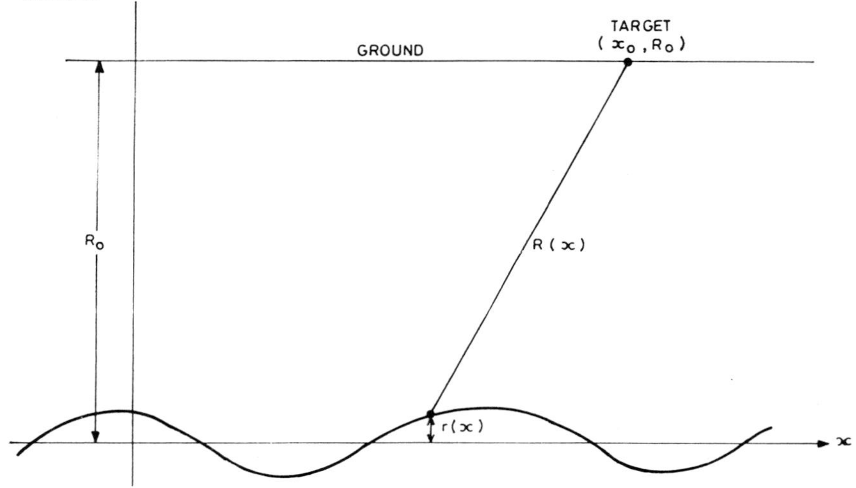

r (x) ≡ PLATFORM TRAJECTORY
Fig. 1 SAR Imaging Geometry

The effects on image focus of different platform trajectories are best illustrated using parabolic trajectories, since it is the quadratic phase variation resulting from such motions which is the fundamental cause of image defocus. Consider a point target, positioned at the origin, being imaged by a SAR which has a constant across-track acceleration, a, so that $x_o = 0$, and $r(x) = \frac{ax^2}{2}$. Then the received signal becomes

$$s(x) = \exp\ [\frac{ik}{R_o}\ (1 - aR_o)x^2]. \tag{2.4}$$

Nominal processing to form an image involves correlating with a reference function which takes the form

$$u(x) = \exp\ \frac{ik}{R_o}\ x^2. \tag{2.5}$$

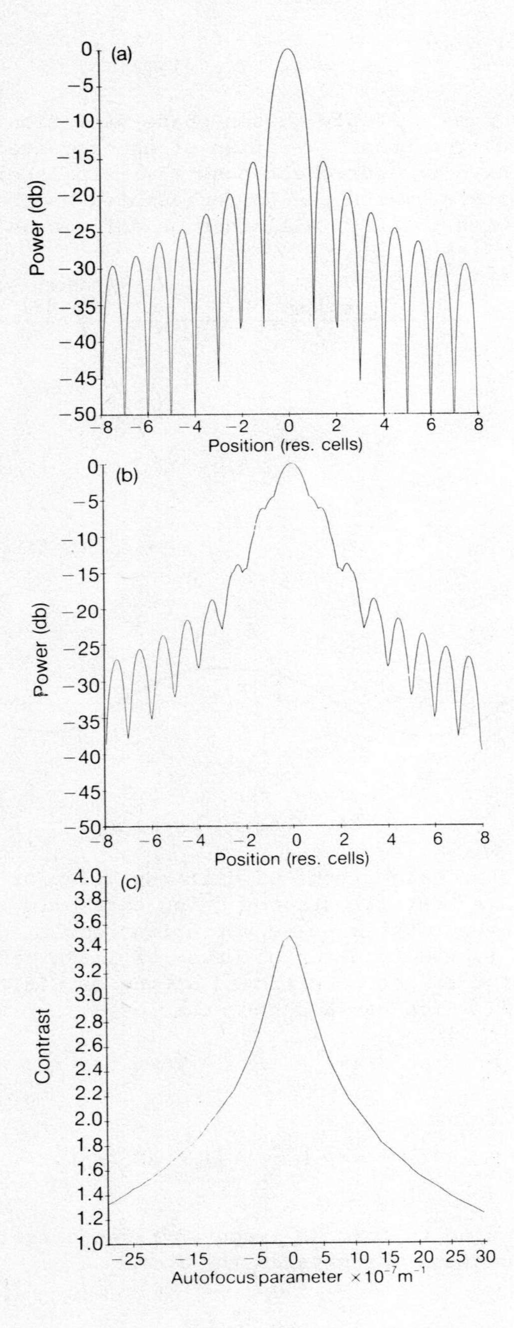
(a)
Power (db)
Position (res. cells)
(b)
Power (db)
Position (res. cells)
(c)
Contrast
Autofocus parameter ×10⁻⁷m⁻¹

Fig. 2

If the SAR platform in fact travelled in a straight line, i.e. $a=0$, then the square-law detected result of this correlation would be the point spread function (PSF) shown in Figure 2(a). However, for a non-zero value of the across-track acceleration, a typical PSF which would result is shown in Figure 2(b). It should be noted that the width of the main lobe of the second PSF is greater than that of the first. This will be true whatever the value of the non-zero across-track acceleration and is the characteristic of the PSF which causes defocussing of the SAR image. However, if the value of the across-track acceleration is estimated as $\hat{a}$, then a modified reference function,

$$u(x) = \exp\left[\frac{ik}{R_0}(1-\hat{a}R_0)x^2\right], \qquad (2.6)$$

can be used which will produce the PSF shown in Figure 2(a) if $\hat{a}=a$, whatever the across-track acceleration. The image has been focussed.

In practice, the value of the across-track acceleration will not be known and must be determined automatically if high quality imagery is to be routinely produced. This process is known as autofocus and can be achieved by performing a numerical search through values of $\hat{a}$, termed the autofocus parameter, until that value which focusses the image is found to a suitable degree of accuracy. The problem lies in finding a measure which indicates how well the image, i.e. the scene convolved with the PSF, has been focussed. The width of the main lobe of the PSF is an obvious candidate but in a SAR image perfect point targets will not occur very frequently and so cannot be used to regularly update the autofocus parameter. Thus some measure is required which can be applied to the image of a general scene.

Finley and Wood [2] use the image contrast, defined as the ratio of the standard deviation to the mean of the pixel values, as a measure of image focus. Figure 2(c) shows the variation of image contrast with autofocus parameter for a point target imaged by a SAR platform with no across-track motion. It can be seen that the contrast peaks when the autofocus parameter is zero, i.e., when the image is focussed. This shows that contrast is a reasonable candidate as a measure of focus, since its variation behaves in the desired manner for a single point target. This is true whatever value of the across track acceleration is used. However, for a homogeneous scene the contrast measure does not give an indication of focus and so the measure cannot be used for a general scene. However, Finley and Wood [2] assumed that, for any features within a scene which start off with a high contrast value, the contrast will be maximized when the image is in focus, i.e., high contrast features will behave in a similar manner to a single point target. Since such features are much more common than isolated point targets the autofocus method based on this assumption can be used to obtain regular updates of the autofocus parameter as the SAR platform progresses in the azimuth direction. The validity of this assumption will be investigated in the next Section.

3. PATHOLOGICAL TARGET ARRANGEMENTS

Will it always be the case that the maximum contrast value for a high contrast feature will occur when the image is in focus? This question can be answered conclusively if a counter-example can be found.

Consider 2N+1 point targets with complex amplitudes a_n at positions x_n, n=1, 2, ..., 2N+1. The signal received from this scene will be

$$s_c(x) = \sum_{n=-N}^{N} a_n \exp \frac{ik}{R_o} (x-x_n)^2. \tag{3.1}$$

It is desired that the image contrast resulting from this signal should peak for some significantly non-zero value of the autofocus parameter. This can be achieved if the signal mimics the signal which would have been received by a SAR moving along a parabolic trajectory imaging a scene for which the maximum image contrast does occur when the image is in focus. The simplest approach to this is to require the signal to mimic the signal received from a single point target imaged by a SAR moving with an across-track acceleration a, namely,

$$s_a(x) = \left\{\exp \frac{ik}{R_o} (1-aR_o)x^2\right\} . \tag{3.2}$$

Equating (3.1) and (3.2) gives the requirement

$$\sum_{n=-N}^{N} a_n \exp\left\{-\frac{2ik}{R_o} xx_n\right\}\exp\left\{\frac{ik}{R_o} x_n^2\right\} \simeq \exp \{-ik\, ax^2\}, \tag{3.3}$$

it being assumed that even approximate equality will be sufficient for the desired behaviour. It can be observed that the LHS of Equation (3.3) has the form of a discrete Fourier transform and so values for a_n and x_n can be obtained by expressing the RHS as a discrete Fourier transform and equating coefficients.

Let

$$f(x) = \exp\{-ikax^2\}, \qquad |x| \leq L/2 . \tag{3.4a}$$

Then, ignoring unimportant multiplicative constants,

$$F(w) \simeq \exp\{-i \frac{\omega^2}{4ak}\}, \qquad |\omega| \leq akL, \tag{3.4b}$$

where L = synthetic aperture length.

Let

$$g(x) = f(x) * \sum_{n=-N}^{N} \delta(x - nL). \tag{3.5a}$$

Then

$$G(\omega) = F(\omega) \sum_{n=-N}^{N} \delta\,(\omega - \frac{2\pi n}{L}). \tag{3.5b}$$

Thus,

$$g(x) = \int_{-\infty}^{\infty} G(\omega) \exp(i\omega x)\, d\omega$$

$$= \sum_{n=-N}^{N} F\,(\frac{2\pi n}{L}) \exp\,(2\pi i n \frac{x}{L}). \tag{3.6}$$

However, $|\omega| \leq akL$, which implies $|n| \leq \frac{akL^2}{2\pi} = \frac{aL^2}{\lambda} = \frac{a}{a_d}$,

where $a_d = \frac{\lambda}{L^2}$ = the depth of acceleration, which is the required accuracy of the autofocus parameter.

Therefore,

$$g(x) = \sum_{|n| \leq a/a_d} \exp\{\frac{i}{4ak}(\frac{2\pi n}{L})^2\}\exp\{2\pi i n\frac{x}{L}\}. \tag{3.7}$$

Comparison of Equations (3.7) and (3.3) gives, after some simplification,

$$a_n = \exp\{2\pi i \frac{n^2}{4} \frac{a_d}{a} (1-aR_o)\}, \tag{3.8a}$$

$$x_n = n\, \ell_a, \tag{3.8b}$$

$$N = [\frac{a}{a_d}], \tag{3.8c}$$

where $\ell_a = \frac{\lambda R_o}{2L}$ = azimuth resolution.

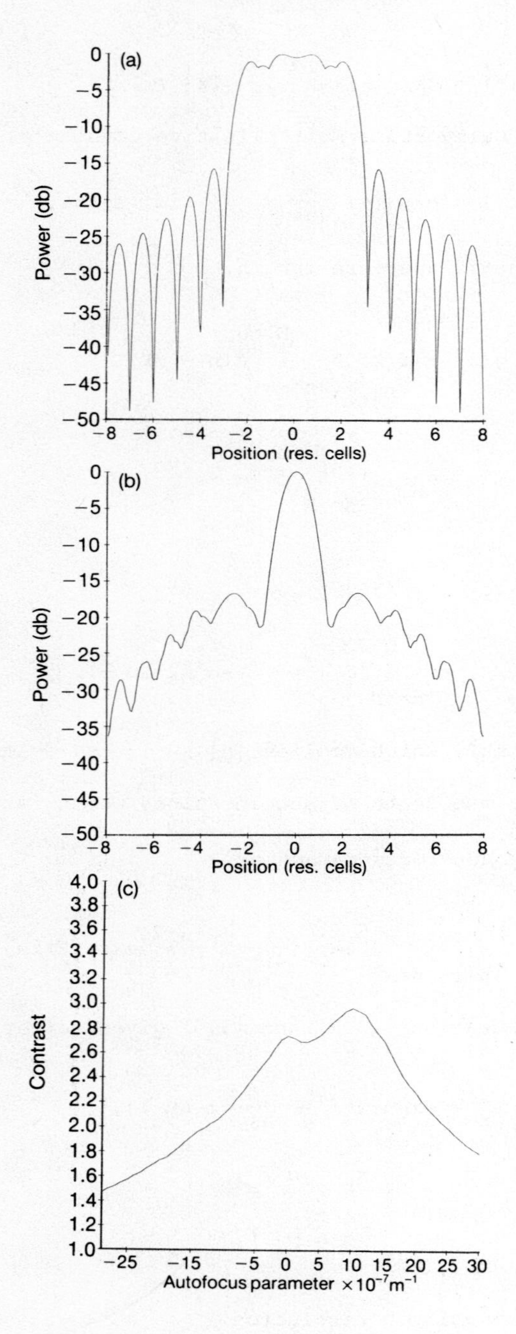
(a)
Power (db)
Position (res. cells)
(b)
Power (db)
Position (res. cells)
(c)
Contrast
Autofocus parameter ×10⁻⁷m⁻¹

Fig. 3

Two examples will illustrate that the behaviour of the image contrast for the target arrangement defined above differs from that assumed for the contrast optimisation algorithm. The SAR parameters to be used are,

$$\lambda = 0.03 \text{ m}$$

$$R_o = 30\,000 \text{ m}$$

$$L = 300 \text{ m}$$

$$\ell_a = 1.5 \text{ m}$$

$$a_d = 3.33 \times 10^{-7} \text{ m}^{-1}.$$

(It should be noted that throughout this discussion acceleration has been used to describe the second <u>spatial</u> derivative of position; hence the units m^{-1}).

(i) Let $a = 3a_d = 10^{-6}$ m^{-1}.

Equation (3.4c) gives N = 3 but, since this is a borderline case between two values of N, the lower value of N = 2 can be used. Thus the scene consists of an arrangement of five point targets with unit amplitude, each separated from the next by the resolution distance. This scene is imaged by a SAR travelling in a straight line, thus to focus correctly the autofocus parameter should be taken to be zero. Figure 3(a) shows the image resulting when the image is focussed correctly and Figure 3(b) shows the image which has maximum contrast. Figure 3(c) shows that the variation of contrast has peaked at the value which was chosen and not at the correct value, despite the local maximum at zero. However, comparing the focussed image, Figure 3(a), with the defocussed image, Figure 3(b), it is difficult to imagine what measure would correctly classify them.

(ii) Let $a = 6a_d = 2 \times 10^{-6}\ m^{-1}$.

In this case the scene consists of an arrangement of eleven point targets with unit amplitude each separated from the next by the resolution distance. The focussed image is shown in Figure 4(a), the defocussed image is shown in Figure 4(b), and the variation of contrast with autofocus parameter is shown in Figure 4(c). The same comments as for the first example apply. Indeed, the false contrast peak is even more pronounced in this case and the defocussed image looks even more like a point target response.

Thus the question posed at the beginning of this section can be answered. It is not always the case that the maximum contrast value for a high contrast feature will occur when the image is in focus.

4. CONSEQUENCES FOR SAR AUTOFOCUS

In example (i), a scene consisting of a symmetrical arrangement of five equally spaced, unit amplitude targets was shown to produce an image which was not in focus when the contrast was a maximum. The two independent phases of the outer targets relative to the central target were determined using Equation (3.8a). It is interesting to observe the behaviour of the contrast peak as these phases are allowed to vary through 360°.

Figure 5 shows the behaviour of the contrast peak for various possible phase combinations. It can be seen that three district regions of the phase space exist, a region where the contrast peaks when the image is in focus and two regions, occupying almost one third of the total region, where the contrast peaks when the image is defocussed. Mathematical analysis of even this simple case rapidly becomes extremely complicated and intractable, so that we have been unable to characterise the different regions (though the rotational symmetry of the figure is readily explained). The behaviour of the contrast measure for more general target arrangement is not known, and currently no treatment to elucidate its behaviour seems obvious.

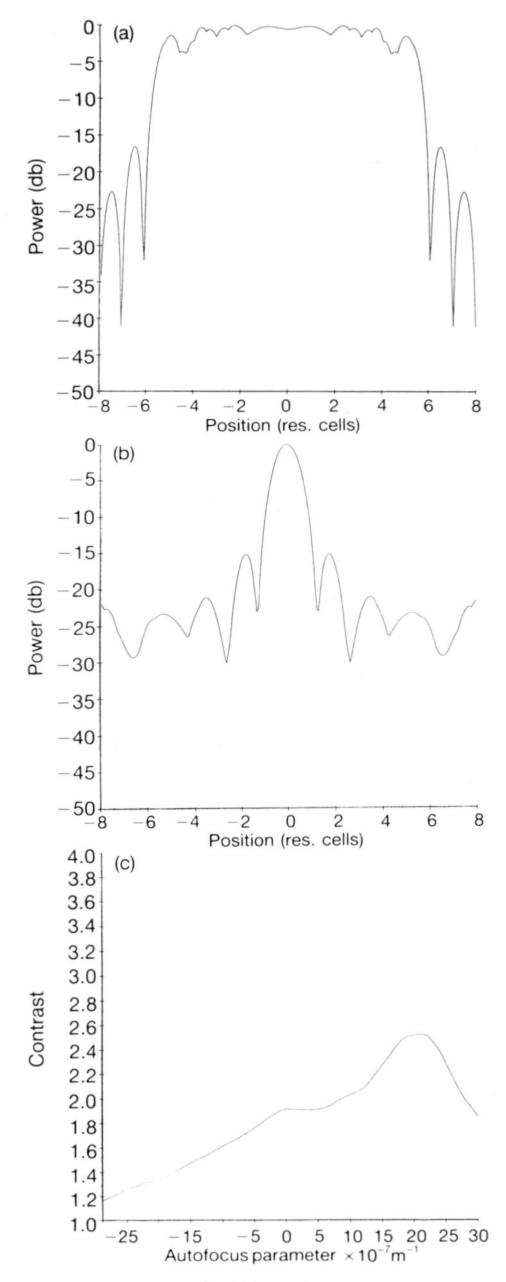
(a)
Power (db)
0
−5
−10
−15
−20
−25
−30
−35
−40
−45
−50
−8
−6
−4
−2
0
2
4
6
8
Position (res. cells)
(b)
Power (db)
Position (res. cells)
(c)
Contrast
4.0
3.8
3.6
3.4
3.2
3.0
2.8
2.6
2.4
2.2
2.0
1.8
1.6
1.4
1.2
1.0
−25
−15
−5
0
5
10
15
20
25
30
Autofocus parameter ×10⁻⁷m⁻¹

Fig. 4

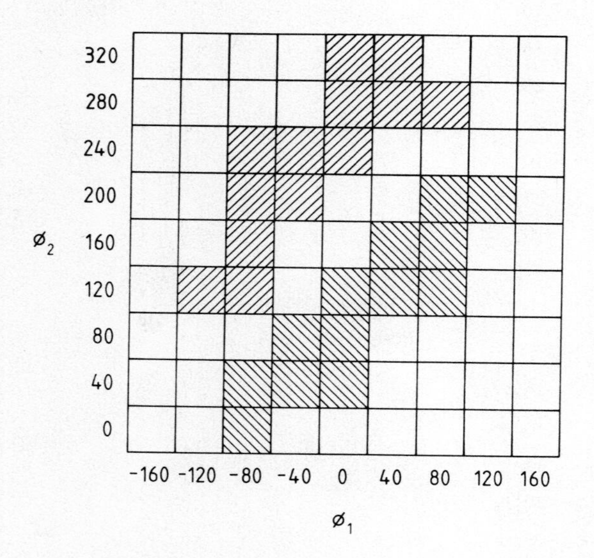

Fig. 5 Behaviour of Contrast Peak for Five Point Targets with Varying Relative Phases.

We therefore have the unsatisfactory situation that the contrast optimisation method of autofocus, which has been successfully used over a number of years, can be shown to fail for scenes containing no dominant scatterers or certain arrangements of dominant scatterers. It is not as yet possible to characterise the arrangements which make the autofocus fail, or to quantify the frequency with which they occur in natural scenes (if at all). It is known that in application contrast optimisation requires averaging of contrast over a number of range lines, and that individual range lines may fail to yield the correct autofocus parameter, suggesting that failures as a result of target arrangements genuinely occur. The number of lines required for averaging is not based on any knowledge of the failure rate, and could be reduced if we better understood the conditions of failure. This paper only exposes the problem, without furnishing a solution.

Progress using simple analytic methods appears barred, and ways round the problem have yet to be found. One possible approach, however, is to change the problem by considering whether contrast is the image measure most likely to relate to focussing. Image entropy may be a more relevant measure, and current work is investigating this possibility.

5. REFERENCES

[1] Kovaly, J.J., (1976) Synthetic Aperture Radar, Artech House.

[2] Finley, I.P. and Wood, J.W., (1985) An Investigation of SAR Autofocus, RSRE Memorandum 3790, Royal Signals and Radar Establishment, Gt. Malvern, Worcs.

6. ACKNOWLEDGEMENT

This work was supported by the Procurement Executive, Ministry of Defence, UK.

IMAGE DATA COMPRESSION COMBINING MULTIRESOLUTION, FEATURE ORIENTATION AND ARITHMETIC ENTROPY CODING

M. Todd and R. Wilson
(University of Warwick)

ABSTRACT

Image data compression is the art of representing an image accurately with as small a data volume as possible. This paper describes a new image data compression algorithm, based on multiresolution interpolation controlled by local feature orientation. The new coder is one of a class of pyramidal coders employing what may be seen as non-causal predictive methods. It is shown to be effective at rates below 1 bit per pixel and to be capable of producing decoded images with no gross distortions at compression ratios of up to 100. The coder is described in the context of a new class of image models - pyramid models - and its performance on a number of test images presented and compared with other methods. The paper is concluded with a discussion of the implications of this work for image modelling and data compression theory.

1. INTRODUCTION

In its raw form digital monochrome image data usually consists of a 2-dimensional rectangular array of N by M picture elements called *pixels*. Each pixel is a data sample representing the intensity or *luminance* of the image at the point in the image defined by the co-ordinates (x,y) $0 \leq x < N, 0 \leq y < M$ of the pixel.

Digital monochrome images with approximately equivalent quality to TV transmissions require an array of around 512 by 512 pixels with 8-bits of information (256 gray-levels) for each pixel. Thus approximately two million bits are required to represent a single monochrome image. The total number of bits required by the image divided by the total number of pixels in the image is known as the bit rate in bits per pixel (bpp) for a single-frame image.

Data compression is possible because most images contain a large degree of structure, giving rise to redundancy, and algorithms can be developed to use this structure to represent the image in a more compact form. There are two basic categories for techniques which compress image data, the first being those which retain an exact representation of the original image by eliminating as much of the statistical redundancy as possible. This is known as noiseless coding. The second is to permit errors between the compressed image and the original. The acceptability of these errors will be related to their effect on the visual perception of the image. Thus a certain amount of error will be allowed between the coded image and the original, and there will be a trade-off between the compression factor that can be achieved and the quality of the image. The minimum acceptable quality (and therefore the maximum compression factor) will depend on the application of the coding scheme.

Conventional image data compression systems fall broadly into two classes: transform coders and predictive coders. The former employ a fast orthogonal transform to approximate the eigenvector transformation of the image autocorrelation function, an approach which can be shown by rate-distortion arguments to be optimal in a mean squared error sense [6]. Because of the non-stationary nature of image data and on computational grounds, such schemes generally operate on small blocks within the image, and are non-causal within the image plane. The latter can again be rigorously justified in terms of a minimum mean squared error distortion criterion [6] but are in general causal within the image plane.

In recent years, there have been a number of developments in image data compression which seek to exploit known properties of the human visual system (e.g. [1,2,8,9,12]). Among these properties, two have been found particularly significant: the use of multiple scales of image representation is known to occur in the visual system and has found many applications in computer vision; the orientation selective properties of simple cells in the primary visual cortex (e.g. [5]) are well documented and this idea has also been exploited in a range of image processing applications including data compression (e.g. [9],[8]). While such methods have been shown to be effective in practice, they still lack the rigorous mathematical basis of more traditional methods, such as predictive or transform coding (e.g. [6]).

The aim of this paper is therefore twofold. First, a general predictive framework is presented for multiresolution image coders based on a new class of image models which are causal not in the image plane, but in the scale dimension. This represents a generalisation of the model discussed in [4] and used for image restoration. Then a new coder which falls within this general class will be described and its performance evaluated. In addition to its use of multiresolution methods, the new coder is distinguished by its application of the local orientation estimation procedure used successfully in [8].

The results achieved with this coder serve to illustrate the potential of methods derived from the new models and to suggest ways in which classical linear signal models can usefully be supplemented by models which are perhaps more closely related back to the content of natural images and to the perceptual machinery humans use to process them.

2. MULTIRESOLUTION IMAGE MODELLING

It has perhaps not been widely appreciated that multiresolution image processing can also be model-based. Indeed the simple linear recursive model of equation (2.1)

$$S(n) = A(n)S(n-1) + B(n)W(n) \qquad 0 < n \leq N \tag{2.1}$$

in which $S(\cdot)$ and $W(\cdot)$ are 2-d vectors

$$S(n) = \left[S_{xy}(n)\right] \qquad 0 \leq x,y < M \tag{2.2}$$

and $A(\cdot)$, $B(\cdot)$ are linear operators, i.e.

$$S_{xy}(n) = \sum_{p,q} A_{xypq}(n)S_{pq}(n-1) + B_{xypq}(n)W_{pq}(n) \tag{2.3}$$

is a suitable basis for a wide class of multiresolution image models, when the vectors $W(n)$ are samples from e.g. a unit variance normal white noise process, so that

$$E\left[W_{xy}(n)W_{pq}(m)\right] = \delta_{nm}\delta_{xp}\delta_{yq} \tag{2.4}$$

and subject to the initial condition $A(0) = 0$ giving,

$$S(0) = B(0)W(0) \tag{2.5}$$

for such a model, the resulting image is just

$$S = S(N) \tag{2.6}$$

Now of course such recursion also serves as the basis of many image plane causal models (e.g. [7]). The multiresolution models, however, have several features not shared by the classical recursive models. The most obvious is that whereas for an image plane causal model, the number of recursions needed to generate the image is $N=M^2$, the image dimension, for a multiresolution model it is typically $N \approx \log_2(M)$. Furthermore, the operators $A(\cdot)$,$B(\cdot)$ have a very different structure for the two classes of model. Specifically, for $n>1$, the operators $A(n)$, $B(n)$ have a *periodic* structure, i.e. $\exists m_n$ such that

$$A_{(x+im_n)(y+jm_n)(p+im_n)(q+jm_n)}(n) = A_{xypq}(n) \tag{2.7}$$

and similarly for B(n). However, as in the classical models, they also have a *locality* property, in that for some $r_n>0$.

$$A_{xypq}(n) = 0 \quad |x-p|>r_n \text{ or } |y-q|>r_n \tag{2.8}$$

and similarly for B(n). Moreover, in the models considered here, there is a passage from *global to local* structure as n increases in that (i) $m_n > m_{n+k}$ and $r_n > r_{n+k}$ for k>0 and (ii) for each n there is an index set $\Lambda_n = \{(x_{n0},y_{n0}),(x_{n1},y_{n1}),\ldots,(x_{nL_n},y_{nL_m})\}$ such that

$$B_{xypq}(n) = 0 \quad (p,q)\notin \Lambda_n \tag{2.9}$$

and for n>1,k>0, $\text{card}(\Lambda_n) = L_n < L_{n+k}$. It is this combination of properties which give the models their *tree* or *pyramid* structure (e.g. [2],[11]). Of the models within this general category, those which have been most thoroughly investigated are the quad-tree models (e.g. [4],[11]), for which fast minimum m.s.e. estimates have been found. Although they have been used successfully in predictive data compression they are not ideal for this task, partly because of blocking artefacts and partly because the signals on levels n<N above the image are not directly accessible, but have to be estimated from the image data. Similar remarks apply of course to more general pyramid models, with the added complication that the m.m.s.e. estimate of some level S(n) say from S(N) cannot be put in a fast recursive form [12]. The class of models considered below, the interpolative models, avoid these problems completely and are ideally suited to data compression applications.

A linear interpolative multiresolution image model is one of the form of equations (2.1)-(2.6), but with the additional constraint that for each point (x,y) there is some $n \leq N$ s.t.

$$\sum_{r,s}|B_{xyrs}(n)| \sum_{p,q}|B_{xypq}(m)| = \delta_{nm}\left(\sum_{r,s}|B_{xyrs}(n)|\right)^2 \tag{2.10}$$

and

$$A_{xyrs}(m) = \delta_{xr}\delta_{ys} \quad \text{if } m>n \tag{2.11}$$

The significance of these equations is that they give the model an interpolative structure - each point in the final image can be expressed as the sum of an innovation term and a linear interpolation from other points in the image

$$S_{xy}(N) = \sum_{p,q} \alpha_{xypq} S_{pq}(N) + \nu_{xy} \tag{2.12}$$

where $\alpha_{xypq} = 0$ if $S_{pq}(N)$ depends either directly or indirectly on $S_{xy}(n)$: there are no cycles of linear dependence in the model. In other words, given the image, it is possible to find any $S(n)$, $n<N$, by means of an equation of the form

$$S(n) = D(n)S(N) \tag{2.13}$$

$D(n)$ is most readily constructed using the recursive formula

$$D(n) = (I-C(n))A(n)D(n-1) + C(n) \quad n \geq 0 \tag{2.14}$$

where I is the identity and $C(n)$ is defined by

$$C_{xypq}(n) = \begin{cases} \delta_{xp}\delta_{yq} & \text{iff } \sum_{p,q} |B_{xypq}(n)| \neq 0 \\ 0 & \text{else} \end{cases} \tag{2.15}$$

It is this property that makes linear interpolative models useful in data compression. In addition to this, such models share with other multiresolution structures an inherent non-stationarity, implied by the periodicity m_n, and an ability to express structures over a range of scales from global to local. Finally, it is simple to generalise the models to a class of parametric models of the form

$$S(n) = A(n,\theta(n))S(n-1) + B(n,\theta(n))W(n) \tag{2.16}$$

where $\theta(n)$ is a vector of model parameters normally of dimension $(M/m_n)\times(M/m_n)$, i.e. with a resolution which matches that of the signal on level n. If an effective procedure exists for estimating $\theta(n)$ from the image, then a recursion of the form of equation (2.16) can be used to give an added degree of local structure to the model.

Among the linear interpolative models, those with a *block* structure are perhaps the most useful. The pyramidal or tree structure comes from successive subdivisions of the blocks into sub-blocks and so on until the block size is (1x1) i.e. a single pixel. One of the first applications of this idea uses recursive binary nesting (RBN) [1] in which the block sizes are divided into two for the passage from level n to level (n+1) (Fig. 1). RBN models are most easily described for images of dimension $(2^N+1)\times(2^N+1)$, for which the operators $A(n), B(n)$ have a structure which is periodic, for $n > 1$.

$$A_{(x+i2^{N-n+1})(y+j2^{N-n+1})(p+i2^{N-n+1})(q+j2^{N-n+1})}(n) = A_{xypq}(n) \quad 0 \le i,\ j < 2^{n-1}$$

$$B_{(x+i2^{N-n+1})(y+j2^{N-n+1})(p+i2^{N-n+1})(q+j2^{N-n+1})}(n) = B_{xypq}(n) \quad n >$$

(2.17)

and which can therefore be defined in terms of the coefficients in the top left hand block: the set β_n of the indexes (x,y).

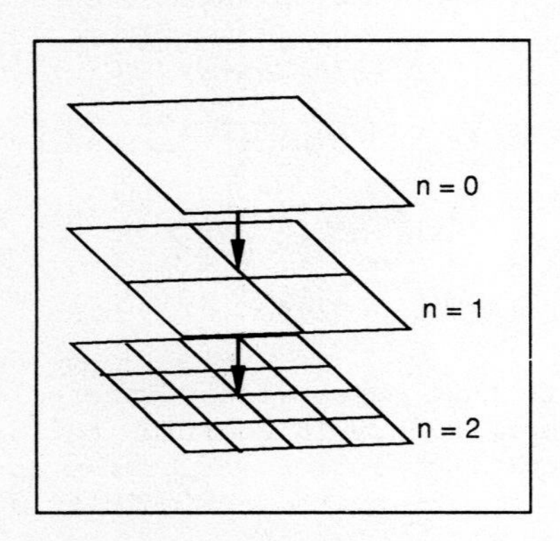

Fig. 1 Recursive Binary Nesting

$$\beta_n = \bigcup_{0\le x,y\le 2^{N-n+1}} (x,y) \tag{2.18}$$

Indeed, defining the block perimeter set π_n by

$$\pi_n = \bigcup_{(x,y)\in\beta_n} \{(0,y)\cup(2^{N-n+1},y)\cup(x,0)\cup(x,2^{N-n+1})\} \tag{2.19}$$

and the centre cross set χ_n by

$$\chi_n = \bigcup_{(x,y)\in\beta_n} \{(x,2^{N-n}) \cup (2^{N-n},y)\} - \pi_n \tag{2.20}$$

then if $(x,y)\in\beta_n$

$$A_{xypq}(n) = 0 \quad \text{if } (x,y)\notin(\beta_n-\pi_n) \text{ or } (p,q)\notin\pi_n \tag{2.21}$$

and

$$B_{xypq}(n) = 0 \quad \text{if } (x,y)\notin\chi_n \text{ or } (p,q)\notin\beta_n \tag{2.22}$$

In effect, the images $\delta(n)$ are *successive approximations* to the final image $S(N)$, based on linear interpolation within edge-sharing blocks of dimension $(2^{N-n}+1)x(2^{N-n}+1)$. The parametric model which has proved most successful in applications is defined in terms of a vector $\theta(n)$ of dimension $2^n x 2^n$ of block orientations: each component $\theta_{xy}(n)$ is an angle in the range $(0,\pi)$, specifying the orientation which the interpolation $A(n)$ should take within the block (Fig. 2). This model therefore expresses both the multiresolution and orientation selective properties of the visual cortex [5].

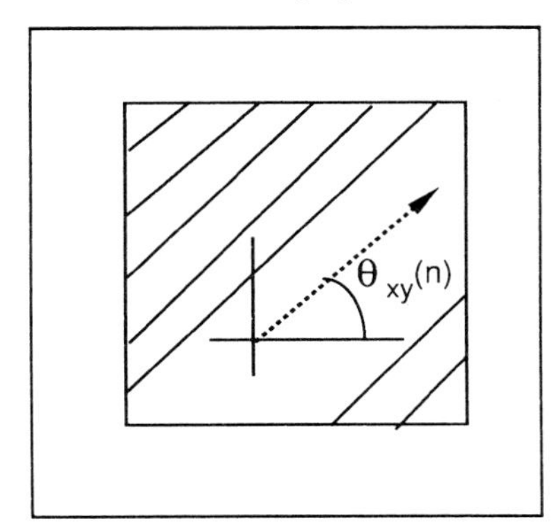

Fig. 2 Oriented Interpolation

3. A PREDICTIVE CODER

The general recursive structure of these models naturally suggests the use of a predictive coder, just as it does in the 1-d case. Indeed it follows from equations (2.1)-(2.5) that the minimum mean squared error predictor for $S(n)$ in terms of $\{S(n-k), 0<k\leq n\}$ is just (e.g. [6]).

$$\hat{S}(n|n-1)=A(n)S(n-1) \tag{3.1}$$

The addition of a quantiser into the loop results in operational equations of the form

$$\hat{S}_q(n|n-1)=A(n)S_q(n-1) \quad N \geq n \geq 1 \tag{3.2}$$

$$E(n)=S(n)-\hat{S}_q(n|n-1) \tag{3.3}$$

$$S_q(n)=Q(E(n))+\hat{S}_q(n|n-1)=S(n)+V_q(n) \qquad (3.4)$$

where $Q(\cdot)$ is a quantisation function, which may be a scalar quantiser or a vector quantiser (both types have been investigated) and $V_q(n)$ is the vector of quantiser noise components. The similarity of these equations to those for the standard predictive coder should be immediately apparent. Indeed there is no formal difference save that $N=\log_2 M$ in the present case. It should also be noted that only those error components corresponding to the innovations at level n are transmitted, that is

$$(Q(E(n)))_{xy} = 0 \text{ if } \sum_{p,q} |B_{xypq}(n)|=0 \qquad (3.5)$$

The orientation dependence of the coder can be emphasised by writing, in place of equation (3.2)

$$\hat{S}_q(n|n-1)=A(n,\theta_q(n))S_q(n-1) \qquad (3.6)$$

where $\theta_q(n)$ is the vector of quantised block orientations. Finally, it may be noted that the successive approximation process in the coder may be terminated at a level $n<N$ if, for example some error criterion for the $(i,j)^{th}$ block at level n is satisfied. In the tests described below, a maximum error criterion was used, viz

$$d_{ij}(n) = \max_{\substack{i2^{N-n}\leq x\leq (i+1)2^{N-n} \\ j2^{N-n}\leq y\leq (j+1)2^{N-n}}} |S_{xy}(N)-S_{qxy}(n)| \leq d \qquad (3.7)$$

where $d>0$ is a threshold. In other words, these coders allow a simple form of threshold coding to be applied: any block for which the error criterion is satisfied is not subdivided. The address information which this produces must of course be transmitted along with the prediction errors.

4. TEST RESULTS

A number of coders based on the above models have been tested. The simplest was a variation on the original RBN coder [1] in which the block interpolation $A(n)$ was just a bilinear interpolation from the four corner points. The prediction errors were quantised using a vector quantiser and a maximum error threshold criterion used. The results from this coder were satisfactory, but did not represent a significant improvement on the basic RBN scheme. The orientation selective coder, on the other hand has produced significantly better results on a number of 8-bit (512x512) pixel test images. The block orientations were estimated using the unbiased estimator described in [8] and the orientation quantised to one of 31 angles

$$\theta_i = i\frac{\pi}{31} \quad 0 \leq i \leq 30 \tag{4.1}$$

The test images were pre-filtered with an anisotropic filter based on the orientation estimate $\theta(N)$. Arithmetic coding with a dynamic probability estimation was employed for all transmitted components and addresses [10]. For this coder, the interpolator $A(n)$ requires the luminances at all points on the block perimeter (π_n of equation (2.19)). This involves 1-d coding of the prediction errors for each block edge. A variety of 1-d coders were tested, including a discrete cosine transform. It was found however that a 1-d version of the RBN algorithm performs as well as other methods. This was the method selected for the tests whose results are shown in Figs. 3-6 and Tables 1,2. Figs. 3 and 4 show the original 512 by 512 8-bits per pixel test images "GIRL" and BOATS". Figs. 5 and 6 show the results of coding the test images respectively at rates of 0.08bpp and 0.26bpp. Although some errors are visible at these rates, there are none of the gross distortions which are typical of most block coders. In Tables 1 and 2, BPP is the bit rate in bits per pixel, MSE is the mean squared error at that bit rate, and PSNR is the peak signal to noise ratio at that bit rate. These tables show that acceptable results can be obtained with these images at rates below 0.5 bpp, comparing favourably with the results reported for either the traditional transform or predictive coders or the more esoteric systems described in e.g.[9].

Fig. 3 "GIRL Original"

Fig. 4 "BOATS Original"

Fig. 5 "GIRL 0.08 bpp"

Fig. 6 "BOATS 0.26 bpp"

GIRL		
BPP	MSE	PSNR
0.08	167.78	+25.88
0.12	119.16	+27.37
0.20	82.33	+28.98
0.29	64.34	+30.05

Table 1

BOATS		
BPP	MSE	PSNR
0.06	398.61	+22.13
0.13	273.22	+23.72
0.26	102.93	+28.01
0.31	87.50	+28.71
0.43	70.90	+29.62

Table 2

5. CONCLUSIONS

Conclusions may be drawn from this work on several levels. First it has been shown that a common framework exists for multiresolution image processing, based on a new subclass of the general recursive linear model of equation (2.1). Some basic properties of these models have been discussed (cf[12]) and a form well suited to image coding - the linear interpolative form - introduced. Parametric forms of the models allowing varying degrees of global and local non-uniformity were also presented and applied successfully to the problem of image data compression. In the second place, by establishing a firmer mathematical footing for these methods, it becomes possible to exploit much of the existing mathematical apparatus of signal theory within the multiresolution framework. This will have important practical consequences for the development of effective procedures for signal estimation and analysis as well as for prediction (see also [3][4]). New applications of rate-distortion theory may also be anticipated. Lastly, this work demonstrates conclusively that image processing algorithms do not have to be either a) based on near sighted extensions of optimal 1-d techniques or b) completely heuristic: it is possible to combine mathematical technique with knowledge of the problem of seeing to create more effective solutions to problems than can be obtained by considering only one half of the partnership.

ACKNOWLEDGEMENTS

This work has been supported by BTRL Martlesham and SERC. Thanks are due both to Dr. C. Nightingale of BTRL and Dr. H. Knutsson of Linköping University whose help has been invaluable.

REFERENCES

[1] Beaumont, J.M., (1988) The RIBENA Algorithm - Recursive Predictive Still Picture Coding, British Telecom Memo RT4343/88/14.

[2] Burt, P. and Adelson, E., (1983) The Laplacian Pyramid as a Compact Image Code, *IEEE Trans.* COM-31, Nr. 4, pp. 532-540.

[3] Calway, A. and Wilson, R., (1988) A Multiresolution Descriptor for Nonstationary Image Processing, Proc. Conf. IMA Mathematics in Signal Processing.

[4] Clippingdale, S.C. and Wilson, R.G., (1987) Quad-Tree Image Estimation: A New Image Model and its Application to Minimum Mean Squared Error Image Restoration, Proc. 5th Scand. Conf. on Image Anal. pp. 699-706.

[5] Hubel, D.H. and Wiesel, T.N., (1979) Brain Mechanisms of Vision, Sci. American, pp. 130-144.

[6] Jain, A.K., (1980) Image Data Compression: A Review, Proc. IEEE, Vol. 68, pp. 366-406.

[7] Jain, A.K., (1981) Advances in Mathematical Models for Image Processing, Proc. IEEE pp. 502-528.

[8] Knutsson, H., Wilson, R. and Granlund, G., (1983) Anisotropic Nonstationary Image Estimation and its Applications: Part I-Restoration of Noisy Images, Part II-Predictive Image Coding, IEEE Trans. COM-31, Nr 3, pp. 388-397, pp. 398-406.

[9] Kunt, M., Ikonomopoulos, A. and Kocher, M., (1985) Second-Generation Image-Coding Techniques, Proc. IEEE Vol. 73, Nr 4, pp. 549-573.

[10] Mitchell, J.L. and Pennebaker, W.B., (1987) Software Implementation of the Q-coder, IBM Research Report, RC 12660.

[11] Wilson, R., (1984) Quad-Tree Predictive Coding: A New Class of Image Data Compression Algorithms, Proc. IEEE Conf. on A.S.S.P., San Diego, CA.

[12] Wilson, R. and Clippingdale, S.C., (1988) A Class of Non-Stationary Image Models and their Applications, Proc. Conf. IMA Mathematics in Signal Processing.

III SPECTRAL ESTIMATION AND STATISTICAL TECHNIQUES

MODEL-BASED SPECTRUM ESTIMATION

J.G. Proakis
(Department of Electrical and Computer Engineering, Northeastern University, Boston, USA)

ABSTRACT

This paper provides a survey of commonly used model-based methods for spectral estimation from a given record of data. Autoregressive (AR), moving average (MA), and autoregressive moving average (ARMA) models are considered for the estimation of the power spectrum and the bispectrum. Also described are eigen-decomposition methods for estimating the parameters of sinusoidal signals, namely, the Pisarenko harmonic decomposition method, the MUSIC algorithm, and the ESPRIT algorithm.

1. INTRODUCTION

The basic problem that we consider in this paper is the estimation of the power density spectrum of a signal from observation of the signal over a finite time interval. The finite record length of the data sequence is a major limitation on the quality of the power spectrum estimate. When dealing with signals that are statistically stationary, the longer the data record, the better the estimate that can be extracted from the data. On the other hand, if the signal statistics are nonstationary, the length of the data record that can be used is limited by the rapidity of the time variations in the signal statistics. Ultimately, our goal is to select as short a data record as possible that will allow us to resolve spectral characteristics of different signal components contained in the data record that have closely spaced spectral peaks and valleys.

Classical power spectrum estimation methods are based on the periodogram which may be efficiently computed by means of the Fast-Fourier Transform (FFT) algorithm (see Oppenheim and Schafer (1975), and Proakis and Manolakis (1988)). The Bartlett method, the Welch method and the Blackman and Tukey method are three examples of classical power spectrum estimation methods which are basically nonparametric in nature, in the sense that no assumptions are made concerning the physical process that generated the data. In general, these methods require long data records in order to yield the necessary frequency resolution that is required in many applications. Furthermore, these methods suffer from spectral leakage effects due to windowing that are inherent in finite-length data records. Often the spectral leakage masks weak signals that are present in the data.

Suppose we observe a finite-duration sequence $x(n)$, $0 \leq n \leq N-$ from which we may compute the time-average autocorrelation sequence

$$r_{xx}(m) = \frac{1}{N - |m|} \sum_{n=0}^{N-|m|-1} x(n)x(n+m), \quad |m| = 0,1\ldots,M$$

where $M < N$. We note that $E(r_{xx}(m)) = \gamma_{xx}(m)$ so that $r_{xx}(m)$ is an unbiased estimate of the true autocorrelation sequence $\gamma_{xx}(m)$. Alternatively, we may compute a biased estimate of $\gamma_{xx}(m)$ by normalizing the sum of lag products by N instead of $N - |m|$. Such an estimate has a lower variance than the unbiased estimate.

From one point of view, the basic limitation of the nonparametric methods is the inherent assumption that the autocorrelation estimate $r_{xx}(m)$ is zero for $m > M$. This assumption severely limits the frequency resolution and the quality of the estimate that is achieved. From another viewpoint, the inherent assumption in the periodogram estimate is that the data are periodic with period M. Neither one of these assumptions is realistic.

In this paper we describe power spectrum estimation methods that do not require such assumptions. In fact, these methods extrapolate the values of the autocorrelation for lags $m \geq N$. Extrapolation is possible if we have some a priori information on how the data were generated. In such a case a model for the signal generation may be constructed with a number of parameters that can be estimated from the observed data. From the model and estimated parameters we can compute the power density spectrum implied by the model.

In effect, the modelling approach eliminates the need for window functions and the assumption that the autocorrelation sequence is zero for $|m| \geq N$. As a consequence, parametric (model-based) power spectrum estimation methods provide better frequency resolution than do the FFT-based, nonparametric methods and avoid the problem of leakage. This is especially true in applications where short data records are available due to time-invariant or transient phenomena.

2. LINEAR SYSTEM MODEL

The parametric methods considered in this paper are based on modelling the data sequence x(n) as the output of a linear system characterized by a rational system function of the form

$$H(z) = \frac{B(z)}{A(z)} = \frac{\Sigma_{k=0}^{q} b_k z^{-k}}{1 + \Sigma_{k=1}^{p} a_k z^{-k}} \tag{2.1}$$

The corresponding difference equation is

$$x(n) = - \sum_{k=1}^{p} a_k \, x(n - k) + \sum_{k=0}^{q} b_k w(n - k) \tag{2.2}$$

where w(n) is the input sequence to the system and the observed data, x(n), represents the output sequence.

In power spectrum estimation, the input sequence is not observable. However, if the observed data are characterized as a stationary random process, then the input sequence is also assumed to be a stationary random process. In such a case the power density spectrum of the data is

$$\Gamma_{xx}(f) = |H(f)|^2 \Gamma_{ww}(f)$$

where $\Gamma_{ww}(f)$ is the power density spectrum of the input sequence and H(f) is the frequency response of the model.

Since our objective is to estimate the power density spectrum $\Gamma_{xx}(f)$, it is convenient to assume that the input sequence w(n) is a zero-mean white noise sequence with autocorrelation

$$\gamma_{ww}(m) = \sigma_w^2 \delta(m)$$

where σ_w^2 is the variance (i.e., $\sigma_w^2 = E[|w(n)|^2]$). Then the power density spectrum of the observed data is simply

$$\Gamma_{xx}(f) = \sigma_w^2 |H(f)|^2 = \sigma_w^2 \frac{|B(f)|^2}{|A(f)|^2} \tag{2.3}$$

A discrete-time random sequence can be represented by (2.2) if its power density spectrum satisfies the Paley-Wiener condition [see Papoulis (1984)]

$$\int_{-1/2}^{1/2} \log \Gamma_{xx}(f)\, df < \infty$$

In the model-based approach, the spectrum estimation procedure consists of two steps. Given the data sequence $x(n)$, $0 \leq n \leq N-1$, we estimate the parameters $\{a_k\}$ and $\{b_k\}$ of the model. Then from these estimates, we compute the power spectrum estimate according to (2.3).

The random process $x(n)$, generated by the pole-zero model in (2.1) or (2.2), is called an autoregressive-moving average (ARMA) process of order (p,q) and it is usually denoted as ARMA (p,q). If $q = 0$ and $b_0 = 1$, the resulting system model has a system function $H(z) = 1/A(z)$ and its ouput $x(n)$ is called an autoregressive (AR) process of order p. This is denoted as AR(p). The third possible model is obtained by setting $A(z) = 1$, so that $H(z) = B(z)$. Its output $x(n)$ is called a moving average (MA) process of order q and denoted as MA(q).

Of these three linear models the AR model is by far the most widely used. The reasons are twofold. First the AR model is suitable for representing spectra with narrow peaks (resonances). Second, the AR model results in very simple linear equations for the AR parameters. On the other hand, the MA model, as a general rule, requires many more coefficients to represent a narrow spectrum. Consequently, it is rarely used alone as a model for spectrum estimation. By combining poles and zeros, the ARMA model provides a more efficient representation from the viewpoint of the number of model parameters to represent the spectrum of a random process.

The decomposition theorem due to Wold (1938) asserts that any ARMA or MA process may be represented uniquely by an AR model of possibly infinite order, and any ARMA or AR process may be represented by a MA model of possibly infinite order. In view of this theorem, the issue of model selection reduces to selecting the model that requires the smallest number of parameters which are also easy to compute. Usually, the choice in practice is the AR model. The ARMA model is used to a lesser extent.

Before describing methods for estimating the parameters in an AR(p), MA(q), and ARMA(p,q) models, it is useful to establish the basic relationships between the model parameters and the autocorrelation sequence $\gamma_{xx}(m)$. In addition, we relate the AR model parameters to the coefficients in a linear predictor for the process $x(n)$.

3. RELATIONSHIPS BETWEEN THE AUTOCORRELATION AND THE MODEL PARAMETERS

There are several relationships that exist between the autocorrelation sequence $\gamma_{xx}(m)$ and the system model parameters. These relationships are established below.

First the z-transform of $\gamma_{xx}(m)$ is defined as

$$\Gamma_{xx}(z) = \sum_{m=-\infty}^{\infty} \gamma_{xx}(m) z^{-m} \tag{3.1}$$

where the region of convergence of $\Gamma_{xx}(z)$ is the annulus $r_1 < |z| < r_2$, which includes the unit circle. Since

$$\Gamma_{xx}(z) = \sigma_w^2 H(z) H(z^{-1}) \tag{3.2}$$

it follows that

$$\gamma_{xx}(m) = \sigma_w^2 \sum_{n=0}^{\infty} h^*(n) h(n+m) \tag{3.3}$$

Secondly, it is easily demonstrated that the relationship

$$\Gamma_{xx}(z) = \sigma_w^2 H(z) H(z^{-1}) = \sigma_w^2 \frac{B(z)B(z^{-1})}{A(z)A(z^{-1})} \equiv \sigma_w^2 \frac{D(z)}{C(z)} \tag{3.4}$$

implies that

$$\begin{aligned} c_m &= \sum_{k=0}^{p-|m|} a_k a_{k+m}, \quad |m| \le p \\ d_m &= \sum_{k=0}^{q-|m|} b_k b_{k+m}, \quad |m| \le q \end{aligned} \tag{3.5}$$

where

$$C(z) = \sum_{m=-p}^{p} c_m z^{-m}$$
$$D(z) = \sum_{m=-q}^{q} d_m z^{-m} \tag{3.6}$$

An explicit relationship between the autocorrelation sequence $\gamma_{xx}(m)$ and the parameters $\{a_k\}$, $\{b_k\}$ can be obtained from (2.2). If we multiply (2.2) by $x^*(n - m)$ and take the expected value of both sides of the equation, we obtain

$$E[x(n)x^*(n-m)] = -\sum_{k=1}^{p} a_k E[x(n-k)x^*(n-m)] + \sum_{k=0}^{q} b_k E[w(n-k)x^*(n-m)]$$

Hence (3.7)

$$\gamma_{xx}(m) = -\sum_{k=1}^{p} a_k \gamma_{xx}(m-k) + \sum_{k=0}^{q} b_k \gamma_{wx}(m-k)$$

But the crosscorrelation $\gamma_{wx}(m)$ may be expressed as

$$\begin{aligned}\gamma_{wx}(m) &= E[x^*(n)w(n+m)] \\ &= E[\sum_{k=0}^{\infty} h(k)w^*(n-k)w(n+m)] \\ &= \sum_{k=0}^{\infty} h(k)E[w^*(n-k)w(n+m)] \\ &= \sigma_w^2 h(-m)\end{aligned}$$

where, in the last step, we have used the fact that the sequence $w(n)$ is white, i.e., $E[w^*(n)w(n+m)] = \sigma_w^2 \delta(m)$. Therefore,

$$\gamma_{wx}(m) = \begin{cases} 0 & m > 0 \\ \sigma_w^2 h(-m) & m \leq 0 \end{cases} \tag{3.8}$$

With the result (3.8) substituted into (3.7) we obtain the desired relationship

$$\gamma_{xx}(m) = \begin{cases} -\sum_{k=1}^{p} a_k \gamma_{xx}(m-k) & m > q \\ -\sum_{k=1}^{p} a_k \gamma_{xx}(m-k) + \sigma_w^2 \sum_{k=0}^{q-m} h(k) b_{k+m} & 0 \leq m \leq q \\ \gamma_{xx}^*(-m) & m < 0 \end{cases} \tag{3.9}$$

The relationships in (3.9) provide a formula for determining the model parameters $\{a_k\}$ by restricting our attention to the case $m > q$. Thus the set of linear equations

$$\begin{pmatrix} \gamma_{xx}(q) & \gamma_{xx}(q-1) & \cdots & \gamma_{x}(q-p+1) \\ \gamma_{xx}(q+1) & \gamma_{xx}(q) & \cdots & \gamma_{xx}(q-p+2) \\ \vdots & \vdots & & \vdots \\ \gamma_{xx}(q+p-1) & \gamma_{xx}(q+p-2) & \cdots & \gamma_{xx}(q) \end{pmatrix} \begin{pmatrix} a_1 \\ a_2 \\ \vdots \\ a_p \end{pmatrix} = - \begin{pmatrix} \gamma_{xx}(q+1) \\ \gamma_{xx}(q+2) \\ \vdots \\ \gamma_{xx}(q+p) \end{pmatrix} \tag{3.10}$$

may be used to solve for the model parameters $\{a_k\}$ by using estimates of the autocorrelation sequence in place of $\gamma_{xx}(m)$ for $m \geq q$. This problem is discussed below.

Another interpretation of the relationship in (3.10) is that the values of the autocorrelation $\gamma_{xx}(m)$ for $m > q$ are uniquely determined from the pole parameters $\{a_k\}$ and the values of $\gamma_{xx}(m)$ for $0 \leq m \leq p$. Consequently, the linear system model automatically extends the values of the autocorrelation sequence $\gamma_{xx}(m)$ for $m > p$.

If the pole parameters $\{a_k\}$ are obtained from (3.10) the result does not help us in determining the MA parameters $\{b_k\}$, because the equation

$$\sigma_w^2 \sum_{k=0}^{q-m} h(k) b_{k+m} = \gamma_{xx}(m) + \sum_{k=1}^{p} a_k \gamma_{xx}(m-k), \quad 0 \leq m \leq q \tag{3.11}$$

depends on the impulse response $h(n)$. Although the impulse response can be expressed in terms of the parameters $\{b_k\}$ by long division of $B(z)$ with the known $A(z)$, this approach results in a set of nonlinear equations for the MA parameters.

If we adopt an AR(p) model for the observed data, the relationship between the AR parameters and the autocorrelation sequence is obtained by setting $q = 0$ in (3.9). Thus we obtain

$$\gamma_{xx}(m) = \begin{cases} -\sum_{k=1}^{p} a_k \gamma_{xx}(m-k) & m > 0 \\ -\sum_{k=1}^{p} a_k \gamma_{xx}(m-k) + \sigma_w^2 & m = 0 \\ \gamma_{xx}^*(-m) & m < 0 \end{cases} \tag{3.12}$$

In this case, the AR parameters $\{a_k\}$ are obtained from the solution of the Yule-Walker or normal equations

$$\begin{pmatrix} \gamma_{xx}(0) & \gamma_{xx}(-1) & \cdots & \gamma_{xx}(-p+1) \\ \gamma_{xx}(1) & \gamma_{xx}(0) & \cdots & \gamma_{xx}(-p+2) \\ \vdots & \vdots & & \vdots \\ \gamma_{xx}(p-1) & \gamma_{xx}(p-2) & \cdots & \gamma_{xx}(0) \end{pmatrix} \begin{pmatrix} a_1 \\ a_2 \\ \vdots \\ a_p \end{pmatrix} = - \begin{pmatrix} \gamma_{xx}(1) \\ \gamma_{xx}(2) \\ \vdots \\ \gamma_{xx}(p) \end{pmatrix} \tag{3.13}$$

and the variance σ_w^2 can be obtained from the equation

$$\sigma_w^2 = \gamma_{xx}(0) + \sum_{k+1}^{p} a_k \gamma_{xx}(-k) \tag{3.14}$$

The equations (3.13) and (3.14) are usually combined into a single matrix equation of the form

$$\begin{pmatrix} \gamma_{xx}(0) & \gamma_{xx}(-1) & \cdots & \gamma_{xx}(-p) \\ \gamma_{xx}(1) & \gamma_{xx}(0) & \cdots & \gamma_{xx}(-p+1) \\ \vdots & \vdots & \vdots & \\ \gamma_{xx}(p) & \gamma_{xx}(p-1) & \cdots & \gamma_{xx}(0) \end{pmatrix} \begin{pmatrix} 1 \\ a_1 \\ \vdots \\ a_p \end{pmatrix} = \begin{pmatrix} \sigma_w^2 \\ 0 \\ \vdots \\ 0 \end{pmatrix} \tag{3.15}$$

Since the correlation matrix in (3.13), or in (3.15) is Toeplitz, it can be efficiently inverted by use of the Levinson-Durbin algorithm.

Thus all the system parameters in the AR(p) model are easily determined from knowledge of the autocorrelation sequence $\gamma_{xx}(m)$ for $0 \leq m \leq p$. Furthermore, (3.12) may be used to extend the autocorrelation sequence for $m > p$, once the $\{a_k\}$ are determined.

Finally, for completeness, we indicate that in a MA(q) model for the observed data, the autocorrelation sequence $\gamma_{xx}(m)$ is related to the MA parameters $\{b_k\}$ by the equation

$$\gamma_{xx}(m) = \begin{cases} \sigma_w^2 \, \Sigma_{k=0}^{q} \, b_k b_{k+m} & 0 \leq m \leq q \\ 0 & m > q \\ \gamma_{xx}^*(-m) & m < 0 \end{cases} \tag{3.16}$$

This result is easily established from (3.9) by setting $a_k = 0$ for $k = 1,2,\ldots,p$ and substituting $\{b_k\}$ for $h(k)$. Furthermore, from (3.5) and (3.6) it is clear that $\gamma_{xx}(m) = \sigma_w^2 d_m$, and therefore the spectrum of the MA process is simply

$$\Gamma_{xx}^{MA}(f) = \sigma_w^2 \sum_{m=-q}^{q} d_m e^{-j2\pi fm} \tag{3.17}$$

Since $\gamma_{xx}(m) = 0$ for $m > q$, this spectrum has the same form as the (estimated) periodogram spectrum.

4. RELATIONSHIP OF AR PROCESS TO LINEAR PREDICTION

The parameters in an AR(p) process are intimately related to a predictor of order p for the same process. The relationship is established in this section.

Let $x(n)$ be a sample sequence of a stationary random process and let us consider a linear predictor of order m for the data point $x(n)$. Thus we have

$$\hat{x}(n) = -\sum_{k=1}^{m} a_m(k)x(n-k) \tag{4.1}$$

where $a_m(k), k = 1,2,\ldots,m$ are the prediction coefficients for the mth-order predictor. We call $\hat{x}(n)$ in (4.1) the forward predictor, since it predicts forward in time. The forward prediction error is defined as

$$\begin{aligned} f_m(n) &= x(n) - \hat{x}(n) \\ &= x(n) + \sum_{k=1}^{m} a_m(k)x(n-k) \end{aligned} \tag{4.2}$$

and its mean-square value is

$$\begin{aligned} \xi_m^f &= E[\,|f_m(n)|^2] \\ &= E[\,|x(n) + \sum_{k=1}^{m} a_m(k)x(n-k)|^2] \end{aligned} \tag{4.3}$$

Since ξ_m^f is a quadratic function of the predictor coefficients, its minimization leads to the set of linear equations

$$\gamma_{xx}(\ell) = -\sum_{k=1}^{m} a_m(k)\gamma_{xx}(\ell - k) \qquad \ell = 1,2,\ldots,m \tag{4.4}$$

The minimum mean-square prediction error is

$$\min[\xi_m^f] \equiv E_m^f = \gamma_{xx}(0) + \sum_{k=1}^{m} a_m(k)\gamma_{xx}(-k) \tag{4.5}$$

If we compare (4.4) and (4.5) to the equations in (3.13) and (3.14) we observe that the prediction coefficients in a linear predictor of order p are identical to the parameters of an AR(p) model. Furthermore, the minimum mean-square prediction error $E_p^f = \sigma_w^2$.

The Levinson-Durbin algorithm may be used to solve (4.4) for the linear predictor coefficients. This algorithm is initialized by solving for the coefficient in the first order (m = 1) predictor

$$a_1(1) = \frac{-\gamma_{xx}(1)}{\gamma_{xx}(0)}$$

$$E_1^f = [1 - |a_1(1)|^2]\gamma_{xx}(0) \tag{4.6}$$

Then the prediction coefficients for the higher-order predictors are given recursively as

$$a_m(k) = a_{m-1}(k) + a_m(m)a^*_{m-1}(m - k) \tag{4.7}$$

where

$$a_m(m) = -\frac{\gamma_{xx}(m) + \Sigma_{k=1}^{m-1} a_{m-1}(k)\gamma_{xx}(m - k)}{E_{m-1}^f} \tag{4.8}$$

$$E_m^f = (1 - |a_m(m)|^2)E_{m-1}^f$$

$$= \gamma_{xx}(0)\Pi_{k=1}^{m}[1 - |a_k(k)|^2] \tag{4.9}$$

Note that the recursion in (4.7) involves the complex conjugate of $a_{m-1}(m - 1)$, which is appropriate for the case in which the autocorrelation $\gamma_{xx}(m)$ is complex valued. We define the prediction coefficients $a_m(m)$ as $a_m(m) = K_m$, where K_m is the mth reflection coefficient in the equivalent lattice realization of the predictor.

If we substitute the recursion for $a_m(k)$ given in (4.7) into (4.2), we obtain an order-recursive equation for the forward prediction error, as

$$f_m(n) = x(n) + \sum_{k=1}^{m-1} a_{m-1}(k)x(n-k) +$$

$$a_m(m) \times \left[x(n-m) + \sum_{k=1}^{m-1} a^*_{m-1}(m-k)x(n-k)\right] \quad (4.10)$$

$$f_m(n) = f_{m-1}(n) + K_m g_{m-1}(n-1)$$

where $g_m(n)$ is the error in an mth-order backward predictor, defined as

$$g_m(n) = x(n-m) + \sum_{k=1}^{m} a^*_m(k)x(n-m+k) \quad (4.11)$$

It is easy to show that the mean-square value of the backward prediction error,

$$\xi^b_m = E[|g_m(n)|^2] \quad (4.12)$$

when minimized with respect to the prediction coefficients yields the same set of equations as (4.4), and

$$\min[\xi^b_m] \equiv E^b_m = E^f_m \quad (4.13)$$

Furthermore, if we substitute the recursion in (4.7) into (4.11), we obtain the order-recursive equation for the backward error as

$$g_m(n) = g_{m-1}(n-1) + K^*_m f_{m-1}(n) \quad (4.14)$$

The pair of order-recursive relations

$$\begin{aligned} f_m(n) &= f_{m-1}(n) + K_m g_{m-1}(n-1) \\ g_m(n) &= g_{m-1}(n-1) + K^*_m f_{m-1}(n) \end{aligned} \quad (4.15)$$

define a lattice filter, as illustrated in Fig. 1, which is a direct consequence of the Levinson-Durbin recursion. The minimization of $\xi^f_m = E[|f_m(n)|^2]$ and $\xi^b_m = E[|g_m(n)|^2]$ with respect to the reflection coefficients $\{K_m\}$ yield the results

$$K_m = \frac{-E[f_{m-1}(n)g^*_{m-1}(n-1)]}{E[|g_{m-1}(n-1)|^2]} \quad (4.16)$$

and

$$K_m^* = \frac{-E[f_{m-1}^*(n)g_{m-1}(n-1)]}{E[|f_{m-1}(n)|^2]} \quad . \tag{4.17}$$

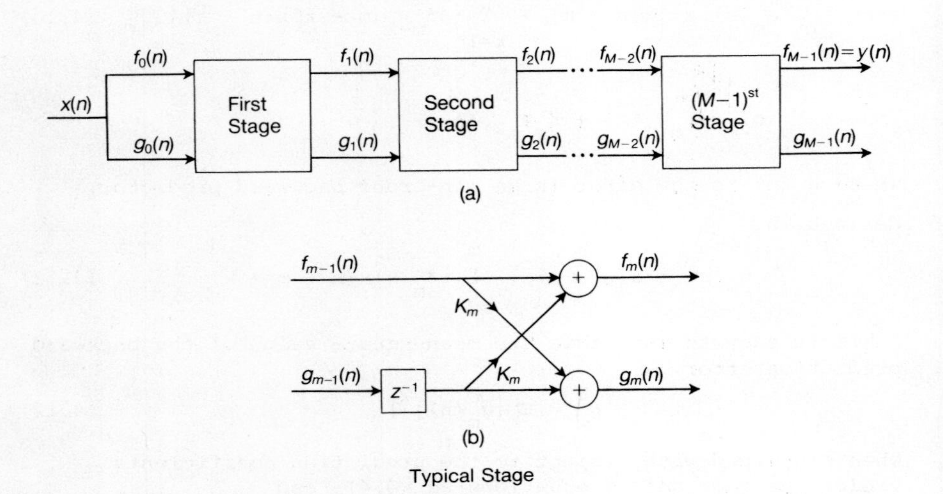

Fig. 1 Lattice Filter Structure

Since the denominator terms in (4.16) and (4.17) are equal, as indicated by (4.13), it follows that the reflection coefficients may also be expressed as

$$K_m = \frac{-E[f_{m-1}(n)g_{m-1}^*(n-1)]}{\sqrt{E[|g_{m-1}(n-1)|^2]E[|f_{m-1}(n)|^2]}} \tag{4.18}$$

From this form it is apparent that the reflection coefficients in the lattice filter are the negative of the (normalized) correlation coefficients between the forward and backward errors in the lattice. It is also apparent from (4.18) that $|K_m| \leq 1$. Since $|K_m| \leq 1$, it follows that the minimum mean-square value of the prediction error, which is given recursively as

$$E_m^f = (1 - |K_m|^2)E_{m-1}^f$$

or equivalently, as

$$\sigma_{wm}^2 = (1 - |K_m|^2)\sigma_{wm-1}^2 \tag{4.19}$$

is a monotically decreasing sequence and can be used in the AR model for determining the order (number of poles p) that provides a good fit to the data.

The parameters $K_m = a_m(m)$, $m = 1,2,\ldots,p$, which are a by-product of the Levinson-Durbin algorithm, are the reflection coefficients of the lattice filter that is equivalent to the linear FIR predictor. The condition $|K_m| \leq 1$ implies that the roots of

$$A_p(z) = 1 + \sum_{k=1}^{p} a_p(k) z^{-k} \tag{4.20}$$

lie inside the unit circle [see Papoulis (1984)]. Therefore, the AR filter model is stable.

The filter with the system function A(z) given by (4.20) is called the forward prediction error filter. Its input is x(n) and its output is the forward prediction error $f_p(n)$. If x(n) is actually an AR process, then A(z) is the inverse filter or the noise whitening filter to x(n). That is, $f_p(n)$ is a white noise sequence. Similarly, the backward prediction error filter with system function

$$B_p(z) = \sum_{k=0}^{p} \beta_p(k) z^{-k} \tag{4.21}$$

where $\beta_p(k) = a_p^*(p - k)$ is also a noise whitening filter when the process x(n) is an AR(p) process.

With the background established above, we will now describe the power spectrum estimation methods for AR(p) and ARMA(p,q) models.

5. THE YULE-WALKER METHOD FOR THE AR MODEL PARAMETERS

In the Yule-Walker method we simply estimate the autocorrelation from the data and use the estimates in (3.13) to solve for the AR model parameters. In this method it is desirable to use the biased form of the autocorrelation estimate,

$$r_{xx}(m) = \frac{1}{N} \sum_{n=0}^{N-|m|} x^*(n) x(n + m) \tag{5.1}$$

to ensure that the autocorrelation matrix is positive semidefinite. The result will be a stable AR model. Although stability is not a critical issue in power spectrum estimation, it is conjectured that a stable AR model best represents the data.

The Levinson-Durbin algorithm given by (4.5) through (4.9) with $r_{xx}(m)$ substituted for $\gamma_{xx}(m)$ yields the AR parameters. The corresponding power spectrum estimate is

$$P_{xx}^{YW}(f) = \frac{\hat{\sigma}^2_{wp}}{\left|1 + \Sigma_{k=1}^{p} \hat{a}_p(k) e^{-j2\pi fk}\right|^2} \tag{5.2}$$

where $\hat{a}_p(k)$ are estimates of the AR parameters obtained from the Levinson-Durbin recursions and

$$\hat{\sigma}^2_{wp} = \hat{E}_p^f = r_{xx}(0)\Pi_{k=1}^{p}[1 - |\hat{a}_k(k)|^2] \tag{5.3}$$

is the estimated minimum mean-square value for the pth-order predictor.

In estimating the power spectrum of sinusoidal signals via AR models, Lacoss (1971) showed that spectral peaks in an AR spectrum estimate are proportional to the square of the power of the sinusoidal signal. On the other hand, the area under the peak in the power density spectrum is linearly proportional to the power of the sinusoid. This characteristic behaviour holds for all AR model-based estimation methods.

6. THE BURG METHOD FOR THE AR MODEL PARAMETERS

The method devised by Burg (1968) for estimating the AR parameters may be viewed as an order-recursive least-squares lattice method based on the minimization of the forward and backward errors in linear predictors, with the constraint that the AR parameters satisfy the Levinson-Durbin recursion.

To derive the estimator, suppose that we are given the data $x(n), n = 0,1\ldots,N-1$, and let us consider the forward and backward linear prediction estimates of order m, which are given as

$$\begin{aligned} \hat{x}(n) &= -\sum_{k=1}^{m} a_m(k)x(n-k) \\ \hat{x}(n-m) &= -\sum_{k=1}^{m} a_m^*(k)x(n+k-m) \end{aligned} \tag{6.1}$$

and the corresponding forward and backward errors $f_m(n)$ and $g_m(n)$ given by (4.2) and (4.11), respectively, where $a_m(k)$, $0 \le k \le m-1, m = 1,2$

are the prediction coefficients. The total squared error is

$$\mathcal{E}_m = \sum_{n=m}^{N-1} [\,|f_m(n)|^2 + |g_m(n)|^2] \tag{6.2}$$

This error is to be minimized by selecting the prediction coefficients, subject to the constraint that they satisfy the Levinson-Durbin recursion given by

$$a_m(k) = a_{m-1}(k) + K_m a^*_{m-1}(m - k) \quad 1 \leq k \leq m-1, \quad 1 \leq m \leq p \tag{6.3}$$

where $K_m = a_m(m)$ is the mth reflection coefficient in the lattice filter realization of the predictor. Recall that when (6.3) is substituted into the expressions for $f_m(n)$ and $g_m(n)$, the result is the pair of order-recursive equations for the forward and backward prediction errors given by (4.15).

Now, if we substitute from (4.15) into (6.2) and perform the minimization of $\mathcal{E}_m$ with respect to the complex-valued reflection coefficient K_m, we obtain the result

$$\hat{K}_m = \frac{-\sum_{n=m}^{N-1} f_{m-1}(n) g^*_{m-1}(n - 1)}{\frac{1}{2}\sum_{n=m}^{N-1}[\,|f_{m-1}(n)|^2 + |g_{m-1}(n - 1)|^2]} \quad m = 1,2,\ldots,p \tag{6.4}$$

The term in the numerator of (6.4) is an estimate of the crosscorrelation between the forward and backward prediction errors. With the normalization factors in the denominator of (6.4), it is apparent that $|K_m| < 1$, so that the all-pole model obtained from the data is stable. The reader should note the similarity of (6.4) with the statistical counterparts given by (4.16) through (4.18).

We note that the denominator (6.4) is simply the least-squares estimate of the forward and backward errors, E^f_{m-1} and E^b_{m-1}, respectively. Hence (6.4) may be expressed as

$$\hat{K}_m = \frac{-\sum_{n=m}^{N-1} f_{m-1}(n) g^*_{m-1}(n - 1)}{\frac{1}{2}[\hat{E}^f_{m-1} + \hat{E}^b_{m-1}]} \quad m = 1,2,\ldots,p \tag{6.5}$$

where $\hat{E}^f_{m-1} + \hat{E}^b_{m-1}$ is an estimate of the total squared error E_m.

It can be shown that the denominator term in (6.5) can be computed in an order-recursive fashion according to the relation

$$\hat{E}_m = (1 - |\hat{K}_{m-1}|^2)\hat{E}_{m-1} - |f_{m-1}(N)|^2 - |g_{m-1}(N - m)|^2 \qquad (6.6)$$

where $\hat{E}_m \equiv \hat{E}^f_m + \hat{E}^b_m$ is the total least-squares error. This result is due to Andersen (1978).

To summarize, the Burg algorithm computes the reflection coefficients in the equivalent lattice structure as specified by (6.5) and (6.6), and the Levinson-Durbin algorithm is used to obtain the AR model parameters. From the estimates of the AR parameters, we form the power spectrum estimate

$$P^{BU}_{xx}(f) = \frac{\hat{E}_p}{\left|1 + \Sigma^p_{k=1}\hat{a}_p(k)e^{-j2\pi fk}\right|^2} \qquad (6.7)$$

The major advantages of the Burg method for estimating the parameters of the AR model are (1) it results in high frequency resolution, (2) it yields a stable AR model, and (3) it is computationally efficient.

The Burg method is known to have several disadvantages, however. First, it exhibits spectral line splitting at high signal-to-noise ratios. [See the paper by Fougere et al. (1976)]. By line splitting, we mean that the spectrum of x(n) may have a single sharp peak, but the Burg method may result in two or more closely spaced peaks. For high-order models, the method also introduces spurious peaks. Furthermore, for sinusoidal signals in noise, the Burg method exhibits a sensitivity to the initial phase of a sinusoid, especially in short data records. This sensitivity is manifest as a frequency shift from the true frequency, resulting in a frequency bias that is phase dependent. For more details on some of these limitations the reader is referred to the papers of Chen and Stegen (1974), Ulrych and Clayton (1976), Fougere et al. (1976), Kay and Marple (1979), Swingler (1979a, 1980), Herring (1980), and Thorvaldsen (1981).

Several modifications have been proposed to overcome some of the more important limitations of the Burg method: namely, the line splitting, spurious peaks, and frequency bias. Basically, the modifications involve the introduction of a weighting (window) sequence on the squared forward and backward errors. That is, the least-squares optimization is performed on the weighted squared errors

$$\xi_m^{WB} = \sum_{n=m}^{N-1} w_m(n)[\,|f_m(n)|^2 + |g_m(n)|^2] \tag{6.8}$$

which, when minimized, results in the reflection coefficient estimates

$$\hat{K}_m = \frac{-\sum_{n=m}^{N-1} w_{m-1}(n)g^*_{m-1}(n-1)f_{m-1}(n)}{\frac{1}{2}\sum_{n=m}^{N-1} w_{m-1}(n)[\,|f_{m-1}(n)|^2 + |g_{m-1}(n-1)|^2]} \tag{6.9}$$

In particular, we mention the use of a Hamming window used by Swingler (1979b), a quadratic or parabolic window used by Kaveh and Lippert (1983), the energy weighting method used by Nikias and Scott (1982), and the data-adaptive energy weighting used by Helme and Nikias (1985).

These windowing and energy weighting methods have proved effective in reducing the occurrence of line splitting and spurious peaks, and are also effective in reducing frequency bias.

The Burg method for power spectrum estimation is usually associated with maximum entropy spectrum estimation, which is a criterion used by Burg (1967, 1975) as a basis for AR modelling in parametric estimation. The problem considered by Burg was how best to extrapolate from the given values of the autocorrelation sequence $\gamma_{xx}(m), 0 \leq m \leq p$, the values for $m > p$, such that the entire autocorrelation sequence is positive semidefinite. Since an infinite number of extrapolations are possible, Burg postulated that the extrapolation be made on the basis of maximizing uncertainty (entropy) or randomness, in the sense that the spectrum $\Gamma_{xx}(f)$ of the process is the flattest of all spectra which have the given autocorrelation values $\gamma_{xx}(m), 0 \leq m \leq p$. In particular the entropy per sample is proportional to the integral [see Burg (1975)]

$$\int_{-1/2}^{1/2} \ln\Gamma_{xx}(f)df \tag{6.10}$$

Burg found that the maximum of this integral subject to the (p + 1) constraints

$$\int_{-1/2}^{1/2} \Gamma_{xx}(f)e^{j2\pi fm}df = \gamma_{xx}(m) \qquad 0 \leq m \leq p \tag{6.11}$$

is the AR(p) process for which the given autocorrelation sequence $\gamma_{xx}(m)$, $0 \leq m \leq p$ is related to the AR parameters by the equation (3.12). This solution provides an additional justification for the use of the AR model in power spectrum estimation.

In view of Burg's basic work in maximum entropy spectral estimation, the Burg power spectrum estimation procedure is often called the maximum entropy method (MEM). We should emphasize, however, that the maximum entropy spectrum is identical to the AR-model spectrum only when the exact autocorrelation $\gamma_{xx}(m)$ is known. When only an estimate of $\gamma_{xx}(m)$ is available for $0 \leq m \leq p$, the AR-model estimates of Yule-Walker and Burg are not maximum entropy spectral estimates. The general formulation for the maximum entropy spectrum based on estimates of the autocorrelation sequence results in a set of nonlinear equations. Solutions for the maximum entropy spectrum with measurement errors in the correlation sequence have been obtained by Newman (1981) and Schott and McClellan (1984).

7. UNCONSTRAINED LEAST-SQUARES METHOD FOR THE AR MODEL PARAMETERS

As described in the preceeding section, the Burg method for determining the parameters of the AR model is basically a least-squares lattice algorithm with the added constraint that the predictor coefficients satisfy the Levinson recursion. As a result of this constraint, an increase in the order of the AR model requires only a single parameter optimization at each stage. In contrast to this approach, we may use an unconstrained least-squares algorithm to determine the AR parameters.

To elaborate, we form the forward and backward linear prediction estimate and their corresponding forward and backward errors as indicated in (6.1). Then we minimize the sum of squares of both errors, that is,

$$\begin{aligned}\xi_p &= \sum_{n=p}^{N-1}\left[\left|f_p(n)\right|^2 + \left|g_p(n)\right|^2\right] \\ &= \sum_{n=p}^{N-1}\left[\left|x(n) + \sum_{k=1}^{p} a_p(k)x(n-k)\right|^2 + \left|x(n-p) + \sum_{k=1}^{p} a_p^*(k)x(n+k-p)\right|^2\right]\end{aligned} \tag{7.1}$$

which is the same performance index as in the Burg method. However, we will not impose the Levinson-Durbin recursion in (7.1) for the AR parameters. The unconstrained minimization of ξ_p with respect to the prediction coefficients yields the set of linear equations

$$\sum_{k=1}^{p} a_p(k) r_{xx}(\ell,k) = -r_{xx}(\ell,0) \qquad \ell = 1,2,\ldots,p \tag{7.2}$$

where, by definition, the autocorrelation $r_{xx}(\ell,k)$ is

$$r_{xx}(\ell,k) = \sum_{n=p}^{N-\ell} [x(n-k)x^*(n-1) + x(n-p+\ell)x^*(n-p+k)] \tag{7.3}$$

The resulting residual least-squares error is

$$\xi_p^{LS} = r_{xx}(0,0) + \sum_{k=1}^{p} \hat{a}_p(k) r_{xx}(0,k) \tag{7.4}$$

Hence the unconstrained least-squares power spectrum estimate is

$$P_{xx}^{LS}(f) = \frac{E_p^{LS}}{\left|1 + \Sigma_{k=1}^{p} \hat{a}_p(k) e^{-j2\pi fk}\right|^2} \tag{7.5}$$

The correlation matrix in (7.3), with elements $r_{xx}(\ell,k)$, is not Toeplitz, so that the Levinson-Durbin algorithm cannot be applied. However, the correlation matrix has sufficient structure to make it possible to devise computationally efficient algorithms with computational complexity proportional to p^2. Marple (1980) devised such an algorithm, which has a lattice structure and employs Levinson-Durbin-type order recursions and additional time recursions.

The form of the unconstrained least-squares method described above has also been called the unwindowed data least-squares method. It has been proposed for spectrum estimation in several papers, including the papers by Burg (1967), Nuttall (1976), and Ulrych and Clayton (1976). Its performance characteristics have been found to be superior to the Burg method, in the sense that the unconstrained least-squares method does not exhibit the same sensitivity to such problems as line splitting, frequency bias, and spurious peaks. With this method there is no guarantee that the estimated AR parameters yield a stable AR model. However, in spectrum estimation this is not considered to be a problem.

8. SEQUENTIAL ESTIMATION METHODS FOR THE AR MODEL PARAMETERS

The three power spectrum estimation methods described in the preceding sections for the AR model may be classified as block processing methods. These methods obtain estimates of the AR parameters from a block of data, say $x(n), n = 0,1...,$ $N - 1$. The AR parameter based on the block of N data points is then used to obtain the power spectrum estimate.

In situations where data are available on a continuous basis, we can still segment the data into blocks of N points and perform spectrum estimation on a block-by-block basis. This is often done in practice, for both real-time and non-real time applications. However, in such applications, there is an alternative approach based on sequential (in time) estimation of the AR model parameters as each new data point becomes available. By introducing a weighting function into past data samples, it is possible to deemphasize the effect of older data samples as new data are received.

Sequential estimation methods for AR models have been developed over the past 20 years as a result of adaptive FIR filtering applications. Adaptive FIR filters and adaptive filtering algorithms are intimately related to linear prediction and linear estimation methods. In view of the relationships we have already established between the coefficients in a linear prediction filter and the parameters of the AR model, it is not surprising that the adaptive filtering algorithms are also applicable directly to power spectrum estimation based on the AR model. In particular, sequential lattice methods directly and optimally estimate the prediction coefficients and reflection coefficients in the lattice realization of the forward and backward linear predictors. The recursive equations for the prediction coefficients relate directly to the AR model parameters. In addition to the order recursive nature of these equations, as implied by the lattice structure, we also obtain time-recursive equations for the reflection coefficients in the lattice and for the forward and backward prediction coefficients.

The sequential recursive least-squares algorithms are equivalent to the unconstrained least squares block processing method described in the preceding section. Hence the power spectrum estimates obtained by the sequential recursive least-squares method retain the desirable properties of the block processing algorithm. Since the AR parameters are being estimated continuously in a sequential estimation algorithm, power spectrum estimates may be obtained as often as desired, from once per sample to once every N samples. By properly weighting past data samples, the sequential estimation methods are particularly suitable for estimating and tracking time-variant power spectra resulting from nonstationary signal statistics.

The computational complexity of efficient sequential estimation methods is proportional to p, the order of the AR process. As a consequence, the sequential estimation algorithms are computationally efficient and, from this viewpoint, may offer some advantage over the block processing methods. Some references to sequential estimation methods for spectrum estimation are the papers by Griffiths (1975), Friedlander (1982b) and Kalouptsidis and Theodoridis (1987).

9. SELECTION OF AR MODEL ORDER

One of the most important aspects of the use of the AR model is the selection of the order p. As a general rule, if we select a model having too low an order, we obtain a highly smoothed spectrum. On the other hand, if p is selected too high, we run the risk of introducing spurious low-level peaks in the spectrum. We mentioned previously that one indication of the performance of the AR model is the mean-square value of the residual error, which, in general, is different for each of the estimators described above. The characteristic of this residual error is that it decreases as the order of the AR model is increased. We can monitor the rate of decrease and decide to terminate the process when the rate of decrease becomes relatively slow. It is apparent, however, that this approach may be imprecise and ill-defined, and other methods should be investigated.

Much work has been done by various researchers on this problem and many experimental results have been given in the literature [e.g., the papers by Gersch and Sharpe (1976), Ulrych and Bishop (1975), Tong (1975,1977), Jones (1976), Nuttall (1976), Berryman (1978), Kaveh and Bruzzone (1979), and Kashyap (1980)].

Two of the better known criteria for selecting the model order have been proposed by Akaike (1969, 1974). For the first, called the final prediction error (FPE) criterion, the order is selected to minimize the performance index

$$\mathrm{FPE}(p) = \hat{\sigma}^2_{wp}\left(\frac{N+p+1}{N-p-1}\right) \tag{9.1}$$

where $\hat{\sigma}^2_{wp}$ is the estimated variance of the linear prediction error. This performance index is based on minimizing the mean-square error for a one-step predictor.

The second criterion proposed by Akaike (1974), called the Akaike information criterion (AIC), is based on selecting the order that minimizes

$$\mathrm{AIC}(p) = \ln \hat{\sigma}^2_{wp} + 2p/N \tag{9.2}$$

Note that the term $\hat{\sigma}^2_{wp}$ decreases and hence $\ln \hat{\sigma}^2_{wp}$ also decreases, as the order of the AR model is increased. However, $2p/N$ increases with an increase in p. Hence a minimum value is obtained for some p.

An alternative information criterion, proposed by Rissanen (1983), is based on selecting the order that minimizes the description length (MDL), where MDL is defined as

$$\mathrm{MDL}(p) = N \ln \hat{\sigma}^2_{wp} + p \ln N \tag{9.3}$$

A fourth criterion has been proposed by Parzen (1974). This is called the criterion autoregressive transfer (CAT) function and is defined as

$$\mathrm{CAT}(p) = \left(\frac{1}{N} \sum_{k=1}^{p} \frac{1}{\bar{\sigma}^2_{wk}} \right) - \frac{1}{\bar{\sigma}^2_{wp}} \tag{9.4}$$

where

$$\bar{\sigma}^2_{wk} = \frac{N}{N-k} \hat{\sigma}^2_{wk} \tag{9.5}$$

The order p is selected to minimize CAT(p).

In applying the criteria given above, the mean should be removed from the data. Since $\hat{\sigma}^2_{wk}$ depends on the type of spectrum estimate we obtain, the model order is also a function of the criterion.

The experimental results given in the references cited above indicate that the model-order selection criteria do not yield definitive results. For example, Ulrych and Bishop (1975), Jones (1976), and Berryman (1978), found that the FPE(p) criterion tends to underestimate the model order. Kashyap (1980) showed that the AIC criterion is statistically inconsistent as $N \to \infty$. On the other hand, the MDL information criterion proposed by Rissanen is statistically consistent. Other experimental results indicate that for small data length, the order of the AR model should be selected to be in the range N/3 to N/2 for good results. It is apparent that in the absence of any prior information regarding the physical process that resulted in the data, one should try different model orders and different criteria and ultimately interpret the experimental results.

10. MA MODEL FOR POWER SPECTRUM ESTIMATION

The moving average model is appropriate for signals that have broad peaks or sharp nulls in their spectra. The MA model is not suitable for signals having spectra with sharp (narrowband) peaks because this model provides poor frequency resolution.

As shown in Section 3, the parameters in a MA(q) model are related to the statistical autocorrelation $\gamma_{xx}(m)$ by (3.16). According to (3.4) and (3.5)

$$B(z)B(z^{-1}) = D(z) = \sum_{m=-q}^{q} d_m z^{-m} \tag{10.1}$$

where the coefficients $\{d_m\}$ are related to the MA parameters by the equation (3.5). Clearly, then,

$$\gamma_{xx}(m) = \begin{cases} \sigma_w^2 d_m & |m| \le q \\ 0 & |m| > q \end{cases} \tag{10.2}$$

and the power spectrum for the MA(q) process is

$$\Gamma_{xx}^{MA}(f) = \sum_{m=-q}^{q} \gamma_{xx}(m) e^{-j2\pi fm} \tag{10.3}$$

It is apparent from these expressions that we do not have to solve for the MA parameters $\{b_k\}$ to estimate the power spectrum. The estimates of the autocorrelation $\gamma_{xx}(m)$ for $|m| \le q$ suffice. From such estimates we compute the estimated MA power spectrum, given as

$$P_{xx}^{MA}(f) = \sum_{m=-q}^{q} r_{xx}(m) e^{-j2\pi fm} \tag{10.4}$$

which is identical to the classical (nonparametric) periodogram power spectrum estimate.

If the parameters of the MA model are desired, they can be obtained by performing the spectral factorization

$$D(z) = B(z)B(z^{-1}) \tag{10.5}$$

This can be accomplished by determining the roots of D(z) which have the symmetry property that if p_i is a root then $1/p_i$ is also a root. Hence, $B(z)$ can be formed by selecting q roots for $B(z)$ and their reciprocals for $B(z^{-1})$. Clearly, this assignment is not unique unless the assumption of a minimum phase model is made.

There is an alternative method for determining $\{b_k\}$ based on a high order AR approximation to the MA process. To be specific, let the MA(q) process be modelled by an AR(p) model, where $p >> q$. Then $B(z) = 1/A(z)$, or equivalently, $B(z)A(z) = 1$. Thus, the parameters $\{b_k\}$ and $\{a_k\}$ are related by a convolution sum, which may be expressed as

$$\hat{a}_n + \sum_{k=1}^{q} b_k \hat{a}_{n-k} = \begin{cases} 1, & n = 0 \\ 0, & n \neq 0 \end{cases} \tag{10.6}$$

where $\{\hat{a}_n\}$ are the parameters obtained by fitting the data to an AR(p) model.

Although this set of equations may be easily solved for the $\{b_k\}$, a better fit is obtained by using a least squares error criterion. That is, we form the squared error

$$\xi = \sum_{n=0}^{p} \left[\hat{a}_n + \sum_{k=1}^{q} b_k \hat{a}_{n-k}\right]^2 - 1, \quad \hat{a}_0 = 1, a_k = 0, k < 0$$

which is minimized by selecting the MA(q) parameters $\{b_k\}$. The result of this minimization is

$$\hat{\underline{b}} = - \underline{R}_{aa}^{-1} \underline{r}_{aa} \tag{10.7}$$

where the elements of $\underline{R}_{aa}$ and $\underline{r}_{aa}$ are given as

$$\begin{aligned} R_{aa}(|i - j|) &= \sum_{n=0}^{p-|i-j|} \hat{a}_m \hat{a}_{n+|i-j|}, \quad i,j = 1,2,\ldots q \\ r_{aa}(i) &= \sum_{n=0}^{p-1} \hat{a}_n \hat{a}_{n+1}, \quad i = 1,2,\ldots,q \end{aligned} \tag{10.8}$$

This least squares method for determining for parameters of the MA(q) model is attributed to Durbin (1959). It has been shown by Kay (1988) that this estimation method is approximately maximum likelihood under the assumption that the observed process is Gaussian.

The order q of the MA model may be determined empirically by several methods. For example, the AIC for MA models has the same form as for AR models, i.e.,

$$\mathrm{AIC}(q) = \ln \sigma_{wq}^2 + 2q/N \tag{10.9}$$

where σ_{wq}^2 is an estimate of the variance of the white noise. Another approach, proposed by Chow (1972), is to filter the data with the inverse MA(q) filter and test the filtered output for whiteness.

11. ARMA MODEL FOR POWER SPECTRUM ESTIMATION

The Burg algorithm, and its variations, and the least-squares method described in the previous sections provide reliable high-resolution spectrum estimates based on the AR model. An ARMA model provides us with an opportunity to improve on the AR spectrum estimate, perhaps, by using fewer model parameters.

The ARMA model is particularly appropriate when our data have been corrupted by noise. For example, suppose that the data x(n) are generated by an AR system, where the system output is corrupted by additive white noise. The z-transform of the autocorrelation of the resultant signal may be expressed as

$$\begin{aligned}\Gamma_{xx}(z) &= \frac{\sigma_w^2}{A(z)A(z^{-1})} + \sigma_n^2 \\ &= \frac{\sigma_w^2 + \sigma_n^2 A(z)A(z^{-1})}{A(z)A(z^{-1})}\end{aligned} \tag{11.1}$$

where σ_n^2 is the variance of the additive noise. Therefore, the process x(n) is ARMA(p,p), where p is the order of the autocorrelation process. This relationship provides some motivation for investigating ARMA models for power spectrum estimation.

As we have demonstrated in Section 3, the parameters of the ARMA model are related to the autocorrelation by the equation in (3.9). For lags $|m| > q$, the equation involves only the AR parameters $\{a_k\}$. With estimates substituted in place of $\gamma_{xx}(m)$, we can solve the p equations in (3.10) to obtain $\hat{a}_k$. For high-order models, however, this approach is likely to yield poor estimates of the AR parameters.

A more reliable method is to construct an overdetermined set of linear equations and to use the method of least-squares on the set of overdetermined equations, as proposed by Cadzow (1982). To elaborate, suppose that the autocorrelation sequence can be accurately estimated up to lag M, where $M > p + q$. Then, we may write the following set of linear equations

$$\begin{bmatrix} r_{xx}(q) & r_{xx}(q-1) & \cdots & r_{xx}(q-p+1) \\ r_{xx}(q+1) & r_{xx}(q) & & r_{xx}(q-p+2) \\ \vdots & \vdots & & \vdots \\ r_{xx}(M-1) & r_{xx}(M-2) & \cdots & r_{xx}(M-p) \end{bmatrix} \begin{bmatrix} a_1 \\ a_2 \\ \vdots \\ a_p \end{bmatrix} = - \begin{bmatrix} r_{xx}(q + \\ r_{xx}(q + \\ \vdots \\ r_{xx}(M) \end{bmatrix} \tag{11.2}$$

or equivalently,

$$\underline{R}_{xx}\, \underline{a} = -\underline{r}_{xx} \tag{11.3}$$

Since $\underline{R}_{xx}$ is of dimension $(M - q) \times p$, and $M - q > p$ we may use the least-squares criterion to solve for the parameter vector $\underline{a}$. The result of this minimization is

$$\underline{\hat{a}} = -\left(\underline{R}_{xx}^{t}\underline{R}_{xx}\right)^{-1} \underline{R}_{xx}^{t}\, \underline{r}_{xx} \tag{11.4}$$

This procedure is called the least-squares modified Yule-Walker method. A weighting factor may also be applied to the autocorrelation sequence to deemphasize the less reliable estimates for large lags.

Once the parameters for the AR part of the model have been estimated as indicated above, we have the system

$$\hat{A}(z) = 1 + \sum_{k=1}^{p} \hat{a}_k z^{-k} \tag{11.5}$$

The sequence $x(n)$ may now be filtered by the FIR filter $\hat{A}(z)$ to yield the sequence

$$v(n) = x(n) + \sum_{k=1}^{p} \hat{a}_k x(n-k) \quad n = 0,1,\ldots,N-1 \tag{11.6}$$

The cascade of the ARMA (p,q) model with $\hat{A}(z)$ is approximately the MA(q) process generated by the model $B(z)$. Hence we may apply the MA estimate given in the preceding section to obtain the MA spectrum. To be specific, the filtered sequence v(n) for $p \le n \le N - 1$ is used to form the estimated correlation sequences $r_{vv}(m)$, from which we obtain the MA spectrum

$$P_{vv}^{MA}(f) = \sum_{m=-q}^{q} r_{vv}(m) e^{-j2\pi fm} \tag{11.7}$$

First, we observe that the parameters $\{b_k\}$ are not required to determine the power spectrum. Second, we observe that $r_{vv}(m)$ is an estimate of the autocorrelation for the MA model given by (3.16). In forming the estimate $r_{vv}(m)$, weighting (e.g., with the Bartlett window) may be used to deemphasize correlation estimates for large lags. In addition, the data may be filtered by a backward filter, thus creating another sequence, say $v^b(n)$, so that both v(n) and $v^b(n)$ can be used in forming the estimate of the autocorrelation $r_{vv}(m)$, as proposed by Kay (1980). Finally, the estimated ARMA power spectrum is

$$\hat{P}_{xx}^{ARMA}(f) = \frac{P_{vv}^{MA}(f)}{\left|1 + \Sigma_{k=1}^{p} a_k e^{-j2\pi fk}\right|^2} \tag{11.8}$$

An alternative method for determining the parameters $\{b_k\}$ of the MA(q) process is the least-squares method of Durbin, described in Section 10, where $\hat{\underline{b}}$ is obtained from (10.7).

The problem of order selection for the ARMA(p,q) model has been investigated by Chow (1972) and Bruzzone and Kaveh (1980). For this purpose the minimum of the AIC index,

$$\text{AIC}(p,q) = \ln \hat{\sigma}_{wpq}^2 + \frac{2(p + q)}{N} \tag{11.9}$$

may be used, where $\hat{\sigma}_{wpq}^2$ is an estimate of the variance of the input noise. An additional test on the adequacy of a particular ARMA(p,q) model is to filter the data through the model and test for whiteness of the output data. This would require that the parameters of the MA model be computed from the estimated autocorrelation, using spectral factorization to determine B(z) from $D(z) = B(z)B(z^{-1})$.

Another approach to order determination in ARMA modelling has been proposed by Cadzow (1982). In this paper he demonstrates that the use of a singular value (eigenvalue) decomposition of an extended autocorrelation matrix provides a good estimate for the order of the ARMA model.

For additional reading of ARMA power spectrum estimation, the reader is referred to the papers by Graupe et al. (1975), Cadzow (1981, 1982), Kay (1980), and Friedlander (1982b).

12. ESTIMATION OF THE BISPECTRUM

There are applications in which higher-order spectra (polyspectra) provide additional information about signal and system characteristics which are not available from the power spectrum. For example, one limitation of the power spectrum is that it does not uniquely characterize the phase information contained in the signal. However, a higher-order spectrum, such as the bispectrum, does provide phase information.

Another reason for computing a higher-order spectrum is to exploit the characteristics of Gaussian processes which often contaminate signals. For example, it is well known that all odd (central) moments of a Gaussian process are zero. Hence, by computing the third moment and its Fourier transform (the bispectrum) of a signal contaminated by additive Gaussian noise, the noise is suppressed and only the signal characteristics are preserved.

A third motivation for considering polyspectra is to detect and characterize nonlinear properties of systems that generate signals which depend on phase relationships of their harmonic components.

In this section we shall briefly consider the computation of the bispectrum. This is defined as the Fourier transform

$$B(f_1,f_2) = \sum_{n=-\infty}^{\infty} \sum_{m=-\infty}^{\infty} \gamma(n,m) e^{-j2\pi(f_1 n+f_2 m)} \qquad (12.1)$$

where $\gamma(n,m)$ is the third-order moment of a zero-mean, stationary random process $x(n)$, i.e.,

$$\gamma(n,m) = E\{x(k)x(k+n)x(k+m)\} \qquad (12.2)$$

where the process is assumed to be real.

Given a finite set of observations $x(n)$, $0 \leq n \leq N-1$, for which the mean value has been removed, the bispectrum can be computed either by means of a nonparametric method or by use of a linear system model. In particular, if the estimation method is nonparametric, the computation may be performed by segmenting the data into K segments and computing the sample moments

$$r_{xx}^{(i)}(n,m) = \frac{1}{M} \sum_{k} x^{(i)}(k)x^{(i)}(k+n)x^{(i)}(k+m), \quad i = 1,2,\ldots,K \tag{12.3}$$

Then, we average these moments, i.e.,

$$r_{xx}(n,m) = \frac{1}{K} \sum_{i=1}^{K} r_{xx}^{(i)}(n,m) \tag{12.4}$$

and, finally, we compute the two-dimensional Fourier transform i.e.,

$$\hat{B}(f_1,f_2) = \sum_{m=-L}^{L} \sum_{n=-L}^{L} r_{xx}(n,m)e^{-j2\pi(f_1 n+f_2 m)} \tag{12.5}$$

Nonparametric methods for bispectrum estimation exhibit the same basic limitations as for power spectrum estimation. The major problem is one of frequency resolution. This serves as the major motivation for considering model-based (parametric) methods for computing $B(f_1,f_2)$.

In an ARMA (p,q) model, the observed data sequence $x(n)$ satisfies the difference equation given by (2), where the noise process $w(n)$ is assumed to be third-order stationary with moments $E[w(n)] = 0, E[w(k)w(k+n)] = \sigma_w^2\delta(n)$, $E[w(k)w(k+n)w(k+m)] = \beta\delta(n,m)$. Furthermore, $x(n)$ is statistically independent of $w(m)$ for $n < m$. We observe that $w(n)$ and $x(n)$ are non-Gaussian. Furthermore, $x(n)$ is third-order stationary if $w(n)$ is third-order stationary and the ARMA (p,q) model represents a stable system.

The bispectrum is easily shown to be

$$B(f_1,f_2) = \beta\, H(f_1)H(f_2)H^*(f_1+f_2) \tag{12.6}$$

where $H(z) = A(z)/B(z)$.

MA(q) Model

If we set $a_{k=0}$, $k = 1,2\ldots,p$, so that (2.2) reduces to the MA(q) model, we find that

$$\begin{aligned}\gamma_{xx}(n,n) &= E[x(k)x^2(k+n)] \\ &= \beta \sum_{k=0}^{q} b_k b_{k+n}^2, \quad n = -q,\ldots,0,\ldots,q\end{aligned} \tag{12.7}$$

This relationship suggests that we use a matching technique which is based on computing the estimate of $\gamma_{xx}(n,n)$ from the data and adopting a squared error criterion. Thus, the function to be minimized is

$$\xi = \sum_{n=-q}^{q} \left[r_{xx}(n,n) - \beta \sum_{k=0}^{q} b_k b_{k+n}^2 \right]^2 \tag{12.8}$$

where the minimization is with respect to the q + 2 unknown parameter $\{b_k\}$ and β. The differentiation of ξ with respect to the q + 2 parameters yields a set of (q + 2) nonlinear equations which may be solved iteratively by one of several known algorithms.

AR(p) Model

If the observed process is an AR(p) process, the third moment sequence $\gamma(n,m)$ satisfies the equation

$$\gamma(-n,-m) + \sum_{k=1}^{p} a_k \gamma(k - n, k - m) = \beta\, \delta(n,m),\ n,m \leq 0 \tag{12.9}$$

where $\delta(n,m)$ is the two-dimensional unit sample sequence. If we evaluate the equations in (12.9) for $n = m = 0,1,\ldots,p$ we obtain a set of p + 1 linear equations which can be expressed in matrix form as

$$\underline{\Gamma}\underline{a} = \underline{\beta} \tag{12.10}$$

where Γ is a Toeplitz matrix with elements $\gamma(n,m)$, $\underline{a} = [1,a_1,a_2,\ldots, a_p]^t$ and $\underline{\beta} = [\underline{\beta}\ ,0,0,\ldots,0]^t$. The estimation of the p + 1 parameters $\{a_k\}$ and β can be accomplished by estimating the third-order moments $\gamma(n,m)$ from the data by use of (12.3) and (12.4). Then, from the estimates $\gamma_{xx}(n,m)$ we form the $(p + 1) \times (p + 1)$ matrix $\underline{R}_{xx}$ and use it in place of $\underline{\Gamma}$ in (12.10). Thus we obtain

$$\begin{aligned} \underline{R}_{xx}\underline{\hat{a}} &= \underline{\hat{\beta}} \\ \underline{\hat{a}} &= \underline{\hat{R}}_{xx}^{-1}\underline{\beta} \end{aligned} \tag{12.11}$$

This method is due to Raghuveer and Nikias (1986). Additional methods are given in the paper by Nikias and Raghuveer (1987).

ARMA (p,q) Model

Compared to the procedures for computing the parameters of the MA(q) and AR(p) models, methods for estimating the parameters of an ARMA(p,q) model are much more involved. For example, one available method utilizes a four-step procedure. First, the power spectrum of the data is estimated. From this we can obtain $|H(f)|$ but not the phase. Secondly, the phase of H(f) is reconstructed from the convertional bispectrum estimate of the data using one of the methods proposed by Brillinger (1981), Lii and Rosenblatt (1982) and Matsuoka and Ulrych (1984). Third, the ARMA(p,q) impulse response is generated from H(f) by using an inverse DFT algorithm. Finally, the p + q + 1 model parameters are obtained from the impulse response by use of the Pade approximation. The paper by Nikias and Raghuveer (1987) contains a more detailed description of this method.

13. ESTIMATION OF PARAMETERS OF SINUSOIDAL SIGNALS

In Section 11 we demonstrated that an AR(p) process corrupted by additive (white) noise is equivalent to an ARMA(p,p) process. In this section, we consider the special case in which the signal components are sinusoids corrupted by additive white noise.

First, we note that a real sinusoidal signal can be generated via the second order difference equation

$$x(n) = -a_1 x(n-1) - a_2 x(n-2) \tag{13.1}$$

where $a_1 = 2\cos 2\pi f_k$ and $a_2 = 1$, and, initially, $x(-1) = -1, x(-2) = 0$. This system has a pair of complex-conjugate poles (at $f = f_k$ and $f = -f_k$) and thus generates the sinusoid $x(n) = \cos 2\pi f_k n$, for $n \geq 0$.

In general, a signal consisting of p sinusoidal components satisfies the difference equation

$$x(n) = -\sum_{m=1}^{2p} a_m x(n-m) \tag{13.2}$$

and corresponds to the system with system function

$$H(z) = \frac{1}{1 + \sum_{m=1}^{2p} a_m z^{-m}} \tag{13.3}$$

The polynomial

$$A(z) = 1 + \sum_{m=1}^{2p} a_m z^{-m} \tag{13.4}$$

has 2p roots on the unit circle which correspond to the frequenci of the sinusoids.

Now, suppose that the sinusoids are corrupted by a white noise sequence w(n) with $E[|w(n)|^2] = \sigma_w^2$. Then we observe that

$$y(n) = x(n) + w(n) \tag{13.5}$$

If we substitute $x(n) = y(n) - w(n)$ in (13.2), we obtain

$$y(n) - w(n) = - \sum_{m=1}^{2p} [y(n - m) - w(n - m)]a_m$$

or, equivalently,

$$\sum_{m=0}^{2p} a_m y(n - m) = \sum_{m=0}^{2p} a_m w(n - m) \tag{13.6}$$

where, by definition, $a_0 = 1$.

We observe that (13.6) is the difference equation for an ARMA(2p,2p) process in which both the AR and MA parameters are identical. This symmetry is a characteristic of the sinusoidal signals in white noise. The difference equation in (13.6) may be expressed in matrix form as

$$\underline{Y}^t \underline{a} = \underline{W}^t \underline{a} \tag{13.7}$$

where $\underline{Y}^t = [y(n)\ y(n - 1)\ldots y(n - 2p)]$ is the observed data vector of dimension (2p + 1), $\underline{W}^t = [w(n)\ w(n - 1)\ldots w(n - 2p)]$ is the noise vector, and $\underline{a} = [1\ a_1 \ldots a_{2p}]$ is the coefficient vector.

If we premultiply (13.7) by $\underline{Y}$ and take the expected value, we obtain

$$E(\underline{Y}\underline{Y}^t)\underline{a} = E(\underline{Y}\underline{W}^t)a = E[(\underline{X} + \underline{W})\underline{W}^t]\underline{a} \tag{13.8}$$

$$\underline{\Gamma}_{yy}\underline{a} = \sigma_w^2 \underline{I}\underline{a}$$

where we have used the assumption that the sequence w(n) is zero mean and white, and $\underline{X}$ is a deterministic signal.

The equation in (13.8) is in the form of an eigenequation, that is,

$$(\underline{\Gamma}_{yy} - \sigma_w^2 \underline{I})\underline{a} = 0 \qquad (13.9)$$

where σ_w^2 is an eigenvalue of the autocorrelation matrix Γ_{yy}. Then the parameter vector a is an eigenvector associated with the eigenvalue σ_w^2. The eigenequation in (13.9) forms the basis for the Pisarenko harmonic decomposition method.

Pisarenko Harmonic Decomposition Method

For p randomly-phased sinusoids in additive white noise, the autocorrelation values are

$$\gamma_{yy}(0) = \sigma_w^2 + \sum_{i=1}^{p} P_i$$

$$\gamma_{yy}(k) = \sum_{i=1}^{p} P_i \cos 2\pi f_i k \quad k \neq 0 \qquad (13.10)$$

where $P_i = A_i^2/2$ is the average power in the ith sinusoid and A_i is the corresponding amplitude. Hence, we may write

$$\begin{pmatrix} \cos 2\pi f_1 & \cos 2\pi f_2 & \cdots & \cos 2\pi f_p \\ \cos 4\pi f_1 & \cos 4\pi f_2 & \cdots & \cos 4\pi f_p \\ \vdots & \vdots & \vdots & \\ \cos 2\pi p f_1 & \cos 2\pi p f_2 & \cdots & \cos 2\pi p f_p \end{pmatrix} \begin{pmatrix} P_1 \\ P_2 \\ \vdots \\ P_p \end{pmatrix} = - \begin{pmatrix} \gamma_{yy}(1) \\ \gamma_{yy}(2) \\ \vdots \\ \gamma_{yy}(p) \end{pmatrix} \qquad (13.11)$$

If we know the frequencies f_i, $1 \le i \le p$, we may use this equation to determine the powers of the sinusoids. In place of $\gamma_{xx}(m)$, we use the estimates $r_{xx}(m)$. Once the powers are known, the noise variance can be obtained from (13.10) as

$$\sigma_w^2 = r_{yy}(0) - \sum_{i=1}^{p} P_i \qquad (13.12)$$

The problem that remains is to determine the p frequencies f_i, $1 \le i \le p$, which, in turn, require knowledge of the eigenvector $\underline{a}$ corresponding to the eigenvalue σ_w^2. Pisarenko (1973) observed [see Papoulis (1984) and Grenander and Szegö (1958)] that for an ARMA process consisting of p sinusoids in additive white noise the variance σ_w^2 corresponds to the minimum eigenvalue of $\underline{\Gamma}_{yy}$, when the dimension of the autocorrelation matrix equals or exceeds (2p + 1) x (2p + 1). The desired ARMA coefficient vector corresponds to the eigenvector associated with the minimum eigenvalue. Therefore, the frequencies f_i, $1 \leq i \leq p$ are obtained from the roots of the polynomial in (13.4), where the coefficients are the elements of the eigenvector $\underline{a}$, corresponding to the minimum eigenvalue σ_w^2. Therefore, the Pisarenko harmonic decomposition method proceeds as follows. First we estimate $\underline{\Gamma}_{yy}$ from the data, i.e., we form the autocorrelation matrix $\underline{R}_{yy}$. Then, we find the minimum eigenvalue and the corresponding minimum eigenvector. The minimum eigenvector yields the parameters of the ARMA(2p,2p) model. From (13.4) we can compute the roots which constitute the frequencies $\{f_i\}$. By using these frequencies, we can solve (13.11) for the signal powers $\{P_i\}$ by substituting the estimates $r_{yy}(m)$ for $\gamma_{yy}(m)$.

As we will observe below, the Pisarenko method is based on the use of a noise subspace eigenvector to estimate the frequenci of the sinusoids.

Eigen-decomposition of the Autocorrelation Matrix for Sinusoids in White Noise

In the above discussion we assumed that the sinusoidal signal consists of p real sinusoids. We shall now assume that the signal consists of p complex sinusoids of the form

$$x(n) = \sum_{i=1}^{p} A_i e^{j(2\pi f_i n + \phi_i)} \qquad (13.13$$

where the amplitudes $\{A_i\}$ and the frequencies $\{f_i\}$ are unknown and the phases $\{\phi_i\}$ are statistically independent random variables uniformly distributed on $(0,2\pi)$. Then, the random process x(n) is wide-sense stationary with autocorrelation function

$$\gamma_{xx}(m) = \sum_{i=1}^{p} P_i e^{j2\pi f_i m} \qquad (13.14$$

where, for complex sinusoids, $P_i = A_i^2$ is the power of the ith sinusoid.

Since the observed sequence is $y(n) = x(m) + w(n)$, where $w(n)$ is a white noise sequence with spectral density σ_w^2, the autocorrelation function for $y(n)$ is

$$\gamma_{yy}(m) = \gamma_{xx}(m) + \sigma_w^2 \delta(m), \quad m = 0, \pm 1, \ldots, \pm(M-1) \qquad (13.15)$$

Hence, the M x M autocorrelation matrix for $y(n)$ may be expressed as

$$\underline{\Gamma}_{yy} = \underline{\Gamma}_{xx} + \sigma_w^2 \underline{I} \qquad (13.16)$$

where $\underline{\Gamma}_{xx}$ is the autocorrelation matrix for the signal $x(n)$ and $\sigma_w^2 \underline{I}$ is the autocorrelation matrix for the noise. Note that if select $M > p$, $\underline{\Gamma}_{xx}$ which is of dimension M x M is not of full rank, because its rank is p. However, $\underline{\Gamma}_{yy}$ is full rank because $\sigma_w^2 \underline{I}$ is of rank M.

In fact, the signal matrix $\underline{\Gamma}_{xx}$ may be represented as

$$\underline{\Gamma}_{xx} = \sum_{i=1}^{p} P_i \underline{s}_i \underline{s}_i^H \qquad (13.17)$$

where H denotes the conjugate transpose and $\underline{s}_i$ is a signal vector of dimension M defined as

$$\underline{s}_i = \left[1, e^{j2\pi f_i}, e^{j4\pi f_i}, \ldots, e^{j2\pi(M-1)f_i}\right] \qquad (13.18)$$

Since each vector (outer product) $\underline{s}_i \underline{s}_i^H$ is a matrix of rank 1 and since there are p vector products, the matrix $\underline{\Gamma}_{xx}$ is of rank p. Note that if the sinusoids were real, the correlation matrix $\underline{\Gamma}_{xx}$ will have rank 2p.

Now, let us perform an eigen-decomposition of the matrix $\underline{\Gamma}_{yy}$. Let the eigenvalues $\{\lambda_i\}$ be ordered in decreasing value with $\lambda_1 \geq \lambda_2 \geq \lambda_3 \geq \ldots \geq \lambda_M$ and let the corresponding eigenvectors be denoted as $\{\underline{v}_i, i = 1, \ldots, M\}$. We assume that the eigenvectors are normalized so that $\underline{v}_i^H \cdot \underline{v}_j = \delta_{ij}$. In the absence of noise the eigenvalues $\lambda_i, i = 1, 2, \ldots, p$ will be non-zero while $\lambda_{p+1} = \lambda_{p+2} = \ldots = \lambda_M = 0$. Furthermore, it follows that the signal correlation matrix can be expressed as

$$\underline{\Gamma}_{xx} = \sum_{i=1}^{p} \lambda_i \underline{v}_i \underline{v}_i^H \tag{13.19}$$

Thus, the eigenvectors $\underline{v}_i$, $i = 1,2...,p$ span the signal subspace as the signal vectors $\underline{s}_i$, $i = 1,2,...,p$ do. These p eigenvectors for the signal subspace are called the principal eigenvectors and the corresponding eigenvalues are called the principal eigenvalues.

In the presence of noise, the noise autocorrelation matrix in (13.6) may be represented as

$$\sigma_w^2 \underline{I} = \sigma_w^2 \sum_{i=1}^{M} \underline{v}_i \underline{v}_i^H \tag{13.20}$$

By substituting (13.19) and (13.20) into (13.16) we obtain

$$\begin{aligned} \underline{\Gamma}_{yy} &= \sum_{i=1}^{p} \lambda_i \underline{v}_i \underline{v}_i^H + \sum_{i=1}^{M} \sigma_w^2 \underline{v}_i \underline{v}_i^H \\ &= \sum_{i=1}^{p} (\lambda_i + \sigma_w^2) \underline{v}_i \underline{v}_i^H + \sum_{i=p+1}^{M} \sigma_w^2 \underline{v}_i \underline{v}_i^H \end{aligned} \tag{13.21}$$

This eigen-decomposition separates the eigenvectors into two sets. The set $\{\underline{v}_i, i = 1,2,...,p\}$, which are the principal eigenvectors span the signal subspace, while the set $\{\underline{v}_i, i = p + 1,...,M\}$ which are orthogonal to the principal eigenvectors, are said to belong to the noise subspace. Since the signal vectors $\{\underline{s}_i, i = 1,2,...,p\}$ are in the signal subspace, it follows that the $\{\underline{s}_i\}$ are simply linear combinations of the principal eigenvectors and, also, are orthogonal to the vectors in the noise subspace.

In this context, we see that the Pisarenko method is based on the estimation of the frequencies by using the orthogonality property between the vectors in the noise subspace and the signal vectors. For complex sinusoids, if we select $M = p + 1$ (for real sinusoids we select $M = 2p + 1$), there is only a single eigenvector in the noise subspace (corresponding to the minimum eigenvalue) which must be orthogonal to the signal vector. Thus, we have

$$\underline{s}_i^H \underline{v}_{p+1} = \sum_{k=0}^{p} \underline{v}_{p+1}(k+1)e^{-j2\pi f_i k} = 0, \quad i = 1,2,\ldots,p \tag{13.22}$$

But (13.22) implies that the frequencies $\{f_i\}$ can be determined by solving for the zeros of the polynomial

$$V(z) = \sum_{n=0}^{p} \underline{v}_{p+1}(k+1)z^{-k} \tag{13.23}$$

all of which lie on the unit circle. The angles of these roots are $2\pi f_i$, $i = 1,2,\ldots,p$.

When the number of sinusoids is unknown, the determination of p may prove to be difficult, especially if the signal level is not much higher than the noise level. In theory, if $M > p + 1$ there is a multiplicity $(M - p)$ of the minimum eigenvalue. However, in practice the $(M - p)$ small eigenvalues of $\underline{R}_{yy}$ will probably be different. By computing all the eigenvalues it may be possible to determine p by grouping the $M - p$ small (noise) eigenvalues into a set and averaging them to obtain an estimate of σ_w^2. Then, the average value may be used in (13.9) along with $\underline{R}_{yy}$ to determine the corresponding eigenvector.

MUSIC Method

The multiple signal classification (MUSIC) method is also a noise subspace frequency estimator. To develop this method, let us first consider the "weighted" spectral estimate

$$P(f) = \sum_{k=p+1}^{M} w_k \left| \underline{s}^H(f)\underline{v}_k \right|^2 \tag{13.24}$$

where $\{\underline{v}_k, k = p + 1,\ldots,M\}$ are the eigenvectors in the noise subspace, $\{w_k\}$ are a set of positive weights, and $\underline{s}(f)$ is the complex sinusoidal vector

$$\underline{s}(f) = \left[1, e^{j2\pi f}, e^{j4\pi f}, \ldots, e^{j2\pi(M-1)f}\right] \tag{13.25}$$

Note that at $f = f_i$, $\underline{s}(f_i) \equiv \underline{s}_i$, so that at any one of the p sinusoidal frequency components of the signal we have

$$P(f_i) = 0, \ i = 1,2,\ldots,p \tag{13.26}$$

Hence, the reciprocal of P(f) is a sharply peaked function of frequency and provides a method for estimating the frequencies of the sinusoidal components. Thus,

$$\frac{1}{P(f)} = \frac{1}{\sum_{k=p+1}^{M} w_k \left| \underline{s}^H(f)\underline{v}_k \right|^2} \tag{13.27}$$

Although theoretically 1/P(f) is infinite at $f = f_i$, in practice the estimation errors result in finite values for 1/P(f) at all frequencies.

The MUSIC sinusoidal frequency estimator proposed by Schmidt (1981) is a special case of (13.27) in which the weights $w_k = 1$ for all k. Hence,

$$P_{MUSIC}(f) = \frac{1}{\sum_{k=p+1}^{M} \left| \underline{s}^H(f)\underline{v}_k \right|^2} \tag{13.28}$$

The estimate of the sinusoidal frequencies are the peaks of $P_{MUSIC}(f)$. Once the sinusoidal frequencies are estimated, the power of each of the sinusoids may be obtained by solving (13.11).

ESPRIT Algorithm

ESPRIT (Estimation of Signal Parameters via Rotational Invariar Techniques) is yet another method for estimating frequencies of a sum of sinusoids by use of an eigen-decomposition approach. As we will observe from the development given below, which is due to Roy et al. (1986), ESPRIT exploits an underlying rotationa invariance of signal subspaces spanned by two temporally displace data vectors.

We again consider the estimation of p complex-valued sinusoids in additive white noise. The received sequence is given by the vector

$$\begin{aligned} \underline{y}(n) &= [y(n), y(n+1), \ldots, y(n+M-1)]^t \\ &= \underline{x}(n) + \underline{w}(n) \end{aligned} \tag{13.29}$$

where $\underline{x}(n)$ is the signal vector and $\underline{w}(n)$ is the noise vector. To exploit the deterministic character of the sinusoids, we define the time-displaced vector $\underline{z}(n) = \underline{y}(n+1)$. Thus,

$$\underline{z}(n) = [z(n), z(n+1), \ldots, z(n+M-1)]^t$$
$$= [y(n+1), y(n+2), \ldots, y(n+M)]^t \quad (13.30)$$

With these definitions we may express the vectors $\underline{y}(n)$ and $\underline{z}(n)$ as

$$\underline{y}(n) = \underline{S}\underline{a} + \underline{w}(n)$$
$$\underline{z}(n) = \underline{S}\underline{\Phi}\underline{a} + \underline{w}(n) \quad (13.31)$$

where $\underline{a} = [a_1, a_2, \ldots, a_p]^t$, $a_i = A_i e^{j\phi_i}$, and $\underline{\Phi}$ is a diagonal p x p matrix consisting of the relative phase between adjacent time samples of each of the complex sinusoids, i.e.,

$$\underline{\Phi} = \text{diag}[e^{j2\pi f_1}, e^{j2\pi f_2}, \ldots, e^{j2\pi f_p}] \quad (13.32)$$

Note that the matrix $\underline{\Phi}$ relates the time-displaced vectors $\underline{y}(n)$ and $\underline{z}(n)$, and may be called a rotation operator. We also note that $\underline{\Phi}$ is unitary. The matrix $\underline{S}$ is the M x p Vandermonde matrix specified by the column vectors

$$\underline{s}_i = [1, e^{j2\pi f_i}, e^{j4\pi f_i}, \ldots, e^{j2\pi(M-1)f_i}] \quad i = 1,2,\ldots,p \quad (13.33)$$

Now, the autocovariance matrix for the data vector $\underline{y}(n)$ is

$$\underline{\Gamma}_{yy} = E[\underline{y}(n)\underline{y}^H(n)]$$
$$= \underline{S}\underline{P}\underline{S}^H + \sigma_w^2 \underline{I} \quad (13.34)$$

where $\underline{P}$ is the p x p diagonal matrix consisting of the powers of the complex sinusoids, i.e.,

$$\underline{P} = \text{diag}[|a_1|^2, |a_2|^2, \ldots, |a_p|^2]$$
$$= \text{diag}[P_1, P_2, \ldots, P_p] \quad (13.35)$$

We observe that $\underline{P}$ is a diagonal matrix since complex sinusoids of different frequencies are orthogonal over the infinite interval. However, we should emphasize that the ESPRIT algorithm does not require $\underline{P}$ to be diagonal. Hence, the algorithm is applicable to the case in which the covariance matrix is estimated from finite data records.

The cross-covariance matrix of the signal vectors $\underline{y}(n)$ and $\underline{z}(n)$ is

$$\underline{\Gamma}_{yz} = E[\underline{y}(n)\underline{z}^H(n)] = \underline{SP\Phi}^H\underline{S}^H + \sigma_w^2\underline{\Gamma}_w \tag{13.36}$$

where

$$\underline{\Gamma}_w = E[\underline{w}(n)\underline{w}^H(n+1)]$$

$$= \sigma_w^2 \begin{bmatrix} 0 & 0 & 0 & \cdots & 0 & 0 \\ 1 & 0 & 0 & \cdots & 0 & 0 \\ 0 & 1 & 0 & \cdots & 0 & 0 \\ \vdots & \vdots & \vdots & & \vdots & \vdots \\ 0 & 0 & 0 & \cdots & 1 & 0 \end{bmatrix} \equiv \sigma_w^2\underline{Q} \tag{13.37}$$

The auto and cross-covariance matrices $\underline{\Gamma}_{yy}$ and $\underline{\Gamma}_{yx}$ are given as

$$\underline{\Gamma}_{yy} = \begin{bmatrix} \gamma_{yy}(0) & \gamma_{yy}(1) & \cdots & \gamma_{yy}(M-1) \\ \gamma_{yy}^*(1) & \gamma_{yy}(0) & \cdots & \gamma_{yy}(M-2) \\ \vdots & \vdots & & \vdots \\ \gamma_{yy}^*(M-1) & \gamma_{yy}(M-2) & \cdots & \gamma_{yy}(0) \end{bmatrix} \tag{13.38}$$

$$\underline{\Gamma}_{yz} = \begin{bmatrix} \gamma_{yy}(1) & \gamma_{yy}(2) & \cdots & \gamma_{yy}(M) \\ \gamma_{yy}(0) & \gamma_{yy}(1) & \cdots & \gamma_{yy}(M-1) \\ \vdots & \vdots & & \vdots \\ \gamma_{yy}^*(M-2) & \gamma_{yy}^*(M-3) & \cdots & \gamma_{yy}(1) \end{bmatrix} \tag{13.39}$$

where $\gamma_{yy}(m) = E[y^*(n)y(n+m)]$. Note that both $\underline{\Gamma}_{yy}$ and $\underline{\Gamma}_{yz}$ are Toeplitz matrices.

Based on this formulation, the problem is to determine the frequencies $\{f_i\}$ and their power $\{P_i\}$ from the autocorrelation sequence $\{\gamma_{yy}(m)\}$.

From the underlying model, it is clear that the matrix $\underline{SPS}^H$ has rank p. Consequently, $\underline{\Gamma}_{yy}$ given by (13.34) has (M - p) identical eigenvalues equal to σ_w^2. Hence,

$$\underline{\Gamma}_{yy} - \sigma_w^2 \underline{I} = \underline{SPS}^H \equiv \underline{C}_{yy} \tag{13.40}$$

From (13.36) we also have

$$\underline{\Gamma}_{yz} - \sigma_w^2 \underline{\Gamma}_w = \underline{SP\Phi}^H \underline{S}^H \equiv \underline{C}_{yz} \tag{13.41}$$

Now, let us consider the matrix $\underline{C}_{yy} - \lambda\, \underline{C}_{yz}$, which can be written as

$$\underline{C}_{yy} - \lambda\, \underline{C}_{yz} = \underline{SP}(\underline{I} - \lambda \underline{\Phi}^H)\underline{S}^H \tag{13.42}$$

Clearly, the column space of $\underline{SPS}^H$ is identical to the column space of $\underline{SP\Phi}^H\underline{S}^H$. Consequently, the rank of $\underline{C}_{yy} - \lambda \underline{C}_{yz}$ is equal to p. However, we note that if $\lambda = \exp(j2\pi f_i)$, the ith row of $(\underline{I} - \lambda\underline{\Phi}^H)$ is zero and, hence, the rank of $[\underline{I} - \underline{\Phi}^H \exp(j2\pi f_i)]$ is $p - 1$. But $\lambda_i = \exp(j2\pi f_i)$, $i = 1,2,\ldots,p$ are the generalized eigenvalues of the matrix pair $(\underline{C}_{yy}, \underline{C}_{yz})$. Thus, the p generalized eigenvalues $\{\lambda_i\}$ that lie on the unit circle correspond to the elements of the rotation operator $\underline{\Phi}$. The remaining M - p generalized eigenvalues of the pair $\{\underline{C}_{yy}, \underline{C}_{yz}\}$ which correspond to the common null space of these matrices, are zero, i.e., the (M - p) eigenvalues are at the origin in the complex plane.

Based on the above mathematical relationships we can formulate an algorithm (ESPRIT) for estimating the frequencies $\{f_i\}$. The procedure is as follows:

1. From the data, compute the autocorrelation values $r_{yy}(m)$, $m = 0,1,2,\ldots,M$ and form the matrices $\underline{R}_{yy}$ and $\underline{R}_{yz}$ corresponding to estimates of $\underline{\Gamma}_{yy}$ and $\underline{\Gamma}_{yz}$.

2. Compute the eigenvalues of $\underline{R}_{yy}$. For $M > p$, the minimum eigenvalue is an estimate of σ_w^2.

3. Compute $\hat{\underline{C}}_{yy} = \underline{R}_{yy} - \hat{\sigma}_w^2 \underline{I}$ and $\hat{\underline{C}}_{yz} = \underline{R}_{yz} - \hat{\sigma}_w^2 \underline{Q}$ where $\underline{Q}$ defined in (13.37).

4. Compute the generalized eigenvalues of the matrix pair $\{\hat{C}_{yy}, \hat{\underline{C}}_{yz}\}$. The p generalized eigenvalues of these matrices that lie on (or near) the unit circle determine the (estimated) elements of Φ and, hence, the sinusoidal frequencies. The remaining M - p eigenvalues will lie at (or near) the origin.

One method for determining the power in the sinusoidal components is to solve the equation in (13.11) with $r_{yy}(m)$ substituted for $\gamma_{yy}(m)$.

Another method is based on the computation of the generalized eigenvectors $\{\underline{v}_i\}$ corresponding to the generalized eigenvalues $\{\lambda_i\}$. We have

$$\left(\underline{C}_{yy} - \lambda_i \underline{C}_{yz}\right)\underline{v}_i = \underline{SP}\left(\underline{I} - \underline{\lambda_i \Phi^H}\right)\underline{S}^H\underline{v}_i = 0 \tag{13.43}$$

Since the column space of $\left(\underline{C}_{yy} - \lambda_i \cdot \underline{C}_{yz}\right)$ is identical to the column space spanned by the vectors $\{\underline{s}_j, j \neq i\}$ given by (13.33), it follows that the generalized eigenvector $\underline{v}_i$ is orthogonal to $\underline{s}_j, j \neq i$. Since P is diagonal, it follows from (13.43) that the signal powers are

$$P_i = \frac{\underline{v}_i^H \underline{C}_{yy} \underline{v}_i}{\left|\underline{v}_i^H \underline{s}_i\right|}, \quad i = 1,2,\ldots,p \tag{13.44}$$

Order Selection Criteria

The eigenanalysis methods described in this section for estimating the frequencies and the powers of the sinusoids also provide information about the number of sinusoidal components. If there are p sinusoids, the eigenvalues associated with the signal subspace are $\{\lambda_i + \sigma_w^2, i = 1,2\ldots,p\}$ while the remaining (M - p) eigenvalues are all equal to σ_w^2. Based on this eigenvalue decomposition, a test can be designed which compares the eigenvalues with a specified threshold. An alternative method also uses the eigenvector decomposition of the estimated autocorrelation matrix of the observed signal and is based on matrix perturbation analysis. This method is described in the recent paper by Fuchs (1988).

Another approach based on an extension of the AIC criterion to the eigen-decomposition method, has been proposed by Wax and Kailath (1985). If the eigenvalues of the sample autocorrelation matrix are ranked so that $\lambda_1 \geq \lambda_2 \geq \ldots \geq \lambda_M$, where $M > p$, the number of sinusoids in the signal subspace is estimated by selecting the minimum value of AIC(p), given as

$$\mathrm{AIC}(p) = (M - p)\ln \left[\frac{\frac{1}{M-p} \Sigma_{i=p+1}^{M} \lambda_i}{\Pi_{i=p+1}^{M} \lambda_i^{-(M-p)}} \right] + p(2M - p) \qquad (13.45)$$

Some results on the quality of this order selection criterion are given in the paper by Wax and Kailath (1985).

14. CONCLUDING REMARKS

We have provided a survey of several model-based spectral estimation methods. AR, MA, and ARMA methods were considered for the estimation of the power spectrum and the bispectrum. We also described eigen-decomposition methods for estimating the frequencies and the powers of sums of sinusoidal signals corrupted by additive noise.

The spectral estimation methods that we have considered did not assume any knowledge of the underlying statistical characteristics of the observed data. When the probability distribution of the observed data is known, it is possible to perform signal-model parameter estimation based on the maximum likelihood criterion. Such as approach yields estimates that are asymptotically efficient.

Usually, the maximum likelihood criterion leads to a set of nonlinear equations which must be solved for the signal parameters. With some approximations, it is sometimes possible to obtain approximate solutions which maintain the important desirable properties of maximum likelihood estimates. The interested reader is referred to the text by Kay (1988) which applies the maximum likelihood criterion to the estimation of signal parameters for model-based spectral estimation.

REFERENCES

Akaike, H., (1969) Power Spectrum Estimation Through Autoregression Model Fitting. *Ann. Inst. Stat. Math.,* vol. 21, pp. 407-419.

Akaike, H., (1974) A New Look at the Statistical Model Identification, *IEEE Trans. Automatic Control,* vol. AC-19, pp. 716-723.

Andersen, N.O., (1978) Comments on the Performance of Maximum Entropy Algorithm, *Prcc. IEEE,* vol. 66, pp. 1581-1582.

Berryman, J.G., (1978) Choice of Operator Length for Maximum Entropy Spectral Analysis, *Geophysics,* vol. 43, pp. 1384-1391.

Brillinger, D.R., (1981) Time Series, Data Analysis and Theory, Holden Day, San Francisco CA.

Burg, J.P., (1967) Maximum Entropy Spectral Analysis, Proc. 37th Meeting of the Society of Exploration Geophysicists, Iklahoma City, Okla. Reprinted in Modern Spectrum Analysis, D.G. Childers, Ed., IEEE Press, New York.

Burg, J.P., (1968) A New Analysis Technique for Time Series Data, NATO Advanced Study Institute on Signal Processing with Emphasis on Underwater Acoustics, Reprinted in Modern Spectrum Analysis, D.G. Childers, Ed., IEEE Press, New York.

Burg, J.P., (1975) Maximum Entropy Spectral Analysis, Ph.D. dissertation, Department of Geophysics, Stanford, Calif.

Cadzow, J.A., (1981) Autoregressive-Moving Average Spectral Estimation: A Model Equation Error Procedure, *IEEE Trans. Geoscience Remote Sensing,* vol. GE-19, pp. 24-28.

Cadzow, J.A., (1982) Spectral Estimation: An Overdetermined Rational Model Equation Approach, *Proc. IEEE,* vol. 70, pp. 907-938.

Chen, W.Y. and Stegen, G.R., (1974) Experiments with Maximum Entropy Power Spectra of Sinusoids, *J. Geophys. Res.,* vol. 79, pp. 3019-3022.

Chow, J.C., (1972) On Estimating the Orders of an Autoregressive Moving Average Process with Uncertain Observations, *IEEE Trans. Automatic Control,* vol. AC-17, pp. 707-709.

Durbin, J., (1959) Efficient Estimation of Parameters in Moving-Average Models, *Biometrika,* vol. 46, pp. 306-316.

Friedlander, B., (1982b) Lattice Methods for Spectral Estimation, *Proc. IEEE,* vol. 70, pp. 990-1017.

Fougere, P.F., Zawalick, E.J. and Radoski, H.R., (1976) Spontaneous Line Splitting in Maximum Entropy Power Spectrum Analysis, *Phys. Earth Planet. Inter.,* vol. 12, 201-207.

Fuchs, J.J., (1988) Estimating the Number of Sinusoids in Additive White Noise, *IEEE Trans. Acoust., Speech. Sig. Proc.,* vol. ASSP-36, pp. 1846-1853.

Gersch, W. amd Sharpe, D.R., (1973) Estimation of Power Spectra with Finite-Order Autoregressive Models, *IEEE Trans. Automatic Control,* vol. AC-18, pp. 367-369.

Graupe, D., Krause, D.J. and Moore, J.B., (1975) Identification of Autoregressive-Moving Average Parameters of Time-Series, *IEEE Trans. Automatic Control,* vol. AC-20, pp. 104-107.

Grenander, O. and Szego, G., (1958) Toeplitz Forms and Their Applications, University of California Press, Berkeley, Calif.

Griffiths, L.J., (1975) Rapid Measurements of Digital Instantaneous Frequency, *IEEE Trans. Acoustics, Speech, and Signal Processing,* vol ASSP-23, pp. 207-222.

Helme, B. and Nikias, C.L., (1985) Improved Spectrum Performance via a Data-Adaptive Weighted Burg Technique, *IEEE Trans. Acoustics, Speech and Signal Processing,* vol. ASSP-33, pp. 903-910.

Herring, R.W., (1980) The Cause of Line Splitting in Burg Maximum Entropy Spectral Analysis, *IEEE Trans. Acoustics, Speech, and Signal Processing,* vol. ASSP-28, pp. 692-701.

Jones, R.H., (1976) Autoregression Order Selection. *Geophysics,* vol. 41, pp. 771-773.

Kalouptsidis, N. and Theodoridis, S., (1987) Fast Adaptive Least-Squares Algorithms for Power Spectral Estimation. *IEEE Trans. Acoustics, Speech and Signal Processing,* vol. ASSP-35, pp. 661-670.

Kashyap, R.L., (1980) Inconsistency of the AIC Rule for Estimating the Order of Autoregressive Models, *IEEE Trans. Automatic Control,* vol. AC-25, pp. 996-998.

Kaveh, M. and Bruzzone, S.P., (1979) Order Determination for Autoregressive Spectral Estimation. Record of the 1979 RADC Spectral Estimation Workshop, pp. 139-145, Griffin Air Force Base, Rome, N.Y.

Kaveh, M., and Lippert, G.A., (1983) An Optimum Tapered Burg Algorithm for Linear Prediction and Spectral Analysis, *IEEE Trans. Acoustics, Speech, and Signal Processing,* vol. ASSP-31, pp. 438-444.

Kay, S.M., (1980) A New ARMA Spectral Estimator, *IEEE Trans. Acoustics, Speech and Signal Processing,* vol. ASSP-28, pp. 585-588.

Kay, S.M., (1988) Modern Spectral Estimation, Prentice Hall, Englewood Cliffs. N.J.

Kay, S.M. and Marple, S.L., Jr., (1979) Sources of and Remedies for Spectral Line Splitting in Autoregressive Spectrum Analysis. Proc. 1979 ICASSP. pp. 151-154.

Lacoss, R.T., (1971) Data Adaptive Spectral Analysis Methods, *Geophysics*, vol. 36, pp. 661-675.

Lii, K.S. and Rosenblatt, M., (1982) Deconvolution and Estimation of Transfer Function Phase and Coefficients for Non-Gaussian Linear Processes, *Ann. Statist.*, vol. 10, pp. 1195-1208.

Marple, S.L., Jr., (1980) A New Autoregressive Spectrum Analysis Algorithm, *IEEE Trans. Acoustics, Speech and Signal Processing*, vol. ASSP-28, pp. 441-454.

Matsuoka, T. and Ulrych, T.J., (1984) Phase Estimation Using the Bispectrum, Proc. IEEE, vol. 72, pp. 1403-1411.

Newman, W.I., (1981) Extension to the Maximum Entropy Method III, Proc. 1st ASSP Workshop on Spectral Estimation, pp. 1.7.1-1.7.6. Hamilton, Ontario, Canada.

Nikias, C.L., and Raghuveer, M.R., (1987) Bispectrum Estimation: A Digital Signal Processing Framework, *Proc. IEEE,* vol. 75, pp. 869-891.

Nikias, C L. and Scott, P.D., (1982) Energy-Weighted Linear Predictive Spectral Estimation: A New Method Combining Robustness and High Resolution, *IEEE Trans. Acoustics, Speech, and Signal Processing,* vol. ASSP-30, pp. 287-292.

Nuttall, A.H., (1976) Spectral Analysis of a Univariate Process with Bad Data Points. via Maximum Entropy and Linear Predictive Techniques, NUSC Technical Report TR-5303, New London, Conn.

Papoulis, A., (1984) Probability, Random Variables, and Stochastic Processes, 2nd ed., McGraw-Hill, New York.

Parzen, E., (1974) Some Recent Advances in Time Series Modelling, *IEEE Trans. Automatic Control,* vol. AC-19, pp. 723-730.

Pisarenko, V.F., (1973) The Retrieval of Harmonics from a Covariance Function, *Geophys. J.R. Astron. Soc.,* vol. 33, pp. 347-366.

Raghuveer, M.R. and Nikias, C.L., (1986) Bispectrum Estimation Via AR Modelling, *Signal Processing,* vol. 9, pp. 35-48.

Rissanen, J., (1983) A Universal Prior for the Integers and Estimation by Minimum Description Length, *Ann. Stat.,* vol. 11. pp. 417-431.

Roy, R., Paulraj, A. and Kailath, T., (1986) ESPRIT-A Subspace Rotation Approach to Estimation of Cisoids in Noise, *IEEE Trans. Acoust. Speech, Sig. Proc.*, vol. ASSP-34, pp. 1340-1346.

Schmidt R.D., (1981) A Signal Subspace Approach to Multiple Emitter Location and Spectral Estimation, Ph.D. dissertation, Department of Electrical Engineering, Stanford University, Stanford, Calif.

Schmidt, R.D., (1986) Multiple Emitter Location and Signal Parameter Estimation, *IEEE Trans. Antennas and Propagation,* vol. AP-34, pp. 276-280.

Schott, J.P. and McClellan, J.H., (1984) Maximum Entropy Power Spectrum Estimation with Uncertainty in Correlation Measurements, *IEEE Trans. Acoustics, Speech, and Signal Processing,* vol. ASSP-32, pp. 410-418.

Swingler, D.N., (1979a) A Comparison Between Burg's Maximum Entropy Method and a Nonrecursive Technique for the Spectral Analysis of Deterministic Signals, *J. Geophys. Res.*, vol. 84, pp. 679-685.

Swingler, D.N., (1979b) A Modified Burg Algorithm for Maximum Entropy Spectral Analysis, *Proc. IEEE,* vol 67, pp. 1368-1369.

Swingler, D.N., (1980) Frequency Errors in MEM Processing, *IEEE Trans. Acoustics, Speech, and Signal Processing,* vol. ASSP-28, pp. 257-259.

Thorvaldsen, T., (1981) A Comparison of the Least Squares Method and the Burg Method for Autoregressive Spectral Analysis, *IEEE Trans. Antennas and Propagation,* vol. AP-29, pp. 675-679.

Tong, H., (1975) Autoregressive Model Fitting with Noisy Data by Akaike's Information Criterion, *IEEE Trans. Information Theory,* vol. IT-21, pp. 476-480.

Tong, H., (1977) More on Autoregressive Model Fitting with Noisy Data by Akaike's Information Criterion, *IEEE Trans. Information Theory,* vol. IT-23, pp. 409-410.

Ulrych, T.J. and Bishop, T.N., (1975) Maximum Entropy Spectral Analysis and Autoregressive Decomposition, *Rev. Geophys. Space Phys.*, vol. 13, pp. 183-200.

Ulrych, T.J., and Clayton, R.W., (1976) Time Series Modelling and Maximum Entropy, *Phys. Earth Planet Inter.*, vol. 12, pp. 188-200.

Wax, M. and Kailath, T., (1985) Detection of Signals by Information Theoretic Criteria, *IEEE Trans. Acoust. Speech Sig. Proc.*, vol. ASSP-33, pp. 387-392.

Wold, H., (1938) A Study in the Analysis of Stationary Time Series, reprinted by Almquist and Wichsells Forlag, Sweden, 1954.

WEIGHT VARIANCE AND STATISTICAL EFFICIENCY IN LEAST SQUARES ADAPTIVE ANTENNAS

J.E. Hudson
(STC Technology Ltd., Essex)

1. INTRODUCTION

The optimum detection procedure for a plane wave in uncorrelated noise consists of a beamforming operation with uniform modulus phase-compensation weights and this ties in with maximum likelihood methods, matched filtering, best SNR etc. When the noise becomes coloured or there are one or more discrete interference sources the situation remains simple in theory but is very complicated in practice. Statistical concepts lead us to compute the likelihood ratio for the data X under signal+noise (S+N) and noise-only (N) hypotheses:

$$p(X|S+N)/P(X|N)$$

and this is one route to the adaptive beamformer [1], in which the weight vector derived is $W = \lambda M^{-1}C$ where M is the covariance of the (real-valued) noise and interference and C is the steering vector which compensates the desired signal time delays, defined as the expected correlation between the desired signal waveform and the X-vector. If M can be measured without corruption by the desired signal then all is well and near-optimum detection is possible. The problem is that in many potential applications the desired signal is present all the time and the interference cannot be estimated in isolation. This paper is concerned with the performance of "optimal" systems under these conditions.

2. LEAST SQUARES SOLUTION

The least squares "Capon" or minimum distortion beamformer is based on the concept that if the desired signal of power π_s has dyadic covariance $\pi_s CC^T$ and the optimum weight vector is known to be

$$\lambda M^{-1}C$$

then when the signal-corrupted matrix is used instead:

$$W = \lambda(M+\pi_s CC^T)^{-1}C$$

straightforward use of the matrix inversion lemma shows that the weight vector is not changed in direction but merely scaled, so the solution is still optimum [2].

However, under practical sampling conditions, the weight vector is actually greatly perturbed from the optimum condition and this is due to two effects: firstly, weight variance or jitter, the topic of the present paper; and secondly, sensitivity to errors in the steering vector C which cause suppression of the desired signal. The latter topic has been treated in detail elsewhere [3]. Signal-induced weight jitter is discussed by Boroson [4] who surveys earlier work and derives exact expressions for the output SNR probability distribution when the signal-present sample covariance matrix is used, and shows that it follows a Beta distribution with a mean of $(K-N+2)/(K+1)$ where K is the number of samples and N is the number of complex weights in the array. His results are specific to all-Gaussian data.

This paper takes the different route of modifying standard least squares parameter covariance theory (eg. Silvey [5]) and offers a different range of solutions which are distribution independent. We start with a conceptual data-vector generator X_i and an associated vector of fixed combination weights W_o which, with the addition of an independent error term e_i, outputs an observation y_i:

$$y_i = X_i^T W_o + e_i \qquad i = 1,\ldots,K \tag{1}$$

The error term is not defined in the usual way as independent noise, rather it is the least-squares estimate of the error as the sample size goes to infinity and is data dependent and problem specific. Using this definition imparts to it all the necessary asymptotic properties of an error signal for least squares purposes.

Eqn. 1 is written in the vector form

$$Y = XW_o + E_o \tag{2}$$

where Y and E_o are K-vectors, X is a KxN data matrix and W_o is an N-vector. The standard least-squares solution to estimate W_o is

$$\hat{W} = (X^T X)^{-1} X^T Y \tag{3}$$

and this is associated with an estimated residual

$$\hat{E} = Y - X\hat{W} = [I - X(X^T X)^{-1} X^T] Y \tag{4}$$

which is the projection of Y onto a subspace orthogonal to the columns of X. This orthogonality of $\hat{E}$ is shown by combining (3) and (4):

$$X^T \hat{E} = X^T Y - X^T X (X^T X)^{-1} X^T Y \equiv 0$$

and, when $K \to \infty$ and $\hat{E} \to E_o$, the implied orthogonality of E_o is a necessary property for an error in the least squares method. The relationship to an adaptive beamformer is that Y represents the main channel signal and is the output of a conventional beam pointing to the desired signal while the columns of X are the auxiliary channel signals and could be the outputs of beams pointing in various other directions or possibly outputs of omnidirectional sensors. The adapted output is the residual vector $\hat{E}$ which, hopefully, contains the desired signal and the least squares condition corresponds to minimum output variance.

3. WEIGHT COVARIANCE

Define

$$\mathrm{Cov}(\hat{W}) = \mathrm{Cov}[(X^T X)^{-1} X^T Y]$$

from (3). Substitution of Y from (2) yields

$$\mathrm{Cov}(\hat{W}) = \mathrm{Cov}[(X^T X)^{-1} X^T (X W_o + E_o)]$$

$$= \mathrm{Cov}[W_o + (X^T X)^{-1} X^T E_o]$$

W_o is obviously fixed and may be ignored, and if E_o is statistically independent of X (it has been shown to be uncorrelated) then, for a given data matrix X,

$$\mathrm{Cov}(\hat{W}) = (X^T X)^{-1} X^T \mathrm{Cov}[E_o] X (X^T X)^{-1} \tag{5}$$

which is a well-known result. Further averaging over the ensemble of data yields

$$\mathrm{Cov}(\hat{W}) = \langle (X^T X)^{-1} X^T \mathrm{Cov}[E_o] X (X^T X)^{-1} \rangle \tag{6}$$

provided such a mean exists.

If E_o, the error sequence, is sequentially uncorrelated, then $Cov[E_o] = \sigma_o^2 I$ and

$$Cov(\hat{W}) = \sigma_o^2 <(X^T X)^{-1}> \tag{7}$$

However, for signals correlated in time, the full expression (6) must be used.

The excess output power due to the weight covariance is

$$P_e = <(\hat{W}-W_o)^T M(\hat{W}-W_o)> = Tr[M.Cov[\hat{W}]]$$

$$= \sigma_o^2 Tr[M.<(X^T X)^{-1}>]$$

Since $<(X^T X)^{-1}>$ converges on M^{-1}/K it can be argued that P_e converges on $\sigma_o^2 Tr[I/K] = \sigma_o^2 N/K$ though there is some approximation here since the mean of $(X^T X)^{-1}$ is found, in practice, to be not well-defined. However $1/P_e$ is well defined and the output SNR of the beamformer, if the optimum output E_o is mostly the desired signal while the excess output is mostly interference, is given by

$$SNR \simeq \sigma^2/P_e = K/N$$

after K samples. Within a small factor, significant only for small K, this expression agrees with the exact result of Boroson for Gaussian signals. It has the advantage that the covariance of the weights is obtained directly which gives additional information on the spatial pattern fluctuation. Moreover expressions can be derived for autocorrelated data via equation (5).

4. STATISTICAL EFFICIENCY

The Cramer-Rao lower bound (CRB) on the estimation, by an N element array, of the amplitude α of a single sample of a sinusoidal source immersed in uncorrelated complex noise of unit analytic variance σ^2 is

$$Var(\hat{\alpha}) \geq \sigma^2/N$$

Also, as is well known, a conventional beamformer is an optimum estimator for a single source under these conditions and it would have an output SNR of $|\alpha|^2 N/\sigma^2$, which is compatible with the CRB variance.

If the equivalent CRB is computed numerically for the case that there are also one or more discrete interference sources present then it is found that very little degradation in accuracy is predicted relative to the no-interference case, provided the interference lies outside the main conventional lobe of the array. Thus we can approximate the multiple-source CRB by the no-interference CRB.

However, when we turn to the "Capon" least-squares beamformer, we find that the ouput SNR calculations based on the weight jitter results are much worse than this value. In fact the statistical efficiency is bounded above by

$$\gamma = \mathrm{var}(\hat{\alpha})_{CRB}/\mathrm{Var}(\hat{\alpha})_{LS} = SNR_{LS}/SNR_{CRB} = \frac{K\sigma^2}{|\alpha|^2 N^2}$$

and only approaches optimality ($\gamma \simeq 1$) if $K \geq |\alpha|^2 N^2/\sigma^2$ which can correspond to a very large number of samples when the desired signal is strong ($|\alpha|^2 >> \sigma^2$). In this sense the Capon beamformer cannot always be regarded as a statistically efficient solution to the interference cancellation problem for small sample numbers.

5. CONCLUSIONS

The weight covariance of a least squares adaptive beamformer, adapted in the presence of the desired signal, has been derived for non-Gaussian signals, both for white-spectrum and autocorrelated signals, and associated SNR computations have been found to give results which are compatible with corresponding SNR probability distributions derived previously by Boroson.

It has also been noted that the least squares adaptive solution will not achieve the Cramer-Rao bound on signal parameter estimation accuracy when adaption occurs in the presence of the desired signal and so it cannot be regarded as an efficient estimator in the statistical sense. Alternative adaptive beamformer criteria can be devised, for example it has been found by simulation in [7] that the process of interference direction finding followed by "open-loop" null-steering can give SNR results which approach the CRB on estimation accuracy for small sample numbers even when the desired signal is present and in retrospect it appears that the ability to do such null steering, which is closely related to applying the Maximum Likelihood method, forms an implicit assumption in the derivation of the CRB.

In practice we can circumvent the problem rather easily by deliberately selecting situations such that the desired signal is absent during the period when the antenna estimates the noise and adapts its weights. These are frozen before the desired signal returns. Such conditions are readily realisable in modern digital communications applications.

6. REFERENCES

[1] Reed, I.S. et al., (1974) Rapid convergence rate in adaptive arrays. IEEE Trans. AES-10, pp. 853-863.

[2] Monzingo, R.A. and Miller, T.W., (1980) Introduction to adaptive arrays, Wiley.

[3] Hudson, J.E., (1981) Adaptive Array Principles, Peter Peregrinus.

[4] Boroson, D.M., (1980) Sample size considerations in adaptive arrays, IEEE Trans. AES-16, pp. 446-451.

[5] Silvey, S.D., (1978) Statistical Inference, Chapman & Hall.

[6] Rife, D.C. and Boorstyn, R.R., (1976) Multiple tone parameter estimation from discrete time observations, Bell Syst. Tech. J. 55, pp. 1389-1410.

[7] Hudson, J.E., (1985) Antenna adaptivity by direction finding and null steering, IEE Proc-H. 132, pp. 307-311.

7. ACKNOWLEDGMENT

The author is grateful to the Directors of STC Technology Ltd. for permission to publish this paper which is largely based on work supported by the Procurement Executive, Ministry of Defence.

STATISTICS OF GENERALIZED EIGENVALUES FOR SIGNAL PARAMETER ESTIMATION

R.N. Madan
(Office of Naval Research, U.S.A.)

and

S.U. Pillai*
(Department of Electrical Engineering, Polytechnic University, New York, U.S.A.)

ABSTRACT

This paper presents an asymptotic analysis of a class of high resolution estimators for resolving correlated and coherent plane waves in noise. These estimators are in turn constructed from certain eigenvectors associated with the covariance matrix generated from a uniform array. This is carried out by observing a well-known property of the signal subspace; i.e. in presence of uncorrelated and identical sensor noise, the subspace spanned by the true direction vectors coincides with the one spanned by the eigenvectors corresponding to all, except the smallest set of repeating eigenvalue of the array output covariance matrix. Further, the bias expressions are used to obtain a meaningful resolution threshold for two equipowered plane waves in white noise, and the result is compared to the one derived by Kaveh et al. [4] for two uncorrelated, equipowered plane waves.

1. INTRODUCTION

In recent times, multiple signal identification utilizing eigenstructure-based techniques has been a topic of considerable research interest in array signal processing. A variety of high resolution techniques that evaluate the directions-of-arrival of incoming planar wavefronts by exploiting certain special eigenstructure properties associated with the sensor array output covariance matrix have been developed.

---

*Dr. R.N. Madan is with the Office of Naval Research, 800 N. Quincy St., Arlington, VA and S.U. Pillai is with the Department of Electrical Engineering, Polytechnic University, 333 Jay St., Brooklyn, New York. This work was supported by the Office of Naval Research under contract N-00014-86-K-0321.

Of these, the relatively new scheme ESPRIT (Estimation of Signal Parameters via Rotational Invariance Techniques) [2]-[3] departs from its predecessors on several important accounts. It utilizes an underlying rotational invariance among signal subspaces induced by subsets of an array of sensors.

To accomplish this, the interelement covariances among the given sensors are used to construct the auto-and cross-covariance matrices and the common noise variance is first evaluated by an eigendecomposition of the auto-covariance matrix. After subtracting the noise variance from the proper elements of the auto- and cross-covariance matrices, the generalized eigenvalues of a matrix pencil formed from the subtracted matrices are computed and they in turn uniquely determine the unknown directions-of-arrival [2]. Compared to the Multiple Signal Classification (MUSIC) technique [1], the ESPRIT scheme is known to reduce the computation and storage costs dramatically. In addition, this method is also shown to be more robust with regard to array imperfections than most of the earlier ones.

Notwithstanding these merits, when estimates of the inter-element covariances are used in these computations, subtracting the estimated noise variance from the auto- and cross-covariance matrices can at times be critical and may result in overall inferior results. To circumvent this difficulty, the TLS-ESPRIT (Total Least Squares ESPRIT) scheme makes use of certain overlapping subarray output- and their cross-covariance matrices simultaneously [3]. Though TLS-ESPRIT is superior in its performance compared to ESPRIT, it is computationally much more complex. However, computational simplicity can be maintained without sacrificing superior performance as described in section II. Results of a first-order perturbation analysis are presented in section III exhibiting the mean and variance corresponding to the new technique, when covariances are estimated from the data directly. These results are then used to derive resolution thresholds associated with two closely-spaced equipowered sources.

2. PROBLEM FORMULATION

Let an uniform array consisting of M sensors receive signals from K narrowband sources $u_1(t), u_2(t), \ldots, u_K(t)$, which are at most partially correlated. Further, the respective arrival angles are assumed to be $\theta_1, \theta_2, \ldots, \theta_K$ with respect to the line of the array. Using complex signal representation, the received signal $x_i(t)$ at the ith sensor can be expressed as

$$x_i(t) = \sum_{k=1}^{K} u_k(t) e^{-j\pi(i-1)\cos\theta_k} + n_i(t). \qquad (1)$$

It is assumed that the signals and noises are stationary, zero-mean circular Gaussian independent random processes, and further the noises are assumed to be independent and identical between themselves with common variance σ^2. Rewriting (1) in common vector notation and with $\omega_k = \pi\cos\theta_k$; $k = 1,2,\ldots,K$, the array output vector $\mathbf{x}(t)$ with size M x 1 can be written as[1]

$$\mathbf{x}(t) = \left[x_1(t), x_2(t), \ldots, x_M(t)\right]^T = \mathbf{A}\,\mathbf{u}(t) + \mathbf{n}(t) \tag{2}$$

where

$$\mathbf{u}(t) = \left[u_1(t), u_2(t), \ldots, u_K(t)\right]^T, \tag{3}$$

$$\mathbf{n}(t) = \left[n_1(t), n_2(t), \ldots, n_M(t)\right]^T \tag{4}$$

$$\mathbf{A} = \sqrt{M}\left[\mathbf{a}(\omega_1), \mathbf{a}(\omega_2), \ldots, \mathbf{a}(\omega_K)\right] \tag{5}$$

and $\mathbf{a}(\omega_k)$ is the normalized direction vector associated with the arrival angle θ_k;i.e.,

$$\mathbf{a}(\omega_k) = \frac{1}{\sqrt{M}}\left[1, \mu_k^{-1}, \mu_k^{-2}, \ldots, \mu_k^{-(M-1)}\right]^T \; ; \quad \mu_k = e^{j\omega_k}. \tag{6}$$

Here $\mathbf{A}$ is an M x K matrix with Vandermomde-structured columns (M>K) of rank K. From our assumptions it follows that the array output covariance matrix has the form

$$\mathbf{R} \triangleq E[\mathbf{x}(t)\mathbf{x}^\dagger(t)] = \mathbf{A}\mathbf{R}_u\mathbf{A}^\dagger + \sigma^2\mathbf{I} \tag{7}$$

where

$$\mathbf{R}_u \triangleq E[\mathbf{u}(t)\mathbf{u}^\dagger(t)] \tag{8}$$

represents the source covariance matrix which remains as nonsingular so long as the sources are at most partially correlated. In that case $\mathbf{A}\mathbf{R}_u\mathbf{A}^\dagger$ is also of rank K and hence, if $\lambda_1 \geq \lambda_2 \geq \ldots \geq \lambda_M$ and $\beta_1, \beta_2, \ldots, \beta_M$ represent the eigenvalues and the corresponding eigenvectors of $\mathbf{R}$ respectively, i.e.,

---

(1) Here onwards $\mathbf{A}^T$, $\mathbf{A}^{*T} \triangleq \mathbf{A}^\dagger$ stand for the transpose and the complex conjugate transpose of $\mathbf{A}$, respectively.

$$\mathbf{R} = \sum_{l=1}^{M} \lambda_l \boldsymbol{\beta}_l \boldsymbol{\beta}_l^{\dagger}, \tag{9}$$

then the above rank property implies that $\lambda_{K+1} = \lambda_{K+2} = \ldots = \lambda_M =$ As a result, we have

$$\boldsymbol{\beta}_i^{\dagger} \mathbf{a}(\omega_k) = 0, \; i = K+1,\ldots,M, \; k = 1,2,\ldots,K. \tag{10}$$

A new outlook can be developed using (10). Since the K true direction vectors $\mathbf{a}(\omega_1), \mathbf{a}(\omega_2), \ldots, \mathbf{a}(\omega_K)$ are linearly independent, they span a K dimensional proper subspace called the signal subspace. Further, from (10), this subspace is orthogonal to the subspace spanned by the eigenvectors $\boldsymbol{\beta}_{K+1}, \boldsymbol{\beta}_{K+2}, \ldots, \boldsymbol{\beta}_M$, implying that the signal subspace spanned by $\mathbf{a}(\omega_1), \mathbf{a}(\omega_2), \ldots, \mathbf{a}(\omega_K)$ coincides with that spanned by the eigenvectors $\beta_1, \beta_2, \ldots, \beta_K$. Using this crucial observation, the eigenvectors $\beta_1, \beta_2, \ldots, {}_K$ in the signal subspace can be expressed as a linear combination of the true direction vectors (columns of $\mathbf{A}$); i.e.,

$$\boldsymbol{\beta}_i = \sum_{k=1}^{K} \tilde{c}_{ki} \mathbf{a}(\omega_k), \; i = 1,2,\ldots,K. \tag{11}$$

Define the M x K signal subspace eigenvector matrix as

$$\tilde{\mathbf{B}} \triangleq \left[\beta_1, \beta_2, \ldots, \beta_K\right]. \tag{12}$$

Using (11)

$$\tilde{\mathbf{B}} = \left[\sum_{k=1}^{K} \tilde{c}_{k1} \mathbf{a}(\omega_k), \; \sum_{k=1}^{K} \tilde{c}_{k2} \mathbf{a}(\omega_k(, \ldots, \; \sum_{k=1}^{K} \tilde{c}_{kK} \mathbf{a}(\omega_k)\right] = \mathbf{A}\tilde{\mathbf{C}} \tag{13}$$

where $\mathbf{A}$ is as defined in (5) and $\tilde{\mathbf{C}}$ is a KxK nonsingular matrix whose $(i,j)^{th}$ element is $\tilde{c}_{ij}/\sqrt{M}$. Further, define two matrices $\tilde{\mathbf{B}}_1$ and $\tilde{\mathbf{B}}_2$ using the first L rows and the 2nd to $(L+1)^{th}$ rows of $\tilde{\mathbf{B}}$ respectively where $K \le L \le M-1$; i.e.

$$\tilde{\mathbf{B}}_1 = \left[\mathbf{I}_L \mid \mathbf{O}_{L,M-1}\right]\tilde{\mathbf{B}} \tag{14}$$

and

$$\tilde{\mathbf{B}}_2 = \left[\mathbf{O}_{L,1} \mid \mathbf{I}_L \mid \mathbf{O}_{L,M-L-1}\right]\tilde{\mathbf{B}}. \tag{15}$$

Then, we have the following interesting result.[2]

Theorem: Let γ_i represent the generalized singular values associated with the matrix pencil $\{\tilde{\mathbf{B}}_1, \tilde{\mathbf{B}}_2\}$. Then

$$\gamma_k = \mu_k, \quad k = 1,2,\ldots,K. \tag{16}$$

Proof: See [6].

Simulation results using this procedure is presented in Fig. 1 for a four-source scene with details as indicated there.

3. PERFORMANCE ANALYSIS

In this section, we examine the statistical behaviour of the estimated generalized eigenvalues in a single-source case and a two-source case for the least favourable configuration $L = K$ as well as the most favourable configuration $L = M - 1$. These results are subsequently used in deriving associated threshold expressions for resolving two closely-spaced sources. For $L = K$, it is shown here that the bias of the estimated generalized eigenvalues is zero and the variance is nonzero within a $1/N$ approximation. The behaviour is unlike the MUSIC scheme where within a $1/N$ approximation, the bias is nonzero and the variance is zero [4,5]. For $L > K$, the situation is considerably more complicated. In particular, it can be shown that for $L = M - 1$ the estimated generalized eigenvalues are no longer unbiased in a two-source scene [6].

Case 1: The Least Favourable Configuration ($L = K$).

A. Single-Source Scene

The mean and variance of the estimated generalized eigenvalue $\hat{\gamma}_1$ for a single source scene are shown to be [6]

$$E[\hat{\gamma}_1] = \gamma_1 + O\left(\frac{1}{N\sqrt{N}}\right) \tag{17}$$

and

$$\mathrm{Var}(\hat{\gamma}_1) = \frac{2M}{N}\left[\frac{1}{\xi} + \frac{1}{\xi^2}\right] + O\left(\frac{1}{N\sqrt{N}}\right) \tag{18}$$

where $\xi = M\,P/\sigma^2$ represents the array output signal-to-noise ratio and $O(1/N\sqrt{N})$ denotes the terms of order less than $1/N$.

---

(2) Here $\mathbf{I}_K$ represents the KxK identity matrix and $\mathbf{O}_{K,J}$ represents the KxJ matrix with all zero entries.

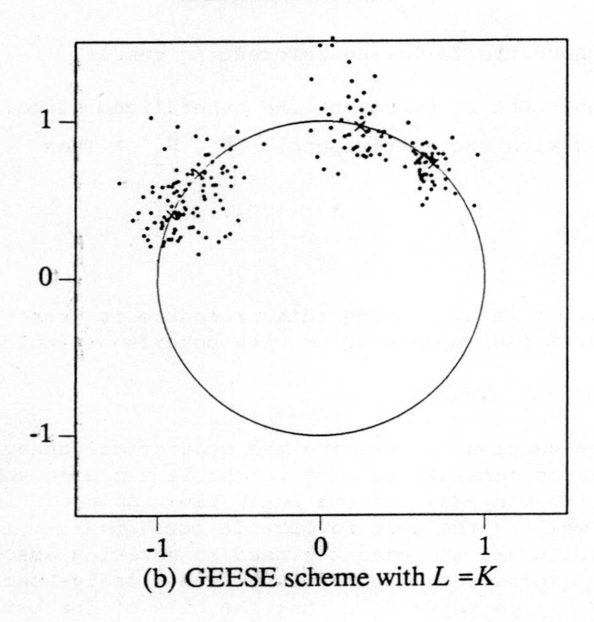

(b) GEESE scheme with $L = K$

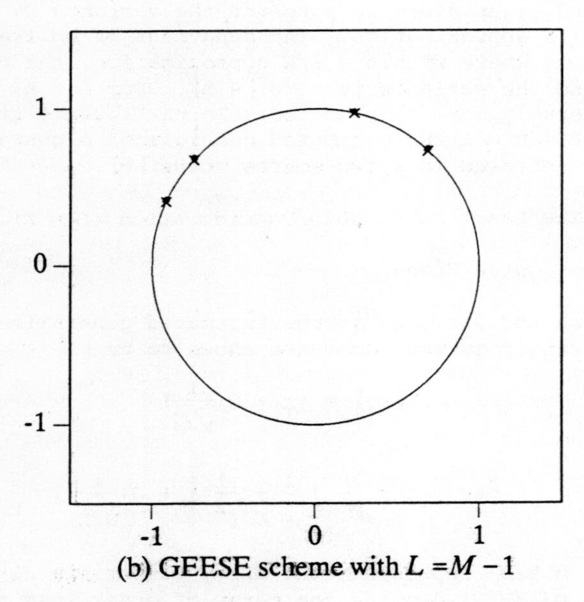

(b) GEESE scheme with $L = M - 1$

Fig. 1 Simulation results for a mixed-source scene. The two sources located at 40°, 50° are uncorrelated and the two sources at 85°, 95° are correlated with a correlation coefficient (0.24,-0.65). A even-element array is used to receive the signals. Input SNR is taken to be 12dB (number of simulations = 50, number of samples = 100).

B. Two-Source Scene

With the help of Appendix A in [5], the mean and variance of the estimated generalized eigenvalues $\hat{\gamma}_i$, $i = 1,2$ in two equipowered uncorrelated source scene are shown to be [6]

$$E[\hat{\gamma}_i] = \gamma_i + O(\frac{1}{N\sqrt{N}}); \quad i = 1,2 \tag{19}$$

and

$$\mathrm{Var}(\hat{\gamma}_i) = \frac{M(2+\cos 2\omega_d)}{N(1-\cos 2\omega_d)}\left[\frac{1}{\xi} + \frac{1}{\xi \quad 1-|\rho_s|^2}\right] + O(\frac{1}{N\sqrt{N}}) \quad i = 1,2 \tag{20}$$

where

$$\rho_s = \mathbf{a}^{\dagger}(\omega_1)\mathbf{a}(\omega_2) = e^{j(M-1)\omega_d}\frac{\sin M\omega_d}{M\sin\omega_d}, \quad \omega_d = \frac{\omega_1-\omega_2}{2}. \tag{21}$$

These expressions can be used to determine the resolution threshold associated with two closely spaced sources. For a specific input SNR, the resolution threshold represents the minimum amount of angular separation required to identify the sources as separate entities unambiguously. From (19) and (20), since the standard derivation of $\hat{\gamma}_i$, $i = 1,2$ is substantially larger than their respective bias, it is clear that the resolution threshold is mostly determined by the behaviour of the standard deviation. In order to obtain the measure of the resolution threshold for two closely spaced sources, consider the situation shown in Fig. 2. Evidently, the sources are resolvable if $\hat{\gamma}_1$ and $\hat{\gamma}_2$ are both inside the cones C_1 and C_2 respectively or equivalently if $|\arg(\hat{\gamma}_i)-\arg(\gamma_i)| < \omega_d$, $i = 1,2$. Exact calculations based on this criterion turns out to be rather tedious. But as computation results in Fig. 1 show, $\hat{\gamma}_1$ and $\hat{\gamma}_2$ are usually within a small circular neighbourhood centred about γ_1 and γ_2. This suggests a more conservative criterion for resolution, i.e., the sources are resolvable if $\hat{\gamma}_1$ and $\hat{\gamma}_2$ are both inside the circles c_1 and c_2 respectively in Fig. 2. In that case, the maximum value of the common radii of these circles is easily shown to be $\sin \omega_d$. Thus, at an SNR satisfying

$$\sqrt{\mathrm{Var}(\hat{\gamma}_i)} = \ell \sin \omega_d \tag{22}$$

where ℓ is some positive integer. Using (22) we finally have the associated threshold SNR to be

$$\xi_{\ell,K} = \frac{c(M,\omega_d)}{N}\left[1 + \left(1 + \frac{2N}{c(M,\omega_d)(1 - |\rho_s|^2)}\right)^{1/2}\right]. \quad (23)$$

where

$$c(M,\omega_d) = \frac{M(2 + \cos 2\omega_d)}{2\ell^2(1 - \cos 2\omega_d)\sin^2\omega_d}.$$

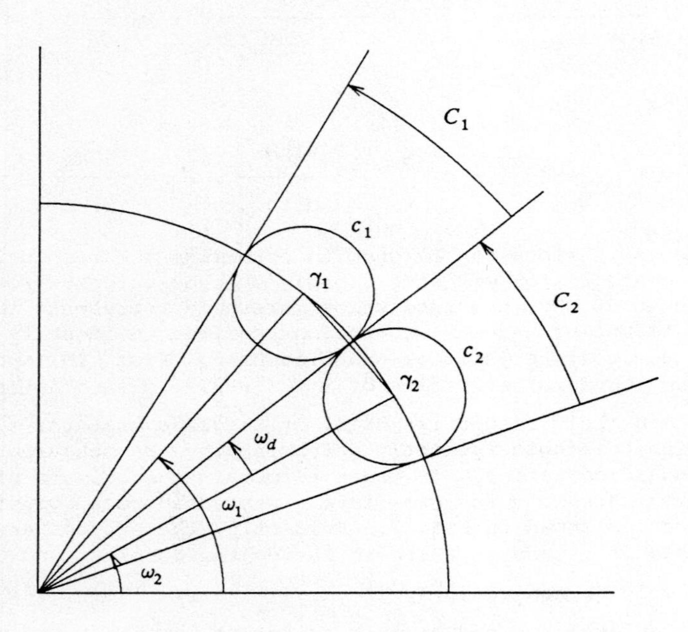

Fig. 2 Resolution threshold analysis

This threshold SNR can also be expressed in terms of the "effective angular separation" parameter Δ^2 defined as [4]

$$\Delta^2 = M^2\omega_d^2/3.$$

For closely-spaced sources, $M^2\omega_d^2 << 1$, and we have

$$\xi_{\ell,K} \approx \frac{M}{6\ell^2 N}\left(\frac{M^4}{2\Delta^4} - \frac{M^2}{\Delta^2}\right)\left[1 + \left(1 + \frac{24\ell^2 N\Delta^2}{M^5}\right)^{1/2}\right]. \quad (24)$$

Notice that calculations for Var $(\hat{\gamma}_i)$ in (20) have been carried out for L = K(=2) case and hence the above threshold expression also corresponds to this least favourable configuration, which only uses part of the available signal subspace eigenvector information in its computations. When higher values of L are used to evaluate $\hat{\gamma}_i$, the corresponding threshold expressions also should turn out to be superior to that in (23). Similar threshold comparisons are carried out in Fig. 3 for the MUSIC scheme and the proposed GEESE scheme. In the case of two uncorrelated sources, by equating the actual bias at the true arrival and middle angles Kaveh et al. has shown the resolution threshold for the standard MUSIC scheme to be [4]

$$\xi_M \approx \frac{1}{N}\left[\frac{20(M-2)}{\Delta^4}\left\{1+\left[1+\frac{N}{5(M-2)}\Delta^2\right]^{1/2}\right\}\right]. \qquad (25)$$

Fig. 3 shows such a comparison using (24) and (25) with $\ell = 2$. The corresponding SNR values are observed to have at least 30 percent probability of resolution.

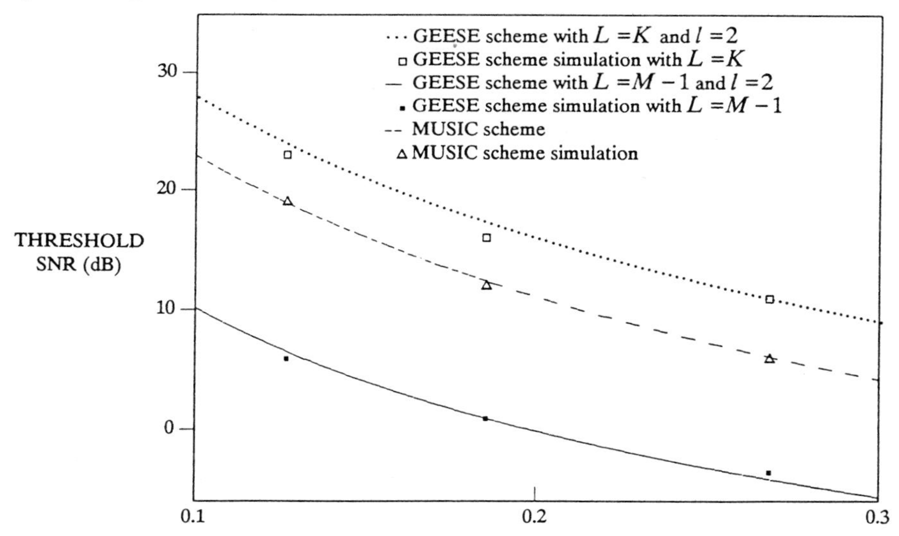

Fig. 3 Resolution threshold vs angular separation for two equipowered sources. A seven element array is used to receive signals in both cases. One hundred snapshots are taken for each simulation. In each simulation, the SNR with 30 percent probability of resolution is taken.

Case 2: The Most Favourable Configuration (L = M - 1).

For L > K, the situation is much more complex and the estimators, $\hat{\gamma}_i$, i = 1,2 are no longer unbiased even within 1/N approximation. The exact bias and variance expressions can be computed. These calculations have been carried out for the most favourable configuration (L = M - 1) in a single-source scene as well as a two-source scene and the results are summarized in [6].

Once again, after a long series of algebraic manipulations, the associated threshold SNR in this case can be shown to be

$$\xi_{\ell,M-1} \approx \frac{M^2 \nu(M,\Delta^2)}{6\ell^2 N\Delta^2}\left[1 + \left(1 + \frac{2\ell^2 N(M-1)^6 \Delta^2}{M^5 (M^2 - 2)}\right)^{1/2}\right], \quad (26)$$

where

$$\nu(M,\Delta^2) \sim \frac{2}{M}\left(\frac{3}{\Delta^4} + \frac{2}{5\Delta^2} - 2\right). \quad (27)$$

For the same source scene and probability of resolution discussed in Fig. 3, new simulation results are presented there for this most favourable configuration. As remarked earlier, the SNR required to resolve two sources in this case is seen to be substantially smaller than that in the former case (L = K). Once again, utilization of all available information in this (L = M - 1) case may be attributed to its superior performance. From these results, it may be reasonably concluded that when all available signal subspace information is exploited, GEESE scheme outperforms the MUSIC scheme.

4. CONCLUSIONS

This paper describes a new technique and its performance analysis for estimating the directions-of-arrival of correlated signals by making use of certain matrices associated with the signal subspace eigenvectors of the array output covariance matrix. This is based on the well-known property that in the case of uncorrelated and identical noise field, the subspace spanned by the true direction vectors is identical to the signal subspace (i.e., the one spanned by the eigenvectors associated with all, except the lowest repeating eigenvalue of the array output covariance matrix). Using a first-order asymptotic analysis, it is shown here that the angle-of-arrival estimator in its least favourable form is unbiased and has nonzero variance in a two-source scene. Although the estimator in its most favourable configuration turns out to be biased, the associated resolution threshold in an equipowered two-source scene is shown to be substantially smaller than that corresponding to the standard MUSIC scheme.

REFERENCES

[1] Schmidt, R.O., (1979) Multiple emitter location and signal parameter estimation, in *Proc. RADC Spectral Est. Workshop*, p. 243-258.

[2] Roy, R., Paulraj, A. and Kailath, T., (1986) ESPRIT - a subspace rotation approach to estimation of parameters of cisoids in noise, in *IEEE Trans. Acoust. Speech, Signal Processing*, vol. ASSP-34, no. 4, pp. 1340-1342.

[3] Roy, R. and Kailath, T., (1987) ESPRIT and total least square, in *Proc. 21st Asilomar Conf.*

[4] Kaveh, M. and Barabell, A.J., (1986) The statistical Performance of the MUSIC and the minimum-norm algorithms in resolving plane waves in noise, *IEEE Trans. Acoust., Speech, Signal Processing*, vol. ASSP-34, no. 2, pp. 331-341.

[5] Pillai, S.U., and Kwon, B.H., (1989) Performance analysis of MUSIC-type high resolution estimators for direction finding in correlated and coherent scenes, *IEEE Trans. Acoust., Speech, Signal Processing*, vol. ASSP-37, no. 8, pp. 1176-1189.

[6] Kwon, B.H., (1989) New high resolution techniques and their performance analysis for angles-of-arrival estimation, Ph.D. dissertation, Polytechnic University.

A BAYESIAN METHOD FOR ADAPTIVE SPECTRUM ESTIMATION USING HIGH ORDER AUTOREGRESSIVE MODELS

A. Houacine
(Institut d'Électronique,
Université des Sciences et de la Technologie d'Alger, Algeria)

and

G. Demoment
(Laboratoire des Signaux et Systèmes, France)

ABSTRACT

Autoregressive (AR) spectrum analysis of non-stationary signals is considered here by assuming a local stationarity of the process studied. To achieve a good spectral resolution when the analysis window has a short length, we use high order AR models whose parameters are estimated with a Bayesian technique to overcome the ill-conditioning of conventional least-squares methods. The method processes the data by blocks and uses smoothness priors or constraints to stabilize the solution. The necessary balance between the fidelity to the data and the fidelity to the priors is tuned through the choice of a regularization coefficient whose value is determined by maximising a marginal likelihood function. A fast implementation is obtained for a particular class of priors with a special fast Kalman filter. The performances of the method are illustrated by examples of spectrum estimation for either short length stationary signals or non-stationary signals.

1. INTRODUCTION

The problem of interest here is the spectral analysis of non-stationary time series. Since the spectrum characteristics are not assummed to be time-invariant, something must be specified, instead, on how fast they vary. Adaptive spectral analysis methods are based on local invariance assumptions on the studied process, which corresponds to slow spectrum distribution variations if compared to the signal amplitude ones. The weakly non-stationary data are then often processed by methods of the stationary case where the filter memory is limited by a rectangular or an exponential window [12]. The necessary specification then appears as a design parameter, the analysis window length N, and preclassification with respect to the degree of non-stationarity has to be done first.

In the situation where only a short data span is available, conventional parametric methods become statistically unreliable, and the resulting spectra suffer from the limitations of these estimators. Let us consider a real random process described by the following AR model:

$$y(n) = z(n) + b(n) \qquad n = 1, 2, \ldots N \tag{1.1}$$

$$z(n) = \sum_{i=1}^{p} a_i \, y(n-i) = \mathbf{a}^t \mathbf{y}_p(n) \tag{1.2}$$

$$\mathbf{a}^t = [a_p, a_{p-1}, \ldots, a_1] \tag{1.3}$$

$$\mathbf{y}_p(n)^t = [y(n-p), \ldots, y(n-1)] \tag{1.4}$$

where $\mathbf{a}$ is the model parameter vector of order p, and b(n) is a real white Gaussian noise with variance σ^2 describing the innovation process. Least-squares methods are widely used to estimate the AR coefficients because they lead to a low computational burden when implemented with fast recursive algorithms [6, 12]. They minimize the following criterion:

$$J_1(\mathbf{a}) = \sum_{n=1}^{N} [y(n) - z(n)]^2 \tag{1.5}$$

The main problem in these methods is the choice of the model order p. The AR model is in general a rough description of the actual signal, and high order models would be necessary to give a good approximation. When the model order becomes relatively high (p equal to N for instance), however, spectrum estimation performed with least-squares methods suffers from the variability of the statistics used. The estimation problem becomes an ill-posed one [13]. The smoothness priors long AR method was developed to mitigate this problem, using regularization principles, in a series of seminal papers by Kitagawa and Gersch [10, 11]. Our method relies on the same basic principles, but differs in the choice of order. In Gersch and Kitagawa's examples, relatively low orders are used (p=20 when N=114 for instance [11]), while their method allows higher order values with, however, a heavier computational cost. But this is a rather arbitrary choice. Why not take orders such as p=25 or p=40? In our method, a maximum order (p=N or N/2 depending on the initial conditions [3]) is used to obtain a maximum potential resolution, and the solution computation is performed recursively by a special fast Kalman filter with a complexity O(p) per recursion.

Our method can be thought of as being mainly an attempt to alleviate the computational problem involved in most regularization techniques and standard Kalman filtering. Computational tractability was the motivation for the choice of an adaptive method with zero-order frequency domain smoothness priors. It allows us to remain within the framework of locally stationary signals and linear models, and to use the corresponding fast estimation algorithms.

Some numerical results are reported here to investigate the feasibility of the method, to assess its performance, and generally to show what kind of results to expect.

2. SMOOTHNESS PRIORS LONG AR MODELS

2.1 Regularization of an ill-posed problem

The regularization of an ill-posed problem is equivalent to adding some *a priori* information about the solution in order to stabilize its computation [2, 13, 14] This can be done by minimizing a modified criterion:

$$J_1(\mathbf{a}) + \mu J_2(\mathbf{a}) \qquad \mu > 0 \qquad (2.1.1)$$

where $J_2(\mathbf{a})$ is a regularizing functional providing some measure of "distance" of the solution to an *a priori*. This can be interpreted as an effort to balance the fidelity of the model to the data, and the fidelity to some priors. The regularization approach provides a family of solutions depending on the priors used and on the μ-values. There is a large variety of $J_2(\mathbf{a})$ functionals [2,14], but, for obvious practical reasons, they are often quadratic to obtain a linear-in-the-data solution. The method introduced by Kitagawa and Gersch consists of using a frequency domain smoothness measure on the solution.

2.2 Frequency domain smoothness priors

Introducing prior information must translate some physical properties of the solution. This cannot be done directly with the AR parameters $\mathbf{a}$. Even though they characterize the process, they are chosen for analytical convenience and are not easy to interpret physically. However, the ordinary least-squares solution is rejected since the unknown spectrum is expected to be significantly smoother, and measures of increasing order k of this function smoothness in the frequency domain are given by:

$$S_k = \int_{-1/2}^{1/2} \left|\frac{\partial^k H(f)}{\partial f^k}\right|^2 df = (2\pi)^{2k} \sum_{m=1}^{p} m^{2k} a_m^2 \quad k = 1, 2, \ldots \qquad (2.2.1)$$

where H(f) is the frequency transfer function of the process whitening filter:

$$H(f) = 1 - \sum_{m=1}^{p} a_m e^{-2i\pi fm} \tag{2.2.2}$$

From the definition in (2.2.1), a large value of S_k means an unsmooth frequency transfer function. In our method, we use only the zero derivative smoothness constraint (k=0) which corresponds to the following regularization functional:

$$J_2(\mathbf{a}) = \sum_{m=1}^{p} a_m^2 \tag{2.2.3}$$

and which becomes in this case the Euclidean norm of the parameter vector $\mathbf{a}$. The fundamental problem now is to choose the μ-coefficient which is a regularization coefficient.

2.3 Bayesian interpretation

The regularized least-squares problem has a simple Bayesian interpretation [1, 8, 11, 14] since minimizing (2.1.1) is equivalent to maximising:

$$L(\mathbf{a}) = \exp\{-\frac{1}{2\sigma^2}(\sum_{n=1}^{N} [y(n) - z(n)]^2 + \sum_{m=1}^{p} \mu a_m^2)\} \tag{2.3.1}$$

The regularized solution $\hat{\mathbf{a}}$ can be considered as the maximum *a posteriori* estimate defined by the following conditional data distribution:

$$f(\mathbf{y}/\mathbf{a}, \sigma^2) \propto \exp\{-\frac{1}{2\sigma^2}(\sum_{n=1}^{N} [y(n) - z(n)^2)\} \tag{2.3.2}$$

and the *a priori* distribution:

$$f(\mathbf{a}/\mu^2,\sigma^2) \propto \exp\{-\frac{\mu}{2\sigma^2}\mathbf{a}^t\mathbf{a}\} \tag{2.3.3}$$

This *a priori* distribution can therefore be considered as the generator of a family of estimates with tuning parameters μ and σ. They are hyperparameters of the problem. The main benefit to be drawn from a Bayesian interpretation is essentially that of methodology for choosing the hyperparameter values [7]. An empirical Bayes technique can be used to estimate these values. Since the model is linear and the distributions are Gaussian, the likelihood of the hyperparameters is also Gaussian and is easily obtained by "integrating the parameters out of the problem" [1]:

$$L(\sigma,\mu/\mathbf{y}) = \int f(\mathbf{y}/\mathbf{a},\sigma)\, f(\mathbf{a}/\mu,\sigma)\, d\mathbf{a} \qquad (2.3.4)$$

The AR parameter estimate is finally computed from maximum likelihood hyperparameter values. The spectrum analysis method is then truly adaptive in the sense that the estimator structure itself depends on the observed data through the priors choice.

The joint density of the observations is computed by applying sequentially Bayes' rule:

$$L(\mu,\sigma/\mathbf{y}) = f[y(1)] \prod_{n=2}^{N} f[y(n)/y(1),\ldots, y(n-1)] \qquad (2.3.5)$$

The conditional density is:

$$f[y(n)/y(1),\ldots, y(n-1)] = \frac{1}{\sqrt{2\pi r(n)}} \exp\{-\frac{e(n)^2}{2r(n)}\} \qquad (2.3.6)$$

where $e(n)$ is the process innovation. The noise variance σ^2 can be estimated explicitly. To see this, introduce the dimensionless variables:

$$r(n)' = r(n)/\sigma^2 \qquad (\sigma^2)' = 1 \qquad (2.3.7)$$

in such a way that the dependence on σ^2 becomes explicit. Then (2.3.6) becomes:

$$f[y(n)/y(1),\ldots, y(n-1)] = \frac{1}{\sqrt{2\pi\sigma^2 r(n)'}} \exp\{-\frac{e(n)^2}{2\sigma^2 r(n)'}\} \qquad (2.3.8)$$

Neither $e(n)$ nor $r(n)'$ depends on σ^2 so that the maximum with respect to σ^2 is attained for:

$$\hat{\sigma}^2 = \frac{1}{N} \sum_{i=1}^{N} \frac{e(i)^2}{r(i)'} \qquad (2.3.9)$$

The logarithm of the likelihood function has the value:

$$P(\mu/\mathbf{y}) = -\frac{1}{2} \sum_{i=1}^{N} [\frac{e(i)^2}{r(i)'} + \ln |r(i)'|] + \text{cste} \qquad (2.3.10)$$

Its dependence with respect to μ cannot be made explicit and its maximum value must be sought by iteration. This problem is suboptimally solved in three steps in our method:

- The set of possible values for μ is reduced to a finite set of discrete *a priori* chosen values.
- For each of these values, the optimal parameter vector $\hat{\mathbf{a}}$, hyperparameter $\hat{\sigma}^2$, and corresponding marginal likelihood are computed from the data. This can be recursively done with a Kalman filter [10].
- The retained model has maximum marginal likelihood on the discrete set.

3. COMPUTATIONAL ASPECTS

3.1 State-space equations for the problem

It is well-known that a standard Kalman filter may compute recursively the solution of the regularized least-squares problem [8]. However, an obstacle prevents its direct use for estimating signal characteristics. The Riccati equations that have to be solved at each recursion to obtain the Kalman gain values are not computationally attractive. To overcome this obstacle, the local stationarity assumption and shift invariance properties of the successive $\mathbf{y}_p(n)$ vectors must be exploited. To do this, let us define an extended data vector corresponding to the whole data block to be processed:

$$\mathbf{y}_m = [y(-p+1), y(-p+2), \ldots, y(0), y(1), \ldots, y(m-p)]^t \quad (3.1.1)$$

and the corresponding extended parameter vector:

$$\mathbf{a}_m(i) = [\mathbf{o}^t_{i-1}, a_p, a_{p-1}, \ldots, a_1, 0, 0, \ldots, 0]^t \quad m \geq 2p \quad (3.1.2)$$

The model equations (1.1) to (1.4) can then be replaced by the following equivalent state-space model:

$$\mathbf{a}_m(i+1) = \mathbf{D}\,\mathbf{a}_m(i) \qquad i = 1, 2, \ldots \quad (3.1.3)$$

$$y(i) = \mathbf{y}^t_m \mathbf{a}_m(i) + b(i) \quad (3.1.4)$$

where $\mathbf{D}$ is the shift matrix:

$$\mathbf{D} = \begin{bmatrix} 0 & 0 & \cdots & \cdots & \cdots \\ 1 & 0 & 0 & \cdots & \cdots \\ 0 & 1 & 0 & \cdots & \cdots \\ \cdots & \cdot & \cdot & & \cdots \\ 0 & \cdots & 0 & 1 & 0 \end{bmatrix} \quad (3.1.5)$$

In this auxiliary model, $\mathbf{D}$ and $\mathbf{y}^t_m$ are constant matrices and $b(i)$ is a stationary noise. We can therefore use Chandrasekhar equations instead of Riccati ones and derive a fast generalized Schur algorithm [3] which presents the advantages of low computational complexity and the possibility of a square-root form if necessary.

3.2 Fast generalized Schur algorithm

The recursive computation of the extended solution can be performed with the following standard Kalman filter:

$$\hat{\mathbf{a}}_m(i+1/i) = \mathbf{D}\,\hat{\mathbf{a}}_m(i/i-1) + \mathbf{k}_m(i)r(i)^{-1}[y(i) - \mathbf{y}_m^t\hat{\mathbf{a}}_m(i/i-1)] \tag{3.2.1}$$

where $\mathbf{k}_m(i)$ is the Kalman gain and $r(i)$ the variance of the innovation $[y(i) - \mathbf{y}_m^t\hat{\mathbf{a}}_m(i/i-1)]$. Chandrasekhar equations propagate the increments of the nominal filter quantities, instead of the quantities themselves, in order to compute auxiliary quantities of lower dimension. A complete deviation of our algorithm is given in [3]. The resulting equations can be compactly written using matrice arrays:

$$\left[\begin{array}{c|c} r(k+1) & \mathbf{O} \\ \mathbf{k}_m(k+1) & \mathbf{B}(k+1) \\ \mathrm{O} & \mathbf{R}^r_{k+1} \end{array}\right] = \left[\begin{array}{c|c} r(k) & \mathbf{y}_m^t\mathbf{B}(k) \\ \mathbf{k}_m(k) & \mathbf{DB}(k) \\ \mathbf{B}(k)^t\mathbf{y}_m & \mathbf{R}^r_k \end{array}\right] \left[\begin{array}{c|c} 1 & -r(k)^{-1}\mathbf{y}_m^t\mathbf{B}(k) \\ \hline -(\mathbf{R}^r_k)^{-1}\mathbf{B}(k)^t\mathbf{y}_m & \mathbf{I}_\alpha \end{array}\right] \tag{3.2.2}$$

The initial conditions are the following:

$$\hat{\mathbf{a}}_m(1/0) = [\mathbf{a}(1/0), 0, \ldots, 0]^t \tag{3.2.3}$$

$$r(1) = \mathbf{y}_m^t\,\mathbf{P}_m(1/0)\,\mathbf{y}_m + \sigma^2 \tag{3.2.4}$$

$$\mathbf{k}_m(1) = \mathbf{D}\,\mathbf{P}_m(1/0)\mathbf{y}_m \tag{3.2.5}$$

$$\mathbf{P}_m(1/0) = \left[\begin{array}{c|c} \mathbf{P}(1/0) & 0 \\ \hline 0 & 0 \end{array}\right] \qquad \mathbf{P}(1/0) = \tau^2\mathbf{I} = \frac{\sigma^2}{\mu^2}\mathbf{I} \tag{3.2.6}$$

The transform matrix:

$$H_k = \left[\begin{array}{c|c} 1 & -r(k)^{-1}\mathbf{y}_m^t\mathbf{B}(k) \\ \hline -(\mathbf{R}^r_k)^{-1}\mathbf{B}(k)^t\mathbf{y}_m & I_\alpha \end{array}\right] \tag{3.2.7}$$

is J-orthogonal. In this algorithm, the Kalman gain $\mathbf{k}_m(k)$ can be interpreted as the covariance of the state or prediction vector and the process innovation. According to the usual classification [6], our algorithm (3.2.2) is of the generalized Schur type, even though equation (3.2.1) is of the generalized Levinson type [4,5]. A comparison can be made with the generalized Schur algorithm [4]:

$$\begin{bmatrix} \sigma_{i+1}^2 & 0 \\ \mathbf{H}_{i+1} & \mathbf{G}_{i+1} \\ 0 & \sigma_{i+1}^2 \Delta_{i+1} \end{bmatrix} = \begin{bmatrix} \sigma_i^2 & -\sigma_i^2 \mathbf{K}_i^t \\ \mathbf{D}\,\mathbf{H}_i & \mathbf{G}_i \\ -\sigma_i^2 \mathbf{K}_i & \sigma_i^2 \Delta_i \end{bmatrix} \begin{bmatrix} 1 & \mathbf{K}_i^t \\ \Delta_i^{-1} \mathbf{K}_i & \mathbf{I}_{\alpha+1} \end{bmatrix} \qquad (3.2.8)$$

which shows that $r(k)^{-1} \mathbf{y}_m^t \mathbf{B}(k)$ can be interpreted as a vector which generalizes the usual scalar reflection coefficients.

4. NUMERICAL EXPERIMENTS

The data for the first example were generated by adding three sine waves of the same amplitude and constant frequencies (their relative frequencies were respectively 0.20, 0.25 and 0.30) in a white Gaussian pseudorandom noise, with a 20 dB signal-to-noise ratio. This example was taken from the literature to illustrate the flexibility and robustness of the method, since a smoothness priors long AR model is not in the class of models that generated the data. The data length was N=64 and we fitted an p=N=64th order model.

In order to illustrate what value of μ can be expected and what ranges of μ are reasonable for preclassification and reduction to a finite discrete set, the log-likelihood function has been computed in Fig. 1 as a function of μ. The estimated regularization parameter $\hat{\mu}$ corresponds to the likelihood maxima.

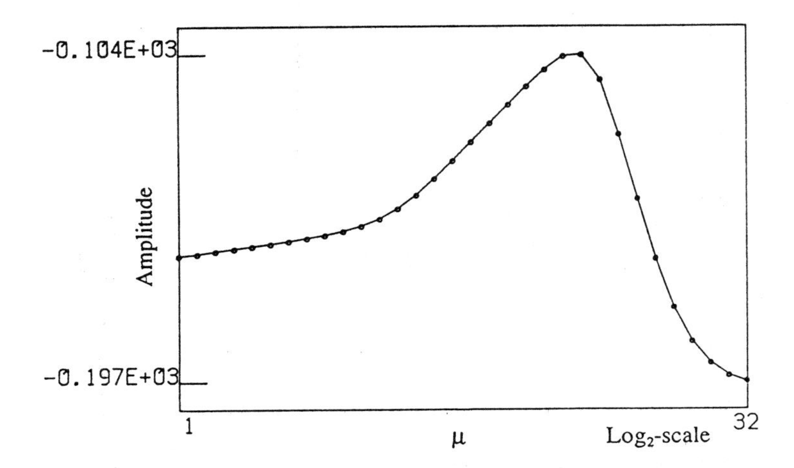

Fig. 1 Log-likelihood function for the test samples as a function of the regularization parameter μ.

The regularization coefficient is the essential tuning parameter of the method. To show what kind of results are obtained, the AR parameters were first computed with different μ-values ranging on 5 decades, and then the spectra were computed. They are shown in Fig. 2.

The data used are able to be represented as an ARMA (6,6) process. The equivalent AR model is of infinite order. By comparing these figures we notice that the Bayesian long AR estimate ($\hat{\mu}=4194$) reproduces the sharp spectral peaks of the signal and yields a smooth spectral estimate between them.

The second example is a synthetic non-stationary signal made of two linearly wobulated sine waves whose trajectories cross each other during the observation time. The data length was 256 and we fitted an p=12th-order model.

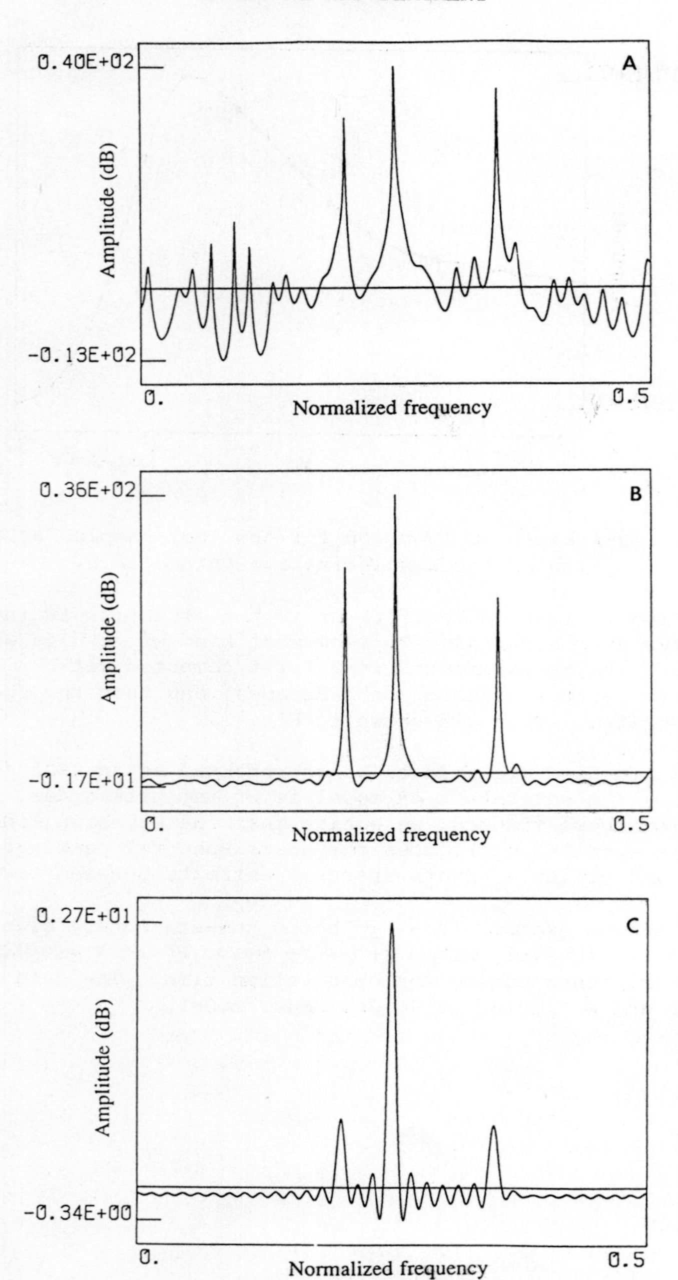

Fig. 2 Estimated power density (in dB) as a function of relative frequency, for different μ-values. (a) $\mu=41.94$, (b) $\mu=\hat{\mu}=4194$ and (c) $\mu=419400$.

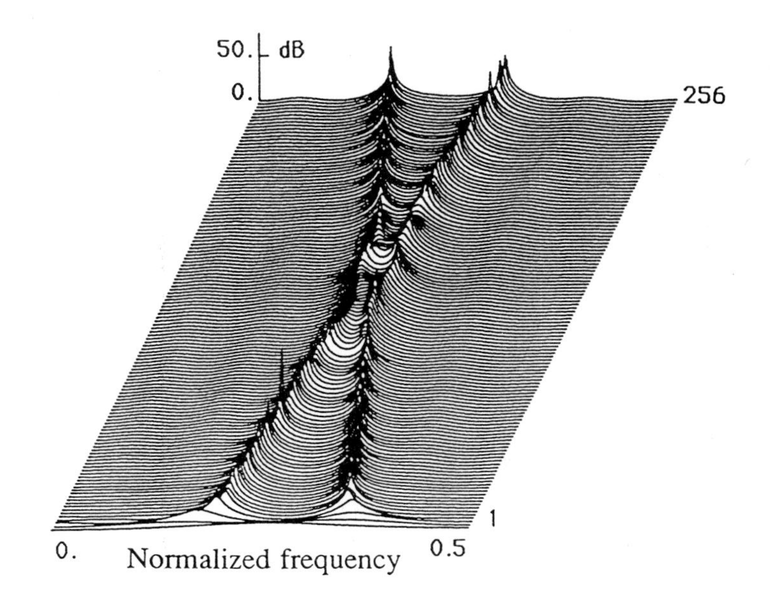

Fig. 3 Estimated instantaneous power density as a function of frequency and time.

The analysis window length was taken as N=12. This result compares favourably with others obtained with various analysis methods such as the Wigner-Ville analysis, conventional AR modelling, AR evolutive methods using Legendre polynomials, or adaptive Pisarenko techniques [3]; and presents the main advantage of a very fast implementation.

5. CONCLUSIONS

In the original smoothness priors approach to the modelling of non-stationary time series, the priors are in the form of difference equation constraints on a few time-varying AR parameters [10]. A different and simpler model is considered here, combining a local stationarity assumption and a long AR modelling with a zero-order frequency smoothness constraint. In this model, the number of AR parameters grows increasingly large with increasing data span N. It might seem questionable that such a model could correctly characterize the signal, but consistency is not a relevant concept here for the AR parameters. The reduction of the data information content is not performed at this first description level, as in conventional parametric methods, but at a second level, defined by the *a priori* distribution, and characterized by two hyperparameters. The parsimony principle is in fact operating at this level! This gives the method much more flexibility than conventional methods. In addition to this, the critical computation of the likelihood of the hyperparameters is greatly facilitated by an original state-space representation and a special fast Kalman filter implementation.

6. REFERENCES

[1] Akaïke, H., (1980) Likelihood and the Bayes procedure, in "Bayesian statistics", Bernando et al. Eds., University Press, Valencia, 146-166.

[2] Cullum, J., (1979) The effective choice of the smoothing norm in regularization, *Math. Comp.* **33**, 149-170.

[3] Demoment, G., Houacine, A., Herment, A. and Mouttapa, I., (1988) Adaptive Bayesian Spectrum Estimation, 4th ASSP Workshop on Spectrum Estimation and Modelling, Minneapolis.

[4] Demeure, C. and Scharf, L., (1987) Vector algorithms for computing QR and Cholesky factors of close-to-Toeplitz matrices, IEEE ICASSP 87, Dallas, 1851-1854.

[5] Friedlander, B., Kailath, T., Morf, M. and Ljung, L., (1978) Extended Levinson and Chandrasekhar equations for general discrete-time linear estimation problems, *IEEE Trans.*, **AC-23**, 653-659.

[6] Guéguen, C., (1985) An introduction to displacement ranks and related fast algorithms, Lecture Notes, NATO Summer School on Digital Signal Processing, Les Houches.

[7] Hall, P. and Titterington, D.M., (1987) Common structure of techniques for choosing smoothing parameters in regression problems, *J.R. Statist. Soc. B*, **49**, 184-198.

[8] Ho, Y.C. and Lee, R.C.K., (1964) A Bayesian approach to problems in stochastic estimation and control, *IEEE Trans.*, **AC-9**, 333-339.

[9] Houacine, A. and Demoment, G., (1987) Approche bayésienne de l'analyse spectrale adaptive: modèles AR longs et filtrage de Kalman rapide, Traitement du Signal, **4**. 389-397.

[10] Kitagawa, G. and Gersch, W., (1985) A Smoothness Priors Time-Varying AR Coefficient Modelling on Nonstationary Covariance Time Series, *IEEE Trans.*, **AC-30**, 48-56.

[11] Kitagawa, G. and Gersch, W., (1985) A Smoothness Priors Long AR Model Method for Spectral Estimation, *IEEE Trans.*, **AC-30**, 57-65.

[12] Marple, S.L., (1987) Digital Spectral Analysis, Prentice Hall, Englewood Cliffs.

[13] Nashed M.Z., (1981) Operator-Theoretic and Computational Approaches to Ill-Posed Problems with Applications to Antenna Theory, *IEEE Trans.*, **AP-29**, 220-231.

[14] Titterington, D.M., (1985) General structure of smoothing techniques in statistics, *Intern. Statist. Rev.*, **53**, 141-170.

A MAXIMUM LIKELIHOOD ALGORITHM FOR TRANSIENT DATA

D.R. Farrier and A.R. Prior-Wandesforde
(Department of Electrical Engineering, University of Southampton)

ABSTRACT

A maximum likelihood algorithm is introduced that is capable of extracting the parameters of an exponentially damped sinusoid with unknown phase buried in Gaussian white noise. The method can be implemented using Fast Fourier Transform (FFT) methods and is shown to perform better than the Kumaresan-Tufts (KT) algorithm and Bresler and Macovski's IQML algorithm. The situation involving more than one damped cosine is also discussed, and an algorithm for solving this problem is proposed.

1. INTRODUCTION

This paper investigates the problem of identifying the parameters of exponentially damped sinusoidal functions buried in Gaussian white noise. Prony's method [1] performs very well when there is very little noise present, but as soon as a realistic amount of noise is added this method fails. Kumaresan and Tufts [2,3] have proposed modifications using Singular Value Decomposition (SVD) which do have superior performance to Prony. We will suggest a maximum likelihood solution that makes use of discrete Fourier transforms. The method is shown to perform better than the Kumaresan and Tufts (KT) algorithm [2] and Bresler and Macovski's IQML algorithm [7]. Kumaresan and Tufts have suggested a variation on their original method [3] which sometimes performs better than their first method but for the examples that we have investigated there is little difference. In this paper the KT algorithm refers to the original method [2].

This new method has several advantages over the KT method; it is easier to program (ie. it does not need complicated SVD routines), the Fast Fourier Transform (FFT) parts of the algorithm could be implemented using parallel structures, and it performs better than the KT method. The disadvantage is that at present it can only cope with one damped sinusoid and although this limitation can be removed it is beyond the scope of the present paper.

Let N noisy data points, y(t), be produced by m damped cosines corrupted by Gaussian, zero mean noise, n(t) i.e.:

$$y(t) = \sum_{k=1}^{m} \rho_k e^{-\beta_k t} \cos(\omega_k t + \psi_k) + n(t) \quad (1)$$

where ω_k, β_k, ψ_k and ρ_k are unknown.

Taking N data points (unity time step) we can form the vectors $\underline{y}$ and $\underline{n}$ such that:

$$\underline{y}^T = [\ y(0)\ y(1)\ \dots\ y(N-1)\ ]^T$$

$$n^T = [\ n(0)\ n(1)\ \dots\ n(N-1)\ ]^T$$

Thus Equation (1) can be rewritten in terms of vectors as:

$$\underline{y} = H\underline{s} + \underline{n}$$

where H is some model of the system, for example:

$$\begin{bmatrix} \cos\psi_1 ; & \dots & ; \cos\psi_m \\ e^{-\beta_1}\cos(\omega_1+\psi_1); & \dots & ; e^{-\beta_m}\cos(\omega_m+\psi_m) \\ e^{-2\beta_1}\cos(2\omega_1+\psi_1); & \dots & ; e^{-2\beta_m}\cos(2\omega_m+\psi_m) \\ \vdots & \dots & \vdots \\ e^{-M\beta_1}\cos(M\omega_1+\psi_1); & \dots & ; e^{-M\beta_m}\cos(M\omega_m+\psi_m) \end{bmatrix} \quad (2)$$

M=N-1, and

$$\underline{s} = [\ \rho_1\ \rho_2\ \dots\ \rho_m]^T$$

The error in the measurements is $\underline{y} - H\underline{s}$ so the squared error will be

$$\gamma = (\underline{y} - H\underline{s})^T(\underline{y} - H\underline{s}) \quad (3)$$

In this equation we wish to minimise the squared error, which for Gaussian noise is also equivalent to maximising the likelihood function. We wish to use these quantities to obtain estimates for the unknown parameters ω_i, β_i, ρ_i and ψ_i. It is easy to show that $\hat{\underline{s}}$, the maximum likelihood estimate for the amplitude vector, is given by:

$$\underline{\hat{s}} = (H^T H)^{-1} H^T \underline{y} \tag{4}$$

and that the maximum likelihood estimate for the unknown parameters (assuming white Gaussian noise) is equivalent to maximising

$$\Phi = \underline{y}^T H (H^T H)^{-1} H^T \underline{y} \tag{5}$$

Although Equation (5) is the correct equation we require to solve, it is a non-linear equation, and there are a number of different approaches which we may take.

2. SINGLE DAMPED SINUSOID

If we look at the case of just one damped cosine (ie. m=1) with arbitrary phase, we see that Equation (5) becomes:

$$\Phi = \frac{(\underline{h}_i^T \underline{y})^2}{\underline{h}_i^T \underline{h}_i} \tag{6}$$

where $\underline{h}_i$ is given by:

$$H = \underline{h}_i = \begin{bmatrix} \cos\psi \\ e^{-\beta}\cos(\omega+\psi) \\ e^{-2\beta}\cos(2\omega+\psi) \\ \vdots \\ e^{-(N-1)\beta}\cos(\{N-1\}\omega+\psi) \end{bmatrix} \tag{7}$$

so that Equation (6) can be written as:

$$\Phi = \frac{\left(\sum_{t=0}^{N-1} y(t)\, e^{-\beta t} \cos(\omega t+\psi) \right)^2}{\sum_{t=0}^{N-1} e^{-2\beta t} \cos^2(\omega t+\psi)}$$

which simplifies to:

$$\Phi = \frac{(a \cos\psi - b \sin\psi)^2}{(f + d \cos 2\psi - c \sin 2\psi)/2} \tag{8}$$

where

$$a = \sum_{t=0}^{N-1} y(t)\, e^{-\beta t} \cos\omega t$$

$$b = \sum_{t=0}^{N-1} y(t)\, e^{-\beta t} \sin\omega t$$

$$c = \sum_{t=0}^{N-1} e^{-2\beta t} \sin 2\omega t$$

$$d = \sum_{t=0}^{N-1} e^{-2\beta t} \cos 2\omega t$$

$$f = \sum_{t=0}^{N-1} e^{-2\beta t}$$

The calculations involved in calculating c, d and f can be reduced by summing them analytically. It should be noted that the parameters c, d and f are independent of the actual data, y(t), thus they only need be calculated once. This can be done before any signal processing, thus making the algorithm quicker. The terms, a and b, do depend on the data, but the calculation of them can be made quicker if we use a Fast Fourier Transform (FFT) on the weighted data $y'(t)$:

$$y'(t) = e^{-\beta t}\, y(t)$$

It is now clear that we could try all possible combinations of ω, β and ψ and choose the global maximum. However, this would be extremely costly and is not recommended. Fortunately, we can eliminate the phase dependence by optimising Equation (8) analytically with respect to $\tan\psi$ (=t). This may be accomplished by first rewriting it as:

$$\Phi = \frac{(a - b\,t)^2}{\left(\frac{f + d}{2}\right) + \left(\frac{f - d}{2}\right) t^2 - c\,t} \tag{9}$$

It is easy to show that the maximum likelihood estimate for $\tan\psi$ occurs when:

$$\tan\psi = \frac{b(d+f) - ac}{a(d-f) + bc} \tag{10}$$

so that we can now use only Equations (9) and (10). This is now only a function of ω and β, since we have removed the dependence on ψ.

Algorithm

(i) Using a FFT algorithm, we plot Φ as a function of ω and α. The values which give a global maximum are approximate maximum likelihood estimates.
(ii) Using a more accurate, local optimisation, (not using an FFT) we then obtain the final estimates. This may be simply achieved by calculating a few sample points, and using quadratic interpolation.

σ	Method	Bias of $\hat{a}_1$	Bias of $\hat{a}_2$	Standard Deviation of $\hat{a}_1$	Standard Deviation of $\hat{a}_2$
0.05	CRLB	-	-	$0.781 \cdot 10^{-2}$	$0.781 \cdot 10^{-2}$
	KT	$-0.70 \cdot 10^{-2}$	$0.89 \cdot 10^{-2}$	$0.134 \cdot 10^{-1}$	$0.122 \cdot 10^{-1}$
	IQML	$0.65 \cdot 10^{-3}$	$-0.49 \cdot 10^{-3}$	$0.117 \cdot 10^{-1}$	$0.110 \cdot 10^{-1}$
	MLA	$-0.41 \cdot 10^{-3}$	$0.45 \cdot 10^{-3}$	$0.116 \cdot 10^{-1}$	$0.110 \cdot 10^{-1}$
0.1	CRLB	-	-	$0.156 \cdot 10^{-1}$	$0.156 \cdot 10^{-1}$
	KT	$-0.24 \cdot 10^{-1}$	$0.32 \cdot 10^{-1}$	$0.295 \cdot 10^{-1}$	$0.238 \cdot 10^{-1}$
	IQML	$0.45 \cdot 10^{-2}$	$-0.38 \cdot 10^{-2}$	$0.252 \cdot 10^{-1}$	$0.210 \cdot 10^{-1}$
	MLA	$0.53 \cdot 10^{-3}$	$0.90 \cdot 10^{-4}$	$0.243 \cdot 10^{-1}$	$0.206 \cdot 10^{-1}$
0.2	CRLB	-	-	$0.312 \cdot 10^{-1}$	$0.312 \cdot 10^{-1}$
	KT	-0.106	0.136	$0.757 \cdot 10^{-1}$	$0.664 \cdot 10^{-1}$
	IQML	$0.55 \cdot 10^{-1}$	$-0.33 \cdot 10^{-1}$	0.261	0.1103
	MLA	$0.25 \cdot 10^{-2}$	$-0.23 \cdot 10^{-2}$	$0.521 \; 10^{-1}$	$0.417 \cdot 10^{-1}$
0.3	CRLB	-	-	$0.469 \cdot 10^{-1}$	$0.469 \cdot 10^{-1}$
	KT	$0.73 \cdot 10^{-2}$	0.354	0.934	0.216
	IQML	0.275	-0.137	0.598	0.252
	MLA	$0.15 \cdot 10^{-1}$	$-0.31 \cdot 10^{-2}$	0.156	$0.710 \cdot 10^{-1}$
0.4	CRLB	-	-	$0.623 \cdot 10^{-1}$	$0.623 \cdot 10^{-1}$
	KT	0.375	0.580	1.473	0.350
	IQML	0.572	-0.221	0.825	0.314
	MLA	0.135	$0.89 \cdot 10^{-2}$	0.586	$0.887 \cdot 10^{-1}$

Table 1 Performance comparison of maximum likelihood algorithm, KT and IQML methods.

Simulations

The proposed method was compared with the backwards KT algorithm [2] and the IQML algorithm [7]. The chosen transient was one damped cosine with $w_i = \cos^{-1}(0.8)$ and $\beta_i = \ln(0.9)$ (ψ_1=0.0 and ρ_1=1.0), with 40 data points and the order of the KT backward polynomial was 8. The standard deviation (σ) of the white noise was varied to see the effect of signal to noise ratio. For each σ the transient was repeated 500 times with different noise samples. This corresponds exactly to the example chosen by Porat and Friedlander [5]. They choose to identify the parameters a_1 and a_2 of the rational transfer function:

$$\frac{b(z)}{a(z)} = \frac{z - 0.72}{z^2 + a_1 z + a_2}$$

where $a_1 = -1.44$ and $a_2 = 0.81$.

The biases and standard deviations of the estimates of $\hat{a}_1$ and $\hat{a}_2$ are shown in Table 1 for the three methods: the KT method, IQML and the proposed method. Also shown in Table 1 are the Cramer-Rao lower bounds on the standard deviation of $\hat{a}_1$ and $\hat{a}_2$ for each case.

3. MULTIPLE SINUSOID CASE

In the case of undamped sinusoids, it is fairly straightforward to follow an approach similar to that given above, to find initial estimates for some of the signals. We then need to:

(a) identify 'major' signals, using the same approach as for a single damped sinusoid.
(b) use a global search for additional signals and place nulls in these positions, using projection matrices.
(c) refine these estimates using a local search.

For the damped sinusoid case, the situation is rather more complex. First, to accurately null signals, we need to estimate both damping and frequency. Second, the 'spectrum' for the damped case exhibits large 'shoulders' which cause several signal spectra to interact strongly. Although some form of 'shading' can be used to reduce this effect, it adversely affects signal-to-noise ratio.

When several damped sinusoids are present, we may still use a set of weighted FFT's for initial estimates, but now the 'shoulders' will interact and it will be more difficult to identify correct frequencies and damping. The problem caused by the shoulders can be reduced by searching for maximum *relative* height (ie. height above the shoulder). This may be accomplished by observing adjacent minima as well as the local maximum.

We will now consider how we may refine our estimates. We have experimented with several approaches, and the two most effective appear to be Gauss Newton iteration, and a projection method. This latter method may also be used for seeking new global maxima and for this situation has the advantage of still requiring FFT structures.

Gauss Newton iteration

Gauss Newton iteration does not have the local convergence properties of a Newton method. However, it does have better stability, and so is preferred. We can define the mean least square error, F, to be:

$$F = \sum_{t=0}^{N-1} e_t^2$$

where

$$e_t = y_t - \sum_{i=1}^{m} \rho_i e^{-\beta_i t} \cos(\omega_i t + \psi_i).$$

If the perturbation vector, $\underline{v}_i$, for the ith damped sinusoid is defined to be

$$\underline{v}_i = [\ \delta\omega_i\ \delta\beta_i\ \delta\psi_i\ \delta\rho_i]^T$$

then we may obtain

$$\frac{\delta F}{\delta \underline{v}_i} = -2 \sum_{t=0}^{M-1} \frac{\delta}{\delta \underline{v}_i} \left(\rho_i e^{-\beta_i t} \cos(\omega_i t + \psi_i) \right) e_t \qquad (11)$$

By defining the error vector to be $\underline{e} = \underline{y} - H\underline{s}$ and J_i to be equal to:

$$J_i = \begin{bmatrix} -\dfrac{\delta}{\delta \underline{v}_i^T} \rho_i \cos\psi_i \\ \vdots \\ -\dfrac{\delta}{\delta \underline{v}_i^T} \rho_i e^{-\beta_i t} \cos(\omega_i t + \psi_i) \\ \vdots \\ -\dfrac{\delta}{\delta \underline{v}_i^T} \rho_i e^{-\beta_i (N-1)} \cos(\omega_i (N-1) + \psi_i) \end{bmatrix}$$

we see that Equation (11) becomes:

$$\frac{\delta F}{\delta \underline{v}_i} = 2\, J_i^T \underline{e} \tag{12}$$

Now if we let:

$$J = [\; J_1 \; J_2 \; .. \; J_m \;]$$

we may apply a Gauss-Newton correction given by:

$$\underline{v} = -\,(J^T J)^{-1} J^T \underline{e} \tag{13}$$

Projection Method

We can partition the matrix H in such a way that the vector $\underline{h}_i$ is one column of H and the matrix A contains the remaining columns i.e.: $H = [A : \underline{h}]$. If we set the matrix P equal to:

$$P = I - A(A^T A)^{-1} A^T$$

the maximum likelihood estimate of the unknown parameters can be shown to be equivalent to maximising.

$$\Phi = \frac{(\underline{h}_i^T P \underline{y})^2}{\underline{h}_i^T P \underline{h}_i} \tag{14}$$

It has been found that we can approximate Equation (14) by:

$$\Phi = \frac{(\underline{h}_i^T P \underline{y})^2}{\underline{h}_i^T \underline{h}_i} \tag{15}$$

without much (if any) loss of accuracy. This approximation is helpful because Equation (15) is quicker to calculate, as it does not require the calculation of $\underline{h}^T P \underline{h}$ for each ω and β. If we maximise Equation (15) in terms of ψ, we obtain a very similar result to that given in earlier.

$$\Phi = \frac{(a \cos\psi - b \sin\psi)^2}{(f + d \cos 2\psi - c \sin 2\psi)/2}$$

where a, b, c, d, f are now given by:

$$a = \sum_{t=0}^{N-1} y'(t)\, e^{-\beta t} \cos\omega t$$

$$b = \sum_{t=0}^{N-1} y'(t)\, e^{-\beta t} \sin\omega t$$

where $\underline{y}' = Py$

and

$$c = 2\underline{h}_c^T \underline{h}_s$$

$$d = \underline{h}_c^T \underline{h}_c - \underline{h}_s^T \underline{h}_s$$

$$f = \underline{h}_c^T \underline{h}_c + \underline{h}_s^T \underline{h}_s$$

Now the most expensive part of the calculation, the inversion of the matrix $A^T A$, can be performed outside any ω and β loop. The calculations may be reduced by using a QR decomposition.

Using this approach, we may try each signal in turn, and perform a *local* maximisation with a small number of frequencies and damping terms, and then use quadratic interpolation to refine our estimates.

If we wish to find new signals which may have been omitted from stage 1, we may use this same technique for a global maximisation, and use FFT techniques. This is the approach which we use in stage 2 of the algorithm.

Multiple Sinusoid Algorithm

The maximum likelihood algorithm that we use can be broken down into three stages, as follows:

Stage 1: The Maximum Likelihood Method proposed for a single signal. To avoid the problems caused by the 'shoulders we use *relative height* to identify the signals.

Stage 2: Reject the root with the lowest spectral height, then use Equation (15) to recalculate a better estimate for this signal.

Stage 3: Refine the initial estimates, using either the Gauss-Newton iteration or projection method, described above.

Simulation using multiple sinusoids

We will consider a 40 data point transient composed of four damped sinusoids having unknown parameters equal to:

$$\omega_1 = 2.9, \quad \omega_2 = 2.0, \quad \omega_3 = 1.3, \quad \omega_4 = 0.6$$

$$\beta_1 = 0.02, \quad \beta_2 = 0.07, \quad \beta_3 = 0.09, \quad \beta_4 = 0.03$$

$$\psi_1 = 1.4, \quad \psi_2 = 0.0, \quad \psi_3 = 0.7, \quad \psi_4 = -1.2$$

The standard deviation of the Gaussian white noise (σ) was varied between 0.1 and 0.6 with the transient being repeated 500 times with different noise samples at each σ.

Table 2 shows the bias and mean square error estimates obtaine for each technique. Also shown are the results obtained using the original Kumerasan and Tufts method [2] (the order of the KT polynomial was 10) and the IQML method of Bresler and Macovski [7]. We may see that the IQML method of Bresler and Macovski performs comparatively well at low noise levels but is less successful at higher noise levels.

σ	Method	Bias	Mean square error
0.10	KT	-0.1499×10^{-1}	0.7066×10^{-3}
	IQML	0.1834×10^{-2}	0.4859×10^{-3}
	G. Newton	0.2780×10^{-3}	0.4748×10^{-3}
	Projection	0.1123×10^{-1}	0.9760×10^{-3}
0.20	KT	-0.5779×10^{-1}	0.4674×10^{-2}
	IQML	0.9316×10^{-2}	0.2767×10^{-2}
	G. Newton	0.3145×10^{-2}	0.2329×10^{-2}
	Projection	0.2031×10^{-1}	0.3497×10^{-2}
0.30	KT	-0.1419	0.1755×10^{-1}
	IQML	0.2517×10^{-1}	0.3297×10^{-1}
	G. Newton	0.2345×10^{-2}	0.1105×10^{-1}
	Projection	0.2925×10^{-1}	0.1318×10^{-1}
0.40	KT	-0.2385	0.4123×10^{-1}
	IQML	0.5046×10^{-1}	0.9438×10^{-1}
	G. Newton	0.2218×10^{-1}	0.3189×10^{-1}
	Projection	0.4416×10^{-1}	0.3716×10^{-1}
0.50	KT	-0.3693	0.8505×10^{-1}
	IQML	0.2342×10^{-1}	0.2050
	G. Newton	-0.7108×10^{-2}	0.9103×10^{-1}
	Projection	0.2918×10^{-2}	0.9334×10^{-1}
0.60	KT	-0.5215	0.1383
	IQML	-0.4612×10^{-2}	0.2398
	G. Newton	-0.1457×10^{-1}	0.1292
	Projection	-0.2270×10^{-1}	0.1283

Table 2 Performance comparison of several maximum likelihood algorithms for four damped sinusoids

4. CONCLUSIONS

We have introduced a maximum likelihood method that performs better than the original Kumaresan and Tufts method [2] and the IQML method of Bresler and Macovski [7] for the estimation of the parameters of transients composed of multiple damped sinusoids buried in Gaussian white noise.

REFERENCES

[1] Hildebrand, F.B., (1965) Introduction to Numerical Analysis, McGraw-Hill, New York, pp. 378-382.

[2] Kumaresan, R. and Tufts, D.W., (1982) Estimating the Parameters of Exponentially Damped Sinusoids and pole-zero modelling in noise, IEEE Trans. ASSP, Vol. 30, pp. 833-840.

[3] Kumaresan, R. and Tufts, D.W., (1983) Estimating the Angles of Arrival of Multiple Plane Waves, IEEE Trans. AES, Vol. 19.1, pp. 134-139.

[4] Pisarenko, V.F., (1973) The Retrieval of Harmonics from a covariance function, Geophs. J.R.astr.Soc., vol. 33, pp. 347-366.

[5] Porat, B. and Friedlander, B., (1986) A Modification of the Kumaresan-Tufts Methods for Estimating Rational Impulse Responses, IEEE Trans. ASSP, Vol. 34.5, pp. 1336-1338.

[6] Rife, D.C. and Boorstyn, R.R., (1974) Single-Tone Parameter Estimation from Discrete-Time Observations, IEEE Trans. IT, Vol. 20, no. 5, pp. 591-598.

[7] Bresler, Y. and Macovski, A., (1986) Exact Maximum Likelihood Parameter Estimation of Superimposed Exponential Signals in Noise, IEEE Trans. ASSP, Vol. 34.5, pp. 1081-1089.

CORRELATION DETECTION USING MULTIPLE-SCALE FILTERS AND SELF-SIMILAR NOISE MODELS

P.G. Earwicker and J.G. Jones
(Special Systems Department, Royal Aerospace Establishment, Hants.)

ABSTRACT

A generalised correlation method for the detection of discrete structured features embedded in fluctuating backgrounds having fractal geometry is described. Measured data are expressed as the sum of an ensemble of discrete features and a Gaussian noise model, where the noise is assumed to be stationary, self-similar and, in two dimensions, isotropic. Such noise is invariant under change of scale as well as under translation and rotation. The detection of a discrete feature of prescribed shape and orientation involves both translation and scale-changing operations, applied to an associated filter, and the extraction of local maxima in operator outputs in a hyperspace whose coordinates represent spatial location and scale. Constraints imposed on the admissible form of the discrete feature by the choice of noise model are described and illustrative examples are presented.

1. INTRODUCTION

In previous work ([4], [9], [10], [11], [12]) a multi-resolution method for analysing geophysical data, including turbulence records and images obtained by remote sensing, has been described in which correlation filters are applied over a range of spatial locations and a range of scales. Measured probability distributions of filter outputs have been shown to exhibit strongly non-Gaussian statistics and to satisfy scaling laws which allow a representation in terms of fractal geometry. The non-Gaussian tails on the probability distributions of the measured data have been interpreted [12] as reflecting the existence in the data of discrete ordered structures. Examples include discrete gusts ([4], [9]) embedded in continuously-fluctuating atmospheric turbulence and discrete edge or bar elements in remote-sensing data.

In one particular form of data analysis to which the proposed form of correlation detection is applicable, a non-Gaussian fractal model f(t) is introduced [6] to represent discrete structured features in the data, and this model is matched to measured data g(t) by the introduction of additive Gaussian noise h(t) such that the equation

$$g(t) = f(t) + h(t) \tag{1.1}$$

is satisfied, where t is in R^{ν}, $\nu = 1$ or 2.

Whilst noise is generally introduced, in the context of signal processing, to represent effects of the measurement process, such as sensor noise, in the above application the noise is introduced primarily to represent residual fluctuations in the data not accounted for by the discrete model f(t). For instance, f(t) may be chosen from a statistical ensemble of functions comprising discrete ramp-shaped gusts whose profile idealizes that of actual measured wind fluctuations and whose statistical distribution incorporates scaling properties of fractal geometry ([5],[7]). In such a situation, where both the measured data g(t) and the model f(t) satisfy fractal scaling laws, the noise term h(t), in equation (1.1), is also required to be fractal.

With this motivation, the standard method of correlation detection [13], which is based on equation (1) where h(t) is taken to be white noise, has been generalised ([2], [3]) to cover the case in which the Gaussian noise h(t) is non-white, but isotropic and self-similar. Such a simplified noise model generally provides an adequate representation of the residual fluctuations in fractal geophysical fields. On the basis of this generalization of correlation detection relationships have been deduced [8] between the form of the correlation filter and the structure of the discrete event, f(t), which is optimally detected. In this paper, some of the results from [3] and [8] are summarized.

2. SELF-SIMILAR PROCESSES AND FIELDS

For a statistically self-similar process or field, fluctuations at different scales are related by an enlargement or compression operation. Specifically, the statistical properties of the graph of the associated function, say z(t), are invariant under an affine scaling in which the t-plane is uniformly stretched by a factor h and the z-axis by a factor h^k. k is referred to as the similarity parameter.

A rigorous treatment of isotropic self-similar Gaussian noise is given in [1], in terms of the theory of distributions.

However, the basic properties may be expressed in terms of appropriately restricted functions h(t) on a Hilbert space.

For t in R^ν, where ν = 1 or 2, the probability functional for the isotropic self-similar noise h(t) is

$$P(h) \propto \exp\left\{-\frac{1}{2\varepsilon^2}(h,Wh)\right\}, \tag{2.1}$$

where the inverse covariance operator W can be expressed in terms of convolution by w:

$$Wh = w * h, \tag{2.2}$$

and the Fourier Transform of w is given by

$$\hat{w}(\omega) = \left\{\phi(\omega)\right\}^{-1} \propto |\omega|^{2k+\nu}. \tag{2.3}$$

Here, $\phi(\omega)$ is the power spectral density.

When the similarity parameter takes the value $k = \nu/2$, white noise is recovered. For $k = \frac{1}{2}$ and $\nu = 1$, on the other hand, Brownian noise is obtained, for which the power spectral density is proportional to $|\omega|^{-2}$. We refer to the two-dimensional field with $\nu = 2$, $k = 0$ as Laplacian noise, as W then takes the form

$$W = -\nabla^2 = -\left(\frac{\partial^2}{\partial t_1^2} + \frac{\partial^2}{\partial t_2^2}\right), \tag{2.4}$$

where ∇^2 is the Laplacian operator. Laplacian and more general fractional Laplacian fields are illustrated in [12].

3. GENERALISED TRANSLATION OPERATORS

A particular instance of correlation detection occurs when a signal f(t) has a waveform of prescribed shape, scale and amplitude but is of unknown location. Cross-correlation [13] between data g(t) and f(t) then involves the translation of the given waveform with respect to g(t). To quantify this process it is convenient to introduce the translation operator T_a which takes a particular function $\bar{f}$ and produces a new function $T_a\bar{f}$ defined by

$$(T_a\bar{f})(t) = \bar{f}(t-a). \tag{3.1}$$

It is easily verified that

$$(T_a g, T_a \bar{f}) = (g, \bar{f}), \tag{3.2}$$

where (,) denotes a Hilbert space inner product. That is, T_a is a unitary operator.

An analogous situation arises when f(t) has prescribed shape and location but is of unknown scale. This problem may be treated by introducing a scaling operator S_h, defined by

$$(S_h \bar{f})(t) = h^k \bar{f}(t/h), \; h > 0 . \tag{3.3}$$

S_h widens $\bar{f}$ by a factor h and increases its magnitude by a factor h^k (compare section 2). S_h may also be shown to be unitary if we take $k = -\frac{1}{2}$, in one dimension, or $k = -1$, in two dimensions.

To generalise these concepts ([2], [3]), an operator U is said to be unitary *with respect to a positive operator* W if, for each g and $\bar{f}$:

$$(Ug, WU\bar{f}) = (g, W\bar{f}). \tag{3.4}$$

By analogy with equation (3.2), such an operator U is referred to as a generalised translation operator. It may be verified in particular that the translation operator T_a, equation (3.1) is unitary with respect to W, as also is S_h, equation (3.3), provided k is chosen in relation to W through equations (2.2) and (2.3).

An important class of generalised translations is that incorporating both spatial translation and scaling, a typical element being $T_a S_h$. The combined operators $T_a S_h$ form a group which is non-commutative, since

$$T_a S_h = S_h T_{a/h} . \tag{3.5}$$

A third operation which can be combined with T_a and S_h to form a three-parameter group is that of multiplication by one of the numbers {1,-1}. This operation commutes with both T_a and S_h. In two dimensions, rotation through an angle can also be incorporated ([2], [3]) to form a four-parameter group.

It may be verified that all the operators introduced in this way are generalised translation operators, satisfying equation (3.4). That is, they are all unitary with respect to the isotropic self-similar noise whose inverse covariance operator is W, as defined in section 2.

4. CORRELATION DETECTION IN SELF-SIMILAR NOISE

In the method of generalised correlation detection ([2], [3]) the function, or 'signal', f(t) in equation (1) is assumed to take the form $U\bar{f}$, where $\bar{f}$ is a function of prescribed location, scale and orientation and U is a unitary operator with respect to the positive operator W defined by equations (2.2) and (2.3).

To simplify the exposition we will assume that the prior distribution $P(U\bar{f})$ is uniform (or alternatively that the noise level is low, $\varepsilon \to 0$). The way in which the equations may be modified to take account of non-uniform prior distributions is described in [6].

Then the process of generalised correlation detection in self-similar noise h(t), with probability functional given by equation (2.1), is based [3] on the equation

$$P(U\bar{f}/g) \propto \exp\left\{-\frac{1}{2\varepsilon^2}(g - U\bar{f},\ Wg - WU\bar{f})\right\}, \qquad (4.1)$$

where $P(U\bar{f}/g)$ is the posterior probability of $U\bar{f}$, given g. The inner product can be expanded:

$$(g - U\bar{f},\ Wg - WU\bar{f}) = (g,Wg) - 2(g,WU\bar{f}) + (\bar{f},W\bar{f}), \qquad (4.2)$$

where it is assumed that $\bar{f}$ has been constrained such that $(\bar{f}, W\bar{f})$ is finite, and we have used the fact that, since the positive operator W is self-adjoint:

$$(U\bar{f},\ Wg) = (Wg,\ U\bar{f}) = (g,WU\bar{f}). \qquad (4.3)$$

Only the middle term in equation (4.2) depends upon U, so to maximise the posterior probability, equation (4.1), with respect to the generalised translation operator U, the inner product $(g,WU\bar{f})$ must be maximised.

5. ILLUSTRATIVE EXAMPLES

When the noise h(t) is Brownian (section 2), it follows from equations (2.2) and (2.3), setting $\nu = 1$ and $k = \frac{1}{2}$, that $W \equiv -d^2/dt^2 = -D^2$. The inner product in equation (2.1) thus becomes

$$\left(h, - \frac{d^2h}{dt^2} \right) = \left(\frac{dh}{dt}, \frac{dh}{dt} \right) = \int \left(\frac{dh}{dt} \right)^2 dt, \qquad (5.1)$$

a quantity we may refer to as the 'gradient energy' of h.

For the group of generalised translations incorporating spatial translation and scaling, the 'generalised correlation function' $(g, WU\bar{f})$, equation (4.3), takes the form

$$\begin{aligned} (g, WU\bar{f}) &= \left(g, - D^2 T_a S_h \bar{f} \right) \\ &= \left(g, - D^2 S_h T_{a/h} \bar{f} \right) \\ &= \left(S_{h^{-1}} g, - D^2 T_{a/h} \bar{f} \right) \\ &= \left(S_{h^{-1}} g, T_{a/h} (-D^2 \bar{f}) \right) , \end{aligned} \qquad (5.2)$$

where in the second line we have used equation (3.5) and in the subsequent line we have used equation (3.4), with $S_{h^{-1}}$ playing the role of U.

The form of the final line is convenient for practical computation, as $(- D^2\bar{f})$ can be computed at the outset to give a fixed filter which is translated $(T_{a/h})$ over scaled data $(S_{h^{-1}}g)$ in steps proportional to the scaling factor h^{-1}. As stated below equation (4.2), it is assumed that $\bar{f}$ has been constrained such that $(\bar{f}, - D^2\bar{f})$ is finite, that is (equation (5.1)) $\bar{f}$ has finite gradient energy.

One elementary signal $\bar{f}$ with this property is that (Figure 1a) whose derivative (i.e. the corresponding function in a 'whitened' space) is a discrete pulse (Figure 1b) of finite energy. Such a signal takes the form (Figure 1a) of a 'smooth-increment' or smooth edge. For this profile the filter weighting function $(- D^2\bar{f})$, shown in Figure 2, can be implemented approximately [8] by convolving a pair of delta functions, of opposite sign, with a smoothing filter whose weighting function is a discrete pulse (as in Figure 1b). The filter whose weighting function is a pair of delta functions of opposite sign is a 'differencing' filter, and its combination with a smoothing operation, to give the weighting function in Figure 2, gives a 'smoothed difference filter', which evaluates a difference of local averages, or average of local differences, as the operations commute. Applications of this filter to the analysis of turbulence records have been described in [4] and [9]. The interpretation of the results in terms of 'smooth increments', as shown in Figure 1a, involves an approximation in which the turbulence background, which generally has a power spectral density close to $1/\omega^{5/3}$, is represented by Brownian noise. An analogous approximation, in which Brownian noise is used over a wide range of scales, is frequently used to generate 'synthetic' background turbulence for aeronautical applications.

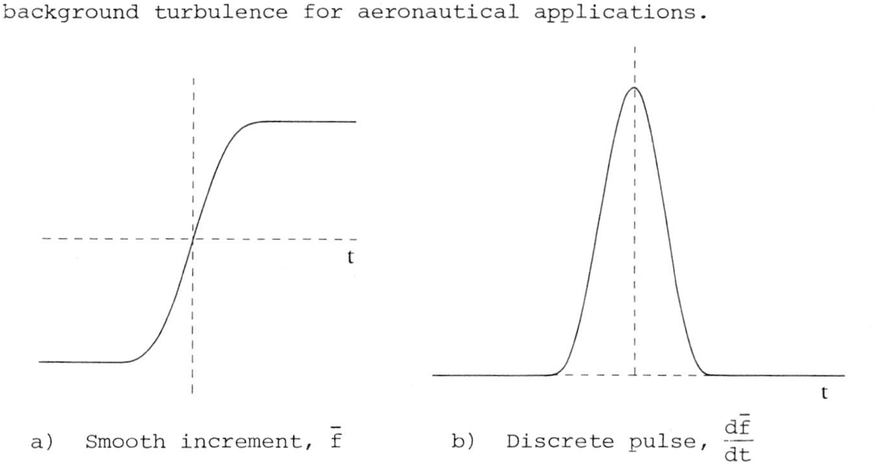

a) Smooth increment, $\bar{f}$ b) Discrete pulse, $\frac{d\bar{f}}{dt}$

Fig. 1 Elementary signal ($\bar{f}$) detectable in Brownian Noise

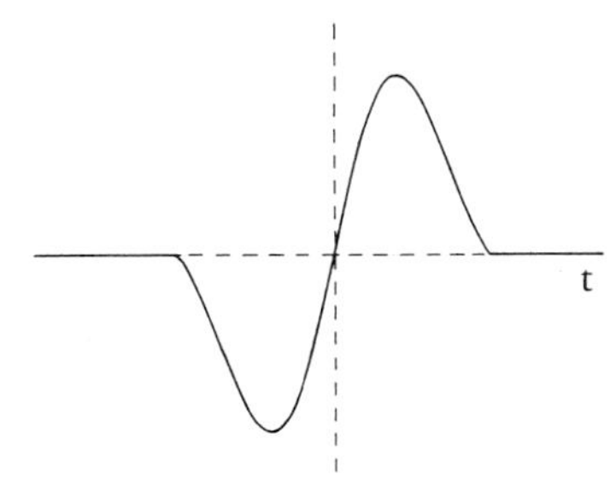

Fig. 2 Smoothed-difference filter weighting function ($-D^2\bar{f}$) with $\bar{f}$ as in Figure 1a

With the choice of parameters $\nu = 1$ and $k = 0$ in equation (2.3), there is obtained '$1/\omega$ noise', widely observed in electronics applications and also in one-dimensional cross sections of remote-sensing data. This requires that the 'generalised energy' $(\bar{f}, |\omega|\bar{f})$ be finite. In the case of Brownian noise, a suitable elementary signal satisfying the generalised-energy constraints was obtained as that whose corresponding function in a whitened space was a discrete pulse. It can be verified [8] that an analogous procedure used with $1/\omega$ noise leads to correlation filters whose weighting functions have slowly-decaying tails (following a fractional power law).

For practical applications, in which data samples are statistically stationary over quite severely constrained regions, such fractional filters are unacceptable and must be replaced by filters which are strongly domain limited. A feasible approach is to employ, in $1/\omega$ noise, the filter already shown to be appropriate for $1/\omega^2$ processes, *viz.* the smoothed-difference filter illustrated in Figure 2. In $1/\omega$ noise the associated signal profile (to be detected) is found [8] by solving the equation

$$z(t) = W\bar{f}(t) \tag{5.3}$$

for $\bar{f}(t)$, with $z(t)$ given by the function in Figure 2 and $W = |\omega|$. The resulting profile is illustrated in Figure 3. Thus, in $1/\omega$ noise, a strongly localised observation in the form of a local maximum in the output of the domain-limited smoothed-difference filter (Figure 2) implies the existence in the input of a 'most probable' waveform having the shape of a smooth increment, or edge, surrounded by slowly-decaying tails.

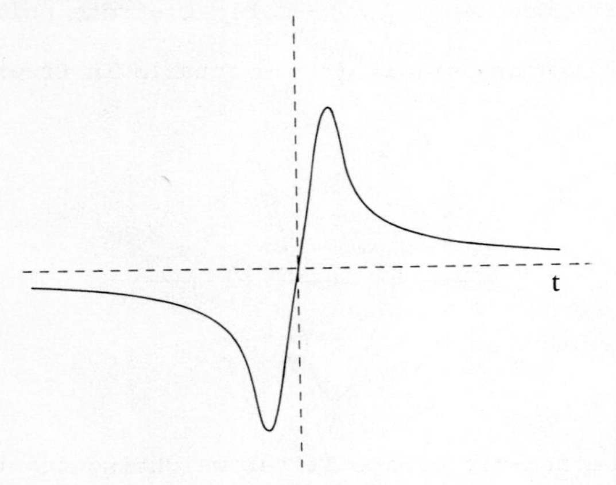

Fig. 3 Discrete profile detected in $1/\omega$ noise by filter in Figure 2.

For other background noises analogous principles apply. For example, when the power-spectral density takes the form $1/\omega^4$ the discrete feature becomes a localised concentration of 'curvature energy' $\int \left(\frac{d^2 h}{dt^2} \right)$ dt. Intermediate spectra such as $1/\omega^3$ pose the same problems of implementation as $1/\omega$ noise and lead to 'elementary' events with slowly-decaying tails.

The choice of elementary profiles to represent discrete events in two dimensions follows the same principles as in one dimension. In [8] the procedure is illustrated with particular reference to the situation where the background noise is isotropic and self-similar with similarity parameter k = 0, i.e. Laplacian noise (equation (2.4)).

6. CONCLUDING REMARKS

A convenient discrete implementation of generalised translation with respect to space and scale begins by initially fixing the scale at h = 1 (equation (3.3)) and performing spatial translations equal to the sampling interval. The scale is then changed, typically to h = 2, by applying the operator $S_{h^{-1}}$ to the data g, in effect compressing it, and then translating the same filter $W\bar{f}$ over the scaled data using the translation operator $T_{a/h}$ which takes reduced incremental steps, as in equation (5.2). The transformations performed in arriving at equation (5.2) have replaced the process of applying a stretched filter to given data by that of applying a fixed filter to compressed data. This has the practical advantage of allowing successive scales to be treated by means of a fixed filter in a simple recursive loop. Examples of this implementation have been presented in [4], [9], [10], [11] and [12].

For application to two-dimensional data, the method is ideally structured for implementation using parallel processing. In a serial implementation, a single filter is translated, with respect to measured data, in successive steps through a hyperspace whose dimensions include spatial location, scale and possibly also rotation. In a parallel implementation a bank of filters spanning the hyperspace is generated as a preliminary step. A given sample of measured data can then be processed by all the filters simultaneously.

REFERENCES

[1] Dobrushin, R.L., (1979) Gaussian and their Subordinated Self-Similar Random Generalised Fields, *Annals of Prob.*, 7, 1-28.

[2] Earwicker, P.G., (1988) Correlation Detection in Self-Similar Noise, Royal Aerospace Establishment Tech. Report 88032.

[3] Earwicker, P.G. and Jones, J.G., (1988) Multiple-Scale Correlation Detection using Self-Similar Noise Models, Submitted for publication.

[4] Foster, G.W. and Jones, J.G., (1987) Measurement and Analysis of Low Altitude Atmospheric Turbulence, AGARD Report 734, Paper No. 2.

[5] Jones, J.G., (1980) Modelling of Gusts and Wind Shear for Aircraft Assessment and Certification, Proc. Indian Acad. Sci. (Eng. Sci.), 3, 1-30.

[6] Jones, J.G., (1986) Statistical Theory for Multi-Resolution Signal Processing and Analysis, Royal Aircraft Establishment Tech. Report 86065.

[7] Jones, J.G., (1989) Statistical-Discrete-Gust Method for Predicting Aircraft Loads and Dynamic Response, Journal of Aircraft, 26, 382-392.

[8] Jones, J.G. and Earwicker, P.G., (1988) Correlation Detection of Discrete Events in Self-Similar Noise Backgrounds, Submitted for publication.

[9] Jones, J.G. and Haynes, A., (1984) A Peakspotter Program applied to the Analysis of Increments in Turbulence Velocity, Royal Aircraft Establishment Tech. Report 84071.

[10] Jones, J.G. and Thomas, R.W., (1985) A Statistical Model for Localised Natural Features in Measured Satellite Imagery, In 'Digital Signal Processing-84', North Holland, Elsevier Science, 638-643.

[11] Jones, J.G., Thomas, R.W. and Earwicker, P.G., (1989) Fractal Properties of Computer-Generated and Natural Geophysical Data, Computers and Geosciences, 15, 227-235.

[12] Jones, J.G., Thomas, R.W. and Earwicker, P.G., (1988) Multi-Resolution Analysis of Remotely-Sensed Imagery, Submitted for publication.

[13] Woodward, P.M., (1953) Probability and Information Theory, with Applications to Radar, Pergamon, Oxford.

SPREAD AND ENTROPY INEQUALITIES FOR WIGNER WEIGHT FUNCTIONS

A.J.E.M. Janssen
(Philips Research Laboratories, The Netherlands)

ABSTRACT

In this paper we give several necessary and, for radially symmetric functions, necessary and sufficient conditions for a function of two variables to be a Wigner weight function. The necessary conditions are in terms of spread and p-norms of the weight function. We also conjecture an entropy inequality for such weight functions. It is shown by examples that none of the necessary conditions is consistently weaker or stronger than any of the others. Hence, each condition represents a particular feature of Wigner weight functions. The present paper is an abridged version of [7]; for all proofs and derivations we refer to [7].

1. INTRODUCTION

A Wigner weight function $K(t,\nu)$ of time t and frequency ν is a real, square-integrable function such that

$$\iint K(t,\nu)W_f(t,\nu)\,dt\,d\nu \geqq 0 \qquad (1.1)$$

for all square-integrable time-functions $f(t)$ ($\int$ stands for integration over R). Here, W_f is the Wigner distribution of f, given by

$$W_f(t,\nu) = \int e^{-2\pi is\nu}\, f(t+\tfrac{1}{2}s)f^*(t-\tfrac{1}{2}s)\,ds. \qquad (1.2)$$

As an important special case of a Wigner weight function we have $K = W_g$ for any square-integrable function g, since we have (Moyal's formula)

$$\iint W_g(t,\nu)W_f(t,\nu)dtd\nu = |(g,f)|^2 \geqq 0 \qquad (1.3)$$

for all square-integrable functions f.

As is well known, Wigner distributions are useful tools for getting insight into the energy distribution over time and frequency of time signals, see e.g. [3], [5]. However, a serious problem with Wigner distributions is that they exhibit, see e.g. [4], [6], negative values to an extent that smearing operations, compatible in some sense with the Heisenberg uncertainty principle, are necessary to obtain an everywhere non-negative time-frequency distribution. From a resolution point of view one is thus interested in Wigner weight functions that smear as little as possible. Therefore, Wigner weight functions that have a low amount of spread

$$\min_{t_0,\nu_0} \iint [(t - t_0)^2 + (\nu - \nu_0)^2] K^p(t,\nu)dtd\nu \qquad (1.4)$$

with p = 1 or 2, or a low amount of entropy,

$$-\iint K^2(t,\nu) \log K^2(t,\nu)dtd\nu \qquad (1.5)$$

are of particular interest.

In this paper we present necessary and, for radially symmetric functions, necessary and sufficient conditions for $K(t,\nu)$ to be a Wigner weight function. The necessary conditions are in terms of spread, entropy and norms of the weight functions. By studying the examples $K(t,\nu) = K_0(2\pi(t^2 + \nu^2))$ with

$$K_0(s) = \left(\frac{1}{s + a}\right)^\nu , \exp(-bs^\alpha) , \left(\frac{1}{s^2 + c^2}\right)^\nu \qquad (1.6)$$

in some detail, we show that none of the necessary conditions is consistently weaker or stronger than any of the others. Hence, each of the necessary conditions represents a particular feature of Wigner weight functions. All prooofs and derivations of our results are omitted here; we refer instead to [7] in which all proofs and further information can be found. There is, however, one key result that is used over and over in [7] which we would like to mention here: a square-integrable $K(t,\nu)$ is a Wigner weight function if and only if there are $c_n \geqq 0$ with $\Sigma_n c_n^2 < \infty$ and orthonormal f_n such that

$$K = \Sigma_n c_n W_{f_n} . \qquad (1.7)$$

Moreover,

$$\iint K^2(t,\nu)dtd\nu = \Sigma_n c_n^2 , \quad \iint K(t,\nu)dtd\nu = \Sigma_n c_n , \qquad (1.8)$$

the last identity being valid when K is integrable.

2. SPREAD OF WIGNER WEIGHT FUNCTIONS

We have the following two results on Wigner weight functions.

Theorem S_1. Assume that $K \in L^2(R^2)$ is a Wigner weight function with $(t^2 + \nu^2)K(t,\nu) \in L^1(R^2)$. When M is a real, positive-definite 2 x 2-matrix, we have $(x = (t,\nu) \in R^2,\ x_0 = (t_0,\nu_0) \in R^2)$

$$\min_{x_0} \iint (x - x_0)^T M(x - x_0)\ K(x)dx \geqq \frac{|M|^{\frac{1}{2}}}{2\pi} \iint K(x)dx. \qquad (2.1)$$

Theorem S_2. Assume that $K \in L^2(R^2)$ is a Wigner weight function with $(t^2 + \nu^2)K^2(t,\nu) \in L^1(R^2)$. When M is a real, positive-definite 2 x 2-matrix, we have $(x = (t,\nu) \in R^2,\ x_0 = (t_0,\nu_0) \in R^2)$

$$\min_{x_0} \iint (x - x_0)^T M(x - x_0)K^2(x)dx \geqq \frac{|M|^{\frac{1}{2}}}{4\pi} \iint K^2(x)dx. \qquad (2.2)$$

The theorems S_1 and S_2 express that Wigner weight functions have a minimal amount of spread. They are generalizations of De Bruijn's results [2], Secs. 14, 15, for the case that K is a Wigner distribution. Theorem S_1 (a version of it) can also be found in [1], Appendix A.

3. NORMS OF WIGNER WEIGHT FUNCTIONS

We have the following result on the norms of Wigner weight functions.

Theorem N_p. Assume that $K \in L^1 \cap L^2(R^2)$ is a Wigner weight function, and that $p \geqq 1$. Then

$$\iint K^{2p}(t,\nu)dtd\nu \leqq \frac{2^{2p-2}}{p} \left(\iint K(t,\nu)dtd\nu\right)^{2p}. \qquad (3.1)$$

Moreover, K is continuous and bounded.

Theorem N_p generalizes the corresponding inequality of Lieb [8] for the case that K is a Wigner distribution. In fact, the proof of Theorem N_p consists of application of Lieb's result together with the representation result (1.7). The two particular cases p = 1 and p = ∞ give

$$\iint K^2(t,\nu)dtd\nu \leqq \left(\iint K(t,\nu)dtd\nu\right)^2 \tag{3.2}$$

with equality if and only if K is a Wigner distribution, and

$$\max |K(t,\nu)| \leqq 2 \iint K(t,\nu)dtd\nu \ . \tag{3.3}$$

These two particular cases readily follow from (1.7), (1.8) and the fact that $|W_f(t,\nu)| \leqq 2$ when f(t) has unit energy.

The bound on the 2p-norm of K as follows from Theorem N_p is one mathematical way to say that a Wigner weight function cannot be highly peaked.

4. ENTROPY OF WIGNER WEIGHT FUNCTIONS

We have the following conjecture on the entropy of Wigner weight functions.

<u>Conjecture</u> E. Assume that $K \in L^2(\mathbb{R}^2)$ is a Wigner weight function. Then

$$-\iint K^2(t,\nu)\log K^2(t,\nu)dtd\nu \geq \tag{4.1}$$

$$-\iint K^2(t,\nu)dtd\nu \log \left[\frac{4}{e}\iint K^2(t,\nu)dtd\nu\right].$$

The conjecture E has been proved by Lieb for the case that K is a Wigner distribution, [8]. In this context it is instructive to mention the following result: when R_f denotes Rihaczek's distribution

$$R_f(t,\nu) = e^{-2\pi i\nu t} f(t)F^*(\nu) \tag{4.2}$$

with F the Fourier transform of f, then

$$- \iint |R_f(t,\nu)|^2 \log |R_f(t,\nu)|^2 dtd\nu \geqq \tag{4.3}$$

$$- \iint |R_f(t,\nu)|^2 \, dtd\nu \log [\tfrac{2}{e} \iint |R_f(t,\nu)|^2 dtd\nu].$$

This shows that the lower bound of the entropy of $|R_f|^2$ exceeds the one for $|W_f|^2$. Since R_f is a serious competitor of W_f as a time-frequency distribution, this is a relevant result.

Lieb's approach to the proof of Conjecture E for Wigner distributions K consists of differentiating the inequality

$$\iint |K(t,\nu)|^{2p} dtd\nu \leqq \frac{2^{2p-2}}{p} \left(\iint K^2(t,\nu)dtd\nu \right)^p, \quad p \geqq 1 \tag{4.4}$$

and setting $p = 1$. However, inequality (4.4) is not valid for general Wigner weight functions.

We mention that the inequality

$$- \iint K^2(t,\nu) \log K^2(t,\nu)dtd\nu \geqq \tag{4.5}$$

$$- \left(\iint K(t,\nu)dtd\nu \right)^2 \log \left[\frac{4}{e} \left(\iint K(t,\nu)dtd\nu \right)^2 \right]$$

which holds when K is a Wigner distribution, fails to hold for general Wigner weight functions.

5. RADIALLY SYMMETRIC WIGNER WEIGHT FUNCTIONS

We consider now $K(t,\nu)$ of the special form

$$K(t,\nu) = K_o(2\pi(t^2 + \nu^2)), \tag{5.1}$$

where the radius function K_o is of the form

$$K_o(s) = \int_0^{\infty} e^{-su} P(u)du \tag{5.2}$$

with $P \in L^2([0,\infty))$. We have the following result.

Theorem R. The K of the above form is a Wigner weight function if and only if

$$\int_0^1 \frac{z^n}{1+z} \left[P\left(\frac{1-z}{1+z}\right) + (-1)^n \frac{1+z}{1-z} P\left(\frac{1+z}{1-z}\right)\right] dz \geqq 0 \quad (5.3)$$

for $n = 0,1,\ldots$.

Corollary. The K of the above form is a Wigner weight function whenever

$$P(u) > \frac{1}{u} \left|P\left(\frac{1}{u}\right)\right| , \quad u \in (0,1]. \qquad (5.4)$$

In [7], Sec. 5, a condition on P is given under which checking of inequality (5.3) is only necessary for $n = 0,1$.

We observe that the function $\exp(-2\pi u(t^2 + \nu^2))$ is a Wigner weight function if and only if $0 < u \leqq 1$. Hence, inequality (5.4) expresses how much weight a point $u > 1$ in the integral (5.2) may have so that K is still a Wigner weight function.

6. EXAMPLES

I. Consider the radius function K_0 given by

$$K_0(s) = \left(\frac{1}{s+a}\right)^{\nu} , \quad a > 0, \ \nu > \tfrac{1}{2}. \qquad (6.1)$$

Theorem R and its extensions yield: there is a function $R(\nu)$, $\nu > \frac{1}{2}$, with $0 < R(\nu) < \nu-\frac{1}{2}$ such that K is a Wigner weight function if and only if $a \geqq R(\nu)$. This function can be expressed in terms of incomplete Γ-functions, but is not easy to handle.

Similarly, Theorems S_1, S_2, N_p and Conjecture E yield functions $S_1(\nu)$, $S_2(\nu)$, $N_p(\nu)$, $E(\nu)$ such that inequality S_1, S_2, N_p, E holds for K if and only if $a \geqq S_1(\nu)$, $S_2(\nu)$, $N_p(\nu)$, $E(\nu)$, respectively. We have for $p \geqq 1$

$$S_1(\nu) \leqq N_1(\nu) \leqq N_p(\nu) \leqq N_\infty(\nu) \leqq S_2(\nu) \leqq E(\nu) \leqq R(\nu), \qquad (6.2)$$

which shows that S_1 gives the weakest condition on a and R gives the strongest condition on a. Also, see Figure 1.

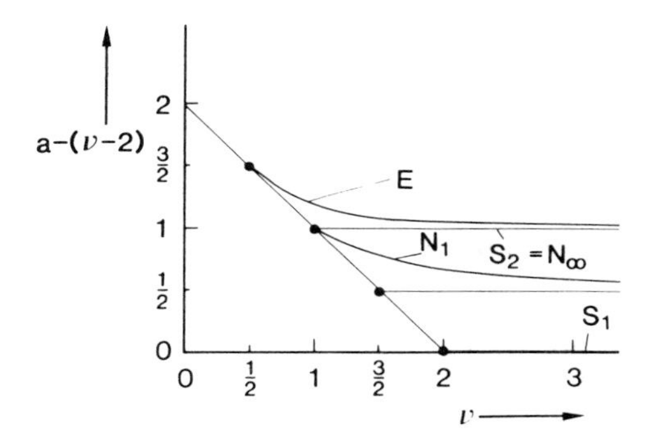

Fig. 1 The bounds S_1, S_2, N_1, N_∞, E on a for Example I as a function of a - (ν-2).

II. Consider the radius function

$$K_o(s) = \exp(-bs^\alpha), \quad b > 0, \ \alpha > 0. \tag{6.3}$$

Then K is a Wigner weight function if and only if $\alpha = 1$ and $0 < b \leqq 1$. As in Example I we find functions $S_1(\alpha)$, $S_2(\alpha)$, $N_p(\alpha)$, $E(\alpha)$ such that inequality S_1, S_2, N_p, E holds for K if and only if $0 < b \leqq S_1(\alpha)$, $S_2(\alpha)$, $N_p(\alpha)$, $E(\alpha)$, respectively. We have $S_1(1) = S_2(1) = N_p(1) = E(1) = 1$, and

$$S_1(\alpha) \geqq N_1(\alpha) \geqq S_2(\alpha) \geqq E(\alpha) \geqq N_\infty(\alpha), \quad 0 < \alpha \leqq 1, \tag{6.4}$$

and

$$S_1(\alpha) \leqq N_1(\alpha) \leqq E(\alpha) \leqq S_2(\alpha) \leqq N_\infty(\alpha), \quad \alpha \geqq 1 \tag{6.5}$$

(note that $S_2(\alpha)$, $E(\alpha)$ in (6.4) and (6.5) are interchanged).

Hence, S_1 gives the weakest condition and N_∞ gives the strongest condition on b when $0 < \alpha \leqq 1$, whereas the state of affairs is reversed when $\alpha \geqq 1$. Also, see Figure 2.

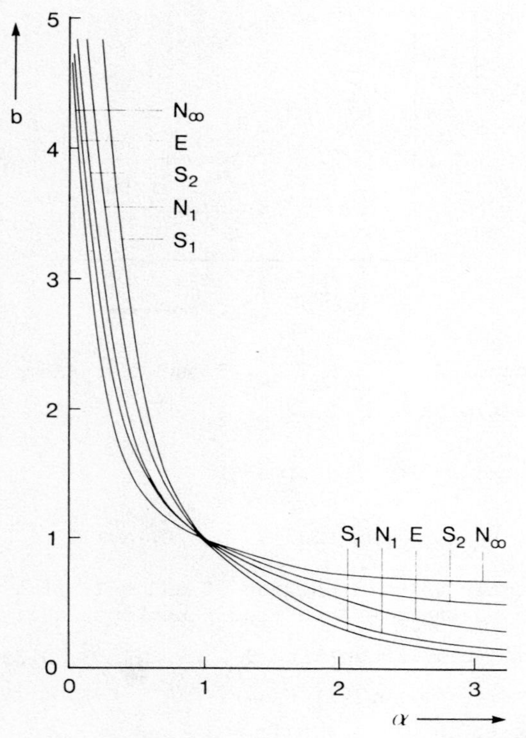

Fig. 2 The bounds S_1, S_2, S_1, N_∞ E on b for Example II as a function of α

III. Consider the radius function

$$K_0(s) = \left(\frac{1}{s^2 + c^2}\right)^\nu, \quad c > 0, \ \nu > \tfrac{1}{4} . \tag{6.6}$$

We do not know a necessary and sufficient condition on (c,ν) such that K is a Wigner weight function. As in Examples I, II we can find functions $S_1(\nu)$, $S_2(\nu)$, $N_p(\nu)$, $E(\nu)$ such that inequality S_1, S_2, N_p, E holds for K if and only if $c \geqq S_1(\nu)$, $S_2(\nu)$, $N_p(\nu)$, $E(\nu)$, respectively. The situation is now more complicated than in the Examples I, II, see Figure 3.

The conclusion from the above examples is that none of the conditions S_1, S_2, N_p, E is consistently stronger or weaker than any of the others.

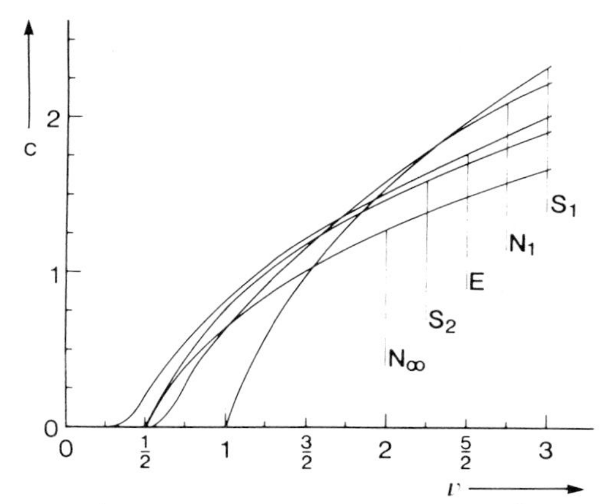

Fig. 3 The bounds S_1, S_2, N_1, N_∞, E on c for Example III as a function of ν.

8. REFERENCES

[1] Bastiaans, M.J., (1983) Uncertainty principle for partially coherent light, J. Opt. Soc. Am. 73, 251-255.

[2] De Bruijn, N.G., (1973) A theory of generalized functions, with applications to Wigner distribution and Weyl correspondence, Nieuw Archief voor Wiskunde 21, 205-280.

[3] Claasen, T.A.C.M. and Mecklenbräuker, W.F.G., (1980) The Wigner Distribution - A tool for time-frequency signal analysis, Part I, *Philips J. Res.* **35**, 217-250.

[4] Hudson, R.L., (1974) When is the Wigner quasi-probability density non-negative?, *Rept. Math. Phys.* **6**, 249-252.

[5] Janse, C.P. and Kaizer, A.J.M., (1983) Time-frequency distributions of loudspeakers: The application of the Wigner distribution, *J. Audio Engrg. Soc.* **31**, 198-223.

[6] Janssen, A.J.E.M., (1988) Positivity of time-frequency distribution functions, Signal Processing 14, 243-252.

[7] Janssen, A.J.E.M., (1988) Wigner weight functions and Weyl symbols of non-negative definite linear operators, submitted to *Philips J. Res.*

[8] Lieb, E.H., (1988) Integral Bounds for Radar Ambiguity functions and Wigner Distributions, preprint.

A MULTIRESOLUTION DESCRIPTOR FOR NONSTATIONARY IMAGE PROCESSING

A. Calway and R. Wilson
(University of Warwick)

ABSTRACT

This paper considers the problem of representing image features within the framework of a new multiresolution image descriptor. A general feature model is defined and the particular case of estimating localised line and edge features is then considered. Results of implementing the estimation scheme on a natural image are presented. In addition, it is shown that the general model can also accommodate the definition of more complex features and is therefore a potential basis for a unified image description.

1. INTRODUCTION

Signal representations which combine both signal and frequency domain information have been of considerable interest in recent years [3][4][5][9] - they provide a means of describing signals which have important features concentrated in both domains, ie they exhibit a degree of nonstationarity. This is particularly applicable to so-called natural signals such as speech, music and optical images.

One such representation which has important novel features was recently introduced in [13][14]. With the aim of overcoming the problems of uncertainty associated with previous methods, this invertible signal descriptor incorporates a multiresolution set of combined representations into a single structure - ranging from the original image to its DFT. Moreover, its hierarchical structure is derived from a generalisation of the pyramidal techniques which have been of considerable use in overcoming the problems of uncertainty encountered in image processing [1][10].

The purpose of this paper is to consider the problem of representing image features within a 2d version of the new descriptor. Towards this end, a general multiresolution image model is defined such that the image is modelled over a set of spatial resolutions, ranging from a per pixel to global basis. Within this model, the particular case of representing and subsequently estimating spatially localised line and edge features is then considered. These features are known to be important in mammalian vision [6][8] and have consequently received considerable attention from a number of workers in image processing and computer vision [2][7].

After defining the structure and notation for the image descriptor, the feature model and estimation scheme are described and results of an implementation on a natural image are presented. The paper concludes with comments on the applicability of the model to more complex features.

2. MULTIRESOLUTION IMAGE DESCRIPTOR

Representing an image $\nu(k,\ell)$ $0 \leq k, \ell < N$ by a 2d vector $\nu_{k\ell} = \nu(k,\ell)$, a descriptor related to the cartesian separable descriptor introduced in [14] is characterised by the invertible linear operator G, ie:

$$u = G\,\nu \quad \nu = G^{-1}u \tag{2.1}$$

The operator G is defined by a set of operators G(k):

$$G = \begin{pmatrix} G(0) \\ \vdots \\ G(k) \\ \vdots \\ G(n) \end{pmatrix} \qquad n = \log_2 N \tag{2.2}$$

while the descriptor coefficients are represented by the vector u:

$$u = \begin{pmatrix} u(0) \\ \vdots \\ u(k) \\ \vdots \\ u(n) \end{pmatrix} \tag{2.3}$$

or in terms of individual components:

$$u_{xypq}(k) = \sum_{ij} G_{xypqij}(k)\, v_{ij} \tag{2.4}$$

$0 \leq x,y < \Omega_k \quad 0 \leq p,q < S_k \quad S_k = 2^k \quad \Omega_k S_k = 2^m N \quad 0 \leq k \leq n \quad 0 \leq m \leq k$

In other words, the vector u consists of a set of $n = \log_2 N$ 4d vectors whose individual components are given by the inner product of the vectors $G_{xypq}(k)$ and the image ν. Consequently it is these vectors which determine the specific nature of the descriptor.

The vectors $G_{xypq}(k)$ each possess the following essential properties:

(i) in the discrete spatial domain, the energy is concentrated into the set $\Lambda = (xS_k, (x+1)S_k - 1) \cap (yS_k, (y+1)S_k - 1)$.

(ii) in the discrete spatial frequency domain, the energy is concentrated into the set $\Gamma_m = ((p - \frac{m}{2})\Omega_k, (p+1+\frac{m}{2})\Omega_k - 1)$ $\cap((q - \frac{m}{2})\Omega_k, (q+1+\frac{m}{2})\Omega_k - 1)$.

(iii) the energies E_Λ and E_{Γ_m}, which are concentrated into the sets Λ and Γ_m respectively, are such that $E_{\Gamma_m} = E_T$ and $\varepsilon_m = E_T - E_\Lambda$, where E_T is the total energy in $G_{xypq}(k)$, ε_0 is a minimum for all possible $G_{xypq}(k)$ and $\varepsilon_0 < \varepsilon_1 \ldots < \varepsilon_k$.

The implications of these properties are as follows. From equation (2.4) and properties (i) and (ii), the individual vectors u(k) represent combined space-spatial frequency descriptions of the image. Furthermore, the vector u is a multiresolution set of such descriptions, with the index k defining the resolution of each level of the set.

This resolution is determined by the criteria in property (iii). The maximally concentrated criterion, ie m=0, ensures that the vectors $G_{xypq}(k)$ satisfy a minimum uncertainty condition which is based upon a measure of energy. Consequently, the resulting image descriptions then possess the maximum possible resolution in one domain given the pre-defined resolution in the other domain. In contrast, for $0<m\leq k$ this condition no longer applies since the spatial frequency resolution is reduced by enlarging the set Γ_0 to Γ_m while still retaining the spatial resolution defined by the set Λ. The resulting descriptions can therefore be regarded as generalised implementations with an increased number of coefficients, eqn.

(2.4). The trade-offs and implications of adopting either the maximally concentrated descriptor or a generalised version are discussed later.

The conditions of minimum uncertainty imposed on the vectors $G_{xypq}(k)$ mean that they are the finite prolate spheroidal sequence (fpss) [13]. However to enable the generalised implementations an additional condition needs to be defined for such sequences. Thus the following definition is adopted:

$$G_{xypq}(k) = B(\Gamma_m) I(\Lambda) e(k,x,y,p,q) \qquad (2.5)$$

where e(.) is the 2d eigenvector corresponding to the largest eigenvalue $\lambda(\Omega_k, S_k)$ of the finite prolate spheroidal equation [13]:

$$B(\Gamma_0) I(\Lambda) e(.) = \lambda(\Omega_k, S_k) e(.) \qquad (2.6)$$

and the operators I(.) and B(.) are truncation and bandlimiting operators respectively:

$$I_{xypq}(\Lambda) = \delta_{xp}\,\delta_{yq} \qquad (x,y) \in \Lambda \qquad (2.7)$$

$$= \quad 0 \qquad \text{else}$$

$$B(\Gamma_0) = F^{+} I(\Gamma_0) F \qquad (2\ 8)$$

with F and F^{+} being the forward and inverse 2d DFT operators.

In the above, equation (2.5) enables a relaxation of the minimum uncertainty condition and thus the definition of a generalised descriptor. The motivation for this relaxation criterion is primarily heuristic although it is derived from signal theory. By allowing the vectors $G_{xypq}(k)$ to occupy a wider bandwidth (ie Γ_0 to Γ_m) the sidelobe energy apparent in the spatial response of the orginal fpss is reduced, at the expense of reducing the amount of energy in the region of interest Λ. It has been found that a trade-off between this sidelobe energy and concentration energy can be adopted to achieve better results when using the descriptor. As a result, a generalised descriptor with m=1 has been found to be an acceptable trade-off.

To conclude this section, the descriptor coefficients represent a multiresolution set of space-spatial frequency descriptions of the image. In its maximally concentrated form each level of the descriptor possesses the maximum possible resolution in one domain given a pre-defined resolution in the other domain. It was in this form that the descriptor was originally introduced in [14]. A generalised version has been defined here to provide a more adaptable descriptor and has been found to have important benefits. The principal properties of the maximally concentrated descriptor are detailed in [14] and these are readily extended to the generalised versions.

3. FEATURE MODEL AND ESTIMATION

3.1 The Model

Having defined the multiresolution structure of the descriptor it is now possible to consider the representation of image features within such a structure. The motivation for doing so is twofold: first, important image features are localised both in space and spatial frequency; second, it is widely accepted that such features exist over a range of spatial scales [15]. Given these two criteria the new descriptor clearly presents a possible basis for a general feature representation.

In order to facilitate this representation it is convenient to first define an image model which is able to incorporate relevant features and is at the same time consistent with the structure of the descriptor. One such model is a particular example of the general multiresolution model described in [12]. This has a linear recursive form,

$$v(k) = A(k)\ v(k+1) + B(k) f(k) \qquad 0 \leq k < n \tag{3.1.1}$$

where in this case $v(k)$ and $f(k)$ are 4d vectors:

$$v(k) = \left(\nu_{xywz}(k)\right) \quad 0 \leq w,z < S_{km} \quad 0 \leq x,y < \Omega_k \quad S_{km} = \frac{S_k}{2^m}$$

and $A(.)$, $B(.)$ are linear operators, i.e: (3.1.2)

$$\nu_{xywz}(k) = \sum_{\alpha\beta\gamma\kappa} A_{xywz\alpha\beta\gamma\kappa}(k)\ \nu_{\alpha\beta\gamma\kappa}(k+1) + B_{xywz\alpha\beta\gamma\kappa}(k)\ f_{\alpha\beta\gamma\kappa}(k) \tag{3 1.3}$$

By defining $f(k)$ to consist of a set of 2d feature vectors $\xi(.)$:

$$f_{xy}(k) = \xi(k,x,y) \tag{3.1.4}$$

this model is able to incorporate the definition of image features over a complete range of resolutions, subject to the initial condition:

$$v(n) = B(n)f(n) \tag{3 1.5}$$

For this model the resulting image ν is just:

$$v = v(0) \tag{3.1.6}$$

The properties of the general model in equation (3.1.1) are fully discussed in [12].

As stated earlier this work is concerned with the representation of localised line and edge features. In this case the feature vectors $\xi(.)$ have the following form:

$$\xi_{wz}(k,x,y) = \sum_i h_{kxyi}(w\cos\theta_i + z\sin\theta_i + C_{kxyi}) + g_{\alpha\beta} \quad \alpha = xS_{km}+w \quad \beta = yS_{km}+z \tag{3.1.7}$$

ie the summation of real 1d functions $h(.)$ in a number of discrete orientations θ_i with offsets C_{kxyi} confined to the region of interest:

$$0 \leq C_{kxyi} < \frac{S_k}{2^m}(\sin\theta_i - \cos\theta_i) \tag{3.1.8}$$

The vector g is a lowpass contribution and is the result of convolving the image $v(0)$ with a Gaussian kernel:

$$g_{xy} = \frac{1}{2\pi\sigma^2} \sum_{pq} e^{\frac{(p^2+q^2)}{2\sigma^2}} \nu_{x-p,y-q}(0) \tag{3.1.9}$$

The model defined in equation (3.1.1) also has the additional constraints:

$$\begin{aligned} B_{xywz\alpha\beta\gamma\kappa}(k) &= \delta_{x\alpha}\delta_{y\beta}\delta_{w\gamma}\delta_{z\kappa} \quad \text{iff} \sum_{wz\alpha\beta\gamma\kappa} \left| A_{xywz\alpha\beta\gamma\kappa}(k) \right| = 0 \\ &= 0 \quad \text{else} \end{aligned} \tag{3.1.10}$$

and

$$\begin{aligned} A_{xywz\alpha\beta\gamma\kappa}(k) &= \delta_{xS_{km}+w,\,\alpha S_{(k-1)m}+\gamma}\;\delta_{yS_{km}+z,\,\beta S_{(k-1)m}+\kappa} \quad \text{iff} \sum_{wz\alpha\beta\gamma\kappa} \left| B_{xywz\alpha\beta\gamma\kappa}(k) \right| = \\ &= 0 \quad \text{else} \end{aligned} \tag{3.1.11}$$

These constraints impose a mutually exclusive condition on equation (3.1.1), ie. a vector $v_{xy}(k)$ is either equal to a new future vector $\xi(k,x,y)$ or is equal to a quadrant of the relevant vector on level k + 1. An example of the structure of this model is illustrated in figure 1.

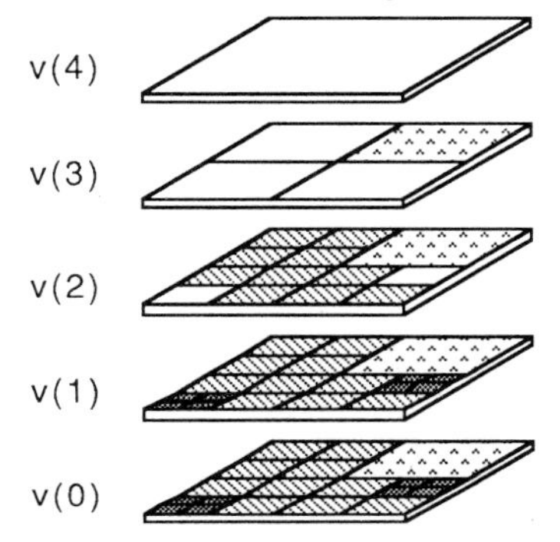

Fig. 1 Multiresolution feature model for N=16 and m=0, shaded areas indicate vectors $v_{xy}(k)$ which contribute to the image v(0).

3.2 Estimation

Now that a suitable model has been defined it is possible to consider the estimation of the line and edge features within its structure, ie f(k). The model constraints in equations (3.1.10) and (3.1.11) mean that these estimates can be derived directly from the original image. Furthermore, given the similarity in structure of the model and the descriptor in section 2, the estimates can be obtained from the descriptor coefficients.

From equations (2.4) and (3.1.2), the vector f(k) has a similar structure to that of the descriptor coefficients u(k). Moreover, from the definition of G(k), the vector $u_{xy}(k)$ is essentially an approximation to a frequency representation of the image region from which the feature vector $\xi(k,x,y)$ is to be estimated, eqns (2.4), (3.1.2) and (3.1.4). Based upon this approximation it is possible to obtain an estimate for the model parameters in equation (3.1.7) and thus an overall estimate for the vector f(k). The validity of the approximation is not discussed here, although it is derived from the completeness of the descriptor [14] and the suitability of the fpss's for spectrum estimation [11].

The estimation scheme is based upon an ideal continuous case. For a continuous image function f(x,y) varying in only one direction, i.e.:

$$f(x,y) = f(x\cos\theta + y\sin\theta + c) \quad (3.2.1)$$

its 2d Fourier transform is given by

$$F(\sigma,\omega) = 2\pi F(\sigma\cos\theta + \omega\sin\theta)\,\delta(\omega\sin\theta - \sigma\cos\theta)\,e^{-j(\sigma\cos\theta + \omega\sin\theta)c} \quad (3.2.2)$$

Using this relationship and assuming the model of equation (3 1.7), the descriptor coefficient vector $u_{xy}(k)$, where:

$$u(k) = G(k)\left[v(O) - g\right] \quad (3.2.3)$$

is assumed to have the following property: in an orientation θ_i on the discrete lattice defined by $u_{xy}(k)$, there exists a linear phase component which is proportional to the offset C_{kxyi}, ie:

$$\varphi_{\theta_i}(r) \approx aC_{kxyi}r + b \qquad a,b - \text{constant.} \quad (3.2.4)$$

and the energy in this orientation is an estimate of the energy in the function $h_{kxyi}(.)$.

The estimation scheme is therefore as follows. From the descriptor coefficient vector defined by equation (3.2.3), an interpolated vector $\tilde{u}(k)$ is given by:

$$\tilde{u}_{xyir}(k) = \sum_{pq} \alpha_{pq}(i,r)u_{xypq}(k) \qquad O \le i,r < S_k \quad (3.2.5)$$

where the weights $\alpha_{pq}(.)$ are chosen so that on the discrete cartesian lattice (p,q) the coefficients $\tilde{u}_{xyi}(k)$ are discrete (bi-linear) interpolated values in an orientation $\theta_i = \frac{i\pi}{S_k}$. A feature estimation vector $\nu(k)$ is then defined:

$$\nu_{xyi}(k) = \tilde{u}_{xyi}^{+}(k)D(k)\tilde{u}_{xyi}(k) \quad (3.2.5)$$

where $^{+}$ indicates conjugate transpose and the operator D(.) is a non-circular shift operator:

$$D_{pq}(k) = \delta_{p+1,q} \qquad O \le q < S_k \qquad O \le p < S_k - 1 \quad (3.2.6)$$

From equations (3.2.2), (3.2.4) and (3.2.5), this feature estimation vector is assumed to have the following form:

$$\nu_{xyi}(k) = \psi(x,y,i)e^{-ja\varsigma_{kxyi}} \tag{3.2.7}$$

where $\psi(x,y,i)$ is related to the energy in $h_{kxyi}(.)$ providing the model of equation (3.1.7) is reasonable and the various approximation errors are reduced by the averaging process implied by the inner product in equation (3.2.5).

4. RESULTS

The feature estimation scheme was implemented on a (256x256) pixel monochrome 'girl' image. The set of feature estimation vectors $\nu(k)$ $0 \leq k \leq 8$ were determined and the results displayed as block line segments on a discrete lattice with offset $C(.)$ and a normalised magnitude $\psi(.)$. The magnitudes of levels 3 and 5 of the descriptor u defined by equation (3.2.3) are shown in figures 2 and 3, where the descriptor is a generalised version with m=1 and the Gaussian kernel of equation (3.1.9) has a variance $\sigma^2 = 32/\pi N$. The levels are shown with the components of the spatial frequency vectors $u_{xy}(k)$ grouped together and displayed on a discrete lattice (x,y). The feature estimation results for $2 \leq k \leq 5$ are shown in figure 4. At the lower levels, where the spatial block size $(S_{km} x S_{km})$ is small, the feature model of equation (3.1.7) appears to be reasonable and the results compare favourably with good edge detection methods. In contrast, and as might be expected, at the higher levels the model becomes too simple: curved shapes occur within the larger blocks while the estimates are straight line approximations.

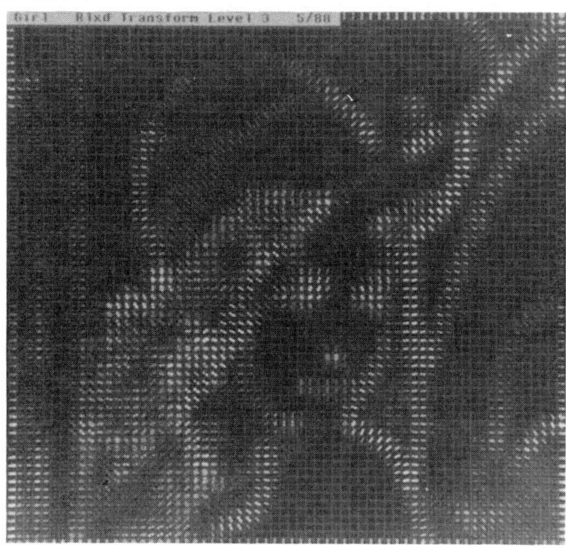

Fig. 2 Coefficient magnitudes for level 3 of the descriptor

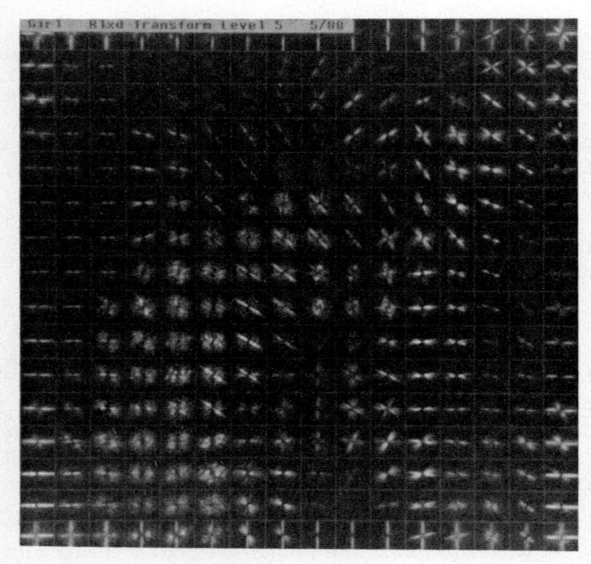

Fig. 3 Coefficient magnitudes for level 5 of the descriptor

Fig. 4 Feature estimation results for levels 2 to 5.

5. CONCLUSIONS

A generalised multiresolution image descriptor has been presented which combines space and spatial frequency information in a hierarchical structure. Within this framework a general feature model was defined and, by way of example, a line and edge estimation scheme was implemented. This showed the feature model to be acceptable at lower levels but too simple at higher levels. At these levels, more complex feature models need to be defined which represent, for example, curves and are able to confirm or contradict the information apparent at the lower levels. It is important to note that such features can be incorporated into the present scheme by re-defining the feature model of equation (3.1.7). Furthermore, due to the structure of the descriptor, the resulting feature vectors can be directly related to their less complex counterparts below. These complex models are the subject of current investigations.

ACKNOWLEDGEMENT

This work has been performed with the support of the UK SERC.

REFERENCES

[1] Burt, P.J. and Adelson, E.H., (1983) The Laplacian Pyramid as a Compact Image Code, IEEE Trans. COM 31, 532-540.

[2] Canny J , (1986) A Computational Approach to Edge Detection, IEEE Trans. PAMI, 8, 6.

[3] Claasen, T.A.C.M. and Mecklenbräuker, W.F.G., (1980) The Wigner Distribution - a Tool for Time-Frequency Analysis II: Discrete Time Signals, Philips Jnl. Res., 35, 276-300.

[4] Daugman, J.G., (1988) Complete Discrete 2-D Gabor Transform by Neural Networks for Image Analysis and Compression, IEEE Trans. ASSP, 36, 7, 1169-1179.

[5] Gabor, D., (1946) Theory of Communication, Proc. IEE, 93, 26, 429-441.

[6] Hubel, D. and Wiesel, T.N., (1979) Brain Mechanisms of Vision Sci. Amer.

[7] Knutsson, H., Wilson, R. and Granlund, G.H., (1983) Estimating the Local Orientation of Anisotropic 2-d Signals, Proc. IEEE ASSP Workshop on Spec. Est., 234-239.

[8] Marr, D., (1982) Vision, Freeman, San Francisco.

[9] Portnoff, M.R., (1980) Time Frequency Representation of Digital Signals and Systems Based on Short-Time Fourier Analysis, IEEE Trans. ASSP, 28, 55-69.

[10] Spann, M. and Wilson, R., (1985) A Quad-Tree Approach to Image Segmentation which Combines Statistical and Spatial Information, Patt. Recog., 18.

[11] Thomson, D.J., (1982) Spectrum Estimation and Harmonic Analysis, Proc. IEEE, 70, 9, 1055-1096.

[12] Wilson, R. and Clippingdale, S.C., (1988) A Class of Nonstationary Image Models and their Applications, Proc. IMA Conf. Maths. in Sig. Process.

[13] Wilson, R. and Spann, M., (1987-88) Finite Prolate Spheroidal Sequences and Their Applications I & II, IEEE Trans. PAMI, 9 & 10.

[14] Wilson, R. and Calway, A., (1988) A General Multiresolution Signal Descriptor and its Application to Image Analysis, Proc. EUSIPCO-88.

[15] Wilson, R. and Granlund, G.H., (1984) The Uncertainty Principle in Image Processing, IEEE Trans. PAMI, 6, 6.

CLASSIFICATION OF POINT PROCESSES USING PRINCIPAL COMPONENT ANALYSIS

N.B. Jones
(Department of Engineering, University of Leicester)

P.J.A. Lago
(Groupo de Matematica Aplicada, Faculdade de Ciencias, Universidade do Porto, Portugal)

and

A. Parekh
(School of Engineering and Applied Sciences, University of Sussex, Brighton)

ABSTRACT

Point processes can be characterised by the Fourier Transform of the covariance density function, the Bartlett spectrum. This paper is concerned with a practical approximation to this spectrum used as a first step in a data reduction procedure. Two methods of classifying these spectra are described, based on the use of principal component analysis. The first uses a single data base of known standards and the second which is new, uses as many data bases of standards as there are classes. Examples are presented in which the point processes are derived from noise-like continuous signals (electromyograms) where events in the point process are derived from the significant peaks in the electromyogram. The methods are shown to be good classifiers of simulated data in the limited study so far carried out.

1. INTRODUCTION

Most signals contain information on the size and the timing of the activities being measured. A point process on the other hand contains just the timing information since it is, by definition, only a sequence of event times.

There are many naturally occurring point processes such as those generated by firing of neurones. However point processes are also generated deliberately as intermediate signals in data reduction schemes. For example this is done to extract timing information from complicated continuous signals which are first reduced to point processes by marking the significant features as events on the time axis.

Examination of time dependency between events in a point process can be conducted using the histogram of intervals or some other time domain method. Alternatively frequency domain techniques can be used. A useful frequency domain description of a point process is the Bartlett spectrum which is the Fourier Transform of the covariance density of the intervals between events [1]. It has been shown [2] that in many practical cases the Bartlett spectrum is well approximated by the Fourier Transform of a binary sequence related to the original point process by a time quantisation and decimation procedure as shown in Fig. 1.

This paper presents a new classification technique for point processes based on the use of principal component analysis of their spectra, where the spectra are defined by the process illustrated in Fig. 1.

The motivation for this work is to find improved ways of discriminating between cases in neurology by examining the time dependence between the major events in the bio-electric process which accompanies muscle contraction.

If needle electrodes are inserted into the muscle being studied, then, for strong contractions, the electrochemical process can be observed as a noise-like electrical signal known as the *"Interference EMG"*. This interference EMG can be reduced to a point process, and hence to a binary sequence, as illustrated in Fig. 2 ready for spectrum analysis using an FFT. The major features of interest in this work are the peaks in the EMG and their relative timing. An algorithm is used to generate the point process by marking significant peaks in the EMG as shown in Fig. 2. This algorithm incorporates two parameters, the *"noise reduction factor"* and the *"peak discrimination factor"*, used to reduce the effect of local noise and of distant bio-electric activity.

Methods are described for classifying the spectra of the point processes derived in this way which seem general enough to be useful for classification of any stationary point processes.

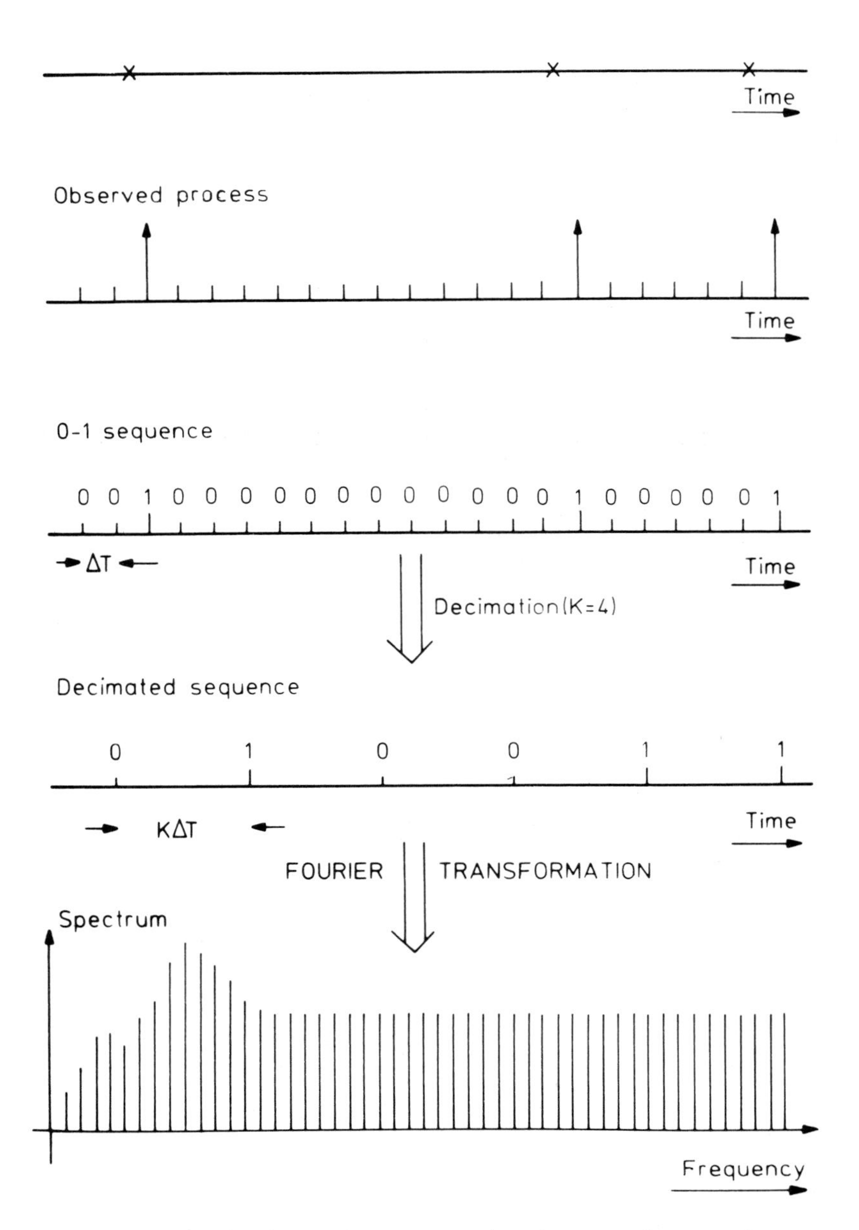

Fig. 1 The calculation of an approximation to the Bartlett Spectrum

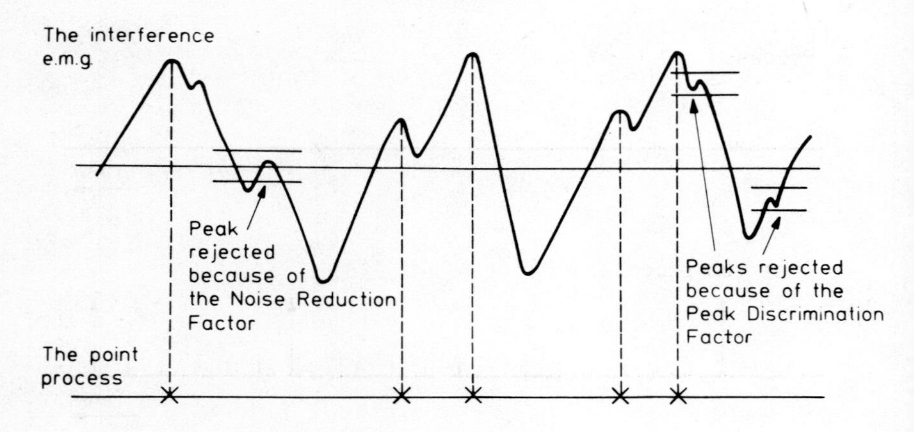

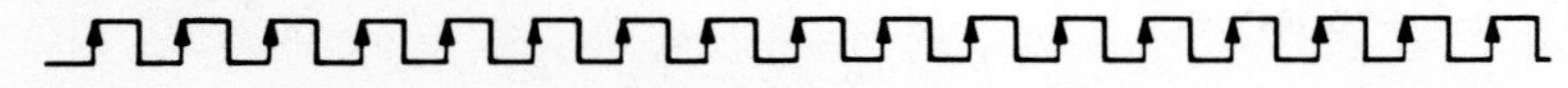

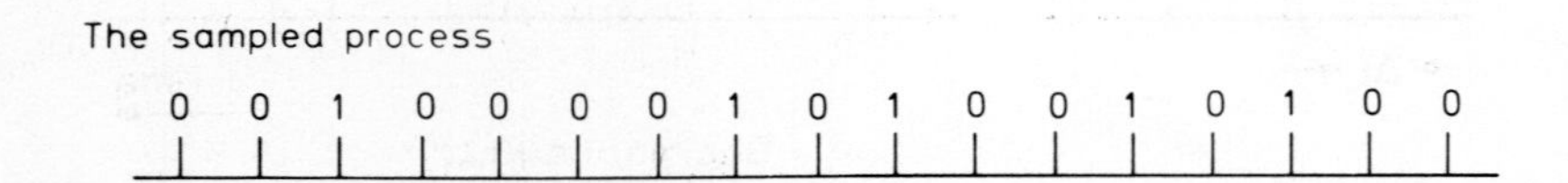

Fig. 2 Forming the point process from the interference EMG

2. ELECTROMYOGRAMS

A muscle contraction is a continuous response to a set of discrete events, the motor neurone firings. Each motor neurone firing causes a particular group of muscle fibres within the muscle to contract and generate force. At the same time an electro-chemical activation wave travels along each group of fibres and produces a component of the EMG known as a *"motor unit action potential"*. One motor neurone and its associated muscle fibres is known as a *"motor unit"* often referred to as a *"unit"*. Each motor unit fires in a quasi-periodic manner and is generally not synchronised with other units. The result is a smooth contraction of the muscle and also the generation of the complex noise-like signal known as the *"interference EMG"*. This noise-like quality arises because of the overlap of non-synchronised motor unit action potentials whose shapes are different because of anatomical differences between the units, differences of the geometry of the units relative to the electrodes and possibly also because of pathology within the units, [3].

Some interference EMGs are shown in Fig. 3. Data in this figure are from pathological cases already diagnosed and are examples of well established disease states. The work presented here arises from an attempt to bring more discriminating power into the diagnosis of less well defined and earlier pathologies.

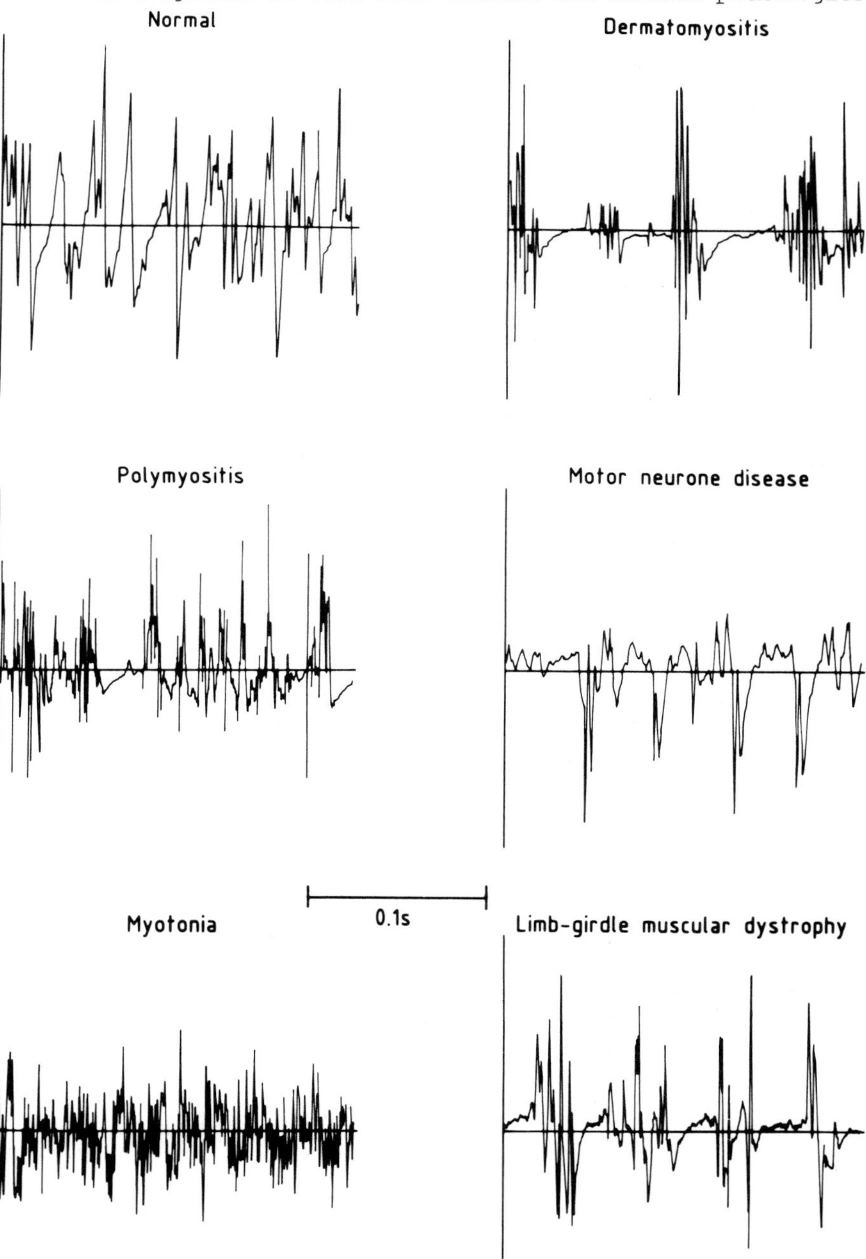

Fig. 3 Interference EMGs

3. TURNING POINTS SPECTRA

The interference EMGs shown in Fig. 3 are much easier to observe than individual motor unit action potentials and even when it is possible to observe the individual potentials it is not certain that the few observed are representative of the whole muscle. In clinical situations it is simple to obtain interference EMGs by asking the patient to contract *"as hard as possible"* but in this situation the signal is complex as shown in Fig. 3.

Nevertheless important diagnostic information exists in the shape of individual potentials but this is obscured when a large number of these potentials interfere and overlap as they do in the interference EMG. Two particularly important features of the shape are, the number of peaks in, and the length of, the action potential. However, it has been shown that direct spectral analysis of the interference EMG is not an efficient method of highlighting this type of information [4] and that it is better to reduce the amount of information in the signal by turning the interference EMG into the point process of the turning points of the EMG and then studying the spectrum of this process using the methods already outlined [2].

The spectra of the turning points derived from the EMGs of Fig. 3 are shown in Fig. 4. It is spectra of this type but derived from simulated EMGs which are used to illustrate the procedure being proposed here.

4. SIMULATIONS

In order to see if useful information can be extracted using the methods indicated it is advantageous to employ simulation of the interference EMG rather than to use real data. This technique allows an a posteriori assessment of how effective the method is in recognising the fundamental properties of the signal which have been built into the simulation.

The simulation is based on methods previously described [5,6]. Briefly, the simulated EMG is built up by the addition of several component quasi-periodic waves, each of which represents an active motor unit. Each motor unit has associated with it a particular action potential which could be from any one of the four classes; normal biphasic, normal triphasic, neurogenic or myopathic. Each active motor unit is therefore a train of action potential of the same basic size and shape (although some random variation is introduced to simulate noise effects) which occur at almost regular intervals. The intervals are in fact varied so that the occurrence times are defined by a Gaussian renewal process. The total EMG can be any combination of up to twenty of such action potential trains.

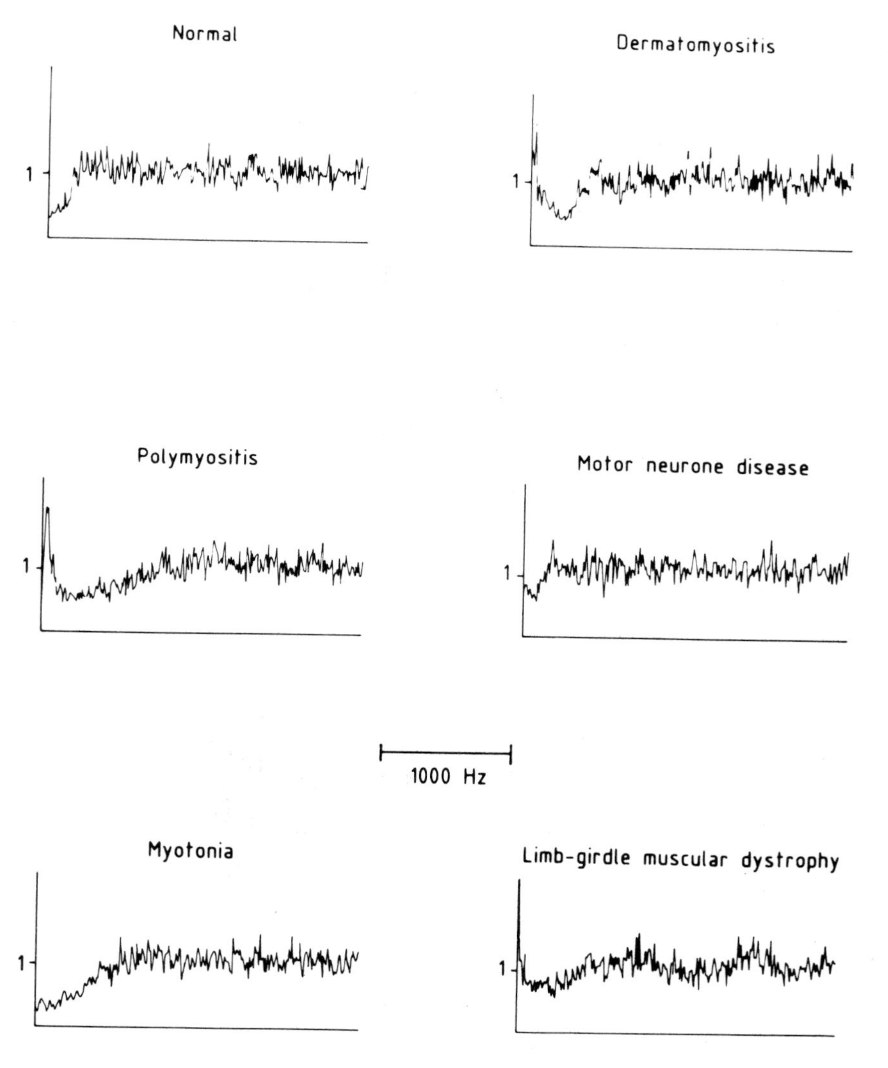

Fig. 4 Turning point spectra

Fig. 5 shows two examples. Both are normal EMGs consisting of 50% biphasic and 50% triphasic action potentials. The first has six active units and the second twenty active units.

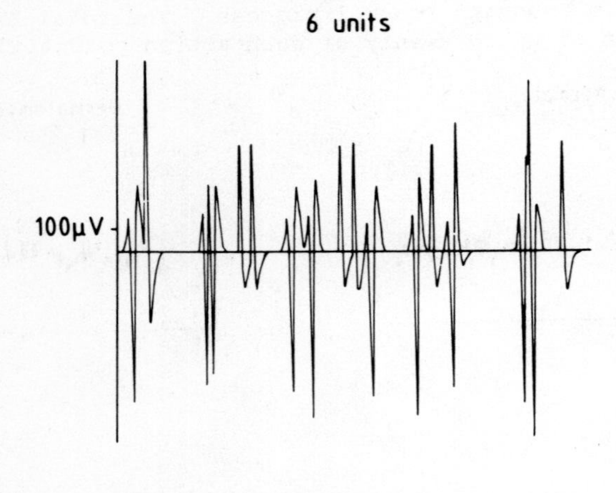

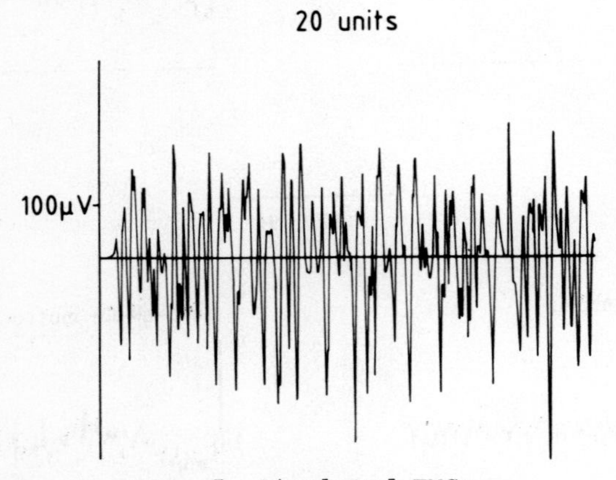

Fig. 5 Simulated EMGs

5. FURTHER DATA REDUCTIONS

In forming the averaged spectra of fig. 4. There has been a reduction of information from the original point process which itself represents a considerable data reduction from the original continuous signal, the EMG. The turning points spectra of Fig. 4 contain information on time dependencies between peaks in the EMG. For the purposes of discriminating between cases it is helpful to reduce the dimensionality of the data still further.

Starting with a spectrum, f(x), defined by say 256 data points there are several ways of reducing the dimensionality, for example by the use of orthogonal series as summarised in the next section.

5.1 Dimensionality reduction using orthogonal series

If f(x) is the spectrum of interest then we can write

$$f(x) = \sum_{n=0}^{N-1} a_n P_n(x) + E_N(x)$$

and, for all f(x) obeying certain rules, the error distribution $E_N(x)$ can be minimised in a mean square sense relative to some weighting function over some interval.

There are two alternatives, either

P_n can be chosen to be a predefined orthogonal set (Generalised Fourier Series analysis) or

P_n is to provide maximum convergence of the series for a sample set of the allowable f(x) (Principal Component Analysis)

In either case f(x) is approximated by $\sum_{n=0}^{N-1} a_n P_n(x)$

where f(x) has high dimensionality (typically M = 256, the number of sample points of f(x)) and its approximating series has low dimensionality (typically N = 3).

6. PRINCIPAL COMPONENT ANALYSIS OF THE EMG TURNING POINT SPECTRUM

6.1 Single reference data base method

In this method all spectra of identified cases used to define the set P_n

New case are then characterised by least squares minimisation of the function

$$f(x) - \sum_{n=0}^{N-1} a_n P_n(x) \quad \text{with respect to } a_n.$$

Now $a_0 \ldots a_{N-1}$ define the point in the reduced dimensionality space which represents the new case.

Fig. 6 shows the mean spectrum M (P_0) and the first three principal component spectra, P_1, P_2 and P_3, of a typical database of spectra derived from simulated EMGs. Figs. 7 and 8 show how an increasing number of myopathic units can be tracked in one and two dimensional space using this method as previously reported [7].

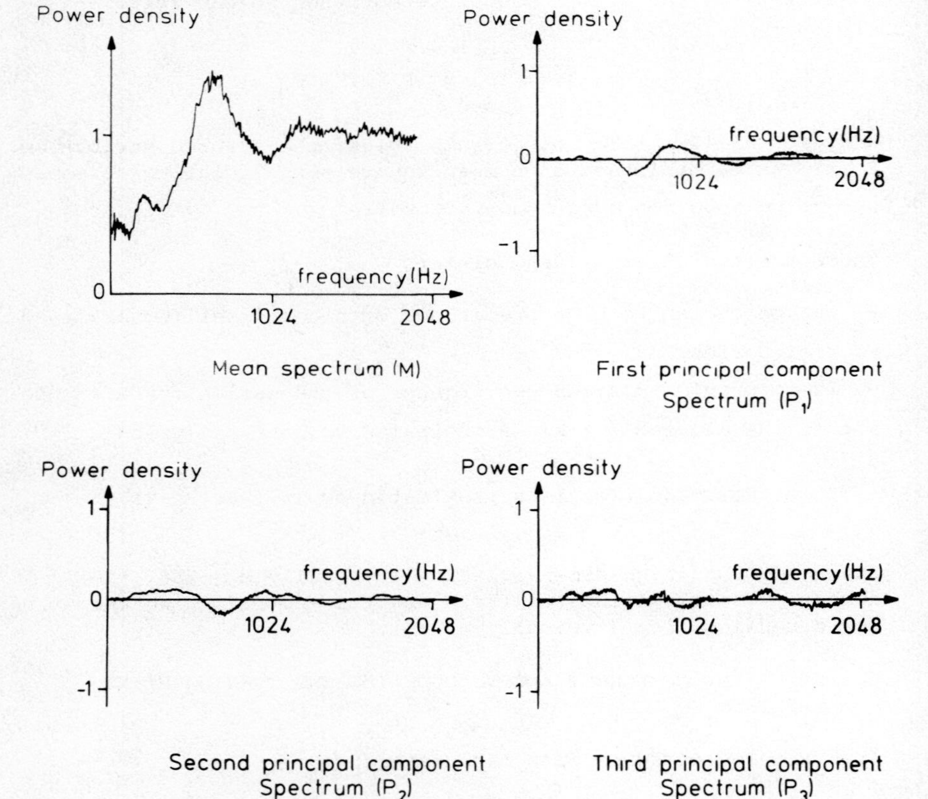

Fig. 6 Mean Spectrum and first three principal component spectra.

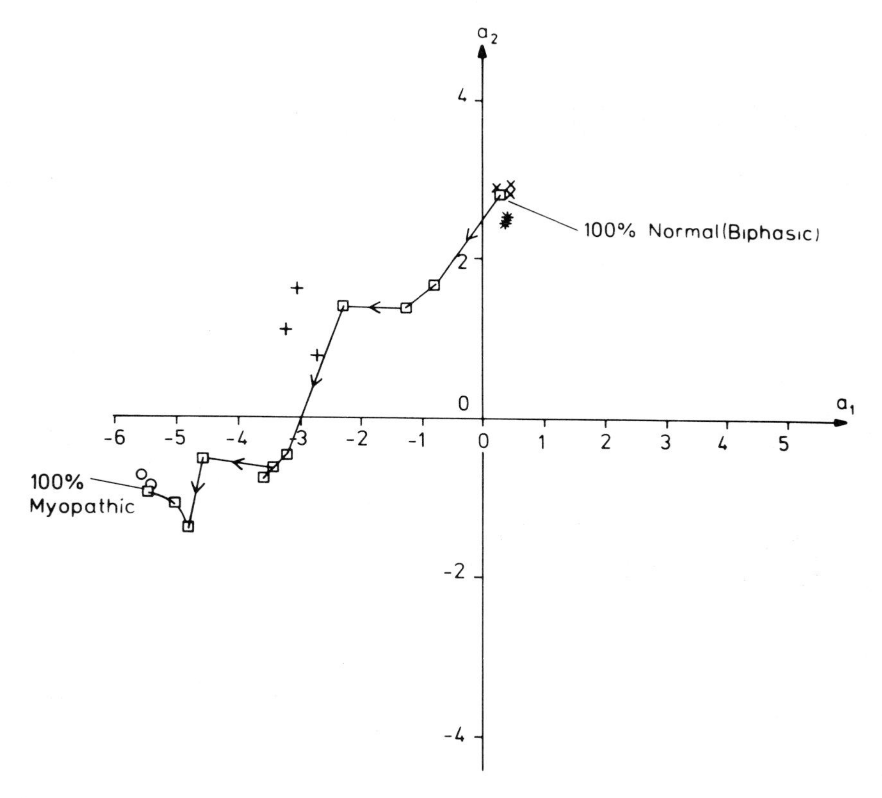

Fig. 7 Increasing percentage of myopathic units in steps of 10% (two dimensions).

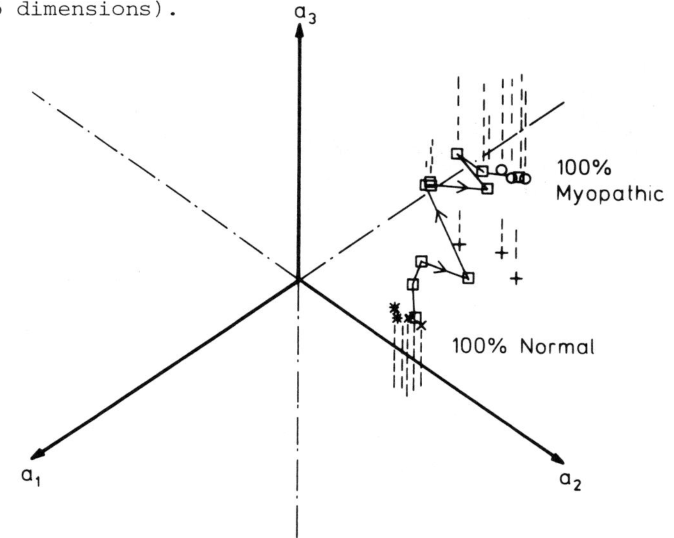

Fig. 8 Increasing percentage of myopathic units in steps of 10% (three dimensions)

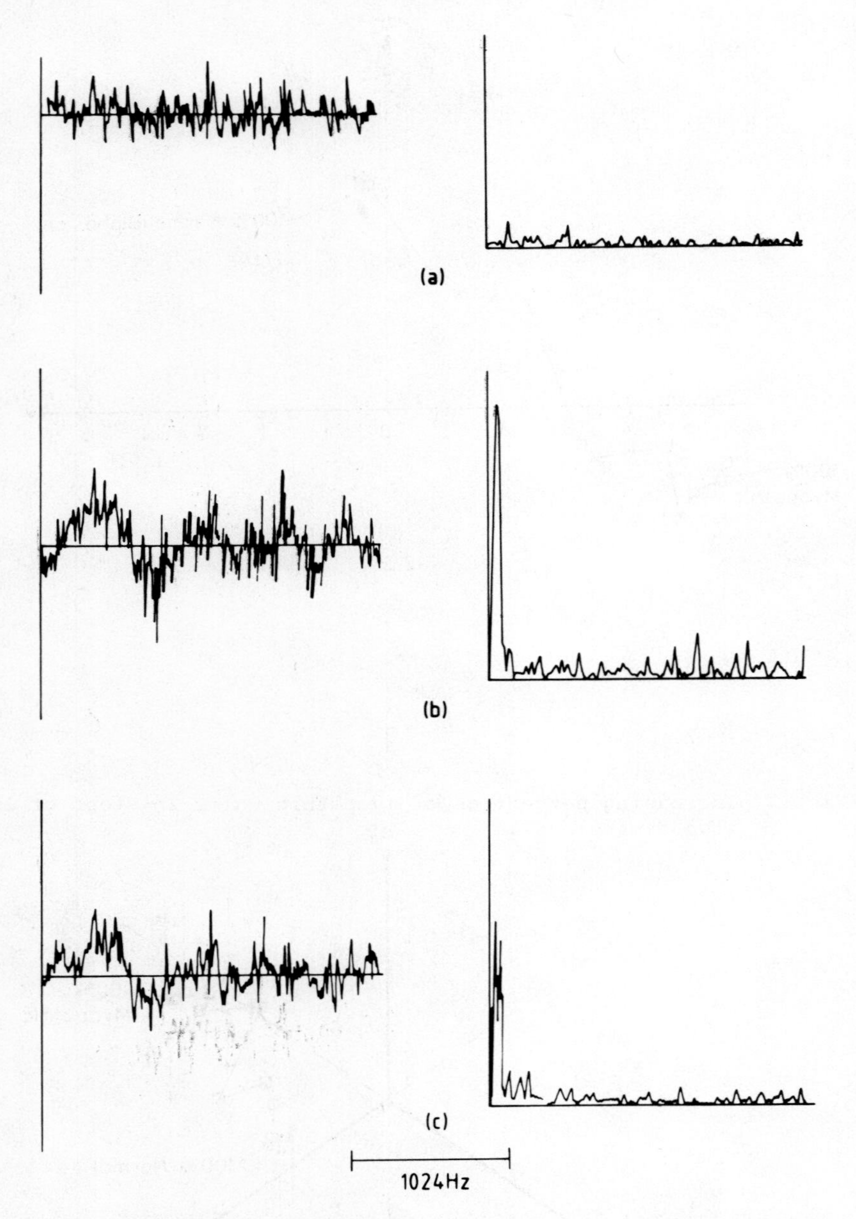

Fig. 9 Residual data and their spectra corresponding to (a) normal, (b) myopathic, and (c) neurogenic pattern class estimates of the test spectrum characterising EMG of 20 normal units.

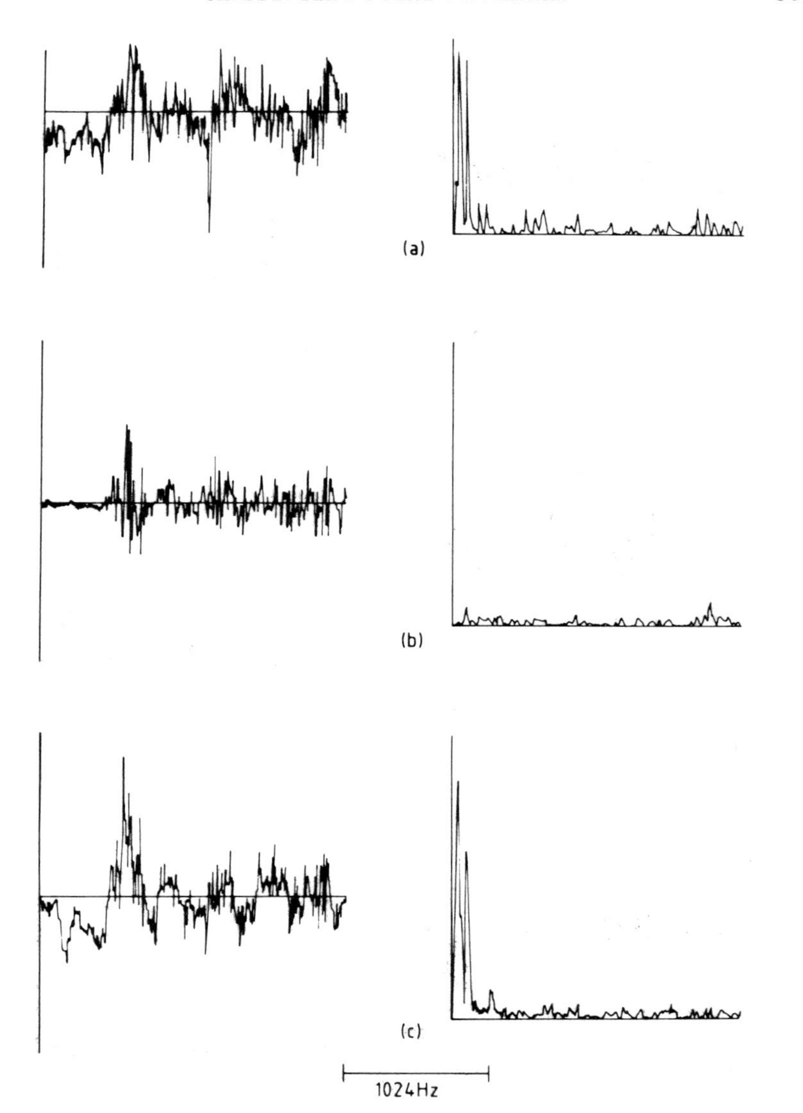

Fig. 10 Residual data and their spectra corresponding to (a) normal, (b) myopathic, and (c) neurogenic pattern class, estimates of the test turning points spectrum of EMG made up of 20 myopathic units.

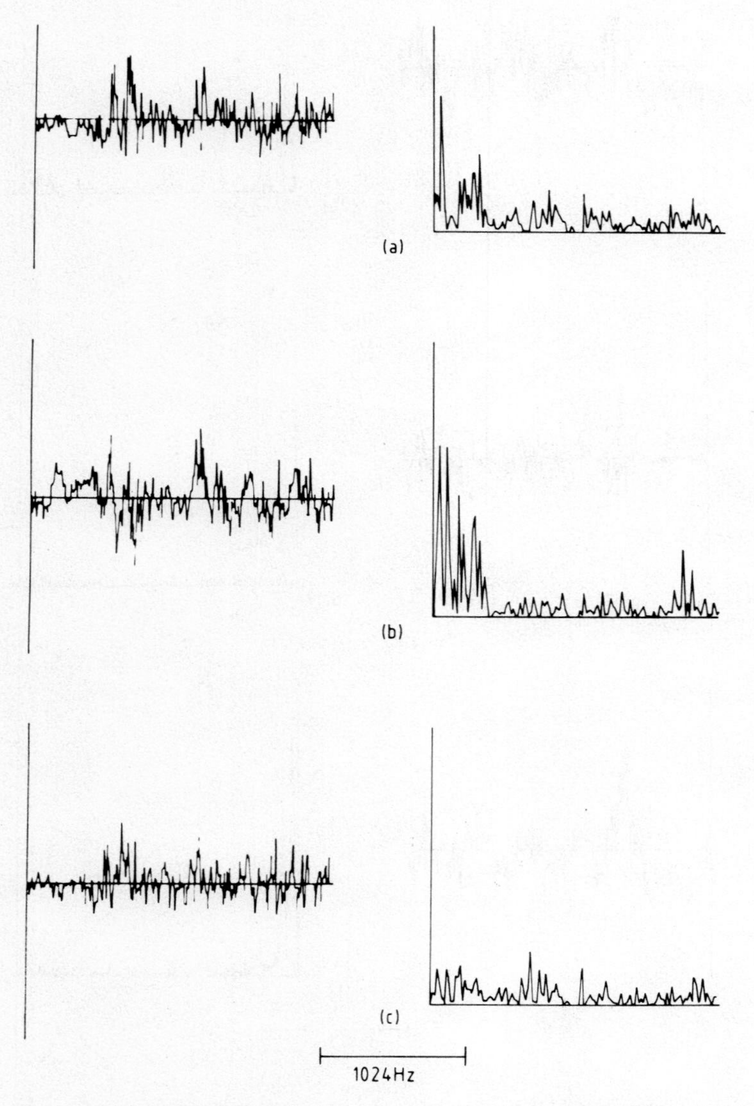

Fig. 11 Residual data and their spectra corresponding to (a) normal, (b) myopathic, and (c) neurogenic pattern class, estimates of the test turning points spectrum of EMG of 20 neurogenic units.

6.2 Multiple reference data base method

In this method all spectra of identified major sub-classes are used to form separate sets of functions P_n for each sub-class.

The a_n are then found for a new case by least square minimisation of the function

$$f(x) - \sum_{n=0}^{N-1} a_n P_n(x) = E_N(x)$$

Finally a new case is identified by either choosing that class which has the smallest value of

$$\sum_{m=1}^{M} (E_N(x_m))^2 \quad \text{and/or the whitest } E_N$$

Figs. 9, 10 and 11 show the residual spectra and *their* spectra for normal, myopathic and neurogenic EMGs respectively. It can be seen that both the smallest and the whitest residuals in each case correspond to the correct classification.

7. SUMMARY AND CONCLUSION

The method proposed for classifying processes is based on principal component analysis of the Bartlett spectrum. In particular it is suggested that, if the point process can be said to belong to one of a limited set of classes of such processes, then *each class* of process should be defined in terms of the first few principal components of the Bartlett spectra. Any new process is then classified to see which set of principal component spectra fits its Bartlett spectrum most closely; *closeness* being measured in terms of the size and whiteness of the residual. The residual is defined as the difference between the actual spectrum and its least squares approximation using weighted sums of the principal component spectra from the particular class being considered.

This pilot study shows that the proposed method is promising and seems likely to be more discriminating than the method using a single database previously published and illustrated again here. It does however consume more computing resources.

The method is illustrated by reference to a problem in neurology: the classification of electromyograms. These waves are reduced to point processes as a means of initial data reduction. In the limited number of cases so far tried, using simulated EMGs as the source of data, all processes have been correctly classified.

ACKNOWLEDGEMENTS

The authors wish to acknowledge the support of the Science and Engineering Research Council, Medelec Ltd., and Instituto Nacional de Investigacao Cientifica.

8. REFERENCES

[1] Bartlett, M.S., (1963) The Spectral analysis of point processes. *J. Roy. Statist. Soc.*, B.25, 664-280.

[2] Lago, P.J. and Jones, N.B., (1983) Turning points spectral analysis of the interference myoelectric activity, *Med. & Biol. Eng. and Computing,* **21**, 333-342.

[3] Basmajian, J.V. and De Luca, C.J., (1985) Muscles Alive, (5th Edition), Williams and Wilkins, London, 65-100.

[4] Jones, N.B. and Lago, P.J., (1982) Spectral analysis and the interference EMG, IEE Proc. 192, 673-678.

[5] Jones, N.B., Lago, P.J. and Parekh, A., (1987) Simulations of the electromyogram and their application in the assessment of new diagnostic tests, Proc. UKSC Conf. on Computer Simulation (R.N. Zobel, Ed.) Concordia, Ghent, 120-125.

[6] Jones, N.B., Lago, P.J. and Parekh, A., (1987) Principal component analysis of the spectra of point processes - an application in electromyography". Mathematics in Signal Processing (T.S. Durani, et al, Eds.) Clarendon Press, Oxford, 147-164.

Autoregressive Spectral Analysis of Point Processes

P.J.A. Lago†, A.P. Rocha†, and N.B. Jones‡

†*Grupo de Matemática Aplicada, Universidade do Porto*
Porto, Portugal

‡*Department of Engineering, University of Leicester*
Leicester, U.K.

---

1 Introduction

Point processes are frequently used for describing experimental data which can be regarded as a series of events randomly occurring in time. The areas of application are numerous; in particular the statistical analysis of point processes is widely used in the characterization of neural data (Cohen 1986). Like any other type of experimental data the analysis of point processes often rely on a frequency domain description and a number of different methods have been proposed for the estimation of the spectrum of a point process (Brillinger 1975, Cox and Lewis 1966, French and Holden 1970, Lago and Jones 1982, Lewis 1970). All these methods share the same characteristic of being nonparametric.

Autoregressive (AR) spectral analysis of time series is now well established and offers some advantages over nonparametric methods (Kay 1988). In principle AR spectral analysis may also be applicable to point processes with similar advantages over the nonparametric approach. However, the AR modelling of the spectrum of a point process may require methods for the estimation of the covariance density which preserve the positive semidefiniteness of that function. Such a property is normally not provided by the methods currently in use for the estimation of the covariance density of a point process (Brillinger 1975, Cox and Lewis 1966, Lewis 1970).

This paper is organized as follows. Section 2 covers some basic concepts and provides a brief introduction to the spectral modelling of point processes with all pole rational models. In Section 3 algorithms giving positive semidefinite estimates for the covariance density of a point process are described. Finally, in Sections 4 and 5 examples of AR spectral analysis of point processes are presented and the potentialities and limitations of the methods discussed in some detail.

2 Spectral analysis of point processes

Two different frequency domain descriptions have been introduced for the study of point processes: the spectrum of the intervals and the spectrum of the counts (Cox and Lewis 1966). The spectrum of the intervals of a point process is the spectrum of the discrete time series made up from the time intervals between consecutive occurrences, therefore the estimation of the spectrum of intervals is not in essence different from the estimation of the spectrum of a discrete time series. The spectrum of the counts of a point process, $G(w)$, is defined as the Fourier transform of the covariance density, $g(t)$,

$$g(t) = \lambda\delta(t) + \lambda(m(t) - \lambda) \tag{1}$$

where $\delta(t)$ is the Dirac delta function, λ is the mean intensity of the process,

$$\lambda = \lim_{\Delta t \to 0+} \text{prob}\{\text{event in }]t, t + \Delta t]\}/\Delta t \tag{2}$$

and $m(t)$ is the conditional intensity function, a symmetric function defined as zero at the origin and for $|t| > 0$ as the conditional probability

$$m(t) = \lim_{\Delta t \to 0+} \text{prob}\{\text{event in }]u + |t|, u + |t| + \Delta t] | \text{event at u }\}/\Delta t. \tag{3}$$

Autoregressive spectral analysis of point processes may be motivated in several ways in particular from spectral modelling (Makhoul 1975). Given a point process with covariance density $g(t)$, consider the discrete sequence $\{g_d(n\Delta t)\}$ obtained from

$$g_d(n\Delta t) = \lambda \Delta t^{-1} \delta_{0n} + \lambda(m(n\Delta t) - \lambda);\ n = 0, \pm 1, \ldots, \tag{4}$$

where δ_{ij} is the Kronecker delta and the spectrum $G_d(\omega)$ defined by

$$G_d(\omega) = \Delta t \sum_{n=-\infty}^{n=+\infty} g_d(n\Delta t) e^{-jn\omega\Delta t}. \tag{5}$$

It follows from eqn (1), (4) and (5) that $G_d(\omega)$ and $G(\omega)$ are related by

$$G_d(\omega) = \lambda + (G(\omega) - \lambda) * \sum_{n=-\infty}^{n=+\infty} \delta(\omega + 2\pi n/\Delta t) \tag{6}$$

where $*$ denotes the convolution operation.

Equation (6) shows that for all practical purposes the spectrum $G_d(\omega)$ may be made a good approximation to $G(\omega)$ by an intelligent choice of Δt. Since for most processes $|G(\omega) - \lambda|$ approaches zero rapidly with increasing values of $|\omega|$, it is clear from eqn (6) that it is always possible to keep aliasing below reasonable limits if Δt is taken small enough to ensure that $|G(\omega) - \lambda| \simeq 0$ for frequencies outside $|\omega| < \pi/\Delta t$. Then we can accept the approximation $G_d(\omega) \simeq G(\omega)$ for $|\omega| < \pi/\Delta t$. Assuming this condition to be satisfied, the modelling of the spectrum $G(\omega)$ may now be stated as the following problem: given $G_d(\omega)$ and a number of poles p it is required to approximate $G_d(\omega)$ by an all pole spectrum $G_a(\omega)$,

$$G_a(\omega) = \frac{V}{|1 + \sum_{n=1}^{p} a_n e^{-jnw\Delta t}|^2} \tag{7}$$

using for error measure, $E(p)$, the integrated ratio of the two spectra $G_d(\omega)$ and $G_a(\omega)$

$$E(p) = \frac{V\Delta t}{2\pi} \int_{-\pi/\Delta t}^{+\pi/\Delta t} \frac{G_d(\omega)}{G_a(\omega)} d\omega. \tag{8}$$

The autoregressive parameters $\{a_n\}$ are the solution of the p linear equations (Makhoul 1975)

$$\sum_{k=1}^{p} a_k g_d(|n-k|\Delta t) = -g_d(n\Delta t),\ n = 1, \ldots, p \tag{9}$$

and V is the minimum error $E(p)$, which can be shown to be equal to

$$E(p) = \Delta t(g_d(0\Delta t) + \sum_{k=1}^{p} a_k g_d(k\Delta t)). \tag{10}$$

We note that the parameters $\{a_n\}$ are equal to those that would be obtained by AR modelling of a discrete time series with covariance function $\{g_d(n\Delta t)\}$. Therefore the criteria of AR order selection for discrete time series may also be used for the selection of the number of coefficients to be retained in the model $G_a(\omega)$ of the spectrum of the counts. The minimum AIC (Akaike 1974) and the minimum CAT (Parzen 1974) have been used in the examples given in Section 4. It is important to realize that the application of these criteria of order selection require the positive semidefiniteness of $g(t)$ to be retained by the estimate of $\{g_d(n\Delta t)\}$. If this condition is verified the AR model spectrum $G_a(\omega)$ has poles inside the unit circle and a nonnegative minimum error measure $E(p)$.

The first step towards the solution of the spectrum modelling problem is the estimation of the covariance density of the process. The estimation of the conditional intensity and of the covariance density functions of a point process is usually based on the statistic (Brillinger 1975, Cox and Lewis 1966)

$$\tilde{m}(t) = N^{-1} \sum_{n=1}^{N-1} \sum_{k=1}^{N-n} \delta(t_{n+k} - t_n - |t|);\ \ 0 < |t| \leq T. \tag{11}$$

where t_k refers to the N observed times of occurrence in the interval $[0, T]$. For all practical purposes smoothing has to be applied to $\tilde{m}(t)$. Using the Daniell's weight function a smoothed estimate for the complete conditional intensity function $v(t)$, $v(t) = \delta(t) + m(t)$, at integer multiples of Δt, may be obtained from

$$\hat{v}_d(n\Delta t) = \Delta t^{-1}(\delta_{0n} + \int_{n\Delta t - \Delta t/2}^{n\Delta t + \Delta t/2} \tilde{m}(u)\, du). \tag{12}$$

From the estimate of the complete conditional intensity function $v(t)$ a smoothed estimate for the covariance density at integer multiples of Δt may be obtained from

$$\hat{g}_d(n\Delta t) = \hat{\lambda}(\hat{v}_d(n\Delta t) - \hat{\lambda}) \tag{13}$$

with $\hat{\lambda} = N/T$.

The method described for the estimation of $g(t)$ is in essence based on histogram construction techniques therefore involving only simple operations of classification and counting. However, since the sequences $\{\hat{v}_d(n\Delta t)\}$ and $\{\hat{g}_d(n\Delta t)\}$ may not be positive semidefinite, these estimates may not be adequate for AR modelling of point processes.

3 Positive semidefinite estimates

The estimation of the covariance density of a point process has strong similarities with the estimation of the covariance $c(t)$ of a time series $\{x(t)\}$ with unknown mean value μ. There are a number of ways of estimating $c(t)$ but in order to guarantee a positive semidefinite estimate one may use

$$\hat{c}(t) = T^{-1} \int_0^{T-|t|} (x(u) - \hat{\mu})(x(u+|t|) - \hat{\mu})\, du\,; \quad |t| \leq T \tag{14}$$

with $\hat{\mu} = T^{-1} \int_0^T x(u)\, du$.

A similar approach may be adopted for the estimation of the covariance density of a point process leading to the unsmoothed estimate (Lago et al. 1989)

$$\tilde{g}_1(t) = T^{-1} \int_{0-}^{T-|t|} (\theta(u) - \hat{\lambda})\,(\theta(u+|t|) - \hat{\lambda})\, du \tag{15}$$

with $\theta(t)$ defined by

$$\theta(t) = \sum_{k=1}^{N} \delta(t_k - t). \tag{16}$$

A smoothed positive semidefinite estimate for the covariance density giving a correct description of the spectrum of the counts at high frequencies may now be obtained from

$$\hat{g}_1(t) = \hat{\lambda}\,\delta(t) + \hat{q}_1(t) \tag{17}$$

where

$$\hat{q}_1(t) = T^{-1}\, w_1(t) * \int_{0+}^{T-|t|} (\theta(u) - \hat{\lambda})\,(\theta(u+|t|) - \hat{\lambda})\, du$$

and $w_1(t)$ is a positive semidefinite function, the simplest being the unit area triangular window over $[-\Delta t, +\Delta t]$. For a given $w_1(t)$, a positive semidefinite estimate for $g(t)$ at integer multiples of Δt may be obtained by sampling $\hat{g}_1(t)$. Then we have,

$$\hat{g}_{1d}(n\Delta t) = \hat{\lambda}\,\Delta t^{-1}\,\delta_{0n} + \hat{q}_1(n\Delta t). \tag{18}$$

Yet another possibility is clearly suggested from inspection of eqn (15), (17) and (18). The estimate $\hat{g}_{1d}(n\Delta t)$ has been obtained by sampling a continuous covariance density function which was previously smoothed with $w_1(t)$. Alternatively, a smoothed discrete estimate for the covariance density may be obtained by smoothing $\theta(t)-\hat{\lambda}$ with a window $w_2(t)$, sampling and the estimation of the discrete covariance function. This approach gives among other possibilities (Lago et al. 1989)

$$\hat{g}_{2d}(n\Delta t) = m^{-1} \sum_{k=0}^{m-|n|} \phi(k)\phi(k+|n|),\ n = 0, \pm 1, \ldots, \pm m-1 \tag{19}$$

with $m = int(T/\Delta t)$, $\phi(n) = (\, w_2(t) * (\theta(t) - \hat{\lambda})\,)_{t=n\Delta t+\Delta t/2}$, $w_2(t)$ the unit area rectangular window with duration Δt and $(\theta(t) - \hat{\lambda})$ defined by eqn (16) over $[0, T]$ and zero otherwise .

Having estimated the covariance density, the complete conditional intensity function may be obtained using the relation $v(t) = \lambda^{-1} g(t) + \lambda$, thus preserving the positive semidefiniteness of the estimate.

Besides positive semidefiniteness the estimates for the covariance density and conditional intensity functions described in this Section can be computed using only simple operations of classification and counting based on the sequence of observed occurrence times, therefore leading to highly efficient computational algorithms. The asymptotic behaviour as $T \to \infty$ of the spectrum of the counts obtained from $\{\hat{g}_{1d}(n\Delta t)\}$ or $\{\hat{g}_{2d}(n\Delta t)\}$ is identical to the spectrum obtained with the covariance density estimated using the histogram construction techniques described in Section 2.

4 Simulation and experimental results

Illustrative examples of autoregressive spectral analysis of point processes are presented in this Section using computer simulated and neurobiological data. The system (9) of the Yule-Walker equations was solved using the Levinson recursion and the number of model parameters p obtained according to the AIC (Akaike 1974) and CAT (Parzen 1974) criteria,

$$\mathrm{AIC}(p) = m \log E(p) + 2(p+1) \tag{20}$$

and

$$\mathrm{CAT}(p) = m^{-1} \sum_{j=1}^{p} \left(\frac{1}{e(j)} - \frac{1}{e(p)}\right) \tag{21}$$

where $m = int(T/\Delta t)$, $E(p)$ indicates the prediction error for the model of order p and $e(j) = (m/(m-j))E(j)$.

Figures 1 and 2 illustrate the normalized spectra of the counts, $G(\omega)/\lambda$, obtained with the methods described in Sections 2 and 3, for a computer simulated renewal Gaussian process. The nonparametric spectrum obtained from eqn (12) and (13) and the Blackman-Tukey method is also illustrated for comparison.

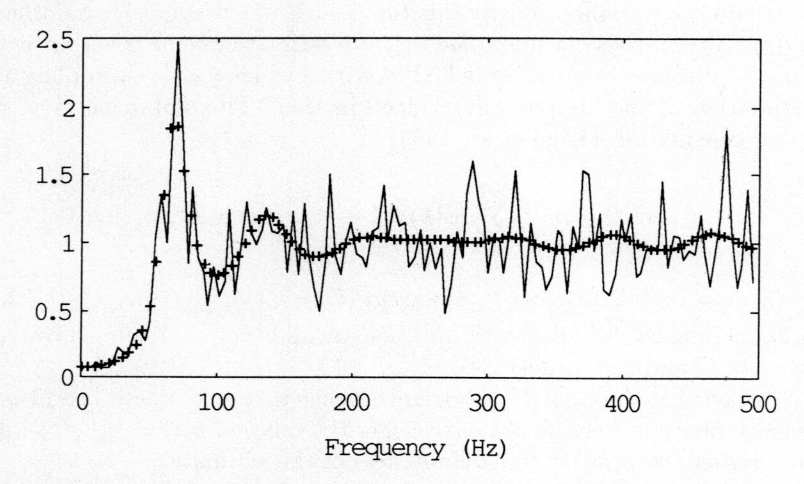

Fig.1. Normalized spectra of the counts, $G(\omega)/\lambda$, of a simulated Gaussian point process ($\lambda = 70$ sec^{-1}, $\sigma/\mu = 0.25$). AR spectrum obtained with the covariance density estimated from eqn (18) and the value for p ($p = 14$) according to the AIC and CAT criteria. Nonparametric spectrum obtained from the covariance density estimated from eqn (13) and the Blackman-Tuckey method (256 lags). The time resolution was $\Delta t = 1$ msec.

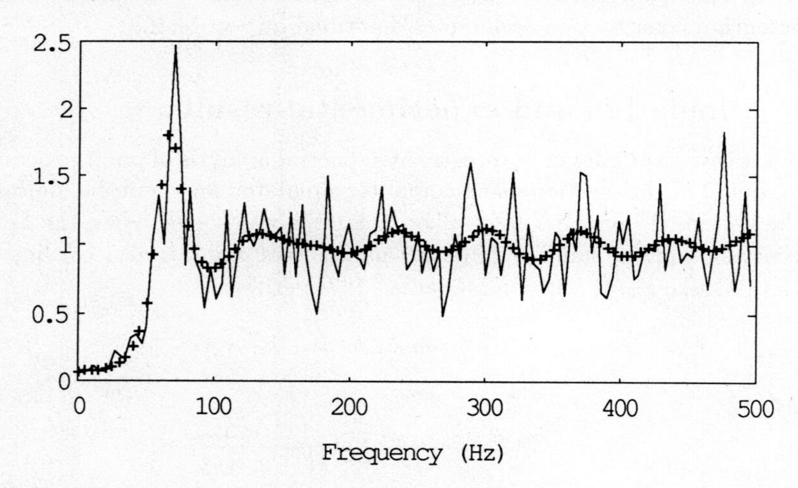

Fig.2. AR normalized spectrum of the counts obtained with the covariance density estimated from eqn (19) and the value for p ($p = 15$) according to the AIC and CAT criteria. Same data as in Fig.1. Nonparametric spectrum obtained with the Blackman-Tuckey method.

The neurobiological data used in the examples herein has been extracted from interference EMG signals using the turning points of the signal as time marks. Details of the method have been given elsewhere (Lago and Jones 1983, Lago et al. 1984). The point process derived from the EMG recorded with concentric needles has typically a mean intensity of 100 to 700 events per second and a spectrum of the counts significantly different from its limit value up to 2 to 8 Khz and close to the mean intensity above that. Examples are given in Fig.3. and 4. Nonparametric estimates are also shown.

These examples and many others have shown that AR spectral modelling gives consistently a good description of the nonparametric spectrum of the counts with a quite significant data reduction. As we would expect, due to the inherent smoothing of AR spectral estimators, the AR spectrum corresponds approximately to a mean curve of the nonparametric spectrum. With the model order selected by the AIC or the CAT criteria the performance of the AR spectral modelling has been found to be equally good in the peaks and also in the valleys of the spectrum. No significant differences were found between the AIC and the CAT criteria for order selection.

The spectrum of the counts of a point process differs significantly from that of a time series since it does not approach zero with increasing frequency. This behaviour introduces some difficulties in the modelling of the high frequency region. Experience with real and with simulated data indicate a slightly better performance of the estimate based on eqn (18) over that

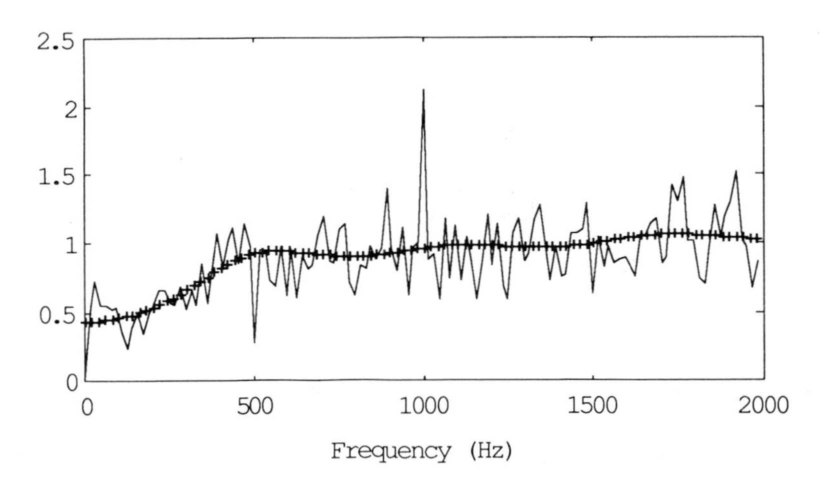

Fig.3. Normalized spectrum of the counts of a point process extracted from the EMG of a normal subject. Nonparametric spectrum obtained with the Blackman-Tuckey method. AR spectrum obtained with the covariance density estimated from eqn (18) and the value for p ($p = 6$) according to the AIC and CAT criteria. ($T = 800$ msec, $\Delta t = 0.25$ msec).

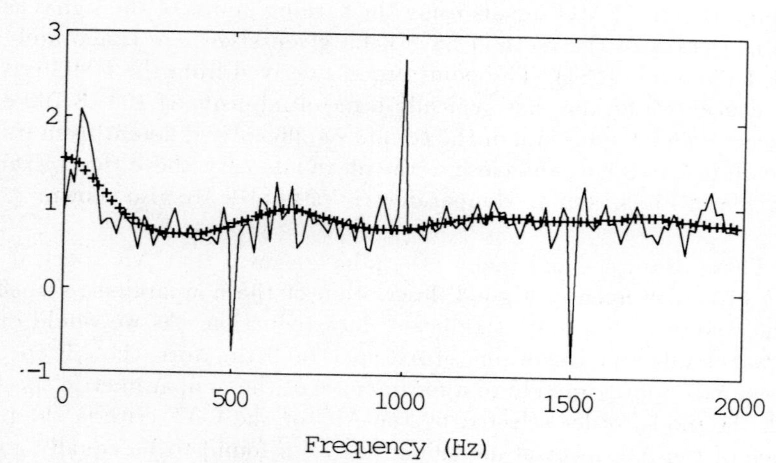

Fig.4. Normalized spectrum of the counts of a point process extracted from the EMG of a patient with Duchenne muscular dystrophy. Nonparametric spectrum from the Blackman-Tuckey method. AR spectrum obtained with covariance density estimated from eqn (18) and optimum model order ($p = 7$) selected by the AIC and CAT criteria. ($T = 800$ msec, $\Delta t = 0.25$ msec).

obtained using eqn (19). On the flat section of the spectrum of the counts (high frequency region) the AR spectrum based on the estimate for the covariance density obtained with the first method often has less ripple. Excluding the high frequency region the two methods give similar results.

5 Overdetermined spectral modelling approach

The difficulty in describing accurately the high frequency region of the spectrum of the counts, which is approximately a constant, is a consequence of the fact that this property has not been explicitly considered in the modelling procedure. AR models are based on very few values of the covariance density and (inherent to AR spectral analysis) the covariance density sequence is implicitly extrapolated beyond that. The natural starting point to improve AR spectral modelling is to use a larger number of covariances.

Using only the first $p + 1$ values for the covariance density, the AR parameters $\{a_n\}$ are the solution of the well-determined set of p linear equations (9), which can be written in matrix form

$$\mathbf{C}_{pp}\mathbf{a}_p = -\mathbf{c}_p, \tag{22}$$

where $\mathbf{C}_{pp}$ indicates the symmetric Toeplitz ($p \times p$) covariance density matrix with elements given by

$$C_{pp}(i,j) = g_d(|i-j|\Delta t), \;\; 1 \leq i \leq p, \; 1 \leq j \leq p,$$

$\mathbf{a}_p$ is the autoregressive parameter vector and $\mathbf{c}_p$ a vector having elements

$$c_p(i) = g_d(i\Delta t), 1 \leq i \leq p.$$

As shown in Section 2 the spectral analysis of a stochastic point process can be reduced to the spectral analysis of a discrete signal. This parallelism suggests that the AR parameters may be obtained from a larger number $t > p$ of equations. In this case (22) is replaced by the overdetermined system of equations which must be solved in the least squares sense,

$$\mathbf{C}_{tp}\mathbf{a}_p = -\mathbf{c}_t, \tag{23}$$

where the elements of the Toeplitz matrix $\mathbf{C}_{tp}$ are $C_{tp}(i,j) = g_d(|i-j|\Delta t)$, $1 \leq i \leq t$, $1 \leq j \leq p$, and $c_t(i) = g_d(i\Delta t)$, $1 \leq i \leq t$.

As illustrated in Fig.5. the spectral estimate obtained from (23) has lower performance and exhibits more ripple in the high frequencies, reflecting the increase in the variance due to the large number of equations used (Kay 1988). This behaviour has been consistently observed in other situations.

Another possibility that has been considered consists in selecting an extended model order $q < t$, much larger than the order p indicated by the AIC/CAT criteria and to use the algebraic properties of the associated $(t \times q)$ covariance density matrix $\mathbf{C}_{tq}$ (Cadzow 1982). If this matrix had a low rank, the singular value decomposition (SVD) of $\mathbf{C}_{tq}$ could be used to obtain an indication for the AR model order and the corresponding extended order model. In this case, a certain number d of singular values would dominate

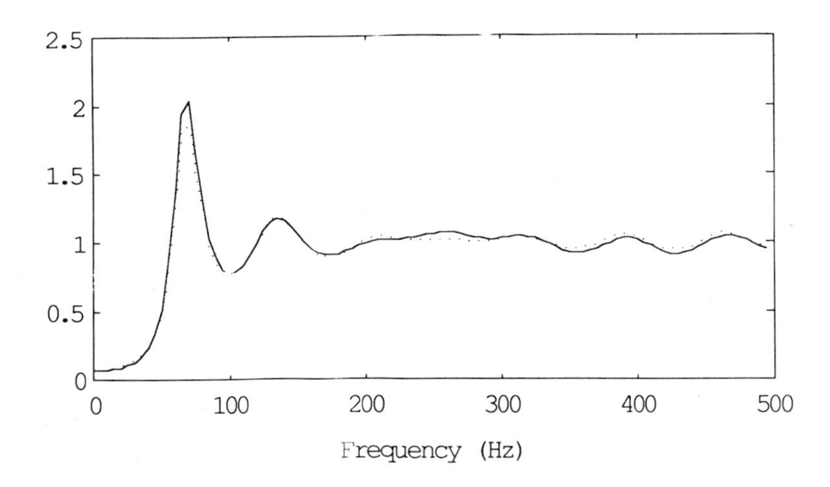

Fig.5. AR normalized spectrum of the counts obtained with the covariance density estimated from eqn (18) and the value for p ($p = 14$) according to AIC and CAT criteria. The continuous line represents the spectrum obtained using least squares from the first $t = 200$ modified Yule-Walker equations (23). Estimate obtained from (22) also shown. Same data as in Fig.1.

the remaining $q-d$ and the AR improved model could be determined from

$$\mathbf{C}_{tq}^{d}\mathbf{a}_{q}=-\mathbf{c}_{t}, \tag{24}$$

where $\mathbf{C}_{tq}^{d}$ indicates the $(t \times q)$ matrix which best approximates matrix $\mathbf{C}_{tq}$ in the Frobenius norm sense (Cadzow 1982). However, for point processes $g(0)$ dominates (equation 4), therefore the covariance density matrix $\mathbf{C}_{tq}$ has always full rank, q, except when displaced Yule-Walker equations are used.

Figure 6 shows the behaviour of the singular value sequence associated with the displaced modified Yule-Walker equations (9) for $n = i, i+1, ..., i+t-1$ with $i > 1$. The rank of the covariance density matrix decreases with increasing values for the displacement, i. Therefore, the singular value sequence associated with the covariance density matrix can have the desired properties if the displacement is chosen big enough. However, experience indicates that the first equations are essential for the determination of the AR model parameters. The results obtained with SVD and displaced Yule-Walker equations have been found both unreliable and poor.

The examples given in this section illustrate the difficulty in improving the AR spectral analysis of point processes and seem to indicate the need for a more drastic step through the use of alternative more complex models such as ARMA(p,p). Figure 7 shows preliminary results of ARMA spectral modelling. In this example the AR parameters were estimated from overdetermined modified Yule-Walker equations and the MA parameters obtained separately from the Fourier transform of the causal part of the covariance density sequence (Cadzow 1982). The results are encouraging, but the choice of model order, being critical, requires further work.

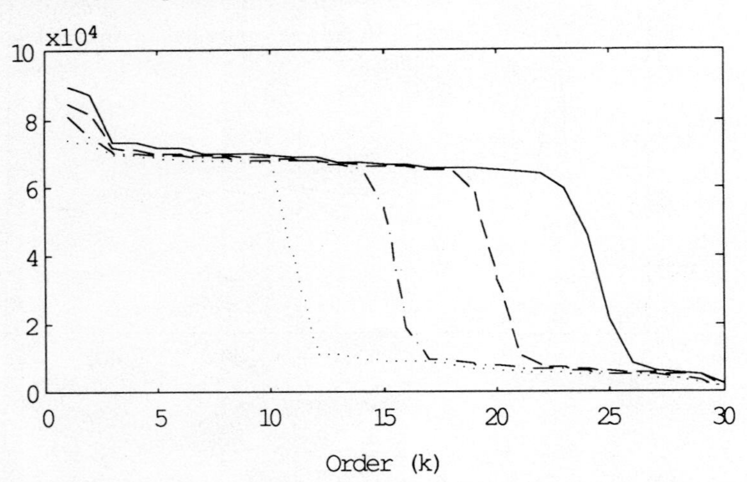

Fig.6. Sequences of singular values associated with displaced modified Yule-Walker equations for $p = 30$, $t = 60$ and displacements $i = 5$ (-), 10 (--), 15 (-·), 20 (···). Same data as in Fig.1.

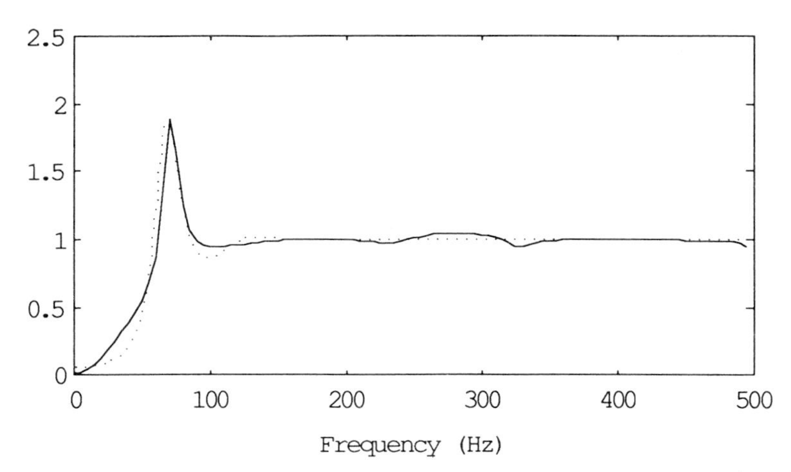

Fig.7. ARMA(p,p) normalized spectrum of the counts obtained with the covariance density estimated from eqn (18). AR parameters from overdetermined modified Yule-Walker equations ($p = 9, t = 20$) and MA parameters estimated from the causal part of the covariance density sequence. The exact spectrum is also illustrated for comparison. Same data as in Fig.1.

6 Concluding remarks

Autoregressive spectral analysis of point processes based on positive semidefinite covariance density estimates gives a number of advantages over the usual nonparametric approach. AR spectral analysis has been found to give a good description of the spectrum of the counts with a quite significant data reduction. With the model order selected by the AIC or the CAT criteria the performance of the AR spectral modelling has been found to be equally good in the peaks and in the valleys, corresponding approximately to a mean curve of the nonparametric spectrum. The experience obtained so far with real and also with computer simulated data indicates that AR modelling compares favourably with nonparametric techniques.

The spectrum of the counts of a point process differs significantly from that of a time series since it approaches the mean intensity with increasing frequency. This behaviour introduces some difficulties in the modelling of the high frequency region. With the error measure adopted, it is unlikely that a significant improvement of AR spectral modelling can be achieved and the use of alternative ARMA models would appear to be the next step. However, the simplicity of AR over ARMA models is sufficiently attractive to justify the search for alternative error measures.

Aknowledgements

P.J.A. Lago acknowledges the assistance of Instituto Nacional de Investigação Científica (INIC), Fundação Calouste Gulbenkian and University of Porto (research contract 36/84). A.P. Rocha acknowledges the INIC grant 23390.

References

Akaike, A. (1974). A new look at the statistical model identification. *IEEE Transactions on Automatic Control*, **19**, 716–23.

Brillinger, D. R. (1975). Statistical inference for stationary point processes. Stochastic processes and related topics (Puri, M. L. ed.). Academic Press, New York.

Cadzow, J. A. (1982). Spectral Estimation: an overdetermined rational model equation approach. *Proceedings of the IEEE*, **70**, 907–39.

Cohen, A. N. (1986). Biomedical Signal Processing. CRC Press, New York.

Cox, D. R. and Lewis, P. A. W. (1966). The statistical analysis of series of events. Chapman and Hall, London.

French, A. S. and Holden, A. V. (1970). Alias free sampling of neuronal spike trains. *Kybernetic*, **8**, 165–71.

Kay, S. M. (1988). Modern spectral estimation. Theory and application. Prentice-Hall, New York.

Lago, P. J. A. and Jones, N. B. (1982). Note on the spectral analysis of neural spike trains. *Medical Biological Engineering and Computing*, **20**, 44–8.

Lago, P. J. A. and Jones, N. B. (1983). Turning points spectral analysis of the interference myoelectric activity. *Medical Biological Engineering and Computing*, **21**, 333–42.

Lago, P. J. A., Jones, N. B. and Rocha, A. P. (1984). Characterization of the complexity of the action potential through spectral analysis. Digital Signal Processing-1984 (Cappellini V., Constantinides, A.G. eds.). Elsevier Science Publishers B.V., Amsterdam, 684–90.

Lago, P. J. A., Rocha, A. P. and Jones, N. B. (1989). Covariance density estimation for autoregressive spectral modelling of point processes. *Biological Cybernetics*, **61**, 3, 195–203.

Lewis, P. A. W. (1970). Remarks on the theory, computation and application of the spectral analysis of series of events. *Journal of Sound and Vibration*, **12**, 353–75.

Makhoul, J. (1975). Spectral linear prediction: properties and applications. *IEEE Transactions on Accoustics Speech and Signal Processing*, **23**, 283–96.

Parzen, E. (1974). Some recent advances in time series modeling. *IEEE Transactions on Automatic Control*, **19**, 723–30.

IV ADAPTIVE FILTERING

ADAPTIVE FILTER THEORY: PAST, PRESENT AND FUTURE

S. Haykin
*(Communications Research Laboratory,
McMaster University, Canada)*

ABSTRACT

In this paper we present the following material:

(a) An assessment of the present status of adaptive filter theory, including finite-precision effects.

(b) Historical notes on the development of this theory.

(c) A discussion of directions for future research in adaptive filter theory.

1. INTRODUCTION

Adaptive filtering has emerged as an important part of signal processing, rich in both theory and application. We say that a filter, be it in hardware or software form is *adaptive* if it satisfies two requirements:

(i) The filter has a built-in mechanism for the automatic adjustment of its coefficients in response to statistical variations of the environment in which the filter operates.

(ii) The coefficient adjustments are made for the purpose of progressively moving the filter (in an iteration-by-iteration or block-by-block manner) toward an optimum performance; here, optimality is defined in some statistical sense.

Adaptive filter theory applies to the processing of a time series as well as a space series. In the temporal case, the filter input may consist of a "vector" of uniformly spaced samples taken from a long data stream. In the spatial case, the filter input may consist of a "snapshot" of elemental outputs derived from an array of uniformly spaced sensors at a particular instant of time. In some practical situations, the uniformity of data samples is not adhered to.

Adaptive filters have been successfully applied in diverse fields, including adaptive equalizers (Qureshi, 1985), echo cancellers (Sondhi and Berkley, 1980), adaptive beamformers for sonar and radar (Owsley, 1985, 1988; Ward, Hargrave and McWhirter, 1986), speech encoders (Markel and Gray, 1976); Jayant and Noll, 1984), time-varying spectrum estimators (Marple, 1987) and system identification (Ljung and Söderström, 1983).

The theory of adaptive filters is well covered in several text books; see for example, Honig and Messerschmitt (1984), Widrow and Stearns (1985), Haykin (1986), Alexander (1986), Belanger (1987), and Haykin (1989).

In Sections 2 through 4 of the paper, we present an assessment of the present status of adaptive filter theory. This is followed by some historical notes in Section 5 on the development of this theory. We finish the paper in Section 6 with a discussion of directions for future research.

2. ADAPTIVE FILTERS USING BLOCK ESTIMATION

The starting point in the development of adaptive filters is the ensemble-averaged or the deterministic (time-averaged) system of normal equations. The development may then proceed using a *block estimation* approach, in which the input data is segmented into blocks, with the length of each block chosen to ensure pseudo-stationarity. Each block of data is used to derive an estimate of the autocorrelation sequence of the input signal or an estimate of the corresponding sequence of reflection coefficients. These parameters are then used to design an adaptive filter, the coefficients of which are varied on a block-by-block basis.

The structures/architectures resulting from the block estimation approach are:

(a) Transversal (tapped-delay-line) filter.

(b) Lattice (sequential) predictor.

(c) Escalator predictor.

(d) Pipeline parallel processor.

(e) Systolic processor.

TABLE I ADAPTIVE FILTERS USING BLOCK ESTIMATION

Structure/architecture	Theoretical tools
Transversal filter	1. Yule-Walker algorithm (assuming estimates of the autocorrelation sequence and incorporating the Levinson-Durbin recursion) 2. Method of least squares
Lattice (sequential predictor	Burg algorithm (incorporating the Levinson-Durbin recursion)
Escalator predictor	Gram-Schmidt algorithm (assuming estimates of the autocorrelation sequence)
Pipeline parallel processor	Schur algorithm (assuming estimates of the autocorrelation sequence)
Systolic (parallel) processor	Cyclic Jacobi algorithm for singular value decomposition (SVD) computation

Table I presents a summary of the algorithms that may be used to design these filtering structures/architectures; for details of these algorithms, see Haykin (1989). It is of interest to note that the Yule-Walker algorithm, the Gram-Schmidt algorithm, and the Schur algorithm are indirect in that they operate on the estimated autocorrelation sequence of the input signal. On the other hand, the method of least squares, the Burg algorithm and the singular value decomposition (SVD) algorithm operate on samples of the input signal directly.

3. ADAPTIVE FILTERS USING RECURSIVE ESTIMATION

A more popular approach for the design of adaptive filters is to use a *recursive estimation* procedure, whereby the filter coefficients are adjusted on an iterative basis (on the arrival of each new sample of input data). This approach is direct in the sense that it eliminates the need for estimating intermediate parameters (e.g., autocorrelations, reflection coefficients). Moreover, in each iteration the filter learns a little more about the incoming signal characteristics, and an improvement to the current coefficient vector of the filter is computed using this new formation.

In this case, the filtering structures of interest are as follows:

(a) Transversal filter(s).

(b) Lattice (sequential) predictor.

(c) Escalator predictor.

(d) Systolic (parallel) processor.

Specifically, Table IIA summarizes three recursive estimation algorithms, namely, the least mean square (LMS) algorithm, the gradient adaptive Lattice (GAL) algorithm, and the gradient adaptive escalator (GAE) algorithm, which represent stochastic gradient approximations of the Wiener filter. They differ from each other in the way in which the approximations to the Wiener filter are made. However, they share a common feature in that their computational complexity increases linearly with the filter order.

Table IIB presents a summary of three recursive estimation algorithms, namely, the recursive least squares (RLS) algorithm, the recursive modified Gram-Schmidt algorithm for least squares estimation, and the recursive QR-decomposition least squares (QRD-LS) algorithm, which owe their origin to the deterministic system of normal equations. In all three algorithms, the

TABLE IIA RECURSIVE ESTIMATION ALGORITHMS USING GRADIENT APPROXIMATIONS OF THE WIENER FILTER

Structure/architecture	Theoretical tools
Transversal filter	Least mean-square (LMS) algorithm, and its leaky version.
Lattice (sequential) predictor	Gradient adaptive lattice (GAL) algorithm.
Triangular (escalator) predictor	Gradient adaptive escalator (GAE) algorithm, based on the classical Gram-Schmidt orthogonalization.

TABLE IIB RECURSIVE ESTIMATION ALGORITHMS BASED ON THE DETERMINISTIC SYSTEM OF NORMAL EQUATIONS

Structure/Architecture	Theoretical tools
Transversal filter	Recursive least squares (RLS) algorithm.
Triangular (escalator) predictor	Recursive modified Gram-Schmidt algorithm for least-squares estimation, involving both time and order updates.
Systolic (parallel) processor	Recursive QR-decomposition least-squares (QRD-LS) algorithm, based on Givens rotations.

TABLE IIC FAST RECURSIVE ESTIMATION ALGORITHMS

Structure/Architecture	Theoretical tools
Transversal filters	1. Fast RLS algorithm. 2. Fast transversal filters (FTF) algorithm.
Lattice (sequential)	Recursive least squares lattice (LSL) algorithm (four different versions).
Systolic (parallel) processor	Fast QRD-LS algorithm.

(a) LMS Algorithm

(i) Due to round-off errors, the steady-state value of the ensemble-averaged coefficient vector of the filter deviates from the infinite-precision (i.e., Wiener) solution by an amount inversely proportional to the step-size parameter of the LMS algorithm. Ordinarily, the step-size parameter is assigned a value small enough to reduce the misadjustment (i.e., deviation from the minimum mean-square value predicted by the Wiener theory). However, a decrease in the step-size parameter of the LMS algorithm operating with finite precision will increase deviation from the infinite-precision performance of the LMS algorithm (Gitlin, Mazo and Taylor, 1973; Caraiscos and Liu, 1984).

(ii) In computer simulation experiments (involving small round-off errors), the effect of this deviation may not be ordinarily observed as it can take on the order of tens of millions of iterations before it becomes noticeable (Cioffi, 1987).

(iii) The effects of round-off errors in the LMS algorithms may be overcome by using leakage techniques (Gitlin, Mazo and Taylor, 1973). This is achieved, however, at the expense of both degraded performance and increased computation, compared to the infinite-precision case.

(b) Standard RLS Algorithm

(i) Round-off errors may cause an inverse matrix (implicit in the standard RLS algorithm) to assume an indefinite form. Even though this effect may not cause overflow, it is nonetheless unacceptable, as this inverse matrix (except for a scaling factor) equals the coefficient-error correlation matrix which, of course, has to be nonnegative definite (Bierman, 1977).

(ii) Due to round-off errors in the coefficient vector up-date computation, the RLS algorithm with no exponential weighting diverges as the number of iterations increases (Ardalan, 1988).

(iii) A decrease in the exponential weighting factor tends to stabilize the RLS algorithm; however, this is achieved at the expense of a more noisy estimate of the desired response due to the combined action of additive noise at the input and round-off errors (Ardalan, 1988). Note that $1-\lambda$ in the RLS algorithm, where λ is the exponential weighting factor, plays a role analogous to the step-size parameter μ in the LMS algorithm.

computational complexity increases as the square of filter order.

Table IIC presents a summary of four fast realizations of recursive least squares. These fast algorithms share a common feature in that they all require a computational complexity that increases linearly with the filter order as with stochastic gradient algorithms - hence, the term "fast". They also provide useful side information about the input data (such as a priori and a posteriori versions of the forward and backward prediction errors). These desirable properties are achieved by exploiting a shifting property of the input data. As such, their use is restricted to the processing of a time series only.

The recursive least-squares lattice (LSL) algorithm listed in Table IIC may assume one of four different forms, depending on the following features:

(i) Whether the algorithm is based on a posteriori or a priori estimation errors.

(ii) Whether the algorithm uses indirect computation of the forward and backward reflection coefficients or a direct procedure (with built-in error feedback)

The gradient adaptive lattice (GAL) algorithm may be viewed as a special case of a recursive LSL algorithm in which a priori and a posteriori estimation errors are made to be the same. Accordingly, the forward and backward reflection coefficients of the GAL algorithm assume a common value (except for a possible complex conjugation in the case of complex-valued data).

It is of interest to note that the LMS and RLS algorithms involve time updates only. On the other hand, all the other algorithms listed in Tables IIA, IIB and IIC involve both time and order updates.

For details of the recursive filtering algorithms summarized in Tables IIA, IIB, IIC, the reader is referred to Haykin (1986); Ling, Manolakis, and Proakis (1986) and Cioffi (1987).

4. FINITE-PRECISION EFFECTS IN ADAPTIVE FILTERS

The algorithms presently available for the design of adaptive filters assume infinite precision. When, however, an adaptive filter so designed is implemented on a digital machine with finite precision, the resulting round-off error effects may cause the performance of the filter to deviate significantly from its theoretical (infinite precision) value. In this section. we briefly review some of these effects in the context of adaptive filters using recursive estimation:

(c) Fast RLS Algorithms

A phenomenon unique to fast RLS algorithms occurs when exponential weighting is used. In particular, round-off errors are increased exponentially, which makes fast RLS algorithms unsuitable for continuous operation (Ljung and Ljung, 1985).

(d) Recursive LSL Algorithm

When the forward and backward reflection coefficients in a recursive LSL algorithm are computed indirectly, the effect of round-off errors may cause the algorithm to become unstable; when, however, they are computed directly, the built-in presence of *error feedback* in their computation tends to stabilize the operation of the algorithm (Ling, Manolakis and Proakis, 1985).

(e) Recursive QR-decomposition LS Algorithm

This algorithm has good numerical properties, which means that it can perform satisfactorily in the presence of round-off errors. A formal proof of the inherent stability of the recursive QRD-LS algorithm, using a finite-precision systolic array implementation, is presented in Leung and Haykin (1989).

For a broad review of finite-precision effects in adaptive filters using recursive estimation, the reader is referred to Cioffi (1987).

5. SOME HISTORICAL NOTES

Adaptive filter theory has an illustrious history, dating back to the pioneering work of Gauss on the method of least squares in the nineteenth century. It has also benefited in great measure from the works of other pioneers, notably, Wiener during the Second World War, and Kalman in 1960.

Other major contributors to the development of adaptive filters, who deserve special mention are:-

(1) Widrow, for inventing the LMS algorithm in 1960, which has withstood the test of time.

(2) Lucky, for demonstrating a practical application of adaptive equalization for the first time in 1965, which contributed to a dramatic increase in reliable data transmission rates over telephone channels.

(3) Kung, for the invention of systolic processing in 1978, and thereby making a novel tool available for the design of a new family of adaptive filters for temporal and spatial applications.

Other noteworthy contributors include Applebaum, for the first demonstration of adaptive arrays in 1966; Björck, for solving the linear least-squares problem by Gram-Schmidt orthogonalization in 1967; Burg, for the design of lattice predictors in 1968; Morf, for paving the way for the development of (exact) least squares lattice algorithms in 1974; Griffiths, for the development of the GAL algorithm in 1977; Falconer and Ljung, for the development of the fast RLS algorithm in 1978; Ahmed and Youn, for the development of the GAE algorithm in 1980; Carayannis, Manolakis and Kalouptsidis in 1983, and Cioffi and Kailath in 1984, for their independent developments of the FTF algorithm; Gentleman and King in 1981, and McWhirter in 1983, for the development of two different versions of adaptive systolic processors; Brent, Luk and Van Loan, for the development of the cyclic Jacobi algorithm for SVD computation in 1985; Ling, Manolakis, and Proakis for the use of the modified Gram-Schmidt algorithm to develop a systolic form of recursive least-squares estimator in 1986; Cioffi, for the development of the fast QRD-LS algorithm in 1987.

6. DIRECTIONS FOR FUTURE RESEARCH

The algorithmic theory of adaptive filters is well-understood. Nevertheless, there remain a number of research problems and new possibilities that require/deserve attention. In this final section of the paper, we highlight five specific issues:

(a) *Efficiency of an Adaptive Filtering Algorithm*

The LMS algorithm is typically viewed as the benchmark against which other adaptive filtering algorithms are compared. The comparison is usually made on the basis of such matters as the rate of convergence, robustness with respect to parameter variations, and computational complexity. The argument that is often levelled against the LMS algorithm is that, although it is computationally simple, nevertheless, the algorithm suffers from a slow rate of convergence and it is sensitive to variations in the eigenvalue spread of the correlation matrix of the input signal. Yet, in spite of these limitations, the LMS algorithm remains highly popular. This is all the more interesting when it is recognized that nowadays computational complexity is no longer the issue that it used to be, thanks to the ever-increasing availability of VLSI technology. There must be other good reasons for the continued interest in practical application of the LMS algorithm. It would therefore appear that we need a more scientific basis for the comparison of adaptive filtering algorithms. One such yardstick may well be the *efficiency* of an adaptive filtering algorithm.

For information measure, we may use Fisher's information matrix that was introduced by Fisher in 1925 especially for the theory of statistical estimation (Fisher, 1925). Other informatic measures include the Kullback-Leibler mean information (Kullback, 1959), and Hajek's J-divergence (Scharf and Moose, 1976), which are closely related to Fisher's information.

Naturally, the efficiency would have to be computed on an iteration-by-iteration basis. Initially, we would expect the efficiency to be low. But as the filter learns more and more about its environment, the efficiency should begin to increase. The form of the efficiency curve will thus be influenced by two factors: structure of the algorithm, and details of the environment responsible for data generation.

(b) *Stabilization of Fast RLS Algorithms*

A variety of ad hoc techniques have been proposed for the stabilization of fast RLS algorithms operating in the presence of round-off errors. The solution of this important problem requires a more scientific approach. In this context, a procedure described in Botto and Moustakides (1987), and Slock and Kailath (1988) deserves special mention. The procedure described herein involves the controlled addition of computationa complexity. The motivation here is similar to that in informatio theory. In the latter case, source encoding is used for the purpose of removing redundancy. This is followed by channel encoding, which is designed to add controlled redundancy, so as to overcome the effect of channel noise. Computational complexit and round-off errors in the design of fast RLS algorithms may thus be viewed as analogous to redundancy and channel noise in the source encoding-channel encoding problem in information theory.

A word of caution is in order here. With the ever-increasing availability of VLSI silicon chips, and therefore the lessening of computational complexity as an issue of concern, there is a real possibility that the stabilization of fast transversal filte algorithms may simply be overtaken by events.

(c) *Adaptation in a Quantized Parameter Space*

Another possibility is to attempt the design of an RLS type of algorithm (and hopefully its fast realizations) by operating directly in a quantized domain. The idea of operating in a quantized parameter space was first described by Gersho in 1968; Gersho's approach is applicable to a gradient search type of adaptation. However, adaptation in a quantized parameter space with an RLS type of algorithm presents monumental theoretical difficulties.

(d) *Bridging the Fields of Adaptive Filtering and Adaptive Control*

The developments of adaptive filtering and adaptive control have been proceeding along almost parallel lines. Yet, the underlying theories of these two fields share several common features. There is the possibility that new results may emerge by building a bridge between the two fields. A particular area of research that may benefit from this cross-fertilization is the study of adaptive infinite impulse response (IIR) filters. Here mention should be made of the monograph by Johnson (1988) that explores the interaction between adaptive filter and control applications in the context of adaptive parameter estimation theory.

(e) *Nonlinear Adaptive Filters*

The adaptive filtering algorithms presently in use are derived from linear estimation theory. The term "linear" refers to the fact that the estimate of the desired response is expressed as a linear combination of elements of the input signal vector. Neural computing, on the other hand, is related to "nonlinear" estimation theory. This form of computing offers several interesting features:

(i) A capability to generalize.

(ii) Fault tolerance.

(iii) Robustness to statistics of the input data.

This would suggest the need for exploring the benefits that may result from the use of nonlinearities in adaptive filters. Although by so doing, we would indeed loose the system of normal equations as a theoretical basis and complicate the mathematical analysis, nevertheless, we would expect to benefit in two ways:

(i) The inclusion of nonlinearity should make it possible for an adaptive filtering algorithm to exploit the information contained in the input data more fully, and thereby enhance the capability of the algorithm.

(ii) The addition of nonlinearity should broaden the areas of application of adaptive filters.

In any event we view nonlinear adaptive filters not as a replacement for but rather as a complement to linear adaptive filters.

It is noteworthy that the use of nonlinearities in the design of adaptive predictors was first explored by Gabor and co-workers in 1960.

ACKNOWLEDGEMENTS

The author is grateful to the Natural Sciences and Engineering Research Council of Canada for financial support, and to Henry Leung for stimulating discussions.

REFERENCES

[1] Ahmed, N. and Youn, D. E. (1980), "On Realization and Related Algorithm for Adaptive Prediction", IEEE Trans. Acoust Speech and Signal Processing, vol. ASSP-28, pp. 494-498.

[2] Alexander, S. T. (1986), "Adaptive Signal Processing: Theory and Applications", (Springer-Verlag).

[3] Ardalan, S. (1988), "Design Issues in the Finite Precision Implementation of Adaptive Filtering Algorithms", Proceedings of the National Communications Forum, vol. 42, No. II.

[4] Belanger, M. (1987), "Adaptive Filters and Signal Analysis", (Marcel Dekker).

[5] Bierman, G. J. (1977), "Factorization Methods for Discrete Sequential Estimation", (Academic Press).

[6] Björck, A. (1967), "Solving Linear Least Squares Problems by Gram-Schmidt Orthogonalization", BIT, vol. 7, pp. 1-21.

[7] Botto, J. L., and Moustakides, G. V. (1987), "Stabilization of Fast Recursive Least-Squares Transversal Filters for Adaptive Filters", Proc. ICASSP 87, Dallas, Texas.

[8] Cioffi, J. M. (1987), "Limited-Precision Effects in Adaptive Filtering", IEEE Trans. Circuits and Systems, vol. CAS-34, pp. 821-833.

[9] Cioffi, J. M. (1987), "The fast QR Adaptive Filter", IEEE Trans. on Acoust. Speech and Signal Processing, submitted for publication.

[10] Fisher, R. A. (1925), "Theory of Statistical Estimation", Proc. Camb. Phil. Soc., vol. 22, pp.700-725.

[11] Gabor, D., Wilby, W. P. L. and Woodcock, P. (1960), "A Universal Non-linear filter, Predictor and Simulator which Optimizes itself by a Learning Process", Proc. IEE (London), vol. 108, pp. 422-428.

[12] Gersho, A. (1968), "Adaptation in a Quantized Paramter Space", Allerton Conference, pp. 646-653.

[13] Gitlin, R. D., Mazo, J. E. and Taylor, M. G. (1973), "On the Design of Gradient Algorithms for Digitally Implemented Adaptive Filters", IEEE Trans. Circuit Theory, vol. CT-20, pp. 125-136.

[14] Haykin, S. (1986), "Adaptive Filter Theory", (Prentice Hall).

[15] Haykin, S. (1989), "Modern Filters", (MacMillan).

[16] Honig, M. L. and Messerschmitt, D. G. (1984), "Adaptive Filters", (Kluwer Academic Publishing).

[17] Jayant, N. S. and Noll, P. (1984), "Digital Coding of Waveforms", (Prentice-Hall).

[18] Johnson, C. R., Jr. (1988), "Lectures on Adaptive Parameter Estimation", (Prentice-Hall).

[19] Kullback, S. (1959), "Information Theory and Statistics, (Wiley).

[20] Leung, H. and Haykin, S. (1989), "Stability of Recursive QRD-LS Algorithms using Finite-Precision Systolic Array Implementation", IEEE Trans. Acoustics, Speech and Signal Processing, to be published.

[21] Ling, F., Manolakis, D. and Proakis, J. G. (1985), "New Forms of LS Lattice Algorithms and an Analysis of their Round-off Error Characteristics", Proceedings ICASSP 85, Tampa, Florida, pp. 1739-1742.

[22] Ling, F., Manolakis, D. and Proakis, J. G. (1986), "A Recursive Modified Gram-Schmidt Algorithm for Least-Squares Estimation", IEEE Trans. Acoustics, Speech and Signal Processing, vol. ASSP-34, pp. 829-836.

[23] Ljung, L. and Söderström, T. (1983), "Theory and Practice of Recursive Identification", (MIT Press).

[24] Markel, J. D. and Gray, A. H., Jr. (1976), "Linear Prediction of Speech", (Springer-Verlag).

[25] Marple, S. L., Jr. (1987), "Digital Spectral Analysis with Applications", (Prentice-Hall).

[26] Owsley, N., (1986) "Sonar Array Processing", in the book "Array Signal Processing", edited by Haykin, S., Chapter 3, (Prentice-Hall).

[27] Owsley, N. (1988), "Systolic Array Adaptive Beamforming", Chapter 6, in the book "Selected Topics in Signal Processing", edited by Haykin, S. (Prentice-Hall).

[28] Qureshi, S. (1985), "Adaptive Equalization", Proc. IEEE vol. 73, pp. 1349-1387.

[29] Scharf, L. L. and Moose, P. H. (1976), "Information Measures and Performance Bounds for Array Processors", IEEE Trans. on Information Theory, vol. IT-22, pp. 11-21.

[30] Shepherd, T. J. and McWhirter, J. G., "Adaptive Systolic Beamforming" in book on "Radar Array Processing", edited by Haykin, S., et al., (Springer-Verlag).

[31] Slock, D. T. M. and Kailath, T. (1988), "Numerically Stable Fast Recursive Least-Squares Transversal Filters", Proc. ICASSP, New York.

[32] Sondhi, M. M., and Berkley, D. A. (1980), "Silencing Echoes in the Telephone Network", Proc. IEEE, vol. 68, pp. 948-963.

[33] Widrow, B. and Stearns, S. D. (1985), "Adaptive Signal Processing", (Prentice-Hall).

THE FAMILY OF FAST LEAST SQUARES ALGORITHMS FOR ADAPTIVE FILTERING

M.G. Bellanger
(T.R.T. France)

ABSTRACT

Three representative algorithms from the family of Fast Least Squares (FLS) algorithms are reviewed. They correspond to the transversal FIR filter, the lattice-ladder structure and the rotation approach respectively. They all solve the least squares problem recursively and are convenient for adaptive filtering.

They lead to three different levels of computational complexity and robustness to round-off errors. Each of them also has special implementation aspects and provides a particular set of signal parameters.

Overall, these three types of algorithms offer a wide and welcome choice to the adaptive filter designer.

1. INTRODUCTION

Fast Least Squares (FLS) algorithms have been worked out for adaptive filters with different kinds of specifications and various structures, and they can handle one-or multidimensional signals [1].

The filter structure may be dictated by the application, but it can also be chosen according to the computational complexity, the particular set of signal parameters which is of interest to the designer or on the basis of robustness against numerical limitations. It must be pointed out that an important aspect of least squares techniques is the round-off error accumulation, which takes place in the transversal filter for example, but apparently not in the lattice-ladder structure or in the rotation method. The purpose of this paper is to review these three major approaches and for each one, assess the computational complexity and comment on the operational organization, the wordlength limitation effects and the set of signal parameters available.

The principle of an adaptive filter is shown in Fig. 1, which gives the various signals involved namely the input x(n), the reference y′(n) and the error e(n). The transversal structure is dealt with first, since the order of growing complexity is retained.

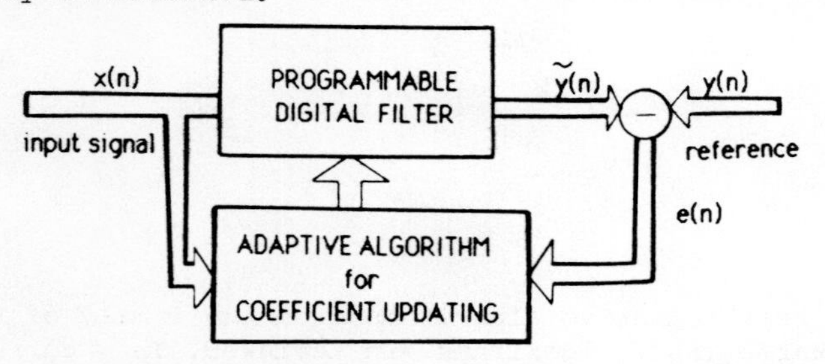

Fig. 1 Principle of an adaptive filter

2. FLS ALGORITHM FOR THE TRANSVERSAL STRUCTURE

A typical FLS algorithm for the transversal structure is given in Fig. 2. A detailed derivation is available in reference [1] and will not be reproduced here. Earlier references include [2, 3, 4].

That approach has two strong points, which are its low computational complexity and the usefulness of the signal parameters involved.

If the filter order is N, the operations listed in Fig. 2 correspond to 8N+4 multiplications and 2 divisions in the adaptation gain updating section and 2 N multiplications in the filter section. The amount of memories is about 6N, to store the coefficients and internal variables. Altogether, the complexity can be viewed as about five times that of the least mean squares (LMS) approach, which is just 2N multiplications.

ALGORITHM F.L.S.1

AVAILABLE AT TIME n.

COEFFICIENTS OF ADAPTIVE FILTER	$H(n)$
- FORWARD PREDICTOR	$A(n)$
- BACKWARD PREDICTOR	$B(n)$
DATA VECTOR	$X(n)$
ADAPTATION GAIN	$G(n)$
FORWARD PREDICTION ERROR ENERGY	$E_a(n)$

NEW DATA AT TIME n:

input Signal : $x(n+1)$; Reference $y(n+1)$

ADAPTATION GAIN UPDATING :

$$e_a(n+1) = x(n+1) - A^t(n)\,X(n)$$
$$A(n+1) = A(n) + G(n)\,e_a(n+1)$$
$$\varepsilon_a(n+1) = x(n+1) - A^t(n+1)\,X(n)$$
$$E_a(n+1) = W\,E_a(n) + e_a(n+1)\,\varepsilon_a(n+1)$$
$$G_1(n+1) = \begin{bmatrix} 0 \\ G(n) \end{bmatrix} + \frac{\varepsilon_a(n+1)}{E_a(n+1)} \begin{bmatrix} 1 \\ -A(n+1) \end{bmatrix} = \begin{bmatrix} M(n+1) \\ m(n+1) \end{bmatrix}$$

$$e_b(n+1) = x(n+1-N) - B^t(n)\,X(n+1)$$
$$G(n+1) = \frac{1}{1-m(n+1)\,e_b(n+1)}\left(M(n+1) + m(n+1)\,B(n) \right)$$
$$B(n+1) = B(n) + G(n+1)\,e_b(n+1)$$

ADAPTIVE FILTER :

$$e(n+1) = y(n+1) - H^t(n)\,X(n+1)$$
$$H(n+1) = H(n) + G(n+1)\,e(n+1)$$

Fig. 2 FLS algorithm for the transversal structure

The sequencing of the operations is reasonably favourable for implementation since weighted sums form the bulk of the processing. A difficulty may arise from the two divisions, in connection with the real time constraints. It is worth mentioning that some flexibility exists in the computation ordering [5].

The adaptation gain updating procedure consists of a real time analysis of the input signal, through linear prediction. The set of forward and backward coefficients obtained is useful in many technical areas, for example for spectral analysis purposes, extraction of information or system modelling and measurement.

The major problem with the FLS transveral adaptive filter is the round-off error accumulation, which takes place, mainly in the backward prediction coefficient updating loops, with a finite precision implementation. In most cases the consequence is instability. Although the problem is still a subject for research, significant progress has been made. First of all, several minor modifications can be made to the basic algorithm, which permit proper working in a number of situations ([6] and chapter 6 in [1]).

The forward prediction error energy updating equation can be modified as follows:

$$E_a(n+1)=WE_a(n)+e_a(n+1)\epsilon_a(n+1)+C \tag{1}$$

The positive constant C prevents the variable $E_a(n+1)$ from reaching zero, which is a crucial point since it is used subsequently as a divider. Moreover it can be shown that the backward prediction coefficient updating recursion is also modified and becomes:

$$B(n+1) = (1-\gamma)B(n)+G(n+1)e_b(n+1) \tag{2}$$

The expectation of the term γ is positive, which yields a leakage effect able to counter the round-off error accumulation. The impact on the forward prediction coefficient updating is difficult to investigate. It can be shown that a leakage effect is also obtained for some signals, for example sinusoids in noise. However further work is needed on this topic.

Another approach is to control the round-off error accumulation by exploiting redundancies in the algorithm, in connection with a recursive least squares procedure. To that purpose a reliable variable has to be chosen. The last element of the gain vector $G_1(n+1)$ is $m(n+1)$, which is related to the backward prediction - variables by the equation:

$$m(n+1) = \frac{\epsilon_b(n+1)}{E_b(N+1)} \tag{3}$$

The two variables $\epsilon_b(n+1)$ and $E_b(n+1)$ can be computed directly, which yields another value, say $m_o(n+1)$. In finite wordlength implementation, a least squares procedure can be worked out and the equations modified in such a way that the energy of the difference between $m(n+1)$ and $m_o(n+1)$ is minimized. That technique has been shown to be successful in a number of cases [7].

Another control variable can be derived from the prediction error ratio and the adaptation gain. Several control variables can also be combined [8].

As concerns the wordlength estimations, in connection with the filter performance specifications, they are difficult to derive and further work is needed. However, rough estimations can be derived from those already available for the LMS algorithm. Consider, for example the wordlength b_c of the prediction coefficients; a first guess can be:

$$b_c = \log\left(\frac{N}{1-W}\right) + \log_2(G_p) + \log_2(a_{max}) \tag{4}$$

where G_p is the prediction gain and a_{max} is the magnitude of the largest coefficient [1]. According to that expression the transversal FLS algorithms require larger wordlengths than LMS algorithms and the difference is $\log_2(N)$.

3. THE CASE OF UPDATING AND DOWNDATING (SLIDING WINDOW)

In most applications of adaptive filters, least squares algorithms are based on cost functions involving infinite exponential weighting of the error sequence. However, in some specific cases, finite memory might yield better performance, for example, when sharp changes are present in the signal, or in image processing, due to the edges. Finite weighting is implemented through sliding windows, and Fast Least Squares (FLS) algorithms can be derived specifically for that case [9, 10].

Given a reference sequence y(n) and an input signal x(n), the cost function considered is:

$$J_{SW}(n) = \sum_{p=n+1-N_o}^{n} [y(p)-X^t(p)H(n)] \tag{5}$$

where N_o is the length of the observation time window, which slides along the time axis. The autocorrelation matrix and crosscorrelation vector estimations are:

$$R_N(n) = \sum_{p=n+1-N_o}^{n} X(p)X^t(p) \tag{6}$$

$$r_{yx}(n) = \sum_{p=n+1-N_o}^{n} y(p)X(p) \tag{7}$$

Using the recurrence relations:

$$R_N(n+1) = R_N(n)+X(n+1)X^t(n+1) - X(n+1-N_o)X^t(n+1-N_o) \tag{8}$$

and:

$$r_{yx}(n+1) = r_{yx}(n)+y(n+1)X(n+1) - y(n+1-N_o)X(n+1-N_o) \tag{9}$$

the following recursive updating of the coefficients is obtained [1]:

$$H(n+1) = H(n)+G(n+1)e(n+1)-G_o(n+1)e_o(n+1) \tag{10}$$

where $e_o(n+1)$ is the backward innovation error:

$$e_o(n+1) = y(n+1-N_o)-X^t(n+1-N_o)H(n) \tag{11}$$

and $G_o(n+1)$ is the backward adaptation gain:

$$G_o(n+1) = R_N^{-1}(n+1)X(n+1-N^o) \tag{12}$$

Both adaptation gains can be updated recursively, in a very similar manner, and the complete algorithm is given in Fig.3.

Clearly, the computational complexity is about twice that of the exponential window algorithm of the previous section.

As concerns performance, the estimation of the mean residual error power can be derived from the same estimation for the exponential weighting case through the equivalence:

$$W = \frac{N_o-1}{N_o+1} \tag{13}$$

where W is the weighting factor.

ALGORITHM F.L.S.S.W.

AVAILABLE AT TIME n:

COEFFICIENTS OF ADAPTIVE FILTER : $H(n)$
COEFF. OF FORWARD PREDICTOR : $A(n)$
COEFF. OF BACKWARD PREDICTOR : $B(n)$
DATA VECTOR : $X(n)$
WINDOW LENGTH : N_0
DELAYED DATA VECTOR : $X(n-N_0)$
ADAPTATION GAIN : $G(n)$
ADAPTATION GAIN FOR DELAYED DATA : $G_0(n)$
FORWARD PREDICTION ERROR ENERGY : $E_a(n)$

NEW DATA AT TIME n:

Input Signal : $x(n+1)$; Reference : $y(n+1)$

ADAPTATION GAIN UPDATING :

$$e_a(n+1) = x(n+1) - A^t(n)\,X(n)$$

$$e_{0a}(n+1) = x(n+1-N_0) - A^t(n)\,X(n-N_0)$$

$$A(n+1) = A(n) + G(n)\,e_a(n+1) - G_0(n)\,e_{0a}(n+1)$$

$$\varepsilon_a(n+1) = x(n+1) - A^t(n+1)\,X(n)$$

$$\varepsilon_{0a}(n+1) = x(n+1-N_0) - A^t(n+1)\,X(n-N_0)$$

$$E_a(n+1) = E_a(n) + e_a(n+1)\,\varepsilon_a(n+1) - e_{0a}(n+1)\,\varepsilon_{0a}(n+1)$$

$$G_1(n+1) = \begin{bmatrix} 0 \\ G(n) \end{bmatrix} + \frac{\varepsilon_a(n+1)}{E_a(n+1)} \begin{bmatrix} 1 \\ -A(n+1) \end{bmatrix} = \begin{bmatrix} M(n+1) \\ m(n+1) \end{bmatrix}$$

$$G_{01}(n+1) = \begin{bmatrix} 0 \\ G_0(n) \end{bmatrix} + \frac{\varepsilon_{0a}(n+1)}{E_a(n+1)} \begin{bmatrix} 1 \\ -A(n+1) \end{bmatrix} = \begin{bmatrix} M_0(n+1) \\ m_0(n+1) \end{bmatrix}$$

Fig. 3 Cont...

...Cont

$$e_b(n+1) = x(n+1-N) - B^t(n)\, X(n+1)$$
$$e_{ob}(n+1) = x(n+1-N-N_0) - B^t(n)\, X(n+1-N_0)$$
$$k = m(n+1) / [\, 1 + m_0(n+1)\, e_{ob}(n+1)\,]$$
$$k_0 = m_0(n+1) / [\, 1 - m(n+1)\, e_b(n+1)\,]$$
$$G(n+1) = \frac{1}{1 - k\, e_b(n+1)} [\, M(n+1) + k\, B(n) - k\, e_{ob}(n+1)\, M_0(n+1)\,]$$
$$G_0(n+1) = \frac{1}{1 + k_0 e_{ob}(n+1)} [\, M_0(n+1) + k_0\, B(n) + k_0 e_b(n+1)\, M(n+1)\,]$$
$$B(n+1) = B(n) + G(n+1)\, e_b(n+1) - G_0(n+1)\, e_{ob}(n+1)$$

ADAPTIVE FILTER:

$$e(n+1) = y(n+1) - H^t(n)\, X(n+1)$$
$$e_0(n+1) = y(n+1-N_0) - H^t(n)\, X(n+1-N_0)$$
$$H(n+1) = H(n) + G(n+1)\, e(n+1) - G_0(n+1)\, e_0(n+1)$$

Fig. 3 Fast Least Squares algorithm for adaptive filtering with a sliding window.

In order to assess the convergence properties, let us assume that, at time n_o, a system to be identified undergoes a step change in coefficients from H_1 to H_2. Then, from the definition of H(n) [10]:

$$H(n) = R_N^{-1}(n) \sum_{n+1-N_o}^{n_o} X(p)y(p) + \sum_{n_o+1}^{n} X(p)y(p) \tag{14}$$

One gets for the exponential window:

$$E[H(n)-H_2] = W^{n-n_o} [H_1-H_2] \tag{15}$$

and for the sliding window:

$$E[H(n)-H_2] = \frac{N_o-(n-n_o)}{N_o} (H_1-H_2); \tag{16}$$

$$n_o \le n \le n_o+N_o$$

The results obtained for the prediction error energy in linear prediction, are shown in Fig. 4; the finite memory effect is clearly apparent.

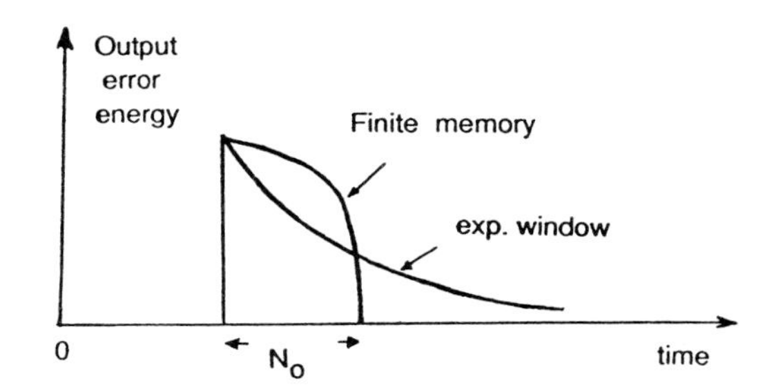

Fig. 4 Responses of the exponential and sliding window algorithms to a step change in input signal

As concerns wordlength limitation effects, they are difficult to investigate in that case. Simulations show that, with some stationary signals, the round-off error accumulation mentioned in the previous section does not occur. It is so for sinusoids in noise for example. However if the input signal is turned on and off regularly for example, then accumulation shows up again, leading to divergence. Further investigation is needed; however, it is clear that the sliding window algorithm is highly sensitive to round-off errors and some control techniques have to be worked out.

The algorithm in Fig. 2 belongs to a group which is based essentially on time recursions. Another group exploits both time and order recursions [11, 12].

AVAILABLE AT TIME n:

REFLECTION COEFFICIENTS : $K_a(n)$, $K_b(n)$

FILTER COEFFICIENTS : $K_f(n)$

BACKWARD PREDICTION ERRORS : $[e_b(n)]$

PREDICTION ERROR ENERGIES : $[E_a(n)]$, $[E_b(n)]$

WEIGHTING FACTOR : W

NEW DATA AT TIME n:

Input signal : $x(n+1)$; Reference : $y(n+1)$

INITIALIZATIONS:

$$e_{ao}(n+1)=e_{bo}(n+1)=x(n+1) \; ; \; e_o(n+1)=y(n+1)$$

$$\varphi_o(n+1)=1 \; ; \; E_{ao}(n+1)=E_{bo}(n+1)=WE_{ao}(n)+x^2(n+1)$$

$0 \leq i \leq N-1$

PREDICTION SECTION:

$$e_{a(i+1)}(n+1) = e_{ai}(n+1) - k_{b(i+1)}(n)\, e_{bi}(n)$$

$$e_{b(i+1)}(n+1) = e_{bi}(n) - k_{a(i+1)}(n)\, e_{ai}(n+1)$$

$$k_{a(i+1)}(n+1) = k_{a(i+1)}(n) + e_{ai}(n+1)\varphi_i(n)\, e_{b(i+1)}(n+1)/E_{ai}(n+1)$$

$$k_{b(i+1)}(n+1) = k_{b(i+1)}(n) + e_{a(i+1)}(n+1)\, e_{bi}(n)\varphi_i(n)/E_{bi}(n)$$

$$E_{a(i+1)}(n+1) = W\, E_{a(i+1)}(n) + e_{a(i+1)}^2(n+1)\varphi_{i+1}(n)$$

$$\varphi_{i+1}(n+1) = \varphi_i(n+1) - \varphi_i^2(n+1)\, e_{bi}^2(n+1)/E_{bi}(n+1)$$

$$E_{b(i+1)}(n+1) = W\, E_{b(i+1)}(n) + e_{b(i+1)}^2(n+1)\varphi_{i+1}(n+1)$$

FILTER SECTION:

$$e_{i+1}(n+1) = e_i(n+1) - k_{fi}(n)\, e_{bi}(n+1)$$

$$k_{fi}(n+1) = k_{fi}(n) + e_{bi}(n+1)\, e_{i+1}(n+1)\varphi_i(n+1)/E_{bi}(n+1)$$

Fig. 5 FLS algorithm for the lattice-ladder structure

4. A LATTICE-LADDER FLS ALGORITHM

Many different combinations of time and order recursions can lead to FLS algorithms for the lattice-ladder structure. The algorithm given in Fig. 5 is obtained when precedence is given to time recursions, which are known for their better robustness to wordlength limitations. The computational complexity amounts to 16N+2 multiplications and 3N divisions (inverse calculations). About 7N memories are required.

A major advantage of that approach is that it provides filters with all orders from 1 to N. Accordingly the operations are arranged in a loop configuration and the programming can be very efficient. For hardware realization a modular cascade structure can be taken, which is quite convenient. Great care has to be exercised in the operation organization, to preserve the indexes and save the relevant variables.

The signal parameters are the reflection or partial correlation coefficients. They can be exploited for signal analysis or feature extraction purposes, but not as widely and easily as the transversal coefficients. From the implementation standpoint, they have a nice property, namely the expectation of their magnitude is bounded by one; the same applies also to the variables $\varphi_i(n)$, which are the ratios of a posteriori to a priori prediction errors. The errors being bounded by the signal values, it turns out that there are no scaling uncertainties in the algorithm of Fig. 5, which turns out to be very robust to round-off errors [13]. A first guess for the wordlength estimation can be obtained from expression (4), taking $a_{max}=1$ and N=1:

$$b_{CL} = \log \left(\frac{1}{1-w}\right) + \log_2 (G_p) \tag{17}$$

As concerns round-off error accumulation, it has been shown theoretically, on an approximate model and to some extent verified experimentally, that the lattice-ladder FLS algorithm is stable [14]. Indeed, it is a critical aspect, which can, in some circumstances, justify the increase in computational complexity.

An important aspect for use in adaptive filtering applications is the availability in each lattice stage of the variable $\varphi_i(n)$, the ratio of a posteriori to a priori prediction errors.

The proper functioning of each stage independently can be checked through that variable, which has to take its value in the interval between zero and unity.

A different type of algorithm, but with similar properties is provided by the rotation method.

5. THE FLS-QR ALGORITHM

In the Q-R decomposition technique the solution to a least squares problem is obtained in two consecutive steps. First an orthogonal matrix is used to transform the data matrix into a matrix in which all the entries are zero except for a submatrix, which takes on a triangular shape. Then the desired coefficients are calculated by solving the new triangular system [15].

The approach can be implemented recursively and the triangularisation process can be performed through a set of rotation operations, which are known for their robustness against finite precision effects.

Let us first show that the rotation matrix $Q(n)$ can be computed recursively. The data matrix $X_N(n)$ is a $(n+1)\text{x}N$ matrix defined by:

$$X_N(n) = \begin{bmatrix} x(n) & x(n-1) & \dots & x(n+1-N) \\ 0 & 0 & \dots & 0 \\ \vdots & \vdots & & \vdots \\ 0 & 0 & \dots & 0 \\ w^{n/2} S_N(n-1) & \dots & \dots & 0 \end{bmatrix}$$

Now, the rotation matrix $Q(n)$ is such that:

$$Q_N(n) X_N(n) = \begin{bmatrix} 0 & \dots & 0 \\ \vdots & & \vdots \\ 0 & \dots & 0 \\ S_N(n) & & \vdots \\ \dots & \dots & 0 \end{bmatrix} \qquad (18)$$

where $S_N(n)$ is a NxN triangular matrix:

$$S_N(n) = \begin{bmatrix} S_{N;N1}(n) & \cdots & \cdots & S_{N;NN}(n) \\ \vdots & & & \vdots \\ \vdots & & & \vdots \\ S_{N;11}(n) & \cdots & \cdots & 0 \end{bmatrix} \tag{19}$$

The data matrix can be rewritten as:

$$X_N(n) = \begin{bmatrix} X(n) & x(n-1) & \cdots & x(n+1-N) \\ & w^{1/2}X_N(n-1) & & \end{bmatrix}$$

Consider the following product:

$$\begin{bmatrix} 1 & 0 \\ 0 & Q_N(n-1) \end{bmatrix} X_N(n) = \begin{bmatrix} x(n) & x(n-1) & \cdots & x(n+1-N) \\ 0 & 0 & \cdots & 0 \\ \vdots & \vdots & & \vdots \\ 0 & 0 & \cdots & 0 \\ \hline \multicolumn{4}{l}{w^{1/2}S_N(n-1) \quad \vdots} \\ \cdots & \cdots & \cdots & 0 \end{bmatrix} \tag{20}$$

If the above matrix is multiplied by the rotation matrix:

$$\begin{bmatrix} \cos\theta_1 & \cdots & -\sin\theta_1 & 0 \\ \vdots & 1 & & \\ \sin\theta_1 & \cdots & \cos\theta_1 & 0 \\ \vdots & & & \vdots \\ 0 & \cdots & \cdots & 1 \end{bmatrix} \tag{21}$$

the term x(n+1-N) is replaced by zero if:

$$\cos\theta_1 \; x(n+1-N) - \sin\theta_1 \; w^{1/2} s_{N,NN}(n-1) = 0 \tag{22}$$

which defines the rotation angle such that:

$$\cos\theta_1 = \frac{w^{1/2} s_{N;NN}(n-1)}{s_{N;NN}(n)}$$

$$\sin\theta_1 = \frac{x(n+1-N)}{s_{N;NN}(n)} \tag{23}$$

where

$$s^2_{N;NN}(n) = w s^2_{N;NN}(n-1) + x^2(n+1-N) \tag{24}$$

Repeating the above operations for x(n+2-N),...,x(n) leads to the recursion:

$$\underset{\sim}{Q}_N(n) = \hat{\underset{\sim}{Q}}_N(n) \begin{bmatrix} 1 & 0 \\ 0 & \underset{\sim}{Q}_N(n-1) \end{bmatrix} \tag{25}$$

where the matrix $\underset{\sim}{Q}_N(n)$ is a product of N rotations similar to (21).

Taking the identity matrix I_N as the initial matrix in the recursive procedure, it appears that the matrix $\underset{\sim}{Q}_N(n)$ is made of rotations and it is orthogonal:

$$\underset{\sim}{Q}_N(n) \underset{\sim}{Q}^t_N(n) = I_N \tag{26}$$

An important property of the rotation matrices is that they keep constant the norms of the vectors. Therefore, the definition equation (18) yields:

$$s^2_{N;NN}(n) = \sum_{p=0}^{n} w^{n-p} x^2(p+1-N) \tag{27}$$

It is the input signal energy and the recursive form is provided by (24).

Once the data matrix has been triangularized, the least squares problem is solved by a set of N substitutions.

Following the principle of other fast least squares algorithms, the fast QR algorithm consists of updating the rotation matrix with extended dimension using forward linear prediction and then obtain the desired matrix with the help of backward linear prediction [16].

The sequence of operations is as follows:

$$\hat{Q}_N(n) \rightarrow \hat{Q}_{N+1}(n+1) \rightarrow \hat{Q}_N(n+1)$$

The corresponding algorithm is given in Fig. 6. In fact it is not $Q_N(n)$ which is involved, but a simplified (N+1)x(N+1) version, denoted $Q_a(n)$ and obtained by deleting all the lines and columns whose elements are only zeros and ones.

For the initialization, all the internal variables are set to zero except for the cosines which are set to one and the prediction error energy $E_a(O)$, which is set to a small positive value, as in the other types of FLS algorithms.

As concerns the operation count, the algorithm requires about 26N multiplications, 2N+1 divisions and 2N+1 square root calculations. Although some simplifications can be made, the computational load is significantly more important than that of the algorithms of the previous sections. However, it must be pointed out that special operators can be designed to carry out the rotations in hardware. Moreover gains can be expected from a reducted internal data wordlength, since most of the internal data are normalized. Further work is needed on that specific topic.

In fact the above algorithm is comparable to a lattice-ladder algorithm with normalization of the variables.

The fact that the operations are essentially rotations leads to a high level of robustness against round-off errors and there are apparently no such divergence effects as encountered in the transversal algorithm of section 2.

AVAILABLE AT TIME n :

Input Data Rotation (N+1)x(N+1) Matrix $Q_a(n)$ [$\cos\theta(i)$; $\sin\theta(i)$; i=1,N]

Backward Data Rotation Matrix $Q_b(n)$ [$\cos\theta_b(i)$; $\sin\theta_b(i)$; i=1,N]

Transformed Input Data N-element Vector : XQ(n)

Transformed Reference N-element Vector : YQ(n)

Forward Prediction Error Energy : $E_a(n)$

Weighting Factor : W

NEW DATA AT TIME n :

Input Signal : x(n+1) ; Reference : y(n+1)

PREDICTION SECTION

$$\begin{bmatrix} \epsilon_{aq}(n+1) \\ \\ XQ(n+1) \end{bmatrix} = Q_a(n) \begin{bmatrix} x(n+1) \\ \\ W^{1/2}\, XQ(n) \end{bmatrix}$$

$$E_a(n+1) = W\, E_a(n) + \epsilon_{aq}^2(n+1)$$

$$\cos\theta_\alpha = [W\, E_a(n) / E_a(n+1)]^{1/2}$$

$$\sin\theta_\alpha = \epsilon_{aq}(n+1) / [E_a(n+1)]^{1/2}$$

$$Q_\alpha = \begin{bmatrix} \cos\theta_\alpha & 0 & -\sin\theta_\alpha \\ 0 & 1 & 0 \\ \sin\theta_\alpha & 0 & \cos\theta_\alpha \end{bmatrix}$$

$$B_1(n+1) = Q_b^t(n) \begin{bmatrix} 1 \\ 0 \\ \vdots \\ 0 \end{bmatrix}$$

$$\begin{bmatrix} U_0 \\ \\ B_2(n+1) \end{bmatrix} = Q_\alpha \begin{bmatrix} Q_a(n) & 0 \\ \\ 0 & 1 \end{bmatrix} \begin{bmatrix} 0 \\ \\ B_1(n+1) \end{bmatrix}$$

Fig. 6 Cont...

...Cont

$$Q_b^t(n+1)\begin{bmatrix}1\\0\\\vdots\\0\end{bmatrix} = B_2(n+1)$$

$$\begin{bmatrix}\gamma_1(n+1)\\ \\ G_{N+1}(n+1)\end{bmatrix} = Q_\alpha \begin{bmatrix}Q_a(n) & 0\\ & \\ 0 & 1\end{bmatrix}\begin{bmatrix}1\\ \\0\end{bmatrix}$$

$$\begin{bmatrix}V_o\\ \\ G_N(n+1)\end{bmatrix} = Q_b(n+1)\; G_{N+1}(n+1)$$

$$Q_a(n+1)\begin{bmatrix}1\\0\\\vdots\\0\end{bmatrix} = \begin{bmatrix}\gamma(n+1)\\ \\ G_N(n+1)\end{bmatrix}$$

$$\epsilon_a(n+1) = \epsilon_{aq}(n+1)\,\gamma(n+1)$$

FILTER SECTION :

$$\begin{bmatrix}\epsilon_q(n+1)\\ \\ YQ(n+1)\end{bmatrix} = Q_a(n+1)\begin{bmatrix}y(n+1)\\ \\ YQ(n)\end{bmatrix}$$

Fig. 6 An FLS-QR algorithm

Like other FLS algorithms, the rotation algorithm performs a real time analysis of the input signal. It can be observed that the rotation procedure corresponds to a projection operation on the signal vectors and the number of large value elements in the transformed vector XQ(n) is, at least for some classes of signals, related to the dimension of the signal space. That property can have interesting practical value and deserves further investigation.

6. CONCLUSION

The main features of the three algorithms which have been reviewed are presented in Fig. 7. Overall, it can be stated that they adequately reflect the characteristics of the family to which they belong. From a user point of view the choice for a given application will depend on what is considered as the most crucial property, computational complexity or numerical stability, operation accuracy or signal parameters. Whether the implementation is carried out in software or hardware may also affect the decision. Worth pointing out also is the need for a fine tuning of the adaptive filter parameters, in order to get a good efficiency. In that respect, the proper choice of the values of the weighting factor W and the initial prediction error energy E_o is crucial.

As concerns research aspects, further work is needed for all three algorithm families, but the rotation method might deserve special attention.

Algorithm	Comput. Complexity			Round-off Error Accumulation	Operation Accuracy	Signal Parameters
	Mult	Div	SqRt			
Transversal	10N	2	0	yes	moderate	very useful
Lattice/ Ladder	16N	3N	0	no	moderate - low	can be useful
Rotation Method	26N	2N+1	2N+1	no	low? **	may be useful **

** Further work is needed

Fig. 7 Main features of the three algorithms

FERENCES

] Bellanger M., Adaptive Digital Filters and Signal Analysis Marcel Dekker Inc, New York, 1987.

] Falconer D., and Ljung L., Application of Fast Kalman Estimation to Adaptive Equalization, IEEE Trans. COM-26, Oct.1978, pp. 1439-1446.

] Carayannis G., Manolakis D., and Kalouptsidis N., A Fast Alg. for LS Filtering and Prediction, IEEE Trans. ASSP-31, Dec.1983, pp. 1394-1402.

] Cioffi J., and Kailath T., Fast Recursive Least Squares Tranversal Filters for Adaptive Filtering, IEEE Trans. ASSP-32, April 1984, pp. 304-337.

] Lawrence V.B, and Tewkesbury S.K., Multiprocessor Implementation of Adaptive Digital Filters, IEEE Trans. COM-31, June 1983, pp. 826-835.

] Fabre P., and Gueguen C., Fast Recursive Least Squares: Preventing Divergence, Proc. IEEE-ICASSP-85, Tampa, Florida, 1985, pp. 1149-1152.

] Botto J.L., Stabilization of Fast Recursive Least Squares Transversal Filters, Proc. IEEE-ICASSP-87, Dallas, Texas, 1987, pp. 403-406.

] Slock D.T.M., and Kailath T., Numerically Stable Fast Recursive Least Squares Transversal Filters, Proceedings of IEEE-ICASSP-88, New York, April 1988, pp. 1365-1368.

] Honig M.L., and Messerschmitt, D.G., (1984) Adaptive Filter Structures, Algorithms and Applications, Kluwer Academic, Boston, 1984.

0] Manolakis D., Ling F., and Proakis J., Efficient Time-Recursive Least Squares Algorithms for Finite Memory Adaptive Filtering, IEEE Transactions, Vol. CAS-34, N°4, April 1987, pp. 400-407.

1] Friedlander B., Lattice Filters for Adaptive Processing, Proc. IEEE, 70, August 1982, pp. 829-867.

2] Turner J.M., Recursive Least Squares Estimation and Lattice Filters, Chapter 5 in Adaptive Filters, Prentice Hall, Englewood Cliffs, New Jersey, 1985.

[13] Ling F., Manolakis D., and Proakis J., Numerically Robust LS Lattice-Ladder Algorithm with Direct Updating of the Coefficients, IEEE Trans, ASSP-34, August 1986, pp. 837-845.

[14] Ljung S., and Ljung L., Error Propagation Properties of Recursive Least Squares Adaptation Algorithms, Automatica 21, 1985, pp. 157-167.

[15] Haykin S., Adaptive Filter Theory, Prentice Hall, Englewood Cliffs, New Jersey, 1985.

[16] Cioffi J.M., High Speed Systolic Implementation of Fast Q.R. Adaptive Filters, Proceedings of IEEE ICASSP-88, New York, April 1988, pp. 1584-1587.

A SQUARE-ROOT FORM OF THE OVERDETERMINED RECURSIVE INSTRUMENTAL VARIABLE ALGORITHM

B. Friedlander
(Signal Processing Technology Limited, USA)

and

B. Porat
(Technion-Israel Institute of Technology)

ABSTRACT

The paper derives a square-root version of the overdetermined recursive instrumental variable (ORIV) algorithm. This square-root version improves the numerical stability of the algorithm, and avoids the problem of loss of positive-definiteness of the inverse covariance matrix. The algorithm uses square-root arrays, and employs both orthogonal and hyperbolic rotations, as necessitated by the nature of the ORIV algorithm.

1. INTRODUCTION

The instrumental variable (IV) method is among the favourite methods for estimating the parameters of dynamical systems [1]. While not very accurate, the IV method is relatively robust, and is less prone to convergence problems than other, more accurate methods.

One way of improving the accuracy of IV algorithms is to use a number of equations greater than the number of estimated parameters. The use of the overdetermined IV method for spectral analysis was advocated by Cadzow [2] and others. Friedlander [3] has developed a recursive version of the overdetermined IV algorithm - the so-called overdetermined recursive instrumental variable (ORIV) algorithm. Recently, the present authors have proposed to use the ORIV algorithm

---

This work was supported by the National Science Foundation under grant no. ISI-8760095.

for the estimation of non-Gaussian ARMA models, using high-order cumulants [4].

The ORIV algorithm of [3] was found, by the present authors and others, to suffer from severe numerical problems. The estimated covariance matrix appearing in the algorithm (the matrix P_t^{-1} - see [3], eq. (10)) is computed as a product of four data matrices, and this may cause this matrix to be ill-conditioned. Furthermore, the inversion of this matrix is done (as in many recursive algorithms) using the matrix inversion lemma, which is known to be numerically unstable. When the matrix P_t loses its positive-definiteness, the algorithm experiences rapid divergence.

The present paper aims at improving the numerical stability of the ORIV algorithm, by deriving a square-root version for it. The square-root algorithm updates the matrix square-root of P_t, and thus avoids the loss of positive-definiteness. A peculiar feature of the ORIV algorithm is the appearance of an indefinite matrix in the square-root array. This necessitates using hyperbolic rotations, as well as orthogonal rotations.

The square-root ORIV algorithm is derived in section 2 of the paper. In section 3, we illustrate the algorithm by an example.

2. THE SQUARE-ROOT ORIV ALGORITHM

The derivation of the ORIV algorithm is given in [3] and we will not repeat it here. We will only outline the underlying model, and summarize the algorithm in a tabular form. We will then proceed to derive the square-root version of the algorithm.

Suppose we are given the following model:

$$Z_t^T Y_t \theta = Z_t^T X_t + E_t, \qquad (2.1)$$

where X_t, Y_t, Z_t are data matrices of dimensions tx1, txN, txK respectively, with K>N. θ is an N-dimensional parameter vector, and E_t is a t-dimensional vector of residual errors.

The least-squares estimate of θ (i.e. the value of θ minimizing the norm of E_t) is given by

$$\hat{\theta}_t = [Y_t^T Z_t Z_t^T Y_t]^{-1} Y_t^T Z_t Z_t^T X_t . \tag{2.2}$$

Suppose that new data x_{t+1}, y_{t+1}^T, z_{t+1}^T become available, and define

$$X_{t+1} = \begin{bmatrix} X_t \\ x_{t+1} \end{bmatrix}; \; Y_{t+1} = \begin{bmatrix} Y_t \\ y_{t+1}^T \end{bmatrix}; \; Z_{t+1} = \begin{bmatrix} \lambda Z_t \\ z_{t+1}^T \end{bmatrix} \tag{2.3}$$

where $0<\lambda\leq 1$ is a user-chosen forgetting factor. The corresponding estimate $\hat{\theta}_{t+1}$ is now given by an equation similar to (2.2), but with the subscript t+1 replacing t throughout.

The ORIV algorithm recursively updates $\hat{\theta}_{t+1}$, using the previous estimate $\hat{\theta}_t$ and several auxiliary matrices. A summary of the ORIV algorithm is given in Table 1.

A major drawback of the ORIV algorithm is its poor numerical behaviour. The matrix $Y_t^T Z_t Z_t^T Y_t$ is often ill-conditioned, and its recursive inversion (using the matrix inversion lemma) may lead to the loss of positive-semidefiniteness. The situation here is much worse than in the standard recursive least-squares (RLS) algorithm (see e.g. [1]), because the matrix to be inverted is a product of four data matrices (rather than two, as in the RLS algorithm).

The square-root ORIV algorithm is based on the triangularization of square-root arrays, using the approach of Morf and Kailath [5]. The algorithm preserves the positive-semidefiniteness of P_t by recursively updating its lower triangular square-root $P_t^{1/2}$. This is a common property of square-root algorithms, and the present algorithm is no exception. However, the ORIV algorithm is different from other recursive algorithms in that one of the auxiliary matrices (namely Λ_{t+1} - See Table 1) is indefinite. Because of that, orthogonal transformations are not sufficient to update the matrix square-roots, and hyperbolic (or J-orthogonal transformations are necessary. To see this, observe that

$$\Lambda_{t+1} = \begin{bmatrix} -z_{t+1}^T z_{t+1} & \lambda \\ \lambda & 0 \end{bmatrix} = \begin{bmatrix} \alpha & 0 \\ -\lambda/\alpha & \lambda/\alpha \end{bmatrix} \begin{bmatrix} -1 & 0 \\ 0 & 1 \end{bmatrix} \begin{bmatrix} \alpha & -\lambda/\alpha \\ 0 & \lambda/\alpha \end{bmatrix} = AJA^T \tag{2.4}$$

where $\alpha = (z_{t+1}^T z_{t+1})^{1/2}$. The matrix J is called the signature of Λ_{t+1}. The fact that the diagonal of J includes the entry -1 results from the indefiniteness of Λ_{t+1}, and leads to the need for hyperbolic transformations, as we shall presently see.

$$w_{t+1} = S_t z_{t+1}$$

$$S_{t+1} = \lambda S_t + y_{t+1} z_{t+1}^T$$

$$\varphi_{t+1} = [\, w_{t+1} \quad y_{t+1}]$$

$$\Lambda_{t+1} = \begin{bmatrix} -z_{t+1}^T z_{t+1} & \lambda \\ \lambda & 0 \end{bmatrix}$$

$$K_{t+1} = P_t \varphi_{t+1} (\Lambda_{t+1} + \varphi_{t+1}^T P_t \varphi_{t+1})^{-1}$$

$$P_{t+1} = \lambda^{-2} [\, P_t - K_{t+1} \varphi_{t+1}^T P_t]$$

$$v_{t+1} = \begin{bmatrix} z_{t+1}^T L_t \\ x_{t+1} \end{bmatrix}$$

$$L_{t+1} = \lambda L_t + z_{t+1} x_{t+1}$$

$$\hat{\theta}_{t+1} = \hat{\theta}_t + K_{t+1} (v_{t+1} - \varphi_{t+1}^T \hat{\theta}_t)$$

Table 1: Summary of a single step of the ORIV Algorithm

Assume that we have found matrix Q of dimensions (N+2)x(N+2) such that

$$Q\begin{bmatrix} J & O \\ O & I_N \end{bmatrix} Q^T = \begin{bmatrix} J & O \\ O & I_N \end{bmatrix}. \tag{2.5}$$

Such a matrix is called J-orthogonal. Furthermore, assume that Q was chosen so that

$$\begin{bmatrix} A & \phi_{t+1}^T P_t^{1/2} \\ O & P_t^{1/2} \end{bmatrix} Q = \begin{bmatrix} L_1 & O \\ M & L_2 \end{bmatrix}, \tag{2.6}$$

where L_1 and L_2 are lower triangular, of dimensions 2x2 and NxN respectively. By (2.5) we will then have

$$\begin{bmatrix} A & \phi_{t+1}^T P_t^{1/2} \\ O & P_t^{1/2} \end{bmatrix} \begin{bmatrix} J & O \\ O & I_N \end{bmatrix} \begin{bmatrix} A^T & O \\ (P_t^{1/2})^T \phi_{t+1} & (P_t^{1/2})^T \end{bmatrix}$$

$$= \begin{bmatrix} L_1 & O \\ M & L_2 \end{bmatrix} \begin{bmatrix} J & O \\ O & I_N \end{bmatrix} \begin{bmatrix} L_1^T & M^T \\ O & L_2^T \end{bmatrix}. \tag{2.7}$$

Hence,

$$AJA^T + \phi_{t+1}^T P_t \phi_{t+1} = \Lambda_{t+1} + \phi_{t+1}^T P_t \phi_{t+1} = L_1 J L_1^T \tag{2.8a}$$

$$P_t \phi_{t+1} = MJL_1^T \tag{2.8b}$$

$$P_t = MJM^T + L_2 L_2^T. \tag{2.8c}$$

From (2.8b) we get

$$M = P_t \phi_{t+1} (L_1^T)^{-1} J, \tag{2.9}$$

hence, substituting in (2.8c),

$$L_2 L_2^T = P_t - MJM^T = P_t - P_t \phi_{t+1} (L_1^T)^{-1} J L_1^{-1} \phi_{t+1}^T P_t$$
$$= P_t - P_t \phi_{t+1} (\Lambda_{t+1} + \phi_{t+1}^T P_t \phi_{t+1})^{-1} \phi_{t+1}^T P_t = \lambda^2 P_{t+1}. \tag{2.10}$$

Therefore, since L_2 is lower triangular, it is proportional to the desired square-root of P_{t+1}, i.e.

$$P_{t+1}^{1/2} = \lambda^{-1} L_2. \tag{2.11}$$

Note also that

$$M L_1^{-1} = P_t \phi_{t+1} (L_1^T)^{-1} J L_1^{-1} = P_t \phi_{t+1} (\Lambda_{t+1} + \phi_{t+1}^T P_t \phi_{t+1})^{-1} = K_{t+1}. \tag{2.12}$$

Equations (2.4), (2.6), (2.11) and (2.12) form the core of the square-root ORIV algorithm. The complete algorithm is summarized in Table 2.

It turns out that the existence of the matrix Q is guaranteed by the fact that the matrix $\Lambda_{t+1} + \phi_{t+1}^T P_t \phi_{t+1}$ has the same signature as Λ_{t+1}, hence equation (2.8a) is non-contradictory. This fact is proven in appendix A.

It remains to describe the details of the hyperbolic transformation Q appearing in (2.6). Consider the array (shown here for N = 4)

$$R = \begin{bmatrix} A & \phi_{t+1}^T P^{1/2} \\ O & P_t^{1/2} \end{bmatrix} = \begin{bmatrix} x & O & x & x & x & x \\ x & x & x & x & x & x \\ O & O & x & O & O & O \\ O & O & x & x & O & O \\ O & O & x & x & x & O \\ O & O & x & x & x & x \end{bmatrix}, \tag{2.13}$$

where the x denote possibly nonzero terms. The entries $\{R_{1,j}; 3 \leq j \leq N+2\}$ can be annihilated using 2x2 hyperbolic rotations, similar to the orthogonal Givens rotations [6]. A general 2x2 hyperbolic rotation corresponding to the signature J is given by

$$\Omega_2^{(H)} = \frac{1}{\sqrt{1-\gamma^2}}\begin{bmatrix} 1 & -\gamma \\ -\gamma & 1 \end{bmatrix}; \quad |\gamma|<1. \tag{2.14}$$

$$w_{t+1} = S_t z_{t+1}$$

$$S_{t+1} = \lambda S_t + y_{t+1} z_{t+1}^T$$

$$\varphi_{t+1} = [\, w_{t+1} \quad y_{t+1}]$$

$$\alpha = (z_{t+1}^T z_{t+1})^{1/2}$$

$$A = \begin{bmatrix} \alpha & 0 \\ -\lambda/\alpha & \lambda/\alpha \end{bmatrix}$$

Find Q such that (see Table 3)

$$R = \begin{bmatrix} A & \varphi_{t+1}^T P_t^{1/2} \\ 0 & P_t^{1/2} \end{bmatrix} Q = \begin{bmatrix} L_1 & 0 \\ M & L_2 \end{bmatrix}$$

$$P_{t+1}^{1/2} = \lambda^{-1} L_2$$

$$K_{t+1} = M L_1^{-1}$$

$$v_{t+1} = \begin{bmatrix} z_{t+1}^T & L_t \\ x_{t+1} & \end{bmatrix}$$

$$L_{t+1} = \lambda L_t + z_{t+1} x_{t+1}$$

$$\hat{\Theta}_{t+1} = \hat{\Theta}_t + K_{t+1}(v_{t+1} - \varphi_{t+1}^T \hat{\Theta}_t)$$

Table 2: A single step of the Square-Root ORIV Algorithm

It can be easily verified that $Q_2^{(H)}$ satisfies the relationship $Q_2^{(H)} J (Q_2^{(H)})^T = J$, as required in (2.5). When $Q_2^{(H)}$ multiplies a 2x2 matrix from the right, the parameter γ can be chosen to annihilate the (1,2) element as follows:

$$\frac{1}{\sqrt{1-\gamma^2}} \begin{bmatrix} x_{11} & x_{12} \\ x_{21} & x_{22} \end{bmatrix} \begin{bmatrix} 1 & -\gamma \\ -\gamma & 1 \end{bmatrix} = \begin{bmatrix} x & 0 \\ x & x \end{bmatrix} \Rightarrow \gamma = \frac{x_{21}}{x_{11}} . \quad (2.15)$$

The special structure of the array R can be used to annihilate the top row of R by N hyperbolic rotations, as depicted in Table 3. Note the main loop annihilates the top row entries from right to left. The lower triangular structure of the lower right block of R, and the zeros at the lower left block, enable a considerable saving in the amount of computations.

The entries $\{R_{2,j};\ 3 \leq j \leq N+2\}$ can be annihilated using 2x2 Givens (orthogonal) rotations. A general 2x2 Givens rotation is given by

$$Q_2^{(O)} = \frac{1}{\sqrt{1+\gamma^2}} \begin{bmatrix} 1 & -\gamma \\ \gamma & 1 \end{bmatrix} . \quad (2.16)$$

When $Q_2^{(O)}$ multiplies a 2x2 matrix from the right, the parameter γ can be chosen to annihilate the (1,2) element as follows:

$$\frac{1}{\sqrt{1+\gamma^2}} \begin{bmatrix} x_{11} & x_{12} \\ x_{21} & x_{22} \end{bmatrix} \begin{bmatrix} 1 & -\gamma \\ \gamma & 1 \end{bmatrix} = \begin{bmatrix} x & 0 \\ x & x \end{bmatrix} \Rightarrow \gamma = \frac{x_{21}}{x_{11}} . \quad (2.17)$$

Table 3 shows the details for annihilating the second row of R by N orthogonal transformations. Again, the special structure of R can be used to save a considerable amount of computations.

```
for j = N+2 down to 3 do
begin
γ = R(1,j)/R(1,1)
R(1,1) = [R(1,1)-γR(1,j)]/√(1-γ²)
for k = 2 and for k = j to N+2 do
begin
α = [R(k,1)-γR(k,j)]/√(1-γ²)
β = [R(k,j)-γR(k,1)]/√(1-γ²)
R(k,1) = α
R(k,j) = β
end
end
for j = N+2 down to 3 do
begin
γ = R(2,j)/R(2,2)
R(2,2) = [R(2,2)+γR(2,j)]/√(1+γ²)
for k = j to N+2 do
begin
α = [R(k,2)+γR(k,j)]/√(1+γ²)
β = [R(k,j)-γR(k,2)]/√(1+γ²)
R(k,2) = α
R(k,j) = β
end
end
```

Table 3: The J-Orthogonal and the Orthogonal Transformations

3. AN EXAMPLE

The square-root ORIV algorithm was programmed in C (on a 8088/8087 personal computer, using double precision) and used to simulate the ARMA parameter estimation algorithm reported in [4]. Previous experience with this algorithm has indicated severe numerical problems. The algorithm in [4] uses high-order sample moments (third or fourth), hence the entries of the matrix P_t have a very large dynamic range. To guarantee stability of the original (non square-root) ORIV algorithm for this application, the computation of P_t had to be restarted every 100 iterations (with double precision arithmetic). Even so, divergence was experienced in about 5% of the runs.

When the same experiments were repeated with the square-root algorithm, the divergence problem disappeared, and the need for periodically restarting P_t was eliminated.

As an example, we illustrate the behaviour of the algorithm for the ARMA process generated by the transfer function

$$\frac{b(z)}{a(z)} = \frac{1-2.05z^{-1}+z^{-2}}{1+0.9801z^{-2}} \tag{3.1}$$

and an exponentially distributed white driving noise. We performed 100 Monte-Carlo runs, with 1000 data points each.

Figure 1 shows the average trajectories of the estimated parameters $\hat{a}_1$, $\hat{a}_2$, $\hat{b}_1$, $\hat{b}_2$, as well as the $\pm 1\sigma$ bounds. No divergence was experienced in this test, and no restarting was needed. We remark that the irregular behaviour of the estimates $\hat{b}_1$, $\hat{b}_2$ during the initial phase is a characteristic of the parameter estimation algorithm, and is not related to its square-root implementation - see [4] for further dicussion.

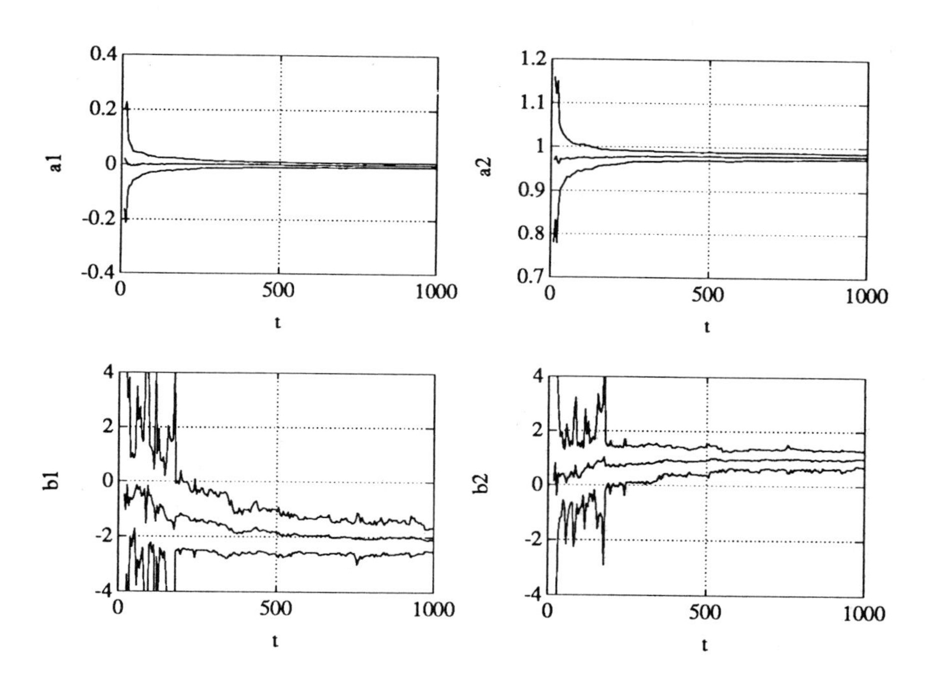

Fig. 1 Monte-Carlo results for the square-root ORIV algorithm

APPENDIX A: The Signature of $\Lambda_{t+1}+\phi_{t+1}^T P_t \phi_{t+1}$

In this appendix we prove that the matrix $\Lambda_{t+1}+\phi_{t+1}^T P_t \phi_{t+1}$ has the signature J, hence its decomposition as $L_1 J L_1^T$ (see equation (2.7)) is valid. Recall from [3] that

$$\phi_{t+1} = [\, S_t z_{t+1} \quad y_{t+1}] \,; \; P_t = (S_t S_t^T)^{-1} ; \; S_t = Y_t^T Z_t . \quad \text{(a.1)}$$

Therefore,

$$\Lambda_{t+1}+\phi_{t+1}^T P_t \phi_{t+1} = \begin{bmatrix} -z_{t+1}^T z_{t+1} & \lambda \\ \lambda & 0 \end{bmatrix} + \begin{bmatrix} z_{t+1}^T S_t^T \\ y_{t+1}^T \end{bmatrix} (S_t S_t^T)^{-1} [\, S_t z_{t+1} \quad y_{t+1}]$$

$$= \begin{bmatrix} -z_{t+1}^T [\, I - S_t^T (S_t S_t^T)^{-1} S_t] \, z_{t+1} & \lambda + z_{t+1}^T S_t^T (S_t S_t^T)^{-1} y_{t+1} \\ \lambda + y_{t+1}^T (S_t S_t^T)^{-1} S_t z_{t+1} & y_{t+1}^T y_{t+1} \end{bmatrix} \quad \text{(a.2)}$$

Now, the matrix $I-S_t^T(S_t S_t^T)^{-1} S_t$ is the projection on the columns of S_t^T. Hence this matrix is positive semidefinite and, furthermore, the (1,1) element of the right-hand side of (a.2) will be strictly negative as long as z_{t+1} is not in the column space of S_t^T. Except for this singular case, the matrix $\Lambda_{t+1}+\phi_{t+1}^T P_t \phi_{t+1}$ will have the form

$$\Lambda_{t+1}+\phi_{t+1}^T P_t \phi_{t+1} = \begin{bmatrix} -a & b \\ b & c \end{bmatrix} = \begin{bmatrix} \sqrt{a} & 0 \\ -\frac{b}{\sqrt{a}} & \sqrt{c+\frac{b^2}{a}} \end{bmatrix} \begin{bmatrix} -1 & 0 \\ 0 & 1 \end{bmatrix} \begin{bmatrix} \sqrt{a} & -\frac{b}{\sqrt{a}} \\ 0 & \sqrt{c+\frac{b^2}{a}} \end{bmatrix} .$$

(a.3)

Since $a>0$ and $c\geq 0$, this decomposition is valid. In the singular case (i.e. when z_{t+1} is in the column space of S_t^T), the square-root recursion cannot proceed in its regular manner, and needs to be modified. This is true, however, for the ORIV algorithm as well, because then the matrix $\Lambda_{t+1}+\phi^T_{t+1}P_t\phi_{t+1}$ will be singular.

REFERENCES

[1] Ljung, L. and Söderström, T., (1983), "Theory and Practice of Recursive Identification", MIT Press.

[2] Cadzow, J.A., (1980), "High Performance Spectral Estimation - A New ARMA Method", IEEE Trans. Acoustics, Speech and Signal Processing, Volume ASSP-28, Number 5, pp. 524-529.

[3] Friedlander, B., (1984), "The Overdetermined Recursive Instrumental Variable Method", IEEE Trans. Automatic Control, Volume AC-29, Number 4, pp. 353-356.

[4] Friedlander, B. and Porat, B., (1989) "Adaptive IIR Algorithms Based on High-Order Statistics", IEEE Trans. Acoustics, Speech and Signal Processing, vol. 37, no. 4, pp. 485-495.

[5] Morf, M. and Kailath, T., (1975), "Square-Root Algorithms for Least-Squares Estimation", IEEE Trans. Automatic Control, Volume AC-20, Number 4, pp. 487-497.

[6] Golub, G.H. and Van Loan, C.F., (1983), "Matrix Computations. The Johns Hopkins University Press.

PERFORMANCE BOUNDS FOR EXPONENTIALLY WINDOWED RLS ALGORITHM IN A NONSTATIONARY ENVIRONMENT

S. McLaughlin, B. Mulgrew and C. F. N. Cowan
(Department of Electrical Engineering, University of Edinburgh)

ABSTRACT

This paper investigates the performance of an exponentially windowed recursive least squares algorithm (RLS) operating as a channel estimator for use in an adaptive equaliser structure. The nonstationary channel considered is represented as a 2nd order Markov process, an approach which is commonly used to model fading dispersive channels such as the HF channel.

In this paper new theoretical results reinforced by simulation are presented which illustrate the optimal performance bounds for the RLS operating in such an environment. The performance of the RLS is characterised for various values of

i) the level of colouration of the input signal,
ii) the level of additive noise in the system,
iii) the degree of nonstationarity.

1. INTRODUCTION

At the present time considerable research effort is being expended to develop adaptive equalisers (as in figure 1) for use in communications systems where the channel is time-varying. The overall aim is to allow communication at data rates greater than are currently possible.

This paper investigates the performance of an exponentially windowed recursive least squares algorithm (RLS), [1], operating as a channel estimator for use in an adaptive equaliser structure. The nonstationary channel considered is represented as a 2nd order Markov process, an approach which is commonly used to model fading dispersive channels such as the high-frequency radio channel.

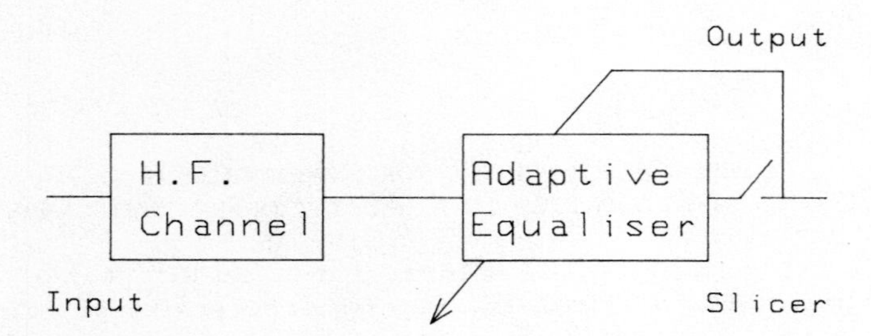

Fig. 1 Adaptive Equalisation of a time-invariant Channel

It has generally been assumed that the RLS algorithm would be suitable for use in time-varying environments because of its fast convergence properties in a stationary environment. However, recently published work [2,3,4] would appear to suggest that in both time-varying and high noise environments the RLS suffers a considerable degradation in performance as demonstrated by a slower rate of convergence and higher minimum mean squared error (MSE).

In this paper the direct modelling (channel estimator) approach is chosen for analysis as opposed to indirect modelling, i.e. equalisation, because the only nonstationarity considered is the time variation of the channel coefficients, since the input signal is stationary and white. This makes the analysis more tractable and isolates the tracking performance of the algorithms. New theoretical results reinforced by simulation are presented here which illustrate the optimal performance bounds for the RLS operating in such an environment. The performance of the RLS is characterised for various values of

i) the level of colouration of the input signal,
ii) the level of additive noise in the system,
iii) the degree of nonstationarity.

The decoupling of the overall error achieved by the RLS in estimating the system into a measurement term and a lag term as in [5], is used to illustrate the degradation in performance due to high additive noise and/or time variations in the system. The relative effect of these errors is shown theoretically and reinforced with simulations. The selection of λ, the exponential windowing factor, to give optimal performance is considered and the trade-off required in its selection is discussed and illustrated.

The simulations which are presented illustrate the performance bounds of the RLS in terms of rate of convergence and minimum achieved MSE given variations in the three parameters detailed above. The results presented will be of use in determining if the RLS is an appropriate choice of algorithm in the design of adaptive equalisers for time-varying channels, given a priori knowledge of the maximum parameter variations possible in the system.

2. RLS MATHEMATICAL ANALYSIS

The problem considered is that of direct modelling of a time-varying system which is characterised by the tapped delay line model of figure 2, where the time-varying taps are generated by filtering random white Gaussian noise through a filter, in this case a 2nd order digital butterworth filter is used with bandwidth very much narrower than the symbol rate. The construction used to carry out the system identification is illustrated in figure 3, this structure was chosen since the tracking behaviour of the algorithms is isolated. That is, the input to the system is also the input to the adaptive algorithm so that only the time-variations of the system under investigation are estimated. In the inverse modelling situation decision errors, inputs which would be both nonstationary and coloured would lead to degradation in the performance of the algorithms for reasons not associated with their tracking performance.

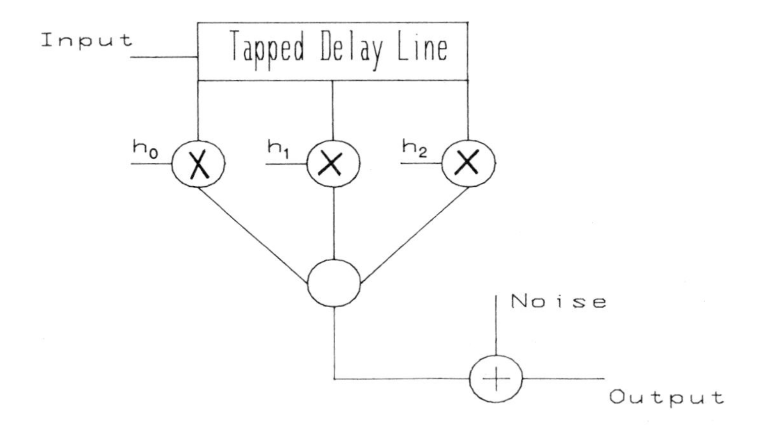

Fig. 2 Channel Model

Using the representation of figure 3 allows the system output at iteration k to be written as,

$$y_k = \underline{x}_k^T \underline{H}_k + n_k \tag{1}$$

where, $\underline{x}_k$ is the input signal vector with the superscript T representing the transpose operator, $\underline{H}_k$ is the tap weight vector of the time varying system and n_k represents the unobservable measurement noise in the system, which in this case is additive white Gaussian noise (AWGN).

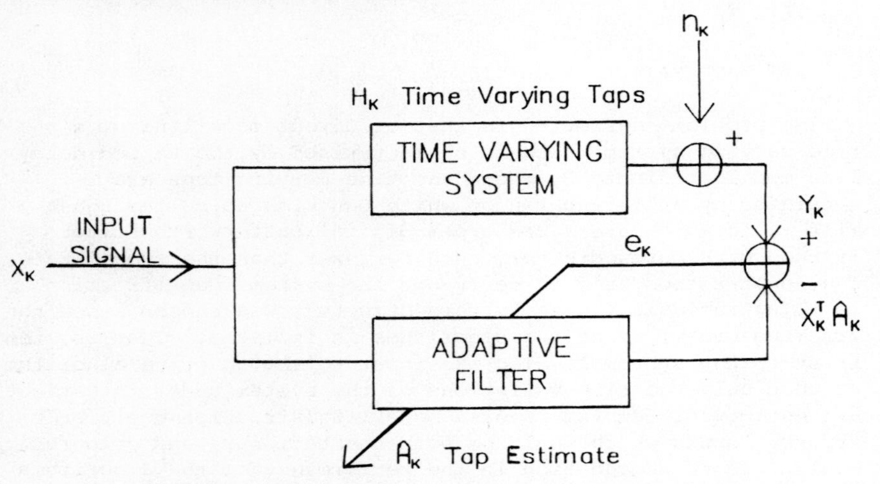

Fig. 3 System Identification of Time-Varying Linear Model

The noise, n_k, and input signal vector $\underline{x}_k$ are assumed to satisfy the following assumptions.

A1:-The sequence $\underline{x}_k$ is stationary with finite moments.

A2:-The sequence n_k is identically distributed and independent of $\underline{x}_k$.

A3:-The time variations of $\underline{H}_k$ are random and independent of $\underline{x}_k$ and n_k.

A4:-The estimate of the autocorrelation matrix, $\hat{R}_k = \sum_{j=1}^{k} \underline{x}_j \lambda^{k-j} \underline{x}_j^T$, of the input signal vector $\underline{x}_j$ can, in the limit, be represented as $\hat{R}_k \approx \Phi(1-\lambda)^{-1}$, where $\Phi = E[\underline{x}_k \underline{x}_k^T]$ and is the exact autocorrelation matrix of the input signal vector $\underline{x}_k$ and initiall· in the case considered here since the input is white $\Phi = I$, where I is the identity matrix. That is,

$$\lim_{k\to\infty} \hat{R}_k = I(1-\lambda)^{-1}$$

This assumes that λ lies close to 1 (normally > 0.9), if this condition is not satisfied then this assumption cannot be considered valid. Assumptions A1 to A3 are identical to those used by Macchi [6,7] in her analysis of the LMS in a nonstationary environment.

It may be argued that these assumptions are not representative of the scenario in which the algorithms have to operate, however they provide a means of analysing the performance of the algorithm and as will be shown later the theoretical predictions obtained from the analysis are in close agreement with simulation results and this in itself is sufficient justification for these assumptions.

If the following variables are defined;

$$\underline{d}_k = \underline{H}_{k+1} - \underline{H}_k \tag{2}$$

and, $\underline{d}_k$ is a measure of the nonstationarity of the channel. Also,

$$\underline{q}_k = \hat{\underline{H}}_k - \underline{H}_k \tag{3}$$

is the tap weight error vector (or misadjustment).

The error e_k would normally be written as,

$$e_k = y_k - \underline{x}_k^T \hat{\underline{H}}_k \tag{4}$$

where $\hat{\underline{H}}_k$ represents the tap weight vector estimate. By using the expressions (1) to (3) it is possible to rewrite (4) as below,

$$e_k = n_k - \underline{x}_k^T \underline{q}_k \tag{5}$$

In this way it is expressed in terms of the unobservable noise in the system and the tap weight error vector.

If the RLS algorithm is written as below,

$$\hat{\underline{H}}_{k+1} = \hat{\underline{H}}_k + \hat{R}_k^{-1} e_k \underline{x}_k \tag{6}$$

where $\hat{R}_k$ is the estimated autocorrelation of the input signal vector $\underline{x}_k$.

As has been noted earlier this representation of the RLS is similar to the structure of the LMS with μ replaced by the inverse of the estimated autocorrelation of the input signal vector.

Using these expressions and by employing some algebraic manipulation a recursive expression for the parameter error vector $\underline{q}_{k+1}$ can be obtained.

$$\underline{q}_{k+1} = (I - \hat{R}_k^{-1}\underline{x}_k\underline{x}_k^T)\underline{q}_k + \hat{R}_k^{-1}n_k\underline{x}_k - \underline{d}_k \tag{7}$$

Expression (7) can be split into two clearly identifiable terms as shown below,

$$\underline{v}_{k+1}^m = (I - \hat{R}_k^{-1}\underline{x}_k\underline{x}_k^T)\underline{v}_k^m + R^{-1}n_k\underline{x}_k \quad ; \quad \underline{v}_1^m = 0 \tag{8a}$$

$$\underline{v}_{k+1}^l = (I - \hat{R}_k^{-1}\underline{x}_k\underline{x}_k^T)\underline{v}_k^l - d_k \quad ; \quad \underline{v}_1^l = 0 \tag{8b}$$

The first term, $\underline{v}_k^m$, can be viewed as a measurement noise term and is present even in the stationary situation. The second term, $\underline{v}_k^l$ can be considered as a lag term and is associated with the time variations of the system. It is present only in the nonstationary case. As a result of assumption A2 & A3 it may be assumed that the two contributions to the error in expression (5) are independent of each other.

In the limit the excess steady state mean square deviation (MSD) is,

$$\lim_{k\to\infty} E|\underline{x}_k^T\underline{q}_k|^2 = \lim_{k\to\infty} E|\underline{x}_k^T v_k^m|^2 + \lim_{k\to\infty} E|\underline{x}_k^T\underline{v}_k^l|^2 \tag{9}$$

That is,

Total MSD = Stationary MSD + Lag MSD

The limits of each term are required to be finite in order to obtain the steady state excess MSE. Since the input signal vector and tap weight error vector are independent and the input signal is stationary it is sufficient to show that the squared norm of the tap weight vector, q_k, is finite in the limit. The analysis for the general case i.e. not utilising assumption (A4) is possible [8] but extremely complex. For the situation considered here it is unnecessary since by utilising assumption (A4) it is possible to substitute $(1-\lambda)I$ for $\hat{R}^{-1}$ and the proof that the tap weight error vector is finite in the limit then follows as in [6]. Once the limits are shown to be finite an expression for the excess steady state MSE can be obtained as follows.

Using the recursions of (8a) and (8b) and taking the measurement noise term first,

$$E[|\underline{v}_{k+1}^{m}|^{2}] = E[|(I-\hat{R}_{k}^{-1}\underline{x}_{k}\underline{x}_{k}^{T})\underline{v}_{k}^{m}+\hat{R}_{k}^{-1}n_{k}\underline{x}_{k}|^{2}] \tag{10}$$

Now if $E[|n_k|^2]=N$ and we replace the inverse of the estimated autocorrelation matrix, $\hat{R}_k^{-1}$ by $(1-\lambda)I$ using assumption (A4), where I is the identity matrix, we obtain

$$E[|\underline{v}_{k+1}^{m}|^{2}] = E|(I - (1 - \lambda)\underline{x}_{k}\underline{x}_{k}^{T})\underline{v}_{k}^{m} + (1 - \lambda)n_{k}\underline{x}_{k}|^{2}] \tag{11}$$

Expanding the expressions and using assumption A2 this becomes,

$$E[|\underline{v}_{k+1}^{m}|^{2}] = E[|(I - (1 - \lambda)\ \underline{x}_{k}\underline{x}_{k}^{T})v_{k}^{m}|^{2}] + (1 - \lambda)^{2}E[|n_{k}\underline{x}_{k}|^{2}] \tag{12}$$

so that,

$$E[|\underline{v}_{k+1}^{m}|^{2}] = E[\underline{v}_{k}^{m\ T}(I - (1 - \lambda)\underline{x}_{k}\underline{x}_{k}^{T})(1 - \lambda)\underline{x}_{k}\underline{x}_{k}^{T})\underline{v}_{k}^{m})] \\ + (1 - \lambda)^{2}NK \tag{13}$$

where K is the order of the system. It now follows from the distributive law that,

$$E[|\underline{v}_{k+1}^{m}|^{2}] = E[\underline{v}_{k}^{m\ T}\underline{v}_{k}^{m}] - 2(1 - \lambda)E[\underline{v}_{k}^{m\ T}\underline{x}_{k}\underline{x}_{k}^{T}v_{k}^{m}] \\ + (1 - \lambda)^{2}E[\underline{v}_{k}^{m\ T}\underline{x}_{k}\underline{x}_{k}^{T}\underline{x}_{k}\underline{x}_{k}^{T}\underline{v}_{k}^{m}] + (1 - \lambda)^{2}NK \tag{14}$$

At this point we can make use of assumption A1 and A3 to obtain,

$$E[|\underline{v}_{k+1}^{m}|^{2}] = E[|\underline{v}_{k}^{m}|^{2}](1 - 2(1 - \lambda) + (1 - \lambda)^{2}\beta) + (1 - \lambda)^{2}NK \tag{15}$$

where $\beta = K - 1 + E[\underline{x}_i^4]$[9].

Thus in the limit the steady state MSD associated with the measurement noise is,

$$\text{Measurement MSD} = KN(1 - \lambda)/(2 - (1 - \lambda)\beta) \tag{16a}$$

Similarly for the lag term,

$$\text{Lag MSD} = D\ /\ ((1 - \lambda)(2 - (1 - \lambda)\beta)) \tag{16b}$$

where $D = E[|\underline{d}_k|^2]$ and is the variance of the time varying tap increment (assuming zero mean). Therefore the steady state MSE achieved by the RLS algorithm is as below;

$$MSE = N + KN(1-\lambda)/(2-\beta(1-\lambda)) + D/((1-\lambda)(2-\beta(1-\lambda))) \tag{17}$$

When the input is no longer white then it is necessary to proceed as follows. Clearly equation (4) which is the RLS tap-weight update equation can be written as,

$$\hat{\underline{H}}_{k+1} = \hat{\underline{H}}_k + (1-\lambda)\Phi^{-1} e_k \underline{x}_k \tag{18}$$

by use of assumption (A4). As a result of the independence assumptions it should be noted that Φ is a diagonal matrix, although this assumption is clearly untrue for the time sequence considered here, a simple unitary rotation would guarantee that Φ would be diagonal and thus allow the analysis to proceed as follows.

If equation (7) is rewritten as,

$$\underline{q}_{k+1} = \underline{q}_k - (1-\lambda)(\underline{q}_k \underline{x}_k + n_k)\ \Phi^{-1}\underline{x}_k - \underline{d}_k. \tag{19}$$

In order to obtain the MSD we require $E[||\underline{q}_\infty||^2]$, so if equation (19) is squared and the expectation taken at the limit then we may proceed as below.

$$E[|\underline{q}_{k+1}|^2] = E[|\underline{q}_k|^2] - 2(1-\lambda)E[|(\underline{q}_k \cdot \underline{x}_k + n_k)(\Phi^{-1}\underline{x}_k)\underline{q}_k|]$$
$$+ (1-\lambda)^2 E[|(\underline{q}_k \underline{x}_k + n_k)^2 (\Phi^{-1}\underline{x}_k)^2|] + E[|\underline{d}_k|^2], \tag{20}$$

and if this is expanded then,

$$E[|\underline{q}_{k+1}|^2] = E[|\underline{q}_k|^2] - 2(1-\lambda)E[|(\underline{q}_k \cdot \underline{x}_k)(\Phi^{-1}\underline{x}_k) \cdot \underline{q}_k|] \tag{21}$$
$$+ (1-\lambda)^2 E[|(\underline{q}_k \cdot \underline{x}_k)^2 (\Phi^{-1} \cdot \underline{x}_k)^2|] + (1-\lambda)^2 NE[|(\Phi^{-1} \cdot \underline{x}_k)^2|] + D.$$

As a consequence of Φ being a diagonal matrix then,

$$E[|\underline{q}_{k+1}|^2] = E[|\underline{q}_k|^2] - 2(1-\lambda)E[|(\sum_{i=1}^{i=K} q_k^i x_k^i)(\sum_{i=1}^{i=K} \frac{1}{\alpha_i}(q_k^i x_k^i))|]$$
$$+ (1-\lambda)^2 E[|(\sum_{i=1}^{i=k}(q_k^i x_k^i)((\frac{x_k^i}{\alpha_i})^2))|] + (\sum_{i=1}^{i=K}\frac{1}{\alpha_i})(1-\lambda)N + D, \tag{22}$$

where x_k^i and q_k^i represent the constituent components in the vectors $\underline{x}_k$ and $\underline{q}_k$ respectively.

The terms α_i represent the eigenvalues of the input signal autocorrelation matrix. If this expression is then taken to the limit then the total MSD is expressed in equation (23) shown below,

$$\text{TotalMSD} = \left(\sum_{i=1}^{i=k} \frac{1}{\alpha_i} \right)(1 - \lambda) \frac{N}{(2 - \beta(1 - \lambda))} + \frac{D}{(1 - \lambda)(2 - \beta(1 - \lambda))}, \tag{23}$$

It is interesting to note that both expressions for the MSD, i.e. in white and coloured input signal conditions allows the effect of the nonstationarity and input signal colouration on the rate of convergence to be assessed, as in [9]. It is clear that only the nonstationarity affects the rate of convergence and this is reinforced by the simulations where the slowing in the rate of convergence of the RLS is clearly demonstrated.

3. DISCUSSION

It has previously been assumed that RLS algorithms would always track time variations of a nonstationary system faster than the LMS algorithm, and Honig [10] demonstrated that the RLS algorithm will always converge faster than the LMS in a stationary environment even when the input is white. The expression obtained for the steady state MSE of the RLS may now be used to evaluate and compare the theoretical performance of the algorithm in a time-varying environment.

Figure 4 illustrates the theoretical steady state MSC for the RLS for constant noise in steps of 10dB (-10dB to 80dB) and for tap variance ranging from -10dB to -80dB. Figure 4b illustrates the performance of the algorithm for constant tap variance in 10dB steps and noise ranging from -10dB to -80dB. In both situations the input is assumed to be white and $\lambda = 0.95$ for the RLS.

As can clearly be seen the algorithm achieves an asymptotic error floor which it cannot improve upon. It is also interesting to note that if the expressions for the predicted steady state MSE, obtained in [6], for the LMS are utilised then the predicted MSE is lower than that for the RLS when the nonstationarity is high (tap variance >-40dB).

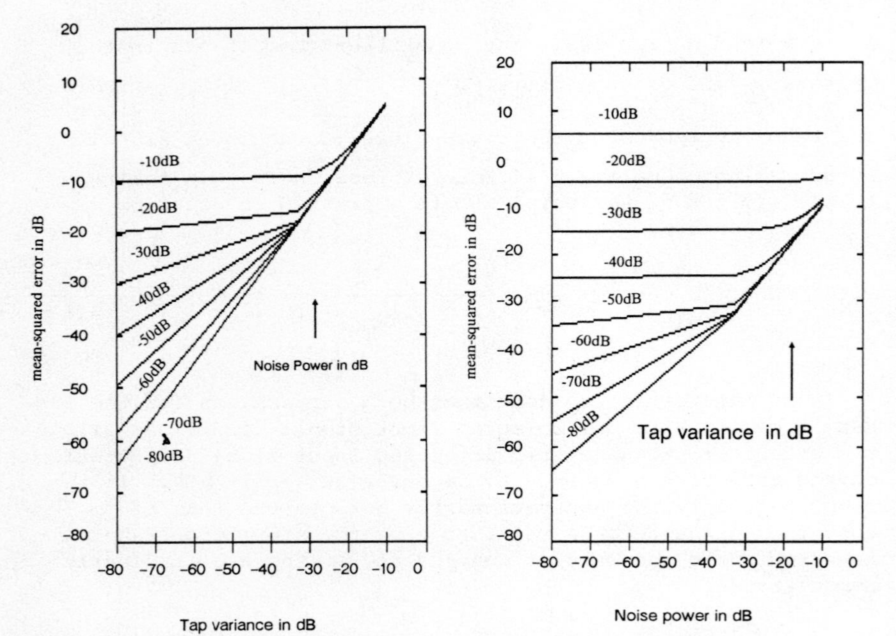

Fig. 4a Theoretical achievable MSE for RLS algorithm in dB plotted against tap variance in dB, for a fixed Noise power ranging from -80dB to -10dB, exponential windowing factor set at 0.95.

Fig. 4b Theoretical achievable MSE for RLS algorithm plotted against Noise power in dB, with tap variance fixed and ranging from -80dB to -10dB, exponential windowing factor set at 0.95.

Figures 5a to 5h illustrate simulated performance of the LMS and RLS algorithms as channel estimators for nonstationarity levels of 25dB and 45dB respectively, signal to noise ratios of 30dB and 50dB and for both white and coloured input signal conditions. The level of colouration being determined by an eigenvalue ratio (EVR) of 16.5. The plots show tap vector norms (or mean squared deviation MSD) against number of iterations. The tap vector norm plots were chosen rather than MSE plots because they illustrate the tracking behaviour of the algorithms more accurately. Table 1 indicates the appropriate values for each figure with MSD_p representing the theoretically predicted MSD and MSD_m representing the measured values from the simulations.

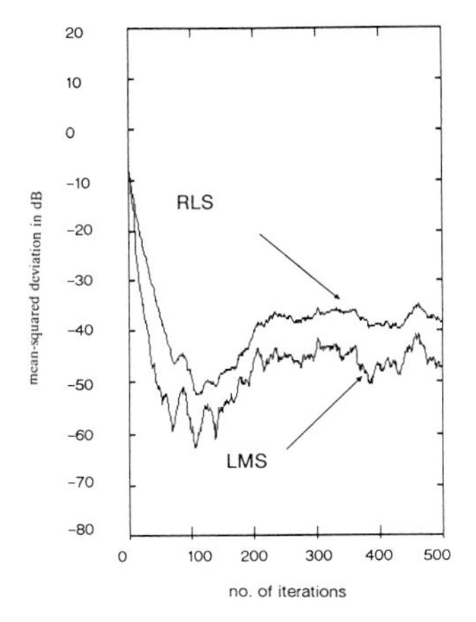

Fig. 5a – Performance comparison of LMS and RLS algorithms as HF Channel estimators with white input.

Fade rate=1Hz (D=-45dB) and Noise power ,N=-50dB, exponential windowing factor for RLS=0.95, LMS step size =0.16666.

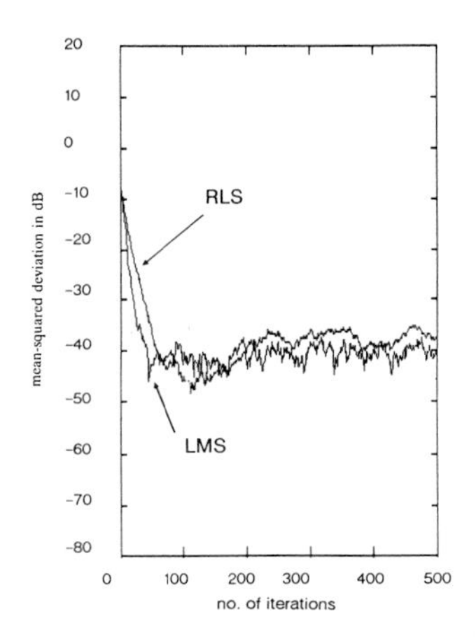

Fig. 5b – Performance comparison of LMS and RLS algorithms as HF Channel estimators with white input.

Fade rate=1Hz (D=-45dB) and Noise power, N=-30dB, exponential windowing factor for RLS=0.95, LMS step size=0.1666.

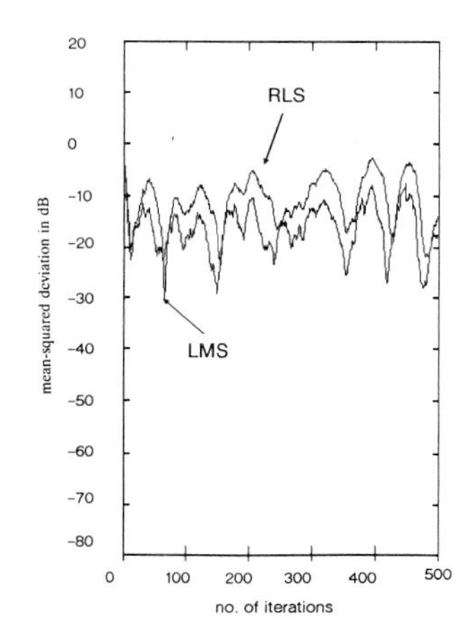

Fig. 5c – Performance comparison of LMS and RLS algorithms as HF Channel estimators with white input.

Fade rate=10Hz (D=-25dB) and Noise power, N=-50dB, exponential windowing factor for RLS=0.95, LMS step size=0.16666.

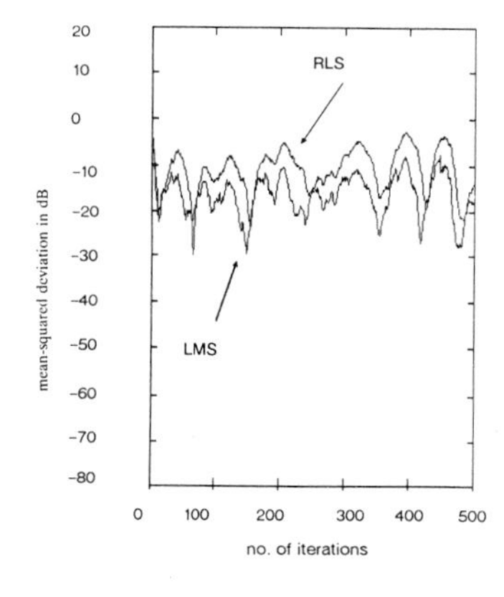

Fig. 5d – Performance comparison of LMS and RLS algorithms as HF Channel estimators with white input.

Fade rate=10Hz (D=-25dB) and Noise power, N=-30dB, exponential windowing factor for RLS=0.95, LMS step size=0.16666.

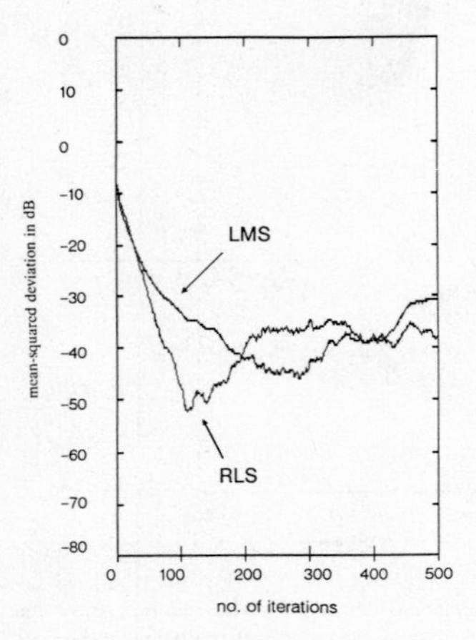

Fig. 5e - Performance comparison of LMS and RLS algorithms as HF Channel estimators for coloured input.

Fade rate=1Hz (D=-45dB) and Noise power, N=-50dB, exponential windowing factor for RLS=0.95, LMS step size=0.1666.

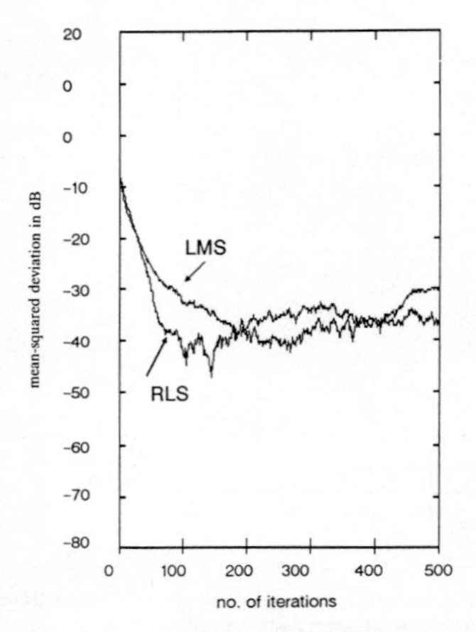

Fig. 5f - Performance Comparison of LMS and RLS algorithms as HF Channel estimators with coloured input.

Fade rate=1Hz (D=-45dB) and Noise power, N=-50dB, exponential windowing factor for RLS=0.95, LMS step size=0.16666.

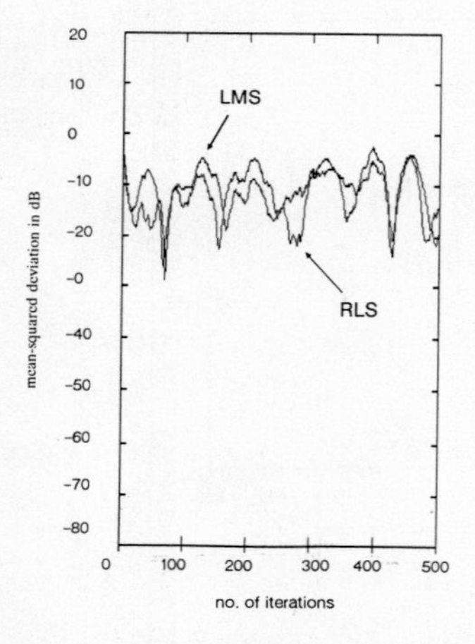

Fig 5g - Performance comparison of LMS and RLS algorithms as HF Channel estimators with coloured input.

Fade rate=10Hz (D=-25dB) and Noise power, N=-50dB, exponential windowing factor for RLS=0.95, LMS step size=0.1666.

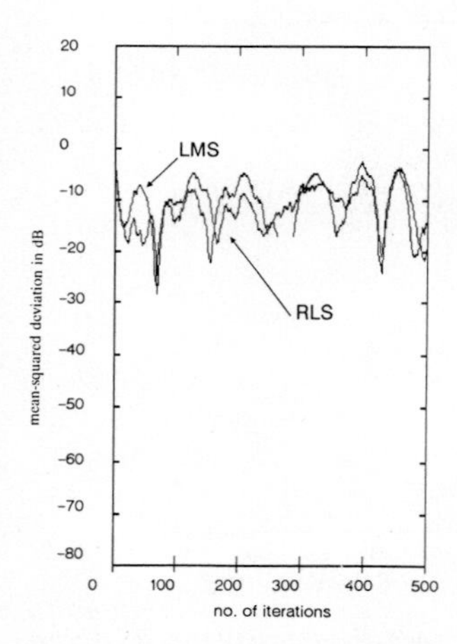

Fig. 5h - Performance Comparison of LMS and RLS algorithms as HF Channel estimators for coloured input.

Fade rate=10Hz (D=-25dB) and Noise power, N=-30dB, exponential windowing factor for RLS =0.95, LMS step size=0.1666.

λ	D	N	EVR	MSD_P	MSD_M
0.95	-45db	-50dB	1	-34.0dB	-35dB
0.95	-45dB	-30dB	1	-33.2dB	-35dB
0.95	-25dB	-50dB	1	-14.6dB	-13.5dB
0.95	-25dB	-30dB	1	-14.6dB	-13.5dB
0.95	-45db	-50db	16.5	-34.4dB	-35dB
0.95	-45dB	-30db	16.5	-32.0dB	-35dB
0.95	-25dB	-30dB	16.5	-14.4dB	-13.5dB
0.95	-25db	-50db	16.5	-14.5dB	-13.5dB

Table 1 Comparison of simulated and theoretical results

The value of μ selected for the LMS used the stability criterion suggested by Feuer and Weinstein in [11], $0 \leq \mu < 1/3\ tr[R]$, in this paper μ was chosen to be at the proposed optimal value, i.e. $\mu = 1/(6\ tr[R])$ which for the white input conditions is $\mu = 1/6K$ where K is the order of the system. This value of μ was chosen to guarantee stability of the LMS but still ensure reasonable tracking performance by the algorithm for the simulations presented. Therefore, for the three tap channel used in the simulations $\mu = 0.05556$.

The simulations represented in figures 5a-5h consider 3 situations.

i) tap variance >> additive noise power,

ii) additive noise power >> tap variance,

iii) additive noise power $\approx$ tap variance.

In all cases when the input is white the LMS performs as well if not better than the RLS.

The predicted values of the RLS are all within 1-2dB of the measured values. In the situations when the input signal is coloured the lack of spectral robustness of the LMS is demonstrated while the RLS is, as expected, relatively unaffected by the signal colouration. The best performance of the RLS (and the theoretical prediction) results when the window length is short, i.e. λ becoming smaller. This is as predicted by the theoretical expression of the RLS. If the expression is differentiated and the measurement and lag terms considered separately it is clear that the contribution of the measurement term becomes smaller as λ approaches 1 and that of the lag term becomes larger as λ approaches 1. This illustrates the trade off required in window selection, i.e. the longer the window length (λ closer to 1) then the better the estimate of the autocorrelation matrix and the shorter the window length (λ getting smaller) then the better tracking speed and thus the smaller the lag error contribution.

The expression $2\mu = (1 - \lambda)$ [1] can be used with the stability criterion shown previously to illustrate the effect of the order of the system. If λ is constrained to lie within the region 0.92-0.999 then clearly if $\lambda = 0.92$ (i.e. $\mu = 0.04$) the LMS could not be guaranteed stable for systems of greater than order 8. However, if $\lambda = 0.98$ was chosen then the LMS would not be guaranteed stable for systems of order greater than 33. Eleftheriou and Falconer [3] applied a similar technique in their work but utilised a less conservative stability criterion for the LMS and considered a channel model represented by a first order Markov process.

It is clear from the results presented in this paper that there is reasonable agreement between the simulation results and theoretical predictions. Work recently published by Clark & Harun [4] and previously by Tront [12] reinforces the results demonstrated in this paper.

4. CONCLUSIONS

It is clear from the results presented in this paper that the tracking performance expected of RLS algorithms in time varying environments is not achieved. Although this paper has only looked at the algorithm's performance in a direct modelling situation it is clear that the tracking performance of the algorithm is an important characteristic of the algorithm which must be clearly separated from other effects such as convergence with coloured inputs.

It would appear that RLS algorithms are not necessarily suitable for use in highly nonstationary environments, such as the HF communication channel. The slow rate of convergence and subsequent degradation in performance of the LMS algorithm with coloured input also makes it unsuited for such applications. Consequently it would appear that some form of nonlinear techniques [13] or new algorithms [14] which estimate the

the parameters generating the nonstationarity may have to be considered to produce adaptation algorithms which can function in such hostile environments.

ACKNOWLEDGEMENTS

This work was funded by the Science and Engineering Research Council of Great Britain.

5. REFERENCES

[1] Cowan, C.F.N. and Grant, P.M., (1985) Adaptive Filters Prentice-Hall.

[2] Cowan, C.F.N., (1987) Performance Comparisons of finite linear adaptive filters, IEE Proceedings Part F, Special issue on Adaptive Filters, Vol. 134, Part No. 3, pp. 211-216.

[3] Eleftheriou, E and Falconer, D.D., (1986) Tracking Properties and Steady State Performance of RLS Adaptive Filter Algorithms, IEEE Trans. ASSP., Vol. ASSP-34, No. 5, pp. 1097-1109.

[4] Clark, A.P. and Harun, R., (1986) Assessment of Kalman-filter Channel estimators for an HF radio link, IEE Proc. Part F, Vol. 133, No. 6, pp. 513-521.

[5] Widrow, B. et al., (1976) Stationary and Nonstationary Learning Characteristics of the LMS Adaptive Filter, IEEE Proceedings, Vol. 64, No. 8, pp. 1151-1162.

[6] Macchi, O., (1986) Optimization of Adaptive Identification for Time-Varying Filters, IEEE Trans. on Automatic Control, Vol. AC-31, No. 3, pp. 283-287.

[7] Eweda, E. and Machhi, O., (1985) Tracking Error Bounds of Adaptive Nonstationary Filtering, Automatica, Vol. 21, No. 3, pp. 293-302.

[8] Davie, A.M., Private Communication.

[9] Gibson, G.J., (1987) A Comparison of Linear Random Search and Least Mean Square Adaptive Algorithms, SPG Internal Report, Dept. of Electrical Engineering, University of Edinburgh.

[10] Honig, M.L. and Messerschmitt, D.G., (1984) Adaptive Filters:Structures, Algorithms and Applications, Kluwer Academic Publishers, Chapter 7, pp. 246-330.

[11] Feuer, A. and Weinstein, E., (1985) Convergence Analysis of LMS filters with Uncorrelated Gaussian Data, IEEE Trans. on Acoustics Speech and Signal Processing, Vol. ASSP-33, No. 1, pp. 222-230.

[12] Tront, R., Performance of Kalman Decision-Feedback Equalisation in HF Radio Modems, M.Sc., University of Victoria, British Columbia.

[13] Gibson, G.J. and Cowan, C.F.N., Communications Data Equalisation with Neural Networks, Submitted to Trans. on ASSP.

[14] McLaughlin, S., Mulgrew, B. and Cowan, C.F.N., (1987) Performance Study of an Extended Kalman Algorithm as a HF Channel Estimator, Proceedings of the 4th International Conference on HF Radio Systems.

FAST QRD-BASED ALGORITHMS FOR LEAST SQUARES LINEAR PREDICTION

I.K. Proudler, J.G. McWhirter and T.J. Shepherd
(Royal Signals and Radar Establishment, Worcestershire)

ABSTRACT

A much simplified and clearer derivation of Cioffi's orthogonal fast Kalman algorithm for least squares linear prediction is presented. The notation used is similar to that adopted in the context of the familiar systolic QR decomposition algorithm.

A fast order-update procedure is then derived. This allows the development of a lattice filter algorithm based entirely on Givens' rotations and requiring O(p) computations for the time-update of the p-th order linear prediction problem.

1. INTRODUCTION

The application of systolic arrays to the problem of least squares minimisation is now well established. A triangular systolic array for implementing the QR decomposition (QRD) algorithm was proposed by Gentleman and King [4] and subsequently adapted by McWhirter [6] to apply to the type of recursive least squares filtering operations which occur in digital signal processing.

Least squares minimisation can be applied to a wide range of digital signal processing problems including adaptive filtering and channel equalisation for noise and echo cancellation. One of the best known application areas is adaptive antenna-array beam-forming. For example, STL Technology Ltd. in the UK [7] and, more recently, the Hazeltine Corporation in the US [5] have implemented the triangular systolic least squares processor in hardware. The triangular systolic array does not assume any particular structure in the data matrix. It may therefore be applied to the most general least squares minimisation problems.

Recursive least squares minimisation may also be applied to the problem of the linear prediction of time series data. In this case, however, the sample data matrix has a Toeplitz structure and as a result it is possible to devise more efficient procedures than the systolic QRD algorithm which would require $O(p^2)$ operations per sample time to update a p-th order prediction problem. Fast algorithms such as the least squares lattice and fast Kalman algorithms take account of the implicit redundancy to reduce the computational load to $O(p)$ operations.

Unfortunately, most of these algorithms exhibit some form of numerical instability and are thus inferior to the QRD algorithm which is based entirely on orthogonal rotations. In a recent benchmark paper[1] [3] Cioffi derived a stable $O(p)$ algorithm for linear prediction by taking account of the Toeplitz data structure within the QRD algorithm. His algorithm is, in effect, an orthogonal form of the fast Kalman algorithm. Cioffi's derivation is, however, unnecessarily complicated and rather difficult to follow. Recognising this fact and also the value of Cioffi's algorithm, Bellanger was motivated to clarify the presentation of his work [1].

In this paper we present a much briefer and greatly simplified derivation of the orthogonal fast Kalman algorithm. This is achieved by using a notation similar to that adopted in the literature in the context of the triangular recursive least squares processor. We then present a new QRD-based least squares lattice algorithm for linear prediction. This follows quite readily given our simplified derivation of the fast Kalman algorithm.

We begin, in section 2, by laying the mathematical framework for the least squares problem in the context of adaptive filtering. In section 3 a simplified derivation of the orthogonal fast Kalman algorithm is presented. This is expanded, in section 4, to produce the orthogonal least squares lattice algorithm.

2. BASIC ADAPTIVE FILTER

In this section we set up the mathematical framework necessary to solve the linear prediction problem. The adaptive filter problem is as follows: choose a weight vector $\underline{w}(n)$ to minimise the weighted mean square error

---

1. This paper has yet to appear in print. See reference [2] for further details.

$$E(n) = ||B(n)\underline{e}(n)|| \tag{2.1}$$

where

$$B(n) = \mathrm{Diag}(\beta^{n-1},\ldots,1) \qquad \underline{e}(n) = \begin{vmatrix} e(n,1) \\ \cdot \\ \cdot \\ e(n,n) \end{vmatrix}$$

$$0<\beta<1$$

$$e(n,m) = \underline{x}^T(m)\underline{w}(n)+y(m) \tag{2.2}$$

$$\underline{x}(m) = \begin{vmatrix} x(m) \\ \cdot \\ \cdot \\ x(m-p+1) \end{vmatrix}$$

Here x(n) is the signal to be filtered and y(n) is the reference signal. The matrix B(n) represents an exponential forgetting factor that allows the algorithm to work for quasi-stationary signals.

From (2.2) we have

$$\underline{e}(n) = X_p(n)\underline{w}(n)+\underline{y}(n) \tag{2.3}$$

where

$$X_p(n) = \begin{vmatrix} \underline{x}^T(1) \\ \cdot \\ \cdot \\ \underline{x}^T(n) \end{vmatrix} = \begin{vmatrix} x(1)\ldots x(2-p) \\ \cdot \qquad \cdot \\ \cdot \qquad \cdot \\ x(n)\ldots x(n-p+1) \end{vmatrix} \quad \text{and} \quad \underline{y}(n) = \begin{vmatrix} y(1) \\ \cdot \\ \cdot \\ y(n) \end{vmatrix}$$

Following [6] introduce the orthogonal matrix $\underset{\sim}{Q}_p(n)$ that triangularizes the weighted data matrix $B(n)X_p(n)$.

Let

$$\underset{\sim}{Q}_p(n)B(n)X_p(n) = \begin{vmatrix} R_p(n) \\ 0 \end{vmatrix} \tag{2.4}$$

where $R_p(n)$ is an upper triangular matrix

so that

$$E(n) = \left\|Q_p(n)B(n)\underline{e}(n)\right\| = \left\|\begin{vmatrix} R_p(n) \\ O \end{vmatrix} \underline{w}(n) + \begin{vmatrix} \underline{u}(n) \\ \underline{v}(n) \end{vmatrix}\right\| \tag{2.5}$$

where

$$\begin{vmatrix} \underline{u}(n) \\ \underline{v}(n) \end{vmatrix} = Q_p(n)B(n)\ \underline{y}(n) \tag{2.6}$$

Note that $Q_p(n)$ has the following decomposition [6]

$$Q_p(n) = \hat{Q}_p(n) \begin{vmatrix} \hat{Q}_p(n-1) & \underline{O} \\ \underline{O}^T & 1 \end{vmatrix} \tag{2.7}$$

i.e.

$$Q_p(n)B(n)X_p(n) = \begin{vmatrix} R_p(n) \\ O \end{vmatrix} = \hat{Q}_p(n) \begin{vmatrix} \beta R_p(n-1) \\ O \\ \underline{x}^T(n) \end{vmatrix} \tag{2.8}$$

From (2.5) it can be seen that the optimum solution occurs when

$$R_p(n)\underline{w}(n) + \underline{u}(n) = \underline{O} \tag{2.9}$$

so that

$$Q_p(n)B(n)\underline{e}(n) = \begin{vmatrix} O \\ \underline{v}(n) \end{vmatrix} \tag{2.10}$$

and

$$e(n,n) = \left[Q_p^T(n) \begin{vmatrix} O \\ \underline{v}_p(n) \end{vmatrix} \right]_n \equiv \gamma_p(n)\alpha(n) \tag{2.11}$$

where $\alpha(n)$ is the last component of the vector $\underline{v}(n)$ and $\gamma_p(n)$ is the lower right-hand element of the matrix $Q_p(n)$. Now from (2.6) we have

$$\begin{vmatrix} \underline{u}(n) \\ \underline{v}(n) \end{vmatrix} = Q_p(n)\ B(n)\underline{y}(n) = \hat{Q}_p(n) \begin{vmatrix} Q_p(n-1) & \underline{O} \\ \underline{O}^T & 1 \end{vmatrix} \begin{vmatrix} \beta B(n-1) & \underline{O} \\ \underline{O}^T & 1 \end{vmatrix} \begin{vmatrix} \underline{y}(n-1) \\ y(n) \end{vmatrix}$$

$$= \hat{Q}_p(n) \begin{vmatrix} \beta u(n-1) \\ \beta \bar{v}(n-1) \\ \bar{y}(n) \end{vmatrix} = \begin{vmatrix} u(n) \\ \beta v(n-1) \\ \bar{\alpha}(n) \end{vmatrix} \qquad (2.12)$$

because of the form $\hat{Q}_p(n)$.

3. "FAST KALMAN" QR ALGORITHM

The linear prediction problem is an adaptive filter problem where the reference sequence is a time shifted version of the data sequence. Depending on the value of this time shift there are two possible problems: the forward problem (subsection 3.2) and the backward one (section 3.1).

The forward and backward p-th order linear prediction problems are linked, however, by the fact that they are based on the same set of data (X_{p+1}). This results in the fact that, via the matrix R_{p+1}, we are able to generate the solution to the backward problem at time (n+1) from the solution to the forward problem at time n. Then by comparison with the solution to the backward problem at time n, we can infer $\hat{Q}_p(n+1)$ and hence the solution to the forward problem at time (n+1). In this way it is possible to develop a time-update algorithm for the p-th order problems that only require O(p) computations.

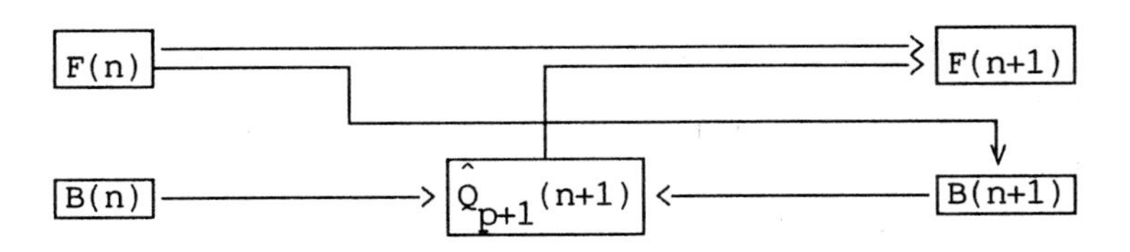

Fig. 3.1 "Fast Kalman" Algorithm

3.1 Backward Linear Prediction

In the p-th order backward linear prediction problem we attempt to estimate x(n-p) from the p samples x(n),...,x(n-p+1). In the following we use the same notation as in Section 2 except for a subscript "b", where applicable, to denote the backward problem.

In this section we derive a connection between the solution to the backward problem and $R_{p+1}(n+1)$.

Let $y_b(n) = x(n-p)$ and prediction error $= e_b(n,n)$

Now by definition (see (2.4) and (2.6))

$$\begin{vmatrix} R_p(n) & \underline{u}_b(n) \\ O & \underline{v}_b(n) \end{vmatrix} = Q_p(n)\ B(n) \begin{vmatrix} X_p(n) & \underline{y}_b(n) \end{vmatrix} \tag{3.1.1}$$

where

$$\underline{y}_b(n) = \begin{vmatrix} x(1-p) \\ \cdot \\ \cdot \\ x(n-p) \end{vmatrix} \tag{3.1.2}$$

hence

$$\begin{vmatrix} X_p(n) & \underline{y}_b(n) \end{vmatrix} = \begin{vmatrix} x(1) & \cdots & x(2-p) & x(1-p) \\ \cdot & \cdots & \cdot & \cdot \\ \cdot & \cdots & \cdot & \cdot \\ x(n) & \cdots & x(n-p+1) & x(n-p) \end{vmatrix} = X_{p+1}(n) \tag{3.1.3}$$

and

$$Q_p(n)\ B(n) X_{p+1}(n) = \begin{vmatrix} R_p(n) & \underline{u}_b(n) \\ O & \underline{v}_b(n) \end{vmatrix} \tag{3.1.4}$$

Consideration of (3.1.4) shows that we have very nearly constructed an upper triangular matrix. If we had succeeded then because of the underlying data matrix (see (3.1.3)) this triangular matrix would be $R_{p+1}(n)$. Clearly to complete the triangularisation all that is required is that all except the first component of $\underline{v}_b(n)$ be annihilated. Let $Q_b(n)$ be the set of rotations that achieves this.

Then

$$Q_b(n) \begin{vmatrix} R_p(n) & \underline{u}_b(n) \\ O & \underline{v}_b(n) \end{vmatrix} = \begin{vmatrix} R_p(n) & \underline{u}_b(n) \\ \underline{O}^T & \varepsilon_b(n) \\ \underline{O} & \underline{O} \end{vmatrix} \tag{3.1.5}$$

i.e.

$$Q_b(n)Q_p(n)B(n)X_{p+1}(n) = \begin{vmatrix} R_p(n) & \underline{u}_b(n) \\ \underline{O}^T & \varepsilon_b(n) \\ \underline{O} & \underline{O} \end{vmatrix} = \begin{vmatrix} R_{p+1}(n) \\ O^T \end{vmatrix} \quad (3.1.6)$$

Equations (3.1.5/6) show that $R_{p+1}(n)$ can be calculated given the solution to the p-th order backward problem. Note in passing that, because of the orthogonal nature of rotations we have that

$$\varepsilon_b^2(n) = ||\underline{v}_b(n)||^2 = ||B(n)\underline{e}_b(n)||^2 \quad (3.1.7)$$

$$\simeq \text{prediction error power.}$$

The calculation of $R_{p+1}(n)$ from the solution to the backward problem via (3.1.5) would require O(n-p) computations. A recursive update procedure can, however, be developed which only requires a fixed amount of calculation. Suppose we have the solution to the backward problem at time n. Consider the update to time (n+1) by addition of a new set of data. We have the following decomposition

$$X_{p+1}(n+1) = \begin{vmatrix} X_p(n+1) & \underline{y}_b(n+1) \end{vmatrix} = \begin{vmatrix} X_p(n) & & \underline{y}_b(n) \\ x(n+1) & \ldots & x(n-p+1) \end{vmatrix} \quad (3.1.8)$$

Hence (see equation (3.1.6))

$$\begin{vmatrix} Q_b(n) & \underline{O} \\ \underline{O}^T & 1 \end{vmatrix} \begin{vmatrix} Q_p(n) & \underline{O} \\ \underline{O}^T & 1 \end{vmatrix} B(n+1)\ X_{p+1}(n+1) =$$

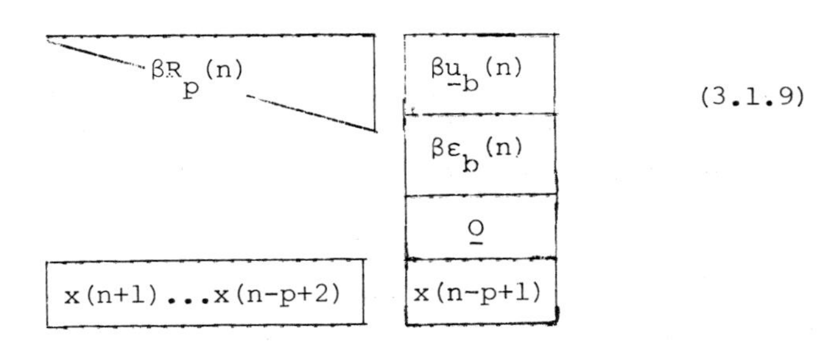

(3.1.9)

In order to generate $R_{p+1}(n+1)$ we need to annihilate the bottom row of the matrix in equation (3.1.9). From equation (2.8) we have

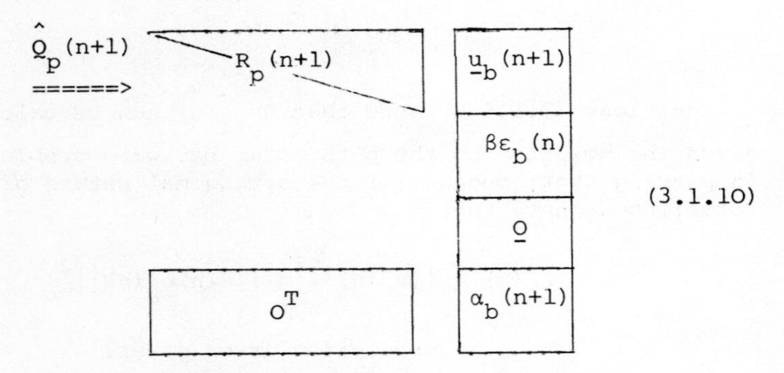

(3.1.10)

Thus we only have to annihilate the element $\alpha_b(n+1)$ in order to create the required triangular matrix. Define $\hat{Q}_b(n+1)$ to be the rotation matrix that achieves this,

i.e.

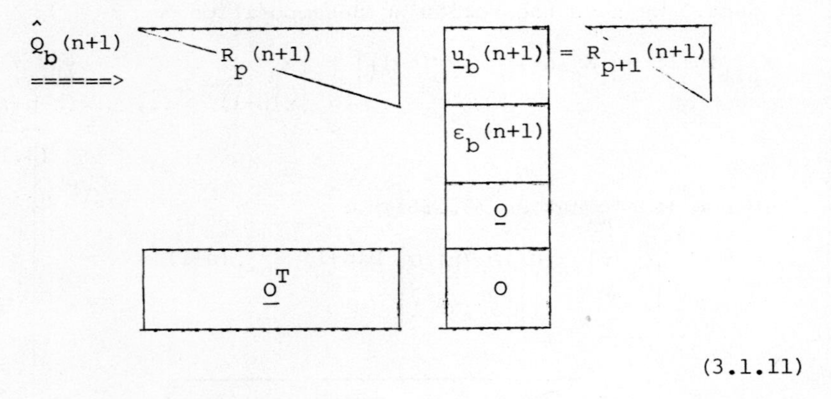

(3.1.11)

Equation (3.1.11) thus defines a method of calculating $R_{p+1}(n+1)$ from the solution to the p-th order backward problem at time n which requires only one rotation.

3.2 Forward Linear Prediction

In the p-th order forward linear prediction problem we attempt to estimate $x(n+1)$ from the p samples $x(n),\ldots,x(n-p+1)$. In the following we use the same notation as in subsection 3.1 substituting a subscript "f", instead of "b", to denote the forward problem.

As in the case of the backward problem, we derive a connection between the solution to the forward problem and $R_{p+1}(n+1)$.

Let $y_f(n) = x(n+1)$ and prediction error $= e_f(n,n)$
By definition.

$$\begin{vmatrix} \underline{u}_f(n-1) & R_p(n-1) \\ \underline{v}_f(n-1) & 0 \end{vmatrix} = \underset{\sim}{Q}_p(n-1)B(n-1)\begin{vmatrix} \underline{y}_f(n-1) & X_p(n-1) \end{vmatrix} \tag{3.2.1}$$

where

$$\underline{y}_f(n-1) = \begin{vmatrix} x(2) \\ \cdot \\ \cdot \\ x(n) \end{vmatrix} \tag{3.2.2}$$

hence

$$\begin{vmatrix} \underline{y}_f(n-1) & X_p(n-1) \end{vmatrix} = \begin{vmatrix} x(2) & x(1) & \cdots & x(2-p) \\ \cdot & \cdot & & \cdot \\ \cdot & \cdot & & \cdot \\ x(n) & x(n-1) & \cdots & x(n-p) \end{vmatrix} \tag{3.2.3}$$

and

$$\begin{vmatrix} x(1) & x(0) & \cdots & x(1-p) \\ \underline{y}_f(n-1) & & X_p(n-1) & \end{vmatrix} = X_{p+1}(n) \tag{3.2.4}$$

If the data is windowed so that $x(n) = 0$ for $n \leq 0$ then (see equations (2.4) and (2.6))

$$\begin{vmatrix} 1 & \underline{0}^T \\ \underline{0} & \underset{\sim}{Q}_p(n-1) \end{vmatrix} B(n)X_{p+1}(n) = \begin{vmatrix} \beta^{n-1}x(1) & \underline{0}^T \\ \underline{u}_f(n-1) & R_p(n-1) \\ \underline{v}_f(n-1) & 0 \end{vmatrix} \tag{3.2.5}$$

Just as in section 3.1 we are aiming to generate R_{p+1}, but now at time n, and thus must convert the matrix in equation (3.2.5) into an upper triangular one. Clearly, this can be achieved by annihilating the vectors $\underline{u}_f(n-1)$ and $\underline{v}_f(n-1)$. We choose to do this in two stages.

Let

$$Q_f^v(n-1) \begin{vmatrix} \beta^{n-1}x(1) & \underline{o}^T \\ \underline{u}_f(n-1) & R_p(n-1) \\ \underline{v}_f(n-1) & O \end{vmatrix} = \begin{vmatrix} \varepsilon_f(n-1) & \underline{o}^T \\ \underline{u}_f(n-1) & R_p(n-1) \\ \underline{O} & O \end{vmatrix} \quad (3.2.6)$$

and

$$Q_f^u(n) \begin{vmatrix} \varepsilon_f(n-1) & o^T \\ \underline{u}_f(n-1) & R_p(n-1) \\ O & O \end{vmatrix} = \begin{vmatrix} R_{p+1}(n) \\ O \end{vmatrix} \quad (3.2.7)$$

so that

$$Q_f^u(n)\ Q_f^v(n-1) \begin{vmatrix} 1 & \underline{o}^T \\ \underline{O} & Q_p(n-1) \end{vmatrix} B(n)\ X_{p+1}(n) = \begin{vmatrix} R_{p+1}(n) \\ O \end{vmatrix} \quad (3.2.8)$$

Equations (3.2.6/7) represent the basic connection between $R_{p+1}(n)$ and the solution to the forward problem (cf. equation (3.1.5/6)). It requires O(n) operations but a faster procedure can be found. Suppose that we have the solution to the forward problem at time n-1, consider updating by the addition of new data. We have the following decomposition for $X_{p+1}(n+1)$:

$$X_{p+1}(n+1) = \begin{vmatrix} x(1) & \underline{o}^T \\ \underline{y}_f(n) & X_p(n) \end{vmatrix} = \begin{vmatrix} x(1) & & \underline{o}^T \\ \underline{y}_f(n-1) & & X_p(n-1) \\ x(n+1) & \dots & x(n-p+1) \end{vmatrix} \quad (3.2.9)$$

Using equations (3.2.5/6) we have

$$\begin{vmatrix} Q_f^v(n-1) & \underline{O} \\ \underline{o}^T & 1 \end{vmatrix} \begin{vmatrix} 1 & \underline{o}^T & O \\ \underline{O} & Q_p(n-1) & \underline{O} \\ O & \underline{o}^T & 1 \end{vmatrix} B(n+1)\ X_{p+1}(n+1) =$$

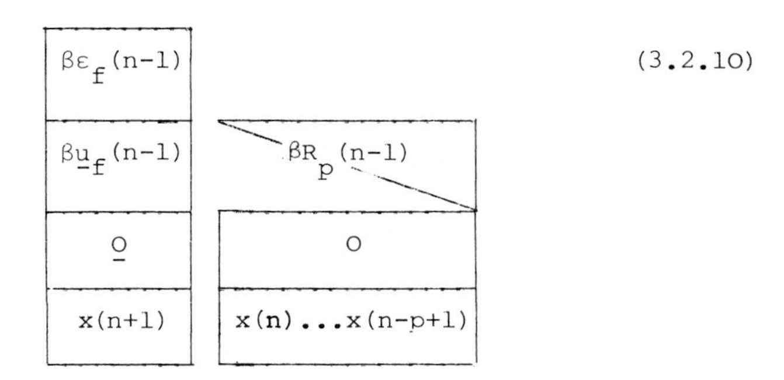

From equation (2.8) we see that

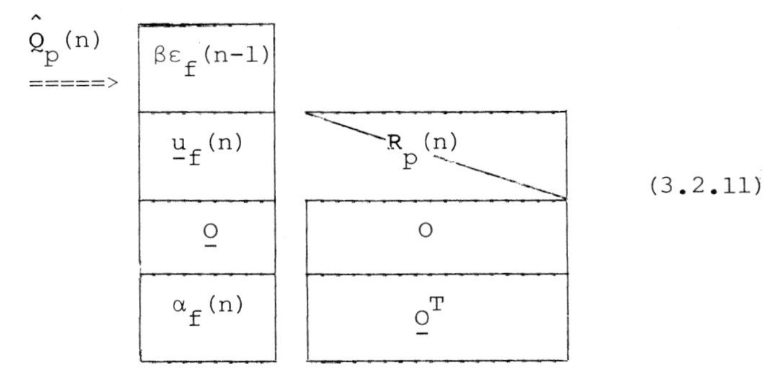

Define a rotation matrix that will annihilate the element $\alpha_f(n)$

i.e.

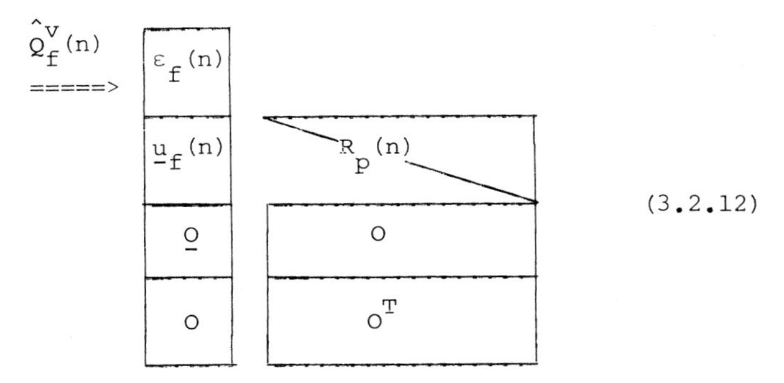

so that (see equation (3.2.7)) in order to calculate $R_{p+1}(n)$

all that is required is a pre-multiplication by $Q_f^u(n+1)$

i.e.

$$\underset{\Longrightarrow}{Q_f^u(n+1)} \begin{vmatrix} R_{p+1}(n+1) \\ \underline{O}^T \end{vmatrix} \qquad (3.2.13)$$

Equations (3.2.12/13) constitute an O(p) algorithm for calculating $R_{p+1}(n+1)$ from the solution to the p-th order forward problem at time n. Along with equation (3.1.11), these equations allow the backward problem at time n+1 to be solved given the solution to the forward problem at time n.

3.3 Algorithm

From sections 3.1 and 3.2 we have the following relations:

$$Q_{p+1}(n+1) = \hat{Q}_b(n+1)\ \hat{Q}_p(n+1) \begin{vmatrix} Q_b(n) & \underline{O} \\ \underline{O}^T & 1 \end{vmatrix} \begin{vmatrix} Q_p(n) & \underline{O} \\ \underline{O}^T & 1 \end{vmatrix} \qquad (3.3.1)$$

$$= Q_f^u(n+1)\ \hat{Q}_f^v(n) \begin{vmatrix} 1 & \underline{O}^T \\ \underline{O} & \hat{Q}_p(n) \end{vmatrix} \begin{vmatrix} Q_f^v(n-1) & \underline{O} \\ \underline{O}^T & 1 \end{vmatrix} \begin{vmatrix} 1 & \underline{O}^T & O \\ \underline{O} & Q_p(n-1) & \underline{O} \\ O & \underline{O}^T & 1 \end{vmatrix} \qquad (3.3.2)$$

Now rather than use the above equations to calculate $\hat{Q}_p(n+1)$ directly, a faster method can be developed based on the idea that $\hat{Q}_p(n+1)$ can be generated from the equation:

$$\left[\hat{Q}_p(n+1)\right]^T \begin{vmatrix} \underline{a}_p(n+1) \\ O \\ \gamma_p(n+1) \end{vmatrix} = \begin{vmatrix} \underline{O} \\ 1 \end{vmatrix} \qquad (3.3.3)$$

where the vector on the left hand side of equation (3.3.3) is by definition the right hand column of the matrix $\hat{Q}_p(n+1)$. Strictly speaking the Q matrix in a QR decomposition of a matrix is not unique and hence equation (3.3.3) does not allow the reconstruction of $\hat{Q}_p(n+1)$. However if the cosines used in the rotations are chosen so that they are all positive then a unique reconstruction is possible. Further because of the orthogonal nature of a rotation and the positive sign of the cosines we have that:

$$\gamma_p(n+1) = +\sqrt{1 - ||\underline{a}_p(n+1)||^2} \qquad (3.3.4)$$

so that all we really need to know is the vector $\underline{a}_p(n+1)$. The calculation is as follows. From equations (2.7) and (3.3.1) we have that:

$$\begin{vmatrix} \underline{a}_{p+1}(n+1) \\ \underline{0} \\ \gamma_{p+1}(n+1) \end{vmatrix} = \hat{Q}_b(n+1) \begin{vmatrix} \underline{a}_p(n+1) \\ \underline{0} \\ \gamma_p(n+1) \end{vmatrix} \qquad (3.3.5)$$

It can be seen from equations (3.1.11) and (3.3.5) that the vector $\underline{a}_p(n+1)$ is just the first p components of $\underline{a}_{p+1}(n+1)$. Hence if $\underline{a}_{p+1}(n+1)$ is known then so is $\underline{a}_p(n+1)$. The vector $\underline{a}_{p+1}(n+1)$ is calculated from equations (2.7) and (3.3.2) as follows:

$$\begin{vmatrix} \underline{a}_{p+1}(n+1) \\ \underline{0} \\ \gamma_{p+1}(n+1) \end{vmatrix} = Q_f^u(n+1)\ \hat{Q}_f^v(n) \begin{vmatrix} 0 \\ \underline{a}_p(n) \\ \underline{0} \\ \gamma_p(n) \end{vmatrix} \qquad (3.3.6)$$

Note that from equation (3.2.12) we have:

$$\hat{Q}_f^v(n) \begin{vmatrix} 0 \\ \underline{a}_p(n) \\ \underline{0} \\ \gamma_p(n) \end{vmatrix} = \begin{vmatrix} h(n) \\ \underline{a}_p(n) \\ \underline{0} \\ \gamma_{p+1}(n+1) \end{vmatrix} \qquad (3.3.7)$$

where h(n) is defined by this equation and that $Q_f^u(n+1)$ is given by (see equation (3.2.13)):

$$Q_f^u(n+1) \begin{vmatrix} \varepsilon_f(n) \\ \underline{u}_f(n) \\ \underline{0} \end{vmatrix} = \begin{vmatrix} r_{p+1,11}(n+1) \\ \underline{0} \end{vmatrix} \qquad (3.3.8)$$

Step by step the algorithm is:

Given $x(n+2)$, $\underline{u}_f(n)$, $\alpha_f(n)$, $\gamma_p(n)$, $\underline{a}_p(n)$ and $\varepsilon_f(n-1)$,

1) Calculate $\hat{Q}_f^v(n)$ from equation (3.2.12):

$$\hat{Q}_f^v(n) \begin{vmatrix} \beta^n \varepsilon_f(n-1) \\ \underline{u}_f(n) \\ \underline{O} \\ \alpha_f(n) \end{vmatrix} = \begin{vmatrix} \varepsilon_f(n) \\ \underline{u}_f(n) \\ \underline{O} \\ O \end{vmatrix} \tag{3.3.9}$$

2) Calculate $h(n)$ from equation (3.3.7):

$$\begin{vmatrix} h(n) \\ \underline{a}_p(n) \\ \underline{O} \\ \gamma_{p+1}(n+1) \end{vmatrix} = \hat{Q}_f^v(n) \begin{vmatrix} O \\ \underline{a}_p(n) \\ O \\ \gamma_p(n) \end{vmatrix} \tag{3.3.10}$$

3) Calculate $Q_f^u(n+1)$ from equation (3.3.8):

$$Q_f^u(n+1) \begin{vmatrix} \varepsilon_f(n) & r_{p+1,11}(n+1) \\ \underline{u}_f(n) & \underline{O} \\ \underline{O} & \underline{O} \end{vmatrix} \tag{3.3.11}$$

4) Calculate $\underline{a}_p(n+1)$ as the first p components of $\underline{a}_{p+1}(n+1)$ using equations (3.3.6) and (3.3.7):

$$\begin{vmatrix} \underline{a}_{p+1}(n+1) \\ \underline{O} \\ \gamma_{p+1}(n+1) \end{vmatrix} = Q_f^u(n+1) \begin{vmatrix} h(n) \\ \underline{a}_p(n) \\ \underline{O} \\ \gamma_{p+1}(n+1) \end{vmatrix} \tag{3.3.12}$$

{NB. $\gamma_{p+1}(n+1)$ is not involved in this rotation and hence need not be known.}

5) Calculate $\gamma_p(n+1)$:

$$\gamma_p(n+1) = +\sqrt{1 - \|\, \underline{a}_p(n+1) \,\|^2} \tag{3.3.13}$$

6) Calculate $\hat{Q}_p(n+1)$ from equation (3.3.3):

$$\left[Q_p(n+1)\right]^T \begin{vmatrix} \underline{a}_p(n+1) \\ \underline{O} \\ \gamma_p(n+1) \end{vmatrix} = \begin{vmatrix} \underline{O} \\ \underline{O} \\ 1 \end{vmatrix} \qquad (3.3.14)$$

7) Update the solution to the forward problem using equations (2.7), (2.12) and (3.2.1):

$$\begin{vmatrix} \underline{u}_f(n+1) \\ \beta\underline{v}_f(n) \\ \alpha_f(n+1) \end{vmatrix} = \hat{Q}_p(n+1) \begin{vmatrix} \beta\underline{u}_f(n) \\ \beta\underline{v}_f(n) \\ x(n+2) \end{vmatrix} \qquad (3.3.15)$$

{NB. $\beta\underline{v}_f(n)$ is not involved in this rotation and hence need not be known.}

8) Forward prediction residual $e_f(n+1,n+1) = \gamma_p(n+1)\ \alpha_f(n+1)$.

{NB. $\underline{u}_f(n+1)$, $\alpha_f(n+1)$, $\gamma_p(n+1)$, $\underline{a}_p(n+1)$ and $\varepsilon_f(n)$ have all been calculated ready for the next time instant.}
The computational load, per time instant, is thus:

i) the calculation and subsequent application of (2p+1) 2x2 rotations.
ii) the calculation of one square-root (other than in the calculation of a rotation).

4. LATTICE FILTER

We note that in the previous section we found a connection between R_{p+1} and the p-th order problems. In order now to solve the (p+1)th problems all we require is the relevant transformed reference sequences. Here we show that they too can be found from the solution to the p-th order problems. Incorporating the usual time-update equations we then develop a fast order-update algorithm.

In the following we use a similar set of notation as used in Section 3 except that we now have to introduce a further subscript to denote the relevant order (p-th or (p+1)th).

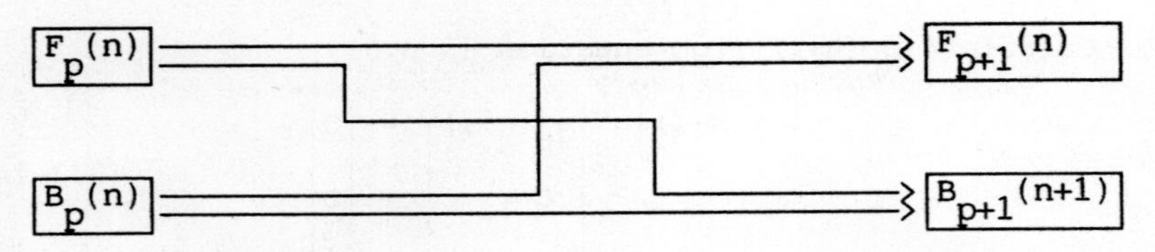

Fig. 4.1 Order iteration

4.1 (p+1)th Order Forward Problem

Now by definition (equations (3.1.1) and (3.2.1))

$$\left|\begin{matrix} \underline{u}_{f,p}(n) & R_p(n) & \underline{u}_{b,p}(n) \\ \underline{v}_{f,p}(n) & 0 & \underline{v}_{b,p}(n) \end{matrix}\right| = \underset{\sim}{Q}_p(n)\ B(n) \left|\begin{matrix} \underline{y}_{f,p}(n) & X_p(n) & \underline{y}_{b,p}(n) \end{matrix}\right| \tag{4.1.1}$$

where

$$\left|\begin{matrix} \underline{y}_{f,p}(n) & X_p(n) & \underline{y}_{b,p}(n) \end{matrix}\right| =$$

$$\left|\begin{matrix} x(2) & x(1) & \dots & x(2-p) & x(1-p) \\ \cdot & \cdot & & & \cdot \\ \cdot & \cdot & & & \cdot \\ x(n+1) & x(n) & \dots & x(n-p+1) & x(n-p) \end{matrix}\right| = \left|\begin{matrix} \underline{y}_{f,p+1}(n) & X_{p+1}(n) \end{matrix}\right| \tag{4.1.2}$$

Note that by virtue of equation (3.1.5) we have

$$Q_b(n) \left|\begin{matrix} \underline{u}_{f,p}(n) & R_p(n) & u_{b,p}(n) \\ \underline{v}_{f,p}(n) & 0 & \underline{v}_{b,p}(n) \end{matrix}\right| = \left|\begin{matrix} \underline{u}_{f,p}(n) & R_p(n) & u_{b,p}(n) \\ \mu_p(n) & \underline{0}^T & \varepsilon_{b,p}(n) \\ \underline{\nu}_p(n) & 0 & \underline{0} \end{matrix}\right|$$

$$\equiv \left|\begin{matrix} \underline{u}_{f,p+1}(n) & R_{p+1}(n) \\ \underline{v}_{f,p+1}(n) & 0 \end{matrix}\right| \tag{4.1.3}$$

where the last step follows by definition and the new variables $\mu_p(n)$ and $\underline{\nu}_p(n)$ are defined by this equation.

Hence it is possible to obtain the solution of the (p+1)th order forward problem at time n from that of the p-th order problems at time n. This order-update requires, however, O(p) operations. The fast update algorithm is developed by applying the standard time-update procedure to equation (4.1.3) Note, however, that we already have developed the rotations necessary for the time update for this problem (see equation (3.1.11)). To complete the fast update procedure for the (p+1)th order all we need therefore is to apply these rotations to the left-hand column of the matrix on the right hand side of equation (4.1.3). Paralleling the development of equation (3.1.11) we have that

$$\hat{Q}_b(n+1)\,\hat{Q}_p(n+1)\begin{vmatrix} Q_b(n) & \underline{0} \\ \underline{0}^T & 1 \end{vmatrix}\begin{vmatrix} Q_p(n) & \underline{0} \\ \underline{0}^T & 1 \end{vmatrix} B(n+1)\begin{vmatrix} \underline{Y}_{f,p+1}(n+1) \end{vmatrix}$$

$$=\hat{Q}_b(n+1)\,\hat{Q}_p(n+1)\begin{bmatrix} \beta\underline{u}_{f,p}(n) \\ \hline \beta\mu_p(n) \\ \hline \beta\underline{\nu}_p(n) \\ \hline x(n+2) \end{bmatrix} = \hat{Q}_b(n+1)\begin{bmatrix} \underline{u}_{f,p}(n+1) \\ \hline \beta\mu_p(n) \\ \hline \beta\underline{\nu}_p(n) \\ \hline \alpha_{f,p}(n+1) \end{bmatrix} = \begin{bmatrix} \underline{u}_{f,p+1}(n+1) \\ \hline \\ \underline{v}_{f,p+1}(n+1) \\ \end{bmatrix} \qquad (4.1.4)$$

where the last step follows by definition. Notice that the last step in both equation (3.1.11) and (4.1.4) (a rotation by $\hat{Q}_b(n+1)$) only requires, as inputs, variables from time n or variables from the p-th order problems at time (n+1). Thus assuming the solutions to the p-th order problems at time n+1 are available, we have the following fast update for the (p+1)th order forward problem.

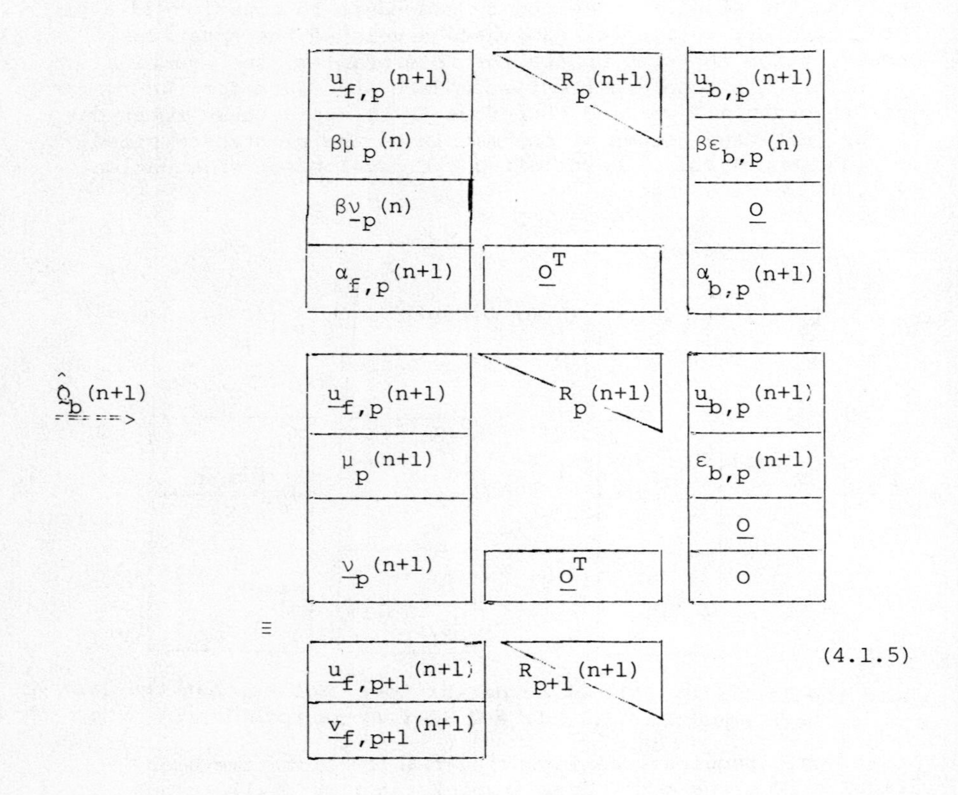

Thus, apart from $\gamma_{p+1}(n+1)$, we have generated a solution to the (p+1)th order forward problem at time n+1 from the solutions to the p-th order forward and backward problems at time n+1. Further, the update requires a fixed number of computations since $\hat{Q}_b(n+1)$ need be applied only to the left and right hand side columns of the above matrix.

Note that $\underline{u}_{b,p}(n+1)$ is not explicitly required for the calculation of $\hat{Q}_b(n+1)$.

4.2 *(p+1)th Order Backward Problem*

The development of the solution to the (p+1)th backward problem proceeds on very similar lines to that in section 3.1.

Again by definition

$$\begin{vmatrix} \underline{u}_{f,p}(n) & R_p(n) & \underline{u}_{b,p}(n) \\ \underline{v}_{f,p}(n) & 0 & \underline{v}_{b,p}(n) \end{vmatrix} = Q_p(n)\ B(n) \begin{vmatrix} \underline{y}_{f,p}(n) & X_p(n) & \underline{y}_{b,p}(n) \end{vmatrix} \quad (4.2.1.)$$

where

$$\begin{vmatrix} \underline{y}_{f,p}(n) & X_p(n) & \underline{y}_{b,p}(n) \end{vmatrix} = \begin{vmatrix} x(2) & x(1) & \ldots & x(2-p) & x(1-p) \\ \cdot & \cdot & & \cdot & \cdot \\ \cdot & \cdot & & \cdot & \cdot \\ x(n+1) & x(n) & \ldots & x(n-p+1) & x(n-p) \end{vmatrix} \quad (4.2.2.)$$

so that

$$\begin{vmatrix} x(1) & \underline{0}^T & 0 \\ \underline{y}_{f,p}(n) & X_p(n) & \underline{y}_{b,p}(n) \end{vmatrix} = \begin{vmatrix} X_{p+1}(n+1) & \underline{y}_{b,p+1}(n+1) \end{vmatrix} \quad (4.2.3.)$$

Now from equation (3.2.12) we have

$$Q_f^v(n) \begin{vmatrix} \beta^n x(1) & \underline{0}^T & 0 \\ \underline{u}_{f,p}(n) & R_p(n) & \underline{u}_{b,p}(n) \\ \underline{v}_{f,p}(n) & 0 & \underline{v}_{b,p}(n) \end{vmatrix} = \begin{vmatrix} \varepsilon_{f,p}(n) & \underline{0}^T & \xi_p(n) \\ \underline{u}_{f,p}(n) & R_p(n) & \underline{u}_{b,p}(n) \\ \underline{0} & 0 & \underline{\lambda}_p(n) \end{vmatrix} \quad (4.2.4)$$

and (equation (3.2.13))

$$Q_f^u(n+1) \begin{vmatrix} \varepsilon_{f,p}(n) & \underline{0}^T & \xi_p(n) \\ \underline{u}_{f,p}(n) & R_p(n) & \underline{u}_{b,p}(n) \\ \underline{0} & 0 & \underline{\lambda}_p(n) \end{vmatrix} = \begin{vmatrix} R_{p+1}(n+1) & \underline{u}_{b,p+1}(n+1) \\ 0 & \underline{v}_{b,p+1}(n+1) \end{vmatrix} \quad (4.2.5)$$

where the new variables $\xi_p(n)$ and $\underline{\lambda}_p(n)$ are defined by these equations. Note however that because of the form of the rotations in $Q_f^u(n+1)$

$$\underline{\lambda}_p(n) \equiv \underline{v}_{b,p+1}(n+1) \quad (4.2.6)$$

Hence it is possible to obtain the solution of the (p+1)th order backward problem at time (n+1) from that of the p-th order problems at time n. Again this order-update requires O(p) computations but by including the time update algorithm we can produce a fast update procedure. With reference to equations

(3.2.12/13) we see that we already know the rotations necessary for the time-update and only have to apply them to the right hand column of the matrix on the right hand side of equation (4.2.4). Paralleling the development of equations (3.2.12/13) we have

$$\hat{Q}_f^v(n)\ \hat{Q}_p(n) \begin{vmatrix} Q_f^v(n) & \underline{0} \\ \underline{0}^T & 1 \end{vmatrix} \begin{vmatrix} 1 & \underline{0}^T & 0 \\ \underline{0} & Q_p(n) & \underline{0} \\ \underline{0} & \underline{0}^T & 1 \end{vmatrix} B(n+2)\ \left|\underline{y}_{f,p+1}(n+2)\right|$$

$$= \hat{Q}_f^v(n)\ \hat{Q}_p(n) \begin{bmatrix} \beta\xi_p(n) \\ \hline \beta\underline{u}_{b,p}(n) \\ \hline \beta\underline{\lambda}_p(n) \\ \hline x(n-p+1) \end{bmatrix} = \hat{Q}_f^v(n) \begin{bmatrix} \beta\xi_p(n) \\ \hline \underline{u}_{b,p}(n+1) \\ \hline \beta\underline{\lambda}_p(n) \\ \hline \alpha_{b,p}(n+1) \end{bmatrix} = \begin{bmatrix} \xi_p(n+1) \\ \hline \underline{u}_{b,p}(n+1) \\ \hline \underline{\lambda}_p(n+1) \end{bmatrix} \qquad (4.2.7)$$

To complete the update we would have to rotate the right hand vector above by the matrix $Q_f^u(n+2)$ but note that we already know $\underline{v}_{b,p+1}(n+2)$ because (see equation (4.2.6))

$$\underline{\lambda}_p(n+1) \equiv \underline{v}_{b,p+1}(n+2)$$

and as $\underline{u}_{b,p+1}(n+1)$ is not explicitly required we need proceed no further. Again we notice that the last step in equations (3.2.12) and (4.2.7) (a rotation by $\hat{Q}_f^v(n+1)$) only requires as inputs variables from time n or variables from the p-th order problems at time (n+1). If the solutions to the p-th order problems are available for time n+1 then we have the following fast update for the (p+1)th order backward problem.

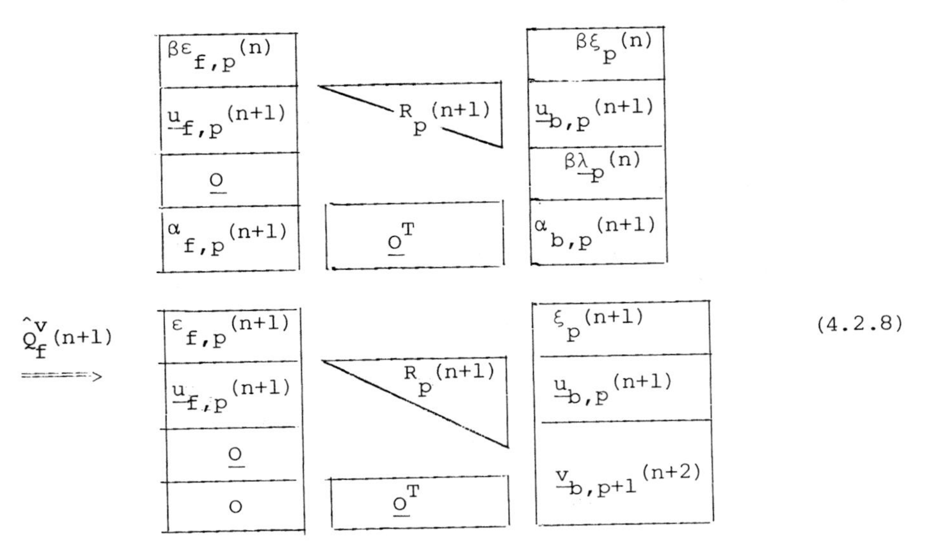

Thus, apart from $\gamma_{p+1}(n+2)$, we have generated a solution to the (p+1)th order backward problem at time n+2 from the solution to the p-th order forward and backward problems at time n+1. Further, the update requires O(1) computations since $\hat{Q}^{v}_{f}(n+1)$ need be applied only to the left and right hand side columns of the above matrix.

4.3 Update of γ *(n)*

In subsections 4.1 and 4.2 we have derived the solution to the (p+1)th order problems from that of the p-th order ones except for the factor γ_{p+1} at the relevant time. This factor is required only in order that the prediction residual can be extracted (see equation (2.11)).

From section 3.1 we notice that (equations (3.1.9/11))

$$Q_{p+1}(n+1) = \hat{Q}_b(n+1)\;\hat{Q}_p(n+1)\begin{vmatrix} Q_b(n) & \underline{O} \\ \underline{O}^T & 1 \end{vmatrix}\begin{vmatrix} Q_p(n) & \underline{O} \\ \underline{O}^T & 1 \end{vmatrix} \qquad (4.3.1)$$

but (see equation (2.7))

$$Q_{p+1}(n+1) = \hat{Q}_{p+1}(n+1)\begin{vmatrix} Q_p(n) & \underline{O} \\ \underline{O}^T & 1 \end{vmatrix} \qquad (4.3.2)$$

hence

$$\hat{Q}_{p+1}(n+1) = \hat{Q}_b(n+1)\,\hat{Q}_p(n+1)\begin{bmatrix} Q_b(n) & \underline{0} \\ \underline{0}^T & 1 \end{bmatrix} \tag{4.3.3}$$

but by definition

$$\hat{Q}_{p+1}(n+1)\begin{bmatrix} \underline{0} \\ 1 \end{bmatrix} = \begin{bmatrix} \underline{a}_{p+1}(n+1) \\ \underline{0} \\ \gamma_{p+1}(n+1) \end{bmatrix} \tag{4.3.4}$$

hence

$$\begin{bmatrix} \underline{a}_{p+1}(n+1) \\ \underline{0} \\ \gamma_{p+1}(n+1) \end{bmatrix} = \hat{Q}_{b,p}(n+1)\begin{bmatrix} \underline{a}_p(n+1) \\ \underline{0} \\ \gamma_p(n+1) \end{bmatrix} \tag{4.3.5}$$

however because of the nature of the rotations in $\hat{Q}_{b,p}(n+1)$ we have

$$\hat{Q}_{b,p}(n+1)\begin{bmatrix} \underline{0} \\ \underline{0} \\ \gamma_p(n+1) \end{bmatrix} = \begin{bmatrix} \underline{0} \\ \Theta_{p+1}(n+1) \\ \underline{0} \\ \gamma_{p+1}(n+1) \end{bmatrix}$$

where

$$\Theta_{p+1}(n+1) = a_{p+1,p+1}(n+1)$$

so that $\underline{a}_p(n)$ need not be calculated explicitly.

Since the above set of rotations is exactly that involved in the updating of the (p+1)th order forward prediction problem we may update $\gamma_p(n)$ at the same time. In fact a similar analysis based on the updating of the (p+1)th order backward problem shows that $\gamma_p(n)$ could also be updated here as well if desired.

4.4 Algorithm

Consideration of equations (4.1.5), (4.2.8) and (4.3.6) shows that not only are $\underline{u}_{b,p}(n)$ and $\underline{u}_{f,p}(n)$ not required for direct residual extraction but also that all except the last elements ($\alpha_{b,p}(n)$ and $\alpha_{f,p}(n)$) of the vectors $\underline{v}_{b,p}(n)$ and $\underline{v}_{f,p}(n)$, respectively, are superfluous.

Given $\alpha_{f,p}(n)$, $\alpha_{b,p}(n)$, $\varepsilon_{f,p}(n-1)$, $\varepsilon_{b,p}(n-1)$, $\gamma_p(n)$, $\mu_p(n-1)$, $\xi_p(n-1)$,

1) Calculate $\alpha_{f,p+1}(n)$, $\varepsilon_{b,p}(n)$, $\gamma_{f,p+1}(n)$, $\mu_p(n)$.

$$\begin{vmatrix} \beta\varepsilon_{b,p}(n-1) & \beta\mu_p(n-1) & 0 \\ \alpha_{b,p}(n) & \alpha_{f,p}(n) & \gamma_p(n) \end{vmatrix} \xrightarrow{\hat{Q}_b(n)} \begin{vmatrix} \varepsilon_{b,p}(n) & \mu_p(n) & \theta_{p+1}(n) \\ 0 & \alpha_{f,p+1}(n) & \gamma_{p+1}(n) \end{vmatrix}$$

NB. $e_{f,p+1}(n,n) = \gamma_{p+1}(n)\,\alpha_{f,p+1}(n)$

This requires the calculation and two applications of a 2x2 rotation.

2) Calculate $\alpha_{b,p+1}(n+1)$, $\varepsilon_{f,p}(n)$, $\xi_p(n)$.

$$\begin{vmatrix} \beta\varepsilon_{f,p}(n-1) & \beta\xi_p(n-1) \\ \alpha_{f,p}(n) & \alpha_{b,p}(n) \end{vmatrix} \xrightarrow{\hat{Q}_f^v(n)} \begin{vmatrix} \varepsilon_{f,p}(n) & \xi_p(n) \\ 0 & \alpha_{b,p+1}(n+1) \end{vmatrix}$$

NB. $e_{b,p+1}(n,n) = \gamma_{p+1}(n)\,\alpha_{b,p+1}(n)$

This requires the calculation and one application of a 2x2 rotation.

CONCLUSION

In our opinion, Cioffi's orthogonal fast Kalman algorithm is a significant new development in the field of least squares linear prediction. We have presented a much simplified derivation of what we believe to be Cioffi's orthogonal fast Kalman algorithm, based on the notation of the QR decomposition algorithm for least squares recursive minimisation. No detailed step by step comparison has yet been undertaken to compare our work with that of Cioffi but, as our algorithm was derived using the same principles, it seems likely that the two algorithms are effectively identical.

The key to the orthogonal fast Kalman algorithm is the interconnection between the forward and backward p-th order linear prediction problems and the matrix R_{p+1}. By broadening this interconnection, we have shown that it is possible to generate the solutions to the (p+1)th order problems in a fixed amount of computation. This leads to a new lattice filter algorithm explicitly based on orthogonal rotations. We have not, as yet, implemented this new algorithm and so can not comment on its numerical properties but being based entirely on orthogonal rotations it is expected to be particularly stable. The results of such a simulation will be presented in due course.

ACKNOWLEDGEMENTS

The authors would like to thank Dr. Adam Bojanczyk of Cornell University for some stimulating discussion on this problem and several colleagues, including F. Ling, M.G. Bellanger and R. Geer, who pointed out a mistake in the original draft.

REFERENCES

[1] Bellanger, M. G., (1988) The FLS-QR Algorithm for Adaptive Filtering, Proc. NATO Advanced Study Institute on Numerical Linear Algebra, Digital Signal Processing and Parallel Algorithms, Leuven, Belgium.

[2] Cioffi, J. M., (1988) High Speed Systolic Implementation of Fast QR Adaptive Filters, Proc. International Conference on A.S.S.P., vol. DIII, 1584-1587, I.E.E.E., New York.

[3] Cioffi, J. M. (198-), Fast QR/Frequency Domain RLS Adaptive Filters, I.E.E.E. transactions on ASSP, to be published.

[4] Gentleman, W. M., Kung, H. T., (1981) Matrix Triangularisation by Systolic Array, Proc. S.P.I.E., 298, Real Time Signal Processing IV, 1981.

[5] Lackey, J. R., Baurle, H. F., Barlie, J., (1988) Application Specific Super Computer, Proc. S.P.I.E., 977, Real Time Signal Processing, 1988.

[6] McWhirter, J. G., (1983) Recursive Least-squares Minimisation using a Systolic Array, Proc. S.P.I.E., 431, Real Time Signal Processing VI, 105-112, 1983.

[7] Ward, C. R., Hargrave, P. J., McWhirter, J. G., (1986) A Novel Algorithm and Architecture for Adaptive Digital Beam-forming, I.E.E.E. transactions, AP-34(3), 338-346, 1986.

FAST NONLINEAR ITERATIVE ALGORITHMS FOR HARMONIC SIGNAL EXTRAPOLATION

A. R. Figueiras-Vidal and J. R. Casar-Corredera
(DSSR, ETSI Telecomunicacion-UPM, Ciudad Universitaria, 28040 Madrid, Spain)

and

D. Docampo-Amoedo and A. Artés-Rodriquez
(DTC, ETSI Telecomunicacion-US, Aptdo. 62, 36280 Vigo (Pontevedra), Spain)

ABSTRACT

A series of gradient algorithms that generalizes the classical ones used in bandlimited harmonic signal extrapolation are obtained from the Moore Penrose pseudoinverse solution to the problem.

The algorithms include adaptive bandwidth reduction by means of applying an adaptive threshold; computationally efficient versions, obtained by using the error sign or by working in the frequency domain, are also considered. The possibility of getting more robustness against added noise combining the iterations with those corresponding to an ideal bandselecting filter are explored. All these schemes are comparatively discussed, and their performances demonstrated by means of simulation examples.

Finally, some suggested further lines of research are given.

1. INTRODUCTION

The application of bandlimited signal extrapolation techniques in spectral analysis, superresolution, signal restoration, etc., justifies the interest that they have received during the last years, and the research activity around them since the first works of Gerchberg [1] and Papoulis [5].

We will focus in this paper on the so called discrete-discrete case (i.e., that obtained when a periodic sequence is assumed), since it is the practical approximation for the other cases. Under this hypothesis, the problem is as follows: to estimate the Nx1 vector x such that

$$SF^{-1}PFx+n=z \qquad (1.1)$$

where z is an Mx1 (M<N) observation vector, n an Nx1 noise vector, F and F^{-1} are NxN DFT and IDFT matrices, P an NxN bandlimitation matrix, and S an MxN time segment selection matrix, respectively. S has only 0 or 1 elements, the last corresponding to the positions of x that are observable as elements of z; P has only nonzero elements, that are also 1, in the main diagonal elements that correspond to frequencies in x (to the signal band).

Assuming that P is known, we will be faced with an estimation problem: to estimate the values of x in equation (1.1), that can be solved by means of a Wiener filter, for example. But, in practi P is unknown: then, our problem is a detection plus estimation problem: to decide the nonzero elements of P and to estimate the values of x under the above decision.

To solve this kind of problem, it is necessary to apply Bayesian decision schemes in a composite hypotheses formulation. This approach is extremely computer consumer and very sensitive to the assumed statistical parameters. Thus, other simpler approaches have been proposed to get reasonable solutions. The oldest is that proposed independently in [1] and [5]: to use

$$x_{m+1}=F^{-1}PF[x_m+S^T(z-SF^{-1}PFx_m)] \quad x_o=0 \qquad (1.2)$$

and to take x_∞ as the solution, simultaneously forcing bandlimitation and compatibility with the observations, P being a conservative bandlimiting operation. A further refinement is introduced in [6], where the application of a rëlatively robust threshold

$$T_m=\min[T_{m-1},\alpha\max\{|X_{m-1}|:|X_{m-1}|>T_{m-1}\}] \qquad (1.3)$$

where $X_m=Fx_m$, serves to introduce an adaptive P_m, the nonzero elements of which are reduced introducing zeros in the positions corresponding to positions in which $|X_{m-1}|<T_{m-1}$. This can be considered as an adaptive detection scheme combined with the previous algorithm.

Although other interesting schemes have been introduced, such as that presented in [3], in which the authors take advantage of the spiky character of the minimum L_1-norm solutions to propose a double algorithm (working in the autocorrelation domain before being used with the data) minimizing spectral L_1 norms, and one proposing L_p minimization of the error [7], they have drawbacks (such as, in the first mentioned example, the possibility of multiple solutions and the sensitivity with respect to the number of used equations and the auxiliary parameters, and the spiky character of the error and not of the signal in the second example) that justify concentrating our interest on Gerchberg-Papoulis' kind of algorithms.

This argument has been reinforced by the results presented in [4]: there, Jain and Ranganeth prove that Gerchberg-Papoulis' algorithm goes to the Moore-Penrose pseudoinverse solution of (1.1):

$$x_{MNLS}=F^{-1}PFS^{T}(SF^{-1}PFS^{T})^{-1}z \tag{1.4}$$

that is a minimum norm (i.e., minimum power), least squares (i.e., forcing a minimum quadratic norm of n) solution of the original equation. It seems clear the interest of this approach when statistical information about the problem is not available.

2. THE BASIC PROPOSED ALGORITHM

It is not difficult to get from (1.4) the general version of the corresponding iterative gradient algorithm, or alternatively by minimizing the error along the data positions:

$$\|Sn\|_2=(z-SF^{-1}PFx)^{T}(z-SF^{-1}PFx) \tag{2.1}$$

arriving to

$$x_{m+1}=x_m+\mu F^{-1}PFS^{T}(z-SF^{-1}PFx_m) \tag{2.2}$$

If we start with a P-bandlimited x_O, $F^{-1}PF$ could be put before x_m as a common factor: we get a generalized ($\mu\neq 1$) Gerchberg-Papoulis' algorithm, equivalent to that proposed in [4].

Combining this with the idea of an adaptive iterative P_m built with the help of threshold (1.3) (or any other equivalent to this) generates the first new algorithm we consider here

$$x_{m+1}=F^{-1}P_mF[x_m+\mu S^{T}(z-SF^{-1}P_mFx_m)] \tag{2.3}$$

Since P_m is nondecreasing, the first $F^{-1}P_mF$ can be suppressed in the practice. Also, this algorithm can be considered as the combination of a direct one

$$x_{m+1}=x_m+\mu S^{T}(z-Sx'_m) \tag{2.4a}$$

with the adaptive reduction of the sinusoidal basis for the x vectors

$$x'_m = F^{-1} P_m F x_m \tag{2.4b}$$

this interpretation justifies the convergence of the algorithms under the convergence conditions for (2.4a) ($0<\mu<2$), and also opens the possibility of combining the idea of adaptive basis reduction with other schemes: it could be used even with the block solution (1.4); but, clearly, it will fail if this solution provides initial high spectral peaks in areas different from the signal band, as it occurs in relatively high noise cases.

Note, on the other side, that we are, in fact, excluding from the basis those elements that offer a reduced scalar product with the solution at each step: a very reasonable way of defining the signal band (as long as the threshold be robust and conservative).

In fact, we have introduced a generalization of Papoulis-Chamzas' algorithm [6] (that corresponds to $\mu=1$), having the advantage, with respect to the previous ones, of the adaptive determination of the signal band, that also provides more robustness against the noise.

This general algorithm works very well except in high noise cases. Figures 1 to 4 show registers corresponding to a first example having $N=256$, $M=51$, $x=1.25\cos(2\pi 12n/256+\pi/6) + 1.5\cos(2\pi 17n/256+\pi/3)$, white, Gaussian, zero mean noise with SNR= 5dB, $\mu=0.5$, $a=1$, $x_0=0$. Fig. 1 is x; Fig. 2, z; Fig. 3, x_{26}; Fig. 4, x_{76} (final estimate). The first threshold is applied on Fx_1 with a value $(3/256)\|x_1\|^2$.

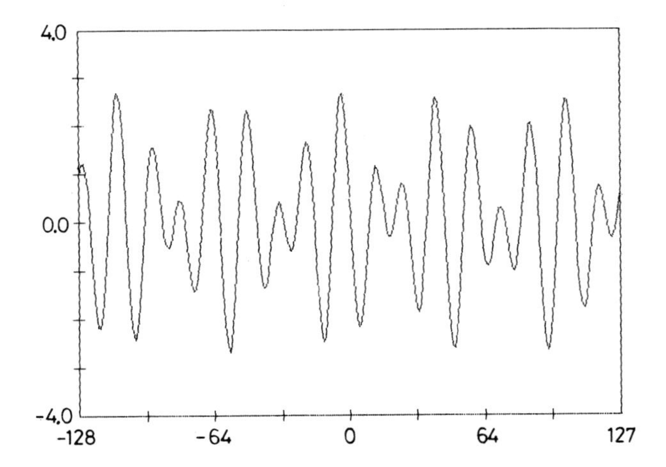

Fig. 1 Signal corresponding to first example;
$x[n] = 1.25\cos(2\pi 12n/256+\pi/6) + 1.5\cos(2\pi 17n/256+\pi/3)$,
$-127 \leq n \leq 128$

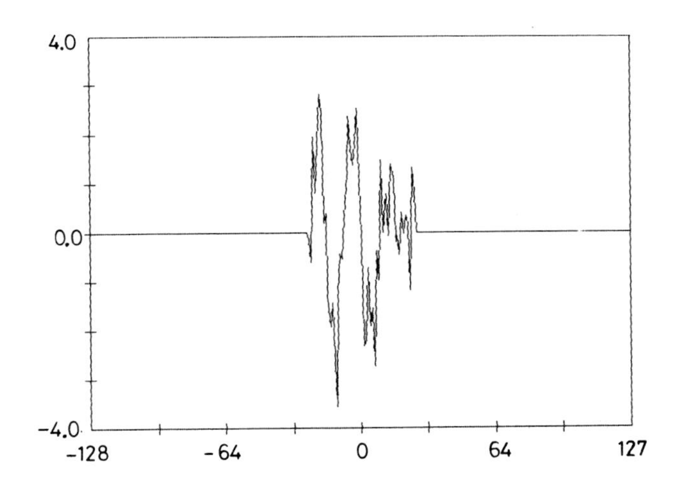

Fig. 2 Data corresponding to first example;
$z[n] = x[n] + r[n]$, $-25 \leq n \leq 25$
$r[n]$: white, Gaussian noise; SNR = 5dB

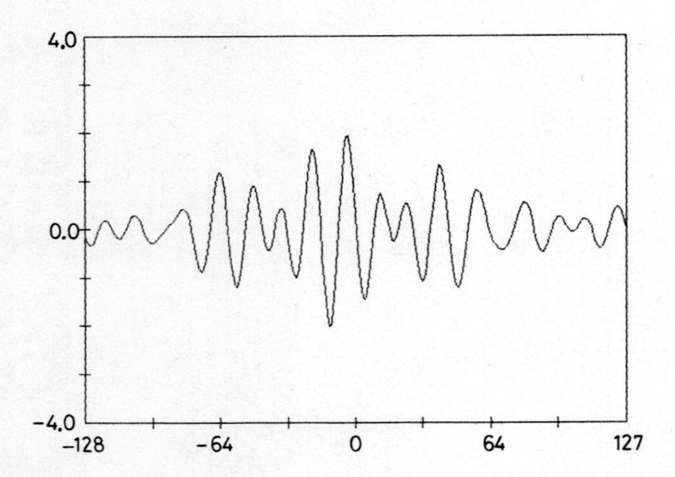

Fig. 3 x_{26} with algorithm (2.3); $x_o = 0$, $\mu = 0.5$, $\alpha = 1$, $T_1 = (3/256)\|x_1$

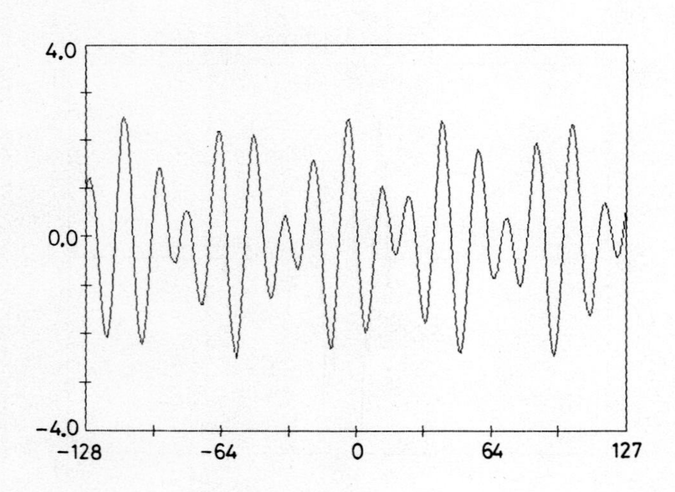

Fig. 4 x_{76} with algorithm (3.1) and the above values for the parameters (final step)

3. SIMPLIFIED VERSIONS

A first reduction of the computational load per step is obtained by introducing the sign version of (2.3)

$$x_{m+1} = F^{-1}P_m F[x_m + \mu S^T \mathrm{sgn}(z - SF^{-1}P_m F x_m)] \tag{3.1}$$

Simulations show that it converges in a slightly greater number of iterations than (2.3); thus reducing the overall number of operations. Figs. 5 and 6 show x_{27} and x_{77} (final iteration) for the same example as previously using (2.4) with $\mu=0.5$, $\alpha=0.98$, $x_0=0$, and first threshold $(2/256)\|x_2\|_2$ applied on x_2.

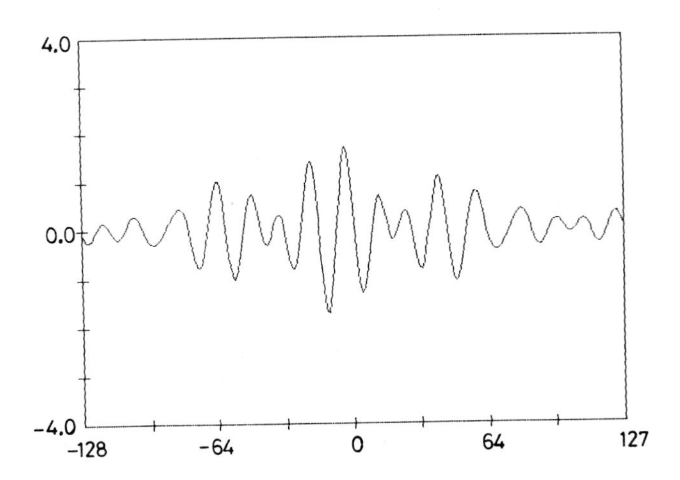

Fig. 5 x_{27} with algorithm (3.1) and the above example; $x_0=0$, $\mu=0.5$, $\alpha=0.98$, $T_2=(2/256)\|x_2\|_2$.

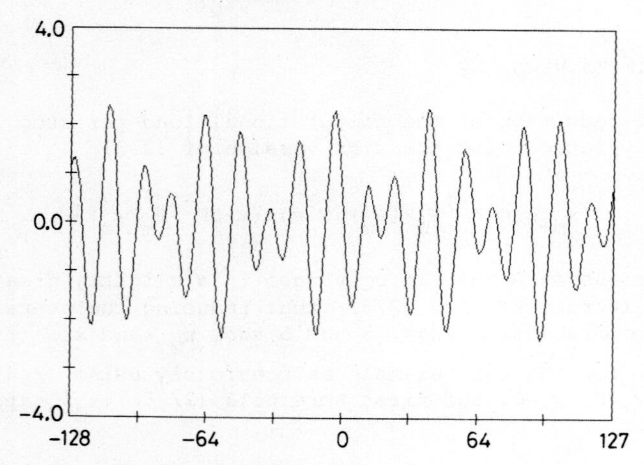

Fig. 6 x_{77} with algorithm (3.1) and the above values for the parameters (final step).

A very interesting approach is that originated by writing (1.1) in the form

$$FS^TSF^{-1}PX+FS^Tn = FS^Tz \tag{3.2}$$

where X=Fx. Using clearly defined spectra, (3.2) becomes

$$CPX+N = Z \tag{3.3}$$

where $C=FS^TSF^{-1}$ is equivalent to apply a circular convolution with the transform of a rectangular window to PX. Working with (3.3) as with (1.1), we arrive at the algorithm

$$X_{m+1} = P_m[X_m+\mu(Z-CP_mX_m)] \tag{3.4}$$

that is the result of applying F to both members of (2.3), but has very clear computational advantages, since X_m has a reduced number of nonzero values. It is obvious that this algorithm provides the same results as the first one: Figs. 7 and 8 show $P_{26}X_{26}$ and $P_{76}X_{76}$ (final step) for the first example working in this domain: X_{26} and X_{76} correspond to Fx_{26} and Fx_{76}, respectively

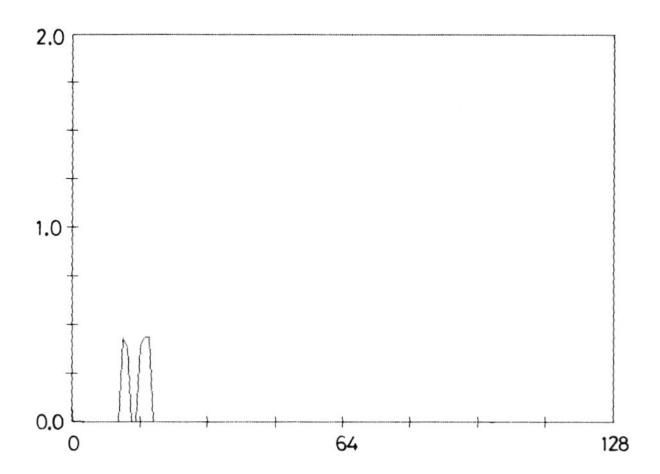

Fig. 7 $P_{26}X_{26}$ for the first example and algorithm (3.4), $x_0=0$, $\mu=0.5$, $\alpha=1$, $T_1=(3/256)\|x_1\|_2$

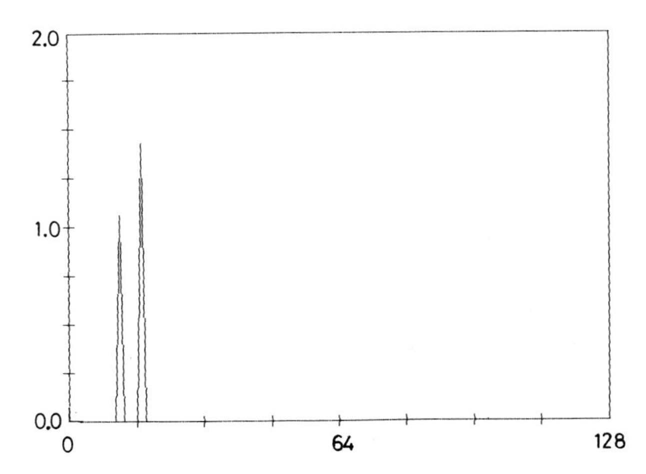

Fig. 8 $P_{76}X_{76}$ for the first example and above and above parameters (final value)

Note that any spiky deconvolution technique could be applied to get a solution of (3.3)

L_1 ("sign") iterative versions are also possible here, but it is clear that the L_1 frequency norm is very computer consuming (it requires to use $E_m/\|E_m\|_2$, E_m being $Z\text{-}FS^TSF^{-1}P_mX_m$).

The L_1 time version is

$$X_{m+1} = P_m[X_m+\mu FS^T \mathrm{sgn}(z\text{-}SF^{-1}P_mX_m)] \tag{3.5}$$

that requires a zero-prunning IDFT (FFT) and a zero-prunning DFT (FFT) of ones and zeros; in any case, it tends to have higher computational cost than (3.4) using direct circular convolution when the bandwidth is narrow.

4. WEIGHTED APPROACHES

It is clear that, extending the data with S^T and considering

$$PX+N' = Z \tag{4.1}$$

where N' is the DFT of the N point error sequence (error on the data interval plus minus the signal extrapolation), the minimum norm least square solution is the P-bandpass filter. This is a reasonable solution when the noise is very high. Its (frequency) iterative algorithm is

$$X_{m+1} = P_m[X_m+\mu(Z-X_m)] \tag{4.2}$$

A weighted combination, putting a weight w on the data interva and 1-w outside it, could serve to get more robustness against the noise. The corresponding algorithm is

$$X_{m+1} = P_m[X_m+\mu\{w(Z-FS^TSF^{-1}P_mX_m)-(1-w)F(I-S^TS)F^{-1}P_mX_m\}) \tag{4.3}$$

w=1 leads to the generalized Papoulis-Chamzas' algorithm, w=1/2 to (4.2). Assuming that we know that the noise power is K times the signal extrapolation power, it seems convenient to weight K times more this part: i.e., to use w=1/(1+K). However, experimental results indicate that a greater value offers better performance.

With the same example we used previously but increasing the noise until SNR=SdB, the generalized Papoulis-Chamzas' algorithm fails in estimating the two sinusoids. However, (4.3) with w=0.85 offers a good result. Fig. 9 shows the register z, Fig. 10 its spectrum, and Figs. 11, 12, and 13, $P_{25}X_{25}$, $P_{75}X_{75}$ (final step) and $F^{-1}P_{75}X_{75}$ for (4.3) with the above value of w and the same parameters as those used in the first example, but $T_5=(2/256)\|x_5\|_2$.

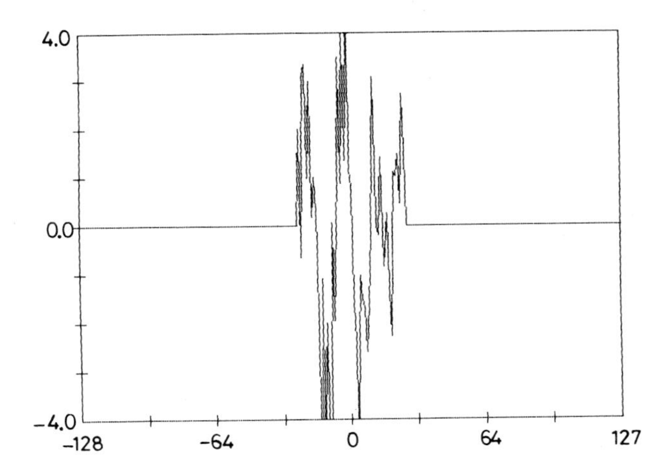

Fig. 9 Data register corresponding to second example:
$z[n] = x[n]+r'[n], -25 \leq n \leq 25$
$r'[n]$: white, Gaussian noise, SNR=2dB

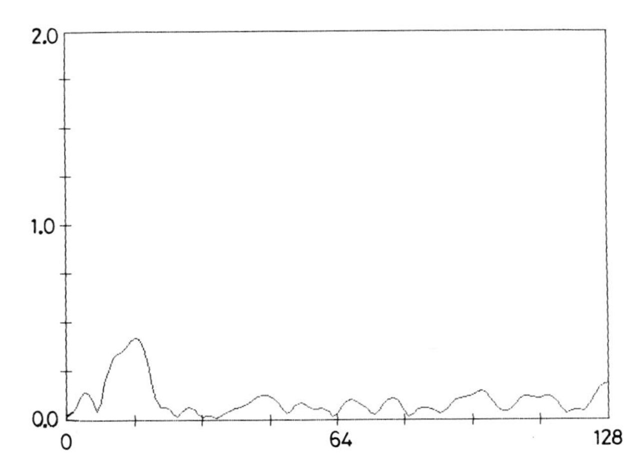

Fig. 10 $Z(k) = \text{DFT}\{z[n]\}$, second example

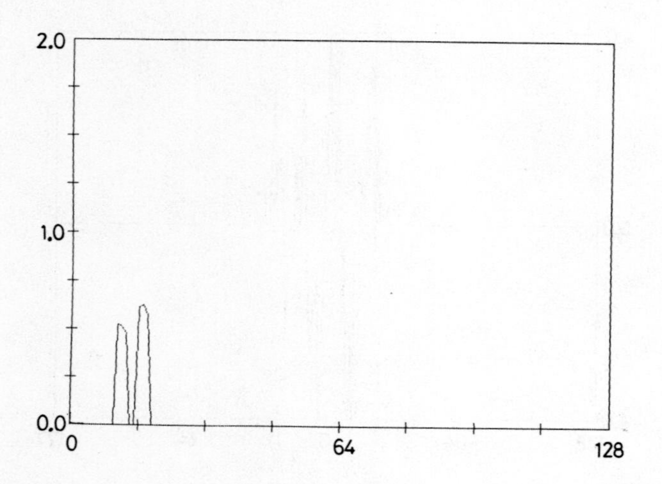

Fig. 11 $P_{25}X_{25}$ for the second example with algorithm (4.3), $w=0.85$, $x_0=0$, $\mu=0.5$, $\alpha=1$, $T_5=(2/256)\|x_5\|_2$

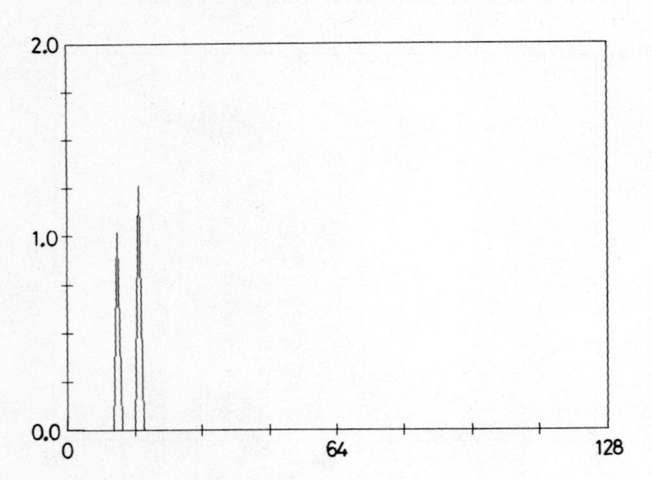

Fig. 12 $P_{75}X_{75}$ for the above example and parameters (final value)

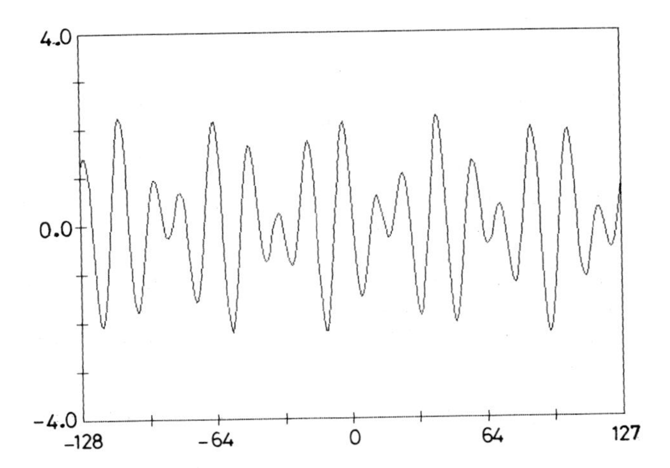

Fig. 13 IDFT $\{P_{75}X_{75}\}$ for the second example and parameters (final time extrapolation).

If we decrease w until 0.8, the weighted algorithm fails.

Using this weighting possibility serves to introduce simple approaches in the direction suggested by Kolba and Parks [2].

Finally, let us mention that one can consider an alternative form of the problem:

$$F^{-1}X = S^{T}z-n' \tag{4.4}$$

and minimize $\|X_m-P_m X_m\|_2$ iterating n'. We get the algorithm

$$n'_{m+1} = n'_m+\mu[S^{T}z-n'_m-F^{-1}P_m(S^{T}z-n'_m)] \tag{4.5}$$

and, if we operate only outside the data segment, we arrive to

$$(I-S^{T}S)x_{m+1} = (I-S^{T}S)x_m+(I-S^{T}S)\mu F^{-1}P_mF[S^{T}z-(I-S^{T}S)x_m] \tag{4.6}$$

that, with μ=1, becomes Papoulis-Chamzas' algorithm. This approach could be explored in a similar way as above: however, it requires more computational effort without improving the results of the previous way.

5. CONCLUSIONS AND FURTHER WORK

A complete set of time and frequency adaptive thresholded gradient-type iterative algorithms based on the Moore-Penrose pseudoinverse for harmonic signal extrapolation have been presented, and their performance and computational load discussed. These algorithms generalize the classical Gerchberg-Papoulis' and Papoulis-Chamzas' versions, and their frequency versions that we introduce are most efficient.

Some generalized versions to take into account noise effects are also developed and their higher performance demonstrated by means of simulation examples. Other possibilities of extending these schemes are shown.

Lines in which further work will provide improvements are:

- alternative thresholding algorithms taking into account statistical knowledge of the problem;
- more efficient iterative algorithms;
- algorithms taking into account statistical information about signal and noise (based on Wiener filtering);
- iterative versions of the last mentioned algorithms, taking advantage of an adaptive thresholding to improve the band definition.

And general open possibilities that we can suggest are

- using other spiky deconvolution techniques in the frequency domain;
- using the basis reduction technique proposed here for other problems having sparse solution (in particular combined with the very effective Projection Onto Convex Sets methods).

REFERENCES

[1] Gerchberg, R. W. (1975) Super Resolution Through Error Energy Reduction, Optical Acta, 21, 709-720,1975.

[2] Kolba, D. P. and Parks, T. W. (1983) Optimal Estimation for Band-Limited Signals including Time-Domain Considerations, IEEE Trans. Acoustics, Speech, and Signal Processing, 31, 113-122,1983.

[3] Levy, S., Walker, C., Ulrych, T. J. and Fullagar, P. K. (1982), A Linear Programming Approach to the Estimation of

the Power Spectra of Harmonic Processes, IEEE Trans. Acoustics, Speech, and Signal Processing, 29, 830-845, 1981.

[4] Jain, A. K. and Ranganath, S. (1981) Extrapolation Algorithms for Discrete Signals with Applications in Spectral Estimation, IEEE Trans. Acoustics, Speech, and Signal Processing, 29, 830-845, 1981.

[5] Papoulis, A. (1975) A New Algorithm in Spectral Analysis and Band-Limited Extrapolation, IEEE Trans. Circuits and Systems, 22, 735-742, 1975.

[6] Papoulis, A. and Chamzas, C. (1979) Detection of Hidden Periodicities by Adaptive Extrapolation, IEEE Trans. Acoustics, Speech, and Signal Processing, 27, 492-500, 1979.

[7] Yargaladda, R., Bednar, J. B. and Watt, T. L. (1985) Fast Algorithms for lp Deconvolution, IEEE Trans. on Acoustics, Speech, and Signal Processing, 33, 174-182, 1985.

AVOIDING TWO POINT BOUNDARY VALUE PROBLEMS IN THE MAXIMUM A PRIORI ESTIMATE OF NOISY DYNAMICAL SYSTEM VARIABLES

A. Graham and J. Smallwood
(Department of Electrical Engineering, University of Southampton)

ABSTRACT

An algorithm for obtaining the maximum a priori estimate of noisy nonlinear dynamical systems is presented and applied to the two dimensional bearings only tracking problem. The algorithm uses a Gauss-Newton iterative scheme implemented by means of Givens rotation methods.

INTRODUCTION

This paper reexamines the important problem of obtaining the maximum a priori (MAP) estimate for discrete, nonlinear, noisy dynamical systems. The MAP estimate is traditionally formulated as a constrained optimisation problem. This in turn is then reformulated as a nonlinear two point boundary value problem (tpbvp) [1], [2]. In this form, calculation of the MAP estimate requires batch processing of the incoming data, and a new tpbvp must be solved as each new observation becomes available. It is generally considered as being impracticable to solve the nonlinear tpbvp in real time, and approximation schemes are usually employed. Typically this leads to some version of the (recursive) Extended Kalman Filter (EKF). Although the EKF is adequate in many nonlinear problems, there are cases where the EKF is known to fail, or at least to perform poorly.

In this paper we show that there are many cases of interest where the MAP estimate can be formulated as an unconstrained optimisation problem. This avoids the need to solve a tpbvp and reduces the dimension of the vector of variables to be estimated, although the problem of having to batch process an increasing number of incoming observations remains. In order to perform the implied minimisation, we discuss a particular descent method based on the Gauss Newton descent direction.

The matrices which arise in the computation of the chosen descent direction are both sparse and have considerable structure. This may be exploited (using Givens rotations) to minimise the computational load at each stage.

As an example the two dimensional bearings only tracking problem is considered and the results compared with the EKF.

1. PROBLEM STATEMENT

The problem we shall consider is identical to that of Cox [1], and a special case of that discussed by Friedland and Bernstein [2]. Consider the noisy dynamical system described by the vector difference equation

$$x(k+1) = \phi(x(k),k)+G(k+1)w(k+1), \quad k=0,\ldots,N-1 \qquad (1.1)$$

where there is an observed signal of the form

$$z(k) = h(x(k),k)+v(k), \quad k=1,\ldots N \qquad (1.2)$$

In the above equations

x is the state vector: $\dim(x)=s$;

w is the random input vector: $\dim(w) = t \leq s$;

z is the observation vector: $\dim(z) = p \leq s$;

G is a matrix of the appropriate dimension and maximal rank;

v is the measurement noise vector;

h and ϕ are vector valued functions.

The explicit dependence of ϕ on k accounts for any deterministic inputs.

We assume that

$$v^N \equiv \{v(1),\ v(2),\ldots,\ v(N)\}$$

and w^N are zero mean independent Gaussian random vector sequences satisfying

$$E\{w(k)w^T(j)\} = Q(k)\delta_{kj}$$

$$E\{v(k)v^T(j)\} = R(k)\delta_{kj}$$

$$E\{v(k)w^T(j)\} = 0$$

where Q and R are positive definite matrices.

In addition we assume that x(0) has an a priori Gaussian distribution given by

$$x(0) \sim N(\hat{x}(0|0),\ P(0))$$

The problem is to estimate the sequence of states x^N given the a priori distribution of x(0) and the measurement sequence z^N.

2. THE MAXIMUM A PRIORI (MAP) ESTIMATE

The MAP estimate is the mode of the conditional probability density function $p(x^N|z^N)$. In [1] and [2] it is shown that the pdf is given by

$$p(x^N|z^N) = C(z^N)\exp(-MAP_N\}$$

where

$$MAP_N = \frac{1}{2}\|x(0)-\hat{x}(0|0)\|^2_{P^{-1}(0)} + \frac{1}{2}\sum_{k=1}^{N}\|(z(k)-h(x(k),k)\|^2_{R^{-1}(k)}$$

$$+\frac{1}{2}\sum_{k=1}^{N}\|w(k)\|^2_{Q^{-1}(k)} \qquad (2.1)$$

subject to constraints, namely the dynamical equations (1.1). (In the above equation we have used the notation that $\|v\|^2_A \equiv v^T Av$.)

Assuming that the pdf is unimodal, the MAP estimate of x^N is obtained by finding the unique stationary point of $MAP_N(x^N, w^N)$. This is in general a problem in nonlinear optimisation, subject to nonlinear constraints. Such problems are known to be hard to solve. The usual method in this case (as in [1] and [2]) is to affix the constraints via Lagrange multipliers (which form a sequence λ^N of s-vectors) and solve the resulting unconstrained minimisation problem. Such an approach leads naturally to a nonlinear two point boundary value problem in the $2(N+1)$-vector $X^T \equiv [(x^N)^T, (\lambda^N)^T]$, (the sequence w^N being easily eliminated in favour of λ^N). Obtaining the numerical solution to a tpbvp is in general a formidable proposition, and the question arises as to whether such a procedure can be avoided.

Providing the matrix $G(k+1)$ is of constant (maximal) rank for all k, then (1.1) may be solved for $w(k+1)$ to give t independent equations

$$w(k+1) = G^{\#}(x(k+1)-\phi(x(k),k))$$

where $G^{\#}$ is the (in general pseudo-) inverse of G. The above equations may be substituted into (2.1) to eliminate $w(k)$. The remaining (s-t) constraint equations then have the form

$$A(x(k+1)-\phi(x(k),k)) = 0$$

where A is the rank (s-t) matrix $I_s - GG^{\#}$. (A annihilates G.) By means of a (possibly k-dependent) transformation of x^N, the objective function (2.1) to be minimised may be written in the form

$$MAP_N = \frac{1}{2}\|x(0)-\hat{x}(0|0)\|^2_{P^{-1}(0)} + \frac{1}{2}\sum_{k=1}^{N}\|z(k)-h(x(k),k)\|^2_{R^{-1}(k)}$$

$$+\frac{1}{2}\sum_{k=0}^{N-1}\|x^{(1)}(k+1)-\phi^{(1)}(x(k),k)\|^2_{Q^{-1}(k+1)} \qquad (2.2)$$

subject to the constraints

$$x^{(2)}(k+1)-\phi^{(2)}(x(k),k)) = 0 \qquad (2.3)$$

where

$$x^T = [x^{(1)T}, x^{(2)T}]$$

and $x^{(1)}(k)$, $x^{(2)}(k)$ have dimension t and (s-t) respectively.

Assuming that it is possible to carry out the above procedure for a given problem, we have at least succeeded in reducing the dimension of the constraint space.

It is always possible, formally at any rate, to solve the constraints (2.3) to give

$$x^{(2)}(k+1) = \psi(x^{(2)}(0), x^{(1)k}) \qquad (2.4)$$

and to substitute the above equations into (2.2). This yields a problem in unconstrained minimisation in the vector $X^T = [(x^{(2)}(0))^T, (x^{(1)N})^T]$. The dimension of X is $[(s-t)+t(N+1)] = s+Nt$.

One may well ask at this stage whether a) the explicit solution of the constraints in the form (2.4) can actually be performed for any problems in practice, and b) whether the resulting dimension of the problem is not so great as to be computationally intractable. The answer to a) is that there are certainly cases, e.g. for linear dynamics and no system noise (t=0) [3], [4] where this procedure has been carried out. Indeed the general nonlinear problem in the absence of system noise is discussed in [5].

In the sequel we shall consider the case where the system noise has the same dimension as the state vector x(k)(t=s). In this case the constraints (2.3) are an empty set, thus avoiding the problem of having to solve them in the form (2.4)), but maximising the dimension of the problem to (N+1)s.

It turns out that there is a great deal of structure in the sparse matrices which occur in obtaining the MAP estimate which may be exploited to reduce the computational load, and goes some way to alleviating objection b).

Finally it is to be noted that the procedure described in the sequel for obtaining the MAP estimate could very easily be modified to the case t<s, providing the constraints could actually be solved in the form (2.4).

3. CALCULATION OF THE MAXIMUM A PRIORI ESTIMATE VIA A GAUSS NEWTON ITERATIVE SCHEME UTILISING GIVENS ROTATIONS

The minimisation of the expression (2.2) for MAP_N is a problem in nonlinear unconstrained optimisation. Since the objective function is smooth, any descent method could be used as an iterative scheme for performing the minimisation. (See [6] for a comprehensive discussion.) That is, the value X^{N*} which minimises MAP_N may be computed via an iteration scheme

$$x^N_{i+1} = x^N_i + \alpha p$$

where α is a step length chosen to ensure convergence and p is a suitable descent direction. Since (2.2) is in the form of a sum of nonlinear least squares, then the use of the Gauss-Newton descent direction suggests itself. We may write

$$MAP_N = \frac{1}{2}f^Tf$$

for some vector $f(x^N)$. Defining $J(x^N)$ as the Jacobian of f, i.e.

$$J = \nabla f$$

then the Gauss-Newton step is obtained by solving

$$J^TJ_p = -J^Tf \tag{3.1a}$$

for p. This is equivalent to solving the linear least squares problem

$$\min_p \| J_p + f \|_2^2 \tag{3.1b}$$

From (2.2) it follows that the vector f is given by

$$f^T = [\,(x(N)-\varphi(x(N-1),N-1))^T C^T(N) \ldots (x(1)-\varphi(x(0),0))^T C^T(1)$$

$$(x(0)-\hat{x}(0|0))^T D^T (z(N)-h(x(N)))^T \sigma^T(N) \ldots (z(1)-h(x(1)))^T \sigma^T(1)]. \tag{3.2}$$

where the matrices D, $\sigma(k)$ and C(k) are the Cholesky factors of $P^{-1}(0)$, $R^{-1}(k)$ and $(GQG^T)^{-1}(k)$ respectively. The vector f has dimension $s(N+1)+pN$.

The Jacobian J of f is given by

$$J = \frac{\partial f}{\partial x^N} = \begin{bmatrix} C(N) & F(N-1) & O & . & . & O & O \\ O & C(N-1) & F(N-2) & . & . & O & O \\ O & O & C(N-2) & . & . & O & O \\ . & . & . & . & . & . & . \\ . & . & . & . & . & . & . \\ O & O & O & . & . & F(1) & O \\ O & O & O & . & . & C(1) & F(O) \\ O & O & O & . & . & O & D \\ K(N) & O & O & . & . & O & O \\ O & K(N-1) & O & . & . & O & O \\ O & O & K(N-2) & . & . & O & O \\ . & . & . & . & . & . & . \\ . & . & . & . & . & . & . \\ O & O & O & . & . & K(1) & O \end{bmatrix} \tag{3.3}$$

where

$$F(k) \equiv -C(k+1) . \frac{\partial x(k+1)}{\partial x(k)} \equiv -C(k+1)\Phi(k+1,k) \tag{3.4}$$

and

$$K(k) \equiv -\frac{\partial h}{\partial x(k)} \sigma^T(k) \equiv -H^T(k)\sigma^T(k) \tag{3.5}$$

The minimisation problem (3.1) for obtaining the Gauss-Newton step is best solved via the complete orthogonal (Q-R) factorisation of J, in which J is upper triangularised by the application of a sequence of orthogonal matrices [7]. The standard method of performing this factorisation is via the well-known Householder transformations. There is however another method using Givens (plane) rotations. The latter method, although simple to use, is less popular than the former since it requires approximately twice as many floating point operations when applied to dense matrices. When applied to sparse matrices however, Givens rotations can be considerably faster than Householder transformations.

The matrix J is certainly sparse, and since D and C(k) are upper triangular, it follows that J is itself already upper triangular apart from the entries K(k). The number of Givens rotations required can easily be calculated by noting that one rotation is required for each element that is to be zeroed, and that in this example, each rotation matrix in general changes up to 2s-1 elements immediately to the right of, and in the same row as, the zeroed element. Thus only $N(N+3)ps/2$ rotations are required as opposed to the $(N+1)s\{Np+[(N+1)s-1)/2]\}$ that would be needed if J were dense. The use of Givens rotations is thus recommended in this case.

4. APPLICATION TO BEARINGS-ONLY TRACKING PROBLEMS

For simplicity, we confine our discussion to the two-dimensional bearings only tracking problem, which is illustrated in Figure 1. The differential equations governing the dynamics of the problem can be written in the first order form

$$\dot{x} = Fx+u+Gw \tag{4.1}$$

where

$$x^T = [\dot{r}_x, \dot{r}_y, r_x, r_y]$$

is the state vector comprising the velocity and range of the target relative to the observer,

$$u^T = [-a_x, -a_y, 0, 0] = [-\underline{a}^T, \underline{0}]$$

where $\underline{a}$ is the known acceleration of the observer,

$$\underline{w}^T = [w_1, w_2]\ ;\ E\{\underline{w}(t)\} = 0,\ E\{\underline{w}(t)\underline{w}^T(\tau)\} = \underset{\sim}{Q}'(t)\delta(t-\tau) \tag{4.2}$$

is a continuous white noise vector, with noise 'strength' matrix $Q'(t)$,

$$F = \begin{bmatrix} O_2 & O_2 \\ I_2 & O_2 \end{bmatrix}$$

and

$$G^T = [\, I_2 \quad O_2]$$

Equations (4.1) can be integrated over a unit time step to give

$$x(k+1) = \Phi(k+1,k)x(k)+U(k+1)+W(k+1) \tag{4.3}$$

where

$$\Phi(k,j) = \Phi(k-j) = \begin{bmatrix} I_2 & O_2 \\ (k-j)I_2 & I_2 \end{bmatrix} \tag{4.4}$$

is the system transition matrix,

$$U(k+1) = \int_k^{k+1} \Phi(k+1-\tau)u(\tau)d\tau \tag{4.5}$$

is the deterministic driving term and

$$W(k+1) = \int_k^{k+1} \Phi(k+1-\tau)G(\tau)w(\tau)d\tau$$

is the random input vector. It follows from (4.2) that the sequence W(k+1) form a discrete, Gaussian, white noise vector process, with zero mean and covariance matrix

$$Q(k+1) \equiv E\{W(k+1)W^T(k+1)\} = \int_k^{k+1} \Phi(k+1-\tau)G(\tau)Q'(\tau)G^T(\tau)\Phi^T(k+1-\tau)d\tau$$

The above can be evaluated numerically for any noise strength matrix $Q'(\tau)$. We shall assume for simplicity that

$$Q'(t) = q^2 I_2$$

where q is constant, in which case

$$Q(k+1) = q^2 \begin{bmatrix} I_2 & -\tfrac{1}{2}I_2 \\ -\tfrac{1}{2}I_2 & {}^1/_3 I_2 \end{bmatrix}$$

It is easy to check that

$$C = \frac{1}{q} \begin{bmatrix} 2I_2 & -3I_2 \\ 0_2 & \sqrt{3}I_2 \end{bmatrix} \tag{4.6}$$

is the Cholesky factor of Q^{-1}.

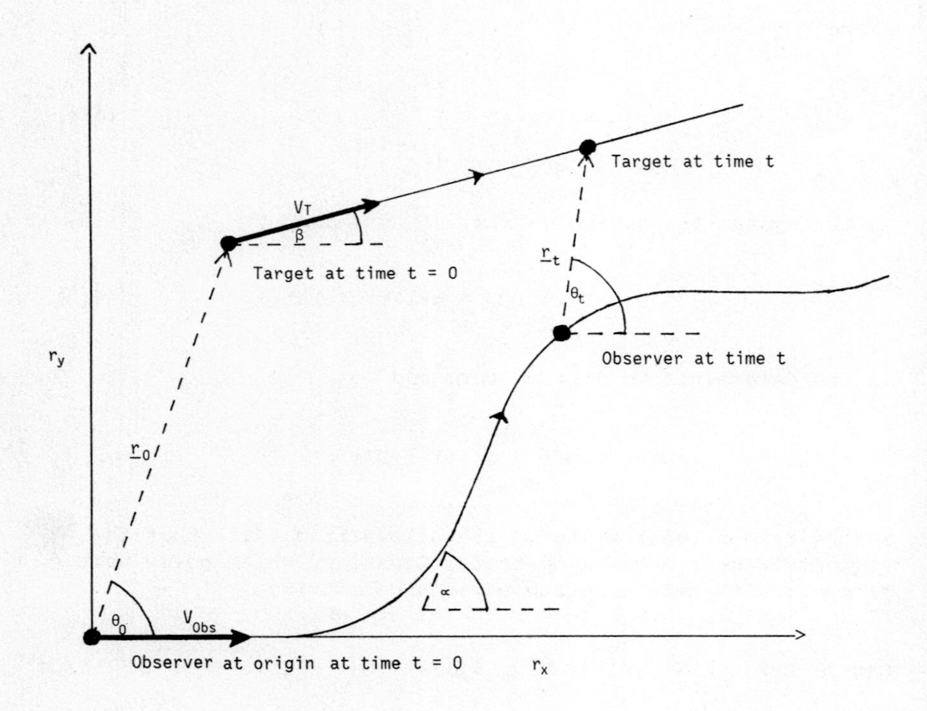

Fig. 1 Two Dimensional Bearings Only Tracking

Note that although the continuous white noise vector w driving the differential equation (4.1) only has dimension two, on integration it generates a discrete random input vector W of the same dimension as the state vector, thus fulfilling the conditions for the algorithm of the previous section to be valid.

The (discrete, scalar) observation process is given by

$$z(k) = \tan^{-1}(r_y(k)/r_x(k))+v(k) \tag{4.7}$$

where we have assumed that the observation rate is constant and defines the unit time step. Since there is only a scalar measurement, it follows that $\sigma(k)$, the Cholesky factor of the inverse measurement noise covariance matrix, is a scalar, which we assume to be constant.

In the Cartesian coordinates we have chosen, the dynamics are linear, but the observation process is nonlinear. Further details and discussion of the system dynamics of two dimensional bearings only tracking can be found in [8], for example.

It follows from (4.7) that the vectors K(k) defined in (3.5) are given by

$$K(k) = \sigma/r^2(k)[0,0,r_y(k),-r_x(k)]. \tag{4.8}$$

where

$$r = (r_x^2+r_y^2)^{\frac{1}{2}}.$$

$P^{-1}(0)$ is usually chosen to be diagonal, hence the calculation of D is trivial.

Collecting together the results in equations (4.3)-(4.8) we may now construct the matrix J and vector f (defined by (3.4) and (3.3) respectively) needed to compute the Gauss Newton step. We see that since the dynamical equations are linear, the only dependence of J upon the state variable sequence x^N occurs in the 2N non zero elements of K(k).

5. SIMULATION RESULTS AND COMPARISONS

The scenario considered is as described in figure 1. In particular note that the target is assumed to be moving with constant but unknown velocity, and the observer follows a path consisting of a sequence of constant speed arcs. The observer must manoeuvre in order to make the process observable, as is well known [8]. The particular parameter values used for this simulation were

initial target range r_0 = 2660 m;
initial target bearing θ_0 = 22.4 deg

target speed V_T = 330 m s^{-1};
target speed angle β = 0

manoeuvre angle α = 30 deg;
manoeuvre turning rate = 15 deg s^{-1}

observer speed V_{Obs} = 990 m s^{-1}.

The above values describe (in the absence of process noise) a collision trajectory, between two high speed vehicles with interception after the observer has executed two manoeuvres at t = 4s. The process noise covariance Q was calculated with q = 100 m $s^{-3/2}$. The observation rate was 20 Hz, with rms measurement noise 0.128 deg.

The initial estimate and covariance were set to

$$\hat{x}_0^T = [-990, 0, 16000\sin z_0, 16000\cos z_0]$$

$$P_0 = [200^2, 200^2, 5000^2, 5000^2]$$

The MAP estimator was compared with the Modified Gain Extended Kalman Filter (MGEKF), which is known to perform better than the EKF [9]. Simulated data were used to accumulate the rms range errors over 250 Monte Carlo runs. These errors are compared with the Cramer-Rao Lower Bound (CRLB) for noise free dynamics [10] and are shown in figure 2. For the scenario considered, the MAP estimator outperforms the MGEKF in the sense that it converges at a much faster rate, although the asymptotic performance of both filters is similar.

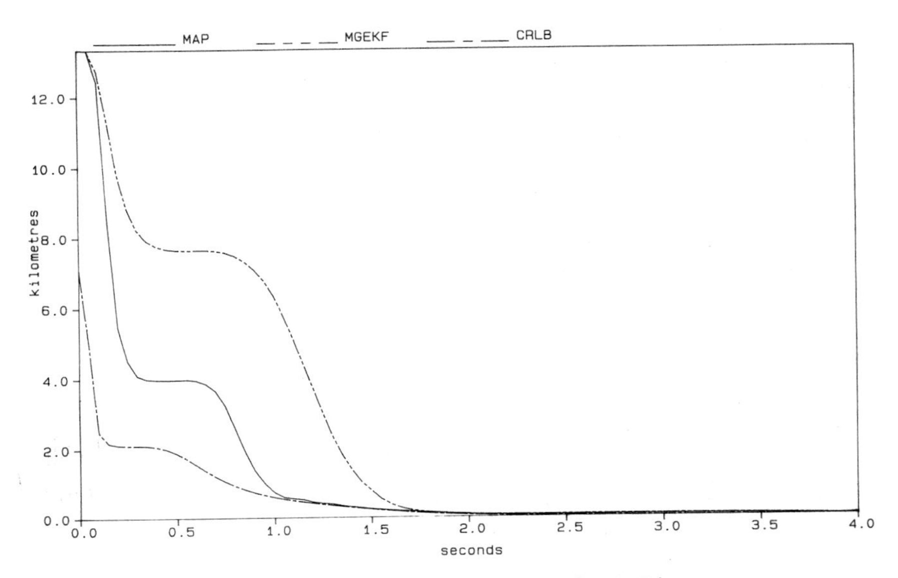

Fig. 2 Rms Range Error Against Time

CONCLUSIONS

We have formulated the MAP estimate as an unconstrained nonlinear optimisation problem and, in the case where the system noise has the same dimension as the state vector, developed an algorithm for its solution by means of a Gauss-Newton iterative scheme. This scheme makes efficient use of Givens rotation techniques to minimise the computational load. We note that the algorithm could quite straightforwardly be modified to deal with the case where the dimension of the system noise is less than that of the state vector. The only proviso is that the constraints (dynamical equations) can be explicitly written in the form (2.4).

The algorithm was applied to a particular scenario in the general two dimensional bearings only tracking problem, and the MAP estimate was seen to compare favourably with that generated by the MGEKF. Other scenarios not reported here have also been investigated. In many cases the MAP estimate is superior to that obtained from the MGEKF, particularly in examples involving slow moving observer/target pairs (i.e. ships) under conditions considered in [8] for example. In general, however, the relative performance of the MAP and the MGEKF is dependent on the particular observer/target motion. This is a subject under current investigation.

The major drawback of the MAP algorithm is the cost. In the absence of system noise the estimator condsidered in this paper reduces to the maximum likelihood (ML) estimator discussed by various authors [3-5]. The time taken (using a single processor) for the MAP algorithm considered here to compute an estimate is (approximately) a factor N greater than that taken by the corresponding ML algorithm in the noise free case, where N is the number of observations. Windowing of the observations would ease this problem, and efficient windowing algorithms are under investigation.

REFERENCES

[1] Cox, H., (1964), "On the estimation of state variables and parameters for noisy dynamic systems", IEEE Trans. Automat. Contr., AC-9, pp. 5-12.

[2] Friedland, B. and Bernstein, I., (1966), "Estimation of the state of a non-linear process in the presence of non-Gaussian noise and disturbances", *J. Franklin Inst.* **281**, pp. 455-480.

[3] Nardone, S.C., Lindgren, A.G. and Gong, K.F., (1984), "Fundamental properties of conventional bearings-only target motion analysis", IEEE Trans. Automat. Contr., AC-29, 9, pp. 775-787.

[4] Chang, C.B., (1980), "Ballistic trajectory estimation with angle-only measurements", IEEE Trans. Automat. Contr., Vol. AC-25, pp. 474-480, June.

[5] Chang, C.B. and Tabaczynski, J.A., (1984), "Application of state estimation to target tracking", IEEE Trans. Automat. Contr., AC-19, 2, pp. 98-109.

[6] Gill, P.E., Murray, W. and Wright, M.H., (1981), "Practical Optimisation", AP.

[7] Golub, G.H. and Van Loan, C.F., (1983), "Matrix Computations", North Oxford Academic.

[8] Aidala, V.J., (1979), "Kalman filter behaviour in bearings-only tracking applications", IEEE Trans. Aerosp. Electron. Syst., Vol. AES-15, pp. 29-39, January.

[9] Song, T.L. and Speyer, J.L., (1985), "A Stochastic Analysis of a Modified Gain Extended Kalman Filter with Applications to Estimation with Bearings Only Measurements", IEEE Trans. Automat. Contr., AC-30, pp. 940-949.

[10] Taylor, J.H., (1979), "The Cramer-Rao estimation error lower bound computation for deterministic nonlinear systems", IEEE Trans. Automat. Contr., Vol. AC-24, pp. 343-344, April.

ADAPTIVE CANCELLATION OF NONLINEAR ECHO IN DATA COMMUNICATION SYSTEMS

J. Chen, J. Vandewalle, D. Vandeputte and M. Vandeurzen,
(Dept. Electric Eng., Katholieke Universiteit Leuven, Belgium)

and

D. Sallaerts
(Alcatel Bell, Microelectronics Dept., Belgium)

ABSTRACT

This paper presents the adaptive cancellations of nonlinear echo in data transmission systems. Two methods for the nonlinear echo cancellation, the Wiener scheme and the Volterra series expansion are investigated by simulations. An adaptive nonlinear echo canceller with Volterra expansion method is designed, which cancels the echo 15 dB better than an adaptive linear echo canceller for the ISDN network.

1. INTRODUCTION

Lots of telecommunication applications nowadays require full duplex data transmission. To realize this, there are several possibilities. First, one can use a four-wire line. This is a rather expensive solution. Another method is using frequency division multiplexing (FDM) on a two-wire line of the existing telephone network. This is cheaper, but data transmission speed is limited.

One can try to combine the advantages of both methods. Fig. 1 gives the scheme of this realization. Hybrids, H, are used to couple the signal from the local sender to the far-end. The problem here is that a near-end echo n_j and a far-end echo n'_j are generated in the hybrids because of the impedance mismatches. The adaptive echo canceller (EC in Fig. 1) is generally considered to be a good solution [3,4,7].

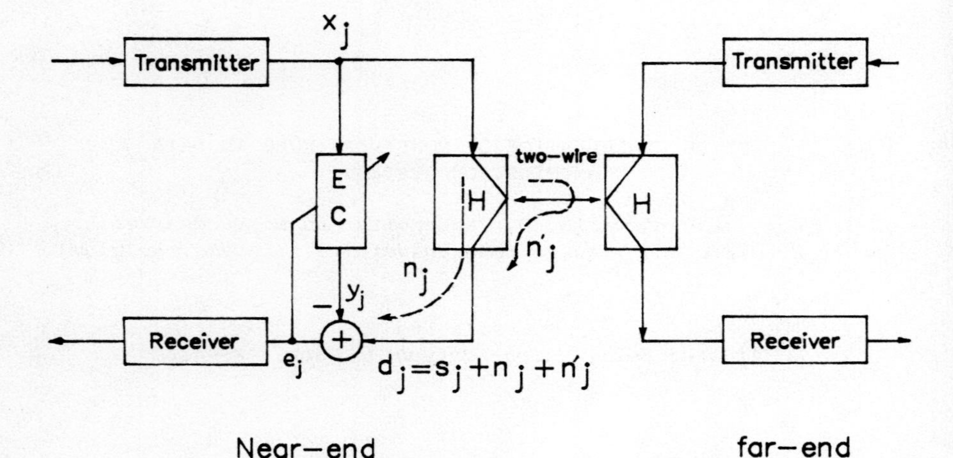

Fig. 1 Block diagram of a data communication system

In Fig. 1 s_j is the useful signal from the far-end. x_j is the signal to be transmitted to the far-end. The adaptive echo canceller imitates the function of the echo path and constructs a replica of the echo. This replica, y_j is then subtracted from the incoming signal d_j. The adaptive echo canceller is usually composed of an adaptive filter of which the filter weights are adaptively adjusted by certain algorithm, usually the LMS (least mean square) algorithm.

In some applications where the level of the echo is lower than the interested signal, a linear echo canceller (i.e , a linear approximation of the function of the echo path) is sufficient. In the two-wire full duplex data transmission system, however the level of the echo signal may be 30 to 40 dB higher than that of the interested signal. Therefore if a signal-to-noise of 20 dB after the echo cancellation is required, the echo canceller should reduce the echo level by 50 to 60 dB. In order to have such high echo reduction, the nonlinear effect of the echo path should be better taken into account.

In a practical data transmission system, following nonlinear distortions can be identified:

- The most important causes of the nonlinearity in the signal are the analog to digital converter (ADC) and the digital to analog converter (DAC).
- Transmitted pulse symmetry.
- Saturation in transformers, which introduces only a slight nonlinearity.

The purpose of our work is to find an adaptive nonlinear echo canceller which will give us about 8 dB more echo cancellation than an adaptive linear echo canceller. The paper is structured as follows: We present in the next section the theory and algorithms for the adaptive nonlinear echo canceller. The LMS algorithm, the Wiener scheme and the Volterra series expansion method are given. The simulation model is illustrated in section 3 and the simulation results are shown in section 4. In section 5 some implementation considerations are presented.

2. THEORY AND ALGORITHMS

2.1 The LMS (Least mean square) algorithm

The adaption algorithm commonly used in an adaptive echo canceller is the well known LMS algorithm. We will briefly state the main points of this algorithm. A detailed discussion can be found in [10].

The purpose of the adaptive FIR (finite impulse response) filter in Fig. 2 is to identify the unknown system, "echo path". For doing so, the filter weights are adapted by the LMS algorithm as follows:

$$H_{j+1} = H_j + 2\mu e_j X_j \tag{1}$$

where, μ is the step-size controlling the convergence and the stability of the algorithm. e_j is the error between the filter output y_j and the desired signal d_j. H_j and X_j are respectively the filter weight vector and the filter input vector at time j, both with the dimension of N (N is the filter order), defined as,

$$H_j^T = [h_{1j}, \ldots, h_{Nj}] \tag{2}$$

$$X_j^T = [x_j, \ldots, x_{j-N+1}] \tag{3}$$

It can be proved [11] that the convergence of the LMS adaptive filter is always guaranteed as long as the following condition holds,

$$0 < \mu < 1/\lambda_{max} \tag{4}$$

where, λ_{max} represents the largest eigenvalue of the autocorrelation matrix R defined as,

$$R = E[X_j X_j^T] \tag{5}$$

In the above equation it is assumed that the input signals are wide-sense stationary. Under this assumption, it can also be shown that after sufficient long adaptations, the mean value of the filter weight vector converges to its optimal values, W* defined by

$$W^* = R^{-1}P \tag{6}$$

where, P is the cross correlation vector defined as:

$$P = E[d_j X_j] \tag{7}$$

Eq(6) is often called Wiener filter.

Assume the eigenvalues of the matrix R are more or less of the same order, the time constant of the LMS adaptive filter can be expressed as,

$$\tau = \frac{N}{4\mu tr(R)} \tag{8}$$

where, tr(R) is the trace of R.

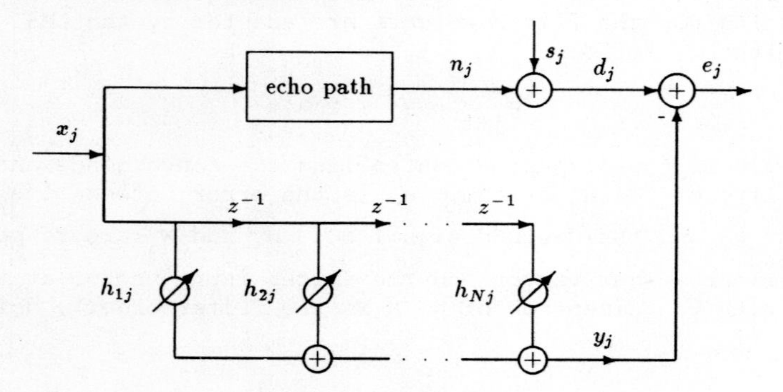

Fig. 2 Echo cancellation with an adaptive FIR filter

2.2 *Nonlinear echo cancellation with Wiener scheme*

An adaptive nonlinear echo canceller with Wiener scheme was proposed in [2] and is shown in Fig. 3. In this method, it is assumed that the echo path is composed of a linear FIR filter with N taps followed by a nonlinear memoryless function d(.). To identify the echo path, an adaptive echo canceller consisting of an adaptive FIR filter and a memoryless function g(.) is constructed.

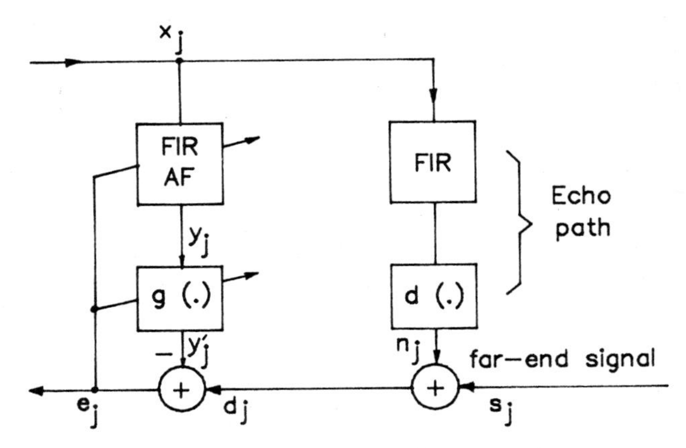

Fig. 3 Wiener scheme for nonlinear echo cancellation

The output of the adaptive filter y_j is equal to

$$y_j = \sum_{k=1}^{N} h_{kj} x_{j-k+1} \tag{9}$$

and the output of the nonlinear processor y_j' is expressed as,

$$y_j' = g(y_j) = \sum_{k=1}^{M} a_{kj} y_j^{k-1} \tag{10}$$

where, a_{kj} (k=1,..,M) are parameters of the nonlinear processor.

The adaptations of the coefficients of the echo canceller can be derived as follows:

$$h_{k,j+1} = h_{kj} + 2\mu_h e_j x_{j-k+1} q_j, \quad k = 1,2,\ldots,N \tag{11}$$

$$a_{k,j+1} = a_{kj} + 2\mu_a e_j y_j^{k-1}, \quad k = 1,2 \ldots,M \tag{12}$$

where,

$$q_j = \sum_{k=1}^{M} a_{kj} (k - 1) y_j^{k-2} \tag{13}$$

is the derivative of the instantaneous estimation of $g(y_j)$, and μ_h and μ_a are step-sizes of the adaptations.

Comparing this method with the conventional LMS algorithm one can observe two differences: First a nonlinear processor is included and its parameters are adaptively adjusted according to Eq. (12); Second, the adaptation of the FIR filter weights is modified. The quantity q_j is added to the modification term of Eq. (11). This cross term makes the behaviour of the echo canceller more complicated and the analysis more difficult. More detailed discussion on that can be found in [2].

2.3 Nonlinear echo cancellation with Volterra series expansion

The Volterra expansion [5] is a popular description of nonlinear circuits and systems with mild nonlinearities. It has been applied for the nonlinear echo cancellation [1,6,8]. The discrete Volterra series for a nonlinear function can be written as

$$y_j = h^{(0)} + \Sigma_{k=0}^{\infty} h_k^{(1)} x_{j-k} + \Sigma_{k_1=0}^{\infty} \Sigma_{k_2=0}^{\infty} h_{k_1,k_2}^{(2)} x_{j-k_1} x_{j-k_2}$$

$$+ \ldots + \Sigma_{k_1=0}^{\infty} \ldots \Sigma_{k_n=0}^{\infty} h_{k_1 k_2 \ldots k_n}^{(n)} x_{j-k_1} x_{j-k_2} \ldots x_{j-k_n} + \ldots \qquad (14)$$

where $h^{(0)}$ is the dc output component. y_j and x_j represent the jth sample of the output and input signals, respectively, $h_{k_1 k_2 \ldots k_n}^{(n)}$ is the weight for the product of n input samples for the $k_1 \ldots k_n$ combination of indices. The superscripts (n) stand for the order of the Volterra series.

For the echo cancellation, a truncated version of the Volterra series is sufficient:

$$y_j = h^{(0)} + \sum_{k=1}^{N} h_k^{(1)} x_{j-k+1} + \sum_{k_1=1}^{N} \sum_{k_2=1}^{N} h_{k_1 k_2}^{(2)} x_{j-k_1+1} x_{j-k_2+1} + \ldots \qquad (15)$$

The above equation can be further simplified by considering the symmetric properties of the nonlinear impulse response, i.e., $h_{k_1 \ldots k_n}^{(n)}$ is the same for every permutation of given indices $k_1 \ldots k_n$. Then one attains,

$$y_j = \Sigma_{k=1}^{N} h_k^{(1)} x_{j-k+1} + \Sigma_{k_1=1}^{N} \Sigma_{k_2=k_1}^{N} h_{k_1,k_2}^{(2)} x_{j-k_1+1} x_{j-k_2+1}$$

$$+ \Sigma_{k_1=1}^{N} \Sigma_{k_2=k_1}^{N} \Sigma_{k_3=k_2}^{N} h_{k_1,k_2,k_3}^{(3)} x_{j-k_1+1} x_{j-k_2+1} x_{j-k_3+1} + \ldots \qquad (16)$$

In the above equation the dc component is omitted.

The nonlinear echo canceller can be realized by the Volterra series shown in Eq. (16), where the x_j is the input signal to the adaptive echo canceller and y_j is its output to be subtracted from the echo combined far-end signal d_j.

It should be noted that the first order Volterra term in Eq. (16) is the same as the linear echo canceller. The second order, third order,...represent the nonlinear parts of the echo. The highest order used in the nonlinear echo canceller must be matched with that of the nonlinearities in the echo path. For example, if the impulse response of the echo path has up to third order harmonics, a third order Volterra series expansion must be implemented for the echo canceller.

The length N in Eq(16) stands for how many previous (including current) input samples are taken into account for calculating the output. It can be observed that N for the first order term should be chosen equal to the effective length of the echo path impulse response. The value of N for other terms is associated to the level of the nonlinearity. As N gets larger, the computation increases dramatically. Thus it is reasonable to select different values of N for the first term and the others:

$$y_j = \sum_{k=1}^{N} h_k^{(1)}(j)x_{j-k+1} + \sum_{k_1=1}^{M} \sum_{k_2=k_1}^{M} h_{k_1,k_2}^{(2)}(j)x_{j-k_1+1}x_{j-k_2+1} + \ldots \tag{17}$$

where $h_{k_1 \ldots k_n}^{(n)}$ are replaced by the adjustable weights $h_{k_1 \ldots k_n}^{(n)}(j)$ (j is the time index). The larger the value of M is chosen, the more the nonlinearities can be compensated.

The filter weights are adapted by the LMS algorithm as follows:

$$h_{k_1 \ldots k_n}^{(n)}(j+1) = h_{k_1 \ldots k_n}^{(n)}(j) + 2\mu_n e_j x_{j-k_1+1}x_{j-k_2+1}\cdots x_{j-k_n+1} \tag{18}$$

Since the Volterra series is a general (not special) representation of a nonlinear function, the nonlinear distortions of the echo path can be cancelled, using the above method as long as the order (n) is high enough and the value of M is large enough.

Another important feature of the Volterra expansion method is that for the elimination of nonlinear distortions, only extra nonlinear taps are required to be added into the existing linear echo cancellation. This is very practical for the purpose of improving an existing system.

3. MODEL DESCRIPTION

In simulations, a simplified model of the real data transmission system is used. It is shown in Fig. 4. The input x_j is either ternary (4B3T) or quaternary (2B1Q) signal. The echo path consists of a linear part (modelled by a transmission line), of an DAC and ADC nonlinearity and of a high pass filter (HP). The nonlinear behaviour of both the ADC and DAC converters is modelled by the following equation:

$$f(x) = x + bx^2 + cx^3 \tag{19}$$

where, x is assumed to be the input of the converter. b and c are variables determined by the measured powers of the second and third order harmonics in the real system.

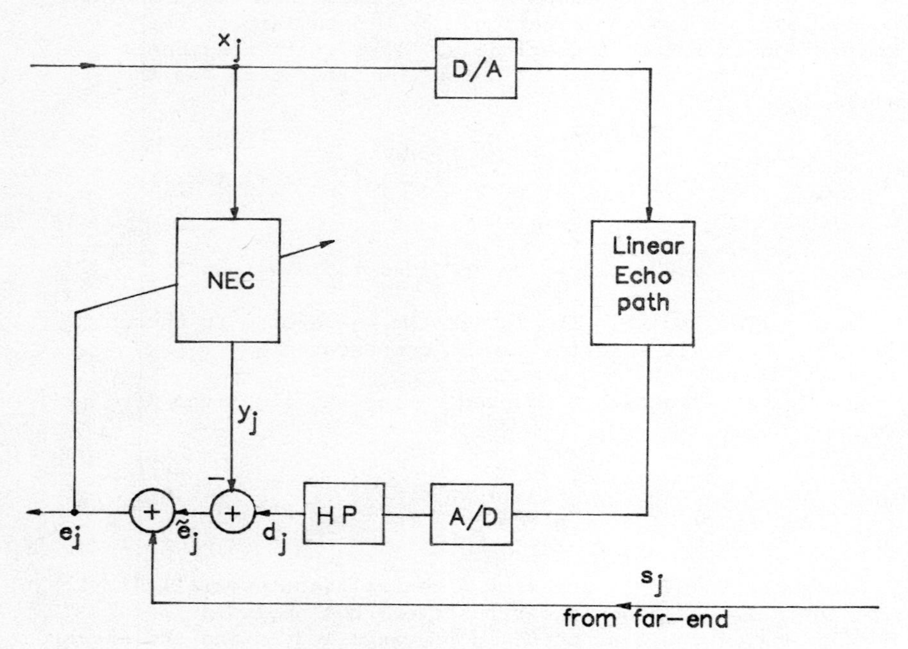

Fig. 4 Simulation model

In order to calculate the echo return loss enhancement (ERLE) consistently for different powers of the far-end signal, the far-end signal s_j is applied after the subtraction of the echo and its replica. This is allowed because the far-end signal only affects the residual error signal used for the adaptation of the echo canceller. In our simulations, this far-end signal can be switched on and off. The echo canceller is based either on the Volterra structure or on the Wiener Scheme.

The ERLE is defined as follows (see Fig. 4).

$$\mathrm{ERLE} = 10\log E\left[\frac{d_j^2}{e_j^2}\right] \tag{20}$$

4. SIMULATION RESULTS

4.1 Simulations with Wiener scheme

The performance of the nonlinear echo canceller based on the Wiener scheme has been thoroughly investigated by simulations [9]. The experiments with exclusion of the high pass filter and the DAC in Fig. 4 resulted in better performance than the Volterra expansion method. The experiments with inclusion of the high pass filter and the DAC, however gave no better performance than a linear echo canceller. This shows that the nonlinear echo canceller based on the Wiener scheme gives good results only when the structure of the echo path is similar to that of the echo canceller itself, i.e., a linear FIR filter and a nonlinear function without memory. The separation between the "memory" (in the linear FIR filter) and the nonlinearity (in the nonlinear function without memory) is the principal cause of this limitation.

4.2 Simulations with Volterra expansion

In order to see the performance of the nonlinear echo canceller with the Volterra series expansion method, we use the following parameters which correspond with the real situation in the ISDN network:

- The order of the Volterra series expansion is 3.
- The attention of the second and third order harmonic distortions of the ADC and DAC are 57 dB and 65 dB, respectively.
- The system is working with quaternary data.
- The attenuation of the far-end signal is 40 dB.

4.2.1 Influence of the length M

As we mentioned before, the computational complexity of the nonlinear echo canceller is determined by the order of the Volterra expansion and the length M for the nonlinear summation, see Eq. (17). Once the order (n) is fixed ((n) = 3), the computational complexity depends only on the value of M. As M gets larger, the computational complexity increases dramatically. However, in principle, better performance can be obtained. Therefore, a good compromise must be made in choosing the value of M.

A series of experiments have been conducted with different values of M and with the exclusion of the far-end signal.

The results are presented in table 1, where the first column represents the length M, second column the total number of taps required by the third order nonlinear echo canceller, third column the adaptation number at which the NEC (Nonlinear echo canceller) converges, fourth column the mean square error (MSE), $E[\tilde{e}_j^2]$ in dB and the last column is the most important one which shows the gain per extra nonlinear tap.

The first row M = 0 means that the linear echo canceller is applied since no quadratic and higher order terms are present. As M gets larger we see that the total number of taps involved increases dramatically, the convergence speed slows down and the performance gets better. But from the last column we can see that increasing M from 0 to 2 produces a gain/tap of 0.756 and from 2 to 3, a gain/tap of 0.528. Starting from M = 3 only little improvement can be obtained by increasing the value of M. This can also be seen from the second column from the right: only 0.8 dB improvement in the echo cancellation is obtained by increasing M from 6 to 7 (with extra taps of 35). The table serves as a good tool for the selection of M.

M	Total taps	Convergence	MSE(dB)	gain (dB)/tap
0	34	5000	-60.76	
2	41	6000	-66.15	0.756
3	50	8500	-70.90	0.528
4	64	10000	-72.50	0.114
5	84	12000	-73.95	0.073
6	111	17500	-75.05	0.041
7	146	22000	-75.85	0.029

Table 1 Influence of the length M on the nonlinear echo canceller

4.2.2 Influence of the far-end signal

Recall that the adaptation of the filter weights is expressed:

$$h_{k,j+1} = h_{kj} + 2\mu e_j x_{j-k+1} \tag{21}$$

Rewrite it as follows (see Fig. 4):

$$h_{k,j+1} = h_{kj} + 2\mu\tilde{e}_j x_{j-k+1} + 2\mu s_j x_{j-k+1} \qquad (22)$$

where, $\tilde{e}_j$ is the residual between the echo and its replica, s_j is the far-end signal.

From Eq. (22) we can see that 1) the far-end signal s_j produces a noise term during the whole adaptive process, 2) the term $2\mu\tilde{e}_j x_{j-k+1}$ is a factor which controls the weight vector to reach its optimal value, but introduces noise in the filter weights in the steady-state after the convergence. To reduce the effects of these two noise terms, one could choose a smaller value of the step-size μ, in doing so, however, the adaptation process would hopelessly be slowed down.

In order to have better performance (lower mean square error) and a faster convergence speed, we propose the following: First, a training sequence is applied. The system works without the far-end signal and with a relatively large value of the step-size (e.g., $\mu = 0.01$). After this training period, the far-end signal is introduced and a smaller value of the step-size (e.g., $\mu = 0.0001$). is applied to reduce the influence of the far-end signal and to keep the echo canceller adaptive to the relatively slow varying characteristics of the echo path. The result is shown in Fig. 5, which presents the ERLE in dB versus the number of the adaptations. Curve (a) results from a linear echo canceller. Curve (b) results from the nonlinear echo canceller, where the first 50000 adaptations represent the training period and after 50000 adaptations, the far-end signal is applied. The improvement (about 15 dB) of the nonlinear echo canceller in the echo reduction can be clearly seen. The influence of the far-end signal can also be observed even with the smaller value of μ. It has been shown that after the training period, with μ equals 0.01 or larger the nonlinear echo canceller gives no better performance than a linear one.

The performance of the nonlinear echo canceller without sending a training sequence was also investigated as follows: the echo canceller started with full-duplex with $\mu = 0.01$ for the first 20000 adaptations and with $\mu = 0.0001$ after 20000 adaptations. The result showed that an improvement of about 8 dB in echo cancellation could be achieved.

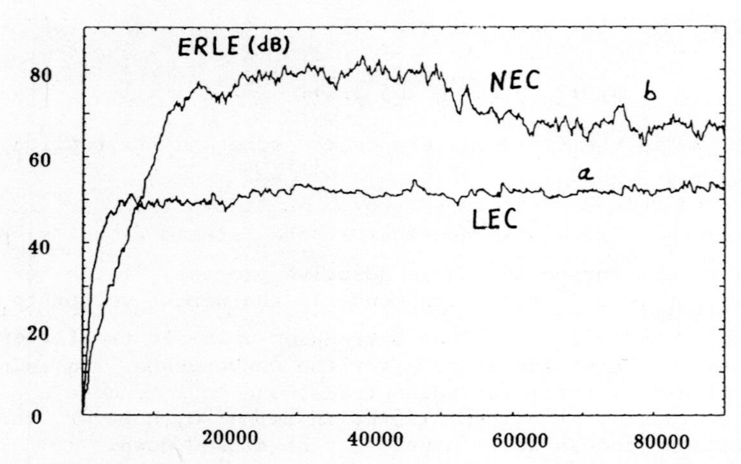

Fig. 5 Performance comparison of an adaptive nonlinear echo canceller with Volterra expansion and an adaptive linear echo canceller.

5. IMPLEMENTATION CONSIDERATIONS

5.1 Required wordlength for the filter weights

The effect of finite wordlength of the filter weights is investigated with different bits under the same conditions as in section 4.

The results have shown that with 15 bits, an improvement of about 12 dB for the full-duplex is obtained, with 13 bits, an improvement of about 10 dB is obtained and with 11 bit, the performance of the NEC becomes unacceptable: nearly no improvement can be achieved. From the results it is observed that the finite wordlength representation of the weights has stronger influence on the nonlinear echo cancellation than on the linear echo cancellation.

5.2 Computational complexity

It has been proved that the performance of the NEC with Volterra expansion method is related to the length M and the order (n) of the Volterra series used to represent the nonlinearities of the echo path. It has also been mentioned that the computational complexity is determined by these two parameters M and (n). Thus, a compromise must be made in the design of an NEC with the Volterra expansion.

The order (n) of the Volterra expansion should be chosen equal or higher than the highest order of the harmonics measured in the real data transmission system. The length M for the nonlinear terms should be large enough to cancel sufficiently the nonlinear distortions. For a fixed value of (n) = 3, a set of simulations have been done with a different number of bits values of M and the results are shown in Fig. 6, where the computational complexity is also illustrated. Figure 6a shows the the total operations required versus the M, where the operations required increase exponentially. The MSEs resulting different values of M are presented in figure 6.b (also seen on table 1). As M gets larger the MSE of the NEC becomes smaller, i.e., the performance becomes better.

Figure 6 and table 1 are very helpful in the design of the NEC. A compromised value of M can be chosen between the acceptable computational complexity and the specified performance of the NEC according to this figure.

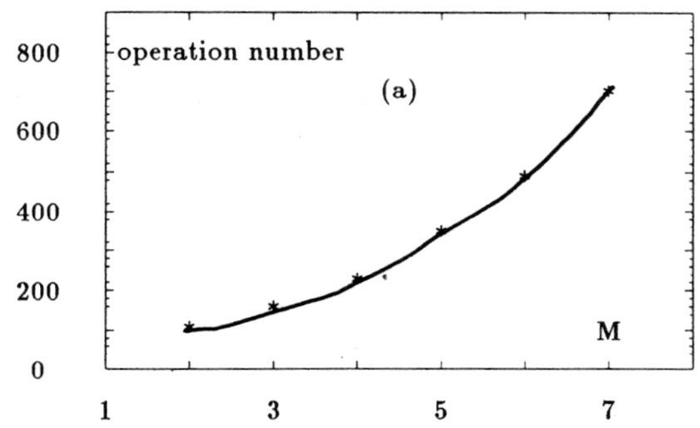

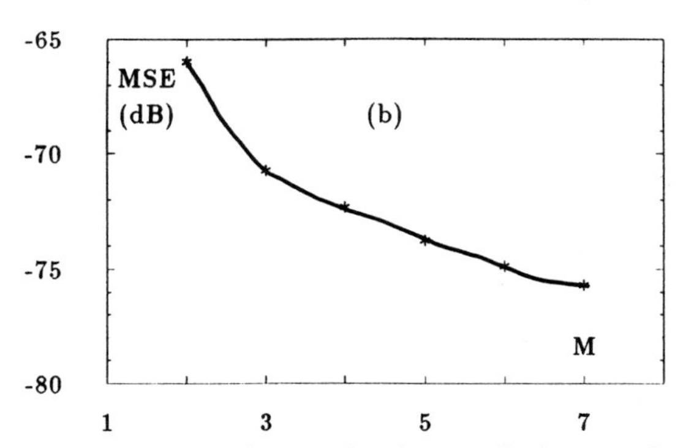

Fig. 6 The computational complexity and the performance of the nonlinear echo canceller. a) Required operations vs. length M; b) Obtained MSE vs. length M.

6. CONCLUSION

In this paper the nonlinear echo cancellers with the Wiener scheme and the Volterra series expansion have been presented. The data transmission system in the ISDN network is modelled. Based on that model the performances of the nonlinear echo canceller with the Wiener scheme and the Volterra series expansion have been investigated. It is shown that the Volterra series expansion method is more suitable for the real system than the Wiener scheme. With suitable values of the step-size μ it is proved that the nonlinear echo canceller with Volterra expansion can achieve 15 db more echo reduction with a training period and 8 dB more echo reduction without the training period than a linear echo canceller. Some design considerations are also given in the paper.

REFERENCES

[1] Agazzi, O., Messerschmitt, D.G. and Hodges, D.A , (1982) Nonlinear echo cancellation of data signals, IEEE Trans. Comm. vol. COM-30, 2421-2433.

[2] Caser-Corredera, J.R., et al., (1985) Data echo nonlinear cancellation, Int. Conf. on Acoustic, Speech and Signal Processing, ICASSP, 1245-1247.

[3] Kelly, Jr., et al., (1970) Self-adaptive echo-canceller, U.S. patent, No. 3,500,000.

[4] Messerschmitt, D.G., (1984) Echo cancellation in speech and data transmission, IEEE J. on Selected Areas in Communications, vol. SAC-2, 283-297.

[5] Schetzen, M., (1980) The Volterra and Wiener Theory of Nonlinear System, John Wiley, New York.

[6] Sicuranza, G.L. and Ramponi, G., (1986) Adaptive nonlinear digital filters using distibuted arithmetic, IEEE Trans. Acoustic Speech and Signal Processing vol. ASSP-34, 518-526.

[7] Sondhi, M.M., (1967) An adaptive echo canceller, bell Syst. Tech. J., vol. 46, 487-511.

[8] Thomas, E.J., (1971) Some considerations on the application of the Volterra representation of nonlinear networks to adaptive echo cancellers, Bell Syst. Tech. J., vol. 50, 2797-2805.

[9] Vandeputte, D., and Vandeurzen, M., (1988) Ontwerp van enn adaptieve niet-lineaire echo-canceler voor toepassing op ISDN-lijnen, Master thesis, Dept. Electrotechnieke, KUL, Belgium.

[10] Widrow, B., et al., (1975) Adaptive noise cancelling: Principles and applications, Proc of IEEE 63, 1692-1716.

[11] Widrow, B., et al., (1976) Stationary and nonstationary learning characteristics of the LMS adaptive filter, Proc. of IEEE, 64, 1151-1162.

V LINEAR ALGEBRA

SINGULAR VALUE DECOMPOSITION: A POWERFUL CONCEPT AND TOOL IN SIGNAL PROCESSING

J. Vandewalle and D. Callaerts[1]
(ESAT-laboratory, Department of Electrical Engineering, Katholieke Universiteit Leuven, Belgium)

ABSTRACT

The increased computational capabilities of modern computers, workstations and VLSI technology on the one hand and the low cost of sensors and data acquisition equipment on the other hand, make it more and more attractive and feasible to use the Singular Value Decomposition (SVD) in signal processing. In this context it is argued that the SVD may become a widespread tool in signal processing, similar to the role played by FFT in digital signal processing in the seventies and eighties.

At the conceptual level the Singular Value Decomposition (SVD) and the Generalized SVD (GSVD) provide a unifying framework and a numerically robust approach for the formulation and computation of new concepts, such as oriented energy and oriented signal-to-signal ratios, angles between spaces and canonical correlation analysis. In our research group these concepts have been very instrumental in order to devise new algorithms for a wide variety of problems in signal processing and identification.

1. INTRODUCTION

Digital signal processing of multichannel, multisensor or 2D images, often includes filtering, signal separation, error correction, compensation, interpolation, decimation, transformation, monitoring, data compression, feature extraction,... Many of these tasks require least squares approximation, low rank approximation, norm and condition number calculation ([12]). Although SVD was already well-known

[1] Supported by the Belgian I.W.O.N.L.

in numerical analysis as a reliable concept for these computations ([11], [13]), it has only recently been rediscovered in signal processing ([2],[3],[5],[10],[20]). In most signal processing applications, the SVD provides a unifying framework, in which one can describe at once the conceptual formulation of the problem, the practical application and an explicit solution that is guaranteed to be numerically robust. In this way, the SVD has become a fundamental vehicle for the formulation and derivation of new concepts such as angles between subspaces, oriented energy, oriented signal-to-signal ratio, canonical correlation analysis,... ([6],[7],[15],[16],[20]) and for the reliable computation of the solutions to problems such as total linear least squares, realization and identification of linear state space models, source separation by subspace methods, etc. It is expected that SVD will become a standard tool on the workbench of a designer of signal processing systems, like FFT was since the mid seventies.

The benefits of using the (generalized) singular value decomposition are most pronounced in those signal processing applications:

- where essentially rank decisions and the computation of the corresponding subspaces determine the complexity and parameters of the model
- where numerical reliability is of crucial importance and the potential loss of numerical accuracy (caused e.g. by the squaring of matrices) cannot be tolerated.
- where a conceptual framework, such as the geometric notion of oriented signal-to-signal ratio, may provide unrevealed additional insight, such as in factor-analysis-like problems.
- where robustness analysis, conditioning and sensitivity optimization are crucial, and are linked together with geometrical insight and interpretation, for which the (G)SVD may provide meaningful quantification (condition numbers, principle angles,...).

Reliable routines for the SVD have been worked out by Golub ([8],[9]) in the sixties and are now available in many libraries like NAG and packages like MATLAB. These avoid the dangerous squaring of matrices. Moderately sized (order of magnitude 50 - 7(SVDs are now feasible on workstations and PCs. Moreover the computational burden may be reduced, since in many digital signal processing applications only part of the SVD information is needed or a prior information on the SVD can be incorporated.

Moreover parallelization and vectorization research on systolic, hypercube, multiprocessor, or dedicated architectures may lead to efficient SVD solvers for arbitrary or structured matrices.

In this paper we first describe in section 2 the role of SVD in the design of digital signal processing systems in order to situate the overall picture to the reader. In section 3 and 4 the (generalized) singular value decomposition and its basic properties are described along with concepts like oriented energy and oriented signal-to-signal ratios. The usefulness of the concepts in the design of digital signal processing algorithms is shown by describing signal separation principles, based on strength or relative strength of signals in section 5. Some practical applications will further motivate the reader. It is concluded that SVD is a useful concept and should be an effective tool on the workbench of a designer of digital signal processing systems.

2. THE SVD AS A USEFUL TOOL IN THE DESIGN OF DIGITAL SIGNAL PROCESSING SYSTEMS

Designers of digital signal processing systems for medical, telecommunication, automation, automotive, consumer or fabrication applications, usually are given a variety of signals measured with a certain precision (say 10%) and are requested to extract in real time useful information from these data. Because of the limited accuracy of the measurements, the data should be handled with care, while on the other hand the speed is rather crucial because of the real time nature of the application. In the design process a number of tools like FFT, are offered from applied mathematics, with which the designer can experiment in order to find a signal processing algorithm which meets the specifications. These tools can be used and evaluated at a high (software) level in the top-down design. Moreover, depending on the needs, these tools can be tuned or adapted to the specific constraints of the application, in order to obtain a cost-effective and reliable signal processing system. FFT has been such a tool in signal processing since the mid seventies and SVD is expected to become an additional one in the near future.

In recent years one can witness an enormous expansion of computational power at all levels: supercomputers, workstations, minicomputers and microcomputers, VLSI, etc. In addition the measurement power is increasing by cheap sensors, transducers and data acquisition equipment. However in practice it is much more difficult to increase the accuracy of the measurements than to increase the volume of the measurements. Hence the need for methods and software which can extract more accurate information from the measured data is more acute even at the expense of Kflops of computations. In this situation we believe that the singular value decomposition (SVD) is a very important tool which allows one to take profit of the expansion of the computational power in order to improve the accuracy. SVD is well known in linear algebra for its solid numerical qualities. However it is an important concept in digital signal processing. On the other hand its widespread use is still hampered by the computational burden. Hence for SVD, like for FFT, special efficient algorithms (software), coprocessors and application specific integrated circuits (ASICs) (hardware) are needed.

3. SVD AND ORIENTED ENERGY OF A VECTOR SEQUENCE

In this section we will introduce the concept of oriented energy of a vector sequence and show that the SVD of the corresponding matrix constitutes the main tool to compute extremal values for this concept.

Definition 1: <u>Oriented energy</u>: ([6],[7],[15])

Consider a sequence of p-vectors $\{a_i\}$, $i = 1,\ldots,q$ and arrange them as the columns of a $p \times q$ matrix A. Then $E_e[A]$ is called the energy of the vectorset in the direction of unit vector $e \in \mathcal{R}^p$ and is defined as:

$$E_e[A] = \sum_{i=1}^{q} (e^t a_i)^2 = \|e^t A\|^2 \qquad (1)$$

with $\|\cdot\|$ the Euclidean norm.

In words, the oriented energy of a vector sequence $\{a_i\}$ in some direction e is a sum of squared projections of the vectors a_i onto direction e. More general, one can define the oriented energy of a vectorset in a subspace $Q \subset \mathcal{R}^p$ as

$$E_Q[A] = \sum_{i=1}^{q} \|P_Q(a_i)\|^2 \qquad (2)$$

with $P_Q(a_i)$ the orthogonal projection of the vectorset onto subspace Q.

Figure 1 shows the polar plot of the oriented energy of a set of vectors in two dimensions. Such a polar plot is found by drawing a vector with length $E_e[A]$ in the direction of a variable unit vector e, and this for all possible directions e in the vectorspace R. The figure then shows in general a p-dimensional surface with p directions of extremal energy: one direction of maximal oriented energy, one of minimal energy and p - 2 directions with a saddle point.

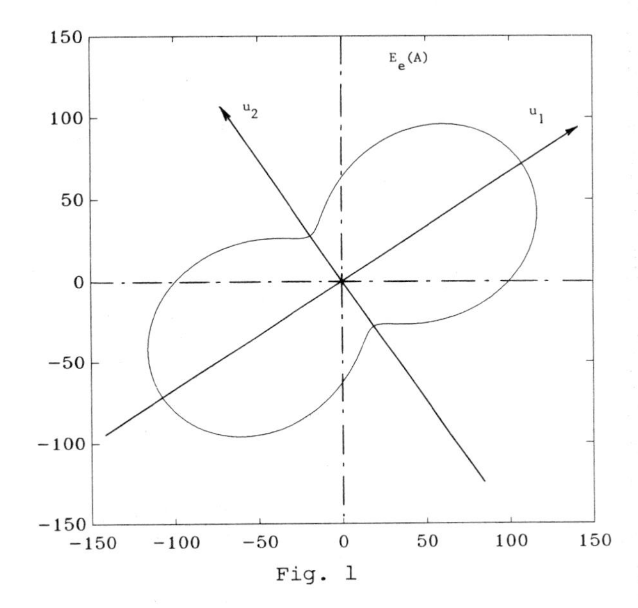

Fig. 1

We will now present the basic SVD-theorem, for our purpose restricted to real matrices. Furthermore we preferred the "economical" factorization, where the dimensions are reduced as much as possible and therefore the pseudo-diagonal matrix is replaced by a square diagonal one.

Theorem 1: Autonne-Eckhart-Young theorem ([9])

For any real p x q matrix A, there exists a real factorization

$$A_{p,q} = U_{p,p} \Sigma_{p,p} V^t_{p,q} \quad \text{(for } p < q\text{)}$$

in which the matrix U is orthogonal ($UU^t = I_p = U^tU$), the matrix V contains p orthonormal columns ($V^tV = I_p$) and $\Sigma_{p,p}$ is a real diagonal matrix with nonnegative diagonal elements, called the singular values σ_i of the matrix A.

Proofs of the existence and uniqueness of the SVD of a matrix are found in [9] and references therein. Some properties, useful in the further outline of this contribution, are mentioned here without proof.

- The number of singular values, different from zero, equals the algebraic rank of the matrix A.

- Any matrix A can be written as the sum of r = rank(A) rank one matrices (dyadic decomposition)

$$A = \sum_{i=1}^{r} u_i \, \sigma_i v_i^t \qquad (3)$$

where $u_i (v_i)$ is the i-th column of U(V) in the SVD of A and σ_i is the i-th singular value.

- Norms of the matrix A can be computed knowing its singular values, since

$$\|A\|_F^2 = \sum_{i=1}^{p} \sum_{j=1}^{q} a_{ij}^2 = \sigma_1^2 + \ldots + \sigma_r^2 \qquad (4)$$

$$\|A\|_2 = \sigma_1$$

- The best rank k approximation of a matrix $A_{p,q}$ with known SVD is given by the partial dyadic decomposition

$$A_k = \sum_{i=1}^{k} u_i \sigma_i v_i^t \quad \text{with } k < r \qquad (6)$$

then $\forall \, B_{p,q}$ with rank (B) = k:

$$\min \|A - B\|_2 = \|A - A_k\|_2 = \sigma_{k+1} \qquad (7)$$

$$\min \|A - B\|_F^2 = \|A - A_k\|_F^2 = \sigma_{k+1}^2 + \ldots + \sigma_r^2 \qquad (8)$$

- Each u_j, column of the matrix U in the SVD of A has the following property

$$\forall \, x \in \mathcal{R}^p, \text{ with } x^t u_i = 0, \qquad (i = 1,\ldots,j-1)$$

$$\frac{\|x^t A\|^2}{\|x\|^2} \leq \|u_j^t A\|^2 = \sigma_j^2 \qquad (9)$$

with σ_j the j-th singular value of A.

From this last property we establish the link between the oriented energy concept and the SVD in the following theorem

Theorem 2: The relation between oriented energy and SVD:

Consider a p x q matrix A with SVD as defined in Theorem 1. Then

$$E_{u_i}[A] = \|u_i^t A\|^2 = \sigma_i^2 \tag{10}$$

$$\forall\, e = \sum_{i=1}^{p} \gamma_i u_i, \qquad E_e[A] = \sum_{i=1}^{p} \gamma_i^2 \sigma_i^2 \tag{11}$$

In words, the SVD of a p x q matrix A looks for directions in the vectorspace $\mathcal{R}^p$, for which the oriented energy of A is extremal. For a more detailed analysis and more properties of the concept of oriented energy, we refer to [6],[7],[15],[17], [18],[20].

4. ORIENTED SIGNAL-TO-SIGNAL RATIO AND GSVD

We can now define the oriented signal-to signal ratio (S-S ratio) of two vector sequences as the ratio of oriented energies, as follows

Definition 2: Oriented signal-to-signal ratio: ([6],[7],[20])

The oriented signal-to signal ratio $R_e[A,B]$ of two sets of p-vectors $\{a_i\}$ and $\{b_j\}$, ($i = 1,\dots q$ and $j = 1 \dots,k$), in the direction of unit vector $e \in \mathcal{R}^p$, is defined as:

$$R_e[A,B] = \frac{E_e[A]}{E_e[B]} = \frac{\|e^t A\|^2}{\|e^t B\|^2} \tag{12}$$

A similar plot as Figure 1 can now be constructed by drawing a vector with length $R_e[A,B]$ in the direction of a variable unit vector e, and this for all possible directions e in the vectorspace $\mathcal{R}^p$. In Figure 2 this is done in two dimensions (p = 2). Figure 2.a shows the oriented energy polar plot of two vectorsets $\{a_i\}$ and $\{b_i\}$, while Figure 2.b is the polar plot for the oriented S-S ratio in all directions. Note that there are again 2 directions (in general p) of extremal S-S ratio, but in contrast with the oriented energy, these directions are not orthonormal in this case. Figure 3 shows the S-S ratio polar plots that arise for different configurations of the oriented energies of matrices A and B.

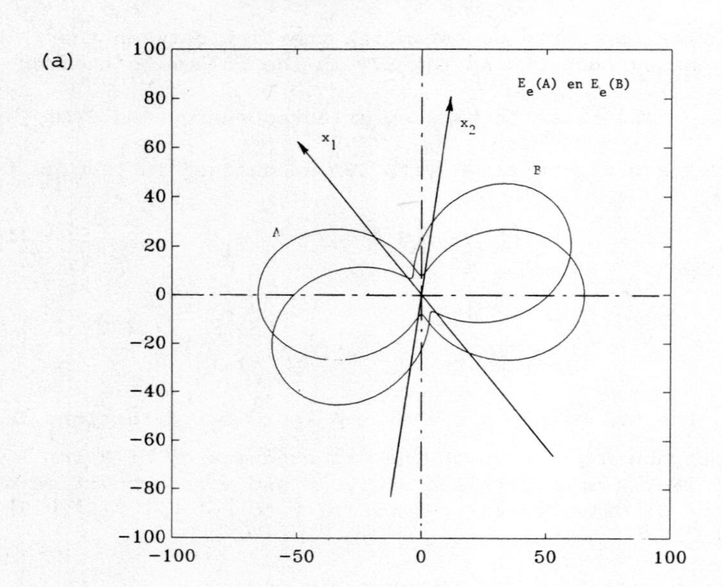
(a)
$E_e(A)$ en $E_e(B)$
x_1
x_2
A
B
100
80
60
40
20
0
−20
−40
−60
−80
−100
−100
−50
0
50
100

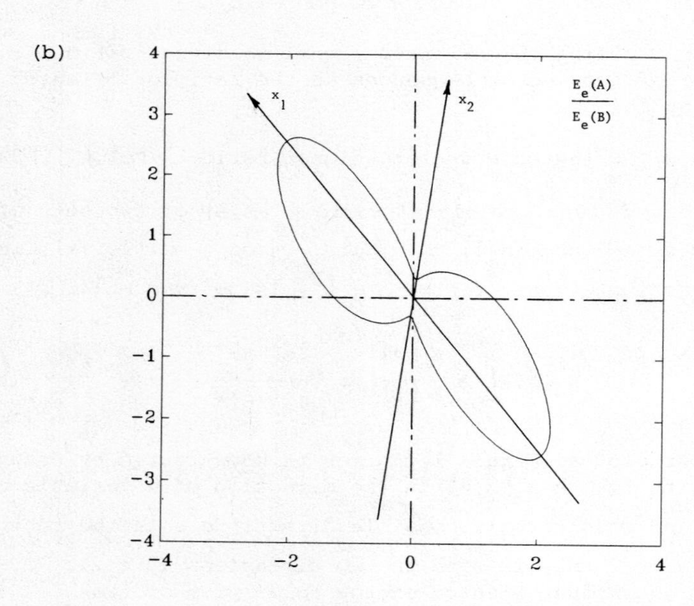
(b)
$\frac{E_e(A)}{E_e(B)}$
x_1
x_2
4
3
2
1
0
−1
−2
−3
−4
−4
−2
0
2
4

Fig. 2

(a)

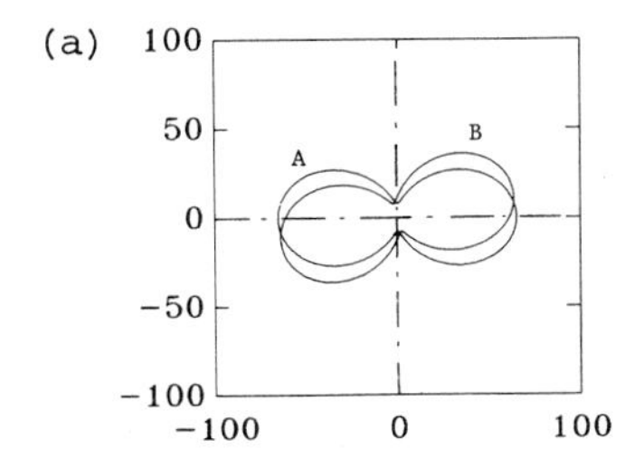
100
50
0
−50
−100
−100
0
100
A
B

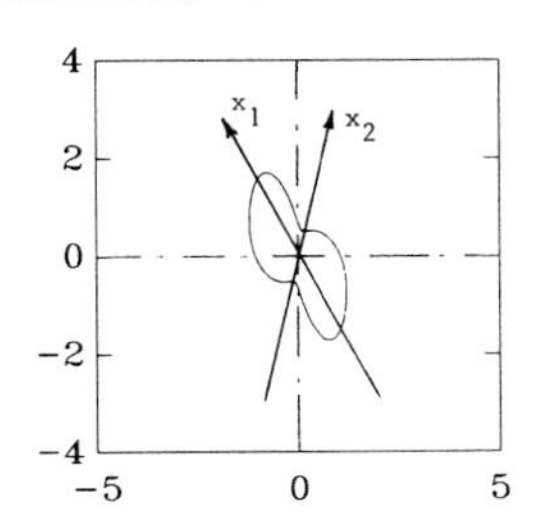
4
2
0
−2
−4
−5
0
5
x1
x2

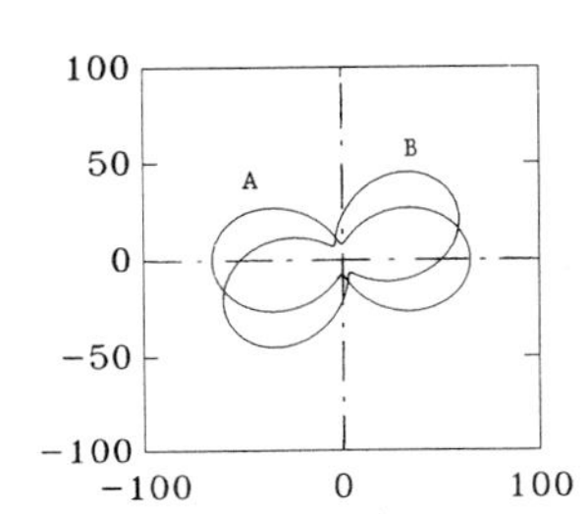
100
50
0
−50
−100
−100
0
100
A
B

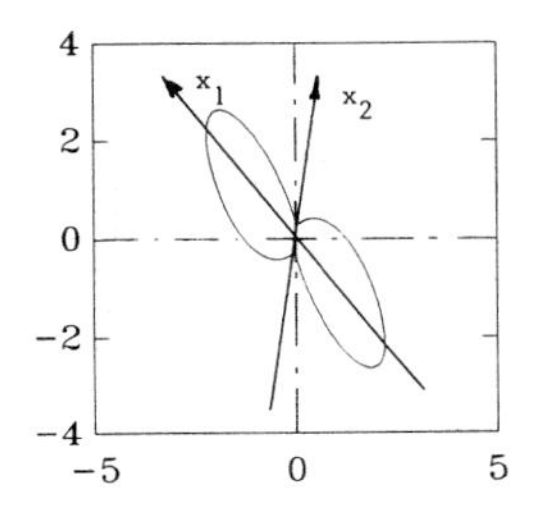
4
2
0
−2
−4
−5
0
5
x1
x2

(b)

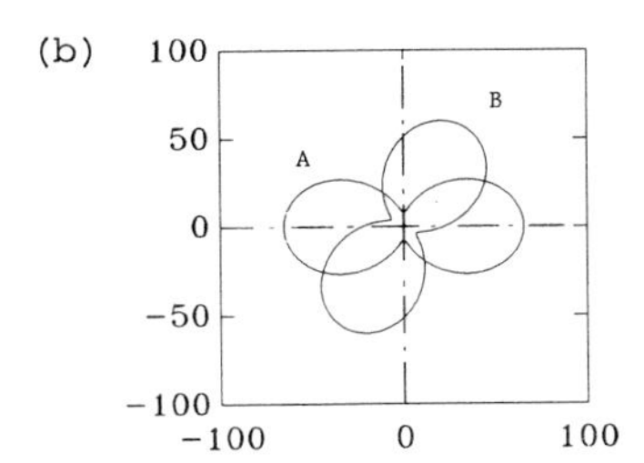
100
50
0
−50
−100
−100
0
100
A
B

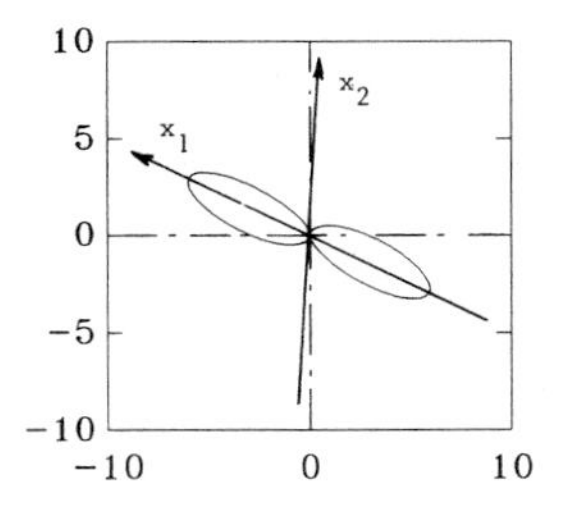
10
5
0
−5
−10
−10
0
10
x1
x2

(c)

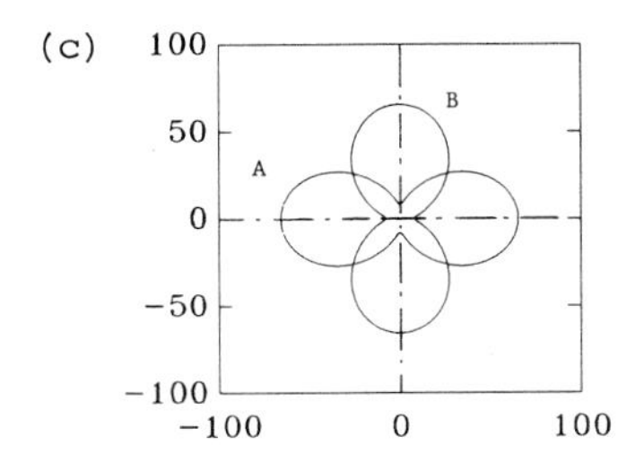
100
50
0
−50
−100
−100
0
100
A
B

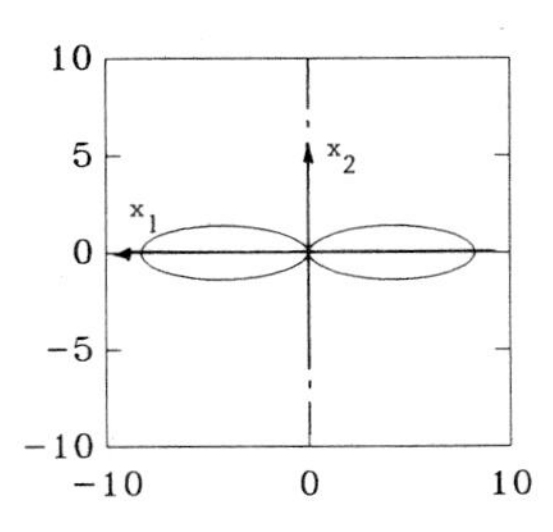
10
5
0
−5
−10
−10
0
10
x1
x2

Fig. 3

A very important property of the S-S ratio is its invariance through a basis transformation in $\mathcal{R}^p$, characterized by the non-singular p x p matrix T ([6],[7]). Indeed, one can verify that

$$R_e[A,B] = R_{e'}[T\,A, T\,B] \quad \text{for } e' = \frac{T^{-t}e}{\|T^{-t}e\|} \tag{13}$$

At this point we present the basic Generalized SVD theorem, again in the most "economical" notations, as follows

Theorem 3 : The Generalized SVD: ([9])

Let A be a p x q and B a p x k matrix ($p < q$ and $p < k$), then there exist matrices U_A (q x p) and U_B (k x p), both with p orthonormal columns, and a non-singular p x p matrix X such that

$$\begin{aligned} A &= X^{-t} D_A U_A^t \\ B &= X^{-t} D_B U_B^t \end{aligned} \tag{14}$$

where $D_A = \mathrm{diag}(\alpha_1,\ldots,\alpha_p)$ and $D_B = \mathrm{diag}(\beta_1,\ldots,\beta_p)$, $(\alpha_i, \beta_i \geq 0)$, are square diagonal p x p matrices and

$\frac{\alpha 1}{\beta_1} \geq \frac{\alpha 2}{\beta 2} \geq \ldots \geq \frac{\alpha r}{\beta r}$, $r = \mathrm{rank}(B)$.

Proof see [9]. The elements of the set $\sigma(A,B) = (\frac{\alpha 1}{\beta 1},\ldots,\frac{\alpha r}{\beta r})$ are referred to as the generalized singular values of A and B. Each x_i, column of the matrix X in the GSVD of (A,B) has the following interesting property

$$\forall\, y \in \mathcal{R}^p \text{ with } y \notin \mathrm{span}\,\{x_1, \ldots, x_{i-1}\},$$

$$\frac{\|y^t A\|^2}{\|y^t B\|^2} \leq \frac{\|x_i^t A\|^2}{\|x_i^t B\|^2} = \frac{\alpha_i^2}{\beta_i^2} \tag{15}$$

In analogy with the relation between the oriented energy concept and the SVD, a link between the oriented signal-to-signal ratio and the Generalized SVD (GSVD) exists:

Theorem 4 : The relation between oriented S-S ratio and GSVD :

Consider a p x q matrix A and a p a k matrix B with GSVD as defined in Theorem 3. Then

$$R_e[A,B] = \left(\frac{\alpha_i}{\beta_i}\right)^2 \quad \text{for } e = \frac{x_i}{\|x_i\|} \tag{16}$$

where x_i is the i-th column of matrix X in the GSVD of (A,B).

Proof: Define the linear transformation $T = D_B^{-1} X^t$, then, due to the invariance property of the S-S ratio

$$\begin{aligned} R_e[A,B] &= R_{e'}[T\,A, T\,B] \\ &= R_{e'}[D_B^{-1} D_A U_A^t, U_B^t] \\ &= E_{e'}[D_B^{-1} D_A U_A^t] \end{aligned} \tag{17}$$

for

$$e' = \frac{T^{-t} e}{\|T^{-t} e\|} = \frac{D_B X^{-1} e}{\|D_B X^{-1} e\|}$$

The matrix $C = D_B^{-1} D_A U_A^t$ can be factorized as follows

$$C = U_c\, \Sigma_c\, V_c^t \quad \text{with } U_c = I_p, \quad \Sigma_c = D_B^{-1} D_A, \quad V_c = U_A$$

such that

$$E_{u_{ci}}[C] = \sigma_{ci}^2 = \left(\frac{\alpha_i}{\beta_i}\right)^2 \tag{18}$$

due to the relation between oriented energy and SVD of C. In other words, $E_{e'}[C]$ is extremal for $e' = u_{ci}$, such that $R_e[A,B]$ is extremal for $e = \frac{t_i}{\|t_i\|} = \frac{x_i}{\|x_i\|}$.

□

The GSVD thus provides the extrema of the oriented S-S ratio and also the directions in which those extrema occur: these are the columns of the matrix X, which need not to be orthonormal.

5. SIGNAL SEPARATION PRINCIPLES

Often, in multichannel signal processing a set of measured, recorded, acquired, captured,... signals is given as a linear combination of some original signals called "source" signals, and corrupted by additive noise. The model of such a measured signal $m_i(t)$ is then

$$m_i(t) = \sum_{j=1}^{r} t_{ij} \; s_j(t) + n_i(t) \qquad \text{for } i = 1,\ldots,p \tag{19}$$

with p the number of measured signals and r the number of source signals involved. The coefficients t_{ij} in the model are called transfer coefficients, since they essentially depend on the signal transfer from the source, that generates the signal, to measurement point, that picks up the signal.

For digital signal processing applications it is required that all the signals are sampled, a process by which a continuous-time signal is transformed into a series of digital numbers. If we consider p signals over a certain time-interval with q samples, the model can be written in matrix notation as follows

$$M_{p,q} = T_{p,r} \; S_{r,q} + N_{p,q} \qquad \text{with} \quad r < p << q \tag{20}$$

where the p rows of the datamatrix M contain the q samples of the p recorded signals. In this equation, only M is known and it is the goal of signal separation to obtain an estimate of the original "source" signals in matrix S.

5.1 SVD of the datamatrix M

Suppose that we partition the r source signals into two groups: r_d desired source signals and r_u undesired source signals ($r = r_u + r_d$). We can then write for matrix $S_{r,q}$

$$S_{r,q} = \overset{q}{\begin{pmatrix} S_u \\ S_d \end{pmatrix}} \begin{matrix} r_u \\ r_d \end{matrix} \tag{21}$$

and partition the r columns of the transfer matrix T similarly

$$T_{p,r} = \begin{pmatrix} \overset{r_u}{T_u} & \overset{r_d}{T_d} \end{pmatrix} \; p \tag{22}$$

These grouped columns of T form subspaces of the p-dimensional column space of $M_{p,q}$. Suppose that the SVD of $M_{p,q}$ can be partitioned as follows ($p_o = p - r_u - r_d$)

$$M = \overset{\quad r_u \quad r_d \quad p_o}{(U_1 \; U_2 \; U_3)} \begin{pmatrix} \Sigma_u & 0 & 0 \\ 0 & \Sigma_d & 0 \\ 0 & 0 & \Sigma_o \end{pmatrix} \begin{pmatrix} V_1^t \\ V_2^t \\ V_3^t \end{pmatrix} \begin{matrix} r_u \\ r_d \\ p_o \end{matrix} \qquad (23)$$

Recent work ([17],[18],[19],[21], [22]) showed that:

- the stronger the undesired signals are present in the recorded signals and
- the more orthogonal the subspaces T_u and T_d of the transfer matrix T are,

the better the subspaces described by U_1 and U_2 match with T_u and T_d respectively. These two conditions can be met by recording the signals $m_i(t)$ as follows ($p_1 + p_2 = p$)

- p_1 signals are recorded as a mixture of desired and undesired signals ($p_1 \geq r_d$)
- p_2 signals are recorded such that they pick up a combination of undesired signals only ($p_2 \geq r_u$).

Example 1: Fetal ECG extraction ([1],[4],[17],[18],[19])

We will illustrate this signal separation principle for the separation of the cutaneously recorded maternal and fetal electrocardiogram. It has been verified (and there exists a physical explanation [14]) that sufficiently far from the adult heart its electrical activity can be described by three independent signals only. In the case of a pregnant woman, the obstetrician is interested in the fetal electrocardiogram (FECG), recorded by only placing electrodes on the maternal skin. The main problem is however the very strong and undesired maternal electrocardiogram (MECG) that is omnipresent in these recordings. Suppose that we apply our signal separation principle to this biomedical application and we assume that the number of undesired signal $r_u = 3$ (MECG), the number of desired signals $r_d = 3$ (FECG) (some experiments show that it can be less than three, but this is a safe upper bound) and we suppose for simplicity that there are no other source signals involved.

We now record $p_1 = 3$ signals on the maternal abdomen, close to the fetal heart, such that they contain a mixture of maternal and fetal ECG (see Figure 4, the last three signals). Furthermore we record $p_2 = 3$ signals that pick up MECG only by placing electrodes on the arms and thorax of the mother, far from the fetal heart (see Figure 4, the first three signals). By constructing the matrix M in this way, the SVD finds 6 orthonormal directions of extremal oriented energy. The three directions, associated with the three largest singular values, correspond to a subspace U_M that is a good estimate of the subspace T_u (the maternal subspace). The two following ones correspond to a subspace U_F, that is a good estimate of the fetal subspace and the last direction corresponds to a subspace of dimension 1 for other much weaker source signals. Projection of the datamatrix M onto these orthonormal directions then results in the signals of Figure 5. These signals are orthogonal estimates for the original source signals in S.

5.2 GSVD of selected data intervals

In some signal processing applications it occurs that, over a specific time interval, the undesired signals contribute considerably more in the measured signals than the desired ones (e.g. for the pulse-like ECG signals the maternal QRS-peaks often appear in between two fetal QRS-peaks (see Figure 6)). In that case we can construct from the data matrix M a matrix $A_{p,k}$ as a sequence of such intervals. This matrix contains in the p-dimensional column space of M mainly contributions from the undesired signals only.

From section 4 we known that the GSVD of the matrix pair (M,A) finds non-orthogonal directions x_i of extremal oriented signal (in M)-to-signal (in A) ratio. Let $\frac{\alpha_i}{\beta_i}$ be the generalized singular values of M and A, arranged in non-increasing order. Then the column x_i of X in the GSVD of (M,A) is a vector for which the oriented energy of matrix M is $\frac{\alpha_i}{\beta_i}$ times larger than the oriented energy of matrix A. The projection of the datamatrix M onto the directions x_i, that correspond with the generalized singular values >> 1, then results in the desired signals.

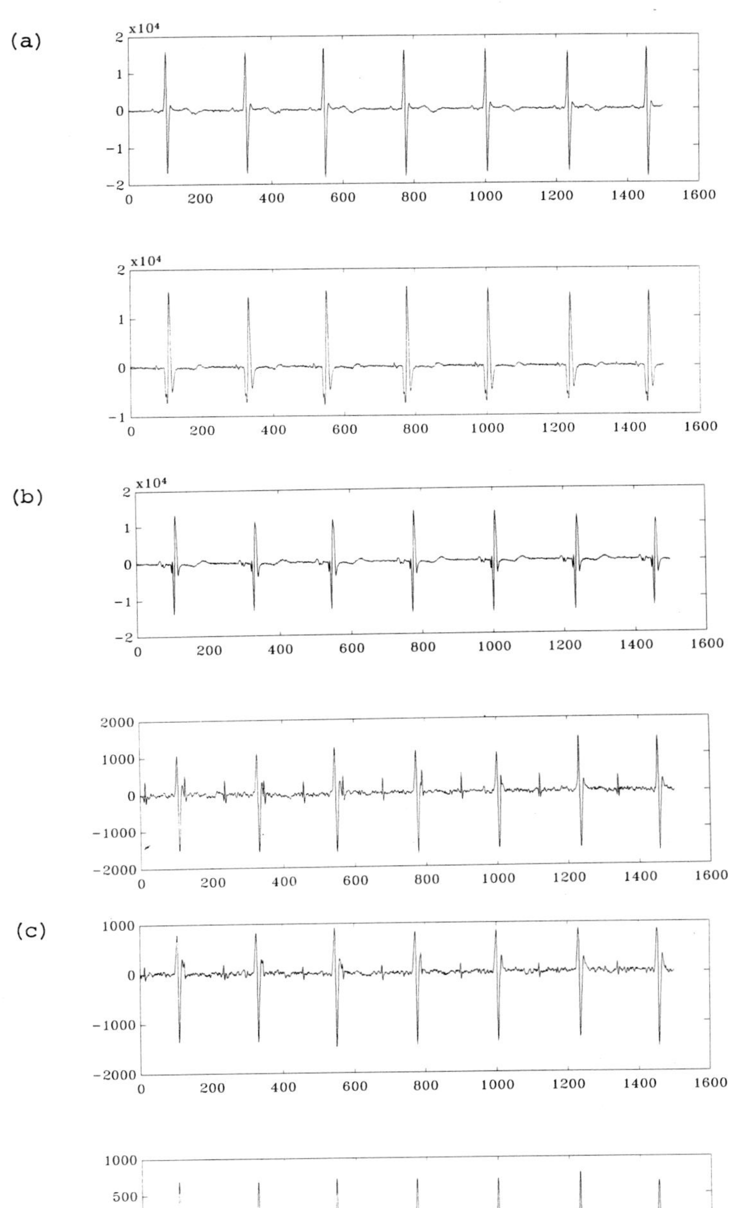
(a)
x10⁴
2
1
0
−1
−2
0 200 400 600 800 1000 1200 1400 1600
x10⁴
2
1
0
−1
0 200 400 600 800 1000 1200 1400 1600
(b)
x10⁴
2
1
0
−1
−2
0 200 400 600 800 1000 1200 1400 1600
2000
1000
0
−1000
−2000
0 200 400 600 800 1000 1200 1400 1600
(c)
1000
0
−1000
−2000
0 200 400 600 800 1000 1200 1400 1600
1000
500
0
−500
−1000
0 200 400 600 800 1000 1200 1400 1600

Fig. 4

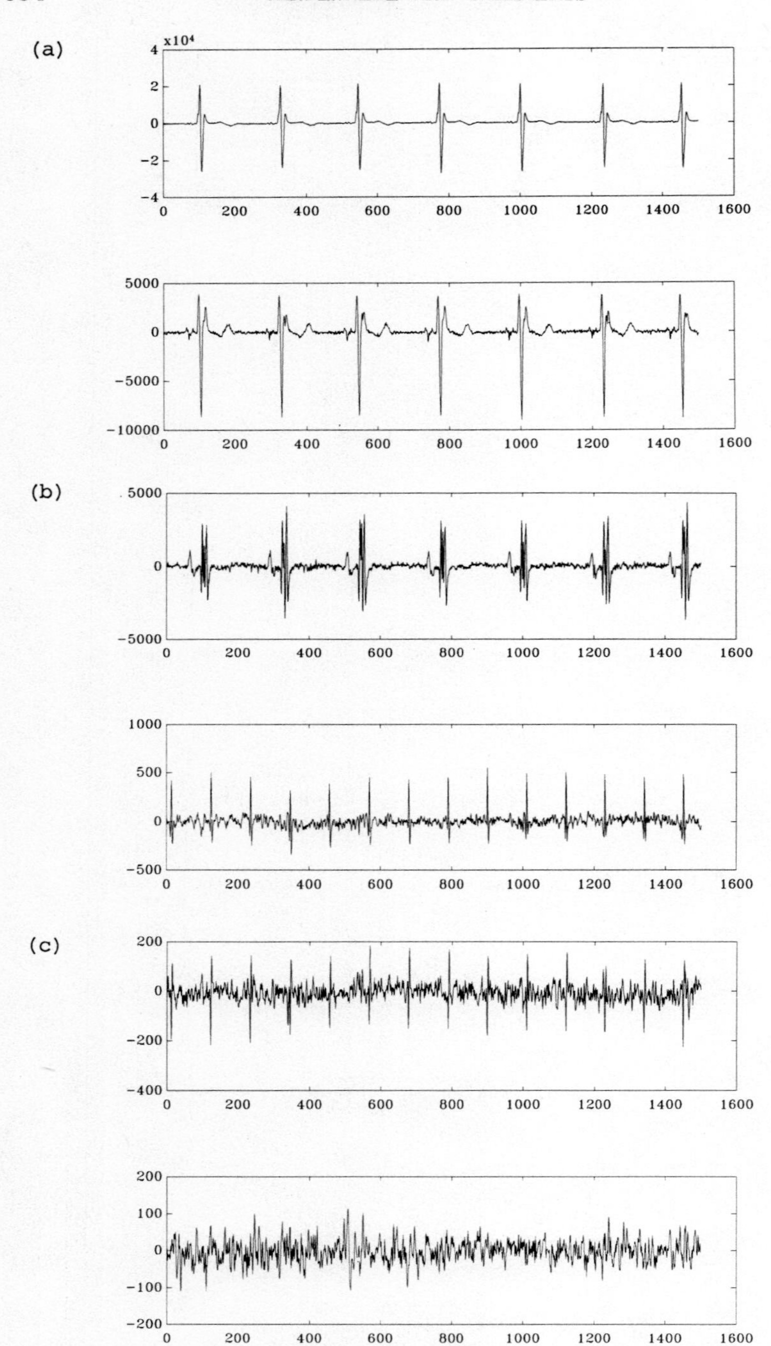
(a)
x10⁴
4
2
0
−2
−4
5000
0
−5000
−10000
(b)
5000
0
−5000
1000
500
0
−500
(c)
200
0
−200
−400
200
100
0
−100
−200
0 200 400 600 800 1000 1200 1400 1600

Fig. 5

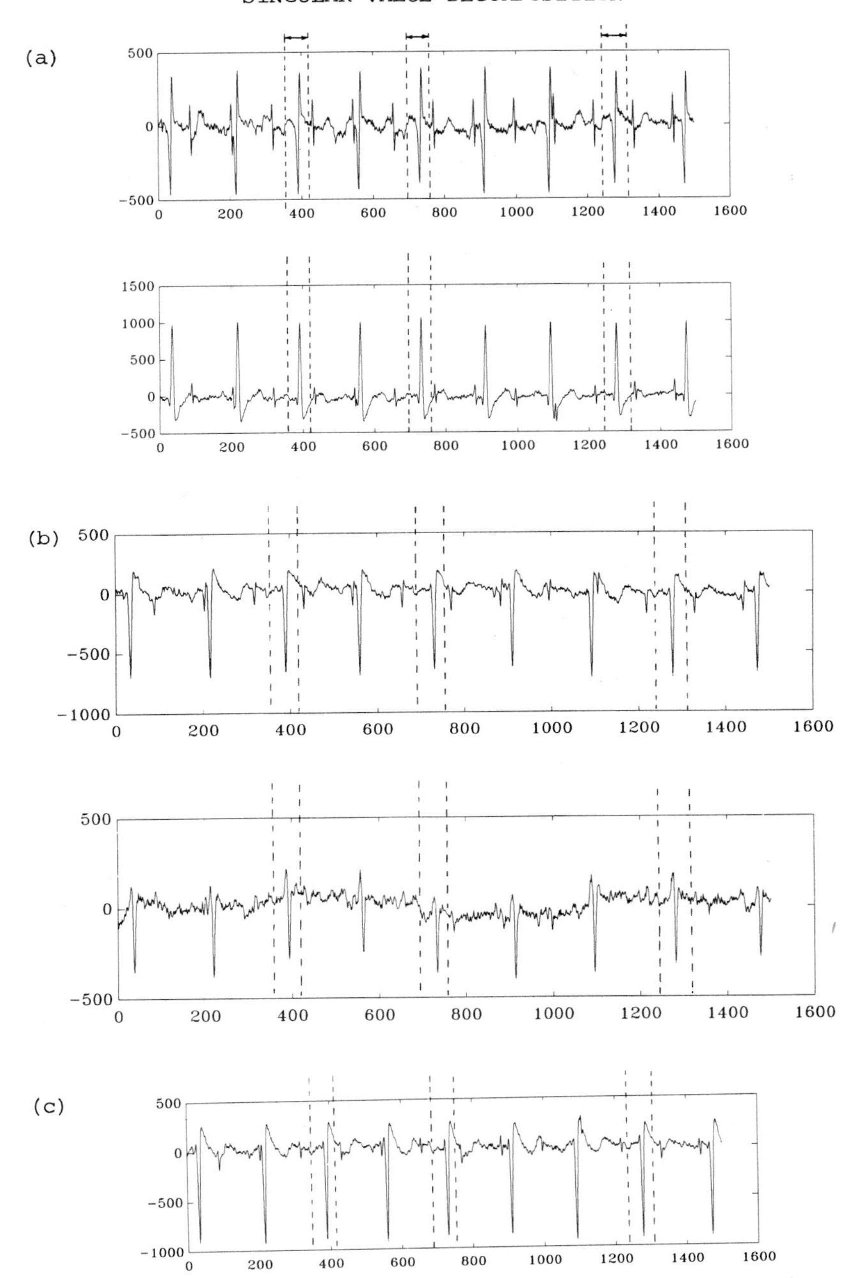
(a)
(b)
(c)
500
0
-500
1500
1000
-1000
0
200
400
600
800
1000
1200
1400
1600

Fig. 6

Example 2: Fetal ECG extraction [4]

Figure 6 shows 5 abdominally recorded potential signals, containing both maternal and fetal ECG. After arranging the signals in a p x q datamatrix M, a matrix M_M is composed as a sequence of several maternal QRS-intervals, not coinciding with fetal complexes (some of those intervals are indicated on Figure 6). The two largest generalized singular values are

$$\frac{\alpha_1}{\beta_1} = 9.986 \quad \text{and} \quad \frac{\alpha_2}{\beta_2} = 5.342$$

and projection onto the corresponding columns x_1 and x_2 of X results in the two MECG-free fetal heart signals of Figure 7.

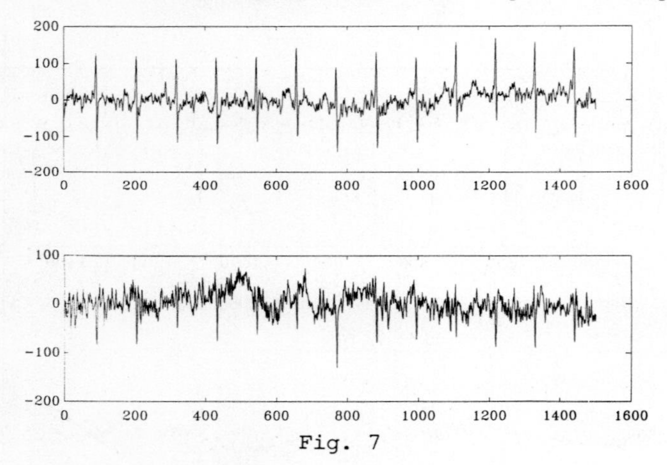

Fig. 7

Example 3: Speech enhancement

At the moment we are doing tests on speech signals, recorded using multiple microphone systems, and corrupted by stationary noise (e.g. a ventilator, a motor running at constant rotation speed, etc.). From the recorded speech signals in datamatrix M, some intervals are selected, where speech is absent and only the noise is present. These intervals are arranged in a special matrix B and again the GSVD is computed of the matrix pair (M,B). Up till now we found an enhancement of the speech-to-noise ratio of about a factor 2.

6. CONCLUSION

In this contribution the use of the (Generalized) Singular Value Decomposition for signal processing and specifically for signal separation purposes is advocated. The SVD and GSVD are not only excellent tools in formulating and describing new geometrical concepts (oriented energy, oriented signal-to-signal ratio), they are also extremely useful in efficiently computing reliable and elegant solutions for many signal problems. Due to the increased computational capabilities of modern computers and the development of dedicated algorithms with incorporation of on-line, adaptive ([2],[3]) and parallel techniques, the computational burden can be reduced substantially. Therefore the SVD should be an effective tool on the workbench of a designer of digital signal processing systems.

REFERENCES

[1] Callaerts, D., Vanderschoot, J., Vandewalle, J., Sansen, W., Vantrappen, G. and Janssens, J., (1986) An adaptive on-line method for the extraction of the complete fetal electrocardiogram from cutaneous multilead recordings, *Journal of Perinatal Medicine,* vol. 14, no. 6, pp. 421-433.

[2] Callaerts, D., Vanderschoot, J., Vandewalle, J. and Sansen, W., (1986) An online adaptive algorithm for signal processing using SVD", in Signal Processing III, Amsterdam, The Netherlands: Elsevier Science Publ. (North Holland), EURASIP 86, The Hague, pp. 953-956.

[3] Callaerts, D., Vandewalle, J., Sansen, W. and Moonen, M., (1987) On-line Algorithm for Signal Separation based on SVD", in SVD and Signal Processing: Algorithms, Applications and Architectures, Deprettere, E., Editor, North-Holland, pp. 269-276.

[3] Callaerts, D., De Moor, B., Vandewalle, J. and Sansen, W., (1989) Comparison of SVD-methods to extract the fetal electrocardiogram from cutaneous electrode recordings, Submitted to Medical and Biological Eng. and Comp.

[5] Damen, A.A. and Van Der Kam, J., (1982) The use of the Singular Value Decomposition in Electrocardiography, Med. Biol. Eng. Comput., vol. 20, pp. 473-482.

[6] De Moor, B., Vandewalle, J. and Staar, J., (1987) Oriented Energy and Oriented Signal-to-Signal Ratio Concepts in the Analysis of Vector Sequences and Time Series, in SVD and Signal Processing: Algorithms, Applications and Architectures, Deprettere, E., Editor, North Holland, pp. 209-232.

[7] De Moor, B., (1988) Mathematical Concepts and Techniques for Modelling of Static and Dynamics Systems, PhD. thesis Dept. of Elect. Eng., Katholieke Universiteit Leuven, Belgium.

[8] Golub, G.H. and Reinsch, C., (1970) Singular Value Decomposition and Least Squares Solutions, *Numer. Math.*, Vol. 14, pp. 403-430.

[9] Golub, G.H. and Van Loan, C.F., (1983) Matrix Computations, North Oxford Academy.

[10] Hansen, P.C. and Nielsen, H.B., (1983) Singular Value Decomposition of images, Proc. of 3-th Scand. Conf. on Image Analysis, pp. 301-307, Copenhagen.

[11] Klema, V.C. and Laub, A.J., (1980) The singular value decomposition: its computation and some applications, IEEE Trans. Automatic Control, Vol. AC-25, no. 2, pp. 164-167.

[12] Lawson, C.L. and Hanson, R.J., (1974) Solving least squares problems, Englewood Cliffs: Prentice Hall Series.

[13] Parlett, B.N., (1980) The Symmetric Eigenvalue Problem, Englewood Cliffs: Prentice Hall Series.

[14] Plonsey, R., (1969) Bioelectric Phenomena, New York: McGraw-Hill.

[15] Staar, J., (1982) Concepts for reliable modelling of linear systems with applications to online identification of multivariable state space descriptions, PhD. thesis, Dept. of Elec. Eng., Katholieke Universiteit Leuven, Belgium.

[16] Staar, J. and Vandewalle, J., (1982) Singular Value Decomposition: A reliable tool in the algorithmic analysis of systems, Journal A, Vol. 23, pp. 69-74.

[17] Vanderschoot, J., Vandewalle, J., Janssens. J., Sansen, W. and Vantrappen, G., (1984) Extraction of weak bioelectrical signals by means of Singular Value Decomposition, in Analysis and Optimization of Systems, Lecture Notes in Control and Information Sciences 63, Bensoussan, A., Lions, J.L., Eds. Springer-Verlag, Berlin, pp. 434-448.

[18] Vanderschoot, J., Vantrappen, G., Janssesn, J., Vandewalle, J. and Sansen, W., (1984) A reliable method for fetal ECG extraction from abdominal recordings, in Medical Informatics Europe 84, Lecture Notes in Medical Informatics 24, Roger, F.H. et al, Eds., Springer-Verlag, Berlin, pp. 249-254.

[19] Vanderschoot, J., Callaerts, D., Sansen, W., Vandewalle, J., Vantrappen, G. and Janssens, J., (1987) Two Methods for Optimal MECG Elimination and FECG Detection from Skin Electrode Signals, IEEE Trans. Biomed, Eng., vol. BME-34, No. 3, pp. 233-243.

[20] Vandewalle, J. and De Moor, B., (1987) A variety of applications of Singular Value Decomposition in Identification and Signal Processing, in SVD and Signal Processing: Algorithms, Applications and Architectures, Deprettere, E., Editor, North-Holland.

[21] Vandewalle, J., Vanderschoot, J. and De Moor, B., (1985) Source separation by adaptive singular value decomposition, Proc. IEEE ISCAS Conf. Kyoto 5-7 June 1985, pp. 1351-1354.

[22] Vandewalle, J., Vanderschoot, J. and De Moor, B., (1985) Filtering of vector signals based on the singular value decomposition, 7th European Conf. on Circuit Theory and Design, Prague, pp. 458-461.

DOWNDATING QR DECOMPOSITIONS

Lars Eldén
(Linköping University, Sweden)

ABSTRACT

We discuss the problem of downdating a QR decomposition, when the orthogonal matrix Q is not available. The stability problems associated with downdating are analyzed, and a new algorithm (due to Björck) with better stability properties than previous algorithms is described. Downdating can be used in the computation of a windowed least squares solution, e.g. in a situation where observations are made in real time, and we want to compute a least squares solution based on the m latest observations. Some numerical experiments with different downdating methods for windowed least squares are reported.

1. INTRODUCTION. Consider the linear least squares problem

$$\min_{w} \| Xw - y \| , \tag{1.1}$$

where X is an $m \times n$ matrix, w and y are vectors of comforming dimensions, and the norm is the Euclidean vector norm $\|x\| = (x^T x)^{1/2}$. We assume that the matrix has full column rank, i.e. the rank is equal to n. A very satisfactory method (from the point of view of numerical stability) for solving (1.1) is to compute a QR decomposition

$$Q^T X = \begin{pmatrix} R \\ O \end{pmatrix}, \tag{1.2}$$

where Q is an orthogonal matrix, and R is an upper triangular $n \times n$ matrix. Using the following identity

$$\| Xw - y \|^2 = \| Rw - y_1 \|^2 + \| y_2 \|^2 , \tag{1.3}$$

the solution of (1.1) is then obtained by solving the

triangular system $Rw = y_1$. In (1.3) y_1 and y_2 are the n first and m - n last components of $Q^T y$, respectively. This method can be implemented in different ways, and it has very good stability properties, see [9].

In some situations it is desirable to be able to update and downdate the decomposition, i.e. to add and delete rows in X without having to recompute the whole decomposition (1.2). (In the sequel we shall often speak about adding and deleting rows in X implicitly assuming that the corresponding elements of y are also added and deleted). This is the case e.g. when the rows are observations of some physical quantities measured in real time, and where we want to have an estimate of the vector w based on the m most recent observations. Recomputing the whole decomposition requires $O(mn^2)$ operations, which would be too costly in most situations. There are many important applications in signal processing, where this situation occurs, an exposition is given in [1]. Another application where downdating is useful is when an observation (e.g. an outlier) is deleted from a regression. Note that often the matrix X has a structure that allows it to be stored economically. In [1] X is a Toeplitz matrix, which can be stored using one vector only.

Updating (1.2) is a simple problem from the point of view of numerical stability, and one such algorithm is implemented in LINPACK as the subroutine SCHUD, see [5]. In the rest of this paper we only discuss the downdating problem.

When the orthogonal matrix is formed explicitly and stored, the downdating problem is easy (see [7],[9p.443]), because in this case essentially we can undo the transformations that were previously used for adding the particular row that is now to be deleted. There are situations, however, when it is inefficent to form and store Q. Therefore, in the rest of this paper we only discuss the case when we want to downdate the R factor in the decomposition, given the matrices R and X. This problem is equivalent to the problem of downdating a Cholesky decomposition of the normal equations $X^T X$ given the same data, see e.g. [10], [1], [4].

There are essentially two approaches for downdating (1.2), one based on orthogonal transformations, see [7], [10], and one based on hyperbolic transformations, see [8], [4], [1] (the latter paper also gives an overview of the history and relevant literature). These methods all require $O(n^2)$ operations. The first is briefly described in Section 2. A new approach due to Björck [3] is also presented. This is based on using the original data for the least squares problem as well as the upper triangular matrix R.

The stability of the downdating problem is analyzed in [10]. In Section 3 of the present paper we give an alternative stability analysis, which to some extent clarifies in what situations one can expect stability difficulties. It is shown that the problem of removing outliers is in general ill-conditioned.

Some numerical experiments with downdating and windowed least squares are reported in Section 4. We illustrate the stability of the Linpack algorithm, Björck's algorithm, and also the algorithm based on hyperbolic transformations.

2. TWO DOWNDATING ALGORITHMS

In this section we shall describe the downdating problem in some more detail, and then we briefly describe two downdating algorithms, namely the Linpack algorithm and the new algorithm due to Björck [3], which can be considered as a stabilization of the Linpack algorithm. We shall refer to these algorithms in the discussion of the stability of the downdating problem in the next section.

We first observe that the right hand side in (1.1) can be treated as an extra column in the matrix X. Therefore we only discuss the downdating of the QR decomposition of X.

Let X be the data matrix and assume that

$$X = \begin{pmatrix} x^T \\ \tilde{X} \end{pmatrix}, \tag{2.1}$$

where x^T is the observation that is to be deleted. Further assume that we have computed a QR decomposition

$$Q^T X = Q^T \begin{pmatrix} x^T \\ \tilde{X} \end{pmatrix} = \begin{pmatrix} R \\ 0 \end{pmatrix}, \tag{2.2}$$

where R is an upper triangular $n \times n$ matrix. Then we want to find an upper triangular matrix $\tilde{R}$, such that

$$\tilde{Q}^T \tilde{X} = \begin{pmatrix} \tilde{R} \\ 0 \end{pmatrix}, \tag{2.3}$$

for some orthogonal matrix $\tilde{Q}$ (which is not needed explicitly).

The following description of the downdating problem essentially follows [3]. Adjoin a column e_1 to the data matrix:

$$(e_1 \ X) = \begin{pmatrix} 1 & x^T \\ 0 & \tilde{X} \end{pmatrix}.$$

Using (2.2) we now have

$$Q^T(e_1 \ X) = \begin{pmatrix} q_1 & R \\ q_2 & 0 \end{pmatrix},$$

where $q^T = (q_1^T q_2^T)$ is the first row of Q. Now determine a sequence of rotations

$$Y^T = J_1 \ldots J_{m-1},$$

where J_κ is a rotation in the plane $(\kappa,\kappa+1)$, chosen to zero the $(\kappa+1)$:st component in q. Then

$$Y^T \begin{pmatrix} q_1 & R \\ q_2 & 0 \end{pmatrix} = \begin{pmatrix} 1 & v^T \\ 0 & \bar{R} \\ 0 & 0 \end{pmatrix}, \tag{2.4}$$

and

$$\hat{Q}^T \begin{pmatrix} 1 & x^T \\ 0 & \tilde{X} \end{pmatrix} = \begin{pmatrix} 1 & v^T \\ 0 & \bar{R} \\ 0 & 0 \end{pmatrix},$$

where $\hat{Q} = QY$, and $\bar{R}$ is upper triangular. Equating the first column on both sides we find that $\hat{Q}^T e_1 = e_1$, which implies that the first row of $\hat{Q}$ is equal to e_1. Hence $\hat{Q}$ must have the form

$$\hat{Q} = \begin{pmatrix} 1 & 0 \\ 0 & \hat{Q}_{22} \end{pmatrix},$$

and it follows that $v = x$ and that $\bar{R} = \tilde{R}$ is the downdated upper triangular matrix.

Note that in order to construct the rotations in Y essentially we must know the first row of Q, but since our assumption is that we do not compute Q, we shall have to reconstruct that vector. To clarify the preceding sentence we note that the first $m - n - 1$ rotations in (2.4) do not affect R or q_1:

$$J_{n+1} \ldots J_{m-1} \begin{pmatrix} q_1 & R \\ q_2 & 0 \end{pmatrix} = \begin{pmatrix} q_1 & R \\ \tau e_1 & 0 \end{pmatrix}, \quad \tau = \|q_2\|. \tag{2.5}$$

From (2.2) we see that

$$x^T = (q_1^T q_2^T) \begin{pmatrix} R \\ 0 \end{pmatrix} = q_1^T R.$$

Since $1 = \|q_1\|^2 + \|q_2\|^2$ it follows that q_1 and τ can be computed from

$$R^T q_1 = x, \quad \tau = \sqrt{1 - \|q_1\|^2}. \tag{2.6}$$

We can then construct the rotations $J_1, \ldots, J_n$, and apply them to (2.5) to obtain $\tilde{R}$. This is the Linpack algorithm [5]. From (2.6) we now see that it can suffer from severe cancellation when $\|q_1\|$ is close to 1.

Thus, to downdate we must reconstruct q_1 and τ.

Before we discuss Björck's algorithm we remind the reader that if only the R factor in a QR decomposition of a matrix X is available, then a least squares problem

$$\min_{\omega} \| Xw - s\|,$$

can be solved using the seminormal equations

$$R^T Rw = X^T s.$$

This method is analysed in [2], and it is shown that if combined with one step of iterative refinement using single precision throughout, the method is weakly stable in a certain sense.

The new approach for downdating suggested in [3] is based on the observation that

$$\begin{pmatrix} R & q_1 \\ 0 & \tau e_1 \end{pmatrix} \tag{2.7}$$

is the R factor of a QR decomposition of (X, e_1). Therefore consider the least squares problem.

$$\min_{\omega} \| Xw - e_1 \|. \tag{2.8}$$

Taking q_1 from (2.6) as an initial approximation of the last column in (2.7) we can compute a solution of (2.8) using seminormal equations and one step of iterative refinement, and thus obtain a better approximation of q_1. The algorithm is as follows:

Algorithm DSNEIR (Downdating using Seminormal Equations and Iterative Refinement)

(i) $R^T q_1 = x$; $Rw = q_1$;

(ii) $r := e_1 - Xw$; $R^T \delta q_1 = X^T r$; $R\delta w = \delta q_1$;

(iii) $q_1 := q_1 + \delta q_1$; $w := w + \delta w$;

(iv) $\tau := \|e_1 - Xw\|$;

(v)

$$Y_O^T \begin{pmatrix} q_1 & R \\ \tau & O \end{pmatrix} = \begin{pmatrix} 1 & x^T \\ O & \tilde{R} \end{pmatrix}.$$

The first triangular systems in step (i) and step (v) are the same as in the Linpack algorithm. The extra work compared to the Linpack algorithm is approximately $3mn + 1.5n^2$, so if m is not very much larger than n we can consider this as an $O(n^2)$ downdating method. It is much more complicated than both the Linpack algorithm and the algorithm based on hyperbolic rotations, and is not very convenient to implement on a parallel computer (this is in contrast to the hyperbolic rotation algorithm, which can easily be implemented on e.g. a systolic array).

Since it does more work and uses also the original data matrix X, Algorithm DSHEIR can be expected to be more accurate than methods based solely on the matrix R and the vector x, e.g. the Linpack algorithm. This is confirmed by the numerical experiments in Section 4.

3. SENSITIVITY TO ROUNDING ERRORS

We first show that the general downdating problem can be reduced to the special case of downdating the vector $x = \kappa e_n$ from an upper triangular matrix. Here it is convenient to regard the problem as a Cholesky decomposition downdating problem. The object is to find the Cholesky factor of the matrix

$$\tilde{X}^T\tilde{X} = X^TX - xx^T = R^TR - xx^T,$$

for an arbitrary vector x. Determine an orthogonal matrix V such that

$$V^Tx = \kappa e_n, \qquad \kappa = \|x\|.$$

Using this we can write

$$\tilde{X}^T\tilde{X} = R^TR - \kappa^2 Ve_ne_n^TV^T = V((RV)^T(RV) - \kappa^2 e_ne_n^T)V^T.$$

The matrix RV is not upper triangular but can be made so by premultiplication by an orthogonal matrix T:

$$T^TRV = S.$$

We then have

$$\tilde{X}^T\tilde{X} = V((RV)^T TT^T (RV) - \kappa^2 e_n e_n^T)V^T = V(S^T S - \kappa^2 e_n e_n^T)V^T.$$

Now we have reduced the problem to that of downdating an upper triangular matrix S when the row vector to be removed is equal to κe_n.

We now go back to the original notation and let

$$X = \begin{pmatrix} \tilde{X} \\ \kappa e_n^T \end{pmatrix} = \begin{pmatrix} \tilde{X} & y \\ 0 & \kappa \end{pmatrix},$$

Further we assume that we know the upper triangular matrix R in the $\underset{\sim}{Q}R$ decomposition (1.2) of X. We want to compute a matrix $\tilde{R}$ such that

$$\bar{\underset{\sim}{Q}}^T \begin{pmatrix} \tilde{R} \\ \kappa e_n^T \end{pmatrix} = \begin{pmatrix} R \\ 0 \end{pmatrix}, \tag{3.1}$$

for some orthogonal matrix $\bar{Q}$, because if (3.1) holds them

$$X^T X = R^T R = \tilde{R}^T\tilde{R} + \kappa^2 e_n e_n^T = \tilde{X}^T\tilde{X} + \kappa^2 e_n e_n^T,$$

which means that $\tilde{R}$ is the required downdated triangular matrix and, equivalently, the Cholesky factor of $\tilde{X}^T\tilde{X}$. Partition

$$R = \begin{pmatrix} R_1 & p \\ 0 & \rho \end{pmatrix}.$$

From (3.1) we now see that $\tilde{R}$ has the partitioning

$$\tilde{R} = \begin{pmatrix} R_1 & p \\ 0 & \gamma \end{pmatrix}, \quad \gamma^2 + \kappa^2 = \rho^2. \tag{3.2}$$

Therefore, in the special case the downdating problem is the problem of determining γ, given ρ and κ. We immediately see that if we solve the second equation in (3.2) for γ then cancellation can occur if κ is only slightly smaller than ρ:

$$\gamma = \sqrt{\rho^2 - \kappa^2}. \tag{3.3}$$

Thus there are stability problems if the downdating is based on (3.3). The Linpack algorithm [5] applied to this special case amounts to using (3.3).

Here we see clearly what makes downdating difficult. In the

computation of R we perform the transformation

$$\bar{Q}^T \begin{pmatrix} R_1 & p \\ 0 & \gamma \\ 0 & \kappa \end{pmatrix} = \begin{pmatrix} R_1 & p \\ 0 & \rho \\ 0 & 0 \end{pmatrix}, \quad \rho = \sqrt{\kappa^2 + \gamma^2}.$$

If κ is much larger than γ, then most of the information in γ is lost when ρ is computed in floating point arithmetic. If there occurs an outlier (in the matrix or in the right hand side) in a windowed least squares computation, then as long as it is present in the matrix it will continue to corrupt information in the reliable observations, and this information can not be retrieved when the outlier is to be removed if we only use the data $\{R,x\}$ in the downdating algorithm. This indicates that the Linpack algorithm (and the hyperbolic rotation algorithm) should be quite sensitive to outliers. Algorithm DSNEIR, on the other hand, uses also the original data matrix and is therefore potentially more stable to outliers.

Now let us investigate in some more detail the sensitivity to perturbations in the data. Assume that there is a perturbation in x:

$$\hat{x} = \hat{\kappa} e_n, \quad \hat{\kappa} = \kappa(1+\xi).$$

Since the only element in the triangular matrix $\tilde{R}$ (see(3.2)) that is affected by the downdating is the element γ, for our analysis we can use the formula (3.3) to compute the perturbation in $\tilde{R}$. Let $\hat{\gamma}$ be the perturbed matrix element and assume that $\rho^2 - \hat{\kappa}^2 > 0$. Then

$$\hat{\gamma} = \sqrt{\rho^2 - \hat{\kappa}^2} = \gamma \sqrt{1 - \frac{\kappa^2}{\gamma^2}(2\xi + \xi^2)}. \tag{3.4}$$

The relative error in $\hat{\gamma}$ is

$$\left|\frac{\hat{\gamma} - \gamma}{\gamma}\right| = \left|\sqrt{1 - \frac{\kappa^2}{\gamma^2}(2\xi + \xi^2)} - 1\right| \le \frac{\kappa^2}{\gamma^2}|2\xi + \xi^2|;$$

(the inequality follows from $|1 - \sqrt{1-x}| \le |x|$). For a small perturbation ξ we then have (ignoring second order terms)

$$\left|\frac{\hat{\gamma} - \gamma}{\gamma}\right| \le \frac{2\kappa^2}{\gamma^2}|\xi| = \frac{2\kappa^2}{\gamma^2}\left|\frac{\hat{\kappa} - \kappa}{\kappa}\right|, \tag{3.5}$$

and $2\kappa^2/\gamma^2$ can be considered as a condition number for the downdating problem with respect to perturbations in the data vector x.

We now compare (3.5) to the corresponding result in [10]. The perturbed downdated matrix is denoted $\hat{R}$ and it is assumed that $|\kappa\xi| = \varepsilon\sigma_1$, where σ_1 is the largest singular value of R (we assume throughout that singular values are ordered $\sigma_1 \geq \sigma_2 \geq \ldots \geq \sigma_n$). In [10] the following estimate for the singular values of $\hat{R}$ and $\tilde{R}$ is derived.

$$\left| \frac{\hat{\sigma}_i - \tilde{\sigma}_i}{\tilde{\sigma}_i} \right| \leq 2\varepsilon \frac{\sigma_1^2}{\tilde{\sigma}_i^2} k = \frac{2\sigma_1 \kappa}{\tilde{\sigma}_1^2} \left| \frac{\hat{\kappa} - \kappa}{\kappa} \right| . \tag{3.6}$$

It is now tempting to take an upper bound for (3.6) as a condition number for the downdating problem. Here we would take

$$\left| \frac{2\sigma_1 \kappa}{\tilde{\sigma}_n^2} \right| . \tag{3.7}$$

This would lead to the conclusion that if $\tilde{\sigma}_n$ is small then the downdating problem is ill-conditioned. However, this is not correct, as is seen from the following examples.

Example 3.1 Assume that ε is small and let

$$X = \begin{pmatrix} \tilde{\mathbf{X}} \\ e_n \end{pmatrix} = \begin{pmatrix} 1 & 0 & 0 \\ 0 & \varepsilon & 0 \\ 0 & 0 & 1 \\ 0 & 0 & 1 \end{pmatrix} .$$

The singular values of R are $\sqrt{2}$, 1 and ε. The formula (3.7) gives $2\sqrt{2}/\varepsilon^2$, but from (3.5) we have the condition number 2. It is obvious that the downdating problem is well-conditioned, and the Linpack algorithm will give a good result.

Example 3.2 Let

$$X = \begin{pmatrix} \tilde{X}_T \\ \kappa e_n \end{pmatrix} = \begin{pmatrix} 1 & 0 & 0 \\ 0 & 1 & 0 \\ 0 & 0 & 1 \\ 0 & 0 & \kappa \end{pmatrix} ,$$

and assume that κ is large. This is the situation when there is an outlier in the matrix that now is to be deleted. Here (3.5) gives $2\kappa^2$ (since $\sigma_1 \approx \kappa$ also (3.7) is large). The downdating problem is ill-conditioned (with respect to perturbations in x). The Linpack algorithm will perform badly.

The correct interpretation of the estimate (3.6) is that the downdating problem is ill-conditioned (with respect to perturbations in x) if any singular value σ_i in R is reduced by a significant amount in the downdating (this was pointed out in [10]). In Example 3.2 it is the largest singular value of R that is reduced from $O(\kappa)$ to 1.

The above discussion shows that the problem of computing $\tilde{R}$ using the input data $\{R,x\}$ may be unstable no matter if the algorithm has good stability properties or not (in [10] the following remark is made on the Linpack algorithm: "any inaccuracies in the results cannot be attributed to the algorithm; they must instead be due to the ill-conditioning of the problem").

4. NUMERICAL TESTS

In this section we report a few numerical tests with windowed least squares using the Linpack algorithm, Algorithm DSNEIR, and the hyperbolic rotation algorithm (HYP). The latter algorithm is the version in [4] that is modified for better stability properties with respect to rounding errors. We have also tested a hybrid algorithm, where the Linpack algorithm is used as long as the downdating problem is well-conditioned and where DSNEIR is used only when the downdating problem is ill-conditioned.

The computations were made using Pro-Matlab with IEEE double precision floating point arithmetic.

We generated two test matrices: the first is a random 50 x 5 matrix with an outlier equal to 10^4 added in position (18,3). The random numbers were taken from a uniform distribution in (0,1). We used the window size 8 throughout. In Figure 1 we give the results with Linpack (solid line), DSNEIR (dashed), and HYP (dot-dashed). We also show the condition number of the window matrix, and the condition number of the downdating problem (which can be shown to be approximately equal to $2/\tau^2$, where τ is given by (2.6)).

Figure 1 shows the relative error (in the Euclidean norm) in the solution of the problem in the window. It is seen that as long as the downdating problem remains well-conditioned all three methods give good accuracy. But when the outlier is to be deleted the error in Linpack and HYP rises to a much higher level (in fact as much as can be expected considering the downdating condition number) and it remains there even though the following downdating problems and the matrix are well-conditioned.

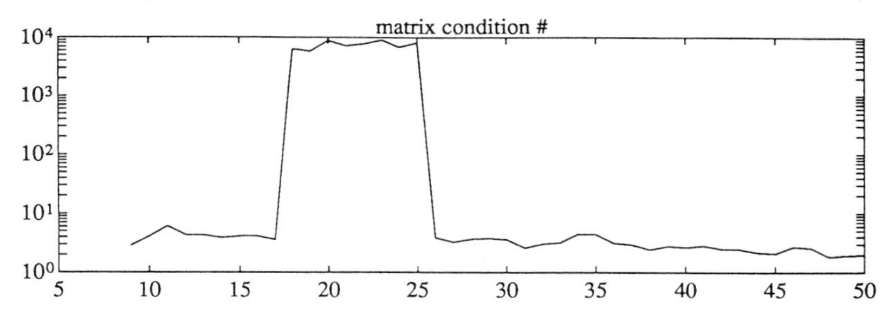

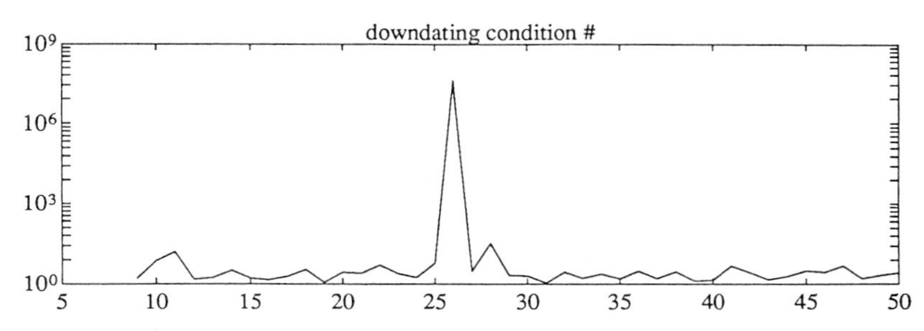

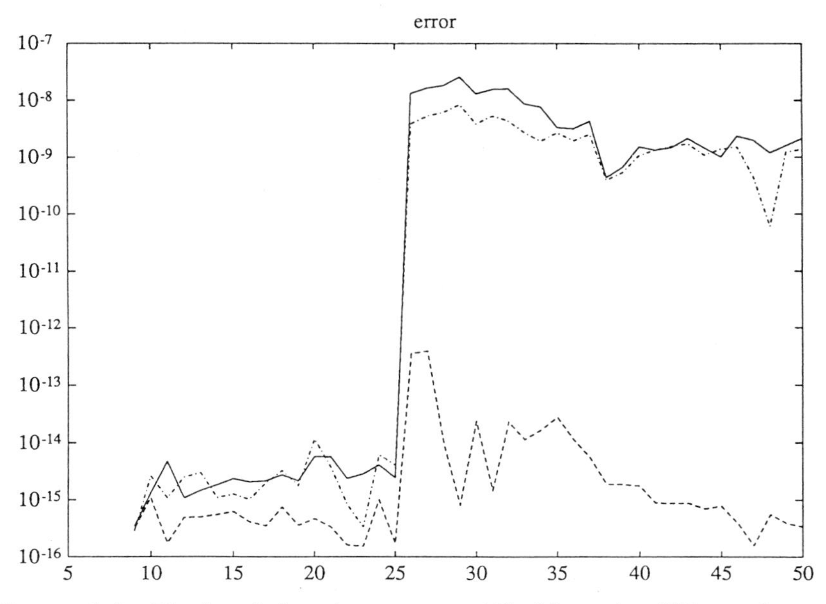

Figure 4.1 Windowed least squares with Linpack, HYP, and DSNEIR. Random matrix with outlier.

In Figure 2 we show the result when the same problem is solved using Linpack, DSNEIR, and a hybrid method (dotted), where Linpack is used except when the downdating problem is ill-conditioned, in which case DSNEIR is used. This indicates that it should be possible to construct a robust and efficient algorithm by combining the two methods.

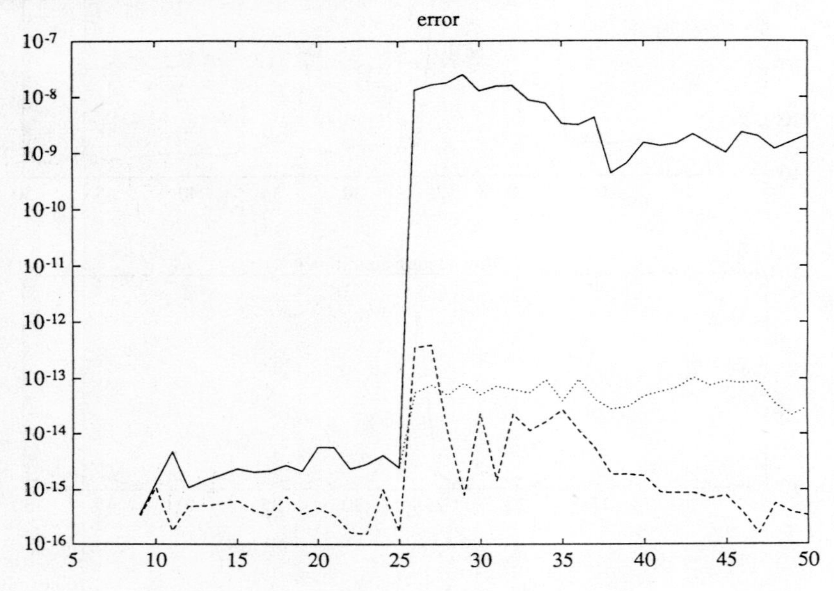

Figure 4.2. Windowed least squares with Linpack, DSNEIR, and hybrid method. Same problem as in Figure 4.1.

The second test problem is a 50 x 5 matrix constructed by taking a 25 x 5 Hilbert matrix as the first 25 rows. Then we turned the Hilbert matrix upside down and put it in the last 25 rows. Finally we added a uniformly distributed perturbation in $(0,10^{-9})$ to the 11 middle rows to prevent the window matrix from becoming too ill-conditioned there. The condition number of the window matrix is shown in Figure 3.

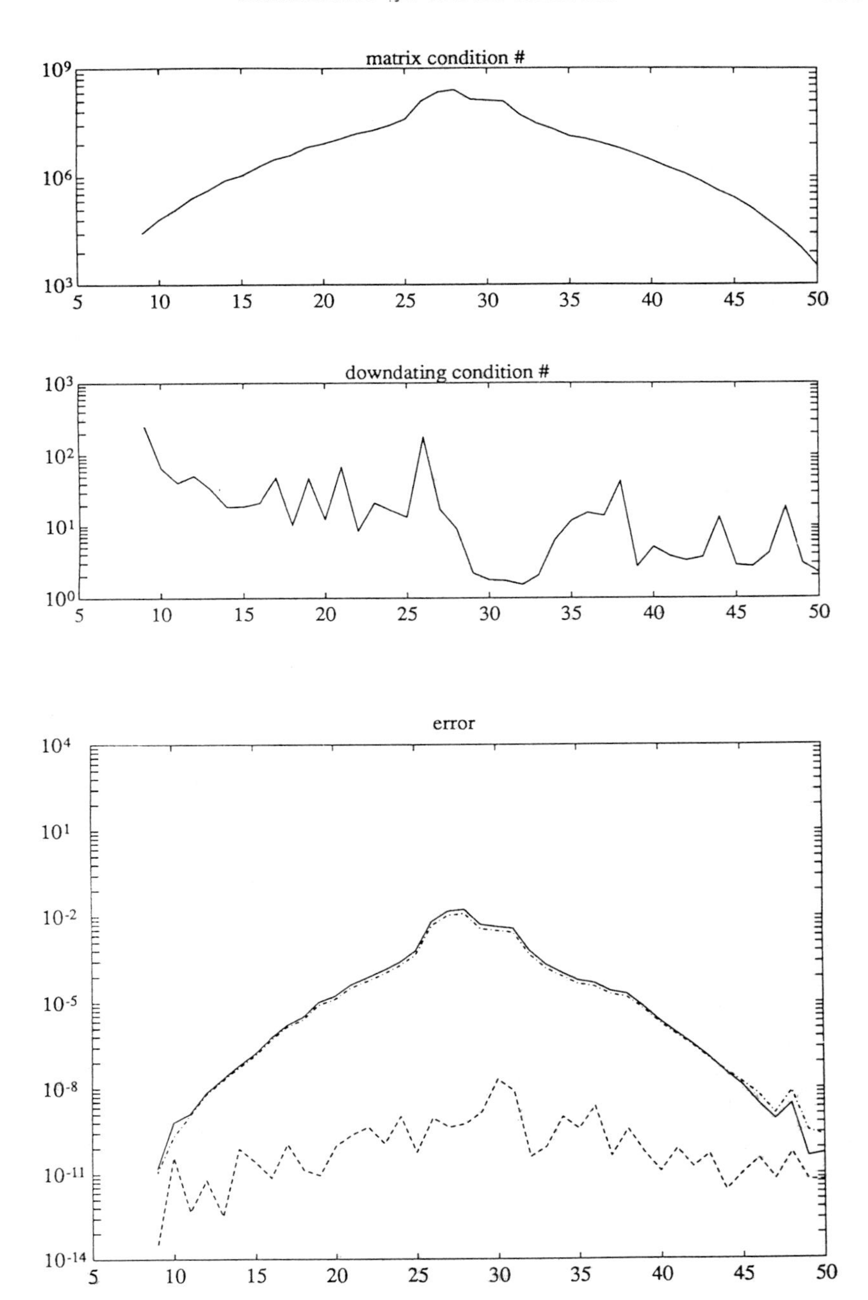

Figure 4.3. Windowed least squares with Linpack, HYP, and DSNEIR. Modified Hilbert matrix.

This problem is solved using Linpack, DSNEIR, and HYP. Since the downdating problem is relatively well-conditioned throughout, the difference between the methods is probably completely due to the fact that the iterative refinement gives very good accuracy in the solution of the linear systems.

REFERENCES

1. Alexander S.T., Pan C.-T., and Plemmons R.J., (1988) Analysis of a recursive least squares hyperbolic rotation algorithm for signal processing, Lin. Alg. Applics., 98, 3-40.

2. Björck Å., (1987) Stability analysis of the method of seminormal equations for linear least squares problems, Lin. Alg. Appl., 88/89, 31-48.

3. Björck Å., (1988) Error analysis of least squares algorithms, NATO Advance Study Institute.

4. Bojanczyk A.W., Brent R.P., van Dooren P. and de Hoog F.R., (1987) A note on downdating the Cholesky factorization, SIAM J. Sci. Stat. Comput., 8, 210-221.

5. Dongarra J.J., Bunch J.R., Moler C.B., and Stewart G.W., (1978) LINPACK User's Guide, SIAM Publications, Philadelphia.

6. Eldén L., and Waldén B., (1989) Downdating QR decompositions with improved stability, Department of Mathematics, Linköping University, Technical report, to appear.

7. Gill P.H., Golub G.H., Murray W., and Saunders M.A., (1974) Methods for modifying matrix factorizations, Mathematics of computation, 28, 505-535.

8. Golub G.H., (1969) Matrix decompositions and statistical computation, in R.C. Milton, J.A. Nelder (eds.), Statistical Computation, Academic Press, New York, 365-397.

9. Golub G.H., and Van Loan C.F., (1983) Matrix computations, North Oxford, Oxford.

10. Stewart G.W., (1979) The effects of rounding errors on an algorithm for downdating a Cholesky factorization, J. Inst. Math. Applics., 23, 203-213.

FAST APPROXIMATION OF DOMINANT HARMONICS BY SOLVING AN ORTHOGONAL EIGENVALUE PROBLEM*

L. Reichel
(Bergen Scientific Centre IBM, Norway)

and

G. Ammar
(Department of Mathematical Sciences, Northern Illinois University, USA)

ABSTRACT

A new computational scheme is described for estimating the dominant harmonics of a stationary time series by the Composite Sinusoidal Modelling or Line Spectrum Pair methods. We compute reflection coefficients γ_j from autocorrelation lags by the Schur algorithm, and associate an orthogonal matrix H' with the γ_j. The arguments of the eigenvalues of H' yield the frequencies of the dominant harmonics, and the first components of the normalized eigenvectors of H' give the amplitudes. We use a divide-and-conquer technique to solve the eigenproblem for H'. Both the Schur and the divide-and-conquer algorithms lend themselves to parallel computation, and therefore are suitable for real-time signal processing, such as speech analysis. The description of our computational scheme suggests a modification of the Composite Sinusoidal Modelling method.

1. INTRODUCTION

Let $\underline{x} = \{x(k)\}_{k=-\infty}^{\infty}$ be a wide-sense stationary real time series, which is ergodic in the autocorrelation. The task of estimating dominant harmonics of such a time series arises in many applications, such as geology, astronomy and speech processing. Recent surveys and comparisons of methods for determining amplitudes and frequencies of dominant harmonics are presented in [3], [13], [14]. In the Composite Sinusoidal Modelling (CSM) method (see [3], [15]), one assumes that $x(k) = y(k)$ for all k, where

---

*Research supported in part by the National Science Foundation under Grant DMS-8704196.

$$y(k) = \sum_{\ell=1}^{n} \rho_\ell \exp(ik\theta_\ell + i\phi_\ell), \quad i := \sqrt{-1},$$

for some (not explicitly known) amplitudes $\rho_\ell > 0$, distinct frequencies $\theta_\ell \in]-\pi,\pi]$, and phases $\phi_\ell \in]-\pi,\pi]$. The first autocorrelation lags of the time series $\underline{y} = \{y(k)\}_{k=-\infty}^{\infty}$ are given by

$$r_{yy}(j) = \sum_{\ell=1}^{n} \rho_\ell^2 \exp(ij\theta_\ell), \quad j = 0, \pm 1, \pm 2, \ldots, \pm(n-1).$$

Let $\tilde{r}_{xx}(j)$, $j = 0, \pm 1, \pm 2, \ldots$, denote computed approximate autocorrelation lags of $\underline{x}$, and assume that the $\tilde{r}_{xx}(j)$ are defined so that the Toeplitz matrix $R = [\tilde{r}_{xx}(j-k)]_{j,k=0}^{n}$ is symmetric and positive definite. The CSM method determines frequencies θ_ℓ and amplitudes ρ_ℓ that satisfy

$$\tilde{r}_{xx}(j) = \sum_{\ell=1}^{n} \rho_\ell^2 \exp(ij\theta_\ell), \quad j = 0, \pm 1, \pm 2 \ldots, \pm(n-1). \tag{1.1}$$

The purpose of the present paper is to address the numerical solution of (1.1). Our approach is to first compute the reflection coefficients γ_j, $1 \le j \le n$, associated with the Toeplitz matrix R by the Schur algorithm; see Section 3. We have $-1 < \gamma_j < 1$ for $1 \le j \le n$, see Lemma 2.1 below, and can associate a Givens reflector $G_j(\gamma_j) \in \mathbb{R}^{n \times n}$ with each γ_j for $1 \le j < n$,

$$G_j(\gamma_j) := \begin{bmatrix} I_{j-1} & & & \\ & -\gamma_j & \sigma_j & \\ & \sigma_j & \gamma_j & \\ & & & I_{n-j-1} \end{bmatrix}, \quad \sigma_j := \sqrt{1-\gamma_j^2}, \tag{1.2a}$$

where I_k denotes the identity matrix of order k. Also introduce the diagonal matrix

$$\tilde{G}_n(\gamma_n) := \begin{bmatrix} I_{n-1} & \\ & -\gamma_n \end{bmatrix}, \tag{1.2b}$$

and the upper Hessenberg matrix

$$H := H(\gamma_1,\gamma_2,\ldots,\gamma_{n-1},\gamma_n) := G_1(\gamma_1)\ldots G_{n-1}(\gamma_{n-1})\tilde{G}_n(\gamma_n). \tag{1.3}$$

If γ_n were of magnitude one, then the matrices $\tilde{G}_n(\gamma_n)$ and H would be orthogonal. Let

$$\gamma_n' \in \{-1,1\}, \tag{1.4}$$

and define the orthogonal matrix

$$H' := H(\gamma_1,\gamma_2,\ldots,\gamma_{n-1},\gamma_n'). \tag{1.5}$$

Let $\lambda_\ell = \exp(i\theta_\ell'), -\pi<\theta_\ell' \leq \pi, 1 \leq \ell \leq n$, be the eigenvalues, and let $\underline{v}_\ell := [\nu_{1\ell},\nu_{2\ell},\ldots,\nu_{n\ell}]^T$ be the corresponding eigenvectors of H'. Assume that the $\underline{v}_\ell$ are normalized so that

$$\underline{v}_\ell^* \underline{v}_\ell = \tilde{r}_{xx}(0),\ \nu_{1\ell}>0,\ 1\leq \ell \leq n,$$

where * denotes transposition and complex conjugation. Then $\theta_\ell := \theta_\ell'$ and $\rho_\ell := \nu_{1\ell}, 1\leq\ell\leq n$, solves (1.1); see Section 2. We therefore can compute a solution of (1.1) by solving an eigenproblem for H'. We solve this eigenproblem by a divide-and-conquer technique (see Section 3), which is suitable for parallel computation. From the arguments of the eigenvalues θ_ℓ', the frequencies Θ_ℓ' of the dominant harmonics are determined by $\Theta_\ell' := \frac{\theta_\ell'}{2\pi}\omega$, where ω is the sampling frequency.

The connection between our computational scheme and frequency estimation by the CSM and Line Spectrum Pair (LSP) methods follow from results in [3]. By setting $\gamma_n' = -1$, we obtain the frequency and amplitude estimates $\{\Theta_\ell',\rho_\ell\}_{\ell=1}^n$ of the CSM method. In the LSP method (see [3], [16], [10]), frequency and amplitude estimates are determined for both $\gamma_n' = -1$ and $\gamma_n' = 1$. Since the matrices $H_{-1} := H(\gamma_1, \gamma_2, \ldots, \gamma_{n-1}, -1)$ and $H_1 := H(\gamma_1, \gamma_2, \ldots, \gamma_{n-1}, 1)$ differ by a matrix of rank one, the divide-and-conquer technique is well suited for computing the estimates of the LSP method. When the frequency and amplitude estimates corresponding to H_{-1} have been determined, it requires only little computational work to determine the estimates corresponding to H_1. Hence, estimates of both the CSM and LSP methods can be computed by solving orthogonal eigenvalue problems. Since both the Schur algorithm and the divide-and-conquer technique are well suited for parallel processing (see Section 3), our approach yields fast numerical schemes that may be attractive for real-time signal processing. Applications of the CSM and LSP methods include speech analysis; see [15], [16], [10]. Illustrative computed examples for the CSM method are presented in [2], where, however, a numerical scheme different from ours is used.

Considering the difference between H and H', defined by (1.3) and (1.5), respectively, suggests a new way to choose γ_n'. In Section 2 we propose the choice

$$\gamma_n' := \begin{cases} \gamma_n/|\gamma_n| & \text{if } \gamma_n \neq 0, \\ -1 & \text{if } \gamma_n = 0. \end{cases} \tag{1.6}$$

Computed examples are presented in Section 4.

2. PROPERTIES OF METHODS FOR APPROXIMATING DOMINANT HARMONICS

Assume that the finite subsequence $\{x(k)\}_{k=0}^{N-1}$ is explicitly known and nonvanishing. Approximations of the autocorrelation lags $r_{xx}(j)$ of $\underline{x}$ can be computed from $\{x(k)\}_{k=0}^{N-1}$ in various ways; see [13]. In the numerical examples of Section 4, we approximate $r_{xx}(j)$ by

$$\tilde{r}_{xx}(j) := \frac{1}{N} \sum_{k=0}^{N-j-1} x(k)x(k+j), \tilde{r}_{xx}(-j) := \tilde{r}_{xx}(j), j = 0,1,2,\ldots . \tag{2.1}$$

Let $R = [r_{j-k}]_{j,k=0}^{n}$ denote the symmetric Toeplitz autocorrelation matrix of order n+1, defined by

$$r_j : = \tilde{r}_{xx}(j), \; j = 0, \pm 1, \pm 2, \ldots, \pm n. \qquad (2.2)$$

By (2.1) - (2.2), it follows that R is positive definite, and we can associate a sequence of reflection coefficients $\gamma_j, 1 \leq j \leq n$, with R. We compute the γ_j by the Schur algorithm, and perturb γ_n. The following lemma relates a perturbation of γ_n to a perturbation of R.

Lemma 2.1. Let $\hat{R}$ be a positive definite symmetric Toeplitz matrix of order n+1. Then its associated reflection coefficients $\hat{\gamma}_j, 1 \leq j \leq n$, satisfy $-1 < \hat{\gamma}_j < 1$. Conversely, an (n+1) - tuple $\{\tilde{r}_0, \tilde{\gamma}_1, \tilde{\gamma}_2, \ldots, \tilde{\gamma}_n\}$, such that $\tilde{r}_0 > 0$ and $1 < \tilde{\gamma}_j < 1$ for $1 \leq j \leq n$, determines a unique positive definite symmetric $(n+1)\mathrm{x}(n+1)$ Toeplitz matrix $\tilde{R}_1$, with the properties that the diagonal element of $\tilde{R}_1$ is $\tilde{r}_0$, and $\tilde{\gamma}_j$, $1 \leq j \leq n$, are the reflection coefficients associated with $\tilde{R}_1$. Moreover, an (n+1) - tuple $\{\tilde{r}_0, \tilde{\gamma}_1, \tilde{\gamma}_2, \ldots, \tilde{\gamma}_n\}$, such that $\tilde{r}_0 > 0, -1 < \tilde{\gamma}_j < 1$ for $1 \leq j < n$, and $\tilde{\gamma}_n = 1$ or $\tilde{\gamma}_n = -1$, determines a unique *singular* positive semi-definite symmetric $(n+1)\mathrm{x}(n+1)$ Toeplitz matrix $\tilde{R}_0$, with the properties that the diagonal element of $\tilde{R}_0$ is $\tilde{r}_0$, and $\tilde{\gamma}_j$, $1 \leq j \leq n$, are the reflection coefficients associated with $\tilde{R}_0$. The matrix $\tilde{R}_0$ has a positive definite leading principal nxn submatrix.

Let the (n+1) - tuple $\{\tilde{r}_0, \tilde{\gamma}_1, \tilde{\gamma}_2, \ldots, \tilde{\gamma}_n\}$ satisfy $\tilde{r}_0 > 0, -1 < \tilde{\gamma}_j < 1$ for $1 \leq j < n$, and $-1 \leq \tilde{\gamma}_n \leq 1$, and let $\tilde{R} = [\tilde{r}_{j-k}]_{j,k=0}^{n}$ be the positive semi-definite symmetric nxn Toeplitz matrix uniquely determined by this (n+1) - tuple in the sense described above. Regard the elements $\tilde{r}_j$ of $\tilde{R}$ as a function of $\tilde{\gamma}_n \in [-1, 1]$. Then the $\tilde{r}_j$ for $0 \leq j < n$ are independent of $\tilde{\gamma}_n$, and

$$\tilde{r}_n = \tilde{r}_n(\tilde{\gamma}_n) = -\tilde{\gamma}_n \tilde{r}_0 \prod_{j=1}^{n-1} (1-\tilde{\gamma}_j^2) + \alpha,$$

for some constant α independent of $\tilde{\gamma}_n$.

Proof. The lemma follows immediately from the recursion formulas (sometimes called Szegö recursions) of the Levinson algorithm; see, e.g., [11], [1] or [6, chap. 5.7]. □

Let $\gamma_n' \in \{-1,1\}$, and let $R' = [r_{j-k}']_{j,k=0}^{n}$ be the singular positive semi-definite symmetric Toeplitz matrix uniquely determined by the (n+1) - tuple $\{r_0, \gamma_1, \gamma_2, \ldots, \gamma_{n-1}, \gamma_n'\}$ in the sense of Lemma 2.1, where r_0 is the diagonal element of R, and $\gamma_j, 1 \le j \le n$, are the reflection coefficients associated with R. We determine the amplitudes ρ_ℓ and frequencies θ_ℓ by requiring

$$\sum_{\ell=1}^{n} \rho_\ell^2 \exp(ik\theta_\ell) = r_k', \quad k = 0, \pm 1, \pm 2, \ldots, \pm n. \qquad (2.3)$$

In view of Lemma 2.1, we have $r_k' = r_k = \tilde{r}_{xx}(k)$ for $-n<k<n$, independently of the choice of γ_n'. The following lemma shows that the value of $r_n' = r_n'(\gamma_n')$ is such that the finite trigonometric moment problem (2.3) is uniquely solvable. The solution of (2.3) depends on γ_n'. A solution of (2.3) also solves (1.1).

Lemma 2.2. Let n>1. Then the finite trigonometric moment problem (2.3) has a unique solution $\{\theta_\ell', \rho_\ell'\}_{\ell=1}^{n}$, such that $\rho_\ell'>0$ and $-\pi<\theta_\ell' \le \pi$ for all ℓ, and $\theta_j' \ne \theta_\ell'$ if $j \ne \ell$.

Proof. We note that R′ is singular, but has a positive definite leading principal nxn submatrix; see Lemma 2.1. By a theorem of Carathéodory [9, chap. 4.1], it follows that (2.3) with k restricted to $1 \le k \le n$ has a unique solution with the stated properties. Part (b) of the proof of Carathéodory's theorem [9, p. 56] shows that this solution also satisfies (2.3) for k=0. □

We determine the solution of (2.3) by solving an eigenproblem for the orthogonal matrix H′.

Lemma 2.3. Let H' be defined by (1.5). Let λ_ℓ, $1 \le \ell \le n$, be the eigenvalues of H', and let $\underline{v}_\ell = [\nu_{1\ell},\ldots,\nu_{n\ell}]^T$ be the corresponding eigenvectors. Assume that the $\underline{v}_\ell$ are scaled so that $\nu_{1\ell} > 0$ and $\underline{v}_\ell^* \underline{v}_\ell = \tilde{r}_{xx}(0)$. Then $\lambda_\ell = \exp(i\theta_\ell')$ and $\nu_{1\ell} = \rho_\ell'$ for $1 \le \ell \le n$, where $\{\theta_\ell', \rho_\ell'\}_{\ell=1}^n$ is the unique solution of (2.3).

Proof. Because R is positive definite, the reflection coefficients γ_j, $1 \le j \le n$, associated with R satisfy $-1 < \gamma_j < 1$; see Lemma 2.1. By (1.2a) we have $\sigma_j > 0$ for $1 \le j \le n$. It follows from (1.5) that the σ_j, $1 \le j \le n$, are the subdiagonal elements of H'. Gragg [7] shows that unitary upper Hessenberg matrices with positive subdiagonal elements relate to positive measures on the unit circle similarly as Jacobi matrices relate to positive measures on an interval. In particular, the eigenvalues λ_ℓ of H' are distinct and the first element $\nu_{1\ell}$ of each eigenvector $\underline{v}_\ell$ is nonvanishing. Therefore, $\underline{\nu}_\ell$ can be scaled as required in the lemma. Then $\nu_{1\ell}^2$ are the weights and λ_ℓ the corresponding abscissas of a Gauss-Szegö quadrature rule on the unit circle. It is shown in [7] that (2.3) is satisfied by $\rho_\ell := \nu_{1\ell}$ and $\theta_\ell := \arg(\lambda_\ell)$, $1 \le \ell \le n$. The unicity of the solution of (2.3) completes the proof. □

In the remainder of this section we consider properties of H' and R' when γ_n' is chosen according to (1.6).

Lemma 2.4. Let H be defined by (1.3) and H' by (1.5), and assume that γ_n' satisfies (1.6). Let U denote the set of nxn unitary matrices, and let $\| \|_U$ denote any unitarily invariant matrix norm on $\mathbb{C}^{n\times n}$. Then

$$\|H-H'\|_U = \min_{W\in U} \|H-W\|_U$$

and

$$\|H-H'\|_2 = 1-|\gamma_n|,$$

where $\| \|_2$ denotes the Euclidean norm.

Proof. By [5, Theorem 1] it suffices to show that H' is a unitary matrix of a polar factorization of H, i.e. we need to show that there is a Hermitian positive semi-definite matrix M, such that $H = H'M$. Assume that $\gamma_n \neq 0$, and let γ_n' be given by (1.6). Then $H = H'\tilde{G}_n(-|\gamma_n|)$ is the desired factorization. Moreover,

$$\|H-H'\|_2 = \|\tilde{G}_n(\gamma_n)-\tilde{G}_n(\gamma_n')\|_2 = |\gamma_n'-\gamma_n| = 1-|\gamma_n|.$$

Now assume that $\gamma_n = 0$ and $\gamma_n' = -1$. Then $H = H'\tilde{G}_n(\gamma_n)$ is a polar factorization of H, and

$$\|H-H'\|_2 = \|\tilde{G}_n(0)-\tilde{G}_n(-1)\|_2 = 1.$$

This completes the proof of the lemma. □

Lemma 2.4 shows that if γ_n' is defined by (1.6), then H' is an orthogonal matrix closest to H with respect to any unitarily invariant matrix norm. We now show how R' depends on the choice of γ_n'.

Lemma 2.5. Let R, R', R_1 and R_{-1} be symmetric positive semi-definite Toeplitz matrices of order $n+1$ with diagonal element r_0. Assume that the first $n-1$ reflection coefficients associated with each of these matrices are $\gamma_1, \gamma_2, \ldots, \gamma_{n-1}$, and let the nth reflection coefficients associated with these matrices be given by Table 1.1, where γ_n' is defined by (1.6).

matrix	nth reflection coefficient
R	γ_n
R'	γ_n'
R_1	1
R_{-1}	-1

Table 1.1

Then

$$\min\{\|R-R_1\|_2, \|R-R_{-1}\|_2\} = \|R-R'\|_2 = (1-|\gamma_n|)r_0 \prod_{j=1}^{n-1}(1-\gamma_j^2). \tag{2.4}$$

Proof. By Lemma 2.1 the symmetric Toeplitz matrix $T = [t_{j-k}]_{j,k=0}^{n} := R-R'$ has elements $t_j = 0$ for $0 \leq j < n$ and $t_n = \varepsilon$, where $\varepsilon := (\gamma_n'-\gamma_n)r_0\prod_{j=1}^{n-1}(1-\gamma_j^2)$. The eigenvalues of T can be determined explicitly, and we obtain $\|T\|_2 = |\varepsilon|$. This shows that

$$\|R-R'\|_2 = \beta|\gamma_n'-\gamma_n| = \beta(1-|\gamma_n|),$$

where $\beta := r_0\prod_{j=1}^{n-1}(1-\gamma_j^2)$. Similarly, one can show that $\|R-R_1\|_2 = \beta(1-\gamma_n)$ and $\|R-R_{-1}\|_2 = \beta(1+\gamma_n)$. This shows (2.4). □

Equation (2.4) shows that among the choices (1.4) of γ_n', the choice (1.6) yields the autocorrelation matrix closest to R with respect to the Euclidean matrix norm.

3. IMPLEMENTATION

We have selected algorithms that lend themselves to parallel computation. The computation of the autocorrelation lags $\tilde{r}_{xx}(j)$, $j = 0,1,\ldots,n$, according to (2.1) parallelizes trivially. Also the computations in the Schur algorithm, described, e.g., in [1], [4], [11], [12], can be organized to allow parallel processing. In order to compute n reflection coefficients from n+1 autocorrelation lags, the Schur algorithm requires $O(n^2)$ arithmetic operations. Kailath [12], [11] discusses how to organize the computations when several processors are available. With n processors the computation of n reflection coefficientsby Schur's algorithm shown below can be carried out in O(n) time units, because each linner loop requires only O(1) time units.

Schur's Algorithm

Input: $\{r_j\}_{j=0}^{n}$, where the $r_j \in \mathbb{R}$ define a positive symmetric Toeplitz matrix $R = [r_{|j-k|}]_{j,k=0}^{n}$; output: $\{\gamma_j\}_{j=1}^{n}$, $\{\sigma_j\}_{j=1}^{n}$, where $\sigma_j := (1-\gamma_j^2)^{1/2}$;

for j: = 0 to n-1 do

$\lfloor\ \beta_j: = r_j;\ \alpha_j: = -r_{j+1};$

for j: = 1 to n do

$\gamma_j: = \alpha_0/\beta_0;$

$\sigma_j^2: = (1-\gamma_j)(1+\gamma_j);$

$\beta_0: = \beta_0\sigma_j^2;$

$\sigma_j: = (\sigma_j^2)^{1/2};$

for k: = 1 to n-j do

$\lfloor\ \tilde{\alpha}_k: = \alpha_k;$

for k: = 1 to n-j do

$\lfloor\ \alpha_{k-1}: = \tilde{\alpha}_k - \gamma_j\beta_k;.$

for k: = 1 to n-j do

$\lfloor\ \beta_k: = \beta_k - \gamma_j\tilde{\alpha}_k;$ □

We note, in passing, that the computation of σ_j^2 in the algorithm is more accurate than if the formula $\sigma_j^2: = 1-\gamma_j^2$ were used. This follows from a straightforward round-off error analysis.

We now outline a divide-and-conquer scheme for the computation of the eigenvalues and the first components of the normalized eigenvectors of the orthogonal matrix H' given by (1.5). Details are presented in [8]. Input to the algorithm is the (2n-1) - tuple $\{\gamma_1,\gamma_2,\ldots,\gamma_{n-1},\gamma_n',\sigma_1,\sigma_2,\ldots,\sigma_{n-1}$

The algorithm proceeds by manipulating this (2n-1) - tuple and the matrix H' does not have to be explicitly formed. We write H' as

$$H' = G_1(\gamma_1)\ldots G_s(\gamma_s)\ldots G_{n-1}(\gamma_{n-1})\tilde{G}_n(\gamma'_n)$$

$$=: \begin{bmatrix} H_1 & \\ & I_{n-s} \end{bmatrix} G_s(\gamma_s) \begin{bmatrix} I_s & \\ & H_2 \end{bmatrix}, \tag{3.1}$$

where $H_1 \in \mathbb{R}^{s \times s}$ and $H_2 \in \mathbb{R}^{(n-s)\times(n-s)}$ are orthogonal matrices. We can write $G_s(\gamma_s)$ as a Householder transformation

$$G_s(\gamma_s) = I - 2\underline{w}\underline{w}^T, \tag{3.2}$$

where $\underline{w} = \underline{e}_s\omega_s + \underline{e}_{s+1}\omega_{s+1} \in \mathbb{R}^n$, $\omega_s = 2^{-1/2}(1+\gamma_s)^{1/2}$, $\omega_{s+1} = -2^{-1/2}(1-\gamma_s)^{1/2}$, and $\underline{e}_j \in \mathbb{R}^n$ denotes the jth axis vector. By (3.1) - (3.2), the matrix H' is orthogonally similar to

$$H'' := \begin{bmatrix} H_1 & \\ & H_2 \end{bmatrix} (I - 2\underline{w}\underline{w}^T) =: \hat{H} - 2\hat{H}\underline{w}\underline{w}^T. \tag{3.3}$$

The eigenproblem for $\hat{H}$ consists of the independent eigenproblems for H_1 and H_2. Let the arguments of the eigenvalues of H_1 be $\hat{\theta}_1 < \hat{\theta}_2 < \ldots < \hat{\theta}_s$, and let the arguments of the eigenvalues of H_2 be $\hat{\theta}_{s+1}, < \hat{\theta}_{s+2} < \ldots < \hat{\theta}_n$. We assume now for simplicity that all the $\hat{\theta}_j$ are distinct. Let $\hat{\delta}_j, 1 \le j \le s$, be the last components of the eigenvectors of length ω_s of H_1, and let $\hat{\delta}_j, s < j \le n$, be the first components of the eigenvectors of length $-\omega_{s+1}$ of H_2. Let $\hat{\delta}_j$ and $\hat{\theta}_j$ be associated with the same eigenvector. Then the arguments of the eigenvalues of H'', and therefore of H', are the zeros in $]-\pi,\pi]$ of

$$\Phi(\theta) := \sum_{j=1}^{n} |\hat{\delta}_j|^2 \cot\left(\frac{\hat{\theta}_j - \theta}{2}\right). \tag{3.4}$$

The first step of the divide-and-conquer algorithm is to split the eigenproblem for H' into trivial eigenproblems of orders two or one, by applying formulas analogous with (3.3) recursively. This step requires no computational work. We then solve these trivial eigenproblems analytically, and from their solution the solutions of larger problems are computed as follows. For each eigenproblem we compute the eigenvalues, as well as the first and last components of the eigenvectors of unit length. These quantities we call the solution of the eigenproblem. Knowledge of the solution of smaller problems enables us to compute the solution of larger ones. Repeatedly, we determine solutions of larger problems from smaller ones already solved until the eigenvalues of H' and the first components of the eigenvectors, normalized according to Lemma 2.3, have been determined. This scheme is suitable for parallel computation because the smaller eigenproblems of equal size can be solved independently, and, moreover, all zeros of functions of the form (3.4) can be computed independently. With n processors we can compute the eigenvalues of H', as well as the first components of the normalized eigenvectors, in $O(n)$ time units. With n^2 processors $O(\log^2 n)$ time units suffice.

4. NUMERICAL EXAMPLES

In the examples of this section, we have used the numerical scheme described in Section 3. The examples suggest that the choice (1.6) of γ_n' is appropriate. However, more computational experience is required. In the examples we generate a "signal" of the form

$$x(k) = \sum_{\ell=1}^{n/2} \rho_\ell(\exp(ik\theta_\ell+i\phi_\ell)+\exp(-ik\theta_\ell-i\phi_\ell))+\varepsilon_\delta(k),\ 0\leq k<N, \tag{4.1}$$

where the $\varepsilon_\delta(k)$, $0\leq k<N$, are uniformly distributed random numbers in the interval $[-\delta,\delta]$ representing "noise". In all examples, we let $N := 100$. We denote the estimated frequencies and amplitudes by θ_ℓ' and ρ_ℓ', respectively, where the θ_ℓ' are the arguments of the eigenvalues of the orthogonal matrix H' defined by (1.5). The ρ_ℓ' are scaled so that $\Sigma_{\ell=1}^{n}(\rho_\ell')^2 = \tilde{r}_{xx}(0)$. If $\{\theta_\ell',\rho_\ell'\}$ is a frequency-amplitude pair with $0<\theta_\ell'<\pi$, then, because H' is real, $\{-\theta_\ell',\rho_\ell'\}$ is another frequency-amplitude pair. The tables of the examples only show frequency-amplitude pairs with $0\leq\theta_\ell'\leq\pi$.

Example 4.1

Let n: = 4, and let θ_ℓ, ϕ_ℓ and ρ_ℓ in (4.1) be defined by Table 4.1. We first let δ: = 0, and obtain $\tilde{r}_{xx}(0) = 1.005$ and $\gamma_n = 0.904$. Table 4.2 shows the computed estimates of the frequencies and amplitudes corresponding with the choice $\gamma_n' = 1$, and Table 4.3 shows the computed estimates for $\gamma_n' = -1$. We recall that in the LSP method the estimates for both $\gamma_n' = 1$ and $\gamma_n' = -1$ are computed, i.e. one computes both Tables 4.2 - 4.3, while in the CSM method the estimates for $\gamma_n' = -1$ are determined, i.e. one determines Table 4.3 only. In the present example the better estimates are obtained for $\gamma_n' = 1$, which is the choice (1.6).

ℓ	θ_ℓ	ϕ_ℓ	ρ_ℓ
1	2.10	0.175	½
2	0.364	-1.22	½

Table 4.1: Signal, n=4.

ℓ	θ_ℓ'	ρ_ℓ'
1	2.10	0.502
2	0.364	0.500

Table 4.2: $\gamma_n' = 1$, $\delta = 0$.

ℓ	θ_ℓ'	ρ_ℓ'
1	π	0.288
2	1.87	0.491
3	0	0.663

Table 4.3: $\gamma_n' = -1$, $\delta = 0$.

We now add noise to the signal by setting $\delta = \frac{1}{4}$. This yields $\tilde{r}_{xx}(0) = 1.026$ and $\gamma_n = 0.867$. We obtain the estimates of frequencies and amplitudes displayed in Tables 4.4 - 4.5. Again $\gamma_n' = 1$ yields the better estimates.

ℓ	θ'_ℓ	ρ'_ℓ
1	2.10	0.501
2	0.359	0.512

Table 4.4: $\gamma'_n = 1$, $\delta = \frac{1}{4}$.

ℓ	θ'_ℓ	ρ'_ℓ
1	π	0.292
2	1.87	0.489
3	0	0.680

Table 4.5: $\gamma'_n = -1$, $\delta = \frac{1}{4}$. □

Example 4.2

Let n: = 4, and let θ_ℓ, ϕ_ℓ and ρ_ℓ in (4.1) be defined by Table 4.6. We let δ: = 0, and obtain $\tilde{r}_{xx}(0) = 0.6489$ and $\gamma_n = -0.280$. Tables 4.7 and 4.8 show the computed estimates of the frequencies and amplitudes for $\gamma'_n = 1$ and $\gamma'_n = -1$, respectively. The estimates for $\gamma'_n = -1$ are better, and this choice of γ'_n satisfies (1.6).

ℓ	θ_ℓ	ϕ_ℓ	ρ_ℓ
1	0.192	0.349	$\frac{1}{2}$
2	0.716	0	$\frac{1}{4}$

Table 4.6: Signal, n = 4.

ℓ	θ'_ℓ	ρ'_ℓ
1	1.67	
2	0.338	0.562

Table 4.7: $\gamma'_n = 1$, $\delta = 0$.

ℓ	θ'_ℓ	ρ'_ℓ
1	π	0.048
2	0.723	0.313
3	0	0.671

Table 4.8: $\gamma'_n = -1$, $\delta = 0$. □

Example 4.3

Let n: = 8, and let θ_ℓ, ϕ_ℓ and ρ_ℓ in (4.1) be defined by Table 4.9. We let δ: $= \frac{1}{10}$, and obtain $\tilde{r}_{xx}(0) = 1.296$ and $\gamma_n = 0.376$. Tables 4.10 and 4.11 show the computed estimates of the frequencies and amplitudes for $\gamma_n' = 1$ and $\gamma_n' = -1$, respectively. The estimates for $\gamma_n' = 1$ are better, and this choice of γ_n' satisfies (1.6).

ℓ	θ_ℓ	φ_ℓ	ρ_ℓ
1	2.50	0.349	½
2	1.92	0.262	½
3	0.908	0.175	¼
4	0.349	0.0873	¼

Table 4.9: Signal, n = 8.

ℓ	θ_ℓ'	ρ_ℓ'
1	2.54	0.505
2	1.94	0.513
3	1.00	0.233
4	0.389	0.274

Table 4.10: $\gamma_n' = 1,\ \delta = \frac{1}{10}$.

ℓ	θ_ℓ'	ρ_ℓ'
1	π	0.202
2	2.45	0.531
3	1.87	0.476
4	0.754	0.289
5	0	0.268

Table 4.11: $\gamma_n' = -1,\ \delta = \frac{1}{10}$. □

5. CONCLUSION

The computation of frequency and amplitude estimates of the CSM and LSP methods by solving an orthogonal eigenproblem is suitable for parallel processing. Our analysis suggests the choice (1.6) of γ_n'. The computed examples show that this choice may yield estimates that are more accurate than those of the CSM method.

ACKNOWLEDGEMENT

We would like to thank Bill Gragg for valuable discussions.

REFERENCES

[1] Ammar, G.S. and Gragg, W.B., (1987), "The Generalized Schur Algorithm for the Superfast Solution of Toeplitz Systems", in Rational Approximation and its Applications in Mathematics and Physics, eds. J. Gilewicz, M. Pindor and W. Siemaszko, Lecture Notes in Mathematics # 1237, Springer, Berlin, pp. 315-330.

[2] Brumme, K., (1987), "Formantbestimmung in der Digitalen Sprachverarbeitung durch LPC und verwandte Verfahren", Diplomarbeit, Department of Applied Mathematics, University of Hamburg, Hamburg, Federal Republic of Germany.

[3] Delsarte, Ph., Genin, Y., Kamp, Y. and Van Dooren, P., (1982), "Speech Modelling and the Trigonometric Moment Problem", Philips J. Res., 37, 277-292.

[4] Delsarte, Ph. and Genin, Y., (1987), "On the Splitting of Classical Algorithms in Linear Prediction Theory", IEEE Trans. Acoust., Speech, Signal Processing, ASSP-35, 645-653.

[5] Fan, K. and Hoffman, A.J., (1955), "Some Metric Inequalities in the Space of Matrices", Proc. Amer. Math. Soc., 6, 111-116.

[6] Golub, G.H. and Van Loan, C.F., (1983), "Matrix Computations", Johns Hopkins University Press, Baltimore, MD.

[7] Gragg, W.B., (1982), "Positive Definite Toeplitz Matrices, the Arnoldi Process for Isometric Operators, and Gaussian Quadrature on the Unit Circle", (in Russian), in Numerical Methods in Linear Algebra, ed. E.S. Nikolaev, Moscow University Press, Moscow, pp. 16-32.

[8] Gragg, W.B. and Reichel, L., (1988), "A Divide and Conquer Algorithm for the Unitary and Orthogonal Eigenproblems, BSC Report 88/31, Bergen Scientific Centre, Bergen, Norway.

[9] Grenander, U. and Szegö, G., (1984), "Toeplitz Forms and their Applications", Chelsea, New York.

[10] Itakura, F., (1975), "Line Spectrum Representation of Linear Predictor Coefficients of Speech Signals", J. Acoust. Soc. Amer., 57, (Supplement # 1), S35.

[11] Kailath, T., (1985), "Linear Estimation for Stationary and Near-Stationary Processes", in Modern Signal Processing, ed. T. Kailath, Hemisphere Publ., Washington, pp. 59-128.

[12] Kailath, T., (1987), "Signal Processing Applications of Some Moment Problems", in Moments in Mathematics, ed. H.J. Landau, Proceedings of Symposia in Applied Mathematics, vol. 37, Amer. Math. Soc., Providence, R.I., pp. 71-109.

[13] Kay, S.M. and Marple, S.L., (1981), "Spectrum Analysis - A Modern Perspective", Proc. IEEE, 69, 1380-1419.

[14] Kung, S.Y., Bhaskar Rao, B.V. and Arun, K.S., (1985), "Spectral Estimation: From Conventional Methods to High-Resolution Modelling Methods", in VLSI and Modern Signal Processing, eds. S.Y. Kung, H.J. Whitehouse and T. Kailath, Prentice-Hall, Englewood Cliffs, N.J., pp. 42-60.

[15] Sagayama, S. and Itakura, F., (1986), "Duality Theory of Composite Sinusoidal Modelling and Linear Prediction", in Proceedings of the 1986 IEEE Intern. Conference on Acoust., Speech and Signal Processing, ICASSP-86, Tokyo, pp. 1261-1264.

[16] Soong, F.K. and Juang, B.-W., (1984), "Line Spectrum Pair (LSP) and Speech Data Compression", in Proceedings of the 1984 IEEE Intern. Conference on Acoust., Speech and Signal Processing, ICASSP-84, San Diego, pp. 1.10.1-1.10.4.

RELIABLE AND EFFICIENT TECHNIQUES BASED ON TOTAL LEAST SQUARES FOR COMPUTING CONSISTENT ESTIMATORS IN MODELS WITH ERRORS IN THE VARIABLES

S. Van Huffel* and J. Vandewalle
(ESAT Laboratory, Department of Electrical Engineering, K.U.Leuven, Heverlee, Belgium)

ABSTRACT

The Total Least Squares (TLS) method has been devised as a more general fitting technique than the ordinary least squares technique for solving overdetermined sets of linear equations $AX \approx B$ when errors occur in all data. If the errors in the measurements A and B are uncorrelated with zero mean and equal variance, TLS is able to compute a strongly consistent estimate of the true solution of the corresponding unperturbed set $A_0X = B_0$. These coefficients are called the parameters of a classical errors-in-variables model.

In this paper, the TLS problem, as well as the TLS computations, are generalized in order to maintain consistency of the parameter estimates in a general errors-in-variables model, i.e. some of the columns of A may be known exactly and the errors in the remaining data matrix may be correlated and not equally sized. For this problem, a computationally efficient and numerically reliable Generalized TLS algorithm GTLS, based on the Generalized SVD (GSVD), is developed. Additionally, the equivalence between the GTLS solution and alternative expressions of consistent estimators, commonly used in linear regression analysis and system identification, is pointed out. These relations allow one to deduce the main statistical properties of the GTLS solution,

* Senior Research Assistant of the Belgian N.F.W.O.
(National Fund of Scientific Research)

1 INTRODUCTION

Every linear parameter estimation problem gives rise to an overdetermined set of linear equations $AX \approx B$. Whenever both A and B are subject to errors, the Total Least Squares (TLS) method can be used for solving this set. Much of the literature concerns the classical TLS problem $AX \approx B$ in which all columns of A are subject to errors, see e.g. [7],[19-21]. If the errors in the measurements A and B are uncorrelated with zero mean and equal variance, then under mild conditions this TLS solution $\hat{X}$ is a strongly consistent estimate of the true solution X of the corresponding unperturbed set $A_O X = B_O$, i.e. $\hat{X}$ converges to X with probability one as the number of equations tends to infinity. However, in many linear parameter estimation problems some columns of A may be error-free. Moreover, the errors on the remaining data may be correlated and not equally sized. In order to maintain consistency of the result when solving these problems, the classical TLS formulation can be generalized as follows (M^{-T} denotes the transposed inverse of matrix M):

<u>Generalized TLS formulation</u>: Given a set of m linear equations in n x d unknowns X:

$$AX \approx B \qquad A \in R^{m \times n},\ B \in R^{m \times d} \text{ and } X \in R^{n \times d} \tag{1}$$

$$\text{Partition} \quad A = [A_1; A_2] \qquad A_1 \in R^{m \times n_1},\ A_2 \in R^{m \times n_2} \text{ and } n = n_1 + n_2 \tag{2}$$

$$X = [X_1^T; X_2^T]^T \qquad X_1 \in R^{n_1 \times d} \text{ and } X_2 \in R^{n_2 \times d} \tag{3}$$

and assume that the columns of A_1 are error-free and that nonsingular error equilibration matrices $R_D \in R^{m \times m}$ and $R_C \in R^{(n_2+d) \times (n_2+d)}$ are given such that the errors on $R_D^{-T}[A_2; B] R_C^{-1}$ are equilibrated, i.e. uncorrelated with zero mean and same variance.

Then, a GTLS solution of eqn (1) is any solution of the set

$$\hat{A}X = A_1 X_1 + \hat{A}_2 X_2 = \hat{B} \tag{4}$$

where $\hat{A} = [A_1; \hat{A}_2]$ and $\hat{B}$ are determined such that

$$\text{Range}(\hat{B}) \subseteq \text{Range}(\hat{A}) \tag{5}$$

$$\| R_D^{-T}[\Delta\hat{A}_2;\Delta\hat{B}] R_C^{-1} \|_F = \| R_D^{-T}[A_2 - \hat{A}_2; B - \hat{B}] R_C^{-1} \|_F \text{ is minimal} \tag{6}$$

The problem of finding $[\Delta\hat{A}_2;\Delta\hat{B}]$ such that eqn (5-6) are satisfied, is referred to as the GTLS problem. Whenever the solution is not unique, GTLS singles out the minimum norm solution denoted by $X = [\hat{X}_1^T;\hat{X}_2^T]^T$.

By varying n_1 from n to zero, this formulation can handle any LS ($n_1 = n; R_C, R_D \sim I$) and generalized LS problem ($n_1 = n, R_C \sim I$), as well as every TLS ($R_C, R_D \sim I$) and GTLS problem. The error equilibration matrices R_D and R_C are the square root of the error covariance matrices $C = E(\Delta^T D^{-1}\Delta)$ and $D = E(\Delta C^{-1}\Delta^T)$ respectively where E denotes the expected value operator and $\Delta_{mx(n_2+d)}$ represents the errors on the noisy data $[A_2;B]$. Often, only C and D are known: in these cases the matrices R_C and R_D are simply obtained from their Cholesky decomposition, i.e. $C = R_C^T R_C$ and $D = R_D^T R_D$, R_C and R_D upper triangular.

Although the name "total least squares" appeared only recently in the literature [7], this method of fitting is certainly not new and has a long history in the statistical literature where the method is known as orthogonal or errors-in-variables regression [5-6]. More recently, the TLS approach to fitting has also attracted interest outside of statistics. In numerical analysis, Golub and Van Loan [7] first studied this problem and presented an algorithm, based on the Singular Value Decomposition (SVD). The efficiency of their computations can be improved with a factor 2 or 3 by computing the SVD only "partially" [20], or iteratively, if a priori information about the solution is available. Here, the inverse iteration method is recommended [22]. We furthermore generalized the algorithm of Golub and Van Loan [7] to all cases in which their algorithm fails to produce a solution, described the properties of these so-called nongeneric TLS problems and proved that the proposed generalization still satisfies the TLS criteria eqn (5-6) if additional constraints are imposed on the solution space [21]. Also in experimental modal analysis, the TLS technique (sometimes called the H_v technique), was studied recently [14], as well as in system identification, where TLS is more commonly known as the eigenvector method [15] or Koopmans-Levin method [3]. If some columns of the data matrix A in the set $AX \approx B$ are error-free, the classical TLS algorithms can be generalized in order to compute the more general TLS estimate

$\hat{X} = [\hat{X}_1^T ; \hat{X}_2^T]^T$ satisfying the TLS criteria eqn (5-6) with $R_C \sim I$ and $R_D \sim I$ [19],[23] (see section 2). In particular, this algorithm is able to compute the Compensated Least Squares (CLS) estimate as derived by Guidorzi [10] and Stoica and Söderström [18]. When the only disturbance of the input-output sequences is given by zero-mean white noise sequences of equal variance, the CLS, GTLS and eigenvector methods all give the same estimate. For a detailed appraisal of the TLS method and its generalizations, see [19],[24].

2 THE GENERALIZED TLS ALGORITHM GTLS

As outlined below, the GTLS algorithm is based on an implicit GSVD method [2], which computes an SVD of a triple matrix product $E^{-1}FG^{-1}$ without explicitly forming the products and without inverting E or G. This guarantees its better numerical performance. Moreover, by first performing a QR factorization [8], only the GSVD of a smaller submatrix is required which makes the GTLS algorithm computationally more efficient than methods described in [3],[18].

Given: An $m \times d$ matrix B and an $m \times n$ matrix $A = [A_1 ; A_2]$ whose first n_1 columns A_1 have full column rank and are error-free, $n = n_1 + n_2$ and $m \geq n + d$.

The error equilibration matrices $(R_D)_{m \times m}$ and $(R_C)_{(n_2+d) \times (n_2+d)}$, as defined in the GTLS formulation.

Step 1: QR and QL factorizations

1.a. Compute the QR factorization of $[A_1 ; A_2 ; B]$:

$$[A_1 ; A_2 ; B] = Q_{AB} \begin{bmatrix} R_{AB} \\ 0 \end{bmatrix} \text{ with } R_{AB} = \begin{bmatrix} R_{11} & R_{12} \\ 0 & R_{22} \end{bmatrix} \begin{matrix} n_1 \\ n_2+d \end{matrix} \quad \text{upper triangular}$$
$$\qquad\qquad\qquad\qquad\qquad\qquad n_1 \quad n_2+d$$

1.b. If $R_D \sim I$ then $E_{22} \leftarrow I$ else compute the QL factorization of $R_D Q_{AB}$:

$$R_D Q_{AB} = Q_{\hat{D}} L_{\hat{D}} \text{ with } L^T_{\hat{D}} = \begin{bmatrix} E_{11} & E_{12} & E_{13} \\ O & E_{22} & E_{23} \\ O & O & E_{33} \end{bmatrix} \begin{matrix} n_1 \\ n_2 + d \\ m-n-d \end{matrix} \quad \text{upper triangular}$$

$$\begin{matrix} n_1 & n_2+d & m-n-d \end{matrix}$$

If $n_I = n$ then begin $Z_2 \leftarrow -I_d$; go to step 2.d. end

1.c. If R_C upper triangular then $R_{\hat{C}} \leftarrow R_C$ else compute the QR factorization of R_C:

$$R_C = Q_{\hat{C}} R_{\hat{C}} \qquad (R_{\hat{C}})_{(n_2+d)\times(n_2+d)} \text{ upper triangular}$$

Step 2: GSVD

2.a. Compute the implicit GSVD

$$U^T E_{22}^{-1} R_{22} R_{\hat{C}}^{-1} V = \text{diag}(\sigma_1, \ldots, \sigma_{n_2+d}) \quad \sigma_{i-1} \geq \sigma_i \quad i=2,\ldots,n_2+d \tag{7}$$

with U and $V = [v_1, \ldots, v_{n_2+d}]$ orthonormal and σ_i the generalized singulär values.

2.b. If not user determined, compute the rank $r(\leq n_2)$ by means of a user-defined rank determinator R_O:

$$\sigma_1 \geq \ldots \geq \sigma_r > R_O \geq \sigma_{r+1} \geq \ldots \geq \sigma_{n_2+d}$$

2.c. If $R_C \sim I$ then $Z_2 \leftarrow [v_{r+1}, \ldots, v_{n_2+d}]$ else solve $R_{\hat{C}} Z_2 = [v_{r+1}, \ldots, v_{n_2+d}]$ by back substitution.

2.d. If $R_D \sim I$ then $\hat{Z} = O$ else solve $E_{22}\hat{Z} = R_{22} Z_2$ by back substitution.

2.e. Solve $R_{11} Z_1 = E_{12}\hat{Z} - R_{12} Z_2$ by back substitution.

If $n_1 = n$ then begin $\hat{X} \leftarrow Z_1$; stop end.

Step 3: GTLS solution $\hat{X} = [\hat{X}_1^T;\hat{X}_2^T]^T$

3.a. If $R_C \neq I_{n+d}$, $d > 1$ and $r < n_2$, orthonormalize $\begin{bmatrix} Z_1 \\ Z_2 \end{bmatrix}$ using a QR factorization:

$$\begin{bmatrix} Z_1 \\ Z_2 \end{bmatrix} = Q_z R_z ; \quad \begin{bmatrix} Z_1 \\ Z_2 \end{bmatrix} \leftarrow Q_z$$

3.b. Perform Householder transformations Q such that

$$\begin{bmatrix} Z_1 \\ Z_2 \end{bmatrix} Q = \begin{bmatrix} W & Y \\ O & \Gamma \end{bmatrix} \begin{matrix} n \\ d \end{matrix} \text{ and } \Gamma \text{ upper triangular} \qquad (8)$$
$$\begin{matrix} n_2-r & d \end{matrix}$$

If Γ nonsingular then solve $\hat{X}\Gamma = -Y$

else lower the rank r with the multiplicity of σ

go back to step 2.c.

END

The following comments are in order:

- Step 1 of the GTLS algorithm, based on the canonical correlation computation procedure of [2], reduces all 3 matrices involved in the GSVD to upper triangular form of equal dimension. In step 2.a. the algorithm PSVD-2 of [2] can readily be applied to find the implicit GSVD. For more details and an analysis of the computational complexity, see [2]. The special case where $R_D \sim I_m$ reduces PSVD-2 to the well-known GSVD algorithms [2-17] for computing the SVD of the product FG^{-1} implicitly. In this case, the GTLS algorithm reduces to the algorithm described in [24]. These algorithms are all based on an implicit Kogbetliantz approach and are suitable for parallel implementation.

- If R_C or R_D are singular, the GSVD algorithm PSVD-2 can be adapted by using matrix adjoints (cf. [17]). Note that in this case the GSVD is not necessarily given by the SVD of of $E_{22}^{\dagger}R_{22}R_{\hat{C}}^{\dagger}$ († denotes the pseudo-inverse). Similarly to [17], the GSVD algorithm could be adapted for the case that $m < n + d$ by adding zero rows and columns, if necessary, to give square matrices of equal dimension. These extensions are not yet fully analyzed.

- If $R_C \sim I_{n+d}$ and $R_D \sim I_m$, the GSVD in step 2 is simply the ordinary SVD of R_{22} so that in this case the GTLS algorithm reduces to the classical TLS algorithms [7-8], [19, sec.1.8.1], [21] for the case that $n_1 = 0$ and [23] for the case that $n_1 \neq 0$. Observe also that the GTLS algorithm solves the ordinary LS problem, using a QR factorization, if all columns of A are error-free ($n_1 = n$), as well as the generalized LS problem described in [16].

- If Γ in eqn (8) is nonsingular (resp., singular), the GTLS solution is called generic (resp., nongeneric). For more details, see [19], [21].

3. PROPERTIES OF THE GENERALIZED TLS SOLUTION

Based on the following important theorem, one can derive the main statistical properties of the GTLS solution:

Theorem 1 Consider eqn (1-2). Denote by σ' (resp., σ) the minimal generalized singular value of the matrix pair $(R_D^{-T} A, R_{Ca}^{*})$ (resp., $(R_D^{-T}[A;B], R_C^{*})$) where $R_C^{*} =$

$$\begin{array}{c} \\ \\ \\ \end{array}\begin{bmatrix} 0 & 0 \\ 0 & R_C \end{bmatrix}\begin{array}{l} n_1 \\ n_2+d \end{array} = \begin{bmatrix} R_{Ca}^{*} & R_{Cab}^{*} \\ 0 & R_{Cb}^{*} \end{bmatrix}\begin{array}{l} n \\ d \end{array}$$

with column dimensions n_1, n_2+d (left) and n, d (right), is upper triangular.

Let σ have multiplicity d and denote $D = R_D^T R_D$, $C_a^{*} = R_{Ca}^{*T} R_{Ca}^{*}$ and $C_{ab}^{*} = R_{Ca}^{*T} R_{Cab}^{*}$. If $\sigma' > \sigma$, the GTLS solution is given by

$$\hat{X} = (A^T D^{-1} A - \sigma^2 C_a^{*})^{-1} (A^T D^{-1} B - \sigma^2 C_{ab}^{*}) \tag{9}$$

Proof: see [24, Theorem 4].

If $D \sim I$, eqn (9) is a well-known expression in linear regression analysis and statistics. Its consistency and other statistical properties have been investigated by Gallo [5]. Gleser [6] studied the special case that $R_C \sim I$ and $n_1 = 0$, corresponding to the classical TLS problem. Using their results and assuming that $\lim_{m\to\infty} \frac{1}{m} A_O^T A_O$ (defined below) exists and is positive definite, it can be concluded that the GTLS solution is a strongly consistent estimate of the true parameters X of the general errors-in-variables model, defined as:

$$B_O = (A_O)_{mxn} X_{nxd} = A_1 X_1 + (A_2)_O X_2; \quad A_2 = (A_2)_O + \Delta A_2 \text{ and } B = B_O + \Delta B \tag{10}$$

X_1 and X_2 are the true but unknown parameters to be estimated, A_1 and $(A_2)_O$ are of full column rank. They consist of constants as well as B_O. A_1 is known but $(A_2)_O$ and B_O not. The observations A_2 and B of the unknown values $(A_2)_O$ and B_O contain measurement errors ΔA_2 and ΔB such that the rows of $[\Delta A_2;\Delta B]$ are independently and identically distributed (i.i.d.) with zero mean and known positive definite covariance matrix $C = R_C^T R_C$, up to a factor of proportionality.

Statistical properties of the GTLS solution for the case that the true values A_O and B_O in eqn (10) are random variables, have been proven by Kelly [12].

Not only in statistics but also in system identification, eqn (9) (with $D \sim I$) is well-known, see e.g. [18], [11], [10]. In particular, eqn (9) arises as a consistent estimator in ARMA modelling.

These models are given by:

$$y(t)+a_1 y(t-1)+\ldots+a_{n_a} y(t-n_a)=b_1 u(t-1)+\ldots+b_{n_b} u(t-n_b) \tag{11}$$

where the $\{u(t)\}$ and $\{y(t)\}$ are the input and output sequences respectively and $\{a_j\}$ and $\{b_j\}$ are the unknown constant parameters of the system. If sufficient observations are taken, eqn (11) gives rise to an overdetermined Toeplitz-like set of equations. Assume that the only disturbances in the observed outputs and inputs (if they can't be measured exactly) are given by mutually independent zero-mean white noise sequences of equal variance, i.e. $R_C \sim I$ and $n_1 = 0$ or n_b. Then, because of the correspondence between the GTLS algorithm and the eigenvector method and compensated LS method, strong consistency of the GTLS solution of this set is proven under the same conditions as described in [1] and [18]. Several authors extended the eigenvector method, described by Levin [15], to multi-input multi-output systems in which the disturbances are not necessarily white provided the covariance matrix $C = R_C^T R_C$ of the correlated noise on the input-output data is known, up to a factor of proportionality. Again, the correspondence between GTLS and these extensions proves that the GTLS solution remains consistent in these cases under the same conditions as described in [13], [9], [11], [4].

REFERENCES

[1] Aoki, M., and Yue, P.C., (1970) On a priori error estimates of some identification methods, IEEE Transactions on Automatic Control, 15, 541-548, 1970.

[2] Ewerbring, L.M., and Luk, F.T., (1989) Canonical Correlations and Generalized SVD: Applications and New Algorithms, Proceedings SPIE 977, Real Time Signal Processing XI, paper 23°, 1988.

[3] Fernando, K.V., and Nicholson, H., (1985) Identification of linear systems with input and output noise: the Koopmans-Levin method, IEE Proceedings, part D, 132, 30-36, 1985.

[4] Furuta, K., and Paquet, J.G., (1970) On the identification of time-invariant discrete processes, IEEE Transactions on Automatic Control 15, 153-155, 1970.

[5] Gallo, P.P., (1982) Consistency of regression estimates when some variables are subject to error, Communications in Statistical Theoretical Methods, 11, 973-983, 1982.

[6] Gleser, L.J., (1981) Estimation in a multivariate "errors in variables" regression model: large sample results, Annals of Statistics, 9, 24-44, 1981.

[7] Golub, G.H., and Van Loan, C.F., (1980) An analysis of the total least squares problem, SIAM Journal on Numerical Analysis, 17, 883-893, 1980.

[8] Golub, G.H., and Van Loan, C.F., (1983) Matrix Computations, The Johns Hopkins Univ. Press, Baltimore, Maryland, 1983.

[9] Grosjean, A., and Fouland, C., (1976) Extensions of the Levin's method (or eigenvector method) for the identification of discrete, linear, multivariable, stochastic, time invariant, dynamic systems, Proceedings 4th IFAC Symposium on Identification and System Parameter Estimation, Tbilisi, USSR, 2003-2010, 1976.

[10] Guidorzi, R.P., (1975) Canonical structures in the identification of multivariable systems, Automatica, 11, 361-374, 1975.

[11] James, P.N., Souter, P., and Dixon., D.C., (1972) Suboptimal estimation of the parameters of discrete systems in the presence of correlated noise, Electronics Letters, 8, 411-412, 1972.

[12] Kelly, G., (1984) The influence function in the errors in variables problem, Annals of Statistics, 12, 87-100, 1984.

[13] Kotta, Ü., (1979) Structure and parameter estimation of multivariable systems using the eigenvector method, Proceedings 5th IFAC Symposium on Identification and System Parameter Estimation, Darmstadt, FRG, 453-458, 1979.

[14] Leuridan, J., De Vis, D., Van der Auweraer, H., and Lembregts, F., (1986) A comparison of some frequency response function measurement techniques, Proceedings 4th International Modal Analysis Conference, Los Angeles, Feb. 3-6, 908-918, 1986.

[15] Levin, M.J., (1964) Estimation of a system pulse transfer function in the presence of noise. IEEE Transactions on Automatic Control, 9, 229-235, 1964.

[16] Paige, C.C., (1985) The general linear model and the generalized singular value decomposition, Linear Algebra and its Applications, 70, 269-284, 1985.

[17] Paige, C.C., (1986) Computing the generalized singular value decomposition, SIAM Journal on Scientific and Statistical Computing, 7, 1126-1146, 1986.

[18] Stoica, P., and Söderström, T., (1982) Bias correction in least squares identification, International Journal of Control, 35, 449-457, 1982.

[19] Van Huffel, S., Analysis of the total least squares problem and its use in parameter estimation, Doct. Dissertation, Dept. Electr. Eng., K.U. Leuven, Belgium, June 1987.

[20] Van Huffel, S., and Vandewalle, J., (1988) The partial total least squares algorithm, Journal of Computational and Applied Mathematics, 21, 333-341, 1988.

[21] Van Huffel, S., and Vandewalle, J., (1988) Analysis and solution of the nongeneric total least squares problem, SIAM Journal on Matrix Analysis and Applications, 9, 360-372, 1988.

[22] Van Huffel, S., and Vandewalle, J., (1988) Iterative speed improvement for solving slowly varying total least squares problems, Mechanical Systems and Signal Processing, 2(4), 327-348, 1988.

[23] Van Huffel, S., and Vandewalle, J., (1989) Comparison of total least squares and instrumental variable methods for parameter estimation of transfer function models, International Journal of Control, 50(4), 1039-1056.

[24] Van Huffel, S., and Vandewalle, J., (1989) Analysis and properties of the generalized total least squares problem $AX \approx B$ when some or all columns in A are subject to error, SIAM Journal on Matrix Analysis and Applications, 10 (3), 294-315.

IMPLEMENTING LINEAR ALGORITHMS FOR DENSE MATRICES ON A HETEROGENEOUS MACHINE

S.C. Tran and D.J. Creasey
(Department of Electronic and Electrical Engineering, University of Birmingham)

ABSTRACT

A common approach in the solution of linear algebra problems is to map the algorithms onto a regular network of processors such as a systolic array. Systolic arrays have been developed for computing eigenvalues and singular values of various type of matrices. However, the practical impact of these algorithms has yet to be considered because of the difficulty of fabricating such devices in silicon. In this paper, an alternative approach for solving eigen-system and linear system equations is described. The method has been implemented, based on earlier work of Lanczos [4], to produce an algorithm that is not only very natural and effective as a parallel algorithm, but is actually faster as a serial algorithm than the standard QR algorithm.

1. INTRODUCTION

It is well known that a matrix written in a frame of reference described by its own eigenvectors will be diagonal [5]. The diagonal form of a matrix and the matrix itself are related by an orthogonal transform [3]. Consequently, any operations performed on the matrix in the diagonal form can be related back to the original system via a similarity transform. Carrying out matrix operations in this new frame of reference simplifies the operations. This theory effectively proves the theory due to Caley and Hamilton [3] on the algebraic properties of matrices.

The expansion of matrices into polynomials using diagonal matrices demonstrates their algebraic properties. The basis of this eigen-decomposition technique lies in the fact that any matrix can be expanded into a Chebyshev polynomial [4]. The eigenvalues are then represented by cosines of different frequencies. The different frequency cosine terms can be identified using normal spectrum-analysis techniques.

2. CHEBYSHEV POLYNOMIAL GENERATOR

Let the matrix A be a real symmetric matrix so that the eigenvalues are all real. It will be shown later that the technique can be also applied to complex eigenvalue problems. From the similarity transform [3], if the matrix U represents the matrix of eigenvectors of A then

$$U.A.U^{-1} = \text{diag}\ (e_1, e_2, e_3 \ldots, e_n)$$

and in general,

$$U.A^k.U^{-1} = \text{diag}\ (e_1^{\ k}, e_2^{\ k} \ldots, e_n^{\ k}) \tag{2.1}$$

It is widely accepted that an arbitrary vector b_o can always be expanded in terms of any orthogonal system [3], i.e.

$$b_o = B_1U_1 + B_2U_2 + \ldots + B_nU_n$$

where B_i are constants and U_i are the eigenvectors of the matrix A. Then by the definition of the principal axes, we obtain

$$A.b_o = B_1e_1U_1 + B_2e_2U_2 + \ldots + B_ne_nU_n$$

and equation (2.1) gives

$$A^k.b_o = B_1e_1^{\ k}U_1 + B_2e_2^{\ k}U_2 + \ldots + B_ne_n^{\ k}U_n \tag{2.2}$$

where the e_i terms are the corresponding eigenvalues. The similarity transform also holds for an arbitrary polynomial which operates on the matrix A, denoted as $P_n(A)$. Therefore by introducing the principal axes of A as a new frame of reference, operation of the matrix polynomial $P_n(A)$ becomes reduced to the algebraic polynomial $P_n(E)$, where now the e_i terms are the successive eigenvalues of the matrix A. Hence,

$$P_n(A) = B_1P_n(e_1)U_1 + B_2P_n(e_2)U_2 + \ldots + B_nP_n(e_n)U_n \tag{2.3}$$

The polynomial $P_n(A)$ is an orthogonal polynomial and it satisfies a general recurrence relation of the form

$$P_{k+1}(A) = (mA-n).P_k(A) - q.P_{k-1}(A) \tag{2.4}$$

The constants m,n and q are predefined constants and they correspond to different sets of polynomials. If the constant vector [m,n,q]=[2,0,1] then equation (2.4) refers to the Chebyshev polynomial.

The Chebyshev polynomials belong to a set of so-called universal polynomials The polynomials express cos(nd) in terms of an polynomial $T_n(x)$, where x=cos(d)

$$T_{n+1}(x) = 2x.T_n(x) - T_{n-1}(x) \tag{2.5}$$

In matrix form, this becomes

$$T_{n+1}(A) = 2.A.T_n(A) - T_{n-1}(A) \tag{2.6}$$

Therefore, if the matrix polynomial is generated using the same recurrence relationship as the Chebyshev polynomials then $P_n(A)$ and $T_n(A)$ will be equivalent. Hence $P_n(e)$ can be replaced with cos(nd). Figure 1 shows a block diagram of a Chebyshev polynomial generator. The implementation is simple and therefore very suitable for VLSI implementation.

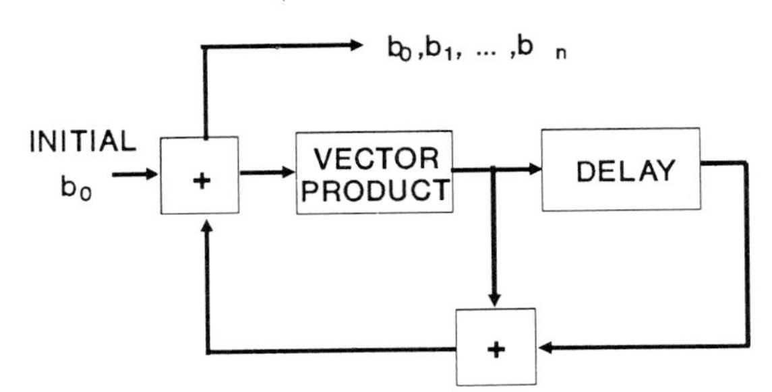

Fig. 1 The Chebyshev polynomial generator

3. THE EIGENVALUE DECOMPOSITION

The Chebyshev polynomials are restricted to the range [1,-1]. Therefore all eigenvalues of A have to be properly normalised. This is achieved by a theorem of Gersgorin which states that the largest eigenvalue of a matrix is never greater than the largest absolute sum of the elements of any rows [5]. Therefore, if the largest sum is defined as s, the eigenvalues of a matrix C = A/s will be confined to the range [1,-1].

Using the recurrence relationship (2.6), a polynomial in C can be generated as

$$b_{k+1} = C.b_k - b_{k-1} \tag{3.1}$$

The polynomial is started by an arbitrary vector b_o. Also from (2.3),

$$P_n(C).b_o = B_1P_n(e_1)U_1 + B_2P_n(e_2)U_2 + \ldots \tag{3.2}$$

If this polynomial is expanded in terms of Chebyshev polynomials then,

$$b_n = B_1.\cos(nd_1).U_1 + B_2.\cos(nd_2).U_2 + \ldots \tag{3.3}$$

where $b_n = P_n(C).b_o$ and the angles $d_1, \ldots, d_n$ are associated with the eigenvalues of C. Figure 2 shows the block diagram representation of the algorithm. Determination of the eigenvalues has now become a harmonic-analysis problem. The angular frequencies d_i can be computed by carrying out a discrete Fourier transform on the sequence b_k, and then relating this back to the eigenvalues using the Chebyshev relation

$$e_i = \cos(d_i) \tag{3.4}$$

For an N-point DFT,

$$\mathrm{DFT}\ (b_k) = 1/2b_o + b_1.\cos(\pi.p/N) + b_2.\cos\ (2.\pi.p/N) + \ldots \tag{3.5}$$

The DFT will identify the eigenvalues by the positions of the peaks w_i in the frequency spectrum, at the points $d_i = Nw_i/\pi$. The eigenvalues of the original matrix A can be computed by using equation (3.4) and rescaling e_i with the normalising factor s. The simulation results for a (6x6) matrix with a full set of eigenvalues and eigenvectors is shown in Figure 3.

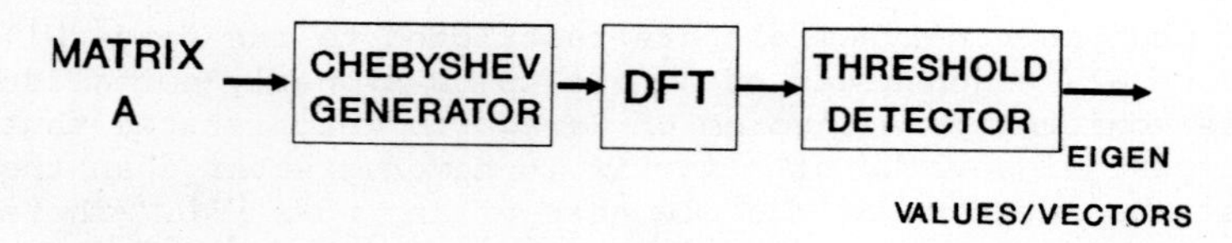

Fig. 2 The eigen-system decomposition algorithm

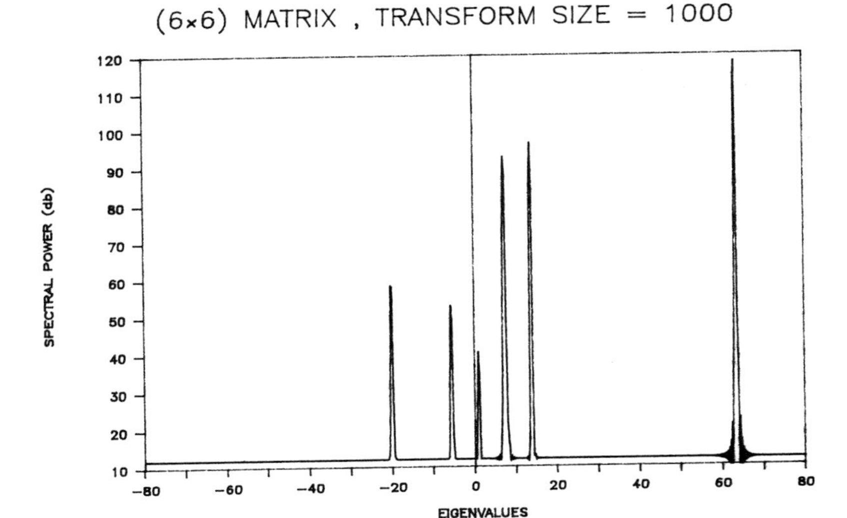

Fig. 3 Simulation results of the eigen-decomposition of a (6x6) matrix

The resolution power of the method is linearly proportional to the number of iterations. While the general shape of the peak which characterises a certain maximum of the function in (3.3) does not change, the ability to resolve adjacent components improves and eventually even very close peaks can be separated. Furthermore, the DFT produces a frequency domain picture of equation (3.3), so that the height of each frequency component will be equal to $B_i U_i$ and hence proportional to the value of the eigenvector U_i. The most surprising feature of this method is that it is entirely free of a dangerous accumulation of rounding errors. Since the signal increases proportionally to the number of iterations, a linear increase in rounding errors would leave the signal-to-noise ratio unchanged. Experiments on small matrices could not detect any damage even after 1000 iterations.

4. LINEAR SYSTEM SOLUTIONS

The application of the Chebyshev polynomials to the eigenvalue problems of matrices with real eigenvalues leads to a method of solving large-scale linear systems without involving matrix inversions.

Consider a set of linear equations in n unknowns

$$a_{11}x_1 + a_{12}x_2 + \ldots\ldots\ldots + a_{1n}x_n + c_1 = 0$$

$$\vdots$$

$$a_{n1}x_1 + a_{n2}x_2 + \ldots\ldots\ldots + a_{nn}x_n + c_n = 0$$

This can be reformulated as a homogeneous set of equations in (n+1) unknowns as follows

$$a_{11}x_1 + a_{12}x_2 + \ldots\ldots + a_{1n}x_n + c_1x_{n+1} = 0$$

$$\vdots$$

$$a_{n1}x_1 + a_{n2}x_2 + \ldots\ldots + a_{nn}x_n + c_nx_{n+1} = 0$$

where $x_{n+1} = 1$. Thus we get an extended (n+1) by (n+1) matrix B.

$$B = \left[\begin{array}{c|c} A & c \\ \hline 0 & 0 \end{array}\right] \tag{4.1}$$

The new system will have the same eigenvalues as the original system plus an extra eigenvalue and eigenvector due to the addition of x_{n+1}. The augmented matrix has a zero row due to the introduction of the extra variable.

The associated eigenvector is defined by the equation $B.U_0=e.U_0=0$. Hence the solution of the original system can be considered as an eigenvalue problem for the augmented system B. The solution to the original system A is then the eigenvector U_0 associated with the zero eigenvalue.

5. ACCURACY

The accuracy of the analysis depends on the width of the windows of the DFT. The DFT is examining a limited sequence of the polynomial and hence a limited number of cycles of each component of the Chebyshev polynomial. This will inevitably lead to inaccuracies in the magnitude of each frequency component and consequently to the determination of the eigenvectors of the system. The eigenvalues, however, are determined by the positions of the maxima. Therefore, the errors in these values are directly proportional to the number of data used in the DFT.

An alternative way of improving the accuracy is by windowing the DFT sequence with different window structures, a selection of which is given in Table 1. The window of the DFT with no weighting applied is a rectangular window producing a response in the frequency domain of sin(x)/x distribution The interference from one peak to another is

$$\sin\,(\pi.w_i)/\pi.(w_i-w_j) \tag{5.1}$$

If one of the Chebyshev components has a frequency which fits an integer number of cycles into the DFT window then there will be a single peak with no tails. If the peak is not at such a frequency, there will be a sin(x)/x distribution around the peak. Windowing techniques may be used to improve the resolution of the system and Figure 4 shows the effects of using different windows in the analysis.

WINDOW	DISCRETE FORM
RECTANGULAR	W(n)=1
TRIANGULAR	W(n)=1-2n/N
HANN	W(n)=cos (pi.n/N)
HAMMING	W(n)=0.54+0.46cos(2.pi.n/N)

Table 1 Various DFT windows

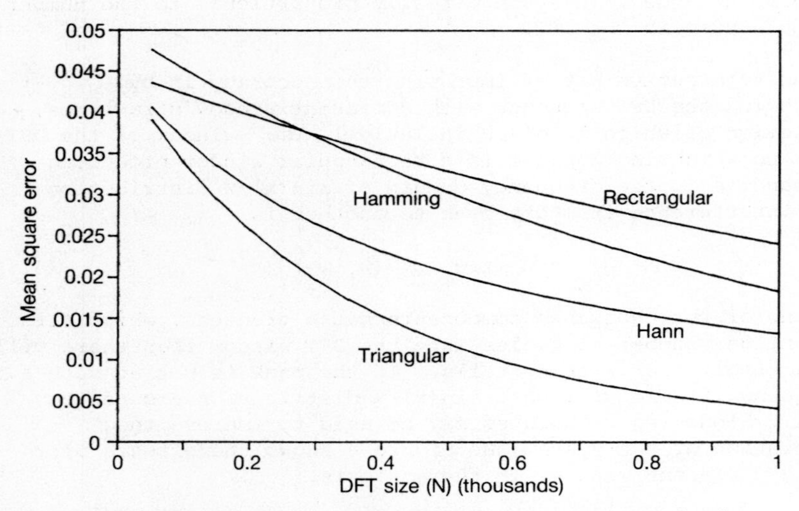

Fig. 4 Effects of DFT windows on eigen-analysis

6. CONCLUSION

In this paper, an alternative technique has been introduced to solve certain linear algebra problems. The method is a hybrid technique which incorporates signal-processing algorithms onto purely algebraic problems. It has been shown that Chebyshev polynomials provide a valuable tool in converting linear systems into frequency-domain problems where it can be conveniently solved by the Fourier-transform technique. The algorithm can be applied to real symmetric matrices but it can be further modified for dense non-symmetric systems by taking into account that the product of a matrix and its transpose is also symmetric.

REFERENCES

[1] Bramley, R.G. and Creasey, D.J., (1988) A New Approach to Signal Processing, Proceedings of the Workshop on Digital Signal Processing, University College London, Torrington Place, London.

[2] Cullum, J. and Donath, W.E., A Block Lanczos Generalisation of the Symmetric S-STEP Lanczos Algorithm, Technical Report RC 4845, Mathematics Department, IBM Research Centre, New York.

[3] Gantmacher, F.R., (1959) The theory of matrices, Chelsea Publishing Company, New York.

[4] Lanczos, C., (1953) Applied analysis, Pitman, London.

[5] Wilkinson, J.H., (1965) The Algebraic Eigenvalue Problem, Oxford University Press.

MATRIX DIAGONALIZATION ALGORITHMS FOR OVERSIZED PROBLEMS ON A DISTRIBUTED-MEMORY MULTIPROCESSOR

H. Park
(Department of Computer Science, University of Minnesota, USA)

ABSTRACT

The singular value decomposition (SVD) and the symmetric eigenvalue decomposition (EVD) are known to be two of the major computations required in many scientific computations. On parallel architectures, Jacobi methods have been established as one of the most popular algorithms for computing the SVD and symmetric EVD due to their excellent parallelism. Often parallel Jacobi algorithms are developed under the assumption that enough processors are available so that the matrix is distributed over the processors by two columns or two by two submatrices. Obviously, there is an upper limit on the problem size in implementing the algorithms with the above assumption on a parallel architecture. We propose algorithms to implement oversized problems on a given parallel architecture with good load balancing and minimal message passing. Performance of the proposed algorithms varies greatly depending on the relation between the problem size and the number of available processors. We give theoretical performance analyses which suggest the faster scheme for a given problem size on a given distributed-memory multiprocessor. We present the implementation results of the algorithms on the NCUBE/seven hypercube.

1. INTRODUCTION

The Jacobi method has been one of the fastest algorithms for parallel computation of the singular value decomposition (SVD) and the symmetric eigen value decomposition (EVD). In this paper, we consider an important practical problem concerning the application of the parallel Jacobi methods to matrices that are oversized relative to the number of available processors. In the ideal environment for developing Jacobi methods on a distributed memory architecture, we have enough processors to distribute the matrix in a certain way : for a matrix of even order n there are n/2 x n/2 processors so that the matrix can be distributed by 2x2 submatrices or n/2 processors so that the matrix is distributed by column pairs over the processors. When the matrix order n is larger than two times the number of available processors, p, then we say that the matrix is oversized. The problem of processing oversized matrices on a given fixed sized array has been dealt with previously in the literature [1,10,11,12]. However, most of the investigation has been confined to matrices of which the orders are such that we can distribute them over the processors by square blocks having an even order or column blocks containing an even number of columns. When the above assumptions are not satisfied, the matrix is padded with zero columns to increase the column dimension in order to satisfy the assumptions. We first review the one-sided Jacobi methods [3,4] tailored for the efficient implementation of Jacobi algorithms on distributed memory architectures. Then, we propose solutions to the oversized problem, and analyze the behaviour of the solutions theoretically. The performance of each method varies greatly depending on the relation between the matrix order n and the number of available processors p. The performance analyses suggest the faster technique for the given n and p. In the last section, we show the implementation results on the NCUBE/seven hypercube, which confirm the validity of the analyses and demonstrate that choosing a good scheme accelerates the performance greatly.

2. ONE-SIDED JACOBI ALGORITHM

Computing the EVD of a symmetric nxn matrix $A \equiv (a_{ij})$ via Jacobi methods, we repeatedly apply plane rotations, to make the matrix converge to a diagonal form with eigenvalues of the matrix A on its diagonal. If we denote the Jacobi rotation in a plane (k,m) as $J(k,m,\theta)$, then the cosine and sine pair $c \equiv \cos\theta$ and $s \equiv \sin\theta$ annihilating the a_{km} and a_{mk} elements satisfy the following relation:

$$\begin{pmatrix} c & s \\ -s & c \end{pmatrix}^T \begin{pmatrix} a_{kk} & a_{km} \\ a_{mk} & a_{mm} \end{pmatrix} \begin{pmatrix} c & s \\ -s & c \end{pmatrix} = \begin{pmatrix} d_1 & 0 \\ 0 & d_2 \end{pmatrix},$$

i.e.,

$$a_{km}(c^2 - s^2) + (a_{kk} - a_{mm})cs = 0. \qquad (2.1)$$

Many Jacobi algorithms differ essentially in the method of choosing the sequence of planes where the transformations take place. This sequence is called an ordering. Jacobi algorithms for the symmetric EVD are summarized in Algorithm 1.

Algorithm 1. Two-sided Jacobi: Given a symmetric nxn matrix A, compute its EVD.

1. Choose an ordering and initialize V as the identity matrix I of order n.
2. Repeat until convergence:
   Sweep through the n(n-1)/2 planes according to the chosen ordering:
   2.1. Determine J via (2.1)
   2.2. Apply J to obtain $A:=J^TAJ$
   2.3. Accumulate J into $V:=VJ$ □

The columns of the resulting matrix V are the computed eigenvectors and the diagonal elements of A are the computed eigenvalues.

For computing the SVD of a square matrix A, we can choose two Jacobi rotations J and K that rotate through the angles θ_1 and θ_2 respectively, satisfying the following relation to annihilate both a_{km} and a_{mk},

$$\begin{pmatrix} c_1 & s_1 \\ -s_1 & c_1 \end{pmatrix}^T \begin{pmatrix} a_{kk} & a_{km} \\ a_{mk} & a_{mm} \end{pmatrix} \begin{pmatrix} c_2 & s_2 \\ -s_2 & c_2 \end{pmatrix} = \begin{pmatrix} d_1 & 0 \\ 0 & d_2 \end{pmatrix}$$

i.e., the relation

$$\begin{cases} \tan(\theta_1 + \theta_2) = (a_{mk} + a_{km})/(a_{mm} - a_{kk}), \\ \tan(-\theta_1 + \theta_2) = (a_{mk} - a_{km})/(a_{mm} + a_{kk}) \end{cases} \qquad (2.2)$$

where $c_i \equiv \cos\theta_i$ and $s_i \equiv \sin\theta_i, 1 \le i \le 2$.

To implement the above algorithms on distributed memory architectures, we initially distribute the matrix by the columns over the processors. The major cost of communications will be incurred if we explicitly update the rows as in stage 2.2 by applying J^T from the left of the matrix A, which requires rotation parameter passing to all the other processors since the matrix is distributed by columns. We can reduce this communication cost by implementing a one-sided variation of the Jacobi scheme [3-6]. In one-sided Jacobi methods for the EVD, instead of applying the rotations from the left side of the matrix explicitly, we accumulate the rotation parameters and retrieve the matrix elements only when necessary.

Algorithm 2. One-sided Jacobi: Given a symmetric nxn matrix $\bar{A} = A$, compute its EVD.

1. Choose an ordering and initialize V as the identity matrix I of order n.
2. Repeat until convergence:
   Sweep through the (n-1)/2 planes according to the chosen ordering:
   2.1. Determine J via (2.1)
   2.2 Apply J to obtain $\bar{A}:=\bar{A}J$
   2.3 Accumulate J into $V:=VJ$
3. Compute the diagonal elements of A □

In stage 2.1, we need to choose the rotation J so that the off-diagonal elements in the (k,m) and (m,k) positions of the matrix A are annihilated after the transformation. At any stage, we do not have the matrix A, but the matrices $\bar{A}$ and V only. But the relation (2.1) for computing the rotation parameters c and s shows that only three elements of A are necessary for computing J, which we can compute through inner products within each processor: $a_{kk} = \langle v_k, \bar{a}_k \rangle$, $a_{mm} = \langle v_m, \bar{a}_m \rangle$, and $a_{km} = \langle v_k, \bar{a}_m \rangle$. Note that although three extra inner products are introduced, this scheme is much faster than the two-sided algorithm on a distributed memory architecture like a hypercube since it avoids message passing for row updating, and is particularly effective if the nodes have vectorization ability.

The schemes we introduce to deal with oversized problems for the EVD work exactly in the same way for the computation of the SVD. For a one-sided SVD computation, there is the well known method by Hestenes [6]. We would like to emphasize that the rotation matrix J in the one-sided Algorithm 2 is identical to J of the two-sided Algorithm 1. If we use Hestenes' rotation parameters for computing the symmetric EVD, the method does not converge to a diagonal form for certain classes of matrices [4]. Interested readers for the parallel SVD computation can refer to [1].

3. A RING ORDERING

We first present a ring ordering developed by Eberlein [3,5]. In the next section, we extend this ordering to deal with the oversized problem. The ring ordering we present is optimal on a ring connected architecture in that it finishes one sweep in a minimum number of stages (n-1 stages if n is even and n stages if n is odd) and each processor sends and receives minimal amount of communication in one sweep. Also, it requires only nearest neighbour communication and is easily generalized for any order n. There exist other orderings with all these properties [5,7,8]. Since the schemes we will introduce can be developed by applying the idea to be presented in this paper on other orderings, we will give a detailed discussion that is based on the ring ordering we present in this section.

We assume that for a matrix of an even order n we have n/2 ring connected processors so that two columns can be stored in each processor. For a matrix of an odd order n, we assume (n - 1)/2 processors. In the following figures, each rectangle represents a processor and two numbers inside a rectangle represent the indices of the two columns allocated to the processor. We illustrate the ring ordering for even n with the case n=8 and p=4 in Figure 1. Two schemes I and II in Figure 1 are used alternately, to generate the index set of each stage.

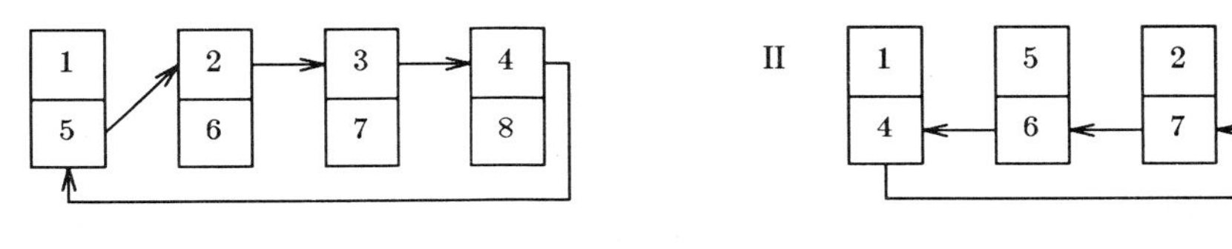

stage 1. (1, 5) (2, 6) (3, 7) (4, 8)
stage 2. (1, 4) (5, 6) (2, 7) (3, 8)
stage 3. (1, 6) (5, 7) (2, 8) (3, 4)
stage 4. (1, 3) (6, 7) (5, 8) (2, 4)
stage 5. (1, 7) (6, 8) (5, 4) (2, 3)
stage 6. (1, 2) (7, 8) (6, 4) (5, 3)
stage 7. (1, 8) (7, 4) (6, 3) (5, 2)

Fig. 1 Ring ordering scheme for n = 8 and p = 4 and one sweep

The ring ordering requires only one send and one receive per stage and finishes one sweep in n - 1 stages. The pattern returns to its original state after 2(n-1) stages.

When n is odd, we can store one extra column in the leftmost processor and apply the same scheme as in the even case to get all possible index pairs in n stages. We illustrate the scheme and one sweep for n = 9 and p = 4 in Figure 2, where the index in the dotted box of the first processor represents the column that is not involved with any computation and the dashed arrow represents the column rearrangement within the first processor.

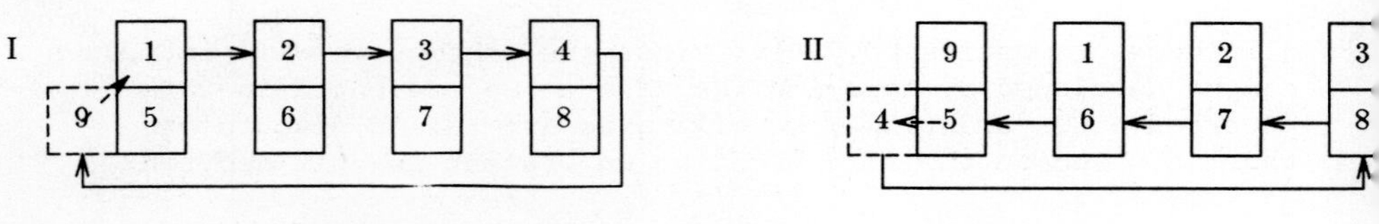

stage 1. 9 (1, 5) (2, 6) (3, 7) (4, 8)
stage 2. 4 (9, 5) (1, 6) (2, 7) (3, 8)
stage 3. 5 (9, 6) (1, 7) (2, 8) (3, 4)
stage 4. 3 (5, 6) (9, 7) (1, 8) (2, 4)
stage 5. 6 (5, 7) (9, 8) (1, 4) (2, 3)
stage 6. 2 (6, 7) (5, 8) (9, 4) (1, 3)
stage 7. 7 (6, 8) (5, 4) (9, 3) (1, 2)
stage 8. 1 (7, 8) (6, 4) (5, 3) (9, 2)
stage 9. 8 (7, 4) (6, 3) (5, 2) (9, 1)

Fig. 2 Ring ordering scheme for n = 9 and p = 4 and one sweep;

4. ALGORITHMS FOR OVERSIZED PROBLEMS

Using some examples, we will explain the notations used in the rest of the paper. The notation [1 2 3 4]*[5 6] denotes the set of all possible pairs between the indices in the set {1 2 3 4 } and the set { 5 6 }, thus it denotes the set of pairs {(1,5)(1,6)(2,5)(2,6)(3,5)(3,6)(4,5)(4,6)} of size 4x2. The notation [1-2-3]*[4-5-6] denotes the union of the set [1 2 3]*[4 5 6] and a set of all possible index pairs out of the set { 1 2 3 } and out of the set { 4 5 6 }. Thus, it is the set of all possible index pairs out of the set { 1 2 3 4 5 6 }, which is the set of pairs,

{(1,4)(1,5)(1,6)(2,4)(2,5)(2,6)(3,4)(3,5)(3,6)(1,2)(1,3)(2,3)(4,5)(4,6)(5,6)

of size $6 \times \frac{5}{2}$. The sets [1 2 3 4]*[5 6] and [1-2-3]*[4-5-6] are ordered sets, where the order is as given in the above examples. We will assume that the number of processors p is an even number since we are particularly interested in the hypercube implementation of the algorithm and it is possible to find a ring of even length in a hypercube. But, the schemes we introduce in this paper are not restricted to the case when p is even, they can be applied to any processor size p. Let $T_{com}(n)$ be the time for completing stages 2.1 to 2.3 of Algorithm 2 for one pair of columns of length n, $T_{msg}(n)$ be the time for sending one vector of length n, and T_s be the start up time for communication.

In this section, we introduce two methods for applying parallel Jacobi algorithms to the oversized matrices. Three criteria we use in developing the methods are: (1) good load balancing should be achieved, (2) one sweep should finish without any repetition of any index pairs in as few stages as possible, and (3) the method should be applied easily for any given n and p with a systematic rule to generate the index pairs in each stage and reorganize them from one stage to the next. The first is the extension method, where we extend the column dimension by padding the matrix with zero columns so that the matrix can be distributed by an equal number of columns over the processors, then apply the ring ordering scheme for even n on column blocks. The extension method can be subdivided according to the ways to distribute the extended matrix over the processors. The second is the residue method. In this method, instead of extending the column dimension like in the extension method, we distribute the columns over the processors by an equal number and store the rest in the first processor. We then apply the ring ordering scheme for odd n. The performance of each scheme varies greatly depending on the relation between the matrix order n and the number of available processors p, which we will show at the end of this section.

4.1 Extension method

In this method, we pad the matrix with zero columns so that we can distribute the matrix over the processors by an equal number of columns. Let us first consider the simplest case when the matrix size is divisible by two times the number of processors. This assumption was made in most of the approaches dealing with the oversized problems [2,10,11,12]. After the matrix is distributed by an equal even number of columns onto the processors, we can divide the columns within each processor into two bins and move the columns around by applying the ring ordering scheme for even n of Section 3 to column blocks. Figure 3 illustrates the case of n = 12 and p = 2.

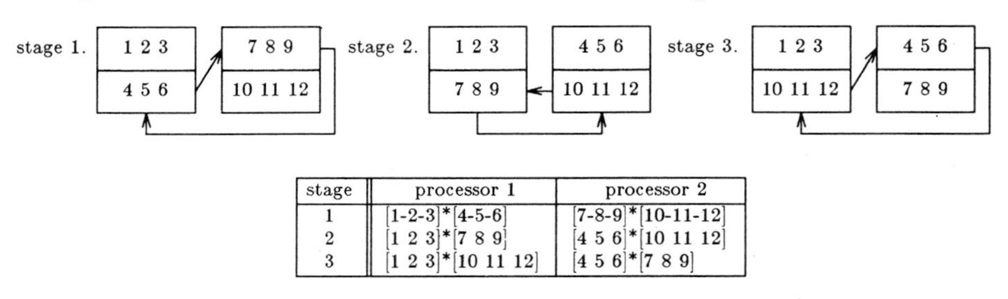

stage	processor 1	processor 2
1	[1-2-3]*[4-5-6]	[7-8-9]*[10-11-12]
2	[1 2 3]*[7 8 9]	[4 5 6]*[10 11 12]
3	[1 2 3]*[10 11 12]	[4 5 6]*[7 8 9]

Fig. 3 Block ring ordering scheme for n = 12 and p = 2 and one sweep

It is clear that every possible column pair is in the same processor at least once throughout 2p - 1 stages. Within each processor, if every possible column pair (i,j) where i is from one bin and j is from the other is processed in each stage, then no pair will be repeated within 2p - 1 stages. The only missing pairs are the ones that can be formed within each bin. They can be recovered if they are processed in one stage of every sweep. Note that one sweep is finished in 2p - 1 stages with minimal communication.

When the matrix size is not divisible by two times the number of processors, 2p, we pad the matrix with zero columns to make its column dimension divisible by 2p. There are many ways to map the properly extended matrix onto the available processors. We consider two mappings: the strip mapping and the wrap mapping [9]. In the strip mapping, the matrix is divided into partitions of the column blocks of the same size and distributed over the processors. In the wrap mapping, the columns are distributed over the processors in wrap around fashion. It is shown that the wrap mapping gives better load balancing in many other computational tasks [9]. Figure 4 shows the case when p = 2 and n = 9 for the strip mapping, where each 0 represents the column that is padded.

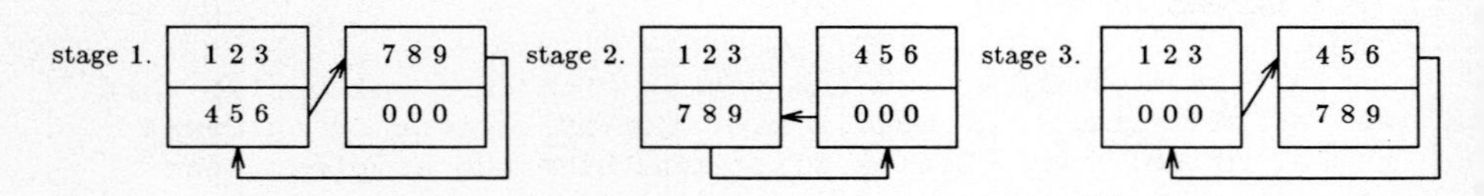

Fig. 4 One sweep of the extension method with strip mapping for n = 9 and p = 2.

In stage 2, while processor 1 spends $3^2 T_{com}(9)$ time for computation, processor 2 is idle, which gives bad load balancing. We can see that the problem may become worse as the matrix size becomes larger. The total computation time required for one sweep of the above example is

$$T_{com}^{strip}(9) \equiv (6 \times \frac{5}{2} + 3x3 + 3x3) T_{com}(9) = 33 T_{com}(9).$$

In the wrap mapping, columns are distributed over the first bin of every processor and then the second bin of every processor, repeatedly until every column is stored. The wrap mapping gives better load balancing in Jacobi algorithms with the pairing scheme of Figure 3, which is shown in Figure 5.

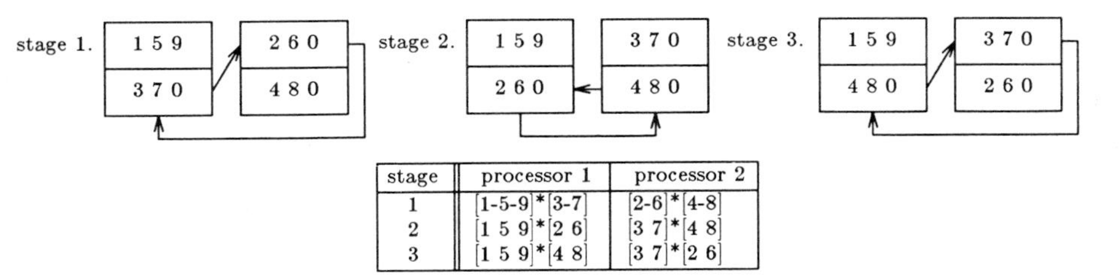

stage	processor 1	processor 2
1	[1-5-9]*[3-7]	[2-6]*[4-8]
2	[1 5 9]*[2 6]	[3 7]*[4 8]
3	[1 5 9]*[4 8]	[3 7]*[2 6]

Fig. 5 One sweep of the extension method with wrap mapping for $n = 9$ and $p = 2$.

The total computation time required for one sweep with the wrap mapping for the above example is

$$T_{com}^{wrap}(9) = (5x \frac{4}{2} + 3x3 + 3x2)T_{com}(9) = 22T_{com}(9).$$

We choose to use the wrap mapping for the extension method for better load balancing.

The extension method can be summarized as follows. Given n and p, we find the smallest nonnegative integer r_e that satisfies the relation

$$n + r_e = p*m_e \tag{4.1}$$

for some positive integer m_e. Then, r_e is the number of zero columns to pad the matrix with so that the matrix is distributed over the processors by equal number of columns, m_e. Note that m_e does not need be an even number. When m_e is an even number, i.e., $m_e = 2k_e$ for some integer k_e, columns are divided into two bins of the same size k_e within each processor. When m_e is an odd number, i.e., $m_e = 2k_e - 1$ for some inter k_e, columns are divided into two bins of size k_e and $k_e - 1$. Then, the ring ordering scheme for even n of Section 3 is applied on column blocks. The maximum work load for each stage when m_e is even is summarized in Table 1.

maximum work load			
stage	$T_{com}(n)$	$T_{msg}(n)$	T_s
1	$k_e(2k_e - 1)$	k_e	1
2	k_e^2	k_e	1
.	.	.	.
.	.	.	.
.	.	.	.
2p - 1	k_e^2	k_e	1

Table 1 Work load for one sweep of the extension method on p processors with $2k_e$ columns each.

Thus, the total time for one sweep of the extension method with $2k_e$ columns in each processor is

$$T^{ee}_{1sweep}(n) \equiv (k_e(2k_e-1)+k_e^2(2p - 2))T_{com}(n)+k_e(2p-1)T_{msg}(n)+(2p -1) \tag{4.2}$$

When m_e is odd, the maximum load for the stages from 2 to 2p - 2 are the same as when m_e is even since there is at least one processor with $2k_e$ columns, which dominates the work load. The maximum work load is summarized in Table 2.

maximum work load			
stage	$T_{com}(n)$	$T_{msg}(n)$	T_e
1	$(2k_e - 1)(k_e - 1)$	k_e	1
2	k_e^2	k_e	1
.	.	.	.
.	.	.	.
.	.	.	.
2p - 2	k_e^2	k_e	1
2p - 1	$k_e(k_e - 1)$	k_e	1

Table 2 Work load for one sweep of the extension method on p processors with $2k_e - 1$ columns each.

Thus, the total time for one sweep of the extension method with $2k_e - 1$ columns in each processor is

$$T^{oe}_{1sweep}(n) \equiv ((k_e-1)(3k_e-1)+k_e^2(2p-3))T_{com}(n) + k_e(2p-1)T_{msg}(n)+(2p-1)T_s. \qquad (4.3)$$

4.2 Residue Method

In the extension method, we pad the matrix with zero columns so that we can distribute the matrix by an equal number of columns over the processors. In the residue scheme, we distribute the columns over the processors by equal numbers without expanding the column dimension and put the remainder in the first processor. The residual scheme can be summarized as follows.

Given n and p, we find nonnegative integers r_r and m_r satisfying the relation

$$n = p*m_r + r_r. \qquad (4.4)$$

Then we distribute m_r columns to each processor except that the first processor gets the remaining r_r columns in addition to its own m_r columns. In each processor, columns are divided into two bins as in the extension method. In the first stage of every sweep, all possible pairs in two bins within each processor are coupled. Additionally, in the first processor, there is one extra bin where r_r extra columns are stored initially, and in the first stage of every sweep, all possible pairs in the extra bin are also processed. In the subsequent stages, all possible pairs in different bins are coupled in each processor. In Figure 6, we show how the column blocks are shuffled when we apply the ring ordering scheme for odd n of Section 3 on the column blocks. The number inside each pair of parentheses represents the number of columns in the bin.

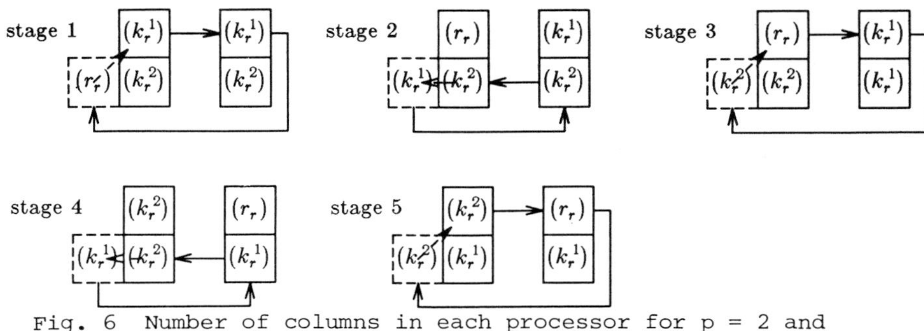

Fig. 6 Number of columns in each processor for p = 2 and $n = p*(k_r^1 + k_r^2) + r_r$.

Let us first assume that $m_r = 2k_r$ for some integer k_r, thus $k_r^1 = k_r^2$ in Figure 6. We can see that the first stage is inevitably badly balanced unless $r_r = 0$ because of the extra work that the first processor has to perform. The maximum work load for each stage for even m_r is summarized in Table 3, from which we can tell that better load balancing can be achieved if m_r and r_r are chosen so that r_r is close to $k_r = m_r/2$.

	maximum work load		
stage	$T_{com}(n)$	$T_{msg}(n)$	T_s
1	$k_r(2k_r - 1)+r_r(r_r-1)/2$	$\max(k_r,r_r)$	1
2	$k_r\max(k_r,r_r)$	$\max(k_r,r_r)$	1
.	.	.	.
.	.	.	.
.	.	.	.
2p + 1	$k_r\max(k_r,r_r)$	$\max(k_r,r_r)$	1

Table 3 Work load for one sweep of the residue method on p processors for $n = p*2k_r + r_r$

Thus, the total time for one sweep of the residue method for $n = p*2k_r + r_r$ on p processors is

$$T_{1sweep}^{er}(n) = (k_r(2k_r - 1)+ r_r(r_r - 1)/2+2pk_r\max(k_r,r_r))T_{com}(n) + (2p + 1)\max(k_r,r_r)T_{msg}(n) + (2p + 1)T_s. \tag{4.5}$$

The maximum work load for each stage for odd $m_r = 2k_r - 1$ is summarized in Table 4, and in this case, $k_r^1 = k_r$ and $k_r^2 = k_r -$ in Figure 6.

maximum work load			
stage	$T_{com}(n)$	$T_{msg}(n)$	T_s
1	$(k_r-1)(2k_r - 1)+r_r(r_r-1)/2$	$\max(k_r,r_r)$	1
2	$(k_r-1)\max(k_r,r_r)$	$\max(k_r,r_r)$	1
3	$\max(k_r^2,(k_r-1)r_r)$	$\max(k_r,r_r)$	1
.		.	.
p + 1	$\max(k_r^2,(k_r-1)r_r)$	$\max(k_r,r_r)$	1
p + 2	$k_r\max(k_r,r_r)$	$\max(k_r,r_r)$	1
.	.	.	.
2p - 1	$k_r\max(k_r,r_r)$	$\max(k_r,r_r)$	1
2p	$k_r\max(k_r-1,r_r)$	$\max(k_r,r_r)$	1
2p + 1	$k_r\max(k_r-1,r_r)$	$\max(k_r,r_r)$	1

Table 4 Work load for one sweep of the residue method on p processors for $n = p*(2k_r-1)+r_r$

Thus, the total time for one sweep of the residue mapping for an odd n on p processors is approximately

$$T^{or}_{1sweep}(n) = ((k_r-1)(2k_r - 1) + r_r(r_r -1)/2 \tag{4.6}$$

$$+ (p - 1)k_r\max(k_r,r_r)+(p + 1)\max(k_r^2,(k_r - 1)r_r))T_{com}(n)$$

$$+ (2p + 1)\max(k_r,r_r)T_{msg}(n) + (2p + 1)T_s.$$

5. IMPLEMENTATION ON THE NCUBE/SEVEN HYPERCUBE

Our implementation results are obtained for the symmetric EVD on NCUBE/seven hypercube. The code was written in Fortran 77. We performed a series of tests on up to 16 processors. The test matrices are generated randomly with a uniform distribution on the interval [-1,1]. The number of sweeps was assigned as log(n) + 2 for a matrix of order n [1]. Because of the limited space, we summarize our results in one Table 5. In Table 5, ncol represents the number of columns each processor gets and the number of extra columns the first processor gets (in brackets) initially. Speedup is computed based on timing of the two-sided Jacobi algorithm with cycle-by-rows ordering run on one node. The speedup and the

p	n	method	ncol	$T_{com}(n)$	$T_{msg}(n)$	T_s	time (sec)	speedup	efficiency (%)
4	79	extension	20	790	70	7	110.192	3.525	88.12
		residue	18[7]	822	81	9	115.696	-	-
4	81	extension	21	925	77	7	139.183	-	-
		residue	18[9]	837	81	9	121.048	3.459	86.48
8	33	extension	5	133	45	15	6.063	-	-
		residue	4[1]	70	34	17	4.487	5.494	68.87
8	34	extension	5	133	45	15	6.451	-	-
		residue	4[2]	71	34	17	4.646	5.526	69.08
8	98	extension	13	757	105	15	150.094	-	-
		residue	12[2]	643	102	17	121.313	6.602	82.53
16	70	extension	5	277	93	31	33.114	8.157	50.98
		residue	4[6]	405	198	33	55.775	-	-
16	98	extension	7	497	124	31	89.614	-	-
		residue	6[2]	304	99	33	60.782	13.837	86.48

Table 5 Theoretical and actual performance of two methods

efficiency in %, which is the speed up divided by the number of processors, are computed only for the faster of the two methods. The coefficients for $T_{com}(n)$, $T_{msg}(n)$, and T_s give us the theoretical estimate of the timing for one sweep of each method using the formula (4.2)-(4.3) and (4.5)-(4.6). Even though the residue method suffers from bad load balancing in the first stage of every sweep, it can perform better than the extension method since the load balancing in the extension method can be worse depending on the relation between n and p. For example, when n = 98 and p = 16, the residue method took only about 68% of the time required for the extension method. Note that the theoretical and experimental timings agree very well. Thus, given n and p, we can use the results of the theoretical analyses to predict the faster method.

REMARKS

We have developed methods for solving oversized problems, which finish one sweep in a minimum number of stages without any repetition of pairs. It would be advantageous to relax this condition and make use of the idle time by repeating some pairs since load balancing is often not achieved. For example, in Figure 5, since processor 2 has only four column pairs to process while processor 1 has six pairs, processor 2 may repeat two pairs, e.g. (3,4) and (3,8), which would not require any extra time. In our implementation, we have fixed the number of sweeps beforehand, which depends only on the matrix order n. But, if we make use of the idle time by repeating some pairs, then it would be more appropriate to do convergence checking periodically: the repeated pairs may change the convergence behaviour of the methods greatly depending on the relation between n and p.

REFERENCES

[1] Brent, R.P., Luk, F.T. and Van Loan, C.F., (1985) Computation of the singular value decomposition using mesh-connected processors, *J. VLSI Computer Systems*, **1**, pp. 242-270.

[2] Bischof, C., (1987) Computing the singular value decomposition on a distributed system of vector processors, Report, TR 87-869, Dept. of Computer Science, Cornell University.

[3] Eberlein, P.J., (1987) On using the Jacobi method on the hypercube, Proc. of the Second Conference on Hypercube Multiprocessors, Ed. M.T. Heath, pp. 605-611.

[4] Eberlein, P.J., (1987) On one-sided Jacobi methods for parallel computation, *SIAM J. Alg. Disc. Meth.*, **8** pp. 790-796.

[5] Eberlein, P.J. and Park, H., (1989) Efficient implementation of Jacobi algorithms and Jacobi sets on distributed memory architectures, Report TR89-10, Dept. of Computer Science, Univ. of Minnesota.

[6] Hestenes, M.R., (1958) Inversion of matrices by biorthogonalization and related results, *J. Soc. Indust. Appl. Math.*, **6**, pp. 51-90.

[7] Luk, F.T. and Park, H., (1989) On parallel Jacobi orderings, *SIAM J. Sci. Statist. Comput.*, **10**, pp. 18-26.

[8] Luk, F.T. and Park, H., (1989) A proof of convergence for the parallel Jacobi SVD algorithms, *IEEE Transactions on Computers* 38, pp. 806-811.

[9] O'Leary, D.P. and Stewart, G.W., (1985) Data-flow algorithms for parallel matrix computations, *Comm. ACM*, **28**, pp. 841-853.

[10] Schreiber, R., (1986) Solving eigenvalue and singular value problems on an undersized systolic array, *SIAM J. Sci. Stat. Comput.*, **7**, pp. 441-451.

[11] Scott, D S., Heath, M.T. and Ward, R.C., (1986) Parallel block Jacobi eigenvalue algorithms using systolic arrays, *Lin. Alg. and its Appl.*, **77**, pp. 345-355.

[12] Van Loan, C., (1985) The block Jacobi method for computing the singular value decomposition, Report, TR 85-680, Dept. of Computer Science, Cornell Univ.

VI PARALLEL ALGORITHMS

SIGNAL PROCESSING COMPUTATIONAL NEEDS: AN UPDATE

J.M. Speiser and H.J. Whitehouse
(Naval Ocean Systems Center, San Diego)

ABSTRACT

Real-time processing imposes a large computational load consisting of a relatively small number of distinct types of computation. This situation has encouraged the continuing development of fast special purpose algorithms, attached processors, and special purpose computers.

The first modern generation of signal processing algorithms and architectures was based on circulant and Toeplitz matrix computation. Its theoretical basis was provided by the convolutions which occur in time-invariant linear systems and the important role of sinusoids as eigenfunctions of every such system The key computational algorithms of this generation were the FTT, the Levinson-Trench algorithm, and their slightly faster descendants. Circulant and Toeplitz methods encouraged the development of attached processor architectures such as the FFT box, the array processor, and the systolic transversal filter. These were the central approaches to meeting signal processing computational needs in the 1965-1980 period, and still play an important role in classical spectral analysis and deconvolution, classical beamforming for uniform linear, circular, and planar arrays, and Fourier-based techniques for image enhancement and data compression.

Although more general matrix-based parallel computation was considered at least as far back as the early 1960s, this field has experienced explosive growth in the 1978-1988 period, and extensive continued development is expected for an additional 5 years. This intense activity is the result of the combination of matrix-based high-resolution signal processing algorithms, high density integrated circuit development making feasible high parallelism in regular designs, and the key concepts of systolic and wavefront processor arrays - which permit the throughput to grow nearly linearly with the number of processors. For a wide range of problems in matrix-based spectrum analysis, beamforming, direction finding, data compression, and pattern recognition, only a few core matrix algorithms for dense matrices are needed to perform the bulk of the computational tasks: matrix-vector or matrix-matrix multiplication, orthogonal triangularization, triangular back-solution, eigensystem solution, singular value decompositions, and solution of the generalized eigenvalue and singular value problems. For all of these tasks, parallel algorithms and architectures are now available which permit throughput to grow linearly with the number of processors - i.e., with constant efficiency independent of the matrix size. These parallel matrix algorithms are being implemented in a variety of ways: on programmable systolic arrays, on hypercube machines, on machines with microprogrammable controller directing a systolic array, and on dedicated hard wired systolic arrays.

A third class of signal processing methods based on time-frequency analysis of nonstationary signals began with the work of Wigner (1932) and Ville (1948) and has seen extensive theoretical development and the beginning of hardware implementation in the 1978-1988 period. The near future will see the combination of time-frequency distribution (TFD) methods with both eigensystem-based signal analysis and wideband ambiguity function and wideband WVD analysis in which Doppler is treated as a time compression or expansion. There will be extensive need for interpolation in wideband TFD analysis, as well as in tomographic signal processing for synthetic aperture and inverse synthetic aperture processing. Each successive generation of signal processing techniques tends to incorporate more detailed signal or noise information into the processing, permitting improved signal analysis for a restricted class of problems. Consequently, each new class of corresponding parallel algorithms and architectures supplements rather than replaces previous methods.

1. INTRODUCTION

This paper provides an updated discussion of the problem of finding a small number of computational building blocks that can address a wide range of computationally intensive signal processing tasks. Emphasis is placed on problems that do not admit the use of structured matrices, since well established efficient architectures are available for the most important structured problems; those involving Fourier transforms and matrices which are circulant or Toeplitz. Well known efficient algorithms for these problems include the FFT and fast Toeplitz solvers such as the Levinson-Trench and Schur methods. In order to consider the additional matrix operations needed for a wider set of tasks, we will examine representative problems in spectrum analysis, beamforming and direction finding, data compression, pattern recognition, and radar imaging.

2. DATA COMPRESSION

Data compression techniques are widely used for speech and television images. For speech, the most popular method is linear predictive coding. For television images, both transform coding and hybrid techniques are used. These methods are based on statistical models for data which have been extensively studied, and use ensemble correlation models. An adaptive data compression using the singular value decomposition was proposed by Harry Andrews [2] [3]. The singular value decomposition (SVD) of a matrix can be viewed as a factorization into the product of a unitary matrix, a diagonal matrix, and another unitary matrix $A = PDQ^H$. The kth column of P is called the kth left singular vector of A. The kth diagonal element of the diagonal matrix, D is called the kth singular value of A, and the kth column of Q is called the kth right singular vector of A. The singular values are real and non-negative, and are assumed to be in decreasing order. This form of the SVD is useful for studying the linear transformation associated with the matrix A, including solving least squares problems. However, for the purpose of data compression, it is more useful to view the SVD as a representation by a sum of rank one matrices, $A = \Sigma_k d_k \mathbf{p}_k \mathbf{q}_k^H$. The sum of the first L terms in this decomposition provides the best approximation of A by a matrix of rank L, in the sense of minimizing the sum of the squared magnitudes of the errors [1]. With this technique, the image is divided into rectangular data blocks, and the current data block is treated as a matrix to be approximated by one of reduced rank. This is similar to transform coding, but uses basis vectors which are derived from the current data block, rather than from a data model or from whole image averages. It should therefore be applicable to data with unknown or time-varying statistics. The reason the method was not widely adopted when it was first introduced was the computation cost of the SVD. However the combination of VLSI devices and parallel architectures for the SVD would now make Andrews' method suitable for a wider class of problems.

3. INTERPRETATION AND APPLICATIONS OF THE SVD

The SVD may be interpreted geometrically as follows: $A = PDQ^H$. P and Q are unitary matrices. They preserve the Euclidean norm of a vector, and may be thought of as generalized rotations. Multiplication by the diagonal matrix D stretches the kth component of a vector by d_k. If the diagonal elements

$d_1 \ldots d_N$ of D are in decreasing order, then d_1/d_N is the condition number of A. It is also the eccentricity of the ellipsoid which is the image of the unit sphere under multiplication by A.

Applications of the SVD include:

- High accuracy least-squares solution with perturbation sensitivity information
- Numerically stable determination of the rank of a matrix
- Approximation of a matrix or linear operator by one of reduced rank
- Data compression for a data source with unknown or time-varying statistics
- High accuracy computation of the eigensystem of $A^H A$ via the SVD of A.

Several extensive introductions to the SVD and its applications are available [1], [4], [5].

3.1 Least-Squares solution via Singular Value Decomposition

The problem to be considered is that of finding the vector $\mathbf{x}$ having minimum Euclidean norm, among those vectors that minimize $\|A\mathbf{x} - \mathbf{y}\|$. Let the matrix A have the singular value decomposition $A = PDQ^H$, where the singular values $d_1 \ldots d_N$ are in decreasing order and exactly r of them are nonzero. Then the matrix A has rank r, and the solution to the least squares problem is $\mathbf{x}_{\mathbf{opt}} = QD_1P^H\mathbf{y}$. The matrix $A^+ = QD_1P^H$ is the Moore-Penrose pseudoinverse of A, and $D_1 = \mathrm{diag}(1/d_1 \ldots 1/d_r, 0, 0 \ldots)$. In practice, only singular values above a threshold determined by roundoff error bounds will be treated as nonzero. The Wieland-Hoffman theorem for singular values [6], p. 287, shows that if a matrix is perturbed, the perturbation of the singular values is no greater than the perturbation of the matrix. If we denote an m by n matrix by A, the perturbation by E, and the kth singular value of a matrix by $\sigma_k(A)$, then

$$\sum_{k=1}^{n} |\sigma_k(A + E) - \sigma_k(A)|^2 \leq \|E\|_F^2$$

If the matrix A or the right hand side vector is only known to within some tolerance, this may also be incorporated into the threshold. In effect the SVD always provides the equivalent of pivoting and rank estimation, as well as permitting the incorporation of uncertainties in the problem formulation.

4. THE GSVD AND AN APPLICATION TO PATTERN RECOGNITION

The Generalised Singular Value Decomposition (GSVD) of Van Loan is a simultaneous reduction of two matrices to diagonal form [6]. If A and B are m by n matrices, with $m \geq n$, then it is possible to write $A = Q_A D_A X$ and $B = Q_B D_B X$, where Q_A and Q_B are unitary, D_A and D_B are diagonal, and X is nonsingular. It then follows that $A^H A - \lambda B^H B = X^H (D_A^2 - \lambda D_B^2) X$. Therefore, the generalized eigenvalues of the matrix pair $(A^H A, B^H B)$ are $(d_A(k)/d_B(k))^2$, where $d_A(k)$ and $d_B(k)$ are the kth diagonal elements of D_A and D_B Numerically stable algorithms for computing the GSVD are now available [7], including a parallel version [8]. The GSVD permits solving the generalized eigensystem for $(A^H A, B^H B)$, without the roundoff error that would be incurred by explicitly computing $A^H A$ and $B^H B$.

When the Hotelling Trace Criterion is used to select measurement vectors for pattern recognition, it is necessary to solve a generalized eigensystem with matrices $S_1 = A^T A$ and $S_2 = B^T B$, where A and B are functions of observed data and S_1 and S_2 may both be singular or nearly so. If the computational word length is short, this computation may best be performed using the GSVD.

5. DIRECTION FINDING

This section provides an overview of selected beamforming and direction finding techniques and their computational requirements. It provides a description of the algorithms and assessment of their range of applicability. It then summarizes the availability of parallel algorithms and architectures for real time implementation of the required computations. Last it recommends areas for future study. With a few noted exceptions, it considers only narrowband processing - i.e. it is assumed that the data snapshots are the Fourier coefficients in one frequency bin for each of the array elements for the current observation interval, and that different frequency bins are processed independently.

5.1 Algorithms Considered

- Classical Beamforming
- MVDR and Capon's Method
- DF by Linear Predictive Spectral Estimation
- MUSIC
- Johnson and DeGraf
- Other MUSIC variants
- Pisarenko Harmonic Retrieval
- ESPRIT
- ESPRIT variants
- Cadzow's Eigenvector Approximation Method
- Maximum Likelihood

5.2 Principal Notation

x	data vector
$x^{(k)}$	data vector for kth snapshot
K	total number of snapshots currently available
M	number of elements in array
B	number of beams
W	total processed bandwidth
$a(u)$	arrival phase shift vector for look direction u. As a function of u, also called the "array manifold"
A	1) general matrix 2) steering phase shift matrix $A = (a(u_1)\ldots a(u_D))$
D	1) number of arriving wavefronts (can be larger than number of sources when multipath arrivals are considered) 2) diagonal matrix in svd or eigenvalue decomposition
s	steering vector, alternate notation for $a(u)$
E	usually denotes statistical expectation
H	used as a superscript to denote Hermitian transpose
P	1) source correlation matrix 2) unitary matrix

Q	1) unitary matrix 2) matrix with orthonormal columns - part of a unitary matrix
U	usually used to denote upper (or right) triangular matrix. In the numerical analysis literature, R is usually used to denote a right triangular matrix. However, since the signal processing literature usually uses R for a correlation matrix, we will usually use U instead, except in the names of numerical algorithms, such as QR decomposition and the QR eigensystem method.
R	spatial correlation matrix. $R = Exx^H$ When it is desired to emphasize the dependence on x, we will use R_x
R_n	1) noise spatial correlation matrix 2) scaled noise correlation matrix
w	weight vector (combined amplitude and phase)
$P D Q^H$	factors in singular value decomposition. P is unitary D is diagonal, and Q is unitary
$Q D Q^H$	factors in eigenvalue decomposition of a Hermitian matrix. Q is unitary, D is diagonal.
n	noise vector
f	1) vector of complex amplitudes for wavefronts arriving at the array 2) frequency
σ^2	noise power: possibly after spatially prewhitening.

6. OVERVIEW AND ALGORITHM ASSESSMENT

6.1 Classical Beamforming

We consider classical beamforming only to establish notation and terminology that will be used in discussing additional algorithms. Standard references for classical beamforming include Ma [9] and Steinberg [10]. For most algorithms we will regard the frequency-domain output of the current data snapshot for the array as a single complex vector, x. The data model used is

$$x = Af + n$$

Classical frequency domain beamforming via phase shift and sum is the frequency domain equivalent of delay and sum time domain, beamforming. For each look direction u, it forms the inner product of the steering phase shift vector $s = a(u)$ with the data vector x. This inner product, $b = [a(u), x]$ is the spatial equivalent of a matched filter. A derivation for general array geometries with simplifications for special array geometries has been provided by Speiser, Whitehouse, and Berg [11]. The function a(u) is sometimes called the "array manifold" [12].

Shading may be used to reduce sidelobe height, at the expense of reduced resolution. The computational load for classical beamforming, exclusive of the temporal Fourier transforms and generation of steering vectors is B * M * W complex multiply-adds per second, where B is the number of beams, M is the number of array elements, and W is the bandwidth. If focussing is performed, B is the number of combined directions with focal depths that are examined. This number can be reduced, of course, for special array geometries that permit the beamforming to be performed via an FFT. These include uniform line and planar arrays, circular and cylindrical arrays, and logarithmic linear and spiral arrays [11].

6.2 MVDR Beamforming

Minimum Variance Distortionless Response (MVDR) beamforming attempts to reject interfering sources while maintaining unit gain and zero phase shift for each look direction [13] [14]. It is sometimes referred to as "Optimum Array Processing" [13]. Unlike classical shaded beamforming, it does not try to create uniformly low sidelobes in the response pattern, but only to form nulls in the direction of interfering sources. For each steering vector s = a(u), it forms b = [w,x], where the weight vector w is chosen to minimize the expected power output, while providing unit gain and zero phase shift for a signal arriving from direction u, i.e. pick w to minimize $E|b|^2$ subject to [w,s] = 1.

$$w_{opt} = \frac{R^{-1}s}{[R^{-1}s,s]}$$

MVDR is not a high resolution method per se, but it is recommended as a baseline for comparison. It is quite robust, useful with any geometry, and very well understood, both in terms of of its performance and parallel algorithms and architectures for real time implementation, including variants incorporating multiple linear constraints [15] [16].

Apart from resolution MVDR does have one limitation: coherent multipath arrivals in the same frequency bin may cause it to effectively lose sensitivity. That is, when the beam is steered to one of the arrivals, even though the weight vector provides unit gain in that direction, the response to a coherent multipath arrival may reduce the signal component of the output.

6.3 Capon's Method

Capon's method estimates the angular spectrum, or power versus arrival direction as $1/[R^{-1}s,s]$ [18], [19], [21]. This is the average power output of an MVDR beamformer with steering vector s. The signal processing literature (including Schmidt's papers) frequently refers to Capon's method as "maximum likelihood". This is a source of confusion, since it does not perform maximum likelihood parameter estimation, and does not have the high resolution of true maximum likelihood spectrum estimation/ direction finding. We will regard Capon Maximum Likelihood as just an auxiliary output easily available from an MVDR beamformer.

6.4 DF by Linear Predictive Spectral Estimation

Linear predictive spectral estimation determines the weights of a prediction error filter for a wide-sense stationary random sequence.

Such a filter converts the sequence into white noise. Since the spectral density of the output of a filter is the product of the input spectral density with the magnitude squared of the filter's transfer function, the input spectral density function is proportional to the reciprocal of the magnitude squared of the filter's transfer function. When the input process has an all-pole spectrum, the optimum prediction based on the infinite past is the same as the optimum predictor using a number of past samples equal to the number of poles. This method is sometimes called Maximum Entropy Spectral Estimation - more specifically the Yule-Walker version of Maximum Entropy Spectral Estimation [21],[17]. It has been proposed as a direction finding technique by utilizing the equivalence of spectral estimation for a wide-sense stationary random process with power versus arrival direction estimation for a uniformly spaced line array receiving signals from uncorrelated sources [20]. Linear predictive spectral estimation can provide high resolution, but has several severe problems and limitations, especially as a direction finding algorithm.

1. It is important to correctly estimate the order of the process. Estimators proposed in the literature, usually based on assumed Gaussian distributions, work well in simulations, but have been observed to grossly overestimate the model order for both acoustic and HF data [27],[26].

2. It is difficult, though possible, to extend the method, to a limited set of two-dimensional arrays.

3. Coherent multipaths violate the assumption of spatial stationarity. Some attempts to correct this have been made by assorted spatial averaging techniques.

4 A recent analysis showed that the Yule-Walker and Pisarenko methods both had poor statistical efficiency (far from the Cramer-Rao bounds when the number of samples was large).

5. Even modified versions of linear predictive estimation can have problems with bias errors and line splitting. Because of these problems and the very restricted set of applicable array geometries, I do not recommend further consideration of linear predictive direction finding.

6.5 MUSIC

The MUSIC method is at this time the most thoroughly studied and best understood of the eigensystem based direction finding techniques. It was independently developed by R. Schmidt [12], [22] in the U.S. and by the team of L. Kopp and G. Bienvenu in France [23]. Schmidt's formulation uses the generalized eigensystem of the data correlation matrix R_x and the scaled noise alone spatial correlation matrix R_n. The Bienvenue and Koff formulation (and many others who have studied MUSIC) assume that the noise is spatially white or the data has been spatially prewhitened, so that $R_n = \sigma^2 I$. The noise power, σ^2 is regarded as an unknown parameter. In this case, the generalized eigensystem reduces to the ordinary eigensystem of R_x alone. In discussing MUSIC it is necessary to distinguish between the behaviour with perfect knowledge of the data correlation matrix and scaled noise alone correlation matrix, versus behaviour with estimated or modelled correlation matrices.

MUSIC requires several assumptions:

1. The number of arriving wavefronts (in the frequency bin being analyzed) is strictly less than the number of elements in the array.

2. The columns of the arrival phase shift matrix, A, are linearly independent.

3. The source correlation matrix, P, is strictly positive definite. (Sources can be coherent - but not perfectly coherent).

4. The noise spatial correlation matrix is nonsingular. (This requirement is unlikely to ever cause a problem).

With perfect knowledge of the data correlation matrix and scaled noise spatial correlation matrix, MUSIC proceeds as follows:

Solve the generalized eigensystem

$$R_x z = \lambda R_n z$$

If the generalized eigenvalues are numbered in decreasing order, the last $M - D$ are equal, and the first D generalized eigenvalues are strictly greater. The span of the generalized eigenvectors corresponding to the last $M - D$ generalized eigenvalues is called the noise subspace. Each vector in the noise subspace is orthogonal to the arrival phase shift vector $a(u_d)$ from each source direction. The number of sources is estimated as $D_1 = M - M_0$ where M_0 is the multiplicity of the smallest generalized eigenvalue.

The arrival directions are then determined by looking for array manifold vectors $a(u)$, orthogonal to the noise subspace. This determines the arrival phase shift matrix A. Then the source correlation matrix **P**, can be computed as

$$P = (A^H A)^{-1} A^H (R_x - \lambda_{min} R_n) A (A^H A)^{-1}.$$

Of course it is preferable to use the pseudoinverse of A, computed by an SVD, rather than forming

$$(A^H A)^{-1} A^H$$

in order to avoid problems of ill-conditioning caused by near collinearity of columns of A when signals arrive from nearly the same direction.

Early simulations by Schmidt, using perfect correlation matrices, showed that MUSIC had far higher resolution than classical beamforming and Capon's maximum likelihood method, and did not have the bias problems of the maximum entropy method.

It is important to note that the MUSIC "DOA spectrum" is simply a function that peaks in directions where the steering vector is nearly orthogonal to the noise subspace. That is, peaks correspond to arrival directions, but the DOA spectrum is not a measure of arrival power versus direction. Furthermore, the processing is highly nonlinear, so there is no simple interpretation of the apparent peak-to-sidelobe ratio.

The previous comments must be modified somewhat when estimated correlation matrices are used, but MUSIC still remains a very strong contender for direction finding. When estimated correlation matrices are used:

1. Instead of a smallest generalized eigenvalue with multiplicity (M - D), there is a spread of small generalize eigenvalues. If we pick too large a dimension for the noise subspace, then signals will be missed. If we pick somewhat too small a dimension for the noise subspace, the method degrades gracefully: Spurious peaks will be found in the DOA spectrum, but these can be eliminated through estimation of the source correlation matrix and eliminating peaks corresponding to zero power.

2. With ideal data and scaled noise correlation matrices, MUSIC can localize in the presence of highly correlated (but not perfectly correlated) multipaths. In simulations using estimated correlation matrices, problems have been reported when the correlation between multipath arrivals was around 0.7 or so.

A survey of work on MUSIC and related eigenvector techniques through 1985, with detailed discussion of source covariance matrix estimation and a signal estimation techniques is available [24].

6.6 Johnson and DeGraf

The method of Johnson and DeGraf is a slight modification of MUSIC. In MUSIC, the generalized eigenvalues are only used in deciding on the dimensionality of the noise subspace, not in forming the DOA spectrum - i.e all the basis vectors for the noise subspace are weighted equally in the DOA spectrum. Johnson and DeGraf use a modified DOA spectrum, in which the eigenvector weighting depends on the corresponding eigenvalue, so that in effect the decision as to the dimension of the noise subspace (or equivalently the number of signals present) is a soft decision [25].

In simulations by the inventors, it was found to provide the same performance as MUSIC when the correct number of signals was chosen for both, but to be more robust when errors were made in picking the number of signals. Studies by Gordon Martin confirmed this behaviour when simulated data was used, but showed better performance for MUSIC when actual hf data from the SARA array was used [26]. In any event, the same software or hardware can easily accommodate both MUSIC and the Johnson & DeGraf method.

6.7 *Other MUSIC variants*

Two MUSIC variants intended to permit use with highly correlated multipath arrivals are only applicable for uniformly spaced line arrays. One technique uses averaging of the correlation matrices from subarrays [28], [29]. In effect this tends to restore the Toeplitz structure that one would expect with uncorrelated sources However, it both reduces the effective aperture and the number of arrivals that can be handled by a factor equal to the number of subarrays used. The second technique is called Root MUSIC. For a uniform line array, the steering vectors are sinusoids, so finding steering vectors orthogonal to the noise subspace is equivalent to finding zeros of polynomials on the unit circle. In Root MUSIC, zeros are rejected if they lie too far from the unit circle, and zeros close to the unit circle are projected onto the unit circle [31]. Root MUSIC is reported to provide better threshold performance than classical MUSIC.

6.8 *Pisarenko Harmonic Retrieval*

Two spectral estimation techniques are both sometimes referred to as "Pisarenko's Method". One is a family of methods based on the eigenvalues and eigenvectors of the correlation matrix. Particular members of this family are chosen by picking a monotone function and its inverse. In this report, we discuss only the Pisarenko special harmonic retrieval method [32]. This method determines the frequencies and amplitudes of sinusoids in additive white noise. Although it preceded (and inspired) MUSIC, it is best understood as a special case of MUSIC, with the following additional assumptions:

1. The noise is white.

2. The data correlation matrix is Toeplitz. In df application this corresponds to uncorrelated sources and a uniformly spaced line array.

3. The noise subspace is treated as having dimension equal to 1.

The Pisarenko method requires finding an eigenvector of a Toeplitz correlation matrix corresponding to the smallest eigenvalue, and then finding sinusoid vectors orthogonal to that eigenvector. Although it can provide high resolution it is not recommended for the following reasons:

1. The Toeplitz assumption makes it inapplicable when even partially coherent multipaths are present.

2. Simulation studies and theoretical analysis have shown that the frequency (or angle in the df application) estimates have large variance or poor statistical efficiency.

3. Gordon Martin's studies using hf data have confirmed the large variance of the direction estimates [26].

6.9 ESPRIT

ESPRIT is a very specialized algorithm. It requires an array consisting of two identical subarrays, with the same displacement vector between each element of the first array and the corresponding element of the second array. Alternatively, the total set of array elements may be regarded as a set of M dipoles, with the same differential displacement vector (length an orientation) for each dipole. The following advantages are claimed for ESPRIT in comparison with MUSIC [34], [35], [36], [31]:

1. ESPRIT eliminates the computationally expensive search through the DOA spectrum.

2. It provides better performance in the presence of highly correlated multipaths.

3. It provides a lower noise threshold.

4. It does not require knowledge of the array manifold.

However, the following disadvantages of ESPRIT should also be noted [33]:

1. It is applicable only to a very limited set of array geometries.

2. Even when a two-dimensional array is used, it only produces the same information as other methods using a line array - i.e. arrival angle with respect to the differential displacement axis.

3. If the length of the differential displacement vector exceeds half a wavelength, then ESPRIT will have grating lobe ambiguities.

For applications other than spectrum analysis and direction finding with a uniformly spaced line array spaced closer than spatial Nyquist, the above problems limit the usefulness of ESPRIT.

6.10 ESPRIT Variants

The original formulation of ESPRIT was based on solving a non-symmetric generalized eigensystem. Two recent alternatives have been proposed: Total Least Squares ESPRIT, and Orthogonal Procrustes ESPRIT [37],[38], [39]. They do not remove any of the disadvantages noted above in connection with the original method. Total Least Squares ESPRIT is claimed to provide slightly better performance at low signal to noise ratio. Both of these methods are conceptually much more complicated than MUSIC and the original ESPRIT formulation.

6.11 Cadzow's Eigenvector fitting method

Cadzow's method is a modification of MUSIC that avoids the requirement that the source correlation matrix be strictly positive definite [40]. That is, perfectly coherent sources are permitted. The method starts with the eigensystem and "DOA spectrum" calculations as in MUSIC. The noise is assumed to be spatially white, or prewhitened if necessary. Then it is observed that each eigenvector of the data correlation matrix is a linear combination of steering vectors from the source directions. Source directions are then found by fitting each eigenvector by linear combinations of steering vectors. Usually only a few steering vectors (corresponding to highly correlated arrivals) are used to fit each eigenvector. Since the steering vectors depend nonlinearly on the arrival directions, this is a nonlinear least squares problem, and must be solved iteratively. In order to reduce the computation time, initial approximations are chosen using the MUSIC arrival directions, generally with small perturbations to make the arrival directions distinct.

6.12 Maximum Likelihood Parameter Estimation

This technique (not to be confused with Capon's so called "maximum likelihood" method) is based on the statistical maximum likelihood parameter estimation method, and assumes that the noise distribution is Gaussian. Maximizing the likelihood ratio for jointly estimating the signal amplitudes and arrival directions is then equivalent to a nonlinear least squares fit of the data vectors by linear combinations of steering vectors - a computational task similar to that required by Cadzow's method, but with a much larger computational load. It is known from the statistical literature that maximum likelihood parameter estimation has excellent large sample properties - asymptotically unbiased, consistent, and asymptotically efficient. Recent empirical studies of maximum likelihood direction finding suggest that the small sample behaviour is very competitive with other high resolution techniques, and that the method is robust in the presence of highly coherent sources [41], [42], [43]. It also extends in a straightforward way to broadband processing. The main difficulty with this technique is keeping the computational load under control. In addition to exact maximisation of the likelihood ratio, several approximate alternatives (Nickel's method and the EM or estimate-maximize method) are available [44], [45], [46].

Maximum Likelihood parameter estimation is an important method for current research, but until the computational requirements are better understood, it is not ready for use in system development.

7. NONSTATIONARY SIGNAL ANALYSIS USING THE WIGNER-VILLE DISTRIBUTION

An increasing set of original processing tasks require an extension of spectral analysis to treat nonstationary functions of a time or space variable. Traditional techniques for the analysis of signals and stationary random processes are based on the fact that the sinusoids are eigenfunctions of every shift-invariant linear operator. This includes both the impulse response of a time-invariant linear filter and the covariance function of a wide-sense stationary random process. The Karhunen-Loeve (K-L) expansion represents an arbitrary random process in terms of the eigenfunctions of its covariance function, and therefore permits an expansion in terms of uncorrelated random variables. However, the K-L expansion does not posess convenient properties for the analysis of signals that have been filtered or modulated.

Recent signal analysis techniques using time-frequency distribution (TFD) functions provide a bridge between the traditional stationary signal analysis methods and the yet to be developed methods for treating general nonstationary problems. The Cohen class of time-frequency distributions provide a time-varying spectrum with many useful properties for a wide class of signals. Different TFDs in the Cohen class are selected via the choice of a weight function. The choice of a weight function determines which useful properties will be possessed by the resulting TFD, and no one choice provides all of the possible desirable properties.

One TFD that has a large set of the possible useful properties, and that has received extensive study in recent years for application to signal processing is the Wigner-Ville Distribution (WVD) [50], [51]. The WVD of a real signal $s(t)$ that has complex analytic signal $g(t)$ is defined as

$$W_g(t,\tau) = \int g(t + \tau/2)g^*(t - \tau/2)e^{-i2\pi f\tau}\, d\tau$$

where $g(t) = s(t) + i\hat{s}(t)$ and $\hat{s}(t)$ denotes the Hilbert transform of $s(t)$. The WVD can also be expressed as

$$W_g(t,\tau) = \int K_g(t,\tau)e^{-i2\pi f\tau}\, d\tau$$

where the kernel $K_g(t,\tau)$ is defined as

$$K_g(t,\tau) = g(t + \tau/2)g^*(t - \tau/2).$$

The WVD of a chirp (a function with constant amplitude and quadratic phase) is distributed on a line through the origin of the time-frequency plane, with slope proportional to the chirp rate. The WVD of the convolution of two signals is the convolution in time of their respective WVDs:

$$W_{g*h}(t,f) = \int W_g(t - u,f)W_h(u,f)du$$

These two properties are crucial to a number of applications of the WVD:

- Location of Sources in the Fresnel Region
- Coherent Integration of Lines Emitted by a Moving Source
- Identification of Stationarity Intervals for a Nonstationary Random Process
- Radar Imaging

Since a narrowband signal or signal component in the Fresnel region of a line array produces a quadratic phase shift as a function of position along the array, such a source can be localized in range and bearing via a spatial WVD in the temporal Fourier transform domain [52].

Estimation techniques using signal subspace concepts can be applied to time-frequency distributions by replacing the Fourier transform of the bilinear kernel by a parametric modelling with respect to the variable τ at each time t [63], [64], [62]. Such a parametric high-resolution modified WVD can be used to enhance the resolution attainable in the spatial localization of a source in the Fresnel region of a line array [66], [53].

8. RADAR IMAGING

A new formulation of tomographic imaging for rotating targets is presented in terms of the Wigner-Ville distribution. This new formulation both simplifies the signal processing required for image generation and is applicable in both mono-static and bi-static situations. The target is assumed to have multiple scatterer within each range-Doppler resolution cell so that the received signal will be a sample function of a Gaussian random variable. When the transmitted waveform is a linear frequency modulated chirp then the expected value of the output of a matched filter receiver is the two-dimensional convolution of the range-Doppler scattering function of the target with the auto-ambiguity surface of the transmitted waveform. Under the additional assumption that the transmitted signal has a Gaussian or rectangular envelope then the receiver can be simplified to a short-time Fourier transform receiver using only a priori information about the envelope and the chirp rate of the transmitter. This simplification eliminates the need for detailed phase information at the receiver and thus bi-static operation is feasible. Since most surfaces are rough at optical frequencies this form of signal processing should be particularly useful with optical sensors such as coherent laser radars (LIDARs).

8.1 Introduction to Range-Doppler Imaging

Range-Doppler imaging (scatterer distribution mapping) is a general class of inverse problem which arises in remote sensing applications in either radar or sonar whenever the sensor is in relative motion with respect to the object or objects under study. In principle the sensors can radiate electromagnetic (radar) or acoustic (sonar) signals. The radiated signals are usually in the form of a modulation on a carrier wave and the bandwidth of the modulation can either be small or large relative to the carrier frequency. The ratio of modulation bandwidth to carrier frequency is called the fractional bandwidth and is usually less than unity in most systems. The fractional bandwidth is usually small in the case of radar and thus motion induced Doppler can be treated as a single frequency shift of the entire modulation spectrum. The fractional bandwidth is often large in the case of sonar and thus motion induced Doppler should be treated as a change in time scale of the modulation signal or equivalently the Doppler shift caused by the motion is proportional to the frequency of the components of the transmitted spectrum. For simplicity we will consider only the small fractional bandwidth case and will use radar terminology and examples. However, the signal processing concepts presented in this section can be extended to the large fractional bandwidth case for use in sonar signal processing.

The problem which will be considered here is the mapping of rigid objects which are in motion with respect to a stationary sensor. Although rigid objects are the primary targets of interest in most remote sensing applications non-rigid scatter distributions such as a flight of birds or a school of fish may be mapped if the overall shape of the swarm does not change too much during the mapping operation. By specializing to rigid objects, it is possible in many applications to incorporate a priori or in some cases a posteriori information about the target and thus convert the measured scatterer distribution to a range and cross-range image of the object. Two types of scattering, deterministic and stochastic, have been considered in the literature. This paper will assume the stochastic model of scattering which is usually valid when the target is rough with respect to the wavelength of the illuminations. Thus the inverse problem is to estimate the range-Doppler scattering function of the target which represents the average power reflected from a resolution cell of the target. If the target is essentially two-dimensional, (i.e. a rotating disk), when viewed along the line of sight (LOS) of the radar then the range-Doppler cells of the scattering function from a one-to-one mapping with range and cross-range cells of the object. If the target is essentially three-dimensional, (i.e. a rotating sphere), then the mapping is no longer one-to-one and in general there may be two or more range and cross-range cells with the same range and Doppler (e.g. the rotation axis of the sphere is perpendicular to the LOS of the radar).

Although the analysis of the tomographic imaging method in this paper is stated in terms of a monostatic radar, (i.e. transmitter and receiver either co-located or closely located), the analysis can be extended to a bi-static radar, (transmitter and receiver at different locations). In the bi-static case the transformation from range and Doppler to range and cross-range is more difficult due to the geometrical terms in the transformation [54]. However, the measurement of the range-Doppler scattering function is still simple when the analysis is performed using the Wigner-Ville distribution instead of the ambiguity surface, (i.e. magnitude squared of the ambiguity function). As will be subsequently shown, the estimation of the scattering function separates into a measurement of the Wigner-Ville distribution of the received signal made at the receiver and the utilization of a priori information about the transmitted signal.

Although the concept presented in this paper is based on a stochastic model of the reflection process, it still works as described even when the target is deterministic and not fluctuating. In this latter case the signal processing proposed in this paper may be simplified by reducing the number of different chirp signals used and thus may be useful at microwave frequencies where longer coherent integration times are required due to the reduced Doppler sensitivity of the lower carrier frequency, compared to the LIDAR case. Also, since the nominal resolution of the range-Doppler cell is proportional to the bandwidth (range resolution) and the duration of coherent integration (Doppler resolution) use at microwave frequencies may require stepped frequency chirp signals instead of continuous chirp signals to satisfy the bandwidth limitations of available components [55].

8.2 WVD Background for Echography

In order to understand the role of the Wigner-Ville distribution in radar/sonar signal processing it is necessary to introduce a model of the received signal which is sufficiently realistic to be practical yet sufficiently simple so that it can be easily understood. Therefore, only scalar signals will be considered. This assumption is true for acoustic waves propagating in a liquid medium and is a simplifying assumption for electromagnetic waves.

The received signal will be modelled as the linear superposition of scaled versions of the transmitted signal delayed by the round trip time and modified by Doppler. Since the transmitted signal has only range and Doppler resolution then the received signal may be simplified by assuming that the scale factor is the product of a constant that depends on the range and a complex random variable which accounts for interference between unresolved target scatterers and propagation multipaths. A reasonable assumption, when the individual contributions are sufficiently numerous and approximately equal in magnitude with uniformly destributed phase, is that the random variable can be modelled as a zero mean Gaussian with variance $\sigma(t,f)$ which depends on the range and Doppler. This function is traditionally called the scattering function and this target model has been described by Van Trees [56]. "Estimation and Modulation," Vol. III, Chapters 9 - 13 [56].

The scattering function will be assumed to be uncorrelated in frequency due to orthogonality of the Doppler shifts for narrow band signals. However, different delays may be correlated depending on the characteristics of the target. In order to be able to include both deterministic as well as stochastic variation from pulse to pulse the random variable will be modelled with an exponential time correlation which can be generated by a recursive filter. Conventional processing only works well when the correlation is high, (i.e. the target scattering is essentially deterministic). Speckle noise is associated with single look SAR images.

8.3 Comparison of Radar Imaging Techniques

Four different analyses of the imaging of radar targets have been proposed in the literature. In the first and simplest the received signal from a spatially distributed target is illuminated at successive frequencies as a function of angle. If the scattering is assumed to be deterministic then the two-dimensional Fourier transform of the received signal gives the range and cross-range components of the scattering distribution provided the target is in the far field of the illumination and the variation in angle is sufficiently small that a rectangular approximation of a polar grid is valid. This type of signal processing is often used in radar ranges where the angular orientation can be measured and the signal to noise ratio is sufficiently large that matched filter processing is not required.

In the second method matched filtering is used with either continuous or stepped frequency chirp signals. Successive range compressed signals from a moving target are aligned in time in order to compensate for the variation in range between successive measurements and then a discrete Fourier transform is computed on the complex matched filter outputs for successive measurements. This Fourier transform computes the Doppler frequency as a function of range and is related to the cross-range of a rigid object if the effective azimuth rotation rate relative to the line of sight (LOS) is known. This method is effective for deterministic signal returns, such as those from metallic aircraft illuminated by microwave frequencies. However when the target is rough relative to the illuminating wavelength then the scattering is stochastic and this type of processing is not appropriate.

In the third method the cross-ambiguity function between the transmitted and the received signal is computed. For deterministic scattering distributions this integral is proportional to the two-dimensional Fourier transform of the product of a complex reflection coefficient with the ambiguity function of the transmitted waveform-but where the proportionality factor is complex and range dependent. This method of range-Doppler imaging has been analyzed by Feig and Grunbaum and leads to a new form of tomographic imaging [57].

In the fourth method the magnitude squared of the cross-ambiguity function between the transmitted and the received signal is computed. For stochastic scatterers the expected value of the magnitude squared of the cross-ambiguity function is proportional to the two-dimensional convolution of the range-Doppler scattering function with the magnitude squared of the auto-ambiguity function of the transmitted signal. Thus for rotation relative to the LOS the effective Doppler of the target becomes coupled to the range of the target through the ambiguity function's magnitude squared response. The latter is sometimes referred to as the ambiguity surface. For narrowband signals of unit energy the ambiguity surface is everywhere positive and has unit volume. Thus it may be considered to be a probability density function which distributes the energy from the scatterer in both range and Doppler according to the particular waveform used by the transmitter. This method of imaging is analogous to the medical method of time-of-flight positron emission tomography (TOFPET) when finite duration chirp signals are used and is analogous to the medical method of computerized axial tomography (CAT) in the limit of infinite duration chirp signals. The analysis of this method of signal processing will be considered next.

8.4 Tomographic radar imaging using ambiguity surfaces

In 1984, Bernfeld proposed the use of a computerized axial tomography (CAT) analogy to process linear fm synthetic aperture (SAR) images [58]. Instead of transmitting repeatedly the same linear fm signal as the radar moves along its path, Bernfeld suggested that chirps of different slopes be transmitted. For a stochastic Gaussian assumption of the scattering process the output of a matched filter receiver may be modelled as the two-dimensional convolution of the range-Doppler scattering function with the auto-ambiguity surface of the transmitted signal. In the limit of an infinite duration linear fm signal and no additive noise the output of the receiver may be considered to be equivalent to a sequence of line integrals through the scattering function, where the orientation of the particular line integral relative to the scattering distribution is determined by the chirp rate of the transmitted signal. The ambiguity surface of an infinite duration linear fm signal may be considered to be a delta function with an orientation in range and Doppler determined by the chirp rate of the signal. Therefore, the convolution of this ambiguity function with the scattering distribution may be considered to be a sequence of tomographic projections of the scattering distribution.

In most sonar and radar applications, infinite duration signals are not practical. Therefore, the tomographic signal processing technique proposed by Bernfield should be modified to accommodate finite duration signals. For a finite duration linear fm signal the ambiguity surface is only an approximation to a line integral. In 1985 Snyder and Whitehouse proposed a signal processing modification based on the medical method of time-of-flight positron emission tomography (TOFPET) [59].

When a linear fm signal with a Gaussian envelope is used as

the transmitted signal in an echo ranging system then the auto-ambiguity surface of the transmitted signal is an asymmetric two-dimensional Gaussian. The inverse problem which results when a number of different orientations of the two-dimensional Gaussian are convolved with an unknown but bounded two dimensional distribution has been studied in the context of reconstructing medical radio nucleide activity distributions [60]. One of the techniques of solving this inverse problem is called the "confidence weighting" algorithm and is a generalization of the CAT technique of back projection and superposition followed by two-dimensional filtering.

A phenomonological description of confidence weighting can be easily constructed to motivate the confidence weighting tomographic algorithm. In the conventional CAT algorithm there is no information in the projection or line integral about the location of the contributions to the line integral. Therefore, the value of the line integral is uniformly distributed along the path of the line integral, (i.e. back projection). In the case of TOFPET or the proposed tomographic radar imaging algorithm the pointspread function of the tomograph or the ambiguity surface provides a priori information about the contributions to the receiver output. Since the normalized ambiguity surface has unit volume and is positive, it may be interpreted as a probability distribution and thus confidence weighting is simply a method analogous to back projection for incorporating the a priori knowledge.

8.5 WVD Interpretation of Imaging

It is well known that the Wigner-Ville distribution (WVD) and the ambiguity function of Sussman are related by a two-dimensional Fourier transform [61]. Thus it is not surprising that radar or sonar detection and estimation can be interpreted in terms of the WVD. However, it is not without some difficulty as will be presented in the following section. But the difficulties are minor compared to the insight provided by reformulating the signal processing equations. In particular, such an interpretation of the signal processing for bistatic operation of the radar or sonar is easier to understand, since the effect of receiver mismatch may be predicted and in the case of linear frequency modulated (LFM) signals, simplified signal processing either in terms of the short time Fourier transform (STFT) or STRETCH processing is possible.

The WVD has many interesting properties [51]. However, only a few will be needed for our purposes. These are:

- Realness
- The shift properties
- Moyal's formula

- The marginal properties
- Relationship to the ambiguity function
- Relationship to the ambiguity surface

Unfortunately, there are some additional properties which would be desirable in a time-frequency distribution, but which are attainable only when other desirable properties are sacrifised. These are:

- Linearity; Must give up realness or resolution.
- Positivity; Must give up resolution, the marginal properties, and the general applicability of Moyal's formula.

Tomographic signal reconstruction techniques restore linearity as well as positivity. High resolution Wigner-Ville analysis shows promise in restoring the resolution [62]. However, Moyal's formula is no longer applicable, and therefore there is a small loss of detectability which is not usually important for imaging applications [65].

The tomographic processing algorithm using the Wigner-Ville distribution can be simply deduced from the corresponding tomographic algorithm using the ambiguity surface. The expected value of the cross-ambiguity surface of the matched filter receiver is the two-dimensional convolution of the scattering function with the auto-ambiguity function of the transmitted waveform [56]. Using the relationship between the ambiguity surface and the Wigner-Ville distribution the cross-ambiguity surface is rewritten as the two-dimensional convolution of the Wigner-Ville distribution of the received signal with the Wigner-Ville distribution of the transmitted signal. In a similar manner the auto-ambiguity surface of the transmitted signal is the two-dimensional convolution of the Wigner-Ville distribution with itself. If the two-dimensional Fourier transform (i.e. ambiguity function) of the transmitted signal is positive then an equivalent formulation is that the expected value of the Wigner-Ville distribution of the received signal is the two-dimensional convolution of the target scattering function with the Wigner-Ville distribution of the transmitted signal.

LASER detection and ranging (LIDAR) technology has significant resolution advantages over conventional microwave radar techniques. The technology has received considerable attention as a means of imaging spaceborne objects using inverse synthetic aperture radar (ISAR) processing techniques using range resolved tomographic reconstruction. These ISAR techniques, however, become ineffective when the object is only viewed for a sm fraction of a rotation, a common occurrence in strategic defence applications. Furthermore, the formidable signal processing requirements inherent in a ISAR imaging application require innovative techniques to minimize the size, weight, and power of the signal processor, particularly for spaceborne systems Recently, Whitehouse and Boashash have suggested that a LIDAR imaging processor which uses tomographic reconstruction techniques to form an image may overcome the limited viewing angle problem associated with range resolved tomographic ISAR processing [66]. In this new technique the chirp rate of the LIDAR transmitted pulse is varied during the illumination period. For these signals, the Wigner-Ville distribution, which simultaneously measures the signal time-delay and Doppler shift, is known to be an approximation to a knife-edge whose orientation in the time-delay Doppler plane is a function of the chirp rate. Thus the output of a Wigner-Ville receiver which is the convolution of this rotated function with the received LIDAR signal and corresponds to calculating a projection of the object's scattering distribution Time of flight positron emission tomography (TOFPET) reconstruction techniques can then be employed to reconstruct an image of the object by confidence weighting, (i.e. convolving the received Wigner-Ville dsitribution with the Wigner-Ville distribution of the illuminating signal). and filtering the superposition of the confidence weighted signals.

9. CONCLUSIONS

The direction finding algorithms we have considered impose a substantial computational load for their real-time implementation. Most of this computational load consists of matrix operations:

- matrix-vector multiplication
- linear equation solution
- orthogonal triangularization
- triangular backsolve
- Hermitian symmetric eigensystem solution
- singular value decomposition
- Hermitian symmetric generalized eigensystem solution

These matrix operations also comprise a substantial part of the computational load for several useful methods for data compression, spectral analsyis, and pattern recognition.

9.1 Research Issues in Parameter Estimation

Two of most promising parameter estimation techniques for further study for application to direction finding are Cadzow's method and the Maximum Likelihood method. MUSIC and MVDR should be viewed as baseline algorithms for comparison with the newer methods. Both Cadzow's eigenvector fitting method and the Maximum Likelihood Method permit high resolution direction finding in the presence of fully coherent multipath, and both are applicable to very general array geometries. However, both require the solution of a computationally expensive nonlinear least squares problem. At present, it is not only difficult to specify a preferred method for solving this nonlinear least squares problem, but even to specify the number of arithmetic operations needed for an acceptable solution. In order to make Cadzow's method and MLM feasible for real-time implementation and examine their robustness, further study is needed for:

- Convergence of Nonlinear Least Squares Algorithms
- Parallelization of Nonlinear Least Squares Algorithms
- NLS solution accuracy needed for Cadzow's Method and MLM
- Effect of element gain, phase, and position errors on Cadzow's Method and MLM

For the special case of radar imaging via estimation of the target scattering function, time-frequency analysis via the application of signal subspace techniques to the Wigner-Ville distribution leads to problems analogous to spectral estimation [30]. For such problems, the spectral analysis of the WVD's bilinear kernel is analogous to direction finding using a uniformly spaced line array, and ESPRIT and its variants are strong candidate methods. These techniques will require parallel implementations for:

- The Nonsymmetric Generalized Eigenvalue Problem
- Total Least Squares
- The Orthogonal Procrustes Problem

9.2 *Parallel Computation*

Surveys of parallel algorithms and architectures for many of the matrix computations discussed in this paper are available [47], [48]. Although nonlinear least squares algorithms use the preceding operations, the resulting iterative algorithms are not as well understood as the classical matrix operations. Additional research is also needed in the area of parallel algorithms for the nonsymmetric eigenvalue problem, since current techniques using parallel Jacobi-like methods for computing the Schur decomposition can have convergence problems for non-normal matrices [49].

ACKNOWLEDGEMENT

One of the authors (HJW) wishes to thank Prof. Donald Snyder of Washington University for many stimulating discussions of tomographic techniques.

REFERENCES

[1] Good, I.J., (1969) Some Applications of the Singular Decomposition of a Matrix, Technometrics, Vol. 11, No. 4, pp. 823-831.

[2] Andrews, H.C. and Patterson, C.L., (1975) Outer Product Expansions and Their Uses in Digital Image Processing, American Math. Monthly, Vol. 82, No. 1, pp. 1-13.

[3] Andrews, H.C. and Patterson, C.L., (1976) Singular Value Decomposition (SVD) Image Coding, IEEE Trans. on Communications, pp. 425-432.

[4] Balance, W.P., (1984) The Singular Value Decomposition: An Introduction, Conference Record of the Eighteenth Asilomar Conference on Circuits, Systems and Computers, held at Pacific Grove, California, IEEE Catalog No. 85CH2220O-4, pp. 148-157.

[5] Nash, J.C., (1979) Compact Numerical Methods for Computers, John Wiley and Sons, New York.

[6] Golub, G.H., and Van Loan, C.F., (1983) Matrix Computations, Johns Hopkins University Press, Baltimore, second edition, 1989.

[7] Paige, C.C., (1986) Computing the Generalized Singular Value Decomposition, *SIAM J. Sci. Stat. Comput.*, Vol. 7, No. 4, pp. 1126-1146.

[8] Luk, F.T., (1985) A Parallel Method for Computing the Generalized Singular Value Decomposition, *Journal of Parallel and Distributed Computing*, Vol. 2, pp. 250-260.

[9] Ma, M T., (1974) Theory and Applications of Antenna Arrays, John Wiley and Sons.

[10] Steinberg, B.D., (1976) Principles of Aperture and Array System Design, John Wiley and Sons.

[11] Speiser, J.M., Whitehouse, H.J. and Berg, N.J., (1978) Signal Processing Architectures using Convolutional Technology, Proc. SPIE Vol. 154, pp. 66-80.

[12] Schmidt, R.O., (1979) Multiple Emitter Location and Signal Parameter Estimation, Proceedings of the RADC Spectrum Estimation Workshop, Rome Aid Development Center, pp. 243-257.

[13] Monzingo, R.A. and Miller, T.W., (1980) Introduction to Adaptive Arrays, John Wiley and Sons.

[14] Owsley, N.L., (1985) Sonar Array Processing, Chapter 3 of Array Signal Processing, edited by S. Haykin, Prentice-Hall.

[15] Schreiber, R. and Kuekes, P.J., (1985) Systolic Linear Algebra Machines for Digital Signal Processing, In VLSI and Modern Signal Processing, edited by S.Y. Kung, H.J. Whitehouse and T Kailath, Prentice Hall, pp. 389-405.

[16] McWhirter J.G., and Shepherd, T.J., Efficient minimum variance distortionless response processing using a systolic array, Proc. SPIE, Vol. 975, paper 975-39, pp. 385-392.

[17] Marple, S.L., (1987) Digital Spectral Analysis, Prentice-Hall.

[18] Capon, J., Greenfield, R. and Kolker, R.J., (1967) Multidimensional Maximum Likelihood Processing of a Large Aperture Seismic Array, Proc. IEEE, Vol. 55, pp. 192-211.

[19] Capon, J., (1969) High Resolution Frequency-Wavenumber Spectrum Analysis, Proc. IEEE, Vol. 57, pp. 1408-1418.

[20] McDonough, R.N. (1979) Application of the Maximum Likelihood Method and Maximum Entropy Method to Array Processing, Chapter 6 of Nonlinear Methods of Spectral Analysis, edited by S. Haykin, Springer-Verlag.

[21] Haykin, S., (Ed.), (1979) Nonlinear Methods of Spectral Analysis, Springer-Verlag.

[22] Schmidt, R.O., (1981) A Signal Subspace Approach to Multiple Emitter Location and Spectral Estimation, Ph.D. Dissertation, Dept. of Electrical Engineering, Stanford University.

[23] Bienvenu, G.S., Influence of the Spatial Coherence of the Background Noise on High Resolution Passive Methods, ICASSP 79 Record, pp. 306-309.

[24] Speiser, J.M., (1985) Progress in Eigenvector Beamforming, SPIE Proceedings, Vol. 564, paper 564-01.

[25] Johnson, D H. and De Graaf, S.R., (1982) Improving the Resolution of Bearing in Passive Sonar Arrays by Eigenvalue Analysis, IEEE Trans. on Acoustics, Speech, and Signal Processing, Vol. ASSP-30, No. 4, pp. 638-647.

[26] Martin, G.E., (1988) Signal subspace processing of experimental radio data, SPIE Proceedings, Vol. 975, paper 975-11, pp. 101-107.

[27] Wagstaff, R.A. and Baggeroer, A., (editors), High Resolution Spatial Processing in Underwater Acoustics, Naval Ocean Research and Development Acitivity, NSTL.

[28] Shan, T.J., Wax, M. and Kailath, T., (1985) On Spatial Smoothing for Direction of Arrival Estimation of Coherent Signals, IEEE Trans. on Acoustics, Speech, and Signal Processing, Vol. ASSP-33, No. 4, pp. 806-811.

[29] Reddi, S.S., (1987) On a Spatial Smoothing Technique Multiple Source Location, IEE Trans on Acoustics, Speech, and Signal Processing, Vol. ASSP-35, No. 5, p. 709.

[30] Whitehouse, H.J., and White, L.B., (1989) Resolution Enhancement in Time-Frequency Signal Processing Using Tomographic Imaging, to appear in ICASSP 89, Glasgow, Scotland.

[31] Roy, R., Paulraj, A, and Kailath, T., Comparative Performance of ESPRIT and MUSIC for Direction-of-Arrival Estimation, ICASSP 87, IEEE Order No. CH2396-0/87, Volume 4, paper 54.12, pp. 2344-2347.

[32] Pisarenko, V.F., (1973) The Retrieval of Harmonics from a Covariance Function, *Geophys. J. R. astr. Soc.*, Vol 33, pp. 347-366.

[33] Speiser, J.M., (1987) Some Observations Concerning the ESPRIT Direction Finding Method, Proc. SPIE, Vol. 826, pp. 178-185.

[34] Paulraj, A., Roy, R. and Kailath, T., (1985) Estimation of Signal Parameters via Rotational Invariance Techniques - ESPRIT, Conference Record of the Nineteenth Asilomar Conference on Circuits, Systems and Computers, pp. 83-89.

[35] Paulraj, A., Roy, R. and Kailath, T., (1986) A subspace rotation approach to estimating directions of arrival, Proc. ICASSP 86, Vol. 4, paper 47.2, pp. 2495-2498.

[36] Paulraj, A., Roy, R, and Kailath, T., (1986) A subspace rotation approach to signal parameter estimation, Proc. IEEE, pp. 1044-1045.

[37] Zoltowski, M.D., Novel Techniques for Estimation of Array Signal Parameters Based on Matrix Pencils, Subspace Rotations, and Total Least Squares, ICASSP 88 Conference Proceedings, Vol. V, pp. 2861-2864.

[38] Zoltowski, M.D., (1988) Generalized Minimum Norm and Total Least Squares with Applications to Array Signal Processing, Advanced Algorithms and Architectures for Signal Processing III, SPIE Vol. 975, paper 975-08, pp. 78-85.

[39] Roy, R. and Kailath, T., (1988) Invariance Techniques and high-resolution null steering, Advanced Algorithms and Architectures for Signal Processing III, SPIE Vol. 975, paper 975-35, pp. 358-367.

[40] Cadzow, J.A., (1988) A High Resolution Direction-of-Arrival Algorithm for Narrow-Band Coherent and Incoherent Sources, IEEE Trans. on Acoustics, Speech and Signal Processing, Vol. 36, No. 7, pp. 965-979.

[41] Bressler, Y. amd Macovski A., (1986) Exact Maximum Likelihood Parameter Estimation of Superimposed Exponential Signals in Noise, IEEE Trans. on Acoustics, Speech and Signal Processing, Vol. ASSP-34, No. 5, pp. 1081-1089.

[42] Weiss, A.J., Wilisky, A.S. and Levy, B.C. (1987) Maximum Likelihood Array Processing for the Estimation of Superimposed Signals, Conference Record of the Twenty-First Asilomar Conference on Signals, Systems, and Computers, held at Pacific Grove, CA, Vol. 2, pp. 845-848.

[43] Ziskind, I. and Wax, M., (1988) Maximum Likelihood Localization of Multiple Sources by Alternating Projection, IEEE Trans. on Acoustics, Speech, and Signal Processing, Vol. ASSP-36, No. 10, pp. 1553-1560.

[44] Nickel, U., (1987) Angle Estimation with Adaptive Arrays and Relation to Superresolution, IEE Proc., Vol. 134, Part H, pp. 77-82.

[45] Feder M. and Weinstein, E., Optimal Multiple Source Location Estimation via the EM algorithm ICASSP 85 Proceedings, Vol. 4, paper 45.6, pp. 1762-1765.

[46] Miller, M.I. and Fuhrmann, D.R., (1987) An EM Algorithm for Direction-of-Arrival Estimation for Narrowband Signals in Noise, Proc. SPIE, Vol. 826, pp. 101-103.

[47] Speiser, J.M., (1986) Signal Processing Computational Needs, Proc. SPIE, Vol. 696, pp. 2-6.

[48] Speiser, J.M., (1987) An Overview of Matrix-Based Signal Processing, Conference Record of the Twenty-First Asilomar Conference on Circuits, Systems, and Computers, held at Pacific Grove, CA, IEEE Computer Society Order No. 760, pp. 284-289.

[49] Eberlein, P.J., (1987) On the Schur Decomposition of a Matrix for Parallel Computation, IEEE Trans. on Computers, Vol. C-36, No. 2, pp. 167-174.

[50] Whitehouse, H.J. and Boashash, B., (1986) Signal Processing Applications of Wigner-Ville Analysis, SPIE Vol. 696, Advanced Algorithms and Architectures for Signal Processing, pp. 156-162.

[51] Mecklenbräuker, W., (1985) A Tutorial on Non-Parametric Bilinear Time-Frequency Signal Representations, in Les Houches, Session XLV, Traitment du Signal / Signal Processing, edited by J.L. Lacoume, T.S. Durrani, and R. Stora, Elsevier Science Publishers, B.V., pp. 277-336.

[52] Breed, B.R., and Posch, T.E., A Range and Azimuth Estimator Based on Forming the Spatial Wigner Distribution, Proc. ICASSP '84, paper 41B9.

[53] Swindlehurst, A.L. and Kailath, T., (1988) Near-field source parameter estimation using a spatial Wigner distribution approach, Advanced Algorithms and Architectures for Signal Processing III, SPIE, Vol. 975, paper 975-09, pp. 86-92.

[54] Arikan, O. and Munson, D.C., Jr., (1988) A tomographic formulation of bistatic synthetic aperture radar, in W.A Porter and S.C. Kak, Editors, Proceedings of the 1988 international conference on advances in communication and control systems, Dept. of Electrical and Computer Engineering, Louisiana State University, Baton Rouge, LA 70803, Baton Rouge, LA, pp. 418-431.

[55] Wehner, D.R., (1987) High Resolution Radar, Artech House.

[56] Van Trees, H.L., (1971) Detection, Estimation, and Modulation Theory, Part III, Radar-Sonar Signal Processing and Gaussian Signals in Noise, John Wiley and Sons, New York.

[57] Feig, E. and Grünbaum, F.A., (1986) Tomographic methods in range-Doppler radar, Inverse Problems, Vol. 2, pp. 185-195.

[58] Bernfeld, M., (1984) Chirp Doppler radar, Proc. IEEE Lett., Vol. 72, No. 4, pp. 540 541.

[59] Snyder, D.L., Whitehouse, H.J., Wohschlaeger, J.T. and Lewis, R.C., (1986) A New Approach to RADAR/SONAR Imaging, Proceedings of the SPIE 30th Annual Technical Symposium, Paper 696-20, San Diego, CA.

[60] Snyder, D.L., Thomas, L.J., Jr., and Ter-Possian, M.M., (1981) A Mathematical Model for Positron Emission Tomography Systems Having Time-of-Flight Measurements, IEEE Trans. on Nuclear Science, Vol. NS-28, pp. 3575-3583.

[61] Szu, H.H., (1982) Two-dimensional optical processing of one-dimensional acoustic data, Optical Engineering, Vol. 21, pp. 804-813.

[62] Boashash, B. and Whitehouse, H.J., (1987) High-accuracy time-frequency signal analysis of fm signals using an autoregressive Wigner-Ville distribution, submitted to IEEE Trans. on Acoustics, Speech and Signal Processing.

[63] Boashash, B. and Whitehouse, H.J., (1987) High Resolution Wigner-Ville Analysis, Onzième Colloque GRETSI, Nice, pp. 205-208.

[64] Boashash, B., Lovell, B. and Whitehouse, H.J., (1987) High Resolution Time-Frequency Signal Analysis by Parametric Modelling of the Wigner-Ville Distribution, Signal Processing Theories, Implementations, and Applications, edited by B. Boashash, Brisbane, Australia.

[65] Flandrin, P., (1988) A Time-Frequency Formulation of Optimum Detection, IEEE Trans. on Acoustics, Speech and Signal Processing, Vol 36, No. 9, pp. 1377-1384.

[66] Whitehouse, H.J. and Boashash, B., (1988) Delay Doppler Radar/Sonar Imaging, presented at EUSIPCO 88, Grenoble, France.

LINEAR ALGEBRA ALGORITHMS ON DISTRIBUTED MEMORY MACHINES

Y. Robert and B. Tourancheau
*(Laboratoire LIP-IMAG
Ecole Normale Supérieure de Lyon, France)*

ABSTRACT

We deal with the implementation of linear algebra algorithms on distributed memory machines. Throughout, we use Gaussian elimination as a target example, and the hypercube as a target machine. We discuss the performance of various allocation functions for different subtopologies of the hypercube and different communication protocols (centralized versus local). We give both experimental and theoretical results. In the last part of the paper, we derive specific algorithms for matrices over finite fields and briefly mention their systolic counterpart.

1. INTRODUCTION

In this paper, we discuss the implementation of linear algebra algorithms on distributed memory machines. Throughout, we use Gaussian elimination as a target example, but the results can be straightforwardly extended to a wide class of algorithms.

Our target machine is a hypercube, which we can also configure as a ring or a 2D-torus. In this paper, we discuss the performances of various allocation functions for these different subtopologies of the hypercube and different communication protocols (centralized versus local). We first give experimental results, and then move to a complexity analysis.

In the last part of the paper, we derive specific algorithms for matrices over finite fields and briefly discuss their systolic counterpart.

This work has been supported by the Research Program C3 of CNRS

To simplify matters when discussing Gaussian elimination, we consider the implementation of a column oriented version without pivoting (we come back to pivoting issues in the following sections). Let A be a square matrix of order n. We have the following algorithm [7]:

```
for k = 1 to n do
    {step k : column k is the pivot column}
    for j = k + 1 to n do
        task T_kj : <   {update column j}

                        a_kj ← a_kj/a_kk

                        for i = k + 1 to n do
                            a_ij ← a_ij - a_ik * a_kj  >
```

Let τ_a be the time for an arithmetics operation. The sequential time is $T_{seq} = 2n^3 \tau_a/3 + O(n^2)$. To design a parallel version on a local memory machine, we distribute the columns of A among the processors. More precisely, we assume that the number of processors p is a divisor of the problem size n, and that each of the p processors holds exactly n/p columns of **the** system matrix in its local memory (equidistribution of the data). Column j of the matrix is allocated to processor alloc(j).

The problem is to determine the allocation function *alloc* which minimizes the total execution time, defined as the sum of the arithmetic time T_a, and of the communication time T_c (which includes idle time).

We start with a brief overview of the main topological properties of hypercubes.

2. HYPERCUBE TOPOLOGY

Hypercubes grow by replication. Given two identical m-dimensional hypercubes, (m-cubes for short, with $p = 2^m$ nodes), we obtain an (m + 1)-cube by linking their corresponding nodes in a one-to-one fashion. Adding that a 0-hypercube consists of one node, we have a recursive definition of hypercubes. It is well known that the nodes of a m-cube can be numbered by m-bits binary numbers, from 0 to 2^m-1, and interconnected so that there is a link between two processors if and only if their binary labels differ by exactly one bit. There are many ways to number the processors likewise, but the most usual technique is the construction of the so-called m-dimensional binary reflected Gray code [19]. We start with the sequence of the two 1-bit numbers $G_1 = \{0, 1\}$. This is a Gray code of dimension 1. Given G_m, denote by $G_m^{(r)}$ the sequence obtained by reversing its order, and by $0G_m$ (resp. $1G_m$) the sequence obtained from G_m by prepending a 0 (resp. a 1) to each element of the sequence: $G_{m+1} = \{0G_m, 1G_m^{(r)}\}$. We have used the Gray code G_4 to label the nodes of the 4-cube illustrated in figure 1.

Perhaps the most important advantage of the hypercubes is that many of the classical topologies such as rings, 2D-torus or 3D torus can be embedded preserving proximity in them. By preserving proximity, we mean that any two adjacent vertices in the original graph are mapped to neighbour nodes in the hypercube [19].

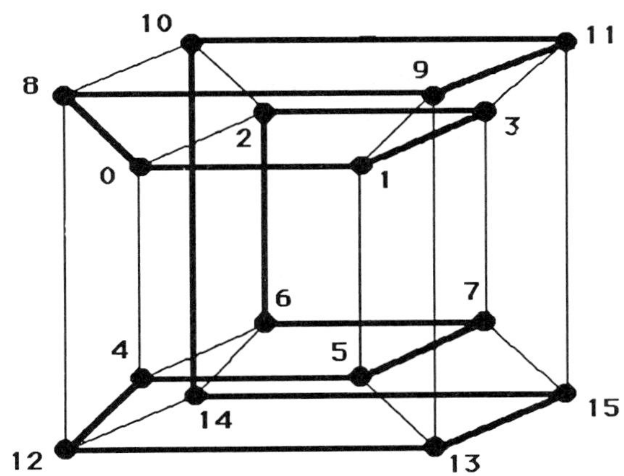

Figure 1 : Gray code numbering of a 4-cube

In the binary reflected Gray code, two consecutive elements differ by exactly one bit, including the last and the first one. Thus a m-dimensional Gray code permits to embed a ring of size 2^m is a m-cube, as illustrated in bold in figure 1 for m = 4: we map the node i in the ring, $0 \leq i \leq 2^m-1$, into the node of the m-cube whose binary label is $g_i^{(m)}$, the i-th element of G_m.

To embed a 2D-torus of size $2^u \times 2^v$ in a m-cube, where $u + v = m$, we simply map the node (i,j) of the grid, $0 \leq i \leq 2^u-1$, $0 \leq i \leq 2^v-1$, into the node of the hypercube whose binary label is $g_i^{(u)} \wedge g_j^{(v)}$, where $\wedge$ denotes concatenation [19]. Observe that the set of nodes obtained by fixing a coordinate and letting the other vary, forms a subcube. In other words, every row or column of processors forms a subcube of the cube.

3. CENTRALIZED ALGORITHMS

We first consider centralized implementations. At each step, the processor which holds the pivot column in its local memory will broadcast it to all the other processors, so that they can use it to update their own internal columns. Hence each step k will consist in two phases:

. a communication phase: a broadcast originated by the processor which holds column k in its local memory, that is the processor i such that alloc(k) = i
. a computation phase: each processor independently updates its internal columns.

We point out that partial pivoting can be straightforwardly introduced in this column-oriented algorithm: at step k, the search for a pivot in column k is a local operation for the pivot processor which does not require any communication. On the contrary, a row oriented version would require an election to determine which processor holds the maximum element (in absolute value) in column k. We obtain the following algorithm [4,6,14,15,18]:

```
Broadcast Algorithm:
{program of processor P_i}

for k=1 to n-1 do
     is alloc(k) = i then
            broadcast pivot column k to all other processors
     else
         receive pivot column k
     endif
     perform the eliminations:
     execute T_kj for all columns j≥k+1 such that alloc(j) = i
 endfor
```

The broadcast operation is usually implemented synchronously. This implies that communications and computations do not overlap within one step of the algorithm, and also that the iteration steps themselves are executed sequentially. There are then two independent kernels to be discussed separately: the implementation of the broadcast function, and the choice of an allocation function which balances the computations among processors.

3.1 Broadcasting

The problem of broadcasting consists in the distribution of the same data from one node to all the other nodes. A simple way to perform a broadcast in a m-cube is to generate a spanning tree of height m for the routing [4,9,17].

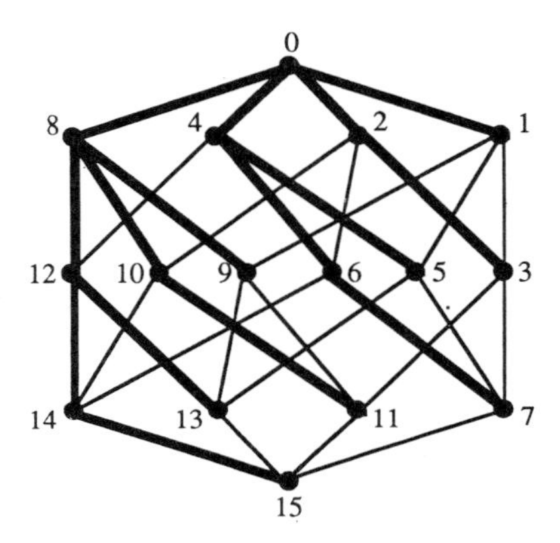

Fig. 2 A spanning tree embedded in a 4-cube

Processors are numbered according to the standard binary reflected Gray code (denoted abs). As shown in figure 2, the spanning tree does not use all the links of the hypercube. The critical path in the spanning tree is of length m. Suppose that the syntax for sending and receiving a message between two neighbours is the following:

```
{receiving}      receive(hypercube_link,receive_buffer,length);
{receiving}      send(hypercube_link,send_buffer,length);
```

where hypercube_link is the dimension along which the message is passed, that is, the index of the bit which differ in the Gray code numbering of the two neighbours.

We have the following broadcast procedure:

```
Broadcast procedure from processor root
#define bit(A,b) ((A>>b)&l){bit(A,b) is the b-th bit of A}
{coming back to the case root = 0000}
pos = abs XOR root;
{index of the first bit equal to one}
first_l=0; while ((bit(pos,first_l)==0)&&(first_l<m))first_l++;
{beginning the neighbour to neighbour communication process}
for(i=m-l; i>0;i--)
     if(i== first_l)
          received(i, recv-buf, length);
     else
          if (i < first_l) send(i,send_buf,length);
```

This algorithm simply describes that the processors receive on the link corresponding to their first bit equal to one, and send on the links corresponding to the preceding zeros.

The time to transfer L words between two adjacent processors is modelled in the literature by $\beta + L\,\tau_C$, where β is the communication start-up and τ_C the elemental transfer time [9,11,18]. The time to broadcast L words using the above procedure is clearly equal to $m\,(\beta + L\,\tau_C)$.

The first optimization consists in pipelining the data in the communication processes [11,17]. The data set is partitioned into several packets and these packets are sent along the links in succession. At step j, the first packet reaches the nodes which are at distance j from the root, the second packet reaches the nodes at distance j-1 and the j-th packet reaches the neighbours of the root. If there are v packets of length L/v_1 the total time will be $(m+v-1)\;(\beta + (L/v)\;\tau_c)$. The optimal number of packets is $v = (L\tau_C(m-1)/\beta)^{1/2}$, leading to a time $((L\tau_C)^{1/2} + ((m-1)\beta)^{1/2})^2$.

Neither the standard broadcast nor its pipelined version makes full use of the hypercube connectivity. We can improve the standard broadcast algorithm by simultaneously generating m different spanning-trees in the m-cube, so as to make full use of the hypercube connectivity. As depicted in figure 3 for m = 4, we rotate the Gray code numbering of the nodes to generate m spanning trees which can be used to schedule m distinct broadcasts in parallel. Note that at a given step, we always communicate on different links for different broadcasts. We split the data into m pieces and transmit them in parallel using the m broadcasts, as stated below in the rotating broadcast procedure. Note that it is not possible to include pipelining in this procedure, because the m generated spanning are not edge-independent. We have to resort to a more complex construction termed the m-Edges disjoint Spanning Binomial Tree (m-ESBT) in the literature [9].

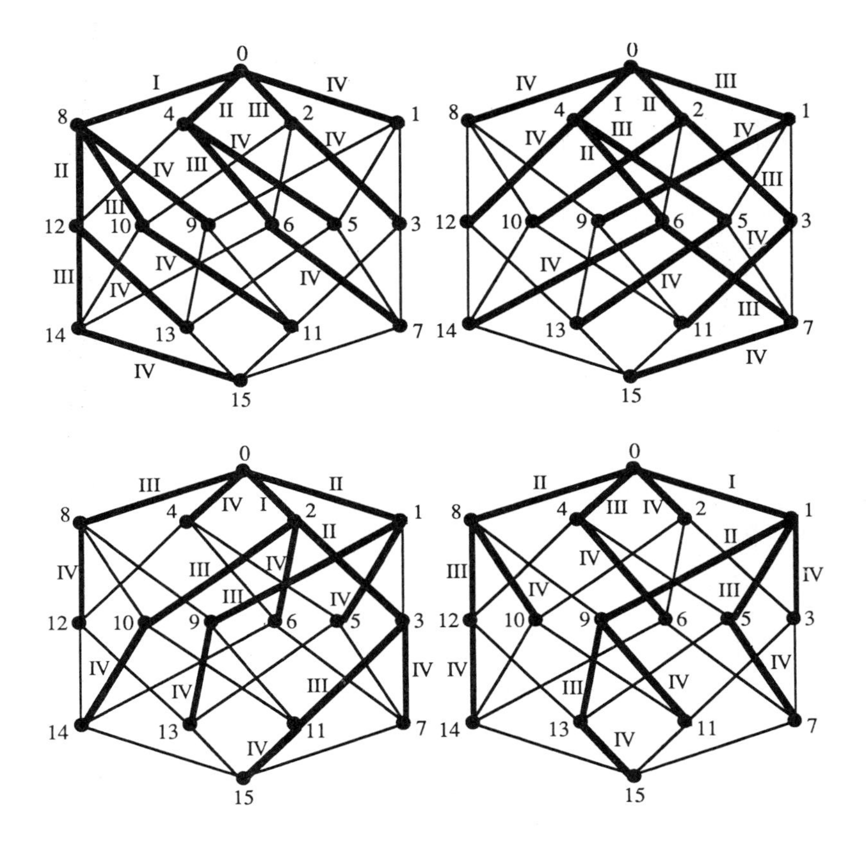

Fig. 3 4 spanning trees embedded in a 4-cube (Roman numbers represent time-steps)

```
Rotating broadcast procedure from processor root
#define bit(A,b) ((A>>b)&1) {bit(A,b) is the b-th bit of A}
#define rot(A,b) (((A>>(m-b))|(A<<b))&(2^m-1)){rotate A of b bits
                                                    leftwards}
{coming back to the case root = 0000}
pos = abs XOR root;
{index of the first bit equal to one and bookkeeping}
length_m = length/m;
send_buffer[0]=send_buffer; recv_buffer[0]=receive_buffer;
for (j=0;j<m;i++)
      first_1[j]=0;
      while ((bit(rot(pos,j),first_1[j] == 0) && (first_1[j]<m))first_1[j]++;
      send_buffer[j+1]=send_buffer[j]+length_m;
      recv_buffer[j+1]=recv_buffer[j]+length_m;
```

```
{beginning the neighbour to neighbour communication process}
for (i=m-1;i>0;i--){loop on time-steps}
     for (j=0;j<m;j++){loop on spanning trees}
          link=(i-j+m)%m;
          if (i == first_l[j])
               receive(link, recv_buffer[j], length_m);
          else
               if (i < first_l[j] send(link,send_buffer[j],length
```

In figure 4, we report the performances of the two broadcast procedures on a FPS T Series hypercube [8], using 16 processors. Each node has a transputer T414 whose 4 links can be activated in parallel. We see that the rotating broadcast is about four times as fast as the standard broadcast, as soon as the length of the message reaches 8K words, which nicely corroborates the timing formulae.

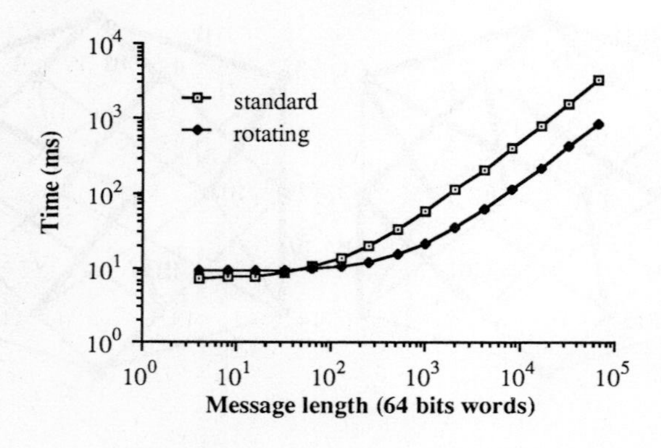

Fig. 4 Performances of the broadcast procedures

3.2 Parallel arithmetic time

Let A be a matrix of size n, and assume that the number of processors p divides n. The problem is to distribute n/p columns of A to each processor so that the arithmetic workload is best balanced. Two allocations functions have been widely used in the literature: the full block repartition, where blocks of n/p consecutive columns are assigned to the processors, and the wrap repartition, where columns of the same index modulo p are assigned to the same processor [4,6,14,15,18]. Here we introduce block-r repartitions, where blocks of r consecutive columns are assigned to the processors in a wraparound fashion. The allocation function by blocks of size r, $1 \leq r \leq n/p$, is defined by alloc(i) = (ceiling(i/r)-1) mod p.

Let τ_a be the time for one arithmetic operation. The weight of a task T_{kj}, $1 \le k < j$, is $W(T_{kj}) = [2(n-k)+1]\tau_a$. Since column j is updated through steps 1 to j-1 during the algorithm, the processor which holds it will perform $W(j) = \Sigma_{1 \le k < j} W(T_{kj}) = (j-1)(2n-j+1)\tau_a$ operations. We see that W(j) is a nondecreasing function of j.

The workload of processor number i is $WL(i) = \Sigma_{alloc(j)=i} W(j)$. The parallel arithmetic time is then $T_a = \max_{0 \le i < p} WL(i)$. We find that

$$T_a = n/(6p)\ [4n^2 + 3rpn - r^2p^2]\tau_a + O(pn)$$

Note that T_a is a non decreasing function of r: the larger the blocks the less balanced the workload distribution. This result is corroborated by table 1 below, where we report the performances for a 1024 matrix, using 16 processors, with various values of the block size r:

Block size	1 (wrap)	2	4	8	16	32	64 (full block)
rotating	67.3	67.7	68.4	70.0	73.1	79.2	90.6
standard	77.4	77.8	78.5	80.0	83.2	89.2	100.7

Table 1: Execution times in seconds for a 1024 matrix and 16 processors using various block sizes and the two broadcast procedures

4. LOCAL PIPELINE ALGORITHMS

In this section, we use two subtopologies of the cube, namely the ring and the grid, to derive pipeline algorithms using only neighbour to neighbour communications.

4.1 Algorithms on a ring of processors

We first configure the hypercube as an oriented ring of p processors numbered from 0 to p-1. Processor P_i receives messages from its predecessor P_{i-1} and sends messages to its successor P_{i+1} (the subscripts are taken modulo p).

At step k, the processor holding the pivot column, say P_i, sends it to its right neighbour P_{i+1}, which in turn sends it to P_{i+2}, and so on until the pivot column reaches the left neighbour of P_i The main advantage of this asynchronous strategy is that communication and computation can be overlapped. A processor can start updating its internal columns as soon as it has received the pivot column and transmitted it to its neighbour. There is no need to wait for all the processors to receive the information before starting the computation. The pipeline ring (PR) algorithm is as follows [3,6,12,14,17]:

```
Pipeline Ring Algorithm:
{program of processor P_i}
for k=1 to n-1 do
    if alloc(k) = i then
            send column k to successor(i)
    else
            receive column k from predecessor(i)
            if successor(i)≠alloc(k) then send column k to successor
    endif
    perform the eliminations:
    execute T_kj for all columns j≥k+1 such that alloc(j) = i
```

To understand the influence of data allocation, we illustrate the communications which occur in the pipeline algorithm for both the full block and the wrap repartition. See figures 5 and 6 where only communications are considered, and are all assumed to be one unit of time. We see that the communication time T_c is less for the full block allocation than for the wrap one, because of the idle time introduced every time that the pivot columns at two consecutive steps are not held by the same processor. This is a general result: the larger r, the smaller T_c.

On the other hand, the arithmetic obeys to the same analysis as in the centralized case: the larger r, the greater T_a. In other words, T_a is minimum for the wrap repartition (r=1), while T_c is minimum for the block repartition (r=n/p). The smaller, r, the more balanced the computations, but the more expensive the communications. As the total time takes into account these two opposite laws, we must be prepared to find a compromise for the choice of the best blocksize r. As evidenced by figure 7, the best value is found to be r=4 for a matrix of order 1024 on the FPS T Series.

	P_0	P_1	P_2	P_3	P_4
Columns	**1,2,3,4**	**5,6,7,8**	**9,10,11,12**	**13,14,15,16**	**17,18,19,20**
t = 1	1	1			
t = 2		1	1		
t = 3	2	2	1	1	
t = 4		2	2	1	1
t = 5	3	3	2	2	
t = 6		3	3	2	2
t = 7	4	4	3	3	
t = 8		4	4	3	3
t = 9			4	4	
t = 10		5	5	4	4
t = 11			5	5	
t = 12		6	6	5	5
t = 13			6	6	
t = 14		7	7	6	6
t = 15			7	7	
t = 16		8	8	7	7
t = 17			8	8	
t = 18				8	8
t = 19			9	9	
t = 21				9	9
t = 22			10	10	9
t = 23				10	10
t = 24			11	11	
t = 25				11	11
t = 26			12	12	
t = 27				12	12
t = 28				13	13
t = 29				14	14
t = 30				15	15
t = 31				16	16

Fig. 5 Communication scheme for Gauss, full block repartition (n=20, p=5, r=4)

	P_0	P_1	P_2	P_3	P_4
Columns	**1,6,11,16**	**2,7,12,17**	**3,8,13,18**	**4,9,14,19**	**5,10,15,20**
t = 1	1	1			
t = 2		1	1		
t = 3			1	1	
t = 4		2	2	1	1
t = 5			2	2	
t = 6				2	2
t = 7	2		3	3	2
t = 8				3	3
t = 9	3				3
t = 10	3	3		4	4
t = 11	4				4
t = 12	4	4			
t = 13	5	4	4		5
t = 14	5	5			
t = 15		5	5		
t = 16	6	6	5	5	
t = 17		6	6		
t = 18			6	6	
t = 19		7	7	6	6
t = 20			7	7	
t = 21				7	7
t = 22	7		8	8	7
t = 23				8	8
t = 24	8				8
t = 25	8	8		9	9
t = 26	9				9
t = 27	9	9			
t = 28	10	9	9		10
t = 29	10	10			
t = 30		10	10		
t = 31	11	11	10	10	12
t = 32		11	11		
t = 33			11	11	
t = 34		12	12	11	11
t = 35			12	12	
t = 36				12	12
t = 37	12		13	13	12
t = 38				13	13
t = 39	13				13
t = 40	13	13		14	14
t = 41	14				14
t = 42	14	14			
t = 43	15	14	14		15
t = 44	15	15			
t = 45		15	15		
t = 46	16	16	15	15	12
t = 47	16	16			
t = 48			16	16	
t = 49		17	17	16	16
t = 50			17	17	
t = 51				17	17
t = 52				18	18
t = 53				18	18
t = 54				19	19

Fig. 6 Communication scheme for Gauss, wrap repartition (n=20, p=5, r=1)

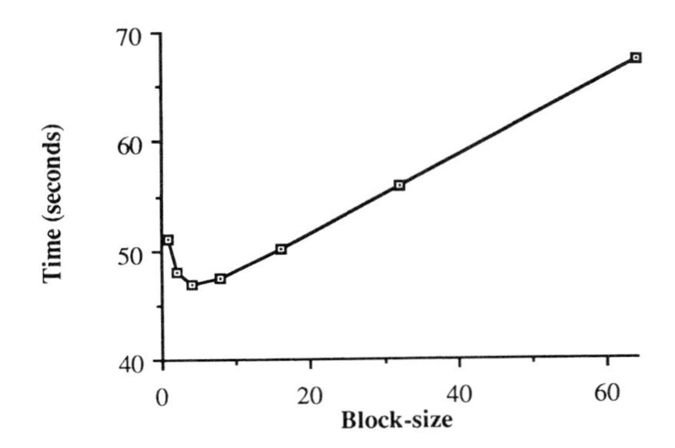

Fig. 7 Execution times for Gauss, various block-sizes ($n=1024$, $p=16$)

4.2 Complexity results on a ring of processors

In the following asymptotic complexity analysis, we assume that the number of processors is proportional to the size of the problem: we let $p = \alpha n$, where $\alpha \in]0,1]$ is a fixed number, and we let n go to infinity. The reason for choosing p proportional to n is as follows: we know from [17] that communications predominate if $p = O(n^2)$ so that we should not use a number of processors greater than the problem size. On the other hand, if p is negligible compared to n, communications have no influence on the asymptotic execution time. The situation under consideration is when p and n are proportional, since the costs of the computations and of the communications are then of the same order $O(n^2)$ since $T_a = O(n^3/p)$ and $T_c = O(n^2)$.

We summarize all the assumptions we make [6,11,16,17]:
.equidistribution of the data: the number of processors p is a divisor of the problem size n, and each of the p processors holds exactly n/p rows of the system matrix in its local memory
.locality assumption: a processor may only modify the data which are stored in that processor
.completion assumption: a processor must complete all computations involving the most recently received message before beginning computations involving the next received message
.oriented communications: a processor can not send data to its right neighbour and receive data from its left neighbour simultaneously
.In the expression $\beta + n\,\tau_c$ which modelizes the time to transfer n words between two adjacent processors, we assume $\beta = 0$ without loss of generality.

Under these assumptions, we have [3,6,16]:

	Block repartition	Wrap repartition
Arithmetic	$\tau_a[n^2/\alpha + O(n)]$	$\tau_a[n^2(4/\alpha+3-\alpha)/6 + O(n)]$
Communication	$\tau_c\,[n^2(1+\alpha) + O(n)]$	$\tau_c[n^2(3+2\alpha-\alpha^2)/2 + O(n)]$

We do not know how to compute T_a for a general allocation function. However, we can compute T_a and T_c for a data repartition by blocks of size r, $1 \le r \le n/p$:

Theorem [16]: On a ring of $p = \alpha\, n$ processors, $0<\alpha<1$, with the allocation function by blocks of size r, $1 \le r \le n/p$, defined by alloc(i) = (ceiling(i/r) - 1) mod p, the execution time for Gauss is

$$T_p = [4/\alpha + 3r - \alpha r^2]n^2\tau_a/6 + [1 + 1/(2r) + \alpha - \alpha^2 r/2]n^2\tau_c + O(n).$$

From the theorem, we see that, given α, T_a is an increasing function of r whereas T_c is a decreasing one, as observed experimentally. Let $\rho = \tau_c / \tau_a$ be the ratio of the elemental costs communication/computation. The efficiency for a block-r repartition is $e_\alpha = 1/[1 + \lambda_1\,\alpha + \lambda_2\,\alpha^2 + \lambda_3\,\alpha^3]$, where

$$\lambda_1 = 3r/4 + 3\rho/2 + 3\rho/(4r),\ \lambda_2 = 3\rho/2 - r^2/4,\ \lambda_3 = -\,3\rho\, r/4$$

Given α and ρ, it is easy to compute the value of r which maximizes the efficiency. To show the good adequation of the theoretical results with the experiments, we let n=480 and use various numbers of processors, from 4 to 16, so that α varies from 1/120 to 1/30. The best experimental value of r is always r=4. We report in table 2 the execution times for r=3, 4 and 5. On the other hand, we estimate $\rho = 14.13$ for the FPS hypercube and messages of average length 240 (half of the matrix size). We report this value in the expression of the efficiency e_α and compute the value of r which maximizes it. This is the "optimal value" reported in table 2.

Block size r	p=4	p=8	p=16
3	35.58	19.97	12.37
4	35.53	19.96	12.26
5	35.71	20.14	12.56
optimal value of r	3.80	3.85	3.97

Table 2: Time (seconds) for block size r and n = 480

Note that for small α, the efficiency can be approximated as

$$e_\alpha = 1/[1 + (3r/4 + 3\rho/2 + 3\rho/(4r))\alpha]$$

(we drop terms in α^2 and α^3). The efficiency is then maximum for $r = \rho^{1/2} \approx 3.76$.

Finally, we state a lower bound to conclude this section:

Proposition [16]: For Gauss on a ring of $p = \alpha n$ processors, $0<\alpha\leq 1$,

$$\tau_a\, 2 n^2/(3\alpha) \leq T_a + O(n)$$

$$\tau_c\, n^2(1+\alpha) \leq T_c + O(n)$$

The full block distribution meets the lower bound on T_c. However, the wrap distribution does not meet the lower bound on T_a. In fact for the wrap distribution, $T_a = \tau_a[T_{bound} + n^2(3-\alpha)/6] + O(n)$, where $T_{bound} = 2n^2/(3\alpha)$. Clearly the bound is not tight for large α, but T_a does not tend to T_{bound} for small α. If we want to achieve a better load balancing for the arithmetic, we can use the reflection mapping [4,6] if $\alpha < 1/2$: processor i, $0\leq i<p$, has rows $2up+i+1$ and $2(u+1)p-i$, $0\leq u\leq n/(2p)-1$, and $T_a = \tau_a[T_{bound} + \alpha n^2/3] + O(n)$. We can also use a bi-reflection mapping if $\alpha < 1/4$: processor i, $0\leq i<p$, has rows $4up+i+1$, $4up+2p-i$, $4up+3p-i$ and $4up+3p+i+1$, $0\leq u\leq n/(4p)-1$, and $T_a = \tau_a[T_{bound} + \alpha n^2/12] + O(n)$. Both the reflection and the bi-reflection mapping tend to T_{bound} for small α.

4.3 Algorithms on a grid of processors

Now we configure the hypercube as a 2D-torus. Processors are labelled P_{ij}, $0 \leq i,j<q$, where $p = q^2$. We consider a matrix of size n, and assume that q divides n. To allocate data to processors, we interleave both rows and columns of the matrix by blocks of size r and map the resulting square blocks onto the grid. The allocation function by blocks of size r, $1 \leq r \leq n/q$, is defined by alloc(i,j) = ((ceiling(i/r) - 1) mod q), (ceiling(j/r) - 1) mod q)). In other words, processor P_{uv} holds the i-th part of column j ivf alloc(i,j) = (u,v). The full wrap (r=1) and the full block (r=n/q) allocation strategies have been discussed in [18].

We derive a pipeline grid algorithm in a similar way as for the ring. At step k, let l = alloc(.,k) the index of the column of processors in the grid which holds the pivot column. First, each processor P_{ul} transmits its part of the pivot column horizontally along the u-th row of the grid in a pipeline fashion. Then, each processor P_{lv} transmits its part of the pivot row vertically along the v-th row of the grid in a pipeline fashion:

```
Pipeline Grid Algorithm:
{program of processor P_ij}
for k=1 to n-1 do
      {horizontal communications}
      if alloc(i,k) = j then
              send-horizontally part of column k to successor(i)
      else
              receive_horizontally part of column k from predecessor
              if successor(i)≠alloc(i,k) then
                    send_horizontally part column k to successor(i)
      endif

  {vertical communications}
  if alloc(k,j) = i then
          send_vertically part of row k to successor(j)
  else
          receive_vertically part of row k from predecessor(j)
          if successor(j)≠alloc(k,j) then
                send_vertically part of row k to successor(j)
  endif
  perform the eliminations:
  execute T_kj for all part of columns j≥k+1 such that alloc(i,j) = i
endfor
```

We can compute the arithmetic workload just as before. For the communications, we note that the critical path corresponds to a virtual ring of size q linking the diagonal processors (hence with an elemental cost twice as before). We refer to [20] for more details and complexity results. Roughly speaking, the results are the same as for the ring algorithm. The smaller r, the better balanced the computations, but the more expensive the communications. The best compromise is illustrated figure 8, where we report the performances for a 1024 matrix, using a 4 x 4 grid, with various values of the block size r:

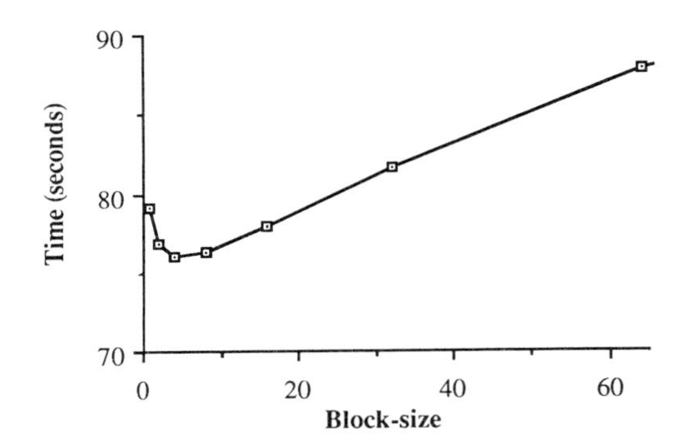

Fig. 8 Execution times for Gauss on the gird, various block sizes (n=1024, p=16)

The main difference between the ring and the grid concerns pivoting issues. If partial pivoting is to be introduced, vertical all-to-all communication must occur between the processors which hold the pivot column. Once the pivot index is computed, we can use a centralized algorithm with horizontal (for broadcasting the parts of the pivot column) and vertical (for broadcasting the parts of the pivot row) one-to-all communications.

5. MATRICES OVER FINITE FIELDS

We briefly deal in this section with the implementation of Gaussian elimination for matrix over finite fields GF(p), and we let p=2 for the sake of simplicity. We refer to [13] for a survey of application domains of modular arithmetic triangularization algorithms.

We can implement the centralized algorithm just as before. However, we have the possibility to store larger matrices, by compressing several row coefficients into one integer word. We report some performances data in figure 9. Note that we use random test matrices, because the timings are data dependent.

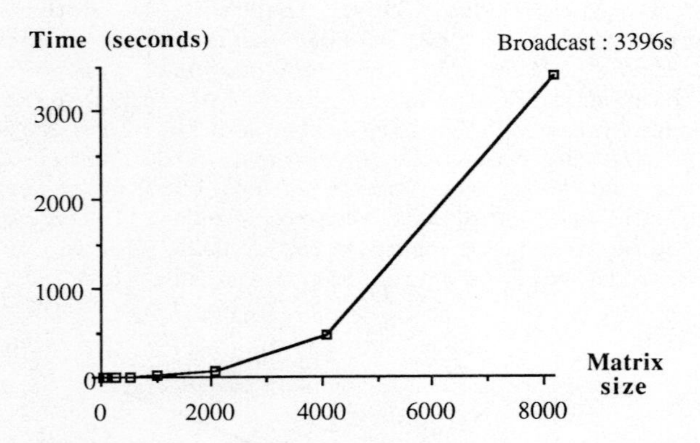

Fig. 9 Broadcast algorithm over GF(2), with 16 processors

Partial pivoting is a *sine qua non* over GF(2), because in average every second coefficient is zero ! But when scanning a column, we do not need any global information: the first non zero element that we find below the diagonal will be chosen as pivot. This property has been used in [10] to derive a systolic architecture for Gaussian elimination with partial pivoting over GF(2). This design can be viewed as the space-time unrolling of the column oriented pipeline ring algorithm.

The basic remark that no global information is needed for the choice of a pivot can be exploited further: we can use several distinct pivots to zero out the elements of a given column simultaneously. To this purpose, we move to a row oriented version where we distribute the rows of the matrix rather than the columns. We first describe a row oriented pipeline algorithm, and then a modification of it called local pivot algorithm.

Let us describe informally step k of the row oriented pipeline algorithm, where alloc(k)=l and processor indices are taken modulo p:

.P_l searches among its rows for a pivot row, sends the pivot row or a failure message (if such a row does not exist) to its right neighbours performs (if necessary) the elimination step updates the array containing the pivot indices and then goes to step k+l.

.P_i, i≠l, receives the message from its left neighbour. If the message contains a pivot row, P_i forwards it to the right neighbour performs the elimination step and then goes to step k+l. If the message does not contain a pivot row, P_i searches among its rows for a pivot row, sends the pivot row or a failure message (if such a row does not exist) to its right neighbour, performs (if necessary) the elimination step, updates the array containing the pivot indices and then goes to step k+1.

```
Row oriented pipeline ring algorithm over GF(2)[2]
{program of processor P_i}

for k = 1 to n do
    if alloc(k) = i then {pivot processor}
        find j such that alloc(j) = i, USED(j) = 0 and a_jk = 1
        if j exists then
            PIVOT(k) = j and USED(j) = 1
            send(j, row(j)) to P_i+1
            for all q such that alloc(q) = i, USED(q) = 0 and
                                        a_qk = 1 do
                row(q,k+1::n) = row(q,k + 1::n) XOR row (j,k+1::n)
        else send(-1) to P_i+1 endif
    else {non pivot}
        receive (PIV, row (PIV)) from P_i-1
        if PIV ≠ -1 then {receiving a pivot}
            if alloc(k) ≠ i+1 then send (PIV, row (PIV))
                                  to P_i+1 endif
            for all q such that alloc(q) = i, USED (q) = 0
                                  and a_qk = 1 do
                row(q,k+1::n) = row (q,k + 1::n) XOR row
                                  (PIV, k+1::n)
        else {no pivot found previously}
            find j such that alloc(j) = i, USED(j) = 0 and a_jk = 1
            if j exists then {take row j as pivot}
                USED(j) := 1;
                if alloc(k) ≠ i+1 then send (j, row (j))
                                to P_i+1 endif
                for all q such that alloc(q) = i, USED(q)=0
                                and a_qk = 1 do
                    row(q,k+1::n) = row(q,k+1::n)
                                    XOR row(j,k+1::n)
                else if alloc(k) ≠ i+1 then send (-1) to
                                    P_i+1 endif
            endif
        endif
    endfor
```

In this algorithm, the arrays PIVOT and USED are distributed among the processors as well as the rows of the matrix. They are used to avoid expensive interchanges of rows. Remark also that the ordering of the rows is unimportant for the algorithm. Hence the distinction between the processor which holds the pivot row and the others is purely artificial. We could choose any processor to initiate the computation, for instance the first one. With this modification, it is not necessary to use a ring: a linear array of processors is sufficient.

The major factor in the communication time arises from the pipelining of the pivot row. Processor p must wait until it receives the information coming from the first processor. Indeed, the basic operation of each processor is to combine two rows with the same number of non-zero elements in order to introduce a new zero. So this row can be any row of the matrix A with a good configuration (i.e. the right number of zeros before the first non-zero element). Hence instead of waiting for receiving the information from its left neighbour each processor could begin the search for a pivot candidate and could use it to eliminate its own rows. Such a candidate, if it exists, will be called a local pivot. However, such a local pivot must be updated using the information coming from the left neighbour. So that, when a processor finds a local pivot, it immediately sends it to its right neighbour in order for it to eliminate its own pivot. If the processor does not find a local pivot, it simply transmits the information coming from the left.

```
Local pivot algorithm over GF(2)[2]
{program of processor P_i}
if i = 1 {program of first processor} then
     for k = 1 to n do
          find j such that alloc(j) = i, USED(j) = 0 and a_jk = 1
          if j exists then {found a pivot}
                PIVOT(k) = j and USED(j) = 1
                send (j, row (j)) to P_i+1
                for all q such that alloc(q) = i, USED(q)
                                         = 0 and a_qk = 1 do
                   row (q,k+1::n) = row(q,k+1::n) XOR row(j,k+1::n)
           else send (-1) to P_i+1
           endif
       endfor
```

```
else {program of processor P_i, i≠1}
     for k = 1 to n do
           find j such that alloc(j) = i, USED(j) = 0 and a_jk = 1
           if j exists then {found a local pivot}
                  PIVOT(k) :=j
                  if i≠p then send (j, row (j)) to P_i+1 endif
                  for all q such that alloc(q) = i, USED (q) = 0
                                                  and a_qk = 1 do
                       row (q,k+1::n) = row(q,k+1::n) XOR row(j,k+1::n)
                  receive (PIV, row(PIV)) from P_i-1
                  if PIV=-1 then {local pivot is final pivot}USED(j) = 1
                  else row(j,k+1::n) = row(j,k+1::n) XOR row
                                             (PIV,k+1::n) endif
              else
                  receive (PIV, row (PIV)) from P_i-1
                  if PIV = -1 and i≠p then send (-1) to P_i+1
                  if PIV ≠ -1 and i≠p then send (PIV, row (PIV)) to P_i+1
              endif
          endfor
      endif
```

Evaluating the execution time of this algorithm is very difficult since it is very data sensitive. Experiments are reported in [2]. Finally, we note that a systolic countepart of the local pivot algorithm has been developed in [1], where a systolic architecture of n(n+1)/2 cells is used to triangularize any matrix of size n in only 2n steps, as opposed to the classical 3n steps of [5] and [10].

6. CONCLUSION

We have presented several variations on a Gaussian thema for distributed memory machines. It is important to point out that many factorization algorithms share the same generic structure:

```
   for k = 1 to n-1
         1. prepare a transformation using column k of A:
               task T_kk:PREP(...,k,...)
         2. apply this transformation to columns k+1 to n
              for j = k+1 to n
                    task T_kj: APPLY(...,k,j,...)
   endfor
```

Hence most of the result presented in this paper can be straightforwardly extended to a wide class of algorithms.

7. REFERENCES

[1] Cosnard, M., Duprat, J. and Robert, Y., (1988) Parallel triangularization in modular arithmetic, in Parallel Processing, M. Cosnard et al. eds., North Holland, 207-220.

[2] Cosnard, M. and Robert, Y., (1987) Implementing the nullspace algorithm over GF(p) on a ring of processors, Second International Symposium on Computer and Information Sciences, Istanbul, 92-110.

[3] Cosnard, M., Tourancheau, B and Villard, G., (1988) Gaussian elimination on message passing architecture, in Supercomputing, E.N. Houstis et al. eds., Lecture Notes in Computer Science 297, Springer-Verlag, 611-628.

[4] Geist, G.A. and Heath, M.T., (1986) Matrix Factorization on a hypercube multiprocessor Hypercube Multiprocessors 1986, M.T. Heath ed., SIAM, 161-180.

[5] Gentleman, W.M. and Kung, H.T., (1981) Matrix triangularisation by systolic arrays", in Proceedings SPIE (Society of Photo-Optical Instrumentation Engineers), vol. 298, Real-time Signal Processing IV, 19-26.

[6] Gerasoulis, A and Nelken, I., (1988) Gaussian elimination and Gauss-Jordan on MIMD architectures, Report LCSR-TR-105, Department of Computer Science, Rutgers University.

[7] Golub, G.H. and Van Loan, C.F., (1983) Matrix Computations, The John Hopkins University Press.

[8] Gustafson, J.L., Hawkinson, S and Scott, K., (1986) The architecture of a homogeneous vector supercomputer, in Proceedings of ICCP 86, IEEE Computer Science Press, 649-652.

[9] Ho, C.T. and Johnsson, S.L., (1986) Distributed routing algorithms for broadcasting and personalized communication in hypercubes, in Proceedings of ICCP 86, IEEE Computer Science Press, 640-648.

[10] Hochet, B., Quinton, P. and Robert, Y., (1987) Systolic solutions of linear systems over GF(p) with partial pivoting, Proc. 8-th Symposium on Computer Arithmetic, IEEE Press, 161-168.

[11] Ipsen, I.C.F., Saad, Y. and Schultz, M.H., (1986) Complexity of dense linear system solution on a multiprocessor ring, Lin. Alg. Appl. 77, 205-239.

[12] Ortega, J.M. and Romine, C.H., (1988) The ijk forms of factorization methods II. Parallel systems, Parallel Computing 7, 149-162.

[13] Parkinson, D. and Wunderlich, M., (1984) A compact for Gaussian elimination over GF(2) implemented on highly parallel computers, Parallel Computing 1, 65-73.

[14] Robert, Y. and Tourancheau, B., (1989) LU and QR factorization on the FPS T Series hypercube, in CONPAR 88, C.H. Jesshope and K.D. Rheinhartz eds., Cambridge University Press, 516-525.

[15] Robert, Y. and Tourancheau, B., (1989) Block Gaussian elimination on a hypercube vector multiprocessor, Revista de Matematicas Aplicadas, 10, 29-69.

[16] Robert, Y., Tourancheau, B and Villard, G., (1989) Data allocation strategies for the Gauss and Jordan algorithms on a ring of processors, Information Processing Letters 31, 1, 21-29.

[17] Saad, Y., (1986) Communication complexity of the Gaussian elimination algorithm on multiprocessors, Lin. Alg. Appl. 77, 315-340.

[18] Saad, Y., (1986) Gaussian elimination on hypercubes, in Parallel Algorithms and Architectures, M. Cosnard et al., eds.,

[19] Saad, Y. and Schultz, M.H., (1988) Topological properties of hypercubes, IEEE Trans. Computers 37, 7, 867-872.

[20] Tourancheau, B., (1988) Mémoires distribuées: répartition des données pour l'algorithme de Gauss, Proc. 3ème Colloque C3, Angoulême, France, Edition du CNRS, 217-230.

LINEAR SYSTOLIC ARRAYS FOR CONSTRAINED LEAST SQUARES PROBLEMS

B. Yang and J. F. Böhme

(Lehrstuhl für Signaltheorie, Ruhr-Universität, 4630 Bochum, FRG)

ABSTRACT

In this paper, we present systolic implementations of various adaptive linearly constrained least squares problems in the context with beamforming and direction finding. Based on some new derived Faddeeva-like equations for matrix computations $D-CA^{-1}B$, we develop efficient update algorithms for both single and multiple constrained beamformers based on the signal-plus-noise spectral density matrix or the noise spectral density matrix. We show that all these algorithms can be implemented by the same linear systolic array.

1. INTRODUCTION

Linearly constrained least squares (LCLS) problems arise in many disciplines of science and engineering. Particular ones are radar and sonar processing, where constrained optimum beamformers are of interest. They are usually adaptive or blockwise adaptive. Because of the high computational amounts of adaptive LCLS problems, high data throughput implementations for real time signal processing are desirable [1]. This leads to the development of stable, highly parallel update algorithms and appropriate array architectures.

Two different approaches have been reported in the literature [2,3] to solve the multiple constrained least squares problem. While Shepherd and McWhirter [2] reduced the constrained least squares problem to a canonical one which is then solved by the QR decomposition of the data matrix, the method of Yang and Böhme [3] is based on rank-one updating and downdating Cholesky factorizations. For the special MVDR (Minimum Variance Distortionless Response) beamforming problem with independent single linear constraints, there exist also two independent works from Yang and Böhme [4] and from McWhirter and Shepherd [5] providing similar solutions, which generalize a previous algorithm proposed by Schreiber [6]. In this paper, we present systolic implementations for some variants of adaptive LCLS problems in the context with beamforming and direction finding. We show the relationships between these adaptive algorithms and the Faddeeva algorithm for matrix computation $D-CA^{-1}B$ [7]. In section 2, we briefly describe a number of constrained optimum beamformers. Section 3 is concerned with some new Faddeeva-like equations. In section 4, we develop efficient update algorithms for the different constrained beamformers. Finally, section 5 presents linear systolic arrays and the CORDIC processor as a suitable processing element.

2. CONSTRAINED OPTIMUM BEAMFORMERS

Assuming that the antenna array consists of N sensors of known but arbitrary geometry, let $x \in C^N$ be the complex column vector of sensor outputs in a particular frequency bin and

$$R = E(xx^H)$$

be its spectral density matrix. "E" and "H" denote expectation and complex conjugate transposition, respectively. The additive noise component of R is denoted by Q which can be, in the cases of active radar and sonar systems, estimated by the signalfree sensor output vector n. The beamformer output is defined as the weighted sum of the sensor outputs

$$y = w^H x,$$

where w is a N-element complex weight vector. Now, we use the MVDR criterion [6, 8, 9] to define different constrained optimum beamformers depending on the underlying spectral density matrix and the number of involved linear constraints.

2.1 Single constrained beamforming based on the signal-plus-noise spectral density matrix

In many applications, only the signal-plus-noise spectral density matrix R is measurable. In this case, we minimize the total beamformer output power

$$\text{minimize } E|y|^2 = w^H Rw \tag{2.1}$$

over the weight vector w subject to a constraint of unity signal response

$$d^H w = 1 \tag{2.2}$$

for a desired look direction. Here, d is the steering vector for this look direction. The beamformer output using the optimum weight vector is given by

$$y = \frac{d^H R^{-1} x}{d^H R^{-1} d} \tag{2.3}$$

2.2 Single constrained beamforming based on the noise spectral density matrix.

It is well known that signal suppression arises when there is a mismatch between the true and the assumed signal spatial characteristic such as imperfect knowledge about the signal direction or by scanning over the look directions [10]. This effect can be avoided if we know the noise spectral density matrix Q and minimize the noise output power

$$\text{minimize } w^H Q w \tag{2.4}$$

subject to (2.2). The beamformer output in this case becomes

$$y = \frac{d^H Q^{-1} x}{d^H Q^{-1} d} \tag{2.5}$$

While equation (2.5) provides a minimum variance unbiased estimate for the signal amplitude, the following beamformer

$$y = \frac{d^H Q^{-1} x}{(d^H Q^{-1} d)^{1/2}} \tag{2.6}$$

with a modified scaling factor for the weight vector corresponds to a consistent estimate for the signal direction [11]. This beamformer is also investigated in this paper.

2.3 Multiple constrained beamforming

Optimum beamformers with multiple linear constraints have been proposed [8,9]. Let C be a NxK full rank constraint matrix (K<N) and g a Kx1 constraint vector. Both optimization problems (2.1) and (2.4) can be formulated subject to the K linear constraints

$$C^H w = g. \tag{2.7}$$

One column of C is usually a steering vector, and the corresponding element in g is the number one resulting in the constraint (2.2). The other columns of C may be derivative constraints to control the mainlobe shape of the beampattern and/or null constraints to place beampattern nulls in specific directions [12]. The solutions of these multiple constrained least squares problems are given by

$$y = g^H (C^H R^{-1} C)^{-1} (C^H R^{-1} x), \tag{2.8}$$

and

$$Y = g^H (C^H Q^{-1} C)^{-1} (C^H Q^{-1} x). \tag{2.9}$$

In adaptive beamforming, the unknown spectral density matrices R and Q are estimated by the rank-one updates

$$R(t) = \mu_1 R(t-1) + \mu_2 x(t) x(t)^H, \tag{2.10}$$

$$Q(t) = \mu_1 Q(t-1) + \mu_2 n(t) n(t)^H, \tag{2.11}$$

where $0<\mu_1, \mu_2<1$ are two forgetting factors to be suitably chosen. In the following, we assume all considered spectral density matrices to be positively definite.

3. SOME BASIC EQUATIONS

Before we start with developing the update algorithms, we derive some equations closely related to the Faddeeva algorithm for matrix computation $D-BA^{-1}C$. These equations provide some insights into the underlying matrix problems of our adaptive beamforming algorithms.

Given the complex matrices A, B, C and D of compatible dimensions, we construct the following composed matrix

$$\begin{bmatrix} A & B \\ C & D \end{bmatrix} \tag{3.1}$$

A is square and non-singular. Faddeeva proposed an elimination method to compute $D-CA^{-1}B$ without any back-substitutions [7]. Based on Gauss eliminations, she added an appropriate linear combination of the rows of [A B] to those of [C D] for annulling the lower left corner of (3.1)

$$\begin{bmatrix} G_{11} & O \\ G_{12} & I \end{bmatrix} \begin{bmatrix} A & B \\ C & D \end{bmatrix} = \begin{bmatrix} G_{11}A & G_{11}B \\ G_{12}A + C & G_{12}B + D \end{bmatrix} = \begin{bmatrix} G_{11}A & G_{11}B \\ O & D-CA^{-1}B \end{bmatrix}. \tag{3.2}$$

In equation (3.2), G_{11} is used to first triangularize A. It is easy to see that after eliminating C with $G_{12}=-CA^{-1}$, the matrix expression $D-CA^{-1}B$ appears in the lower right hand corner of the resulting composed matrix.

However, the above method is not appropriate for systolic implementations since pivoting is required for the triangularization causing global communications in systolic arrays. This problem does not arise if A is triangular. In this case, equation (3.2) simplifes to

$$\begin{bmatrix} I & O \\ G_{12} & I \end{bmatrix} \begin{bmatrix} A & B \\ C & D \end{bmatrix} = \begin{bmatrix} A & B \\ G_{12}A+C & G_{12}B+D \end{bmatrix} = \begin{bmatrix} A & B \\ O & D-CA^{-1}B \end{bmatrix} \tag{3.3}$$

In fact, we can generalize the Faddeeva algorithm to compute more matrix expressions. Let U be any non-singular matrix applied to (3.1) for eliminating C

$$U. \begin{bmatrix} A & B \\ C & D \end{bmatrix} = \begin{bmatrix} U_{11} & U_{12} \\ U_{21} & U_{22} \end{bmatrix} \begin{bmatrix} A & B \\ C & D \end{bmatrix} = \begin{bmatrix} \tilde{A} & \tilde{B} \\ O & \tilde{D} \end{bmatrix} \tag{3.4}$$

Comparing the lower left and right corners of both sides of equation (3.4), we get

$$U_{21}A+U_{22}C = O, \tag{3.5}$$

$$U_{21}B+U_{22}D = \tilde{D}. \tag{3.6}$$

Solving U_{21} from (3.5) and inserting it into (3.6) yields

$$\tilde{D} = U_{22}(D-CA^{-1}B). \tag{3.7}$$

Note, U_{22} is square and non-singular. So we can extract $D-CA^{-1}B$ from $\tilde{D}$ if we know U_{22}.

On the other hand, equation (3.4) can be reformulated

$$\begin{bmatrix} A & B \\ C & D \end{bmatrix} = U^{-1}. \begin{bmatrix} \tilde{A} & \tilde{B} \\ O & \tilde{D} \end{bmatrix} = \begin{bmatrix} \tilde{U}_{11} & \tilde{U}_{12} \\ \tilde{U}_{21} & \tilde{U}_{22} \end{bmatrix} \begin{bmatrix} \tilde{A} & \tilde{B} \\ O & \tilde{D} \end{bmatrix}, \tag{3.8}$$

where $\tilde{U}_{ij}$ (ij=1,2) describe the submatrices of U^{-1}. Similar to (3.5) and (3.6), we get two equations for the lower left and right corners of both sides of (3.8)

$$C = \tilde{U}_{21}\tilde{A}, \tag{3.9}$$

$$D = \tilde{U}_{21}\tilde{B}+\tilde{U}_{22}\tilde{D}. \tag{3.10}$$

This leads to a second expression for $\tilde{D}$ if we replace $\tilde{U}_{21}$ in (3.10) with $C\tilde{A}^{-1}$

$$\tilde{U}_{22}\ \tilde{D} = D - C\tilde{A}^{-1}\tilde{B}. \tag{3.11}$$

From equations (3.7) and (3.11), we see that both $D\text{-}CA^{-1}B$ and $D\text{-}C\tilde{A}^{-1}\tilde{B}$ can be extracted from $\tilde{D}$. While equation (3.7) describes a more general class of methods to compute $D\text{-}CA^{-1}B$ for any four given matrices A, B, C and D, equation (3.11) shows a new method for adaptive computations of the type $D\text{-}C\tilde{A}^{-1}\tilde{B}$. Note that $D\text{-}C\tilde{A}^{-1}\tilde{B}$ depends on the updated values $\tilde{A}$ and $\tilde{B}$. If we interpret equation (3.4) as an update operation, where A and B are computed using the "new sample update" C, then $D\text{-}C\tilde{A}^{-1}\tilde{B}$ can be interpreted as a result corresponding to the recent sample update. In conjunction with the adaptive constrained beamforming problems, we see that all beamformer outputs (2.3), (2.5), (2.6), (2.8) and (2.9) consist of expressions of the form $C\tilde{A}^{-1}\tilde{B}$.

Equations (3.7) and (3.11) are valid for any non-singular matrix U. In the following, we are interested in two special cases.

A. Unitary transformation

If U is an unitary matrix satisfying

$$U^H U = UU^H = I, \tag{3.12}$$

equation (3.4) is characterized by

$$\tilde{A}^H\tilde{A} = A^H A+C^H C. \tag{3.13}$$

It shows strong similarities to equations (2.10) and (2.11) updating the spectral density matrices. In this case, the quantities U_{22} in (3.7) and $\tilde{U}_{22}$ in (3.11) are related by

$$\tilde{U}_{22} = U_{22}^H. \tag{3.14}$$

An unitary transformation can be realized, for example, by using Householder transformations or Givens rotations [13]. The latter is more appropriate for systolic implementations because of its regular structure [14].

B. Hyperbolic transformation:

If U describes a hyperbolic transformation defined by

$$U^H \cdot \begin{bmatrix} I & O \\ O & -I \end{bmatrix} \cdot U = \begin{bmatrix} I & O \\ O & -I \end{bmatrix}, \tag{3.15}$$

where the matrices I and -I have the dimensions of U_{11} and U_{22}, respectively, equation (3.4) is characterized by the downdating operation

$$\tilde{A}^H \tilde{A} = A^H A - C^H C. \tag{3.16}$$

In this case, the relation (3.14) is still valid. A hyperbolic transformation can be realized by hyperbolic rotations.

Now, we discuss how to compute U_{22} and $\tilde{U}_{22}$ for extracting $D-CA^{-1}B$ in (3.7) and $D-\tilde{C}\tilde{A}^{-1}\tilde{B}$ in (3.11) from $\tilde{D}$ for the scalar case. Although it is generally difficult to compute $\tilde{U}_{22}$ as the last diagonal element of U^{-1}, we can easily do it exploiting (3.14) for unitary and hyperbolic transformations. For example, we apply U to the "pinning" vector $\pi=[0,0...,1]^H$ to extract U_{22}

$$U.\pi = \begin{vmatrix} U_{11} & U_{12} \\ U_{21} & U_{22} \end{vmatrix} .\pi = \begin{vmatrix} U_{12} \\ U_{22} \end{vmatrix}. \tag{3.17}$$

Directly, U_{22} can be computed by

$$U_{22} = \prod_{i=1}^{N} c_i, \tag{3.18}$$

where c_i is the last diagonal element of the complex Givens or hyperbolic rotation matrix for eliminating the i-th element of the N-element row vector C in (3.4), cf. [2]. In real data case, c_i is simply the cosine or the hyperbolic cosine coefficient. Note that (3.17) requires to apply the transformation U to an additional column vector, while (3.18) assumes c_i being available and requires N-1 complex multiplications.

4. UPDATE ALGORITHMS

In this section, we derive the update algorithms for the different adaptive constrained beamformers. In most applications, we are only interested in the resulting beamformer output which is used for appropriate post-processing operations. The optimum weight vector is only an intermediate quantity. Thus, its explicit computation is not concerned in this paper.

This section is organized as follows. We first consider the multiple and single constrained beamformers based on the signal-plus-noise spectral density matrix. Then, we discuss how to modify the update algorithms for computing the beamformer output based on the noise spectral density matrix.

4.1 Multiple constrained beamforming based on the signal-plus-noise spectral density matrix

The operations we have to implement are given by the following equations,

$$R(t) = \mu_1 R(t-1) + \mu_2 x(t) x(t)^H, \tag{4.1}$$

$$y(t) = g^H (C^H R(t)^{-1} C)^{-1} (C^H R(t)^{-1} x(t)). \tag{4.2}$$

$R(t)$ is a NxN Hermitian, positively definite matrix, and C is of dimension NxK with rank $K<N$. For each time instant, the sample update $x(t)$ is used both to update the spectral density matrix (4.1) and to compute the beamformer output (4.2). A straight-forward implementation of equations (4.1) and (4.2) by computing the matrix inverse $R(t)^{-1}$ leads to $O(N^3)$ operations. We follow the idea proposed by Schreiber [6] to recursively update the Cholesky factor of $R(t)$ and develop and $O(N^2)$ update algorithm.

Let

$$R(t) = L(t) L(t)^H \tag{4.3}$$

be the Cholesky decomposition, where $L(t)$ is a complex, lower triangular matrix with positive diagonal elements. We define the new variables

$$\tilde{C}(t) = L(t)^{-1} C, \tag{4.4}$$

$$D(t) = \tilde{C}(t)^H \tilde{C}(t), \tag{4.5}$$

$$b(t) = C^H R(t)^{-1} x(t) = \tilde{C}(t)^H L(t)^{-1} x(t) \tag{4.6}$$

The beamformer output (4.2) can then be expressed by

$$y(t) = g^H D(t)^{-1} b(t). \tag{4.7}$$

In equations (4.6) and (4.7), we need to compute two matrix expressions of the form $C\tilde{A}^{-1}\tilde{B}$. This can easily be done using the generalized Faddeeva algorithm of section 3. Let

$$S(t-1) = \begin{bmatrix} \sqrt{\mu_1} L(t-1)^H & \tilde{C}(t-1)/\sqrt{\mu_1} \\ \sqrt{\mu_2} x(t)^H & 0 \end{bmatrix} \tag{4.8}$$

be a composed matrix. We apply an unitary transformation $U(t)$ to $S(t-1)$ for eliminating the row vector $\sqrt{\mu_2}x(t)^H$ while preserving the triangular structure of $L(t-1)$

$$S(t) = U(t).S(t-1) = \begin{bmatrix} L(t)^H & \tilde{C}(t) \\ 0 & z(t) \end{bmatrix} \tag{4.9}$$

Because of the matrix density

$$S(t)^H S(t) = S(t-1)^H U(t)^H U(t) S(t-1) = S(t-1)^H S(t-1),$$

we can show the following equations,

$$L(t)L(t)^H = [\sqrt{\mu_1}L(t-1)][\sqrt{\mu_1}L(t-1)]^H + [\sqrt{\mu_2}x(t)][\sqrt{\mu_2}x(t)]^H, \tag{4.10}$$

$$L(t)\tilde{C}(t) = \sqrt{\mu_1}L(t-1).\tilde{C}(t-1)/\sqrt{\mu_1}, \tag{4.11}$$

$$\tilde{C}(t-1)^H\tilde{C}(t-1)/\mu_1 = \tilde{C}(t)^H\tilde{C}(t) + z(t)^H z(t). \tag{4.12}$$

Considering (4.1), (4.3) and (4.4), we see that the computed $L(t)$ and $\tilde{C}(t)$ in equation (4.9) correspond to the desired updates of $L(t-1)$ and $\tilde{C}(t-1)$.

More interesting, $b(t)$ in (4.7) can be directly extracted from $z(t)$ in (4.9) without performing the matrix computations (4.6). Comparing the Faddeeva-like equations (3.4), (3.11) and (3.14) with (4.8) and (4.9), we see that $z(t)$ can be expressed by

$$z(t) = -(\sqrt{\mu_2}/U_{22}^H(t)).x(t)^H L(t)^{-H}\tilde{C}(t) = -(\sqrt{\mu_2}/U_{22}^H(t)).b(t)^H. \tag{4.13}$$

The scaling factor $\sqrt{\mu_2}/U_{22}^H(t)$ is obtained by appling U(t) on a suitable "pinning" vector as described in (3.17),

$$U(t).\begin{bmatrix} 0 \\ 1/\sqrt{\mu_2} \end{bmatrix} = \begin{bmatrix} \tilde{x}(t) \\ \alpha(t) \end{bmatrix} \tag{4.14}$$

with

$$\tilde{x}(t) = L(t)^{-1}x(t), \quad \alpha(t) = U_{22}(t)/\sqrt{\mu_2}. \tag{4.15}$$

Thus, b(t) is simply given by

$$b(t) = -\alpha(t)z(t)^H. \tag{4.16}$$

Now, we discuss how to compute the beamformer output (4.6) for given b(t). The straightforward way of doing this leads to equation (4.5) requiring $O(NK^2)$ operations and the computation of $D(t)^{-1}$ by additional $O(K^3)$ operations. We propose an efficient method based on recursively updating the Cholesky factor of D(t), which requires $O(K^2)$ operations. Combining equations (4.5) and (4.12), we get the following recursion

$$D(t) = D(t-1)/\mu_1 - z(t)^H z(t) \tag{4.17}$$

Let

$$D(t) = F(t)F(t)^H \tag{4.18}$$

be the Cholesky decomposition of D(t). Defining the new variable

$$\tilde{g}(t) = F(t)^{-1}g, \tag{4.19}$$

the beamformer output (4.7) can be expressed by

$$y(t) = \tilde{g}(t)^H F(t)^{-1} b(t). \tag{4.20}$$

Note that the set of equations (4.17), (4.18) and (4.20) is very similar to that of (4.1), (4.3) and (4.6) except for the minus sign in (4.17). This leads to a downdating operation [15] which can be performed by a hyperbolic transformation.

Again, we construct a composed matrix consisting of F(t-1), $\tilde{g}(t-1)$, z(t) and $\alpha(t)$. We apply a hyperbolic transformation H(t) to it to eliminate the row vector z(t) while preserving the triangular structure of F(t-1)

$$H(t).\begin{vmatrix} F(t-1)^H/\sqrt{\mu_1} & 0 & \sqrt{\mu_1}\tilde{g}(t-1) \\ z(t) & \alpha(t) & 0 \end{vmatrix} = \begin{vmatrix} F(t)^H & * & \tilde{g}(t) \\ 0 & & \beta(t)\gamma(t) \end{vmatrix} \quad (4.21)$$

Based on the "pseudo unitary" [16] property of (3.15), we can show the following equations,

$$F(t)F(t)^H = [F(t-1)/\sqrt{\mu_1}][F(t-1)/\sqrt{\mu_1}]^H - z(t)^H z(t), \quad (4.22)$$

$$F(t)\tilde{g}(t) = F(t-1)/\sqrt{\mu_1}.\sqrt{\mu_1}\tilde{g}(t-1). \quad (4.23)$$

This means that F(t) and $\tilde{g}(t)$ are the desired updates of F(t-1) and $\tilde{g}(t-1)$, respectively. Moreover, we obtain the following expressions for $\beta(t)$ and $\gamma(t)$ corresponding to (3.17) and (3.11),

$$\beta(t) = \alpha(t)/H_{22}(t), \quad (4.24)$$

$$\gamma(t) = -(1/H_{22}^H(t)).z(t)F(t)^{-H}\tilde{g}(t). \quad (4.25)$$

Combining equations (4.24), (4.25), (4.16) and (4.20), we get finally

$$\beta(t)\gamma(t)^H = \tilde{g}(t)^H F(t)^{-1}(-\alpha(t)z(t)^H) = \tilde{g}(t)^H F(t)^{-1} b(t) = y(t) \quad (4.26)$$

The update algorithm can now be summarized as follows.

Initial step: initialize $L(0)^H$, $\tilde{C}(0)$, $F(0)^H$, $\tilde{g}(0)$

For t=1,2,...do

1. Scaling operations:

$$\sqrt{\mu_1}.L(t-1)^H, (1/\sqrt{\mu_1}).\tilde{C}(t-1), (1/\sqrt{\mu_1}).F(t-1)^H, \sqrt{\mu_1}.\tilde{g}(t-1), \sqrt{\mu_2}.x(t)^H \tag{4.27a}$$

2. Unitary transformation:

$$U(t).\begin{bmatrix} \sqrt{\mu_1}L(t-1)^H & \tilde{C}(t-1)/\sqrt{\mu_1} & 0 \\ \sqrt{\mu_2}x(t)^H & 0 & 1/\sqrt{\mu_2} \end{bmatrix} = \begin{bmatrix} L(t)^H & \tilde{C}(t) & \tilde{x}(t) \\ 0 & z(t) & \alpha(t) \end{bmatrix} \tag{4.27b}$$

3. Hyperbolic transformation:

$$H(t).\begin{bmatrix} F(t-1)^H/\sqrt{\mu_1} & 0 & \sqrt{\mu_1}\tilde{g}(t-1) \\ z(t) & \alpha(t) & 0 \end{bmatrix} \begin{bmatrix} F(t)^H & * & \tilde{g}(t) \\ 0 & \beta(t) & \gamma(t) \end{bmatrix} \tag{4.27c}$$

4. Beamformer output computation

$$y(t) = \beta(t)\gamma(t)^H \tag{4.27d}$$

End

Now, we briefly comment on the numerical stability properties of above update algorithm. While the unitary transformation does not cause any problems, the hyperbolic one may be critical. However, recent works [15,16] have shown that, with careful implementation of the hyperbolic rotations, the hyperbolic transformation is stable in a certain "mixed" sense. It works well in the presence of rounding errors, provided that the matrix D(t) is not ill conditioned.

4.2 Single constrained beamforming based on the signal-plus-noise spectral density matrix

The update algorithm presented in the previous section can be considerably simplified if we consider the single constrained beamformer (K=1). In this case, the NxK constraint matrix C becomes the steering vector d, and the Kx1 constraint vector g becomes the number one. Similar to (4.4) and (4.5), we define

$$\tilde{d}(t) = L(t)^{-1}d, \tag{4.28}$$

$$D(t) = |\tilde{d}(t)|^2 \tag{4.29}$$

D(t) is now a scalar quantity. The downdating operations (4.17) and equation (4.7) for computing the beamformer output consist of only simple scalar operations. The hyperbolic transformation (4.21) can thus be saved.

We remark that the above update algorithm can be easily extended to scanning over a desired set of look directions with only O(N) additional operations per additional direction. If we attach to each direction dependent quantity a subscript i=1,2,... indicating different look directions, we can formulate the update algorithm as follows.

Initial step; initialize $L(0)^H$, $\tilde{d}_i(0)$, $D_i(0)$ (i=1,2...)

For t= 1,2,...do

1. Scaling operations:

$$\sqrt{\mu_1}.L(t-1)^H, \quad (1/\sqrt{\mu_1}).\tilde{d}_i(t-1) \quad (i=1,2...), \quad \sqrt{\mu_2}.x(t)^H \tag{4.30a}$$

2. Unitary transformation:

$$U(t).\begin{vmatrix} \sqrt{\mu_1}L(t-1)^H & 0 & \tilde{d}_1(t-1)/\sqrt{\mu_1} & \tilde{d}_2(t-1)/\sqrt{\mu_1} & \cdots \\ \sqrt{\mu_2}x(t)^H & 1/\sqrt{\mu_2} & 0 & 0 & \cdots \end{vmatrix}$$

$$= \begin{vmatrix} L(t)^H & \tilde{x}(t) & \tilde{d}_1(t) & \tilde{d}_2(t) & \cdots \\ 0 & \alpha(t) & z_1(t) & z_2(t) & \cdots \end{vmatrix} \tag{4.30b}$$

3. Beamformer output computations:

$$D_i(t) = D_i(t-1)/\mu_i - |z_i(t)|^2, \tag{4.30c}$$

$$y_i(t) = -\alpha(t)z_i(t)^H/D_i(t) \quad (i=1,2...). \tag{4.30d}$$

End

Note that by computing the quantity $D_i(t)=d_i^H R(t)^{-1} d_i$, the above update algorithm also implements Capon's beamformer [17] calculating the inverse $1/D_i(t)$ of the estimated beamformer output power. A better numerical stability can be achieved, if equation (4.30c) is replaced by (4.29).

4.3 Multiple constrained beamforming based on the noise spectral density matrix

For active radar and sonar systems, we can measure both the signal and noise wavefields. Let n(t) be the noise sensor output vector measured at times $t \in T_n = \{t \mid t_o+1,\ldots,t_o+t_n\}$ and x(t) be the signal-plus-noise sensor output vector measured at times $t \in T_x = \{t \mid t=t_o+t_n+1,\ldots,t_o+t_n+t_x\}$. The multiple constrained beamformer based on the noise spectral density matrix can be formulated as follows. At first, we measure n(t) and estimate the noise spectral density matrix by the rank-one update

$$Q(t) = \mu_1 Q(t-1)+\mu_2 n(t)n(t)^H \quad (t \in T_n) \tag{4.31}$$

Then, we measure x(t) and compute beam signals by using the latest, fixed noise spectral density matrix estimate $Q(t_o+t_n)$

$$y(t) = g^H (C^H Q(t_o+t_n)^{-1} C)^{-1} (C^H Q(t_o+t_n)^{-1} x(t) \quad (t \in T_x) \tag{4.32}$$

Equations (4.31) and (4.32) describe a blockwise adaptive beamformer.

The first step of our update algorithm is very similar to that in 4.1. Let

$$Q(t) = L(t)L(t)^H \tag{4.33}$$

be the Cholesky decomposition. We define $\tilde{C}(t)$, D(t), F(t) and $\tilde{g}(t)$ as in equations (4.4), (4.5), (4.18) and (4.19). Replacing x(t) by n(t) in (4.8), we perform the unitary (4.9) and the hyperbolic (4.21) transformation for $t \in T_n$. The final results at $t=t_o+t_n$ are the updates $L(t_o+t_n)$, $\tilde{C}(t_o+t_n)$, $F(t_o+t_n)$ and $\tilde{g}(t_o+t_n)$.

Now, we compute the beamformer output (4.32) for $t \in T_x$. According to equations (4.6) and (4.20), we can express (4.32) by

$$b(t) = \tilde{C}(t_o+t_n)^H L(t_o+t_n)^{-1} x\ (t) \quad (t \in T_x), \tag{4.34}$$

$$y(t) = \tilde{g}(t_o+t_n)^H F(t_o+t_n)^{-1} b(t) \quad (t \in T_x), \tag{4.35}$$

These are two matrix expressions of the form $CA^{-1}B$. The quantities L, $\tilde{C}$, F and $\tilde{g}$ are given and need not be updated. Thus, we apply the Faddeeva algorithm based on Gauss eliminations (3.3) to compute (4.34) and (4.35), where the triangular structure of $L(t_o+t_n)$ and $F(t_o+t_n)$ is exploited. We summarize the algorithm as follows.

Assuming $L(t_o)^H$, $\tilde{C}(t_o)$, $F(t_o)^H$ and $\tilde{g}(t_o)$ to be known

For $t=t_o+1, t_o+2, \ldots, t_o+t_n$ do (when measuring the noise wavefield)

1. Scaling operations:

$$\sqrt{\mu_1}.L(t-1)^H, \ (1/\sqrt{\mu_1}).\tilde{C}(t-1), \ (1/\sqrt{\mu_1}).F(t-1)^H, \ \sqrt{\mu_1}.\tilde{g}(t-1), \ \sqrt{\mu_2}.n(t)^H \tag{4.36a}$$

2. Unitary transformation:

$$U(t).\begin{bmatrix} \sqrt{\mu_1}L(t-1)^H & \tilde{C}(t-1)/\sqrt{\mu_1} \\ \sqrt{\mu_2}n(t)^H & 0 \end{bmatrix} = \begin{bmatrix} L(t)^H & \tilde{C}(t) \\ 0 & z(t) \end{bmatrix} \tag{4.36b}$$

3. Hyperbolic Transformation:

$$H(t).\begin{bmatrix} F(t-1)^H/\sqrt{\mu_1} & \sqrt{\mu_1}\tilde{g}(t-1) \\ z(t) & 0 \end{bmatrix} = \begin{bmatrix} F(t)^H & \tilde{g}(t) \\ 0 & \gamma(t) \end{bmatrix} \tag{4.36c}$$

End

For $t=t_o+t_n+1, t_o+t_n+2, \ldots, t_o+t_n+t_x$ do (when measuring the signal wavefield)

4. Gauss elimination:

$$G_1(t).\begin{bmatrix} L(t_o+t_n)^H & \tilde{C}(t_o+t_n) \\ x(t)^H & 0 \end{bmatrix} = \begin{bmatrix} L(t_o+t_n)^H & \tilde{C}(t_o+t_n) \\ 0 & v(t) \end{bmatrix} \tag{4.36d}$$

5. Gauss elimination:

$$G_2(t) \cdot \begin{bmatrix} F(t_o+t_n)^H & \tilde{g}(t_o+t_n) \\ v(t) & 0 \end{bmatrix} = \begin{bmatrix} F(t_o+t_n)^H & \tilde{g}(t_o+t_n) \\ 0 & y(t)^H \end{bmatrix} \tag{4.36e}$$

End

Note that the Gauss eliminations (4.36d) and (4.36e) can also be replaced by unitary transformations. Then, we need to perform operations (3.7) and (3.17) to compensate the scaling "$\tilde{U}_{22}$", and we require additional storages for keeping $L(t_o+t_n)$, $\tilde{C}(t_o+t_n)$, $F(t_o+$ and $\tilde{g}(t_o+t_n)$ unchanged, since these data are destroyed while applying unitary transformations.

4.4 Single constrained beamforming based on the noise spectral density matrix

Using the ideas from 4.2 and 4.3, the update algorithm for the single constrained beamformer (2.5) based on the noise spectral density matrix by scanning over a desired set of look directions can be similarly derived.

By measuring the noise wavefield at $t \in T_n$, we perform operations similar to (4.30a) and (4.30b) for updating $L(t)^H$ and $\tilde{d}_i(t)(i=1,2,\ldots)$. Note that in this case, $L(t)$ is the Cholesky factor of $Q(t)$, and $x(t)$ in (4.30a) and (4.30b) is replaced by $n(t)$. The column vector $[0\ 1/\sqrt{\mu_2}]^H$ in (4.30b) can be omitted. Now, when we receive $x(t)$ at $t \in T_x$, a Gauss elimination operation

$$G(t) \cdot \begin{bmatrix} L(t_o+t_n)^H & \tilde{d}_1(t_o+t_n) & \tilde{d}_2(t_o+t_n) & \ldots \\ x(t)^H & 0 & 0 & \ldots \end{bmatrix} \begin{bmatrix} L(t_o+t_n)^H & \tilde{d}_1(t_o+t_n) & \tilde{d}_1(t_o+t_n) & \ldots \\ 0 & v_1(t) & v_2(t) & \ldots \end{bmatrix} \tag{4.37}$$

is carried out. The beamformer outputs are then computed by

$$y_i(t) = -v_i(t)^H / |\tilde{d}_i(t_o+t_n)|^2 \quad (i=1,2,\ldots), \tag{4.38}$$

where equation (4.29) is used for calculating the denominators in (4.38).

The optimum beamformer for signal direction estimation (2.6) can be calculated using the same algorithm if we replace $|\tilde{d}_i(t_o+t_n)|^2$ by $|\tilde{d}_i(t_o+t_n)|$ in (4.38).

5. SYSTOLIC IMPLEMENTATIONS

5.1 Basic operations and processing elements

We have seen that the basic operations we have to implement are complex Givens rotations, complex hyperbolic rotations and complex Gauss eliminations. They are applied to two complex row vectors for eliminating the first element of the second row. That is

$$\begin{bmatrix} a_1 & a_2 & a_3 & \dots \\ b_1 & b_2 & b_3 & \dots \end{bmatrix} \rightarrow \begin{bmatrix} \tilde{a}_1 & \tilde{a}_2 & \tilde{a}_3 & \dots \\ 0 & \tilde{b}_2 & \tilde{b}_3 & \dots \end{bmatrix} \tag{5.1}$$

Note, a_1 is real and positive because it is a diagonal element of the Cholesky factor of a spectral density matrix.

There are several different ways to realize the above complex operations. One of them based on real rotations can be formulated as follows:

complex Givens rotation
$$\begin{bmatrix} \tilde{a}_i \\ \tilde{b}_i \end{bmatrix} = \begin{bmatrix} \cos\theta & \sin\theta \\ -\sin\theta & \cos\theta \end{bmatrix} \begin{bmatrix} 1 & 0 \\ 0 & e^{-j\ell} \end{bmatrix} \begin{bmatrix} a_i \\ b_i \end{bmatrix}, \tag{5.2}$$

complex hyperbolic rotation
$$\begin{bmatrix} \tilde{a}_1 \\ \tilde{b}_i \end{bmatrix} = \begin{bmatrix} \cosh\theta & -\sinh\theta \\ -\sinh\theta & \cosh\theta \end{bmatrix} \begin{bmatrix} 1 & 0 \\ 0 & e^{-j\ell} \end{bmatrix} \begin{bmatrix} a_i \\ b_i \end{bmatrix}, \tag{5.3}$$

complex Gauss elimination
$$\begin{bmatrix} \tilde{a}_i \\ \tilde{b}_i \end{bmatrix} \begin{bmatrix} 1 & 0 \\ -\theta & 1 \end{bmatrix} \begin{bmatrix} 1 & 0 \\ 0 & e^{-j\ell} \end{bmatrix} \begin{bmatrix} a_i \\ b_i \end{bmatrix}. \tag{5.4}$$

We first determine the phase angle ℓ of the element b_1 and scale the second row with $e^{-j\ell}$ to make b_1 real. This corresponds to a real Givens rotation applied to the real and imaginary part of b_i. Then, we compute the "rotation angle" θ from a_1 and $|b_1|$ and apply a real Givens rotation (or real hyperbolic rotation or real Gauss elimination) to the two rows to eliminate $|b_1|$.

The above operations can be carried out by adder-multiplier based processors. An alternative is the use of dedicated processors. Because of the "rotational" operations in (5.2) to (5.4), we propose to use the CORDIC (COordinate Rotation DIgital Computer) processor [18] to realize the processing elements (PEs). Based on a sequence of iterative shift-add operations, the CORDIC processor can perform a variety of functions of the initial values x_o, y_o and z_o (table 1). Here, m and VR are two "programming" parameters, and $\varepsilon=\pm 1$ is a sign parameter. We see that all desired operations can be performed by the CORDIC processor. Moreover, a suitably designed CORDIC processor requires the same computation time for all functions, which is a very useful property for pipelining. It has a regular structure, a high data throughput in a pipeline realization and can be easily programmed. Recently, CORDIC processors have been available in various realizations [19,20,21].

Note that for implementing equation (5.1) with a CORDIC processor, we need to add a programming bit VR to the data vectors. This bit is set to one for the first data pair (a_1,b_1) and zero otherwise. In this way, the CORDIC processor first computes suitable "rotation angles" ℓ and θ from (a_1,b_1) for eliminating b_1 and then apply the corresponding complex "rotation" to the successive data pairs $(a_i,b_i)(i=2,3,\dots)$.

	m=+1	m=0	m=-1
VR=1	$x=\sqrt{x_o^2+y_o^2}$ $z=z_o+\varepsilon\arctan(y_o/x_o)$	$x=x_o$ $z=z_o+\varepsilon y_o/x_o$	$x=\sqrt{x_o^2-y_o^2}$ $z=z_o+\varepsilon \mathrm{artanh}(y_o/x_o)$
VR=0	$x=x_o\cos z_o-\varepsilon y_o\sin z_o$ $y=y_o\cos z_o+\varepsilon x_o\sin z_o$	$x=x_o$ $y=y_o+\varepsilon z_o x_o$	$x=x_o\cosh z_o+\varepsilon y_o\sinh z_o$ $y=y_o\cosh z_o+\varepsilon x_o\sinh z_o$

Table 1 : CORDIC functions

5.2 Systolic arrays

In recent papers [3,4], we have shown that the single and multiple constrained beamformers based on the signal-plus-noise spectral density matrix can be implemented by linear systolic arrays. We show in this section that the same systolic arrays can be used to implement the constrained beamformers based on the noise spectral density matrix.

Figure 1 shows the systolic implementation of the multiple constrained beamformer based on the signal-plus-noise spectral density matrix (4.27). It consists of two rows of PEs. The upper one with PEs P1 performs the scaling operations (4.27a) to prepare the input data for (4.27b) and (4.27c). The lower one contains three different types of PEs, P2, P3 and P4 which implement the unitary transformation (4.27b), the hyperbolic transformation (4.27c) and the multiplication (4.27d) respectively. The right side input data in Figure 1 are arranged such as if we spatially rotate $[L^H \tilde{C}\tilde{x}]$ in (4.27b) and $[F^H * \tilde{g}]$ in (4.27c) by 90 degrees counterclockwise and put them into the systolic array. A corresponding time skew is imposed here to avoid "broadcasting" of data. The left side data vector x has an attached programming bit VR which is not drawn in Figure 1 for clarity. This bit is set to one for the first element of x and zero otherwise. All data are simply delayed by one computation cycle per PE as they propagate from left to right and from top to bottom across the array except for the programming bit VR. It is delayed by two computation cycles per PE P2 or P3 so that VR=1 and the i-th (i=1,...,N) element of x always arrive in the i-th PE P2 at the same time for eliminating the element of x.

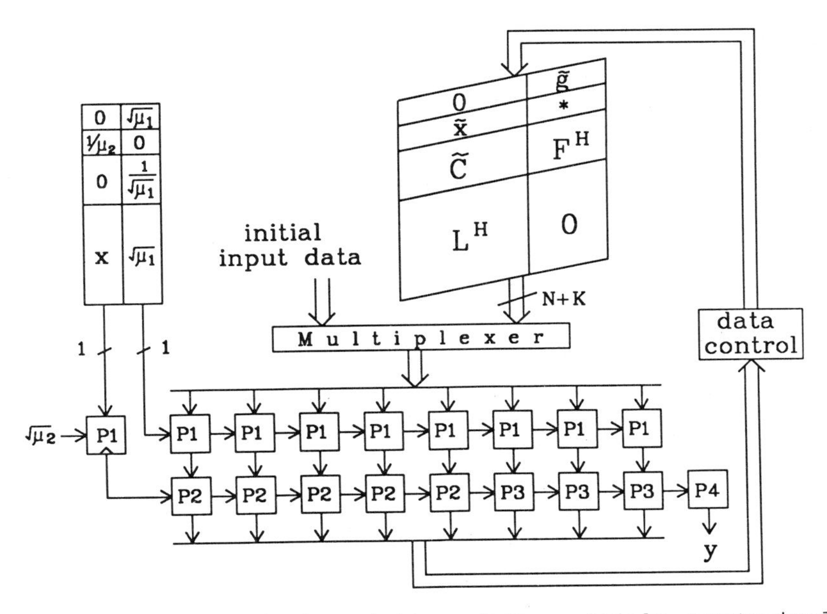

Fig. 1a) Systolic implementation of the multiple constrained adaptive beamformer based on the signal-plus-noise spectral density matrix (N=5, K=3)

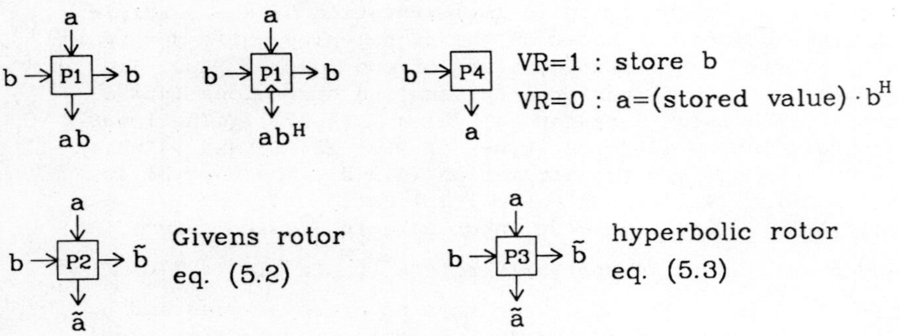

Fig. 1b) Processing elements

Let us assume that, after suitable scaling operations in the PEs P1, the first element of the vector x and the (1,1)-th element of the matrix L^H reach the most left PE P2. Controlled by VR=1, two rotation angles ℓ and θ are computed to eliminate the element of x. These angles are stored in the PE. The same complex Givens rotation (5.2) is then applied to successive input pairs of the most left PE according to VR=0. The results appear at the right and bottom output of the PE. Now, the second transformed element of x, the second diagonal element of L^H and VR=1 reach the second PE P2. A similar complex Givens rotation is performed there to eliminate the second element of x. Repeating this procedure, all N elements of x are successively eliminated in the N PEs P2.

After completing the unitary transformation (4.27b), the resulting vector z appears at the right output of the most right PE P2. It enters the PEs P3 synchonously with F^H. Similarly, complex hyperbolic rotations (5.3), which realize the hyperbolic transformation (4.27c), are performed to eliminate z. Finally, the PE P4 collects the scalar quantities β and γ and performs the multiplication (4.27d) for computing the beamformer output.

Figure 2 shows the systolic implementation of the multiple constrain beamformer based on the noise spectral density matrix (4.36). We see that the array architecture is identical to that in Figure 1 except for the PE P4. The input data pattern on the left contains either the noise sensor output vector z or the signal-plus-noise sensor output vector x. In the former case, the systolic array implements the operations (4.36a) to (4.36c) as in Figure 1 to update L^H, $\tilde{C}$, F^H and $\tilde{g}$. The PEs P2 and P3 act as Givens rotors and hyperbolic rotors defined in (5.2) and (5.3), respectively. In the latter case, when the signal-plus-noise sensor output vector is put in, the systolic array computes the beamformer output by performing the operations (4.36d) and (4.36e). The PEs P2 and P3 are now "Gauss eliminators" carrying out complex Gauss eliminations (5.4).

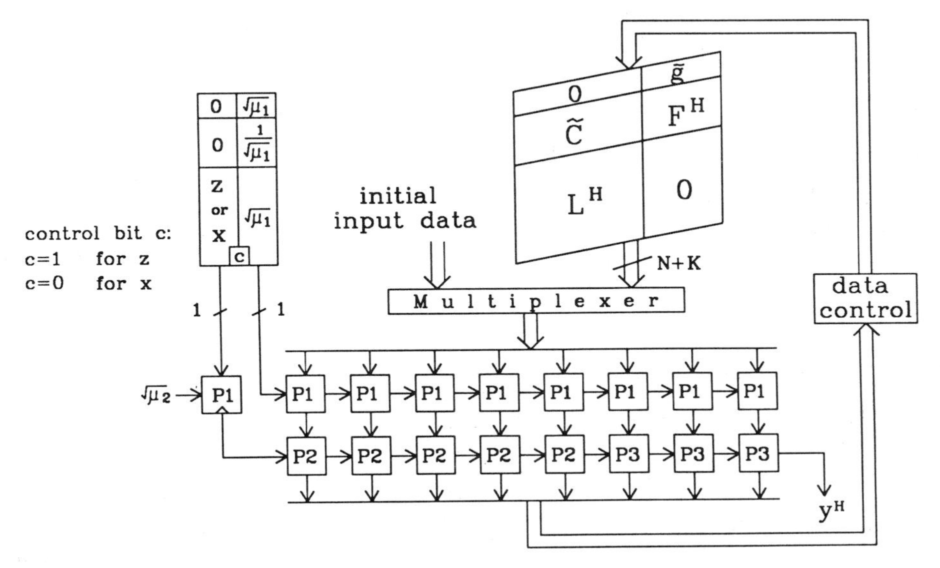

Fig. 2a) Systolic implementation of the multiple constrained adaptive beamformer based on the noise spectral density matrix (N=5, K=3)

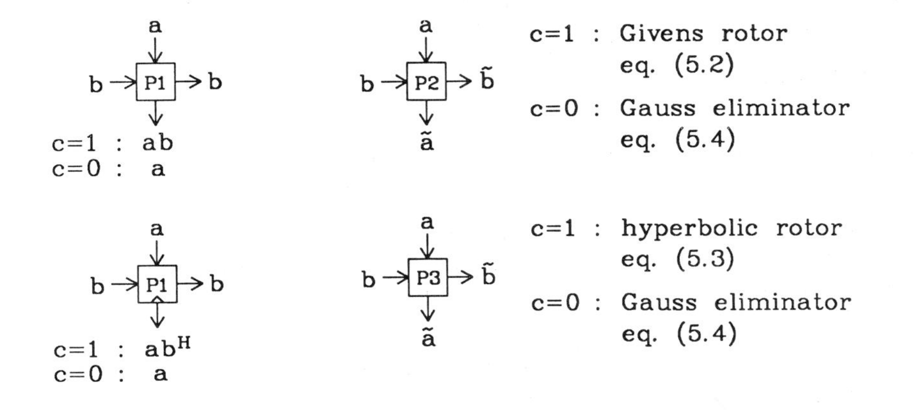

Fig. 2b) Processing elements

These two different modes of the systolic array can be simply controlled by a data driven control bit c which flows through all PEs. By setting c to one or zero, the above systolic array can be arbitrarily switched between adapting the noise wavefield and computing beam signals for the signal wavefield. The requirement on the PEs to be programmable for performing different types of operations does not cause any problems if we use CORDIC processors. It is interesting to note that McWhirter and Shepherd [2] proposed a similar technique to "freeze" adaptions by setting an input data to zero in relation with the square root free algorithm.

For the single constrained beamformers, we can show similar results. The linear systolic array described in [4] can also be used to implement the single constrained beamformer based on the noise spectral density matrix. However, space does not permit us to discuss it in more details.

Finally, we mention that all update algorithms developed in this paper can also be implemented by two-dimensional systolic arrays. Related works can be found by McWhirter and Shepherd [2,5]. By doing this, we can usually achieve a higher efficiency defined by the percentage of the busy time-space spans over the total time-space spans. The reason is that we can map the involved matrix structure exactly onto the structure of the systolic array causing fewer or no idle PEs.

REFERENCES

[1] Speiser, J. M. (1986) "Signal Processing Computational Needs", Proc. SPIE, Advanced Algorithms and Architectures for Signal Processing, vol. 696, pp.2-6.

[2] Shepherd, T. J. and McWhirter, J. G. (1985) "A Pipelined Array for Linearly Constrained Least-Squares Optimisation", Proc. IMA Conf. on Mathematics in Signal Processing.

[3] Yang, B. and Böhme, J. F. (1988) "A Multiple Constrained Adaptive Beamformer and A Systolic Implementation", Proc. EUSIPCO, pp.283-287.

[4] Yang, B. and Böhme, J. F. (1988) "Systolic Implementation of A General Adaptive Array Processing Algorithm", Proc. IEEE ICASSP, pp. 2785-2789.

[5] McWhirter, J. G. and Shepherd, T. J. (1988) "An Efficient Systolic Array for MVDR Beamforming", Proc. IEEE Int. Conf. on Systolic Arrays.

[6] Schreiber, R. (1986) "Implementation of Adaptive Array Algorithms", IEEE Trans. ASSP, vol. 34, no. 5, pp.1038-1045.

[7] Faddeeva, V. N. (1959) Computational Methods of Linear Algebra, pp.90-99, Dover Publications, New York.

[8] Frost, O. L. (1972) "An Algorithm for Linearly Constrained Adaptive Processing", Proc. IEEE, vol.60, no. 8, pp.926-935.

[9] Haykin, S. (1985) Array Signal Processing, Prentice Hall. (see chapter Sonar Array Processing by N. Owsley)

[10] Cox, H. (1973) "Resolving Power and Sensitivity to Mismatch of Optimum Array Processors", J. Accoust. Soc. Amer., vol. 54, no. 3, pp.771-785.

[11] Nickel, U. (1987) "Angle Estimation with Adaptive Arrays and Its Relation to Super-Resolution", Proc. IEE, vol. 134, no. 1, pp. 77-82.

[12] Cox, H., et al, (1987) "Robust Adaptive Beamforming", IEEE Trans. ASSP, vol.35, no.10, pp.1365-1374.

[13] Golub, G. H. and Van Loan, C. (1983) Matrix Computations, John Hopkins Press, Baltimore.

[14] Nash, J. G. and Hansen, S. (1984) "Modified Faddeev Algorithm for Matrix Computation", Proc. SPIE, Real Time Signal Processing, vol. 495, pp.39-46.

[15] Bojanczyk, A. et al, (1987) "Downdating the Cholesky Factorization", Algorithms and Applications on Vector and Parallel Computers, Elsevier Science Publishers B.V. (North-Holland), ed. H. J. J. te Riele, et al, pp.307-323.

[16] Alexander, S. T., et al., (1988) "Analysis of a Recursive Least Squares Hyperbolic Rotation Algorithm for Signal Processing", Linear Algebra and Its Applications, vol. 98, pp.3-40

[17] Capon, J. (1969) "High Resolution Frequency - Wavenumber Spectrum Analysis", Proc. IEEE, vol. 57, pp.1408-1418.

[18] Walther, J. S., (1971) "An Unified Algorithm for Elementary Functions", Proc. IEEE SJCC, pp.379-385.

[19] Schmidt, G., et al, (1988) "Design of 16-Bit Fixed-Point Recursive CORDIC Processors and Evaluation Tools", Proc. EUSIPCO, pp. 1557-1560.

[20] de Lange, A. A. J., et al, (1988) "An Optimal Floating-Point Pipeline CMOS CORDIC Processor", Proc. IEEE ISCAS, pp. 2043-2047.

[21] Yoshimura, H., et al, (1988) "A 50MHz CMOS Geometrical Mapping Processor", Proc. IEEE ISSCC.

A SYSTOLIC SQUARE ROOT COVARIANCE KALMAN FILTER

F.M.F. Gaston and G.W. Irwin
(The Queen's University, Belfast)

1. INTRODUCTION

In the last twenty years the Kalman filter has become a fundamental tool for state estimation and prediction, particularly in the areas of communications, signal processing and modern control. The equations defining the filter are computationally intensive, involving a number of matrix operations for each recursion, and this inevitably restricts the full exploitation of the technique in broad-band, real-time applications. Parallel processing is considered to be a possible solution to this computational bottleneck.

It has been shown that systolic array architectures [1] can significantly reduce the processing time for matrix operations and a number of researchers have applied this approach to Kalman filtering [2,3,4]. The fastest and most efficient full Kalman filter to date, where efficiency is measured by processor utilization, is that of Gaston and Irwin [3] which implements a novel square root information Kalman filter. A full state update can be obtained every O(3n) timesteps, where n is the number of states, and the processor utilization achieved is between 33 and 63%. The algorithm implemented was derived from an approach of Paige and Saunders [5].

In this paper the square root covariance Kalman filter of Morf and Kailath [6] will be derived using the approach of Paige [7]. This will clearly illustrate the relationship between this form of the Kalman filter and least squares. A systolic array implementation of this algorithm will also be described. This implementation obtains a new state update every O(2n) timesteps and achieves a processor utilization of between 44 and 100%, a considerable improvement on the corresponding figures for information filtering.

2. PRELIMINARIES

The linear discrete-time system is described by

$$\underline{x}(k+1) = A(k)\underline{x}(k) + B(k)\underline{u}(k) + \underline{w}(k) \tag{1}$$

$$\underline{z}(k) = C(k)\underline{x}(k) + \underline{v}(k) \tag{2}$$

where $\underline{x}(k)$ is the (nxl) state vector, $\underline{z}(k)$ is the (mxl) measurement vector ($m \leq n$) and $\underline{u}(k)$ is the (pxl) control or deterministic forcing vector. A(k), B(k) and C(k) are known matrices of appropriate dimensions. Also $\underline{w}(k)$ and $\underline{v}(k)$ are zero mean, independent, Gaussian white noise sequences with known covariances W(k) and V(k) respectively.

Further, W(k) is assumed to be a symmetric, non-negative definite matrix while V(k) is assumed to be a symmetric, positive definite matrix and thus the following 'square roots', or Choleski factors, which can be either upper or lower triangular are defined as

$$V = (-V^{1/2})(-V^{T/2}) \quad ; \quad W = (-W^{1/2})(-W^{T/2}) \tag{3}$$

The Kalman filter estimates the state of the system from a sequence of measurements. Several types of state estimate are possible, in particular the predicted state estimate, $\underline{\hat{x}}(k+1/k)$ (the estimate of the state at time k+1 given measurements up until time k) and the filtered state estimate, $\underline{\hat{x}}(k/k)$ (the estimate at time k given measurements up until time k). The error covariance matrices corresponding to these estimates are

$$P(k+1/k) = E\{(\underline{x}(k+1)-\underline{\hat{x}}(k+1/k))(\underline{x}(k+1)-\underline{\hat{x}}(k+1/k))^T\} \tag{4}$$

$$P(k/k) = E\{(\underline{x}(k)-\underline{\hat{x}}(k/k))(x(k)-\underline{\hat{x}}(k/k))^T\} \tag{5}$$

3. LEAST SQUARES ESTIMATION AND SQUARE ROOT COVARIANCE KALMAN FILTERING

In [8] two novel square root information algorithms were derived from a least squares point of view. Here the relationship between least squares and the square root covariance Kalman filter [6] will be illustrated using the approach of Paige [7].

The available information, given by equations 1 and 2 together with the initial conditions, $\underline{\hat{x}}(0)$ and P(0), can be combined into equation 6.

$$\begin{bmatrix} \hat{\underline{x}}(0) \\ \underline{z}(0) \\ -B(0)\underline{u}(0) \\ \underline{z}(1) \\ -B(1)\underline{u}(1) \\ \vdots \\ -B(k)\underline{u}(k) \end{bmatrix} = \begin{bmatrix} I & 0 & 0 & & & \\ C(0) & 0 & 0 & & & \\ A(0) & -I & 0 & & & \\ 0 & C(1) & 0 & & & \\ 0 & A(1) & -I & & & \\ \vdots & & & & & \\ 0 & 0 & 0 & A(k) & -I \end{bmatrix} \begin{bmatrix} \underline{x}(0) \\ \underline{x}(1) \\ \underline{x}(2) \\ \underline{x}(3) \\ \underline{x}(4) \\ \vdots \\ \underline{x}(k+1) \end{bmatrix} + \begin{bmatrix} \underline{e}(0) \\ \underline{v}(0) \\ \underline{w}(0) \\ \underline{v}(1) \\ \underline{w}(1) \\ \vdots \\ \underline{w}(k) \end{bmatrix} \quad (6)$$

where $\underline{e}(0)$, the error associated with the initial estimate, has covariance matrix $P(0)$. Now let

$$\underline{e}(0) = P^{1/2}(0)\tilde{\underline{e}}(0), \quad \underline{v}(k) = -V^{1/2}(k)\tilde{\underline{v}}(k), \quad \underline{w}(k) = -W^{1/2}(k)\tilde{\underline{w}}(k)$$

where $\tilde{\underline{e}}$, $\tilde{\underline{v}}$ and $\tilde{\underline{w}}$ all have unit covariances. Equation 6 then becomes

$$\begin{bmatrix} \hat{\underline{x}}(0) \\ \underline{z}(0) \\ -B(0)\underline{u}(0) \\ \vdots \\ -B(k)\underline{u}(k) \end{bmatrix} = \begin{bmatrix} I & 0 & & \\ C(0) & 0 & & \\ A(0) & -I & & \\ \vdots & & & \\ 0 & 0 & A(k) & -I \end{bmatrix} \begin{bmatrix} \underline{x}(0) \\ \underline{x}(1) \\ \underline{x}(2) \\ \vdots \\ \underline{x}(k+1) \end{bmatrix} + \begin{bmatrix} P^{1/2}(0) & & & \\ & -V^{1/2}(0) & & \\ & & -W^{1/2}(0) & \\ & & & \ddots \\ & & & & -W^{1/2}(k) \end{bmatrix} \begin{bmatrix} \tilde{\underline{e}}(0) \\ \tilde{\underline{v}}(0) \\ \tilde{\underline{w}}(0) \\ \vdots \\ \tilde{\underline{w}}(k) \end{bmatrix} \quad (7)$$

The equations in 7 are organised sequentially so that $\hat{\underline{x}}(0)$ is the initial state prediction $\hat{\underline{x}}(0/-1)$ and $P(0)$ is the initial state error covariance matrix, $P(0/-1)$. Row 2 contains the first measurement and this information will therefore correct $\hat{\underline{x}}(0/-1)$ to $\hat{\underline{x}}(0/0)$ while row 3 contains the information to make the next prediction; thus the information is arranged in a predict/correct fashion. If rows 1 and 2 are used then $\hat{\underline{x}}(0/0)$ can be found, while if rows 1,2 and 3 are used then $\hat{\underline{x}}(1/0)$ can also be calculated.

Since equation 7 is simply a set of linear equations, rows can be multiplied and combined without affecting the variables. For example, if equation 7 is premultiplied by

$$\begin{bmatrix} I & 0 & 0 \\ C(0) & -I & 0 \\ \vdots & & \vdots \\ 0 & 0 & I \end{bmatrix}$$

then equation 8 will result

$$\begin{bmatrix} \underline{\hat{x}}(0) \\ C(0)\underline{\hat{x}}(0)-\underline{z}(0) \\ -B(0)\underline{u}(0) \\ \vdots \\ -B(k)\underline{u}(k) \end{bmatrix} = \begin{bmatrix} I & & & & \\ O & & & & \\ A(0) & -I & & & \\ \vdots & & \ddots & & \\ O & & & A(k) & -I \end{bmatrix} \begin{bmatrix} \underline{x}(0) \\ \underline{x}(1) \\ \underline{x}(2) \\ \vdots \\ \underline{x}(k+1) \end{bmatrix} + \begin{bmatrix} P^{1/2}(0) & & & \\ C(0)P^{1/2}(0) & V^{1/2}(0) & & \\ & & -W^{1/2}(0) & \\ \vdots & & & \ddots \\ & & & -W^{1/2}(k) \end{bmatrix} \begin{bmatrix} \underline{\tilde{e}}(0) \\ \underline{\tilde{v}}(0) \\ \underline{\tilde{w}}(0) \\ \vdots \\ \underline{w}(k) \end{bmatrix} \tag{8}$$

If equation 8 is now premultiplied by

$$\begin{bmatrix} O & I & & O \\ A(0) & O & -I & O \\ \vdots & & & \vdots \\ O & O & & I \end{bmatrix}$$

equation 9 results

$$\begin{bmatrix} C(0)\underline{\hat{x}}(0)-\underline{z}(0) \\ A(0)\underline{\hat{x}}(0)+B(0)\underline{u}(0) \\ \vdots \\ -B(k)\underline{u}(k) \end{bmatrix} = \begin{bmatrix} O & O & & \\ O & I & & \\ \vdots & & \ddots & \\ O & & A(k) & -I \end{bmatrix} \begin{bmatrix} \underline{x}(0) \\ \underline{x}(0) \\ \vdots \\ \underline{x}(k+1) \end{bmatrix} + \begin{bmatrix} C(0)P^{1/2}(0) & V^{1/2}(0) & & \\ A(0)P^{1/2}(0) & & W^{1/2}(0) & \\ \vdots & & & \ddots \\ & & & -W^{1/2}(K) \end{bmatrix} \begin{bmatrix} \underline{\tilde{e}}(\\ \underline{\tilde{v}}(\\ \underline{\tilde{w}}(\\ \vdots \\ \underline{w}(\end{bmatrix} \tag{9}$$

The information in the first row of equation 7 has been combined into rows 2 and 3 to form the first rows of equation 9 without having any effect on the state or the noise variables. Equation 9 can now be used to find the predicted state estimate $\underline{\hat{x}}(1/0)$ in the following way.

Consider the first two rows of equation 9 alone and ignore the timescripts for reasons of clarity.

$$\begin{bmatrix} C\underline{\hat{x}}-\underline{z} \\ A\underline{\hat{x}}+B\underline{u} \end{bmatrix} = \begin{bmatrix} O & O \\ O & I \end{bmatrix} \begin{bmatrix} \underline{x}(0) \\ \underline{x}(1) \end{bmatrix} + \begin{bmatrix} CP^{1/2} & V^{1/2} & O \\ AP^{1/2} & O & W^{1/2} \end{bmatrix} \begin{bmatrix} \underline{\tilde{e}} \\ \underline{\tilde{v}} \\ \underline{\tilde{w}} \end{bmatrix} \tag{10}$$

The last product of equation 10 can be written

$$\begin{bmatrix} CP^{1/2} & V^{1/2} & O \\ AP^{1/2} & O & W^{1/2} \end{bmatrix} Q^T Q \begin{bmatrix} \underline{\tilde{e}} \\ \underline{\tilde{v}} \\ \underline{\tilde{w}} \end{bmatrix}$$

where Q is any orthogonal matrix which triangularises the compound matrix, i.e.

$$\begin{bmatrix} CP^{1/2} & V^{1/2} & O \\ AP^{1/2} & O & W^{1/2} \end{bmatrix} Q^T = \begin{bmatrix} L_1 & O & O \\ L_2 & L_3 & O \end{bmatrix} \tag{11}$$

Let

$$Q \begin{bmatrix} \underline{\tilde{e}} \\ \underline{\tilde{v}} \\ \underline{\tilde{w}} \end{bmatrix} = \begin{bmatrix} \underline{e}' \\ \underline{v}' \\ \underline{w}' \end{bmatrix}$$

and note that $\underline{e}'$, $\underline{v}'$ and $\underline{w}'$ still have unit covariance matrices. Equation 10 becomes therefore

$$\begin{bmatrix} C\underline{\hat{x}}-\underline{z} \\ A\underline{\hat{x}}-B\underline{u} \end{bmatrix} = \begin{bmatrix} O & O \\ O & I \end{bmatrix} \begin{bmatrix} \underline{x}(O) \\ \underline{x}(1) \end{bmatrix} + \begin{bmatrix} L_1 & O & O \\ L_2 & L_3 & O \end{bmatrix} \begin{bmatrix} \underline{e}' \\ \underline{v}' \\ \underline{w}' \end{bmatrix} \tag{12}$$

Assuming that L_1 is non-singular $\underline{e}'$ can be expressed as

$$\underline{e}' = L_1^{-1}(C\underline{\hat{x}}-\underline{z}) \tag{13}$$

Also, from equation 12

$$A\underline{\hat{x}}+B\underline{u} = \underline{x}(1)+L_2\underline{e}'+L_3\underline{v}' \tag{14}$$

Rearranging equation 14 and substituting for $\underline{e}'$ from equation 13 gives

$$\underline{x}(1)+L_3\underline{v}' = A\underline{\hat{x}}+B\underline{u}+L_2L_1^{-1}(\underline{z}-C\underline{\hat{x}}) \tag{15}$$

$\underline{x}(1)$ is the state of the system at time k=1, and $L_3\underline{v}'$ is the noise term associated with the state estimate given measurements up until k=0. In other words

$$\underline{x}(1)+L_3\underline{v}' = \underline{x}(1)+\underline{e}(1) = \hat{\underline{x}}(1/0) \tag{16}$$

$$L_3 = P^{1/2}(1/0) \tag{17}$$

and

$L_2L_1^{-1}$ is the Kalman gain.

This is intuitively correct and can be checked by 'squaring' both sides of equation 11 and after some simple matrix manipulation comparing the results with the standard covariance Kalman filter equations [9].

The general form of the square root covariance Kalman filter is

$$Q\begin{bmatrix} P^{T/2}(k/k-1)C^T(k) & P^{T/2}(k/k-1)A^T(k) \\ V^{T/2}(k) & 0 \\ 0 & W^{T/2}(k) \end{bmatrix} = \begin{bmatrix} V_e^{T/2} & V_e^{-1/2}C(k)P(k/k-1)A^T(K) \\ 0 & 0^{T/2}(k+1/k) \\ 0 & 0 \end{bmatrix} \tag{18}$$

where

$$V_e^{1/2}V_e^{T/2} = C(k)P(k/k)C^T(k)+V(k)$$

and

$$\hat{\underline{x}}(k+1/k) = (A(k)\hat{\underline{x}}(k/k-1)+B(k)\underline{u}(k)) + A(k)P(k/k-1)C^T(k)V_e^{-1}[\underline{z}(k)-C(k)\hat{\underline{x}}(k/k-1)] \tag{19}$$

It has been assumed that L_1, i.e. $V_e^{1/2}$, is non-singular. Examination of the expression above for V_e and knowing that V(k) is positive definitive at all times should reveal that this assumption is always true.

Equations 18 and 19 define the standard form of the square root covariance Kalman filter and they can now be mapped onto a systolic array architecture. The above derivation of the Kalman filter shows how equation 18 gives the Kalman gain which is then used equation 19. This relationship aids the development of a systolic array architecture for the square root covariance Kalman filter.

4. SYSTOLIC IMPLEMENTATION

Figure 1 shows the systolic array implementation and data flow for this Kalman filter formulation. Figure 2 gives the detail of the cell operations. The architecture has four separate sections; a trapezoidal section, a triangular section, a linear section and one single addition cell.

Consider equation 18, an orthogonal decomposition is required to triangularise the prearray. This can be implemented on the trapezoidal and triangular sections using any orthogonal decomposition, but for the purposes of this discussion Givens rotations will be used. The linear section simply passes the data through unchanged. It should be noted that the order in which the rows of data are fed into the systolic array has no effect on the final result and therefore the rows containing $V^{T/2}(k)$ and $W^{T/2}(k)$ can be preloaded and then only the rows containing $P^{T/2}(k/k-1)C^T(k)$ and $P^{T/2}(k/k-1)A^T(k)$ need to be fed in.

The parts of the postarray, which end up in the memory of the trapezoidal section, can be used to obtain the next predicted state estimate according to

$$\underline{\hat{x}}(k+1/k) = A(k)\underline{\hat{x}}(k/k-1)+B(k)\underline{u}(k)+L_2L_1^{-1}[\underline{z}(k)-C(k)\underline{\hat{x}}(k/k-1)] \tag{20}$$

Yeh [10] implemented the Fadeev algorithm [11] which showed that $D+CA^{-1}B$ can be calculated using simple row operations of the compound matrix

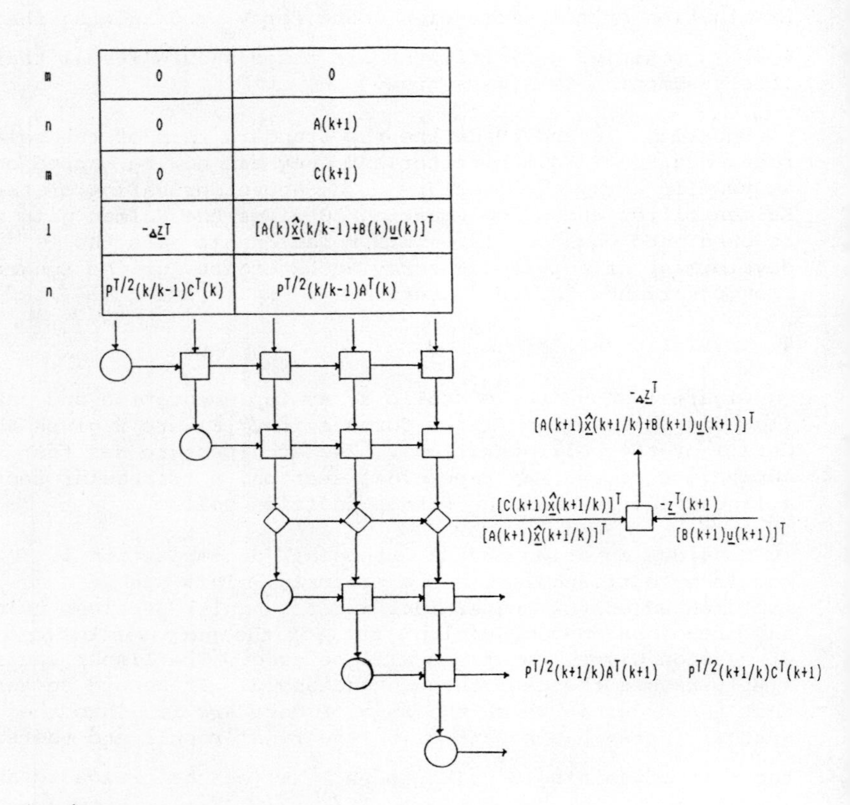

$\Delta \underline{z} = \underline{z}(k+1) - C(k+1)\hat{\underline{x}}(k+1/k)$

Fig. 1 Systolic array implementation of the square root covariance Kalman filter.

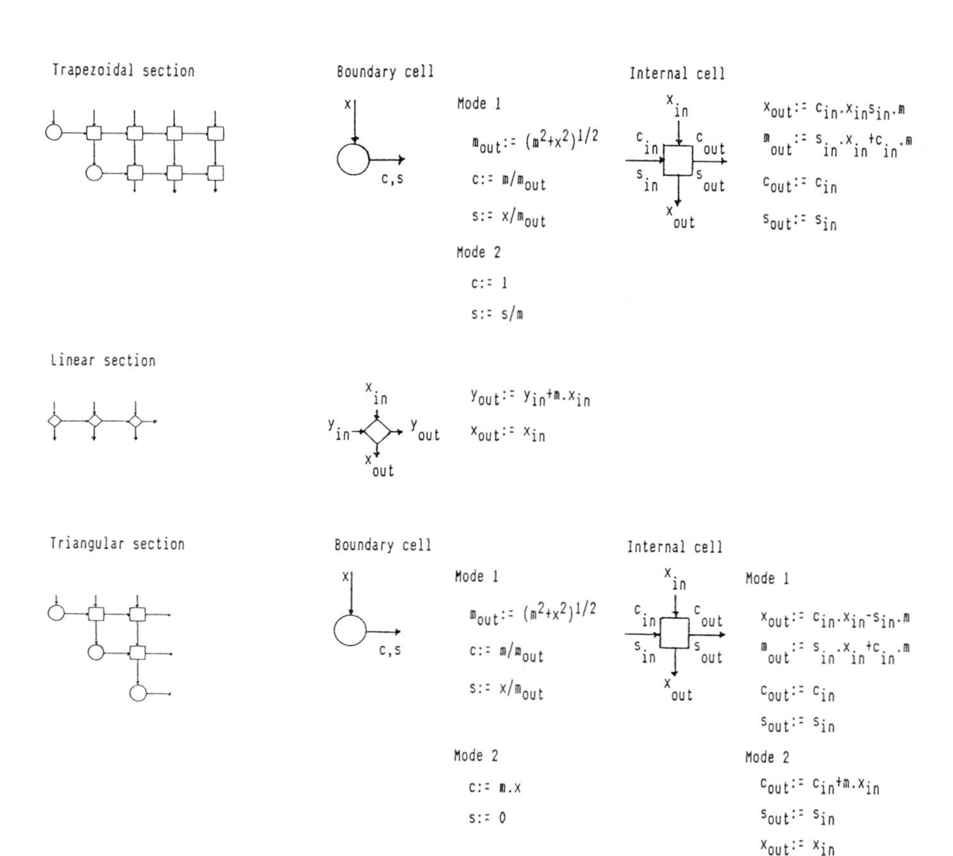

Fig. 2 Cell operations for the architecture of the square root covariance Kalman filter shown in figure 1.

$$\begin{bmatrix} A & B \\ -C & D \end{bmatrix}.$$

The transpose of the right hand side of equation 20 is of this form, thus

$$\underbrace{[A(k)\underline{\hat{x}}(k/k-1)+B(k)\underline{u}(k)]^T}_{D} + \underbrace{[\underline{x}(k)-C(k)\underline{\hat{x}}(k/k-1)]}_{C}\underbrace{L_1^{-T}}_{A^{-1}}\underbrace{L_2^T}_{B}$$

The A and B terms have already been loaded and therefore the C and D terms can be passed in.

The vectors $-[\underline{z}(k)-C(k)\underline{\hat{x}}(k/k-1)]^T$ and $[A(k)\underline{\hat{x}}(k/k-1)+B(k)\underline{u}(k)]^T$ are fed in immediately after the rows containing $P^{T/2}(k/k-1)C^T(k)$ and $P^{T/2}(k/k-1)A^T(k)$. The cells in the trapezoidal section now change their mode of operation to complete the operation described by Yeh. Once the vectors have passed through the trapezoidal section $\underline{\hat{x}}(k+1/k)$ will be captured by the linear section of the array.

For the next iteration the updates, $\underline{\hat{x}}^T(k+1/k)$ and $P^{T/2}(k+1/k)$, must both be post-multiplied by $A^T(k+1)$ and $C^T(k+1)$. The architecture now has all of its memory full. The linear and triangular sections now operate as matrix-matrix multipliers and $C(k+1)$ and $A(k+1)$ are passed in immediately after the vector terms and the results $\underline{\hat{x}}^T(k+1/k)C^T(k+1)$, $\underline{\hat{x}}^T(k+1/k)A^T(k+1)$, $P^{T/2}(k+1/k)C^T(k+1)$ and $P^{T/2}(k+1/k)A^T(k+1)$ are obtained. The matrix terms can be fed straight back into the next prearray while the vector terms need to be combined with $-\underline{z}^T(k+1)$ and $(B(k+1)\underline{u}(k+1))^T$ respectively in the addition cell.

5. CONCLUSIONS

The relationship between least squares and square root covariance Kalman filtering was illustrated and the square root covariance Kalman filter implemented on a systolic array architecture which achieved O(2n) timesteps between full state updates and a processor utilization of between 44 and 100%.

REFERENCES

[1] Kung, H.T., (1982), "Why Systolic Architectures", Computer, Volume 15, pp. 37-46.

[2] Jover, J.M. and Kailath, T., (1986), "A parallel architecture for Kalman filter measurement update and parameter estimation", Automatica, Volume 22, Number 1, pp. 43-57.

[3] Gaston, F.M.F. and Irwin, G.W., (1988), "A systolic square root information Kalman filter", Proc. IEEE Int. Conf. on Systolic Arrays, San Diego, California, pp. 643-652.

[4] Sung, T-Y. and Hu, Y-H., (1986), "VLSI implementation of real-time Kalman filter", Proc. ICASSP, pp. 2223-2226.

[5] Paige, C.C. and Saunders, M.A., (1977), "Least squares estimation of discrete linear dynamic systems using orthogonal transformations", *SIAM J. Numer. Anal.*, Volume 14, Number 2, pp. 180-193.

[6] Morf, M. and Kailath, T., (1975), "Square root algorithms for least squares estimation", IEEE Trans. Auto. Contr., Volume AC-20, Number 4, pp. 487-497.

[7] Paige, C.C., (1985), "Covariance matrix representations in linear filtering", special issue of Contempory Mathematics on "Linear algebra and its role in systems theory", pp. 309-321.

[8] Gaston, F.M.F. and Irwin, G.W., (1989), "The systolic approach to square root information Kalman filtering", *Int. J. of Control.*, Volume 15, Number 1, pp. 225-228.

[9] Anderson, B.D.O. and Moore, J.B., (1979), "Optimal Filtering", (Prentice-Hall).

[10] Yeh, H.G., (1986), "Kalman filtering and systolic processes", Proc. ICASSP, pp. 2139-2142.

[11] Fadeev, D.K. and Fadeeva, V.N., (1963), "Computational methods of linear algebra", W.H. Freeman and Co.

A SYSTOLIC TOEPLITZ LINEAR SOLVER

D.J. Evans and G.M. Megson*
(Department of Computer Studies,
Loughborough University of Technology)

ABSTRACT

This paper briefly reviews some important parallel processing algorithms and systolic architectures for real time signal processing.

Then, the solution of Toeplitz systems of equations is presented on a systolic array with O(n) basic inner product cells requiring O(n) computation time. The design is highly concurrent in the sense that the actual computation is performed in parallel and separate instances of the same global problem can be pipelined to produce high throughput.

1. INTRODUCTION

For many important real time signal processing tasks it is well known that the major computational requirements can be reduced to a common set of basic matrix operations for which a variety of simple linear, rectangular and hexagonal systolic array architectures and algorithms are available which provide modular parallelism, local interconnection, regular data flow and high efficiency.

However the increasing demands of modern signal processing operations require additional prior information concerning the signal structure, etc. resulting in the need for extensive utilisation of linear algebra and eigensystem evaluation software packages with the inevitable increased computational loads.

Further for applications in which the sampling rate approaches the computational cycle time, parallel architectures

*Oriel College, Oxford.

are required for real time implementation of tasks often requiring an operations count proportional to the cube of the number of points. The promising characteristics of the systolic array parallel architecture makes it ideallly suitable for these applications.

2. TOEPLITZ LINEAR SYSTEMS

The solutions of partial differential equations with periodic boundary conditions using finite difference techniques leads to an X-band form matrix, (Megson and Evans, [85]), i.e.,

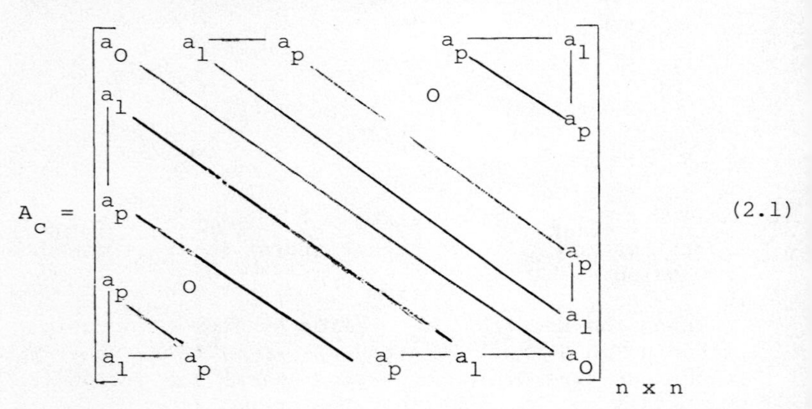

(2.1)

Factorising this matrix produces fill-in along the last p rows of the L factor and the last p columns of the U factor and a hexagonal systolic array must treat A_c as a full matrix. Consequently the solution of Toeplitz systems such as circulant, symmetric and skew-symmetric forms like (2.1) have received much scientific applications attention. The most successful designs to date from a systolic view point are the pipelined lattice processor (PLP) S.Y. Kung and Hu [83]) and a linear array of Brent and Luk [83]. Both designs have O(n) time and the latter is also sensitive to band structure and can solve both symmetric and skew symmetric forms. The PLP consisting of three tiers of n inner product cells each, and a LIFO structure of $O(n^2)$ cells, is based on Levinsons's algorithm and requires O(2n) time. The linear array is based on the Bareiss algorithm and consists of $\lceil \frac{n}{2} \rceil$ super cells each with a control algorithm and requires O(4n) time with six ips per cell and is essentially soft-systolic.

In contrast our technique is based on new factorisation methods developed in Chen [85] and Audish and Evans [85a,b] and permits the pipelining of solutions to more than one system through the array. The previous schemes cannot pipeline successive instances and for a dedicated Toeplitz solver our method should improve throughput significantly.

We consider the solution of the nxn system,

$$A_c x = f, \tag{2.2}$$

where $x=(x_1,x_2,\ldots,x_n)^t$ is unknown and f is the known right hand side. Chen [85] shows that if A_c is strictly diagonally dominant it can be factorised into the form,

$$A_c = \beta_0^{-1}\,\tilde{L}\,\tilde{L}^T, \tag{2.3}$$

where,

$$\tilde{L} = \begin{bmatrix} \beta_0 & & & & \beta_p & \cdots & \beta_1 \\ \beta_1 & \ddots & & & & \ddots & \vdots \\ \vdots & & \ddots & & O & & \beta_p \\ \beta_p & & & \ddots & & & \\ & \ddots & & & \ddots & & \\ & O & \ddots & & & \ddots & \\ & & & \beta_p & \cdots & \beta_1 & \beta_0 \end{bmatrix} \tag{2.4}$$

(2.2) is then solved by the coupled systems,

$$\text{a)}\quad \tilde{L}y = d, \qquad \text{b)}\quad \tilde{L}^T x = y, \tag{2.5}$$

where $d=\beta_0 f$.

Now put,

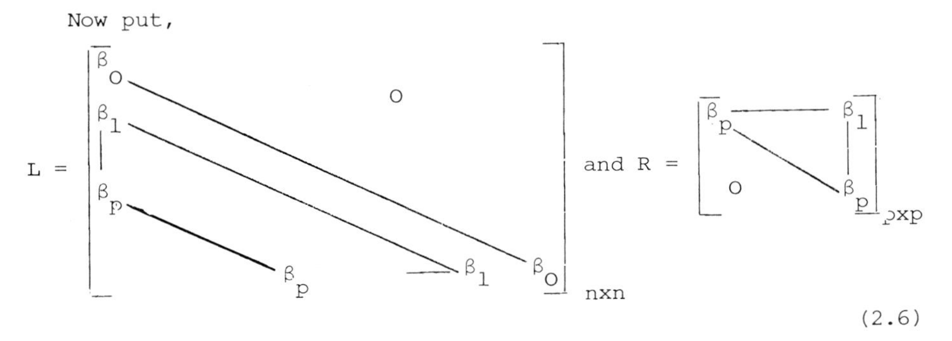

(2.6)

then,

$$\tilde{L} = L + \begin{bmatrix} I_p \\ O \end{bmatrix} R[O^T \; I_p], \tag{2.7}$$

where I_p = pth order identity matrix, and O the (n-p)*p null matrix. We now apply the well known rank annihilation formula (Golub and Van Loan [87]) to yield,

$$\tilde{L}^{-1} = L^{-1} - L^{-1}\begin{bmatrix} R \\ O \end{bmatrix}\{I+[O^T \; I_p]L^{-1}\begin{bmatrix} R \\ O \end{bmatrix}\}^{-1}[O^T \; I_p]L^{-1}, \tag{2.8}$$

and the coupled system (2.5) is solved explicitly, viz.,

$$\text{a) } y = \tilde{L}^{-1}d \text{ and b) } x = \tilde{L}^{-T}y. \tag{2.9}$$

The method extends easily to the simple banded Toeplitz matrix A_t where,

$$A_t = A_c - \begin{bmatrix} I_p \\ O \end{bmatrix} U[O^T \; I_p] - \begin{bmatrix} O \\ I_p \end{bmatrix} U^T[I_p \; O^T], \tag{2.10}$$

and the corresponding linear system,

$$A_t x = f, \tag{2.11}$$

is given by,

$$x - A_c^{-1}\begin{bmatrix} I_p \\ O \end{bmatrix} U[O^T \; I_p]x - A_c^{-1}\begin{bmatrix} O \\ A_p \end{bmatrix} U^T[I_p \; O^T]x = A_c^{-1}f, \tag{2.12}$$

or

$$x = y + B_1 U x^{(3)} + B_3 U^T x^{(1)}, \tag{2.13}$$

with,

$$y = A_t^{-1}x,\; B_1 = A_c^{-1}\begin{bmatrix} I_p \\ O \end{bmatrix}, \text{ and } B_3 = A_c^{-1}\begin{bmatrix} O \\ I_p \end{bmatrix},$$

where $x^{(1)}=(x_1,\ldots,x_p)^T$, $x^{(2)}=(x_{p+1},\ldots,x_{n-p})^T$ and $x^{(3)}=(x_{n-p+1},\ldots,x_n)^T$ and U is a pth order matrix like R but with elements $a_1,\ldots,a_p$. The solution is then determined by premultiplying

(2.13) by (I_p, O^T) and (O^T, I_p) to produce the system,

$$\begin{aligned}(I_p - M_{1p}U^T)x^{(1)} - M_{11}Ux^{(3)} &= y^{(1)}\\ -M_{11}U^T x^{(1)} + (I_p - M_{1p}^T U)x^{(3)} &= y^{(3)}\end{aligned} \qquad (2.14)$$

where M_{11} and M_{1p} are the pth order submtrices of A_c^{-1} at the northwest and northeast corners. We then find $x^{(1)}$ and $x^{(3)}$ using (2.14) and x with (2.13). This latter scheme is more computationally complex than existing systolic schemes for A_t matrix structures but demonstrates that the solution of A_t is simply a linear combination of the solution to (2.2) and the first and last p-columns of A_c^{-1}.

3. A PIPELINED SOLVER

A solution to (2.2) can be constructed by a simple pipeline arrangement illustrated in Fig. (1), and consists of a triangular inverter for finding L^{-1} and L^{-T}, a rank annihilation pipeline for (2.8) and a matrix vector array for the coupled systems (2.9). The important component is the inverter and its operation requires some explanation.

The inverter itself comprises two back to back triangular inverters, one on the left producing upper triangular inverses (or L^T) and one on the right for lower triangular inverses. These two components are operated in mutually exclusive fashion so no conflicts occur in later stages of the pipe and make use of the special form of A_c. Since A_c is a symmetric circulant matrix so is A_c^{-1} and is uniquely defined by its first column hence L^{-1} us found by,

$$L\gamma = (1,0,0,\ldots,0)^T, \qquad (3.1)$$

and can be computed on a p-cell backsubstitution array in $T = n+p$ cycles, producing the solution sequence $\gamma_0 0 \gamma_1 0 \ldots 0 \gamma_{n-1}$. However, the rank annihilation array accepts input in standard diagonal format,

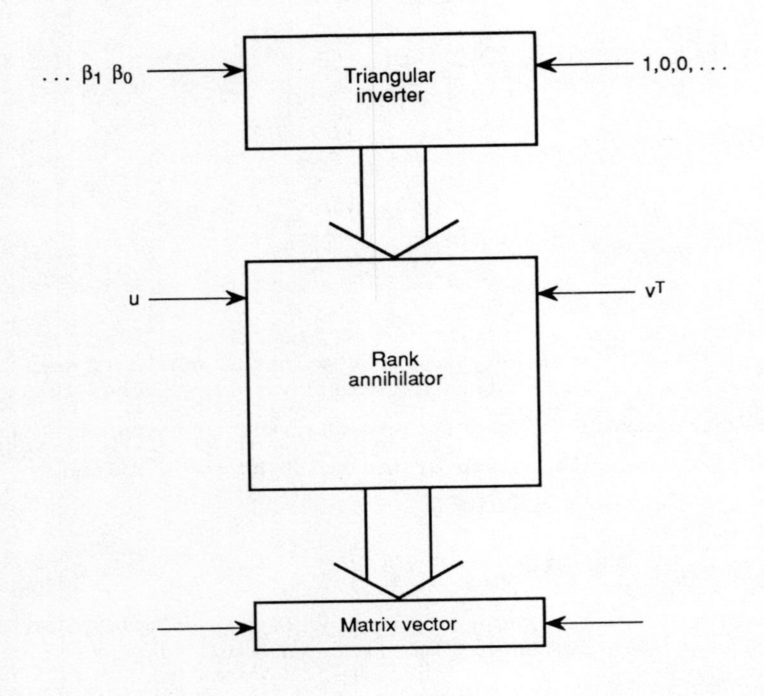

Fig. 1 Pipelines Toeplitz solver

e.g.,

$$\begin{array}{c}
\gamma_0 \\
0 \; - \; \gamma_1 \\
0 \; - \; \gamma_0 \; - \; \gamma_2 \\
0 \; - \; 0 \; - \; \gamma_1 \; - \; \gamma_3 \\
0 \; - \; 0 \; - \; \gamma_0 \; - \; \gamma_2 \; - \; \gamma_4 \\
0 \; - \; 0 \; - \; \gamma_1 \; - \; \gamma_3 \\
0 \; - \; \gamma_0 \; - \; \gamma_2 \\
0 \; - \; \gamma_1 \\
\gamma_0
\end{array} \downarrow \tag{3.2}$$

for L^{-1} and n=5, and a transposed form for L^{-T}. Each inverter component is a bi-linear array, shown in Fig. (2). The top tier is a modified backsubstituter which preloads the constant diagonal values β_i, i=0(1)p, and the bottom tier contains n cells to generate the non-zero portion of the above diamond input. As the 'don't care' slots ('-') can be replaced by the neutral elemen zero, and the components are operated in mutually exclusive fashic The off-state of a component generates the zero side of the input automatically.

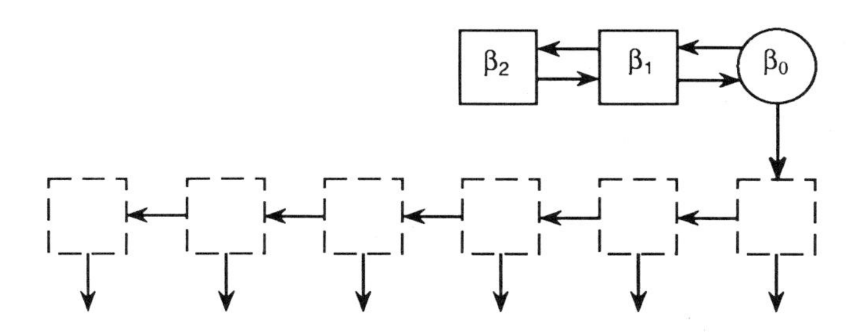

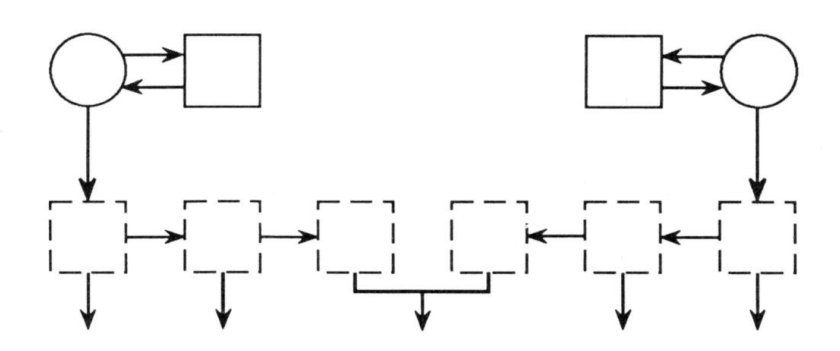

Fig. 2 Triangular inverter

Second tier generating cells consist of loadable registers and simple control logic as defined below

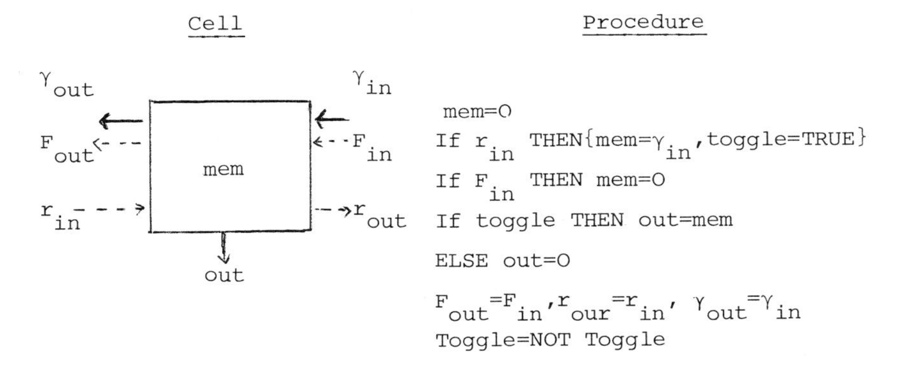

As the backsubstituter produces the γ_i values, they are pumped from right to left along the second tier. Associated with and travelling one cycle in front of, γ_0 is a control value (F).

This forward control signal on the right to left journey resets cells by putting mem=0, and when F drops off the left end of the array it is immediately input as a return signal (r). By virtue of the γ_i spacing and its lead on γ_0, r meets each γ_i, i=0(1)n-1, as it moves left to right loading them into generating cells, and locking the cell into an alternating output cycle (see Fig 3). On reaching the rightmost cell, r loads γ_{n-1} which is output only once. Consequently on leaving the array r is pumped back in as F to reset cells forming the remaining input pattern of (3.2). The control travels in cyclic fashion forming a central ring and indicates that the next solution sequence can be fed right to left along the second tier while the current inverse is still being output. If we allow 2(p+1) additional cycles to load the new parameters into the first tier and output the first result $\bar{\gamma}_0$, the reset control is 2(p+1) cycles in front, causing erroneous loading of the next inverse coefficients. Hence the overlapping of different instances on the same inverter component requires a modified control arrangement.

Theorem 1:

The lower triangular inverter of a symmetric Toeplitz matrix L of bandwidth p requires T=3n+2(p+1) cycles to generate a diagonal input format.

Proof:

We require p+1 cycles to load the parameters representing L, and a further (p+1) cycles for the first result to emerge from tier 1 of Fig. (2). This first result requires n cycles to filter through the second tier to its correct position. As the element corresponds to the diagonal and hence longest data sequence in (3.2) containing 2n elements, T=2n+n+2(p+1) follows immediately.

Corollary 1: Successful matrix outputs of the same inverter components are separated by at least 2(p+1) cycles.

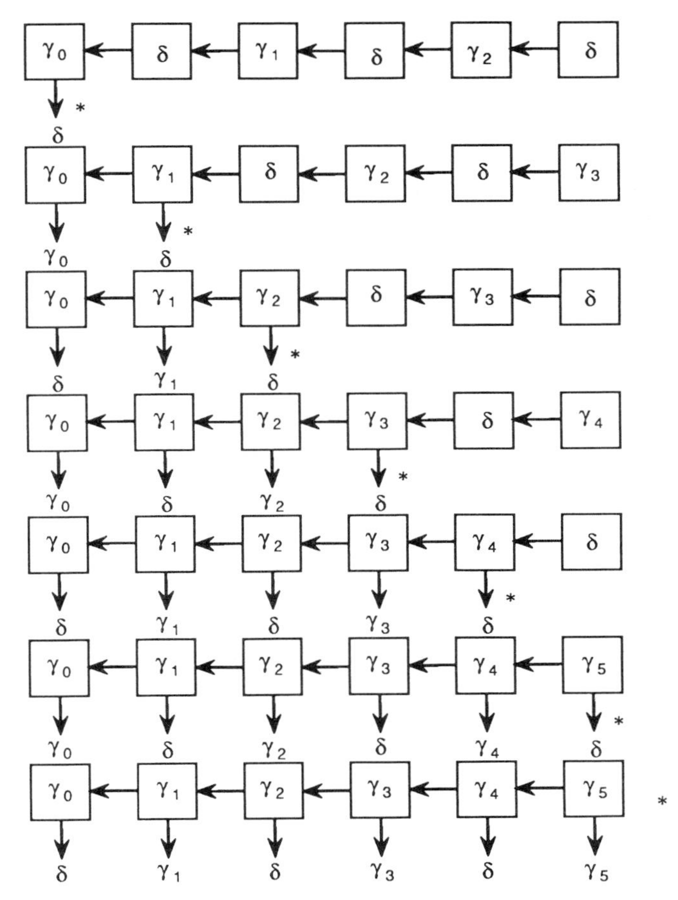

Fig. 3 Snapshots of inverse output generation * = control signal

Proof:

Computation in tier 1 is complete after 2(n+p+1) cycles, and tier two cells begin to switch off. Allow 2(p+1) cycles for parameter loading and the computation of the first result of the next instance on tier 1, switch off controls are 2(p+1) cycles in front of the new data. As the leftmost cell of tier 2 is the first and last to output, successive matrix diamond patterns must be separated by 2(p+1) cycles. This completes the description of the inverter.

We can now consider the operation of the pipeline in Fig. (1). Initially the parameters β_i, i=0(1)p are loaded into the right component of the inverter and after n+2(p+1) cycles L^{-1} begins to emerge. These values are pipelined onto the rank annihilator which for simplicity is assumed to be a RANK-1 pipe with a ring capability. Converting L^{-1} to $\tilde{L}^{-1}$ requires p column updates to distinct columns and the vectors u_i and v_i in the rank annihilation procedure can be precomputed. It follows that $\tilde{L}^{-1}$ is computed by p cycles around the ring requiring p(4n+2) cycles (Evans & Megson, 1987) and the fact that initial synchronization and output are overlapped with other pipe segments. Finally $\tilde{L}^{-1}$ is pipelined onto the matrix vector array to produce (2.9a). Now using the left component of the inverter L^{-T} and its first update computations can be overlapped with the last modification of $\tilde{L}^{-1}$ producing an additional time of (p-1)(4n+2) cycles to complete $\tilde{L}^{-T}$. Feeding y back into the matrix vector array to synchronise with $\tilde{L}^{-T}$ produces the result (2.9b).

Theorem 2:

The solution of k(nxn) circulant symmetric matrix systems of the form $A_c x=f$ with semi-bandwidth p+1 is computed on a Toeplitz solver in T=(6-4k)n+8kp(n+1)+2 cycles.

Proof:

$$T = T_1 + T_2 + T_3 \tag{3.3}$$

where T_1 = initialization and output latency delays

where T_2 = total length of input/output sequence

T_3 = additional delays spent cycling in rank annihilator.

Now T_1=2(n+p+1) as the inverter requires n+2(p+1) to produce the first element of the first L^{-1}, and the first output is delayed by n cycles on its way out of the matrix vector array.

There are k systems and allowing for the inverter delay, each one is represented by two diamond forms like (3.2) of length 2n separated by 2(p+1) cycles. One diamond represents L^{-1} the other L^{-T}. Thus, a single system has input length 4n+2(p+1) generated by the inverter. Furthermore to retain synchronisation each system must be separated by 2(p+1) cycles. Thus the total input length to the rank annihilator is,

$$T_2 = 2k(2n+p+1) + 2(k-1)(p+1) \tag{3.4}$$

Now for a semi-bandwidth p+1 the rank annihilator performs p modifications to L^{-1} and L^{-T} of each system. Cycles of the rank annihilator introduce additional delays effectively lengthening the input incident on the matrix vector array. For the first system as described above this delay is p(4n+2)+(p-1)(4n+2)= (2p-1)(4n+2). Using the fact that the production of $\tilde{L}^{-T}$ of the ith system can be overlapped with $\tilde{L}^{-1}$ of the (i+1)th system for subsequent solutions add a delay 2(p-1)(4n+2) each hence,

$$T_3 = (2p-1)(4n+2)+2(k-1)(p-1)(4n+2), \qquad (3.5)$$

forcing the summation (3.3) and some algebraic manipulation produces the theorem time.

Corollary 2: The solution of a single nxn circulant symmetric matrix system $A_c x=f$ of semi-bandwidth (p+1) requires T=2(4p+1) (n+1) cycles using the Toeplitz solver.

Proof: Use k=1 in Theorem (2).

Further improvements to these timings are possible by using a RANK-2 pipeline and by noticing that for p>1 the 2(p+1) delay associated with the inverter can be overlapped with the last update in the rank annihilator. An alternative scheme is to try and interleave the computation of $\tilde{L}^{-1}$ and $\tilde{L}^{-T}$ using the fact that a cycle length is 4n+2 cycles and data length is 2n; thereby halving the delay associated with the rank annihilation. But from Corollary (1) the input diamonds of L^{-1} and L^{-T} are separated by 2(p+1) cycles giving a combined input length of 2n+2(p+1) cycles. Hence even with p=1 interleaving is not possible.

Now p=1 is an interesting problem because A_c becomes a circulant tridiagonal and only a single update is required to L^{-1} and L^{-T}. Consequently terms in (3.4) and (3.5) associated with cycling disappear, improving throughput and decreasing computation time. These attributes can be retained for general A_c bandwidths by considering an alternative factorisation method due to Audish and Evans [85b]. The idea is to factorise A_c such that,

$$A_c = Q_1, Q_2, \ldots, Q_p, Q_p^T, \ldots, Q_1^T \quad , \qquad (3.6)$$

where,

$$Q_i = \begin{bmatrix} 1 & & & \alpha_i \\ \alpha_i & 1 & & O \\ & \ddots & \ddots & \\ O & & \alpha_i & 1 \end{bmatrix}, \quad i=1(1)p \qquad (3.7)$$

substituting for A_c in (2.2) allows x to be computed using the coupled systems,

$$\left.\begin{array}{l} i=1, Q_1 y_1 = f \\ 1<i\leq p,\ Q_i y_i = y_{i-1}, \\ p<i\leq 2p,\ Q^T_{2p-i+1} y_i = y_{i-1}, \end{array}\right\} \qquad (3.8)$$

where y_i, $i=1(1)2p$ are auxiliary vectors and $x=y_{2p}$. By a simple extension of the solution method in (2.7)-(2.9), (3.8) reduces to 2p matrix-vector multiplications. It follows that (2.2) is solved by interpreting A_c as 2p special circulant problems of semi-bandwidth $\bar{p}+1$ (with $\bar{p}=1$). Hence

Theorem 3:

The solution of $A_c x=f$ where A_c is an nxn symmetric circulant matrix of semi-bandwidth r requires, $T=2(2r+3)n+8r+2$ using the Toeplitz solver, and the factorisation (3.6).

Proof:

Using the factorisation above we have 2r problems with semi-bandwidth $\bar{p}=1$. So the inverter latency is $n+2(\bar{p}+1)=n+4$, and a single pass through the rank-1 pipe costs $4n+2$ cycles. For 2r problems, the input length is $2r(2n)+(2r-1)2(\bar{p}+1)=4rn+4(2r-1)$. All these values pass through the matrix vector array generating the same number of outputs which filter out of the array with an additional n cycles delay. Summing these delays gives $T=2(2r+3)n+8r+2$. This answer is verified by applying Theorem (2) with k=r and $p=\bar{p}+1$. Note that k=r not 2r because the proof of Theorem (2) assumes each system computes L^{-1} and L^{-T} but the solutions in (3.8) are a special case using only L^{-1}. Consequently 2r problems can be compressed into the space and time of r problems by using the spare L^{-T} places. Now solving k problems of semi-bandwidth r+1 using the old factorisation is equivalent to solving rk problems of semi-bandwidth 2 with the new factorisation, and Theorem (2) yields the speed-up inequality

$$(6-4k)n+8kr(n+1)+2 > (6-4kr)n+8kr(n+1)+2 \quad (3.9)$$

It follows that for $r>1$ the new factorisation is faster. Furthermore the method can be applied to different problems of varying bandwidth. For example, the time to solve $\bar{k}$ problems of semi-bandwidth r_i+1, $i=1(1)\bar{k}$ is given by Theorem (2) with $k = \sum_{i=1}^{\bar{k}} r_i$ $p=1$.

Throughout the discussions so far we have assumed that successive matrix inputs arriving at the matrix vector array in Fig. (1) synchronise. For a $p>1$ column rank annihilation strategy this is trivial to arrange using (2.9) because there is plenty of time for y in (2.9a) to filter out of the array and then be pumped back to synchronise with $\tilde{L}^{-T}$ in (2.9b). For (3.8) the arrival of successive y_i, $i=1(1)2p$ is a time critical problem. The typical structure of two solutions Q_i and Q_{i+1} is given by,

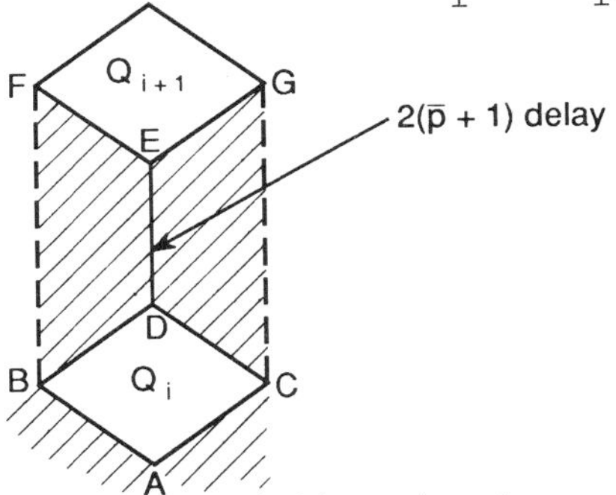

Fig. 4 The Q_i and Q_{i+1} solution structure

where the shaded regions are zero elements used solely for synchronisation and AB=AC=BD=EF=n as the matrix vector array has 2n-1 cells. In general the computation of $y_i=Q^{-1}y_{i-1}$ starts when the element at A enters the array. After a further n cycles elements along BC have entered implying that the last component of y_{i-1} has been input and the first component of y_i has been output. Now the elements of BDEF have the property of never modifying results (i.e., neutral computation). Consequently the synchronisation of known elements of y_i with Q_{i+1} can be overlapped with computation of unknown y_i values, by a simple feedback loop with a delay $2(\tilde{p}+1)+1=5$ cycles as $\tilde{p}=1$. Note that the loop must be switchable to allow the input of the initial vector (y_1) of a new problem. A similar argument verifies that successive rank annihilations can start in the same way, and we conclude that no additional delays are required over those incorporated into the theorems.

REFERENCES

[1] Megson, G. M. and Evans, D. J. (1986) Soft Systolic Pipelined Matrix Algorithms, 171-180, Parallel Computing 85, eds. Feilmeier, Joubert & Schendel, Elsevier Science Pub.

[2] Kung, S. V. and Hu (1983) A Highly Concurrent Algorithm and Pipelined Architecture for Solving Toeplitz Systems, IEEE Trans. on Acoustics, Speech & Signal Processing, Vol. ASSp-31, 66-75.

[3] Brent, R. and Luk, F. (1984) A Systolic Array for the Linear Time Solution of Toeplitz Systems of Equations, Jour. VLSI & Comp. Syst. 1, 1-22.

[4] Chen, M. (1985) On the Solution of a Class of Toeplitz Systems, Report YALE/DCS/RR-417.

[5] Audish, S. E. and Evans, D. J. (1985) On the Parallel Solution of Certain Circulant Banded Linear Systems, Int.J. Comp.Math., 18, 83-90.

[6] Audish, S. E. and Evans, D. J. (1985) A Parallel Circulant Linear Solver, C.S. Rep.238, L.U.T.

[7] Golub, G. H. and Van Loan, C. F. (1983) Matrix Computations, John Hopkins, Univ. Press.

[8] Evans, D. J. and Megson, G. M. (1987) Matrix Inversion by Systolic Rank Annihilation, Int.J.Comp.Math., 21, 319-358

THE PERFORMANCE OF A PARALLEL SUPER-RESOLUTION ALGORITHM FOR SYNTHETIC APERTURE RADAR IMAGES

G.C. Pryde and L.M. Delves
(Centre for Mathematical Software Research, University of Liverpool)

S.P. Luttrell
(R.S.R.E. Malvern)

ABSTRACT

In a previous paper [3] we have described an effective super-resolution algorithm for the detailed analysis of portions of a SAR image. That paper also discussed indirect implementation techniques for the algorithm, with the aim of reducing the very high cost of the algorithm when applied to realistic size images. Despite the large speedups made, even the indirect algorithm remains expensive. We therefore describe here some initial experiments with parallel versions of the original (direct) algorithm, on both the AMT DAP and on transputer arrays. The results show that the algorithm can be very effectively implemented on either architecture.

1. INTRODUCTION

Many image processing algorithms are computationally demanding; and many such applications have a requirement for rapid or even real time processing. There has therefore been considerable interest in the use of parallel computers for image processing [1,4,8], and especially the use of SIMD architectures such as the DAP [1,8]. The ability of the DAP to work efficiently at the short wordlengths required by many such problems, and to provide sufficient processors to support a realistic processor-to-pixel assignment, has been demonstrated in a number of studies [1,8]. However, the peak Mflop rating of the current DAP 510 systems is between 5-10 Mflops; transputer arrays of much higher theoretical peak power are available. It is not yet clear whether this additional peak power makes such MIMD machines better image processing engines than the DAP. For "traditional" image processing applications, with relatively low noise level data, numerous single or comparative studies have been carried out [1,4,8].

However, these studies are not directly relevant to image processing algorithms which are not of traditional "local pixel processing" type, and for these little or no work on either MIMD or SIMD architectures seems to have been carried out. Such algorithms are common in the field of SAR (Synthetic Aperture Radar) Image processing [2,3,5,6,9,11]. In this field, the noise inherent in the image is multiplicative, with a 1:1 signal:noise ratio; such a high noise level has led to the development of new processing techniques which are non-local in nature and which tend to be even more computationally intensive than more traditional techniques. We study here a particular SAR problem: that of image super-resolution, and show that:

a) parallelism can be used to give large reductions in the otherwise high runtimes for this problem;

b) the direct super-resolution algorithm described in [3] can be implemented effectively on both the DAP and on transputer arrays.

2. THE SUPER-RESOLUTION PROBLEM

The basic Super-Resolution problem is :- given an image g in the form of a set of nxn (generally complex) pixel values, and knowledge of the imaging system (represented by the operator T), super-resolve the image by a factor R, i.e. find a set of RnxRn values f consistent with the image g and imaging function T. In practice we pack the quantities f and g into vectors (of length R^2n^2 and n^2 respectively). The imaging operation is then represented as multiplication by an $n^2 \times R^2n^2$ matrix T :-

$$g = T\,\underline{f} + \underline{n}$$

where $\underline{n}$ represents additive instrument noise. One approach to this problem is given by the "Stochastic Inverse" reconstruction method described in [3]; the basic steps in this method are as follows:

1) Solve $\quad [TWT^+ + N]\underline{h} = g \qquad (2.1)$

for $\underline{h}$

where W is defined as $< \underline{f}\,\underline{f}^+ >$

and N as $< \underline{n}\,\underline{n}^+ >$

2) Form $\quad \underline{f}_r = WT^+\,\underline{h}$

3) Update W (details given in [3]) and repeat until convergence.

The simplest method of obtaining a reconstruction is clearly to form the matrix $L = TWT^{+} + N$, solve the equation $L\underline{h} = \underline{g}$ using a linear equation solver, and finally calculate $\underline{f} = WT^{+}\underline{h}$. On serial machines this is (for any reasonably sized problem) far too computationally intensive, and a variety of techniques for speeding up the reconstruction process have been developed [3]. These have proved very effective, and speedups of a factor of over one hundred relative to a straightforward implementation of the algorithm, are reported in [3].

However, even the speeded up algorithm remains computationally expensive, with times for a single iteration of the order of seconds on a microVax, even for a small problem with an 8 by 8 image and super-resolution factor R = 4. We therefore consider here the extent to which the super-resolution algorithm is parallelisable, on an SIMD AMT DAP, and on an MIMD local memory transputer array. To obtain a direct initial comparison of the "easily available" power of these machines, we implement on each a single iteration of the "direct" algorithm: that is equations (2.1) are formed and solved directly, and the update step 3) for W is omitted. A further simplification is also made: the super-resolution problem is essentially complex - both the initial image data and final reconstruction data being complex valued. This poses no particular problem when coded serially in FORTRAN 77. Unfortunately neither Fortran Plus (on the DAP) nor OCCAM (on transputer arrays) support complex arithmetic. We restrict ourselves here to a real-valued version of the algorithm.

3. SERIAL IMPLEMENTATION

For comparison purposes a serial version of the super-resolution algorithm to solve a purely real-valued problem was implemented in VAX FORTRAN 77 on a MicroVAX II. As N increases the execution time is dominated by the time to evaluate L, which is of order N^6R^2. The following timings (in seconds) were obtained.

Image Size N	Super Resolution Factor R		
	2	4	6
4	0.7	2.3	4.8
6	5.4	19.2	42.7
8	25.4	93.1	207.7

4. THE DAP ALGORITHM

The current AMT DAP 510 has a 32 by 32 processor array. For a typical image size of 8x8, the final linear system formed is of size 64x64, and the problem is one of partitioning into DAP-size chunks rather than struggling to use as many processors as possible. For the direct approach being used here, the software tools required are straightforward. A linear equation solver for large systems (BIGSOLVE) forms part of the DAP subroutine library. Apart from this some basic shifting and matrix arithmetic facilities for the partitioned arrays enable successive reconstructions to be produced relatively straightforwardly, and (most importantly) with a high degree of parallelism. Hence, producing an efficient direct implementation on the DAP is straightforward.

We do not have direct access to DAP hardware. For testing purposes the performance of the DAP510 was simulated using the DAP-ADA package developed at Liverpool [12]. This is a package of Ada declarations and operators which provides functionally modelled after the array extensions provided in DAP-Fortran; DAP-Ada programs can be readily transliterated on a line-by-line basis into DAP-Fortran. More importantly for our purposes, DAP-Ada contains a pseudo-timer which reports the time which the equivalent DAP-Fortran code would have taken on a "genuine" DAP. The original timings were those appropriate for the mainframe DAP; we have updated these to reflect the operations timings recently released by AMT for the DAP510 (given to within 25% (worst case) [10]. A number of extensions to DAP-Ada were also made to cope with the large size arrays required. We also had to implement a DAP-Ada version of BIGSOLVE, but we ignore this need in assessing "ease of use".

For problems of a realistic size (i.e. $N \geq 6$, $R \geq 4$) the processing time on the DAP is still dominated by the time to form the product $L = T * WT^+ . T$ and WT^+ are of size $N^2 x N^2$, with each element being of length R^2. The method chosen for implementation on the DAP was to take the loop over R^2 outside so that we now require the sum of R^2 products of matrices of scalars. These matrices are partitioned into blocks of size 32x32. To form the product of matrices containing MxM blocks requires order M^3 operations on individual blocks.

On the DAP 32x32 matrices can be multiplied in true matrix fashion in 32 operations (each involving one multiply and one add). Thus the time to form L involves $R^2 \times M^3 \times 32 \times$ (multiply time + add time). The quoted multiply and add times are [10] 0.0001557s and 0.0001050s respectively. Thus the predicted time for forming L is about 0.008 $R^2 M^3$. For the 'standard' problem size of N = 8, R = 4 this evaluates to about 1 s.

More accurate estimates were obtained by coding the direct algorithm in DAP-Ada. The following simulated timings were obtained for one complete cycle of the reconstruction process, i.e. form L, solve (2.1) for $\underline{h}$, and form a new $\underline{f}$ and W, together with the speedups relative to the μVAX timings.

Image Size N	Run Time / s			Speedup Relative To μVAX
	Super Resolution Factor R			(Independent of R)
	2	4	6	
4	0.05	0.19	0.42	12.1
6	0.41	1.48	3.28	12.8
8	0.42	1.50	3.29	62.0
10	3.16	11.61		
12	6.11			

The speedups are independent of R since on both machines the methods have cost proportional to R^2.

5. THE MIMD ALGORITHM

For the transputer array, we again study a direct parallelisation of the direct implementation for comparison purposes; and again, this involves primarily the use of existing library routines (from the parallel library described in [13]) for matrix multiplication and the solution of systems of linear equations.

Occam code was written to implement the direct algorithm on a transputer array. The hardware used was originally the RSRE Protonode at Liverpool, containing 16 T414 Transputers. Later, an ITEM box containing 9 T800 Floating Point Transputers, and a prototype Supernode with 16 T800s became available; the results from these T800 arrays are reported here. Both arrays were driven by a single Transputer on a B004 board hosted by a PC. The configuration used was dictated by the fact that the matrix multiply library routine was written for a square array, and the linear equation solver for a linear chain of processors. It was therefore necessary to use a square array of processors and take advantage of the fact that a linear path can easily be made within this array.

Since matrix multiplication is by far the most expensive part of the Super-Resolution process (followed by solving the linear system), it is sufficient to perform the rest of the calculations in serial on the B004 Transputer. The serial section of OCCAM code thus does the initial setting up of all arrays, and performs any necessary i/o. It contains calls to previously written driver routines for matrix multiplication and linear equation solution.

As in the SIMD case the vector-valued nature of the matrix elements was dealt with by having an outer loop over R^2, performing this many matrix multiplications and summing the results. Once again the matrix multiplication stage dominates the total time taken; some of the times measured for this part of the operation are shown in the next table.

Matrix Multiplication times in seconds

Matrix Size	No. of T800s		
	1	4	9
36	0.142	0.058	0.042
64	0.731	0.246	0.152
100	2.676	0.817	0.461
128	5.520	1.618	0.864

	Relative Speedups (Efficiency)		MegaFlop Rates		
Matrix Size	No. of T800s		No. of T800s		
	4	9	1	4	9
36	2.46 (61%)	2.97 (37%)	0.64	1.57	2.35
64	2.97 (74%)	4.82 (54%)	0.70	2.08	3.37
100	3.27 (81%)	5.80 (64%)	0.75	2.45	4.35
128	3.41 (85%)	6.39 (71%)	0.76	2.59	4.85

The general pattern of these results is as we would expect. We have to output $2xN^2$ data items, perform N^3 multiplications and additions, and receive back N^2 results. The data output is bottlenecked at the start; all the data must pass out to the array on one link, and this dominates the communications overhead which is (roughly) dependent only on the matrix size. Therefore for a fixed N, as the number of processors increases the proportion of time spent calculating (as opposed to communicating) drops, hence the efficiency drops. Conversely as N rises with P fixed, the number of floating point operations increases a factor of N faster than the number of data items, and the efficiency increases.

6. COMPARISON OF RESULTS

Figure 1 shows a comparison of the measured times for one complete iteration of the super-resolution algorithm, on the microVax; the DAP 510, and on a T800 array of size from 4 to 64. The image size in this figure is 8 by 8; the super-resolution factor R = 4. The 64-transputer times are extrapolations.

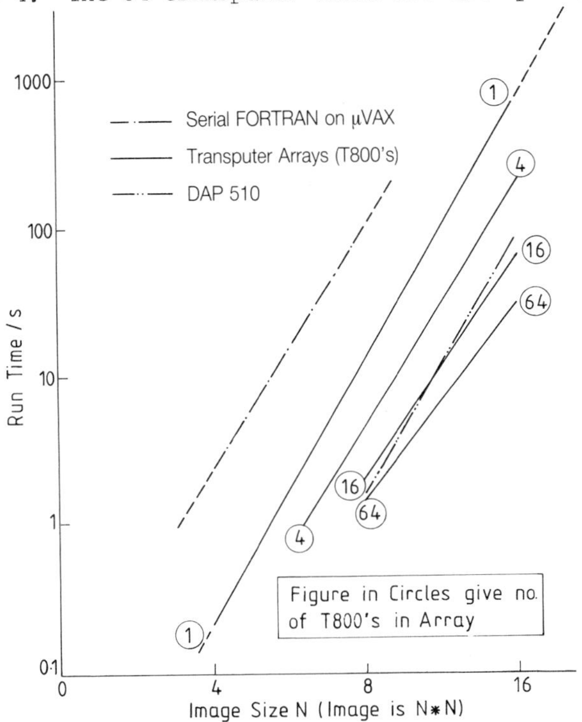

Fig. 1 Super-Resolution Run Times vs Image Size for various processors; super-resolution factor R = 4. The figures given in circles relate to transputer arrays with the specified number of processors.

We see that the timings for 16 transputers are indistinguishable from those for the DAP 510. Both are much better than those for the microVax; the problem clearly parallelises well, and this is shown up also by the steady reduction in time as the array size increases. Both machines were also easy to program; this reflects primarily the availability of numerical libraries on the two machines. On a cost-for cost basis, the transputer array wins comfortably; a sixteen transputer array can be bought for £20,000. Further, the problem size taken is smaller than is desirable in "production" mode; a further advantage of the transputer array is the ability to scale the power by increasing the array size (though we note that larger DAP arrays are now available from AMT).

These conclusions must be tempered by the reminder that we have only implemented the crudest form of the super-resolution algorithm. The fast serial algorithms described in [3] can also, we believe, be parallelised; and we have so far made no efforts to develop new algorithms with parallel architectures in mind.

ACKNOWLEDGEMENTS

This work was carried out with support from the Procurement Executive, Ministry of Defence.

REFERENCES

[1] Arnot, N. R., Wilkinson, G. G. and Burge, R. E. (1982) Applications of the ICL DAP for Two-Dimensional Image Processing, Computer Physics Communications 26 pp 455-457 (VAPP I Proceedings)

[2] Delves, L. M., McQuillan, R. T. and Oliver, C.J. A One-Dimensional De-Speckling Algorithm for SAR Images, Inverse Problems to be published.

[3] Delves, L.M., Pryde, G.C. and Luttrell, S.P. A Super-Resolution Algorithm for SAR Images, Inverse Problems to be published.

[4] Harp, J. G., Roberts, J. B. G. and Ward, J. S. (1985) Signal Processing with Transputer Arrays (TRAPS), Computer Physics Communications 37 pp 77-86 (VAPP II Proceedings)

[5] Luttrell, S. P. (1985) Prior Knowledge and Object Reconstruction using the Best Linear Estimate technique, Optica Acta 32 pp 703-716.

[6] Mammone, R. J. and Eichmann, G. (1982) Super-resolving Image Restoration using Linear Programming, Applied Optics 21 pp 496-501.

[7] Reddaway, S. F., Bowgen G. and van den Berghe, C. S.. High Performance Linear Algebra on the AMT DAP 510, AMT Technical Report.

[8] Simpson, P. and Merrifield, B. C. (1985) Real Time Signal Processing Applications of a Distributed Array Processor, Computer Physics Communications 37 pp 133-140 (VAP II Proceedings)

[9] White, R. G. (1987) A Model based approach to the low-level segmentation of noisy imagery Int J. Remote Sensing to be published.

[10] van den Berghe, C. S., FORTRAN-PLUS Operation timings, AMT Technical Report.

[11] Zala, C. A., Barrodale, I. and Kennedy J. S. (1986) High Resolution Signal and Noise Field Estimation using the L1 Norm, preprint, University of Victoria.

[12] Delves, L. M. and McCrann, M. (1986) DAP-Ada User Manual, University of Liverpool.

[13] Delves, L. M. and Brown, N. G. (1988) A Parallel Library for Transputer Arrays, Final Report of Esprit 1 Project P1085 (Supernode), Brussels, November.

2-D SYSTOLIC SOLUTION TO DISCRETE FOURIER TRANSFORM

K.J. Jones
(Plessey Avionics Ltd., Havant, Hants.)

ABSTRACT

A number of systolic architectures have appeared over the past few years for performing the Discrete Fourier Transform (DFT) and Fast Fourier Transform (FFT) algorithms, using both linear and orthogonal processing networks. This paper shows how a rectangular array of N CORDIC (Co-ORdinate DIgital Computer) processing elements can be used to carry out an efficient 2-D systolic implementation of the N-point DFT, offering highly attractive throughput rates in relation to other N-processor solutions, such as the conventional linear systolic array.

1. INTRODUCTION

A number of processing architectures have recently been discussed [9] for performing the DFT and FFT algorithms. These are based upon the original systolic processing ideas of Kung [4], who defines a systolic architecture as being an array or network of processing elements, each capable of performing some simple operation such as multiplication/accumulation, which synchronously computes and passes data through the system. Architectures such as these are particularly attractive for very-large-scale integration (VLSI), owing to the associated simple and regular communication structures.

The conventional linear systolic array of Kung, for example, which provides the benchmark for the results presented here, uses N processing elements to 1-dimensionally pipeline a length-N DFT with O(N) time-complexity. The more recently developed orthogonal processors for performing the DFT and FFT algorithms offer superior throughput to the linear array, via 2-dimensional pipelining and/ or pipelining of parallel data streams, at the expense of an increased number of processing elements. For large N, therefore, the hardware requirements of these orthogonal processors can prove prohibitive, even in VLSI.

This paper discusses a word-level systolic implementation of the DFT algorithm, which uses a rectangular processor array of just processing elements to 2-dimensionally pipeline the small DFTs that result from the arbitrary factor [1] and prime factor [2] decompositions. It is shown, in particular, that if N has factors of N_1 and N_2, the time-complexity can be expressed as a function of N_1+N_2, rather than N, which can mean great improvements in throughput over the conventional linear systolic array, for the same number of processing elements.

A very simple 2-state processing element is developed for performing the complex arithmetic operations that constitute the DFT computation for the 2-factor decomposition. This is based upon the use of Horner's rule [3] from polynomial algebra for representing the Fourier coefficient expansion, with the associated storage/processing requirements being minimised by means of hardware-efficient CORDIC arithmetic [8].

2. MATRIX-PRODUCT FORMULATION OF DFT ALGORITHM

The Fourier matrix, W_N, can be viewed, up to a scaling factor, as a unitary operator upon the vector space of complex N-tuples, $\mathbb{C}_N$, so that the vector output, $\underline{X}_N$, of a length-N DFT can be simply related to the vector input, $\underline{x}_N$, where $\underline{x}_N$, $\underline{X}_N \in \mathbb{C}_N$, via the express

$$\underline{X}_N = W_N . \underline{x}_N \tag{2.1}$$

Suppose the length N of this DFT can be written as:

$$N = N_1 \times N_2 \tag{2.2}$$

where N_1 and N_2 are arbitrary factors (i.e. with or without common factors). Then by application of the lexicographical index mapping [1] to the input and output data vectors, the single DFT of length N can be decomposed into a partial DFT, followed by a pointwise matrix multiplication to account for the twiddle factors, followed by a second partial DFT.

The pointwise matrix multiplication can be simply eliminated, however, by making the factors N_1 and N_2 to be relatively prime, whereupon the Fourier matrix, W_N, can be factored as:

$$W_N = W_{N_1} \otimes W_{N_2} \tag{2.3}$$

where $\otimes$ denotes the Kronecker product. This factorisation allows the DFT output vector, $\underline{X}_N$, to be written as:

$$P_2 . \underline{X}_N = (W_{N_1} \times W_{N_2}) . P_1 . \underline{x}_N \qquad (2.4)$$

the prime factor decomposition, where P_1 and P_2 are permutation matrices obtained from Good's index mapping [2], this in turn based upon the Chinese Remainder Theorem [7].

Denoting $P_1 . \underline{x}_N$ by x_{N_1,N_2} and $P_2 . \underline{X}_N$ by X_{N_1,N_2} enables equation (2.4) to be rewritten as:

$$X_{N_1,N_2} = W_{N_1} . x_{N_2,N_2} . W_{N_2} \qquad (2.5)$$

which can be simply evaluated via the two matrix-products:

$$Y_{N_1,N_2} = x_{N_1,N_2} . W_{N_2} \qquad (2.6)$$

followed by:

$$X_{N_1,N_2} = W_{N_1} . Y_{N_1,N_2} \qquad (2.7)$$

where equation (2.6) performs the N_1 DFTs of length N_2 corresponding to the first partial DFT, and equation (2.7) the N_2 DFTs of length N_1 corresponding to the second. Note that for the arbitrary factor solution the exponential twiddle factors:

$$\mathrm{EXP}\{-i2\pi(n_1-1)(n_2-1)/N\} \qquad (2.8)$$

are simply applied to the matrix output of equation (2.6), prior to computation of the second partial DFT.

The computation of the original DFT has thus been reduced, essentially, to that of two matrix-products, as shown in the processing scheme of Figure 1, so that efficient computation of the DFT, in particular for the 2-factor decomposition (with or without twiddle factors), rests upon the availability of a computationally efficient matrix multiplier.

3. DFT COMPUTATION VIA SYSTOLIC MATRIX MULTIPLIER

An important processing element for systolic DFT computation is that of the complex inner product step processor, which performs the set of recurrences:

$$c_{pq}^{(k)} = c_{pq}^{(k-1)} + a_{pr} \times b_{rq} \qquad k=1,2,\ldots,N \qquad (3.1)$$

In addition, matrix-matrix multiplication, where one of the two matrices is square (and thus corresponding to a Fourier matrix), can be efficiently performed via the systolic processor [5] shown in Figure 2, where each element of the processor performs the recurrence of equation (3.1) above.

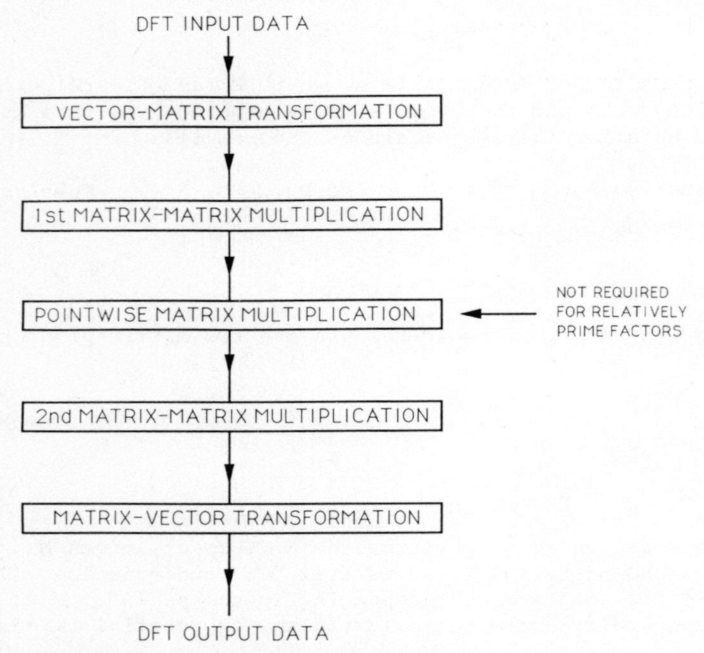

Fig. 1 DFT Processing Scheme for 2-factor Decomposition

To compute the product of two matrices of dimensions $N_1 \times N_2$ and $N_2 \times N_2$ via this processor, c_{11} is initialised to zero at time-step zero; c_{12} and c_{21} are initialised to zero by time-step 1; etc, so that by time-step $k+N_2$, all of the complex inner products corresponding to the position of the k'th antidiagonal are accumulated in-place. The accumulation of the last term is therefore completed after:

$$T_1 = (N_1-1) + (N_2-1) + N_2 = N_1 + 2N_2 - 2 \qquad (3.2)$$

time-steps. Similarly, computation of the product of two matrices of dimensions $N_1 \times N_1$ and $N_1 \times N_2$ can be completed after:

$$T_2 = (N_2-1) + (N_1-1) + N_1 = 2N_1 + N_2 -2 \qquad (3.3)$$

time-steps, via the same processor configuration.

From these results, it is clear that the first and second stages of the computation of the prime factor DFT, as given by

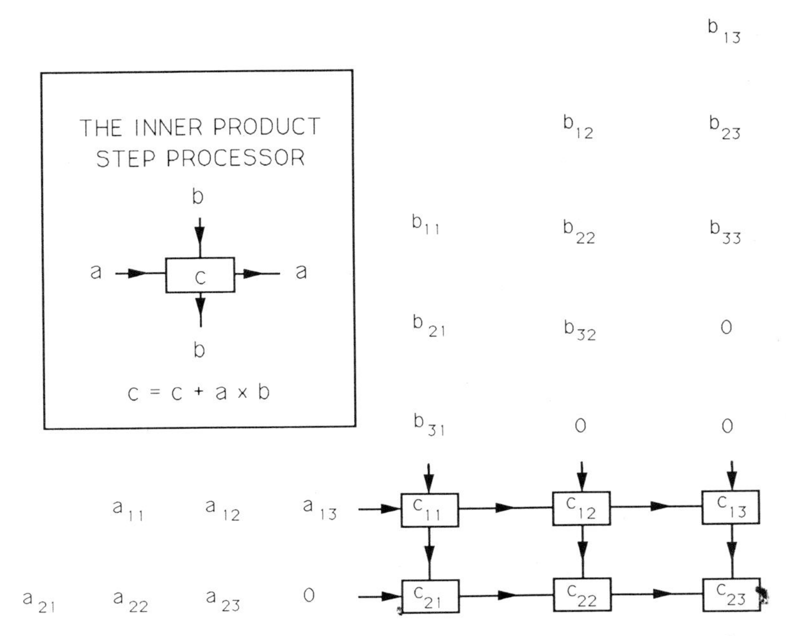

Fig. 2 Systolic Matrix Multiplier for 2 x 3 Dimensional Product

equations (2.6) and (2.7), can be carried out in T_1 and T_2 time-steps, respectively, via this systolic matrix multiplier, using just N complex inner product step processing elements. The pointwise matrix multiplication necessary for the arbitrary factor solution requires just one additional time-step.

Therefore, the two matrix-products that comprise the DFT computation can be carried out in a total of T_R time-steps, where:

$$T_R = T_1 + T_2 = 3.(N_1+N_2) - 4 \tag{3.4}$$

which for sufficiently large N compares very favourably with the T_L time-steps, where:

$$T_L = 2N - 1 \tag{3.5}$$

of the conventional, N-processor, linear systolic array.

Upon completion of the first matrix-product, however, the contents of the rectangular array must be transferred to the vertical data buffer, from where it is input, in the skewed form of Figure 2, to the second matrix-product. This data transfer can be efficiently performed by pipelining the data out of the array via vertical data paths, resulting in an additional N_1 time-steps. A similar number of time-steps is required to empty the contents of the array, containing the computed DFT output, after the second matrix-product.

The total computation time for the prime factor DFT (excluding the index mappings) can therefore be written as:

$$T_R = 5N_1 + 3N_2 - 4 \tag{3.6}$$

time-steps. It is now seen how this figure can be reduced by appropriate implementation of the processing element.

4. 2-STATE CORDIC PROCESSING ELEMENT

For an efficient VLSI implementation, it should be noted that as the elements of the Fourier matrices are complex exponentials, it is possible to carry out the complex multiplications as simple phase rotations with bit-recursive CORDIC arithmetic.

This technique involves the implementation of a co-ordinate rotation via a combination of computational and table look-up techniques. The key to the technique is the decomposition of the rotation angle into the sum of several smaller rotation angles, each of which can be simply implemented by additions and shifts.

This is normally followed by the application of a scaling factor to correct for the ensuing magnification. However, as discussed by Dewilde, Deprettere and Nouta in [6], the scaling can be efficiently performed in VLSI by interleaving with the iterations of the co-ordinate rotation.

Given that the scaling operation follows that of the rotation, the same hardware can be used for both operations, as shown in the simple CORDIC circuit of Figure 3, reducing the hardware requirements to 2 parallel shifters and 2 parallel adders per processing element. A switch is used to alternate the function of the circuit between that of rotation and scaling.

$XR^{(0)}$, $XI^{(0)}$ - COMPLEX INPUT

$XR^{(2L)}$, $XI^{(2L)}$ - ROTATED & SCALED OUTPUT

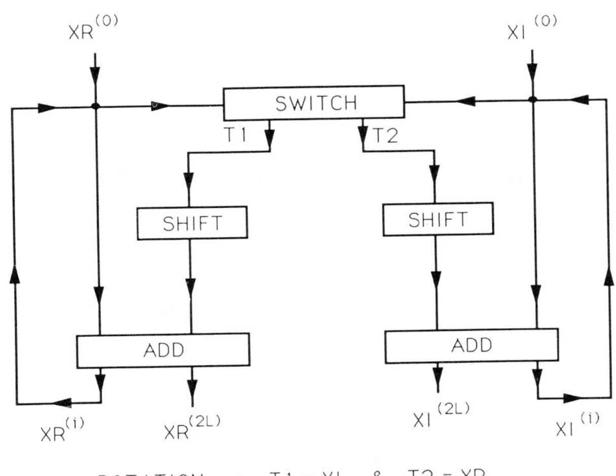

ROTATION ⇒ T1 = XI & T2 = XR

SCALING ⇒ T1 = XR & T2 = XI

2L ITERATIONS REQUIRED FOR L-BIT PRECISION

Fig. 3 Hardware Efficient CORDIC Circuit for Rotation & Scaling

The CORDIC approach has obvious attractions for the computation of the DFT, as the N terms of the n'th row (column) of a Fourier matrix, W_N, can be simply generated via rotation with the single phase element:

$$\varphi_n = -2\pi(n-1)/N \tag{4.1}$$

where the initial term of each row (column) is given by 1.

The significance of this property is evident when the n'th Fourier coefficient is recursively computed from the data by Horner's rule for polynomial evaluation:

$$X(n) = (\ldots((x(1).\alpha + x(N)).\alpha + x(N-1)).\alpha + \ldots + x(2)).\alpha \tag{4.2}$$

where

$$\alpha = \mathrm{EXP}\{i\varphi_n\} \tag{4.3}$$

as it suggests that for the prime factor algorithm, each processing element has only to store the encoded versions of two phase elements - one corresponding to the first matrix-product, and one corresponding to the second - in order to generate the required Fourier coefficients. The arbitrary factor algorithm requires an additional encoded phase term to account for rotation by the twiddle factors.

Note that as $\alpha^{0} = \alpha^{N} = 1$, the term x(1) is fed first, rather than last, into the above recurrence equation, ensuring the function of this 2-state processing element is the same for all N recursions - namely that of "addition/rotation".

Thus, to implement this recurrence, it is necessary that for the first matrix-product, the left-most element of each row-vector of data enter (→) the array first, whilst for the second matrix-product, the top-most element of each column-vector of partial DFT output must enter (↓) the array first. Both requirements are met by appropriately modifying the input index mapping:- apply a single rotation (←) to each row-vector of permuted data, followed by a single rotation (↑) to each of the resulting column-vectors.

5. REDUCED-COMPLEXITY PROCESSING SCHEME

An additional attraction of using Horner's rule for each DFT coefficient expansion, is that the requirement for skewing the data into the array is removed, so that each matrix-product reduces, via the rectangular array, to the parallel computation of matrix-vector products, where the encoded phases of the Fourier matrix data are pre-stored within the processing elements. The resulting computational savings are given by N_1-1 time-steps for the first partial DFT, and N_2-1 time-steps for the second.

Control information, in the form of a bit-sequence of O's and 1's, can be broadcast to the array to indicate when the first state ends and the second state begins. This enables the processing elements to be simply programmed, as shown in Figures 4a and 4b, using the simple additions and shifts of CORDIC arithmetic to perform the required phase rotations.

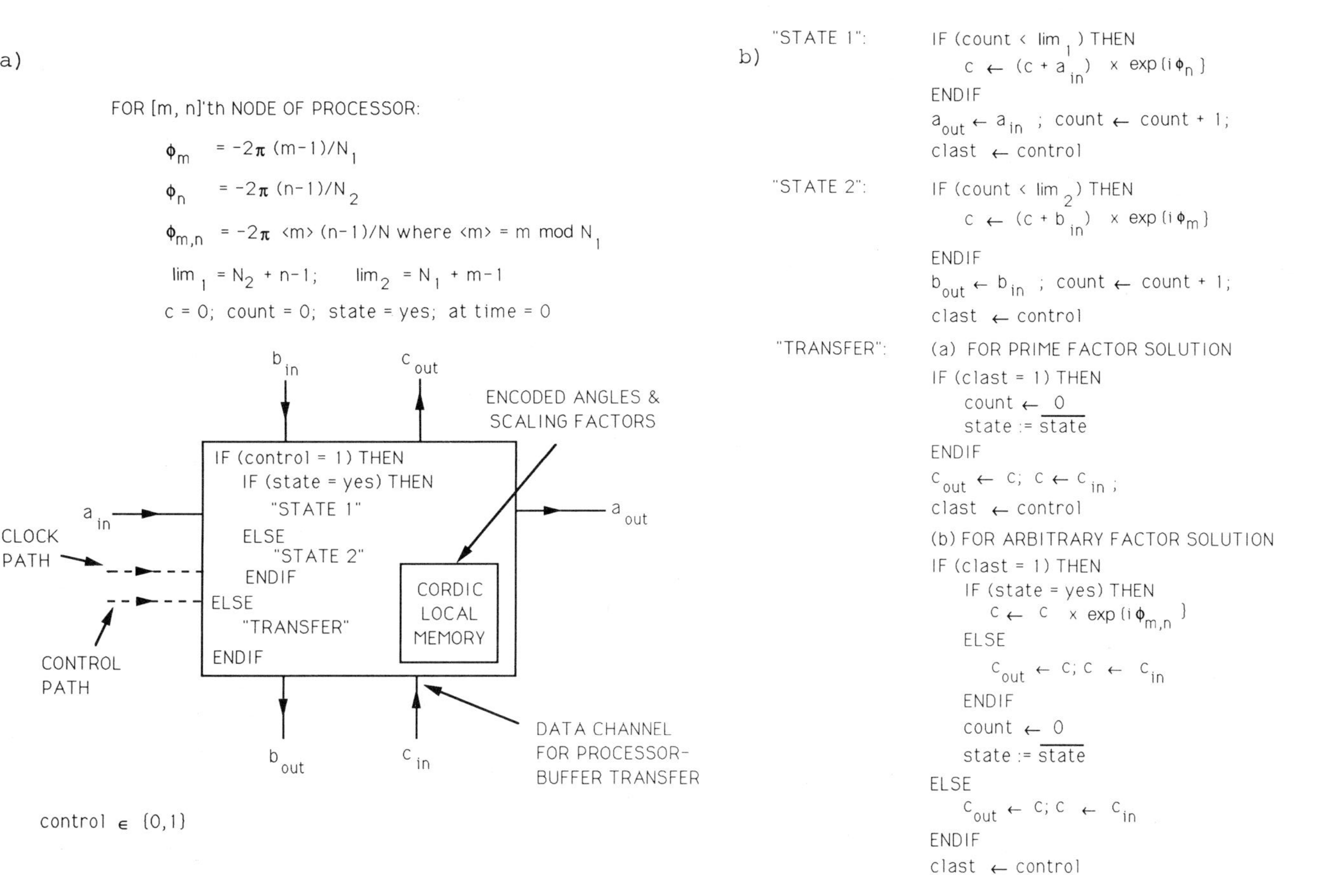

Fig. 4 a) 2-State Processing Element, b) State Functions

The second state of the processing element is therefore simply obtained from the first state (and vice versa) by inter-changing the roles of the inputs, "a" & "b", and by interchanging the CORDIC phase elements corresponding to its column and row addresses. The flow of data into such a processor is described in Figure 5 for the simple case of the 6-point prime factor DFT.

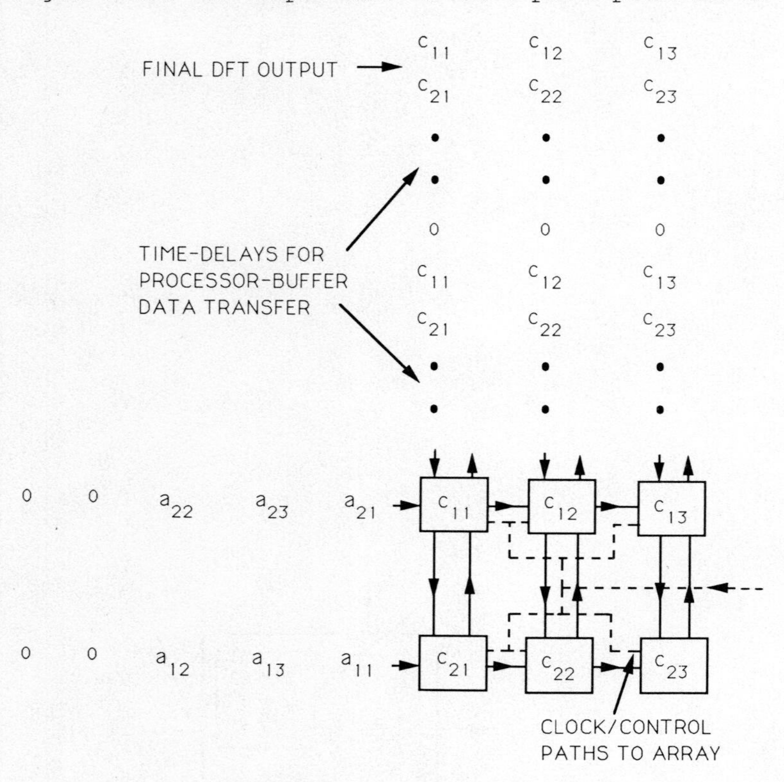

Fig. 5 Processing Scheme for 6-Point Prime Factor DFT

6. THROUGHPUT & IMPLEMENTATIONAL CONSIDERATIONS

The elimination of the skewing requirement, mentioned above, reduces the total computation time of equation (3.6) to just:

$$T_R = 4N_1 + 2N_2 - 2 \tag{6.1}$$

time-steps, so that, for the example:

$$N = 1025 = 25 \times 41 \tag{6.2}$$

the expressions for T_R and T_L reduce to:

$$T_R = 180 \quad \& \quad T_L = 2049 \tag{6.3}$$

time-steps, giving an 11:1 reduction in computing time via the prime factor and, indeed, arbitrary factor systolic solutions.

For the computation of each successive DFT by the rectangular processor array, the input data vector must be transformed into matrix form in order that the data be correctly loaded into the array. The linear array, however, processes the input data and outputs the results in their natural order, so that subsequent output vectors can be obtained in N time-steps. This means that for the 1025-point example given, a reduction of nearly 6:1 in computing time is still achieved with the 2-D systolic solution in obtaining subsequent output vectors.

Finally, in order to analyse the numerical properties of the 2-factor DFT implementation discussed in this paper, it should be noted that there are in fact two levels of recursion taking place:- a word-level recursion, as expressed via Horner's rule, for implementation of each Fourier coefficient expansion - and a bit-level recursion, as expressed via the CORDIC algorithm, for implementation of each phase rotation.

At the bit-level, with the CORDIC iteration defined as:

$$\text{"iteration"} = \text{"rotation"} + \text{"scaling"} \tag{6.4}$$

it can be shown that the number of such iterations controls the precision to which the rotation angle is resolved, with the angular error being bounded by:

$$\delta_K < \text{TAN}^{-1}\{2^{-(K-1)}\} \tag{6.5}$$

and K iterations producing K bits of precision in the rotated output. This accuracy, however, necessitates the use of an additional $\text{LOG}_2\{K\}$ guard bits for the internal CORDIC arithmetic operations.

At the word-level, each rotation is preceded by a complex addition, so that after the first partial DFT, the data length will have extended by up to $\text{LOG}_2\{N_2\}$ bits, followed by a possible $\text{LOG}_2\{N_1\}$ bits after the second - the CORDIC operation will not itself extend the maximum length of the data being rotated.

Therefore, if the original data is represented by just K bits, and the CORDIC operation by K iterations, these extra bits will be redundant and can be dispensed with by application of an appropriate truncation/rounding scheme to the intermediate results.

Alternatively, by using sufficient storage for accumulation of the rotated data, and sufficient iterations and storage for the CORDIC operations, the requirement for such a scheme can be eliminated. For the DFT implementation discussed here, this involves a total accumulator wordlength of L bits, where:

$$L = K + LOG_2\{N_1\} + LOG_2\{N_2\} \tag{6.6}$$

together with L iterations and $LOG_2\{L\}$ guard bits for the CORDIC operation.

A simple compromise might be to truncate/round, corresponding to the available wordlength, just the first partial DFT output.

7. SUMMARY AND CONCLUSIONS

This paper has shown how a rectangular array of N processing elements can be used to carry out an efficient systolic implementation of the N-point DFT, via the arbitrary factor and prime factor decompositions, for the particular case where N is expressed as a product of just two factors.

It differs from the conventional systolic DFT approach, in that the input and output data must undergo vector-matrix and matrix-vector transformations, respectively, in order that the 2-D formulation of the DFT be fully exploited by the rectangular processor array.

Despite the indexing requirements, for sufficiently large N, the 2-D systolic solution offers the attraction of very high throughput rates, compared to the conventional linear systolic array, for the same number of processing elements.

The application of Horner's rule facilitates the construction of a very simple 2-state processing element, for the recursive computation of the small DFTs that result from the arbitrary factor & prime factor decompositions, with high-precision output achieved via the use of hardware-efficient CORDIC arithmetic.

8. REFERENCES

[1] Brigham, E. (1974) "The Fast Fourier Transform", Prentice-Hall, Englewood Cliffs, 1974.

[2] Good, I. (1958) "The Interaction Algorithm and Practical Fourier Analysis", J. Roy.Stat.Soc., B-20, 361-372, 1958.

[3] Knuth, D. (1981) "The Art of Computer Programming", Vol. 2, Addison-Wesley, 1981.

[4] Kung, H. (1979) "Let's Design Algorithms for VLSI Systems", Caltech Conference on VLSI, 65-90, Jan. 1979.

[5] Kung, S. (1984) "On Supercomputing with Systolic/Wavefront Array Processors", IEEE Proc. 72, 867-884, 1984.

[6] Kung, S. et al (EDS), (1985) "VLSI and Modern Signal Processing", Prentice-Hall, 1985.

[7] Niven, I. and Zuckerman, H. (1980) "An Introduction to the Theory of Numbers", New York, Wiley, 1980.

[8] Volder, J. (1959) "The CORDIC Trigonometric Computing Technique", IRE Trans. Elec. Comp. EC-9,330-334, Sept. 1959.

[9] Willey, T., Chapman, R., Yoho, H., Durrani, T. and Preis, D. (1985) "Systolic Implementations for Deconvolution, DFT and FFT", IEE Proc. pt. F, 132, 466-472, Oct. 1985.

PARALLEL DFT ALGORITHMS FOR A DISTRIBUTED ARRAY OF PROCESSORS

R. C. Green
(Active Memory Technology)

and

J. J. Soraghan
(Department of Electronic Engineering, University of Strathclyde)

1. INTRODUCTION

This paper discusses the implementation of the DFT on a distributed array of processors such as the AMT family of DAPs. The DAP is described in Section 2 together with programming considerations. Section 3 describes the mapping choices for a range of discrete Fourier transform (DFT) problems and presents the corresponding parallel algorithms. The performance for these are given for 32-bit floating point and 16-bit fixed bit arithmetic using the DAP-510. In Section 4 the complex "twiddle" factor contribution to the overall cost is analysed. It is shown that this ranges from 11% to 35%. Section 5 investigates the implementation of the Prime Factor Algorithm (PFA) and the Winograd DFT on the DAP. In this study it is assumed that the DAP is operating with a medium degree of parallelism. Following on from this study the algorithms, designed to permute data arrays that have dimensions that are relatively prime are presented. A complexity analysis of these algorithms is included. A comparison is made of the implementation of the DFT on the DAP-610 using the normal FFT approach and the new permutations. It is shown that the latter method may reduce up to 77.8% off the "twiddling" time.

2. THE DISTRIBUTED ARRAY PROCESSOR (DAP)

Figure 1 shows a schematic of the massively parallel DAP 510 (Distributed Array Processor). The heart of the machine is a 2-d array of simple 1-bit processing elements 1024 in total (PEs). Each PE within the array has connections to its nearest neighbours, the other PEs in its row and column, and to its own local memory. Memory addressing is under the control of the MCU (Master Control Unit) so a memory access by the PEs can be considered as the processing of a bit plane. Within an individual PE there are essentially four registers: the Q, C,A, and D. The Q (accumulator)

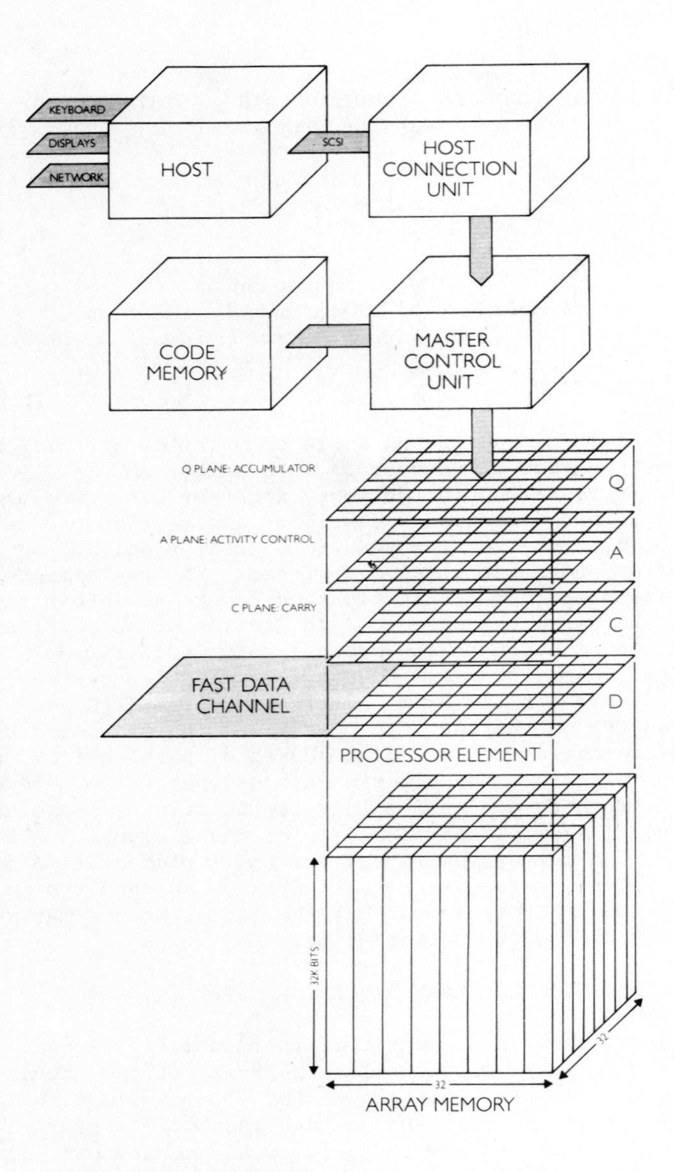

Fig. 1 The DAP 510 computer

and the C (carry) registers are used to develop the arithmetic. Whereas the A and D registers have more specialised purposes the A or activity register controls the writing of results to memory, so while all PEs execute the same instruction not all results need be written back to memory. The D register on the other hand provides the array with a route to the real world. Planes of memory can be loaded into the D place of registers then, asynchronously to the subsequent operations can be fed to the Input/Output devices at a sustained rate of 50 MBytes per second. At this I/O rate there is only a 4% overhead for the 500 series machines and a 1% overhead for the 600 series machines, which have 4096 PEs. To summarise the DAPs potential for signal processing, it can perform high-speed arithmetic, input/output, and data movement; in short all the functions needed for high-speed real-time signal processing.

Floating-point, fixed-point, Boolean and user-defined number formats are supported by DAP computers. Access to these number formats is essentially by two programming languages: Fortran-plus and APAL. Fortran-plus is an extended Fortran, having array and vector data types, with a comprehensive set of functions and operators to process arrays and vectors. Most notably there are extensions that allow the user to utilise the nearest neighbour and broadcast features of the DAP. Alternatively the user may choose to program the DAP using APAL; the DAP's assembly language.

While floating-point and fixed-point arithmetic and their use in signal processing algorithms is well understood, the potential benefits of user defined-arithmetic are less well understood. As the DAP builds up its arithmetic functions from bit-serial operations, arithmetic of arbitrary precision can be performed. In particular the arithmetic precision can be tailored to the precise requirements of the algorithm. For example, high performance FFTs can be programmed by tapering the arithmetic precision to the word growth through the algorithm.

3. DFT ON THE DAP

The degree of parallelism offered by a DAP containing P PEs is variable and ranges from 1 (Scalar Mode) to P (Maximum Machine Parallelism). The design of efficient parallel DFT algorithms begins with a proper choice of data mapping.

Given N DFTs each of length M and assuming that N > P then clearly $[N/P]_F$ transforms may be simultaneously performed using a maximum parallelism of P. For greatest efficiency it is advantageous for N to be an exact multiple of P. (The $[.]_F$ is the 'integer floor' function). This type of problem is referred to as the ideal parallel problem wherein an optimum sequential algorithm is simultaneously performed within each PE. In the case of the DFT set it is defined as the Vertical Transform (VT) and is depicted in Fig. 2(a).

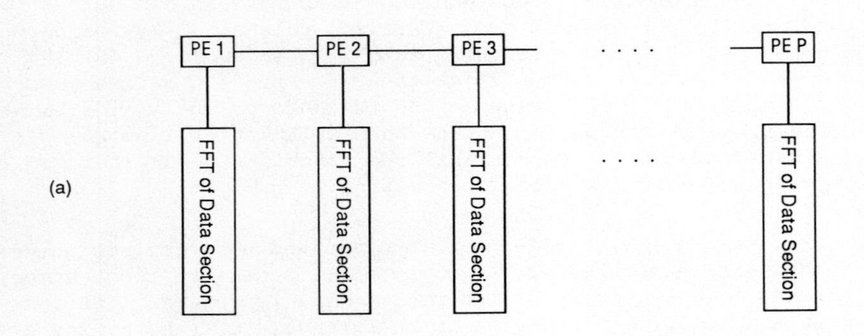

Fig. 2(a) Vertical Transform (VT) Similar FFT applied to all P, PEs.

When N < P then the N transforms may be simultaneously performed using a parallelism of N. For optimum efficiency it is advantageous that N divides P exactly. This type of algorithm (where the data is spread across/down PEs) is defined as the Horizontal/Vertical Transform (HVT) and is shown in Fig. 2(b). Depending on N and the data length M the algorithm may consist of a pure Horizontal Transform (HT) (r=1) in Fig. 2(b)) or a HVT (r > 1 in Fig. 2(b)).

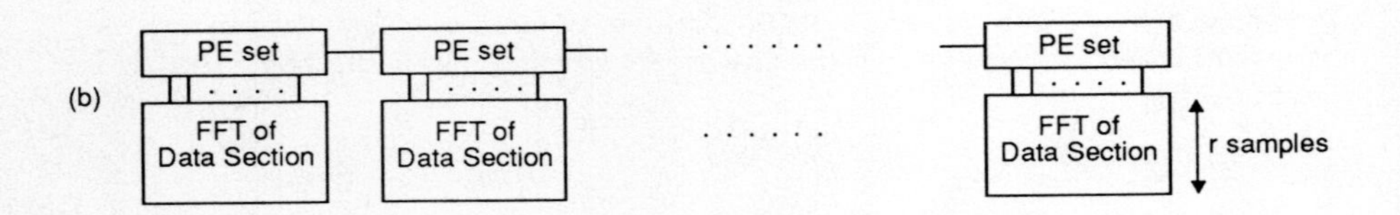

Fig. 2(b) r = 1 = > Horizontal Transform (HT)
r > 1 = > Horizontal/Vertical Transform (HVT)
Similar FFT applied to each set of PEs

When N and P are powers of 2 and N < P then for N,M-point DFTs the DAP is firstly partitioned into N sections each containing P/N PEs. Each data set of length M is mapped under the P/N PEs down a depth of (NM/P) into the DAP store. The 1-d data sets are thus mapped into a natural 2-d structure and a parallel version of the familiar ROW/TWIDDLE/COLUMN DFT algorithm [1] may be used to implement the sets of 1-d transforms. In the general case the resulting DAP algorithm would consist of a VT, twiddling and finally a series of HTs.

The difference between a VT and a HT is that in the latter data communication is required between the PEs. Hockney [2] has shown that where possible, it is more economical to perform the complete transform using VTs and rotation of data if the data sets extend into memory by more than two samples i.e., if (NM/P) > 2.

In this paper it is assumed that the DFTs may be performed using only VTs.

An expression for the complexity of N,M-point DFTs on the DAP is developed below.

Let
$$r = \frac{NM}{P} \tag{1}$$
and
$$q = \left\lfloor \log_r \left(\frac{P}{N}\right) + 1 \right\rfloor_F$$

then following the above discussion q VTs will be performed with a rotation of data and 'twiddling' of data occurring between each VT. Having performed the maximum number of 'r' point VTs the length of the remaining transform is

$$\frac{M}{r^q} \tag{2}$$

and r^{q+1}/M of these will need to be performed per PE section. The combined cost C(N,M) may be written as:

$$C(N,M) = qC_{VT}(r) + (q-1+S)[C_{TW}(r-1) + rC_{RO}(\frac{M}{r})]$$
$$+ \frac{r^{q+1}}{M}[C_{VT}(\frac{M}{r^q})] \tag{3}$$

where $\quad S = 1$ if $(M/r^q) > 1$

$\quad\quad = 0$ otherwise

$C_{VT}(X)$ is the cost of performing a VT (optimum sequential algorithm in each PE) of length X.

$C_{TW}(X)$ is the cost of performing X twiddles on the data (X complex multiplications)

and $C_{RO}(X)$ is the cost per complex sample (in secs) of rotating data in X PEs down the memory store.

An expression for this latter cost is [3]:

$$C_{RO}(X) = 4W_d \sum_{i=0}^{K} (5+2^i) \text{ cycles per complex sample of } W_d \text{ bit precision} \quad (4)$$

where $K = \text{Log}_2 X$ and the cycle time of the DAP 510 is 100ns.

A performance table for various sized DFT sets is shown in Table 1. The performance index used in the figure is the execution time per transform. The throughput for both 32-bit floating-point and 16-bit fixed-point arithmetic is incorporated in the table. As the parallelism of the problem increases the throughput increases as expected.

N	64	256	1024
M	(FlP, FxP)	(FlP, FxP)	(Flp, FxP)
32	(.13, .047)	(.059, .013)	(.043, .011)
128	(.56, .13)	(.336, .085)	(.276, .074)
512	(2.0, .6)	(1.76, .46)	(1.46, .376)
1024	(4.02, 1.11)	(3.84, .99)	(3.9, .93)
2048	(8.94, 2.54)	(8.48, 2.17)	

Table 1 Performance Table for both 32-bit Floating-Point (FlP) and 16-bit Fixed-Point (FxP) arithmetic. The figures shown represent the throughput or the cost per transform in msecs for the DAP 510.

4. INCREASING THE THROUGHPUT

In many problems the DAP will be used with a medium degree of parallelism. This may be due to the nature of the problem (non-stationary) or due to a limitation of DAP memory. In order to enhance the throughput for DFT problems of this type equation (3) is studied.

Initially assume $r^q = M$, then from equation (3) the cost of performing N transforms may be written as:

$$C(N,M) = q\, C_{VT}(r) + (q-1)[C_{TW}(r-1) + r\, C_{RO}(\frac{M}{r})] \qquad (5)$$

The Twiddle factors are complex matrix-by-matrix multiplications whereas the multiplications incorporated within the VTs are scalar-by-matrix multiplications. The contribution of the 'Twiddles' to the overall cost is approximated below. Defining η as the ratio of 'Twiddle Cost' to overall cost, it may be written as:

$$\eta = \frac{(q-1)C_{TW}(r-1)}{qC_{VT}(r) + (q-1)[C_{TW}(r-1) + rC_{RO}(\frac{M}{r})]}$$

$$\eta = \frac{1}{\frac{q}{q-1}\frac{C_{VT}(r)}{C_{TW}(r-1)} + 1 + \frac{r\, C_{RO}(\frac{M}{r})}{C_{TW}(r-1)}} \qquad (6)$$

Taking an expression for the optimised radix-2 sequential FFT algorithm and assuming that a complex multiplication requires 4 real multiplications and 2 real additions then equation (6) may be written as:

$$\eta = \frac{1}{\frac{q}{q-1}\frac{r}{r-1}[\frac{(2+3\alpha)}{(4+2\alpha)}\log_2 r - 1 + \frac{1}{r}] + 1 + \frac{rC_{RO}(\frac{M}{r})}{(r-1)(4+2\alpha)t_r}} \qquad (7)$$

where α is the ratio of real addition to real multiplication execution time, and t_r is the cost of performing a real multiplication.

Letting r=4, q=2 and $\alpha=2/3$ (32-bit floating-point) gives $\eta=27\%$ and for r=4, q=2 and $\alpha = 2/3$ yields $\eta = 15\%$ approx. In this case η increases to 19% for $\alpha = 1/10$.

In general as r or the precision increases then η decreases. Through direct substitution of appropriate values into expression (3) it may be observed that η ranges from 35% (low r) to 11% (large r) for floating point arithmetic.

5. FFT ALTERNATIVES

One of the advantages of FFT alternatives such as the Prime Factor Algorithm (PFA) or Winograd's DFT (WFTA) [4] is that the comp 'Twiddle' factors analysed above are replaced by noiseless permutations of data elements. When the DAP is operating with a medium degree of parallelism then unlike sequential algorithms, the permutations involved in the PFA/WFTA may not be performed using simple indexing.

The two data permutations encountered in the PFA and WFTA are:

(A) The simple permutation (SP).

(B) The Chinese Remainder Theorem permutation (CRTP) each of which is defined as follows.

Given a 1-D set of numbers x_n, $n=0,1,\ldots M-1$ and that $M = Lr$ where $(L,r)=1$ [i.e. L and r are relatively prime] then the SP is obtained using the following index substitution:

$$n = [\, L.i + r.j \,] \text{ modulo } M \tag{8}$$

for $i = 0,1,2\ldots r-1$ and $j = 0,1,2\ldots L-1$

The CRTP is obtained using the following substitution for the index n:

$$n = [\, L.i.g + r.j.p \,] \text{ modulo } M \tag{9}$$

for $i = 0,1,2\ldots r-1$ and $j = 0,1,2\ldots L-1$

The integers g and p are obtained as solutions to the following Diophantine [4] equations:

$$L.g = 1 \text{ modulo } r \quad \text{and } r.p = 1 \text{ modulo } L \tag{10}$$

The above permutations may be extended to any number of dimensions, once all the dimensions are relatively prime.

The M-point PFA with $M = \Pi_{i=1}^{D} p_i$ and $(p_i, p_j) = 1 \; i \neq j$, involves the following tasks:

(a) An SP to permute the 1-D sequence into a D-dimensional sequence

(b) $\left(\frac{M}{p_i}\right)$ short length DFTs for each dimension 'i', i=1,D

(c) A CRTP to permute the output sequence to its natural order.

In an M-point WFTA tasks (a) and (c) are similar to those of the PFA. For task (b) Winograd devised an efficient nesting scheme which minimises the total number of multiplications required in the resulting algorithm [4].

To construct efficient DAP SP and CRTP algorithms the following constraints are imposed given N sets of M-point DFTs [5](M = Lr,GCD(L,r) = 1):

(a) L must be a power of two and r = L-1

(b) If either the Natural to CRTP or its inverse map is required then the data must be mapped under L PEs.

(c) If either the Natural to SP or its inverse map is required then the data must be mapped under r PEs.

For example consider the mapping of 4 sets of 12 elements onto a hypothetical 4x4 DAP. The following mappings result for 1 data set:

PEs	1	2	3	4	1	2	3	4	1	2	3	4
S												
T	0	1	2	3	0	1	2		0	1	2	3
O	4	5	6	7	3	4	5		4	5	6	7
R	8	9	10	11	6	7	8		8	9	10	11
E					9	10	11					
	Natural				Map for SP				Map for CRTP			

Performing the permutations yields:

PEs	1	2	3	4	1	2	3	4	1	2	3	4
S												
T	0	1	2	3	0	4	8		0	9	6	3
O	4	5	6	7	3	7	11		4	1	10	7
R	8	9	10	11	6	10	2		8	5	2	11
E					9	1	5					
	Natural				SP				CRTP			

Notice that both the SP and CRTP are toroidally ordered as indicated by the dashed lines. Both maps may therefore be obtained from the Natural map by cyclically shifting each column, in a northern direction i by V(i), i=0,1,2,3 where V = [0,1,2,0] for the SP and V = [0,2,1,0] for the CRTP.

A flow diagram of both algorithms is shown in Figure 3.

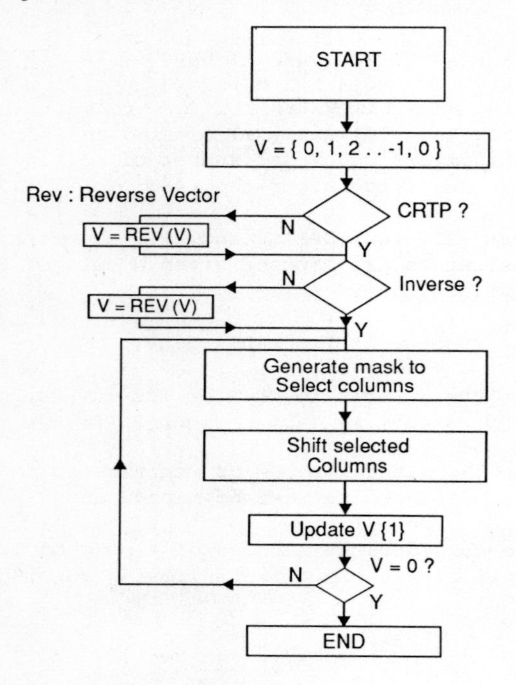

Fig. 3 The SP and CRTP Algorithms

The SP(CRTP) to Natural or Natural to SP(CRTP) is performed according to the inverse Boolean Flag.

The cost in terms of number of exchanges required by the SP algorithm may be written as:

$$C_{SP}(L) = L/2+(L/2+L/4)+(L/2+L/4+L/8)+ \ldots +(L/2+L/4+L/8 + \ldots + 2+1)$$

$$C_{SP}(L) = \sum_{i=1}^{\log_2 L} i2^{i-1} \tag{11}$$

The complexity of the CRTP is complicated by the fact that the depth in memory is not a power of 2. The complexity depends on the actual memory depth. For example consider r=15, the total cost, in terms of exchanges, may then be written as:

$$C_{CRTP}(15) = 2(7) + 2(5) + 4(3) + 7(2)+14(1) \qquad (12)$$

The figures in brackets represent the degree of cyclic shifts performed at a particular stage. If S represents the degree of required shift at a particular stage and S divides r then S elements are exchanged r/S-1 times to implement the cyclic permutation. If S does not divide r then an extra layer of exchanging is required as seen for stage 1 (S=7) and stage 4 (S=2) above.

Table 2 compares the costs of using twiddles to both the combined SP/CRTP and single SP costs on the 100nsecs DAP-610 for 32-bit floating-point and 16-bit fixed-point arithmetic. The latter comparison is included because in many DSP applications and in the case of Synthetic Aperture Radar Processing in particular, the CRTP is not required and thus a single SP will suffice. It is seen from the Table 2 that a 50% and 77.8% time saving is obtained in the case of 32-bit floating-point for the combined SP/CRTP and single SP respectively. The corresponding figures for 16-bit fixed-point are 16.8% and 62.4%.

Arithmetic	(tm: cost of 1 DAP complex multiply)	Permutations Cost (n bit planes EXCHANGED costs {5+.4n} microsecs)	
	64 tm	CRTP+SP	SP
32-bit floating-point. (tm=.8)	51.2	25.6	11.42
16-bit fixed-point (tm=.304)	19.46	16.39	7.32

Table 2. Comparison of twiddle costs to both CRTP+SP costs and SP cost for 64 sets of 4032 complex DFTs using 32-bit floating point and 16-bit fixed point arithmetic. All times are in msecs. for the AMT DAP-610.

6. CONCLUSIONS

A Parallelism ranging from 1 to P is available from the AMT DAP with P PEs. However efficient parallel algorithm designs depend on proper mapping selection. Throughput for a given algorithm thus depends on the degree of data parallelism inherent within the application. Combined PFA/WDFT/FFT algorithms may be mapped on to the DAP by making several restrictions. These restrictions are necessary for the design of efficient permutation algorithms. Such algorithms can reduce the cost of twiddle factor multiples by up to 77%. For this paper we have assumed the use of 32 bit floating-point and 16 bit fixed-point arithmetic. As the DAP is bit-serial word parallel machine the word-length may be tailored to suit the particular application. In situations where the data word-length is low, for example image or radar data, the throughput may be significantly improved by using "tapered" precision within the algorithms.

REFERENCES

[1] Nussbaumer, H. J. (1982), "Fast Fourier Transforms and Convolution Algirithms", Springer Verlag, New York.

[2] Hockney, R. W. and Jesshope, C. R. (1981), "Parallel Computers", Adam Hilger Ltd., U.K.

[3] Flanders, P. M. (1988) Private communication.

[4] Blahut, R. E. (1985), "Fast Algorithms fcr Digital Signal Processing", Addison-Wesley.

[5] Soraghan, J. J. (1988), "New Data Movement Algorithms for Processor Arrays", In Proc. ICASSP, New York.

PARALLEL WEIGHT EXTRACTION FROM A SYSTOLIC ADAPTIVE BEAMFORMER

T.J. Shepherd and J.G. McWhirter
(Royal Signals and Radar Establishment, Malvern, Worcestershire)

and

J.E. Hudson
(Standard Telecommunications Laboratories, Harlow, Essex)

ABSTRACT

A novel method is described whereby all elements of the least squares weight vector may be extracted in parallel from a systolic adaptive beamforming network. This constitutes a significant improvement over the traditional "back-solving" and "weight-flushing" techniques which could only be used to extract the weight vector one element at a time. The new method requires a triangular array of "slave" processing cells which update the inverse Cholesky factor of the correlation matrix using the same rotations as those computed within the triangular array to update the Cholesky factor itself.

1. INTRODUCTION

The use of systolic arrays in performing least squares minimisation for adaptive beamforming has been the subject of recent extensive investigation. In this application, signals from several antenna channels are processed so as to minimise undesired additive interference (or "jamming") present in either a given signal channel, or in a linear combination of channels (or "beam"). The unwanted interference is minimised after adjustment of a vector of complex weights that constitutes a set of coefficients for the optimum linear combination of channels.

An important operation central to the least squares minimisation used in the adaptive beamforming has proved to be QR decomposition of the input data channel matrix. This process can be performed iteratively in a numerically well-conditioned fashion, and is amenable to rapid computation via parallel and pipelined (systolic) processing arrays. One such array, due to Gentleman and Kung [1], comprises a triangular arrangement of processing cells, reflecting the triangular structure of the matrix $\underline{R}$ formed by the procedure.

In the original work of Gentleman and Kung the required least squares weight vector was obtained by first extracting the elements of the matrix $\underline{R}$ from the array, and then solving a set of linear equations off-line by means of back-substitution. This method interrupts data flow within the triangular array, and requires (2p-1) clock cycles to compute a p-element weight vector. McWhirter[2] later demonstrated that if only the noise-reduced signal was required (the least squares residual), then this was available at every clock cycle directly from the triangular array without the need to extract the weight vector. It is then possible to show that the vector elements are sequentially obtainable at the residual output as the impulse response of the system when adaptivity of the array is temporarily suspended[3]. This second method of obtaining the weights, known as "weight flushing", requires no additional hardware, but consumes p clock cycles to output the vector, and also interrupts normal input data flow for the same time.

In the following we shall describe how a complete least squares weight vector may be obtained at each clock cycle without affecting input data flow. The method will be illustrated here for the cases of both constrained and unconstrained (or "canonical") least squares minimisation.

2. STATEMENT OF PROBLEM

In "sidelobe cancellation" adaptive beamforming a linear combination of p auxiliary channel signals is used to cancel the interference from a primary channel containing the corrupted signal. Specifically, this consists of minimising the mean-square of the quantity

$$e(t_q) = \underline{x}^T(q)\underline{w} + y(t_q), \tag{2.1}$$

where $y(t_q)$ is the complex value of the primary channel signal at time t_q $(q=1,2,\ldots,n)$, and $\underline{x}(q)$ is a p-element vector "snapshot" of auxiliary channel values at t_q. $\underline{w}$ is the p-element weight vector. (We shall use superscripts T and H to denote the operations of matrix transposition and hermitian conjugation, respectively).

The mean-square of e(n) is minimised in the least squares sense when the L_2 norm of the vector

$$\underline{e}(n) = \underline{X}(n)\underline{w}(n) + \underline{y}(n), \tag{2.2}$$

is minimised with respect to the elements of $\underline{w}(n)$. Here $\underline{e}(n)$ (the residual vector) and $\underline{y}(n)$ are n-element vectors containing the entries $\{e(t_q)\}$ and $\{y(t_q)\}$, $(q=1,2,...,n)$, respectively, and $\underline{X}(n)$ is an nxp data matrix with rows $\{\underline{x}^T(q)\}$, $(q=1,2,...,n)$. We refer to the determination of $\underline{w}(n)$ as the "unconstrained" or "canonical" least squares problem. (Note that we are including no data deweighting in this exposition).

Constrained least squares, on the other hand, is employed in the MVDR (Minimum Variance Distortionless Response) adaptive beamforming problem. Here a constant antenna array gain is maintained for a beam defined by a given vector $\underline{c}$, and the linear combination of p channels found which otherwise minimises the interference. There is no primary channel. Expressed as a least squares problem, the task is to minimise the L_2 norm of the complex residual vector

$$\underline{e}_c(n) = \underline{X}(n)\underline{w}_c(n), \tag{2.3}$$

subject to the constraint

$$\underline{c}^T\underline{w}_c(n) = \mu , \tag{2.4}$$

where μ is the desired antenna gain.

We shall now describe how $\underline{w}(n)$ and $\underline{w}_c(n)$ may be computed at each value of n.

3. QR DECOMPOSITION

The QR decomposition of the matrix $\underline{X}(n)$ may be defined by the equation

$$\underline{Q}(n)\underline{X}(n) = \begin{bmatrix} \underline{R}(n) \\ \underline{O} \end{bmatrix} , \tag{3.1}$$

where $\underline{Q}(n)$ is a unitary nxn matrix, and $\underline{R}(n)$ is pxp upper triangular. (We assume n>p, and that $\underline{X}(n)$ is of rank p). Applying equation (3.1) to the unconstrained problem specified by equation (2.2) it is easily shown that the norm of $\underline{e}(n)$ is minimised when $\underline{w}(n)$ satisfies the triangular set of equations

$$\underline{R}(n)\underline{w}(n) + \underline{u}(n) = \underline{O} , \tag{3.2}$$

where $\underline{u}(n)$ comprises the first p elements of $\underline{Q}(n)\underline{y}(n)$,

$$\underline{Q}(n)\underline{y}(n) = \begin{bmatrix} \underline{u}(n) \\ \underline{v}(n) \end{bmatrix} , \qquad (3.3)$$

It may also be shown[4] that the constrained weight vector $\underline{w}_c(n)$ is given by

$$\underline{w}_c(n) = \lambda \underline{R}^{-1}(n)\underline{d}^*(n) , \qquad (3.4)$$

where

$$\underline{d}(n) = \underline{R}^{-T}(n)\underline{c} , \qquad (3.5)$$

and λ is a Lagrange multiplier given by

$$\lambda = \mu\ (\underline{d}^H(n)\underline{d}(n))^{-1} . \qquad (3.6)$$

$\underline{w}_c(n)$ can also be expressed in the more familiar form

$$\underline{w}_c(n) = \lambda \underline{M}^{-1}(n)\underline{c}^* , \qquad (3.7)$$

where $\underline{M}(n)$ is an estimator for the channel cross-correlation matrix

$$\underline{M}(n) = \underline{X}^H(n)\underline{X}(n) = \underline{R}^H(n)\underline{R}(n) . \qquad (3.8)$$

$\underline{R}(n)$ is the Cholesky square root matrix of $\underline{M}(n)$.

The matrix $\underline{R}$ may be computed and updated from t_{n-1} to t_n using a triangular array of the type proposed by Gentleman and Kung[1], as shown in figure 1. The array is an arrangement of two types of cell: the boundary cell, which computes parameters for a two-dimensional planar, or Givens, rotation, and an internal cell, which applies the rotation to the pair (x,r), where x is the value of data entering the top of the cell, and r is an element of $\underline{R}$ stored and updated within the cell. In the arrangement shown in the figure, each cell initially contains an element of $\underline{R}(n-1)$. The row vector $\underline{x}^T(n)$ enters the array from the north, and in the first boundary cell the Givens rotation parameters computed which annihilate the rotated data from the pair $(x_1(n), r_{11}(n-1))$. These parameters are then passed to the east, where the remaining data are rotated with elements in the first row of $\underline{R}(n-1)$. The rotated data are then passed to the

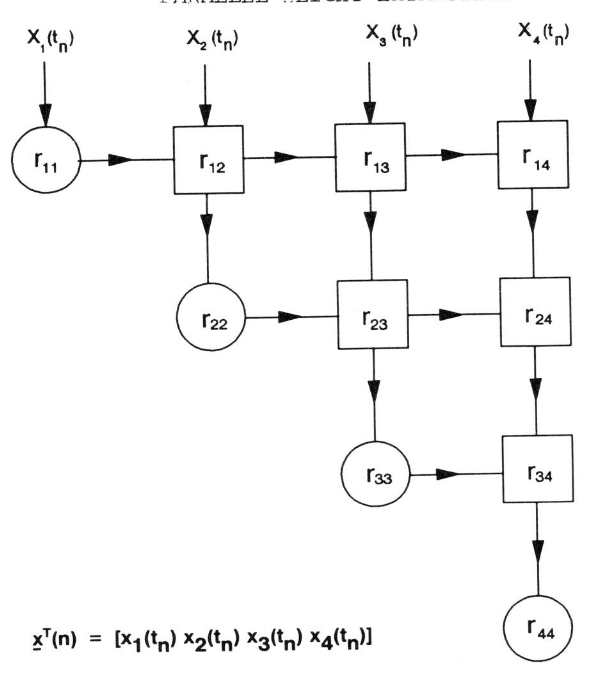

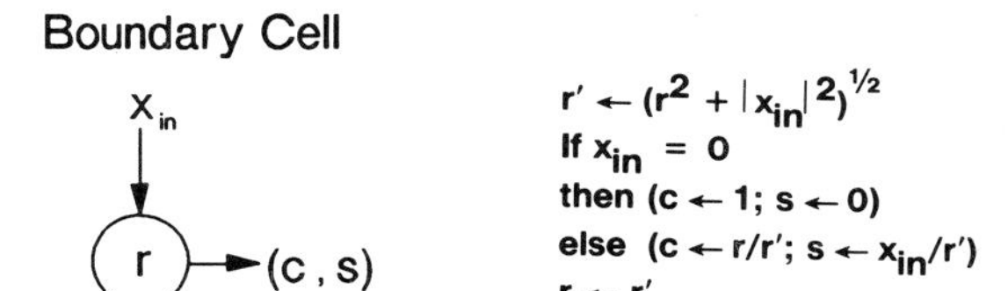

Boundary Cell
X_{in}
r
(c , s)
$r' \leftarrow (r^2 + |x_{in}|^2)^{1/2}$
If $x_{in} = 0$
then ($c \leftarrow 1$; $s \leftarrow 0$)
else ($c \leftarrow r/r'$; $s \leftarrow x_{in}/r'$)
$r \leftarrow r'$

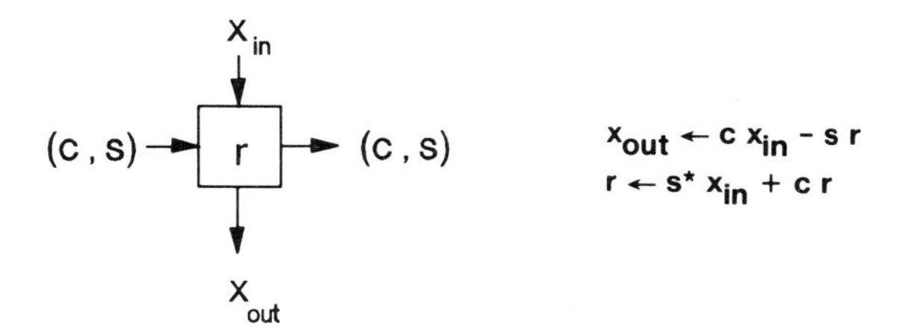

Fig. 1 Gentleman-Kung array for updating the matrix $\underline{R}$. Note that explicit systolic configuration of data is not shown.

next row, where the process is repeated until all data are annihilated, and all elements of $\underline{R}$ updated. For the purpose of clarity we have chosen for the moment to ignore the fully systolic operation of the array, and input data "skew". This will be discussed in section 8. A modification of the Gentleman-Kung array to a system that iteratively computes e(n) or $e_c(n)$ (the nth elements of vectors $\underline{e}(n)$ and $\underline{e}_c(n)$) has been reported in detail by McWhirter[2] and McWhirter and Shepherd[5], and so will not be described here.

The update operation may be defined by the matrix equation

$$\hat{\underline{Q}}(n) \begin{bmatrix} \underline{R}(n-1) \\ \underline{O} \\ \underline{x}^T(n) \end{bmatrix} = \begin{bmatrix} \underline{R}(n) \\ \\ \underline{O} \end{bmatrix} \qquad (3.9)$$

where $\hat{\underline{Q}}(n)$ is the nxn unitary matrix product of p individual Givens rotation matrices employed in a single update. It may be shown by construction that $\hat{\underline{Q}}(n)$ takes the form [2]

$$\hat{\underline{Q}}(n) = \begin{bmatrix} \underline{A}(n) & \underline{O} & \underline{a}(n) \\ \underline{O} & \underline{I} & \underline{O} \\ \underline{b}^T(n) & \underline{O} & \gamma(n) \end{bmatrix}, \qquad (3.10)$$

where $\underline{A}(n)$ is a pxp matrix, $\underline{a}(n)$ and $\underline{b}(n)$ are px1 vectors, $\gamma(n)$ is a (real) scalar, and $\underline{I}$ is the unit matrix.

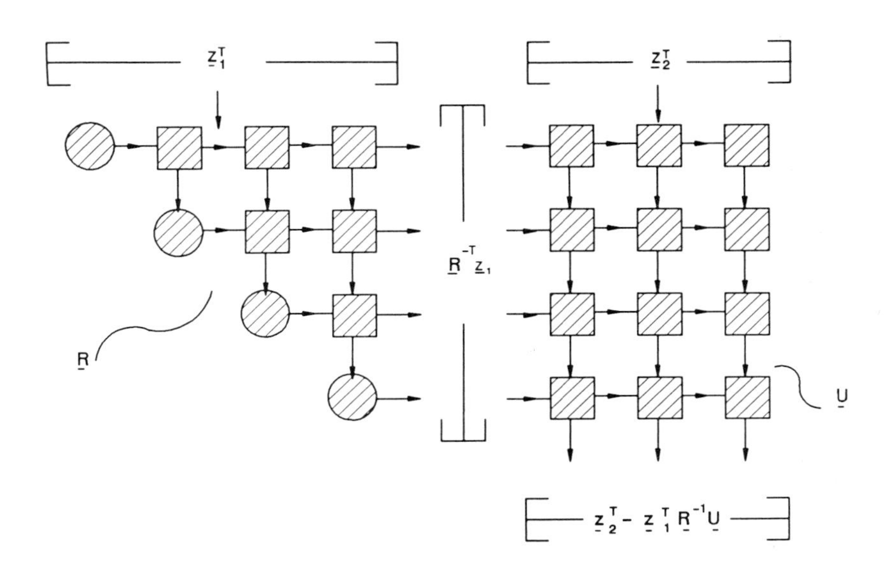

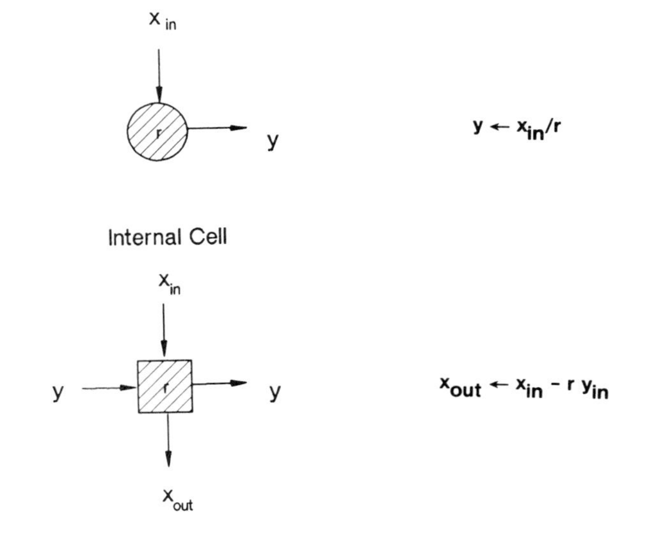

Fig. 2 Trapezoidal fixed network. The triangular and rectangular sections contain elements of $\underline{R}$ and $\underline{U}$, respectively.

4. FIXED NETWORK

Consider the network in figure 2. The whole is a concatenation of a triangular pxp array, containing elements of a matrix $\underline{R}$, and a rectangular array, containing elements of a matrix $\underline{U}$. Component cell functions are given in the figure, and again distinguish boundary from internal cells. It is straightforward to verify that when a row vector $\underline{z}_1^T$ is input to the triangular section from the north, this section outputs the column vector $\underline{R}^{-T}\underline{z}_1$ eastwards to the rectangular section. If a vector $\underline{z}_2^T$ is simultaneously input to the north of the rectangular section, a row vector equal to $\underline{z}_2^T-\underline{z}_1^T\underline{R}^{-1}\underline{U}$ is output from the south of this section, as shown in figure 2. Note that the cells perform a subset of the computations used in the adaptive array of Gentleman and Kung; in fact they may be obtained by suppressing the update of the stored parameters r of the former array. For this reason we have chosen to give the name "fixed networks" to arrays of cells performing the latter operation, and the transformation of the former operation to the latter we call "weight freezing".[6]

5. UPDATE OF $\underline{R}^{-H}$

The update for the matrix $\underline{R}^{-H}(n)$ may be performed alongside that of $\underline{R}(n)$. By writing the identity $\underline{R}^H(n-1)\underline{R}^{-H}(n-1)=\underline{I}$ in the form

$$p\ \begin{matrix} [\ \underline{R}^H(n-1) & \underline{O} & \underline{x}^*(n)\ ] \\ p & n-p-1 & 1 \end{matrix} \begin{bmatrix} \underline{R}^{-H}(n-1) \\ \underline{\zeta} \\ \underline{O}^T \end{bmatrix} \begin{matrix} p \\ n-p-1 \\ 1 \end{matrix} = \underline{I}, \qquad (5.1)$$

where $\underline{\zeta}$ denotes an arbitrary px(n-p-1) matrix, we may insert the identity $\hat{\underline{Q}}^H(n)\hat{\underline{Q}}(n)=\underline{I}$ between the matrices on the left-hand side of equation (5.1), and use equation (3.9) to give the identity

$$[\ \underline{R}^H(n)\quad \underline{O}\quad \underline{O}\ ] \begin{bmatrix} \underline{F}(n) \\ \underline{\zeta} \\ \underline{\xi}^T \end{bmatrix} = \underline{I}\ . \qquad (5.2)$$

Here the right-hand matrix in the product derives from the form of $\hat{\underline{Q}}(n)$ as given in equation (3.10). $\underline{F}(n)$ and ξ denote the results of appropriate submatrix multiplication. Equation (5.2) provides the immediate identification

$$\underline{F}(n) = \underline{R}^{-H}(n) \quad , \tag{5.3}$$

and so we may infer from equation (5.1) that

$$\hat{\underline{Q}}(n) \begin{bmatrix} \underline{R}^{-H}(n-1) \\ \underline{\zeta} \\ \underline{O} \end{bmatrix} = \begin{bmatrix} \underline{R}^{-H}(n) \\ \underline{\zeta} \\ \underline{\xi}^{T} \end{bmatrix} ; \tag{5.4}$$

that is to say that the same Givens rotations that update $\underline{R}(n)$ will also update the matrix $\underline{R}^{-H}(n)$.

Figure 3 shows the array of cells which simultaneously updates the matrices R(n) and $\underline{R}^{-H}(n)$. It consists of the concatenation of a Gentleman-Kung array and an adjacent triangular arrangement of internal cells. The first of these sections contains the elements of $\underline{R}(n)$, and the second the elements of $\underline{R}^{-H}(n)$. If the two sections are configured at time t_{n-1} to contain, respectively, elements of $\underline{R}(n-1)$ and $\underline{R}^{-H}(n-1)$, then insertion of the data vector $[\underline{x}^{T}(n), \underline{O}^{T}]$ updates the cell contents to give $\underline{R}(n)$ and $\underline{R}^{-H}(n)$, as both sections share the same set of Givens rotations.

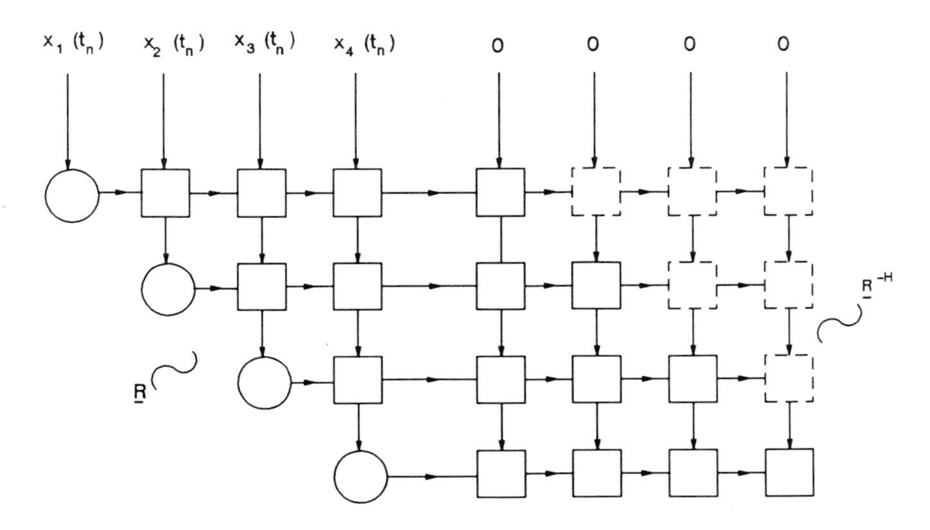

Fig. 3 Network updating $\underline{R}$ and $\underline{R}^{-H}$. The internal cells shown in dotted outline may be included, but are unnecessary.

In relation to this network, we note that the matrix $\underline{R}^{-H}(n)$ can be set up in the right-hand cells at time t_n by "freezing" the left-hand array, and then inputting the series of row vectors that make up the unit matrix. From the discussion in section 4, successive rows of the matrix $\underline{R}^{-T}(n)$ will then emerge from the east. (We have assumed that $\underline{X}(n)$ is full rank and so $\underline{R}^{-H}(n)$ exists). The complex conjugate values of these elements may then be "captured" column-by-column in the right-hand section in a manner described in reference [5].

It should be clear that a full pxp *square* matrix of internal cells will also perform the required update of $\underline{R}^{-H}(n)$; the cells for the upper triangular part of this matrix would, however, only store the value zero, and pass on zero data during adaptation.

Finally, if we can freeze the complete array at some time $t_s > t_n$ and input an arbitrary vector $[\underline{c}^T, \underline{O}^T]$, the discussion in section 4 reveals that the vector $-\underline{c}^T\underline{R}^{-1}(s)\underline{R}^{-H}(s)$ will be output from the south. A little algebra shows that this is just the vector $-[\underline{M}^{-1}(s)\underline{c}^*]^H$. The whole "parallelogram" network can thus be used as an inverse correlation matrix operator, which can both update the inverse matrix $\underline{M}^{-1}$ and perform the multiplication of $\underline{M}^{-1}$ by $\underline{c}^*$. This has obvious application in producing quantities such as the constrained weight vector $\underline{w}_c$ in equation (3.7); in this mode of operation, however, the data flow must be interrupted for one clock cycle each time the matrix product is required. More direct methods for obtaining $\underline{w}$ and $\underline{w}_c$ are given in the following sections.

6. CANONICAL WEIGHT EXTRACTION - METHOD 1

We shall now derive a network which is capable of parallel computation of the vector $\underline{w}(n)$. Suppose initially that the right-hand section of the parallelogram network outlined in Section 5 is frozen at time t_n, and therefore contains the matrix $\underline{R}^{-H}(n)$. Inputting the vector $\underline{u}^*(n)$ from the west, and using the frozen network functions, produces a row vector, output from the south, equal to $-(\underline{u}^H(n)\underline{R}^{-H}(n))$. From equation (3.2), this is precisely the vector $\underline{w}^H(n)$, so that $\underline{w}^T(n)$ is then obtainable simply upon complex conjugation. It is obvious that this quantity may be obtained without the need to interrupt data input if the cells

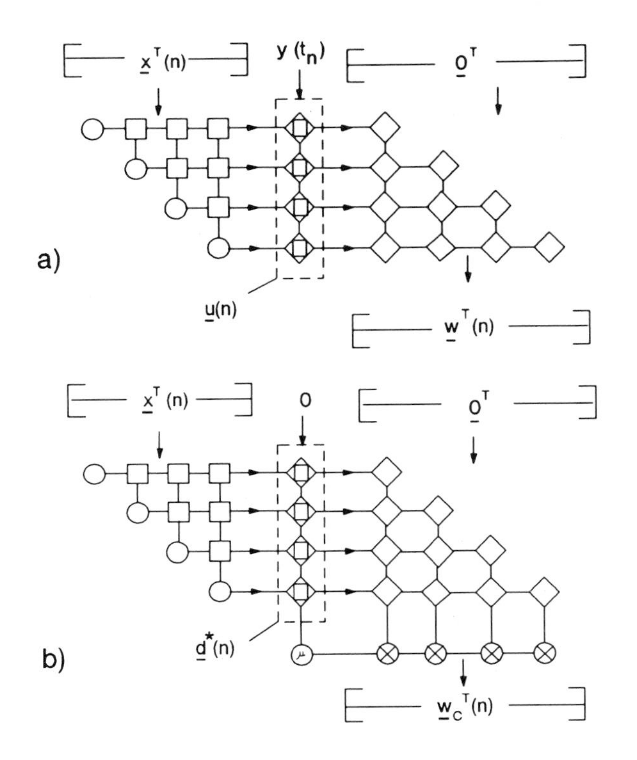

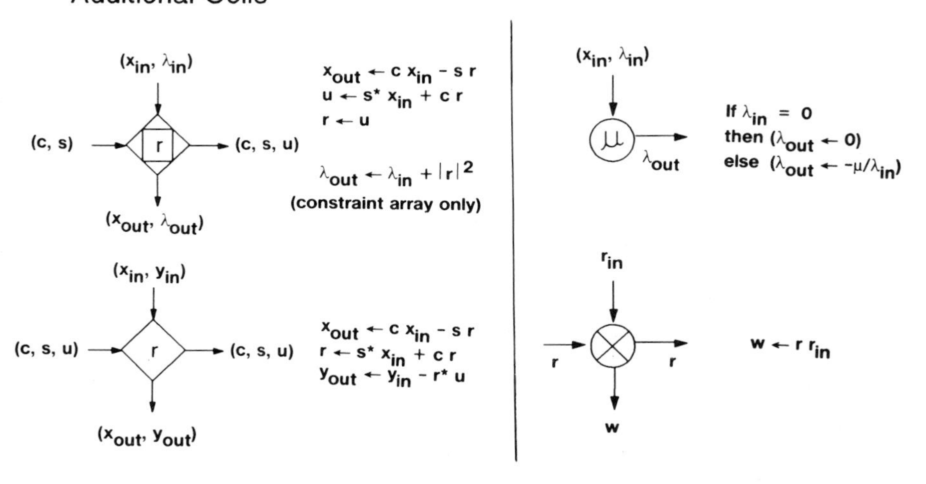

Fig. 4. First method of a) canonical, and b) constrainted, weight vector extraction.

in the right-hand section are simultaneously capable of carrying out Givens rotations *and* performing fixed-cell functions. The vector $\underline{u}^*(n)$ is calculable in an intermediate column of rotation cells, as shown in figure 4(a); these cells must also possess the ability to pass their contents to the east, as well as the Givens rotation parameters computed in the left-hand section.

A simplification is achieved if the fixed-cell functions in the right-hand section employ elements of the matrix $\underline{R}^{-T}(n)$, instead of $\underline{R}^{-H}(n)$. Only the vector $\underline{u}(n)$, and not $\underline{u}^*(n)$, need then be input from the west, and the vector $\underline{w}^T(n)$ emerges from the south: this is the function of the cells shown in figure 4(a).

7. CANONICAL WEIGHT EXTRACTION - METHOD 2

There exists a second method of canonical weight extraction which avoids the simultaneous use of fixed cell functions in the right-hand section of the network. We may show that the parallelogram network of the last section explicitly computes the canonical weight vector $\underline{w}(n)$ at each n. Consider first the QR decomposition of the extended data matrix

$$\overline{\underline{X}}(n) = [\ \underline{X}(n),\ \underline{y}(n)\ ], \tag{7.1}$$

so that

$$\overline{\underline{Q}}(n)\overline{\underline{X}}(n) = \begin{bmatrix} \overline{\underline{R}}(n) \\ \underline{O} \end{bmatrix}, \tag{7.2}$$

where $\overline{\underline{R}}(n)$ is a (p+1)x(p+1) upper triangular matrix, and $\overline{\underline{Q}}(n)$ is the corresponding nxn unitary matrix. It is easy to show that $\overline{\underline{R}}(n)$ has the structure

$$\overline{\underline{R}}(n) = \begin{bmatrix} \underline{R}(n) & \underline{u}(n) \\ \underline{O}^T & r(n) \end{bmatrix} \tag{7.3}$$

where $\underline{R}(n)$ and $\underline{u}(n)$ are the matrices defined in equations (3.1) and (3.3), respectively. (Equation (7.3) may be proved algebraically, but it is obvious when the effect of an enlarged Gentleman-Kung array is considered). r(n) is defined merely as the $(p,p)^{th}$ element of the matrix $\overline{\underline{R}}(n)$. Defining now the extended vectors

$$\tilde{\underline{w}}(n) = \begin{bmatrix} \underline{w}(n) \\ 1 \end{bmatrix} \begin{matrix} p \\ 1 \end{matrix}\ ;\quad \underline{s} = \begin{bmatrix} \underline{O} \\ 1 \end{bmatrix} \begin{matrix} p \\ 1 \end{matrix}\ , \tag{7.4}$$

equation (3.2) may be written in the form

$$\underline{\bar{R}}(n)\underline{\bar{w}}(n) = \underline{s}\ r(n)\ . \tag{7.5}$$

Solving for $\underline{\bar{w}}(n)$, we obtain

$$\begin{aligned}\underline{\bar{w}}(n) &= r(n)\ \underline{\bar{R}}^{-1}(n)\ \underline{s} \\ &= r(n) \times (\text{last row of } \underline{\bar{R}}^{-H}(n)\ )^*\ . \end{aligned} \tag{7.6}$$

Thus $\underline{\bar{w}}(n)$ may be obtained using the network shown in figure 5. It comprises a parallelogram network for the extended matrices $\underline{\bar{R}}(n)$ and $\underline{\bar{R}}^{-H}(n)$, updated as described in section 5. Equation (7.6) shows that $\underline{\bar{w}}(n)$ may be extracted by outputting the complex conjugate of the final row of the right-hand network, and multiplying by $r(n)$, which is locally available from the left-hand network.

8. PARALLEL EXTRACTION OF CONSTRAINED VECTOR

A slight modification to the network outlined in the last two sections provides a system capable of computing the constrained weight vector $\underline{w}_c(n)$. Schreiber[4] has shown how the vector $\underline{d}^*(n)$ in equation (3.5) may be updated using the same Givens rotations that update $\underline{R}(n)$. It follows from this result that if we replace the elements of $\underline{u}(n)$ in either of the networks just described by those of the vector $\underline{d}^*(n)$, then this vector will be automatically updated.

If the network used for Method 1 is used, the transit of $\underline{d}^*(n)$ through the right-hand modified frozen section will produce the vector $-(\underline{R}^{-1}(n)\underline{d}^*(n))^T$ from the south of the right-hand section at each time step. This vector may then be suitably scaled by the factor λ of equation (3.6). λ emerges from the south of the intermediate column in which it is computed, as described in reference [5]. This will complete the computation of $-\underline{w}_c^T(n)$. The resulting network is illustrated in figure 4(b).

A similar modification to the network of section 7 will also furnish the vector $\underline{w}_c(n)$. To see this, we define the matrix

$$\underline{\bar{R}}_c(n) = \begin{bmatrix} \underline{R}(n) & \underline{d}^*(n) \\ \underline{0}^T & r(n) \end{bmatrix} \tag{8.1}$$

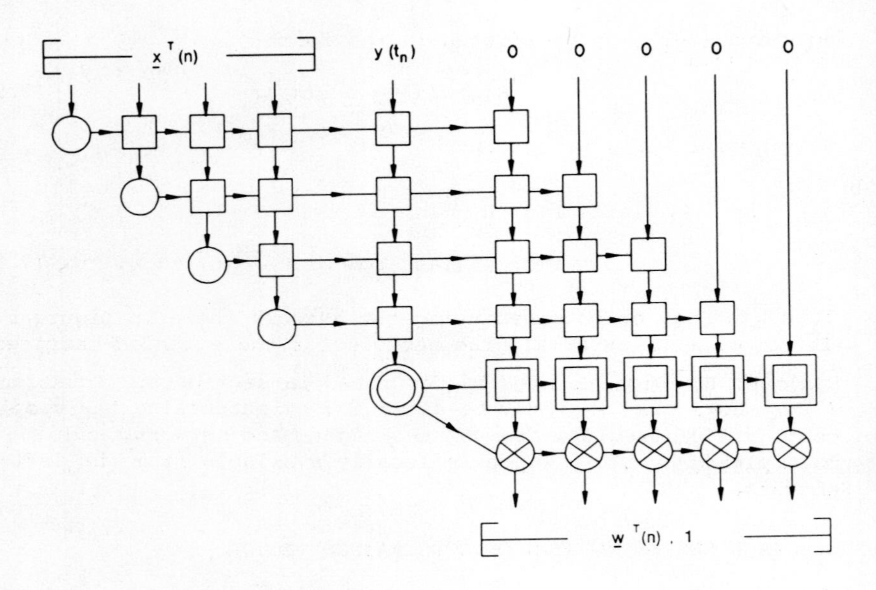

Additional Cells

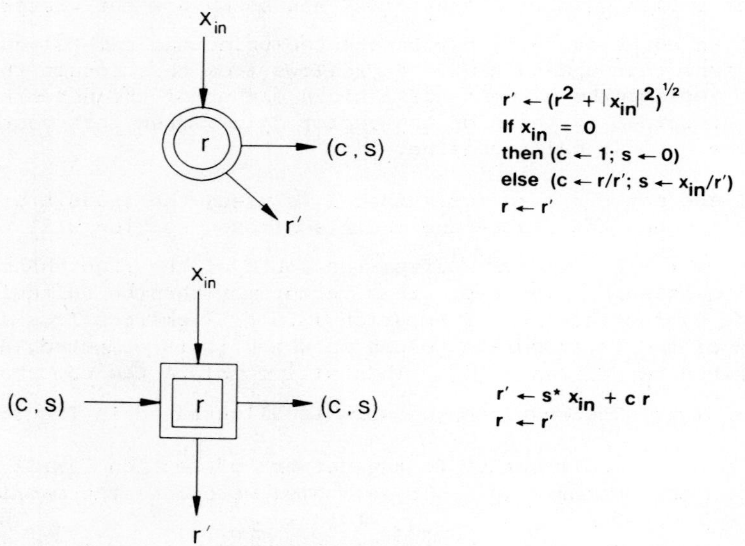

Fig. 5 Parallel extraction of canonical weight vector using second method. The corresponding network for extraction of the constrained weight vector is described in the text.

where r(n) is, for the moment, arbitrary. It is straightforward to show that

$$\underline{\bar{R}}_c^{-H}(n) = \begin{bmatrix} \underline{R}^{-H}(n) & \underline{O} \\ \underline{V}^H(n) & r^{-1}(n) \end{bmatrix} , \tag{8.2}$$

where

$$\underline{V}(n) = -r^{-1}(n)\ \underline{R}^{-1}(n)\underline{d}^*(n) . \tag{8.3}$$

The vector $\underline{V}(n)$ is seen from equation (3.4) to be proportional to the required vector $\underline{w}_c(n)$.

If the elements of $\underline{\bar{R}}_c$ are set up at time t_s in the left-hand section of the network shown in figure 5, then the component matrices $\underline{R}$ and $\underline{d}^*$ will be correctly updated, as previously explained. The parameter r(s) stored in the lowest boundary cell can be set to any desired value at t_s, and will be updated accordingly; the inverse of $\underline{\bar{R}}_c(s)$ will only exist, of course, if r(s) is non-zero. The matrix $\underline{\bar{R}}_c^{-H}(s)$ can then be captured in the right-hand triangular array in the manner described in section 5, and will be correctly updated using the same rotations as for the left-hand triangle. According to the discussion above, the required vector $\underline{w}_c(n)$ is subsequently obtained by outputting the complex conjugate of the final row in $\underline{\bar{R}}_c^{-H}(n)$, and scaling by the factor $-r(n)/|\underline{d}(n)|^2$, which may be computed within the central column containing the vector $\underline{d}^*(n)$. (This is a simple modification to the network drawn in figure 5, and is not explicitly shown here). Again, this method of obtaining $\underline{w}_c(n)$ requires no fixed-cell operations for internal cells in the right-hand section, although it does require the extra computation of the scaling parameter $\lambda^{-1} = |\underline{d}(n)|^2$ within the intermediate column.

9. COMMENTS AND CONCLUSIONS

Novel methods have been described for the parallel computation of least squares weight vectors in both the canonical and single constraint problems, as encountered in adaptive beamforming. They involve the extension of the triangular Gentleman-Kung systolic array by a further triangular network in which the inverse Cholesky factor of the channel correlation matrix estimator is updated.

The array analyses provided have deliberately avoided a description of full systolic operation. It is sufficient to state that all operations described here - including those for frozen arrays - may be rendered fully parallel upon conventional input data skewing[1],[5]. Consequently, elements of a desired weight vector corresponding to a given epoch (t_n, say) will emerge over p successive clock cycles, and within a given cycle elements of vectors corresponding to p different time epochs (t_{n-p-1}, t_{n-p-2}, ..., t_n) will emerge simultaneously.

Lack of space precludes a detailed discussion of issues such as operation count, algorithm modification (eg. square-root-free versions, or inclusion of data deweighting), and numerical stability. These topics will be discussed in future publications.

ACKNOWLEDGEMENTS

The authors wish to thank the Directors of STC Technology Ltd., for permission to publish this paper and Drs. C. Ward and P. Gosling of STC for useful discussions.

REFERENCES

[1] Gentleman, W. M. and Kung, H. T. (1981) "Matrix triangularization by systolic arrays", Proc. SPIE, **298**, Real Time Signal Processing IV, 19.

[2] McWhirter, J. G., (1983) "Recursive least-squares minimisation using a systolic array", Proc. SPIE, **431**, Real Time Signal Processing VI, 105.

[3] Shepherd, T. J. and McWhirter, J. G. (1987), "A pipelined array for linearly constrained least-squares optimisation", in "Mathematics in Signal Processing", (ed. T. S. Durrani et al.), 457, (OUP, Oxford).

[4] Schreiber, R., (1986) "Implementations of adaptive array algorithms", IEEE Trans. **ASSP-34**, No. 5, 1038.

[5] McWhirter, J. G. and Shepherd, T. J., (1988) "An efficient systolic array for MVDR beamforming", in Proc. International Conference on Systolic Arrays (ed. K. Bromley et al.), **11**, (Computer Society Press).

[6] Shepherd, T. J. and McWhirter, J. G. (1987) "A systolic array for linearly constrained least-squares optimisation", in Proc. First International Conference on Systolic Arrays, (ed. A. McCabe and R. Urquhart), 151, (Adam Hilger, Bristol and Boston).

CHECKSUM SCHEMES FOR FAULT TOLERANT SYSTOLIC COMPUTING

R.P. Brent
(Australian National University, Australia)

and

F.T. Luk and C.J. Anfinson
(Cornell University, New York)

ABSTRACT

The weighted checksum scheme has been proposed as a low-cost fault tolerant procedure for parallel matrix computations. To guarantee multiple error detection and correction, the chosen weight vectors must satisfy some very specific properties about linear independence. We will provide a theoretical framework for these properties, and prove that for a distance d+1 scheme, if a maximum of [d/2] errors ensue, the exact number of errors can be determined. We will derive a procedure for correcting the errors. Previous weight generating methods that fulfil the independence criteria have troubles with numerical overflow. We will present a new scheme that generates weight vectors to meet the requirements about independence and to avoid the difficulties with overflow.

1. INTRODUCTION

The importance of solving signal processing problems in real time and the development of VLSI and wafer-scale technology have led to research in systolic arrays and algorithms. There is a need for high-performance digital signal processing systems which are extremely reliable. Algorithm-based fault tolerance has been proposed to meet this reliability need since the most common alternative of duplicating hardware is often too expensive to be practical.

The weighted checksum scheme, originally developed by Abraham and students [6], [7], provides low-cost error protection for applications that include matrix addition, matrix multiplication, and triangular decompositions (see also [8] and [10]). We will present a theoretical framework for the scheme and show how to decide upon the exact number of errors. Techniques for error correction were known only for the cases of one error [7] and two errors [1]. We will present a scheme for correcting [d/2] errors. The previously proposed weights are powers of integers, and thus can become very large. We will propose a technique for generating reasonably sized weights.

This paper is organized as follows. We first review the weighted checksum scheme and discuss an important theorem that guarantees multiple error detection and correction. Then we prove that, if a maximum of [d/2] errors ensue in a distance d+1 code, the errors can be detected and the exact number of errors determined. Furthermore, we show how we can always correct the errors and present a procedure for doing so. Lastly, we discuss the problem of numerical overflow and propose a new method for weight generation that overcomes this difficulty.

2. BACKGROUND

In [1], [6] and [7], a linear algebraic model of the weighted checksum scheme is developed, allowing parallels to be drawn between algorithm-based fault tolerance and coding theory. We briefly review this important background material, and discuss the fault model relevant to our results, as well as several important assumptions and their implications.

2.1 Definitions

We define a $d \times (n+d)$ consistency check matrix H by

$$H = \begin{bmatrix} \Psi_0^0 & \Psi_1^0 & \cdots & \Psi_{n-1}^0 & 1 & 0 & \cdots & 0 \\ \Psi_0^1 & \Psi_1^1 & \cdots & \Psi_{n-1}^1 & 0 & 1 & \cdots & 0 \\ \cdot & \cdot & \cdots & \cdot & 0 & 0 & \cdots & 0 \\ \cdot & \cdot & \cdots & \cdot & \cdot & \cdot & \cdots & \cdot \\ \cdot & \cdot & \cdots & \cdot & \cdot & \cdot & \cdots & \cdot \\ \Psi_0^{d-1} & \Psi_1^{d-1} & \cdots & \Psi_{n-1}^{d-1} & 0 & 0 & \cdots & 1 \end{bmatrix} \qquad (2.1)$$

It is proved in [1] that for $\Psi_i \neq \Psi_j$, where $i \neq j$, every set of d columns of H is linearly indpendent. This matrix is similar to the parity check matrix in coding theory and is said to generate the code. In order to make precise the meaning of distance in the code space, we define a metric upon the domain of H. It is easily checked that the metric satisfies the properties of a distance [4].

Definition 2.1. The code space C of H consists of all vectors which lie in the null space N(H) of H, where $N(H) = \{x : Hx = 0\}$.

Definition 2.2. The distance between two vectors v and w in the domain of H, *dist* (v,w), equals the number of components in which v and w differ.

Definition 2.3. The distance of the code space C is the minimum of the distances between all possible pairs of distinct vectors in N(H); i.e., distance of C = min $\{dist(v,w) : v,w$ in $N(H), v \neq w\}$.

Definition 2.4. Let x be in N(H), and $\hat{x}$ be a possibly erroneous version of x. Define the syndrome vector s by $s = H\hat{x}$, and the correction vector c by $c = \hat{x} - x$.

Note that $Hc = s$ since $Hc = H\hat{x} - Hx = H\hat{x} = s$. For simplicity we will subscript the vectors c and s as follows:

$$c = (c_0, c_1, \ldots, c_{n+d-1})^T$$

and

$$s = (s_0, s_1, \ldots, s_{d-1})^T.$$

We can now rigorize the definitions of error detection and correction. Let α denote the total number of errors that have occurred and define γ by

$$\gamma = \left[\frac{d}{2} \right]. \qquad (2.2)$$

A coding system can *detect* α errors if the syndrome vector s is nonzero whenever $1 \leq \alpha \leq d$. A coding scheme can *correct* α errors if $\hat{x}$ can be corrected to x for $1 \leq \alpha \leq \gamma$. The next result from [1] states the sufficient conditions for error detection and correction.

Theorem 2.1. If every set of d columns of H is linearly independent, then the distance of the code is d+1, a maximum of d errors can be detected, and a maximum of γ errors can be corrected. □

2.2 Fault model and assumptions

As we are primarily interested in multiprocessor systems for real time digital signal processing (e.g., systolic arrays), we assume that a module (a processor or computational unit) makes arbitrary logical errors in the event of a fault and that the system is periodically checked for permanent and intermittent errors. We focus our attention on soft errors, and suppose that the soft error rate is small under normal operating conditions, which is reasonable since it has been reported that the soft error rate for a large VLSI chip (1 cm^2) is 10^{-4} per hour [12].

Two major assumptions are made in this paper. First, for our correction procedure, we assume that no errors occur in the checksums themselves. It should be noted that for error detection this assumption is not needed. We also assume that $1 \leq \alpha \leq \gamma$.

3. ERROR DETECTION

In this section we discuss a matrix factorization that allows us to determine the exact value of α. This a priori knowledge of α will prove important later in the correction procedure.

3.1 The syndrome vector

Consider the syndrome vector s. By assuming that no errors occur in the checksums themselves, we get $c_n = c_{n+1} = \ldots = c_{n+d-1} = 0$. Hence, the jth element of s is given by the formula

$$s_j = \sum_{k=0}^{n-1} \Psi_k^j c_k, \text{ for } j = 0, 1, \ldots, d-1.$$

Clearly there are only α nonzero elements in the set $\{c_0, c_1, \ldots, c_{n-1}\}$, because by definition $c = \hat{x} - x$, and $\hat{x}$ differs from x by α elements. Denote the nonzero elements of c by c_{i_k}, for $k = 1, \ldots, \alpha$. Hence, s_j may be expressed as

$$s_j = \sum_{k=1}^{\alpha} \Psi_{i_k}^{j} c_{i_k}, \text{ for } j = 0, 1, \ldots, d-1. \tag{3.1}$$

We would like to express s_j in terms of known parameters, and at this time α is unknown and γ is known. So, let

$$y_k = c_{i_k} \text{ and } \xi_k = \Psi_{i_k}, \text{ for } k = 1, \ldots, \alpha,$$

and

$$y_k = 0 \text{ and } \xi_k = \Psi_j, \text{ for } k = \alpha+1, \ldots, \gamma,$$

such that $\xi_i \neq \xi_\ell$ for all $i \neq l$, where $i, \ell = 1, \ldots, \gamma$. With these substitutions s_j becomes

$$s_j = \sum_{k=1}^{\gamma} \xi_k^j y_k, \text{ for } j = 0, 1, \ldots, d-1. \tag{3.2}$$

3.2 Find number of errors

Define a $\gamma \times \gamma$ symmetric Hankel matrix by

$$K = \begin{bmatrix} s_0 & s_1 & \cdots & s_{\gamma-1} \\ s_1 & s_2 & \cdots & s_\gamma \\ \cdot & \cdot & \cdots & \cdot \\ \cdot & \cdot & \cdots & \cdot \\ \cdot & \cdot & \cdots & \cdot \\ s_{\gamma-1} & s_\gamma & \cdots & s_{2\gamma-2} \end{bmatrix} \tag{3.3}$$

Theorem 3.1. The matrix K has the factorization $K = VYV^T$, where $Y = \mathit{diag}[y_1, y_2, \ldots, y_\gamma]$ and

$$V = \begin{bmatrix} 1 & 1 & \cdots & 1 \\ \xi_1 & \xi_2 & \cdots & \xi_\gamma \\ \cdot & \cdot & \cdots & \cdot \\ \cdot & \cdot & \cdots & \cdot \\ \cdot & \cdot & \cdots & \cdot \\ \xi_1^{\gamma-1} & \xi_2^{\gamma-1} & \cdots & \xi_\gamma^{\gamma-1} \end{bmatrix}.$$

Furthermore, the matrix has rank α, where α denotes the number of errors that have occurred.

Proof. Multiplying out VYV^T we find that the (i,j) entry is equal to s_{i+j-2}, and thus $(K)_{ij}$. The second part of our claim follows from the observation that $rank(K) = rank(Y)$. □

An accurate, albeit expensive, way to determine the rank of the symmetric matrix K is to compute its eigenvalue decomposition. As will be seen, we will need a nonsingular matrix for the correction procedure to work effectively. The following corollary to Theorem 3.1 will allow us to handle the case of a rank deficient K. Define K_α as the leading $\alpha \times \alpha$ principal submatrix of K:

$$K_\alpha = \begin{bmatrix} s_0 & s_1 & \cdots & s_{\alpha-1} \\ s_1 & s_2 & \cdots & s_\alpha \\ \cdot & \cdot & \cdots & \cdot \\ \cdot & \cdot & \cdots & \cdot \\ \cdot & \cdot & \cdots & \cdot \\ s_{\alpha-1} & s_\alpha & \cdots & s_{2\alpha-2} \end{bmatrix}. \tag{3.4}$$

Corollary 3.1. The $\alpha \times \alpha$ matrix K_α has full rank, and can be factored as $K_\alpha = V_\alpha Y_\alpha V_\alpha^T$, where $Y_\alpha = diag[y_1, y_2, \ldots, y_\alpha]$ and

$$V_\alpha = \begin{bmatrix} 1 & 1 & \cdots & 1 \\ \xi_1 & \xi_2 & \cdots & \xi_\alpha \\ \cdot & \cdot & \cdots & \cdot \\ \cdot & \cdot & \cdots & \cdot \\ \cdot & \cdot & \cdots & \cdot \\ \xi_1^{\alpha-1} & \xi_2^{\alpha-1} & \cdots & \xi_\alpha^{\alpha-1} \end{bmatrix}.$$

4. ERROR CORRECTION

In this section we outline our correction procedure. We will assume that α has been determined such that K_α is nonsingular.

4.1 A polynomial

Define a polynomial $P(z)$ whose roots are the unknown weights ξ_i, for $i = 1,, \ldots, \alpha$. We will show that we can determine the coefficients of the polynomial by solving a linear system involving the matrix K_α.

$$P(z) = \prod_{i=1}^{\alpha} (z - \xi_i) = \sum_{i=0}^{\alpha} a_i z^i, \tag{4.1}$$

where the coefficients of P are given by

$$a_i = (-1)^{i+\alpha} \sum_{j_1<\ldots<j_{\alpha-i}} \xi_{j_1} \cdots \xi_{j_{\alpha-i}}, \tag{4.2}$$

and $a_\alpha = 1$. The next theorem and its two corollaries will lay a foundation for the correction procedure. We find necessary the following definitions. First, define A_{-k} as the matrix A with the kth column deleted and

$$a = (a_0, a_1, \ldots, a_\alpha)^T.$$

Then, let

$$f = (s_\alpha, s_{\alpha+1}, \ldots, s_{2\alpha-1})^T$$

and

$$F = (K_\alpha, f) = \begin{bmatrix} s_0 & s_1 & \cdots & s_{\alpha-1} & s_\alpha \\ s_1 & s_2 & \cdots & s_\alpha & s_{\alpha+1} \\ \cdot & \cdot & \cdots & \cdot & \cdot \\ \cdot & \cdot & \cdots & \cdot & \cdot \\ \cdot & \cdot & \cdots & \cdot & \cdot \\ s_{\alpha-1} & s_\alpha & \cdots & s_{2\alpha-2} & s_{2\alpha-1} \end{bmatrix}.$$

Writing out the jth element of the matrix-vector product Fa, we get

$$(Fa)_j = \sum_{k=1}^{\alpha} \xi_k^j y_k P(\xi_k) = 0, \text{ for } j = 0,1,\ldots,d-1.$$

Theorem 4.1. The coefficients of the polynomial $P(z)$ satisfy the equation

$$Fa = 0. \quad \square \tag{4.3}$$

Corollary 4.1. Let $D_\alpha = \det(K_\alpha)$. For $k = 0,1,\ldots,\alpha-1$, the coefficient a_k can be computed by the formula

$$a_k = (-1)^{k+\alpha} D_k / D_\alpha, \tag{4.4}$$

where $D_k = \det(F_{-k})$. $\square$

Corollary 4.2. Define the polynomial $Q(z)$ as $\sum_{i=0}^{\alpha} (-1)^i D_i z^i$. Then $Q(z)$ has roots $\xi_1, \xi_2, \ldots, \xi_\alpha$. $\square$

4.2 Correction procedure

Thus, in order to find the unknown weights, we need to find the vector a of coefficients of the polynomial $P(z)$. So, we may first solve (4.3) for a, and then find the roots of $P(z)$. Alternatively, we could compute the determinants D_ks and find the roots of $Q(z)$, which will give us the unknown weights by Corollary 4.2. Using the above facts, there are four steps that need to be accomplished for correction.

Procedure

(1) Find *rank* (K). This can be done by finding the eigenvalues of the symmetric matrix K.

(2) Solve for the coefficients $(a_0, a_1, \ldots, a_\alpha)$ of $P(z)$. Note that this can be accomplished by either solving the system (4.3) or by using equation (4.4). This step requires K_α to be of full rank.

(3) Find the roots $(\xi_1, \ldots, \xi_\alpha)$ of $P(z)$ or $Q(z)$. Note that these can be computed as the eigenvalues of a certain companion matrix. This will give us the vector ξ.

(4) Find the vector $y = (y_1 y_2, \ldots, y_\alpha)^T$ by solving the systems $\Xi y = s$, where

$$\Xi = \begin{bmatrix} \xi_1^0 & \xi_2^0 & \cdots & \xi_\alpha^0 \\ \xi_1^1 & \xi_2^1 & \cdots & \xi_\alpha^1 \\ \cdot & \cdot & \cdots & \cdot \\ \cdot & \cdot & \cdots & \cdot \\ \cdot & \cdot & \cdots & \cdot \\ \xi_1^{d-1} & \xi_2^{d-1} & \cdots & \xi_\alpha^{d-1} \end{bmatrix}.$$

Actually Ξ is a nonsingular Vandermonde matrix and there are fast techniques available to solve systems of linear equations involving a Vandermonde matrix [4]. Note that solving the system $\Xi y = s$ is equivalent to solving $Hc = s$ for the nonzero values of c, and that our procedure is very similar to the Reed-Solomon code [11]. Our procedure details the mathematical analysis of the problem, and can use further improvement as a numerical algorithm. The procedure will not be prohibitively expensive to implement. Furthermore, one may use a systolic array, or some other parallel machine which can solve the eigenvalue problem rapidly [2], [13]; since $d \ll n$, the size of the eigenvalue problem we need to solve is much smaller than that of the original problem.

5. AVOIDING OVERFLOWS

We present a new technique for generating weight vectors which satisfy the conditions of Theorem 2.1, and avoid the weight overflow problem which plagues previous schemes. Big weights introduce large roundoff errors that cannot be distinguished from true errors caused by faults (see [9] for a discussion.) Our scheme is dependent on the values of n and d, but as the step of generating the weight vectors is pre-processing, it will not slow down the execution of the algorithm. Our method uses modular arithmetic, but not in the same way as suggested by Abraham et al. [6], [7].

5.1 Modular arithmetic

Consider the values of n and d to be known and fixed. The size of the input matrix determines n, and d is selected by the user according to a model which decides on the expected number of errors. Choose a prime number p such that $p > n+d$. It is well known that for any prime p there exists a finite field of p elements [5]. Let α be a primitive element of the finite field of p elements. Recall that a primitive element β of a finite field with q elements is an element with order $q-1$, i.e., the smallest integer m for which $\beta^m = 1 \bmod q$ is $m = q-1$. It is easy to see that powers of β generate all the nonzero elements of the field.

Let B be the following matrix

$$B = (b_{ij}) = (\alpha^{(i-1)(j-1)}), \text{ for } 1 \le i \le d \text{ and } 1 \le j \le n+d. \quad (5.1)$$

Partition B as

$$B = [V|M] \quad (5.2)$$

Here V has dimension dxn and W has dimension dxd. Now, in order to get the desired form of H, we multiply the matrix B *mod* p by the matrix W^{-1} *mod* p. That is

$$H = W^{-1}[V|M] \bmod p = [W^{-1}V \bmod p|I]. \quad (5.3)$$

To make the above procedure valid, we must prove the following. First, we need to show that W is nonsingular so that the multiplication in (5.3) is valid. Then we need to prove that every set of d columns of the resulting matrix H is linearly independent, so that we can guarantee the detection of a maximum of d errors. A proof of the next theorem can be found in [3].

Theorem 5.1. The matrix B *mod* p, where B is given by equation (5.2), satisfies the condition that every set of d columns is linearly independent. □

Corollary 5.1. The dxd matrix W *mod* p, where W is defined in equation (5.2), is nonsingular. □

Corollary 5.2. The matrix H, given by equation (5.3), satisfies the hypotheses of Theorem 2.1. □

We note that the choice of W is by no means unique. We could choose any d columns of B and permute the matrix into the form given by equation (5.2).

5.2 An example

Suppose we are given a problem where $n = 8$ and d is selected to be 4. Thus, by fulfilling the requirements of Theorem 2.1 we will be able to detect a maximum of 4 errors in a 12-element encoded vector. We take $p = 13$, and $\alpha = 2$ is a primitive root of p. The matrix B *mod* p has the form

$$B \textit{ mod } p = \begin{bmatrix} 1 & 1 & 1 & 1 & 1 & 1 & 1 & 1 & 1 & 1 & 1 & 1 \\ 1 & 2 & 4 & 8 & 3 & 6 & 12 & 11 & 9 & 5 & 10 & 7 \\ 1 & 4 & 3 & 12 & 9 & 10 & 1 & 4 & 3 & 12 & 9 & 10 \\ 1 & 8 & 12 & 5 & 1 & 8 & 12 & 5 & 1 & 8 & 12 & 5 \end{bmatrix}.$$

Now we need to find the inverse of W *mod* p. If we use the Gauss-Jordan method for finding matrix inverses, we can do the inversion *mod* p since the procedure consists of only elementary row operations. We then perform the multiplication $W^{-1}V$ *mod* p. Alternatively, we can find the LU factors of W via Gaussian elimination *mod* p, which also involves only elementary row operations, and then compute each column of the product $W^{-1}V$ *mod* p using the LU factors. While computing $W^{-1}V$ *mod* p is expensive, we once again point out that it is all pre-processing; it will not add extra running time to the algorithm. Hence, we can find the inverse of W *mod* p,

$$W^{-1} \bmod p = \begin{bmatrix} 8 & 5 & 6 & 8 \\ 6 & 11 & 0 & 12 \\ 5 & 0 & 9 & 7 \\ 8 & 10 & 11 & 12 \end{bmatrix}.$$

Multiplying out $W^{-1}B \bmod p$, we get the desired matrix

$$H = \begin{bmatrix} 1 & 2 & 12 & 4 & 7 & 6 & 1 & 10 & 1 & 0 & 0 & 0 \\ 3 & 7 & 12 & 11 & 12 & 12 & 9 & 5 & 0 & 1 & 0 & 0 \\ 8 & 6 & 12 & 5 & 2 & 8 & 7 & 11 & 0 & 0 & 1 & 0 \\ 2 & 12 & 4 & 7 & 6 & 1 & 10 & 1 & 0 & 0 & 0 & 1 \end{bmatrix}.$$

5.3 Comparison with previous work

We examine how our method differs from these previous schemes. For means of comparison we consider fixed point arithmetic since it is used in [6] and [7]. Let r denote the word length. Huang and Abraham [6] suggested that we compute the actual checksum values modulo 2^r. In Jou and Abraham [7], a similar technique is proposed where the checksum column is computed modulo 2^r and the weighted checksum column is computed modulo N, where N is the largest prime less than 2^{r+1}. They proved that for this choice of weights, the scheme detects errors as desired. One difference between the methods is that our new one is all pre-processing while theirs adds extra time to the algorithm as we must compute the checksums using modular arithmetic. Furthermore, the prime that is selected by Jou and Abraham is somewhat larger than the prime we come up with. For example, in [7], if the word length $r = 32$ then the prime $N = 8589934583$. Our scheme is relatively independent of the word length; it depends upon the values of n and d. For $n = 1000$, a large value in light of the dimensions required by current signal processing problems, and $d = 50$, the prime $p = 1051$, which is much smaller than the Jou-Abraham value of N.

We next compare the sizes of the elements in the matrix H for various weight generating schemes. Suppose that $n = 500$ and $d = 10$. Then the range of values for i and j is $i = 1,\ldots,10$ and $j = 1,\ldots,500$. Jou and Abraham proposed the set of weights $w_j^{(i)} = 2^{(i-1)(j-1)}$. While easily implemented as shifts, this set of weights becomes extremely large very quickly, resulting in overflows. For i = 10 and j = 500, we get $w_j^{(i)} = 2^{4491}$. Another suggestion was to let $w_j^{(i)} = j^{i-1}$ [8], which will get large very rapidly, although not as quickly as the prévious technique. Using the same example, with i = 10 and j = 500, we have $w_j^{(i)} = 500^9$. For our new scheme, the smallest prime p satisfying p>n+d = 510 is p = 521. Therefore, every element $w_j^{(i)}$ will be bounded above by 521. Thus, we see that for this moderately sized problem, our new scheme generates a parity-check matrix whose elements are likewise of moderate size.

6. CONCLUSIONS

In this paper we have proved that, given a consistency check matrix H which defines a distance d+1 code, we can determine the exact number of errors and, if the total number of errors that have occurred lies between 0 and γ, then we can correct all errors. We have also presented a new method for generating the weighted checksum matrix whose elements are bounded in size by the smallest prime p such that p>n+d. Since we usually consider $d \ll n$,p is approximately O(n). To generate the matrix H involves only pre-processing steps, and thus will take no additional time in the algorithm.

7. ACKNOWLEDGEMENTS

This work was supported in part by the Joint Services Electronics Program under contract F49620-87-C-0044 at Cornell University. Cynthia Anfinson was also suported by a U.S. Army Science and Technology Graduate Fellowship.

8. REFERENCES

[1] Anfinson, C.J. and Luk, F.T., (1988), "A linear algebraic model of algorithm-based fault tolerance", IEEE Trans. Comput., Vol. C-37, No. 12, pp. 1599-1604.

[2] Brent, R.P. and Luk, F.T., (1985), "The solution of singular-value and symmetric eigenvalue problems on multiprocessor arrays", *SIAM J. Sci. Statist. Comput.*, **6**, pp. 69-84.

[3] Brent, R.P., Luk, F.T. and Anfinson, C.J., (1989), "Choosing small weights for multiple error detection", Proc. SPIE, Vol. 1058, High Speed Computing II, pp. 130-136.

[4] Golub, G.H. and Van Loan, C.F., (1983), "Matrix Computations", The Johns Hopkins University Press, Baltimore, Maryland.

[5] Herstein, I.N., (1975), "Topics in Algebra", Second Edition, John Wiley and Sons, New York.

[6] Huang, K.H. and Abraham, J.A., (1984), "Algorithm-based fault tolerance for matrix operations", *IEEE Trans. Comput.*, Vol. C-33, No. 6, pp. 518-528.

[7] Jou, J.Y. and Abraham, J.A., (1986), "Fault-tolerant matrix arithmetic and signal processing on highly concurrent computing structures", Proc. IEEE, Vol. 74, No. 5, Special Issue on Fault Tolerance in VLSI, pp. 732-741.

[8] Luk, F.T., (1985), "Algorithm-based fault tolerance for parallel matrix equation solvers", Proc. SPIE, Vol. 564, Real Time Signal Processing VIII, pp. 49-53.

[9] Luk, F.T. and Park, H., (1988), "An analysis of algorithm-based fault tolerance techniques", *J. Parallel Distrib. Comput.*, Vol. 5, pp. 172-184.

[10] Luk, F.T. and Park, H., (1988), "Fault-tolerant matrix triangulations on systolic arrays", *IEEE Trans. Comput.*, Vol. C-37, No. 11, pp. 1434-1438.

[11] Reed, I.S. and Solomon, G., (1960), "Polynomial codes over certain finite fields", *J. Soc. Ind. Appl. Math.*, Vol. 8, pp. 300-304.

[12] Sarrazin, D.B. and Malek, M., (1984), "Fault-tolerant semiconductor memories", *IEEE Computer, Vol. 17*, No. 8, Special Issue on Fault-Tolerant Computing, pp. 49-56.

[13] Stewart, G.W., (1985), "A Jacobi-like algorithm for computing the Schur decomposition of a non-Hermitian matrix", *SIAM J. Sci. Statist. Comput.*, **6**, pp. 853-864.

SIMULATION OF LUK'S SVD ARRAY USING A TRANSPUTER

G. de Villiers
(Royal Signals and Radar Establishment,
Malvern, Worcestershire)

ABSTRACT

In this paper we discuss the simulation of Luk's SVD array using OCCAM running on a single transputer. The simulation is of a wavefront array rather than a systolic array since this is simpler in OCCAM. We discuss the basic mathematics and then the simulation details. A couple of examples of matrices with known singular values are given. The simulation produced the correct singular values for these matrices.

1. INTRODUCTION AND BASIC MATHEMATICS

The singular value decomposition (SVD) of a rectangular mxn matrix A is given by

$$A = U \Sigma V^{\dagger} \qquad (1.1)$$

where U is a unitary mxm matrix, V is a unitary nxn matrix and Σ is an mxn diagonal matrix whose diagonal entries are real positive numbers or zero. These diagonal entries are known as the singular values of A. For real A, U and V are orthogonal matrices. We shall assume throughout this paper that A is real.

Various methods have been proposed for finding singular values by using an array of processors. Recently one which has received a lot of attention is the triangular array of Luk [2]. The matrix A is fed into this and is turned into an upper triangular matrix using one-sided orthogonal transformations. This upper triangular matrix is then operated on by a sequence of Kogbetliantz transformations to reduce it to diagonal form with its singular values along the diagonal. The QR stage is accomplished in a finite number of steps whereas the SVD stage is iterative, carried on until some convergence criterion has been satisfied.

We now briefly review the mathematical basis for the algorithm using the same convention as Luk [2].

The mathematics for the QR stage is very well known (see for example Kung and Gentleman [1] McWhirter [3]). The basic tool is the 2x2 rotation

$$J(\theta) = \begin{bmatrix} \cos\theta & \sin\theta \\ -\sin\theta & \cos\theta \end{bmatrix} \qquad (1.2)$$

The QR decomposition of A is built up from 2x2 QR decompositions:

$$\begin{bmatrix} a & b \\ c & d \end{bmatrix} = J(\theta)^T \begin{bmatrix} p & q \\ 0 & r \end{bmatrix} \qquad (1.3)$$

i.e.

$$J(\theta) \begin{bmatrix} a & b \\ c & d \end{bmatrix} = \begin{bmatrix} p & q \\ o & r \end{bmatrix} \qquad (1.4)$$

The angle θ is given by

$$c = 0, \ \theta = 0$$

$$c \neq 0, \ \sin\theta = \frac{c}{\sqrt{a^2+c^2}}, \quad \cos\theta = \frac{a}{\sqrt{a^2+c^2}} \qquad (1.5)$$

To see the effect this 2x2 rotation has on the remaining elements of the matrix let us consider a general rotation matrix designed to annihilate a subdiagonal element of a matrix, say

$$\begin{array}{r} \\ \\ i^{th}\text{ row} \\ i+1^{th}\text{ row} \\ \\ \end{array} \begin{bmatrix} 1 & & & & & & 0 \\ & 1 & & & & & \\ & & 1 & & & & \\ \ldots & \ldots & \ldots & c & s & & \\ \ldots & \ldots & \ldots & -s & c & & \\ & & & & & 1 & \\ & & & & & & 1 \\ 0 & & & & & & 1 \end{bmatrix} \qquad (1.6)$$

Denoting this matrix by Rot we have

$$\text{Rot}_{k\ell} = \delta_{k\ell} + \delta_{ki}\delta_{\ell i}(c-1) + \delta_{ki+1}\delta_{\ell i+1}(c-1) + \delta_{ki}\delta_{\ell i+1}s - \delta_{ki+1}\delta_{\ell i}s \qquad (1.7)$$

Operating on A with Rot we have

$$a'_{kn} = \sum_{\ell} Rot_{k\ell} a_{\ell n} \tag{1.8}$$

$$= a_{kn} \quad k \neq i, \text{ or } i + 1$$

$$= ca_{kn} + sa_{k+1n} \quad k = i$$

$$= ca_{kn} - sa_{k-1n} \quad k = i + 1$$

What the preceding mathematics says is that only the rows i and i+1 of the matrix A are transformed to something different. The transformation of these rows is a rotation between their elements of same n, i.e. in the same column.

The basic principle of the QR stage is as follows. The matrix is fed in row by row. The top row of processors converts every arriving row of the incoming matrix into a row with a zero as the first element. This shorter row is then passed down to the second row of processors. The second row of processors converts every arriving row into a row with a zero as the first element and then passes down this shorter row to the third row of processors and so on. All the stored numbers are initialised to zero before the matrix is fed in. The leading diagonal processors perform the 2x2 rotations discussed above and then pass on the rotation parameters to the other processors in their row that these may apply them. The process continues until the final row to be fed in at the top reaches the bottom. The rows of processors stop functioning as soon as the final row of the matrix has passed through them. The triangular matrix R then resides on the array.

The leading diagonal processors now perform 2x2 Kogbetliantz transformations:

$$J(\theta)^T \begin{bmatrix} w & x \\ O & z \end{bmatrix} K(\phi) = \begin{bmatrix} d_1 & O \\ O & d_2 \end{bmatrix} \tag{1.9}$$

(Note the difference in convention. Here we are using a transposed rotation matrix from the left). We follow Luk in using the two stage procedure for the matrix $J(\theta)^T$. First we symmetrise the upper triangular matrix from the left.

$$S(\psi)^T \begin{bmatrix} w & x \\ O & z \end{bmatrix} = \begin{bmatrix} p & q \\ q & r \end{bmatrix} \tag{1.10}$$

The angle is given by

$$x = 0, \ \psi = 0$$

$$x \neq 0$$

$$\rho \equiv \cot \psi = \frac{w \mp z}{x}, \ \sin \psi = \frac{\operatorname{sign}(\rho)}{\sqrt{1 + \rho^2}}, \cos \psi = \rho \sin\psi \qquad (1.11)$$

Next we diagonalise the result

$$K(\varphi)^T \begin{bmatrix} p & q \\ q & r \end{bmatrix} K(\varphi) = \begin{bmatrix} d_1 & 0 \\ 0 & d_2 \end{bmatrix} \qquad (1.12)$$

Following Luk we choose the "outer rotation" solution for φ, $\pi/4 < |\varphi| < \pi/2$

$$\cos \varphi = \frac{1}{\sqrt{1 + t^2}}, \ \sin \varphi = t \cos \varphi \qquad (1.13)$$

where

$$t = -\operatorname{sign}(\rho) \left[|\rho| + \sqrt{1 + \rho^2} \right] \qquad (1.14)$$

and

$$\rho = \frac{r - p}{2q} \qquad (1.15)$$

For the case $q = 0$ we set $\varphi = \pi/2$.

In order to describe the SVD stage we first look at the leading diagonal processors. We number these from top to bottom with the top processor being processor one. The SVD part of the algorithm as stated before, is iterative. It consists of a set of sweeps, each sweep consisting of two steps.

In the first step the odd numbered processors find the appropriate rotation parameters as above. These are passed on to be applied by the remaining processors in the same rows and columns as the odd processors.

In the second step the even processors repeat the performance, passing on the resultant rotations to the remaining processors in the same rows and columns as themselves.

Let us now describe these operations in more detail. We saw in the QR stage that multiplication on the left by a matrix of the form of Rot resulted in the appropriate pair of rows being transformed and the rest of the matrix staying fixed. We recall that between the QR and SVD stages there is a sign change in the 2x2 rotations (following Luk). Hence in the SVD part the effect of a 2x2 rotation on the left designed around the ith and $i+1^{th}$ diagonal elements is

$$
\begin{aligned}
a'_{kn} &= a_{kn} && k \neq i \text{ or } i+1 \\
a'_{kn} &= ca_{kn} - sa_{k+1n} && k = i \\
a'_{kn} &= ca_{kn} + sa_{k-1n} && k = i+1
\end{aligned}
\tag{1.16}
$$

We now consider the effect of multiplying on the right by a matrix of the same form as Rot (note that this will have the same form as the transpose of the matrix we have just applied from the left). Using the expression given for $\text{Rot}_{k\ell}$ we have

$$
\begin{aligned}
a'_{kn} &= \sum_{\ell} a_{k\ell}\text{Rot}_{\ell n} \\
&= a_{kn} \quad n \neq i \text{ or } i+1 \\
&= ca_{kn} - sa_{kn+1} \quad n = i \\
&= ca_{kn} + sa_{kn-1} \quad n = i+1
\end{aligned}
\tag{1.17}
$$

In words, these equations say that elements not in the ith or i+1th columns are unaffected by multiplication on the right by Rot. However, elements in these columns are rotated pairwise, each element being rotated with the one in the same row but neighbouring column.

As we saw earlier the 2x2 transformations in the SVD stage are of the form of (1.9).

If we now view the 2x2 matrix

$$
\begin{bmatrix} w & x \\ o & z \end{bmatrix}
\tag{1.18}
$$

as one of the 2x2 submatrices along the leading diagonal, so that, say w is a_{ii} and z, a_{i+1i+1} the rotation on the left involving ϕ will only affect the ith and $i+1^{th}$ rows. Similarly the rotation on the right involving θ will only affect the ith and $i+1^{th}$ columns.

Hence the rotation parameters cos ϕ and sin ϕ need only be passed along the ith and $i+1^{th}$ rows and the rotation parameters cos θ and sin θ need only be passed up the ith and $i+1^{th}$ columns.

The rotation parameters cos ϕ and sin ϕ are termed the horizontal rotation parameters and cos θ and sin θ are termed the vertical rotation parameters.

After the various rotation parameters have been calculated by the diagonal processors they are passed on to be applied by the remaining processors. The process is continued until all the off-diagonal elements of the matrix are sufficiently small.

This completes the description of the basic mathematics and we turn now to the rest of the paper.

The next section is concerned with the OCCAM simulation of Luk's array.

The final section contains the conclusions.

2. SIMULATION

In OCCAM processors must communicate solely by using channels. Hence all processors sharing matrix elements in Luk's array must be connected by channels. The layout of functioning processors is slightly different for the two stages. Figures 1 and 2 show stages 1 and 2 for a general matrix n even. In these diagrams the curly lines indicate where extra processors are, for large n. The dots at the bottom do likewise. The arrows indicate directions of data flow. Figure 3 shows the combined architecture we simulated for n even. Figure 4 shows the special case n = 4. Note that these architectures would not be appropriate as they stand for multiple transputer implementation since transputers only have four links.

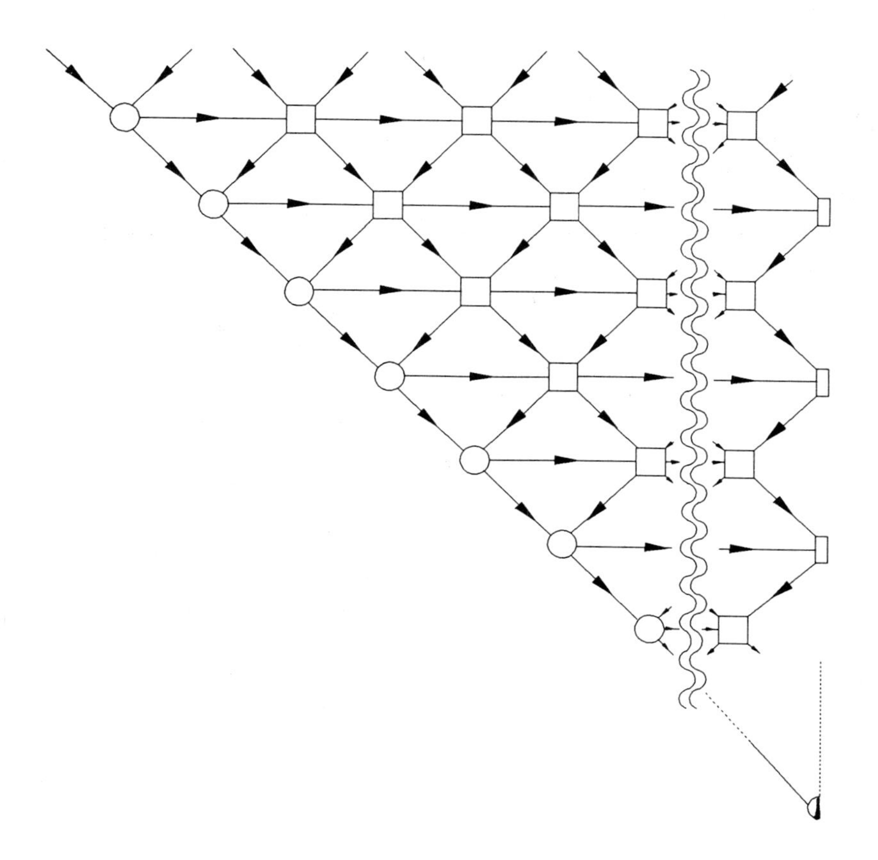

Fig. 1 Stage 1, n even, Luk's Array

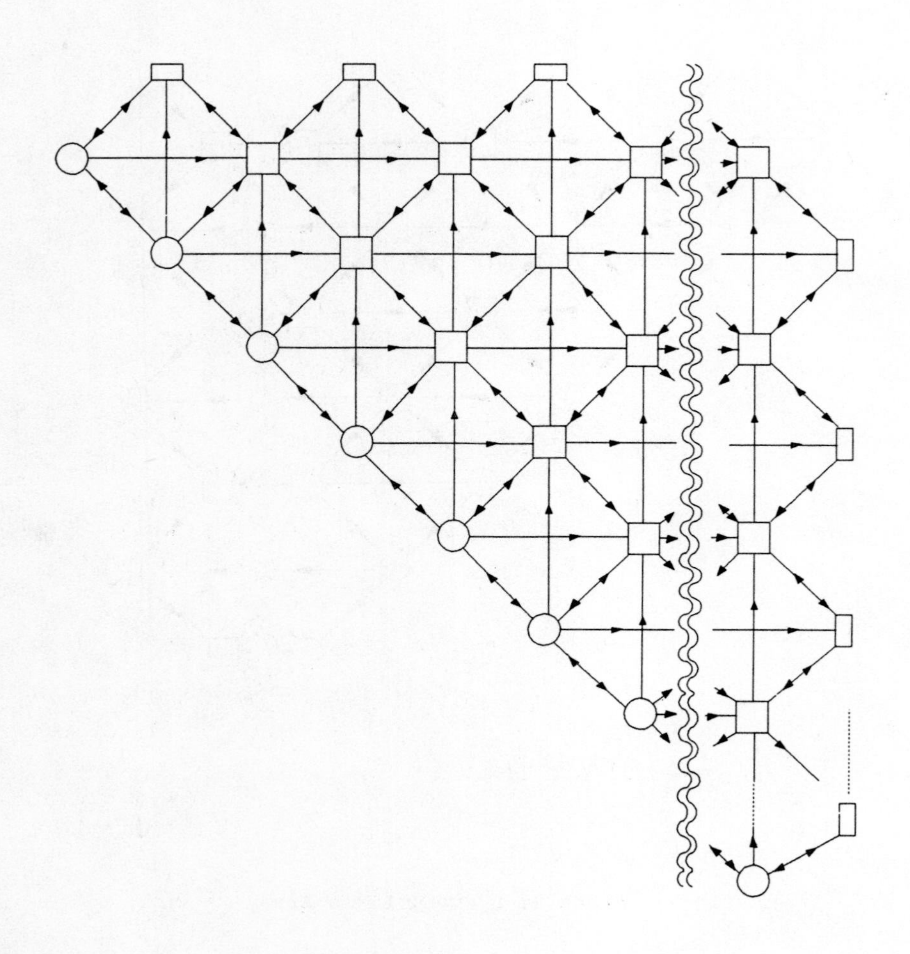

Fig. 2 Stage 2, n even, Luk's Array

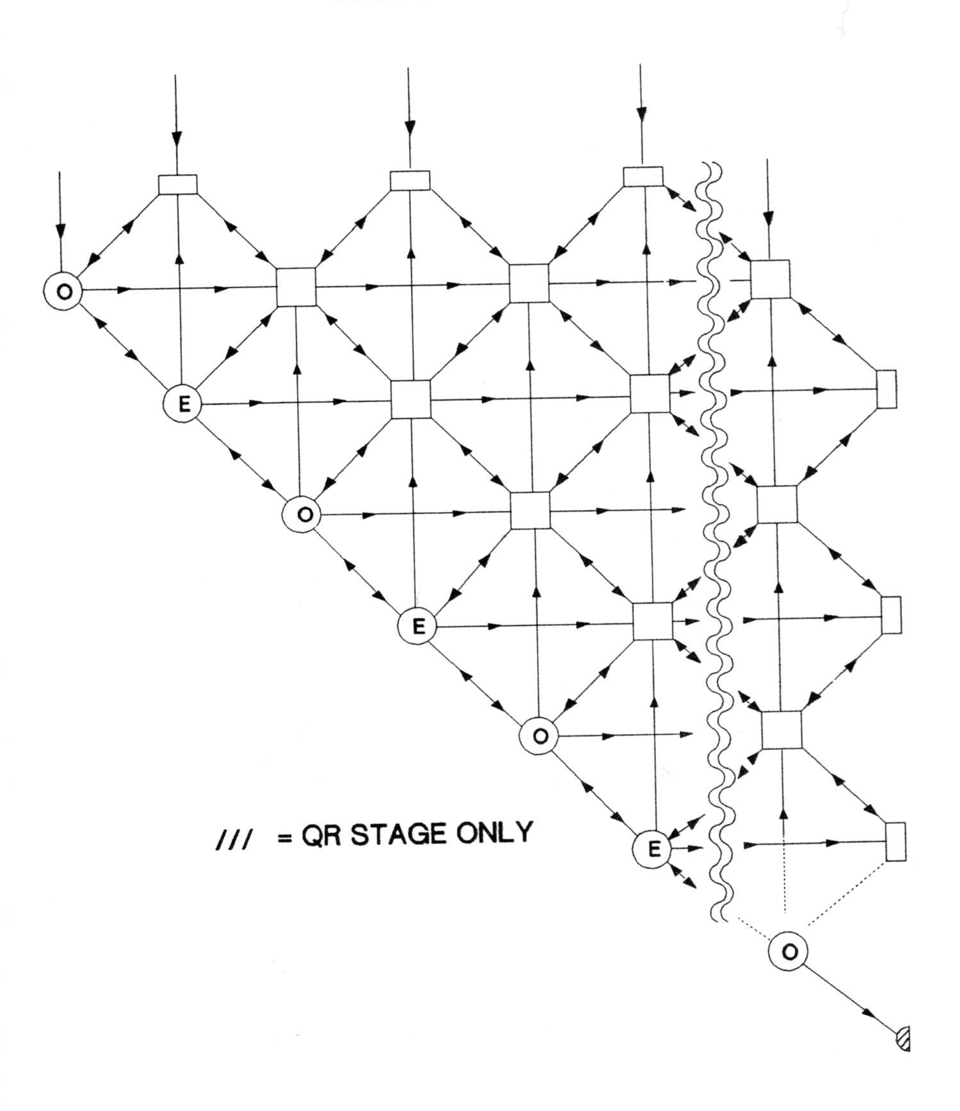

Fig. 3 Unified Architecture for n even

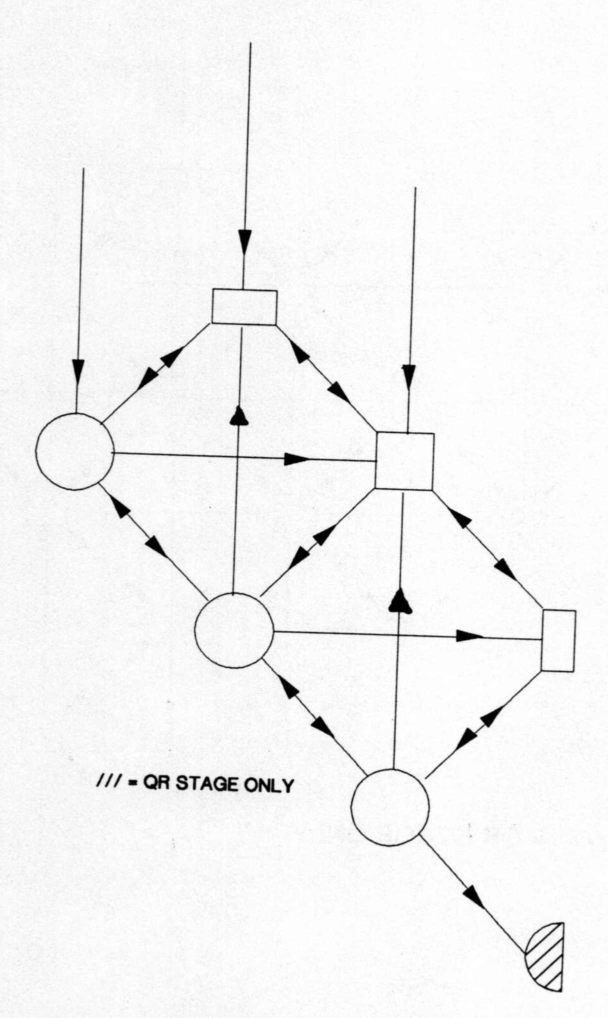

Fig. 4 Unified Architecture for n = 4

We simulated Luk's array in OCCAM using a single transputer. The transputer was on a Transtech Devices TSB04-2 board inside an Olivetti M24Q computer. The TSB04-2 board is basically an INMOS B004 board lookalike consisting of a T4 and associated memory. The OCCAM used was OCCAM2, Beta 2 release and the MSDOS was MSDOS version 3.

Luk's paper was concerned with a systolic array implementation. In our work it was easier to consider a wavefront array implementation. Calculations are done as and when all the necessary numbers are available. The QR stage can be thought of as a wavefront of data sweeping down through the array being processed as it goes. Associated with this are wavefronts of calculations starting on the diagonal and moving across the array to the right.

The SVD stage can be thought of as wavefronts of calculations sweeping upwards and across to the right from the leading diagonal. This stage may start before the QR stage has finished.

In Luk's paper he advocates stopping the iterations in the SVD stage after a fixed number of iterations (sweeps) rather than carrying out a convergence test. In the simulation we have followed this idea since it is by far the simplest solution and is a lot less wasteful of time.

Starting off with matrix A the first problem was how to feed it into the array. Comparing figures 1 and 2 we may see that the top row of processors used in stage 2 is not used in stage 1. Hence by attaching additional channels to the top of these processors they can be used for feeding A in. Extra channels are required at either end to feed in a[i][1] and a[i][n]. Figures 3 and 4 show these details.

The next problem is how to feed the matrix out and display it on the screen. The way this was done was to make the matrix R a global variable, which could then be sent to the screen after it had been diagonalised. A by-product of making R global was that it was much easier to see what was going on in the programme for debugging purposes. A bare outline of the programme is

Feed in matrix A

Triangularise A and write result in global matrix R

Diagonalise R

Write result to screen

A direct consequence of making R global is that one must divide the processors into two groups - the "storage" and "non-storage" processors. By "storage" we mean that the elements of R are only referred to in the code associated with these processors. The matrix elements are passed as local variables to the "non-storage" processors. This construction avoids the possibility of a global variable being referred to in two processors running in parallel. The convention we chose to follow was to have the processors in the same rows as the odd processors (see previous section for the definition of these) as the storage processors and the processors in the same rows as the even processors as non-storage ones.

The mathematics carried out by the processors was put into separately compiled library routines. These routines were a 2x2 QR routine, a 2x2 Kogbetliantz routine and a rotation routine. The channel communications were left in the processes corresponding to the processors.

The top level program structure was

```
PROC svd (CHAN OF ANY Keyboard, Screen)
  ...access to various libraries
  ...declaration of size of global matrices
  SEQ
    ...read in matrix A
    ...channel declarations
    ...processors
    ...output to screen
:
```

The only interesting fold here is the processor one. This contains a PAR followed by the list of different types of processor, most of which are in the PAR loops themselves. The code for a typical processor looks like

```
...variable declarations
SEQ
  SEQ j=1 FOR m
    QR stage
  ....details of setting up of R
  SEQ j=1 FOR 10
    ...SVD stage
```

This gives a sketchy view of the programme. Space forbids more detail.

To test the simulation a couple of rectangular matrices A_1 and A_2 with known singular values were processed. The matrices

$$A_1 = \begin{bmatrix} 0.05 & 0.05 & 0.25 & -0.25 \\ 0.25 & 0.25 & 0.05 & -0.05 \\ 0.35 & 0.35 & 1.75 & -1.75 \\ 1.75 & 1.75 & 0.35 & -0.35 \\ 0.30 & -0.30 & 0.30 & 0.30 \\ 0.40 & -0.40 & 0.40 & 0.40 \end{bmatrix} \quad A_2 = \begin{bmatrix} 22.25 & 31.75 & -38.25 & 65.50 \\ 20.00 & 26.75 & 28.50 & -26.50 \\ -15.25 & 24.25 & 27.75 & 18.50 \\ 27.25 & 10.00 & 3.00 & 2.00 \\ -17.25 & -30.75 & 11.25 & 7.50 \\ 17.25 & 30.75 & -11.25 & -7.50 \end{bmatrix}$$

are used as examples in the NAG library documentation for various SVD routines. The singular values of these matrices, as given in the NAG library documentation are:

$$\sigma_1 : 3.000 \quad 2.000 \quad 1.000 \quad 0.000$$

$$\sigma_2 : 91.000 \quad 68.250 \quad 45.500 \quad 22.750$$

The simulation produced the same answers to the quoted accuracy after 10 sweeps.

3. CONCLUSIONS

We have simulated Luk's SVD array on a single transputer in OCCAM. We have shown that the results of the simulation agree with correct results for a couple of examples given in the NAG library documentation. We have shown that, as far as single transputer simulation goes, the array functions correctly as a wavefront array.

REFERENCES

[1] Gentleman, W.M. and Kung, H.T., (1981) Matrix Triangularisation by Systolic Arrays, SPIE Vol. 298, Real-Time Signal Processing IV 19-26.

[2] Luk, F.T., (1986) A Triangular Processor Array for Computing Singular Values, Linear Algebra and its Applications, 77, 259-273.

[3] McWhirter, J.G., (1983) Recursive Least-Squares Minimisation using a Systolic Array. SPIE Vol. 431 Real-Time Signal Processing VI 105-112.